Polarization-Based Optical Imaging for Clinical Diagnostics

Polarization-Based Optical Imaging for Clinical Diagnostics

Edited by

Yurii Ushenko

Weidong Dou

Oleksandr Ushenko

Oleksandr Dubolazov

Iryna Soltys

WILEY VCH

Editors

Yurii Ushenko
Shaoxing University, Shaoxing, China
Yuriy Fedkovych Chernivtsi National University, Chernivtsi, Ukraine

Weidong Dou
Shaoxing University, Shaoxing, China

Oleksandr Ushenko
Taizhou Institute of Zhejiang University, Taizhou, China
Yuriy Fedkovych Chernivtsi National University, Chernivtsi, Ukraine

Oleksandr Dubolazov
Yuriy Fedkovych Chernivtsi National University, Chernivtsi, Ukraine

Iryna Soltys
Yuriy Fedkovych Chernivtsi National University, Chernivtsi, Ukraine

Cover Design: Wiley
Cover Image: oxygen/Getty Images

Typesetting: Lumina Datamatics, Madurai, India.

Library of Congress Card No.: applied for

British Library Cataloguing-in-Publication Data A catalogue record for this book is available from the British Library.

Bibliographic information published by the Deutsche Nationalbibliothek The Deutsche Nationalbibliothek lists this publication in the Deutsche Nationalbibliografie; detailed bibliographic data are available on the Internet at <http://dnb.d-nb.de>.

Print ISBN: 9783527414697
ePDF ISBN: 9783527857647
ePub ISBN: 9783527857630
oBook ISBN: 9783527857654

Printed and bound by CPI Group (UK) Ltd, Croydon, CR0 4YY

C9783527414697_030926

Table of Contents

Part III Biomedical and Clinical Applications of 3D Jones-Mueller Digital Holographic Mapping

Foreword

Dear Reader!

I am pleased to present to the scientific community a monograph by a team of researchers of Yuriy Fedkovych Chernivtsi National University, which celebrates the 150th anniversary of its foundation in 2025.

The publication summarizes the results of many years of fundamental and applied research in the field of laser polarization biomedical imaging, carried out over the past 25 years.

The results obtained by the authors have significant scientific and practical value and can be useful for specialists, teachers, and students working in the field of life sciences. The presented work makes a significant contribution to the development of modern biomedical technologies aimed at preserving and strengthening human health.

Rector
Ruslan BILOSKURSKIY

About the Editors

Yurii Ushenko is a Professor at School of Mathematics and Information, Department of Physics, Shaoxing University, Shaoxing, Zhejiang, China, since 2024. Head of the Computer Science Department, Chernivtsi National University, Chernivtsi, Ukraine. Head of the Laboratory of Multifunctional Light and Fluorescence Microscopy and Laser Polarimetry. Head (CEO) of the NGO "Bukovinian Cluster of Innovation Technologies named after Joseph Schumpeter," Association "Cluster Bit."

Professor Ushenko has authored over 300 peer-reviewed publications indexed in Scopus, including more than 40 papers in Q1 journals, 20 monographs, and over 70 patents. His work has been published in leading international journals. He serves as a reviewer for several high-impact journals in optics and biomedical photonics and has led numerous nationally funded research projects in laser and biomedical optical technologies.

Weidong Dou is a professor at the Department of Physics of Shaoxing University. He received his PhD in surface and interface physics from Zhejiang University in 2009, then worked as a postdoctoral researcher and research associate at Center of Super-Diamond and Advanced Films (COSDAF) in the City University of Hong Kong during the periods of 2009–2010, 2011–2012, and 2013–2014. His research interests focus on the fabrication and applications of nanomaterials.

Oleksandr Ushenko is a Professor and Head of the Department of Printing, Multimedia and Optical Technologies at the Yuriy Fedkovych Chernivtsi National University, Ukraine and a Research Fellow at the Taizhou Institute of Zhejiang University, China. He received the degree of Doctor of Physical and Mathematical Sciences in optics and laser physics and holds the academic title of Professor. His research interests focus on laser polarimetry, Mueller- and Jones-matrix imaging, polarization singular optics, correlation optics, and biomedical optical diagnostics of biological tissues and fluids.

Professor Ushenko has authored over 300 peer-reviewed publications indexed in Scopus, including more than 40 papers in Q1 journals, 21 monographs, and over 100 patents. His work has been published in leading international journals. He serves as a reviewer for several high-impact journals in optics and biomedical photonics and has led numerous nationally funded research projects in laser and biomedical optical technologies.

Oleksandr Dubolazov holds an MSc in Laser Optics, 2007. PhD (2010) and DSc (2021) in Optics and Laser Physics. He is a Professor of the Department of Printing, Multimedia, and Optical Technologies at Yuriy Fedkovych Chernivtsi National University, Ukraine. His research focuses on laser polarimetry, polarization interferometry, and digital holography, image processing, computer vision.

His academic output includes over 50 patents, more than 10 monographs, and over 200 Scopus-indexed publications. He has led five funded research-and-development projects as a principal investigator.

He serves as a reviewer for several high-impact journals in optics and biomedical photonics and has led numerous nationally funded research projects in laser and biomedical optical technologies.

Iryna Soltys received her PhD in Optics and Laser Physics from Yuriy Fedkovych Chernivtsi National University, Ukraine, in 2014. She is currently an Associate Professor at the Yurii Fedkovych Chernivtsi National University.

Her research interests include biomedical optics, laser polarimetry, Mueller- and Jones-matrix imaging, digital holographic polarimetry, and polarization-based diagnostics of biological tissues and fluids, with applications in medical and forensic science.

Dr. Soltys has authored over 100 peer-reviewed publications indexed in Scopus, including several monographs published by international publishers, and is the inventor or co-inventor of 24 Ukrainian patents. She has served as principal investigator and senior researcher on multiple nationally funded research projects and acts as a reviewer for international scientific journals.

Preface

The book, "*New Techniques of 3D Jones–Mueller Digital Holographic Mapping for Biomedical and Clinical Applications*," contains original materials on the development of a new multifunctional direction in biophotonics – the development of a set of the latest principles, digital methods, and systems for highly sensitive, express laser digital holographic reconstruction of the layer-by-layer micro- and nanostructure of optically anisotropic polycrystalline components of biological tissues and fluids of human organs for early (preclinical) detection and monitoring of the most common, systemic inflammatory conditions (type II diabetes, sepsis, arthritis, and arthrosis), pathological conditions (such as precancer, cancer) and necrotic conditions (such as heart attack, acute coronary insufficiency, and ischemic heart disease), as well as cause and time of death. The main fundamental idea of all collected and analyzed materials in the field of three-dimensional (3D) Jones–Mueller digital holographic mapping is based on a single principle – the search and physical substantiation of a set of diagnostically significant relationships between objective (statistical, correlation, fractal, singular, wavelet, Fourier, etc.) digital markers characterizing layered Jones–Mueller matrix images and reconstructed optical anisotropy maps and a wide range of inflammatory, pathological, and necrotic changes in the polycrystalline architecture of biological tissue and fluid samples. The primary focus of this research is the polycrystalline, optically anisotropic component at both the micro (architectonics) and nano (molecular) levels of biological samples, including various biological tissues and fluids with diverse morphological structures and physiological states. The experimental technique and its instrumental implementations are based on the synthesis of the principles of Stokes polarimetry, polarization interferometry, digital holography, phase scanning, the formalism of the Mueller and Jones matrices, which provide the possibility of obtaining exhaustively complete information on the polarization, polarization-correlation, and singular manifestations of the set of mechanisms of linear and circular birefringence and dichroism of optical anisotropy of the polycrystalline architectonics of biological samples.

To obtain a set of new digital criteria (markers) for the objective evaluation of experimental data from laser 3D Jones–Mueller digital holographic mapping of biological samples, a complex algorithmic data processing approach was employed, utilizing principles of statistical, correlational, fractal, Fourier, and wavelet analysis. The found digital markers were used within the framework of evidence-based medicine to determine the operational characteristics (sensitivity, specificity, accuracy) of the diagnostic power of laser 3D Jones–Mueller digital holographic mapping of biological samples.

The materials of the monograph contain physical, biomedical, and clinical information that is unique for a wide range of modern optical tomographic studies. Here, for the first time, the following materials are presented within the framework of a unified methodological approach:

- A unified phenomenological theory of the methods of polarization interferometry, Mueller and Jones matrix mapping, digital holography, and phase scanning of polarization-inhomogeneous object speckle fields of optically anisotropic architectonics of biological tissues and liquids of various morphological structures and physiological states.
- Descriptions and characteristics of the primary experimental schemes of laser 3D Jones–Mueller matrix mapping of biological samples.
- Descriptions and characteristics of the main algorithms for analytical processing of experimental data of laser 3D Jones–Mueller matrix mapping systems of biological samples.
- Analysis and discussion of the results of experimental studies using 3D Jones–Mueller matrix mapping methods of optically anisotropic architectonics of biological samples in various fields of medicine – oncology, gynecology, cardiology, forensic medicine.

The distribution and promotion of our book will ensure the formation of a modern unified methodological basis for a theoretical, algorithmic, and engineering platform for the practical implementation of a wide range of highly sensitive, express multiparametric laser biophysical methods of optical mapping in early (preclinical) diagnostics of a wide range of the most common and severe pathologies of the human body.

Yurii Ushenko
Weidong Dou
Oleksandr Ushenko
Oleksandr Dubolazov
Iryna Soltys

Acknowledgments

Authors acknowledge the support from the National Research Foundation of Ukraine, Project 2023.03/0174; the Academy of Finland (grant project 358200).

1

Methods and Systems of Optical Mapping in Biomedical Research

Yurii Ushenko[1,2], Weidong Dou[1], Oleksandr Ushenko[2,3], Oleksandr Dubolazov[2], Iryna Soltys[2], Victoria Kyfyak[2], and Olena Vinnychuk[2]

[1] *Shaoxing University, Shaoxing, China*
[2] *Yurii Fedkovych Chernivtsi National University, Chernivtsi, Ukraine*
[3] *Taizhou Institute of Zhejiang University, Taizhou, China*

1.1 Overview of Modern Optical Mapping Techniques

1.1.1 Diffuse, Fluorescence Molecular, Bioluminescence, Photoacoustic, Laser Coherent, and Laser Polarization-sensitive Systems

1.1.1.1 Diffuse Optical Tomography Methods and Systems

Diffuse optical tomography (DOT) represents a noninvasive biomedical imaging modality that generates tomographic reconstructions of the spatial distribution of optical properties, notably absorption and scattering coefficients, within biological tissues. The methodology involves surface illumination of the tissue and the subsequent detection of the emergent scattered radiation field [1–7]. Utilizing near-infrared (NIR) wavelengths ensures significant depth of penetration, where the probe beam is primarily attenuated by oxy- and deoxyhemoglobin. This allows for the precise quantification of hemoglobin concentrations and oxygen saturation levels, facilitating the diagnosis of breast malignancies and pathologies characterized by angiogenesis [8–12]. Furthermore, DOT is used in neuroimaging to monitor cortical activity [13] and to assess cerebral oxygenation in neonatal intensive care settings [14]. These clinical applications leverage the tracking of regional hemodynamic variations and scattering coefficients as critical biomarkers of physiological tissue status.

While DOT is a promising biomedical imaging modality, it faces significant technical limitations primarily due to the intense multiple scattering and absorption of light within biological media. Unlike X-ray computed tomography (CT), which assumes linear, ballistic photon travel, DOT must account for highly diffusive light propagation, which makes standard back-projection algorithms ineffective for image reconstruction. Consequently, DOT requires complex, nonlinear inverse modeling to accurately map optical properties against a background of scattered radiation. Image reconstruction in DOT is fundamentally comprised of two interconnected procedures. Initially, forward analytical modeling is used to characterize light propagation and generate predicted measurement data derived from defined absorption and scattering coefficients. Subsequently, an optimization framework is utilized to minimize the variance between empirical observations and these theoretical predictions. The mathematical foundation of the forward problem is established via the radiative transfer equation (RTE) or its approximations, notably the photon diffusion equation (PDE), the Spherical harmonics (PN) approximation, and Monte Carlo simulations, which facilitate the analytical derivation of surface measurements from biological specimens. The optimization phase involves the iterative refinement of initial scattering and absorption coefficients to attenuate the residuals between predicted and experimental datasets. This objective is achieved through diverse numerical optimization strategies, most notably the Newton–Raphson method, which minimizes the squared error functional, complemented by regularization techniques and the integration of a priori constraints. Such methodologies demonstrate robust efficacy across a spectrum of practical optimization challenges. Consequently, addressing the inverse problem in NIR reconstruction necessitates a synergistic application of forward modeling and optimization algorithms.

In clinical practice, this integrated framework is indispensable for neutralizing the confounding effects of diffuse background signals inherent in biological media. *Current research aimed at optimizing analytical reconstruction in NIR imaging focuses on addressing critical technical constraints, specifically the suboptimal spatial resolution and limited quantitative accuracy inherent in DOT of heterogeneous biological media.* To enhance spatial resolution, advanced methodologies such as analytical sparsity regularization – frequently derived from compressed sensing paradigms – are being implemented. Furthermore, the integration of a priori morphological data regarding internal tissue structures serves as a robust strategy for refining the precision of DOT image reconstruction. Structural constraints are typically derived from multimodal imaging frameworks, such as CT or magnetic resonance imaging (MRI). Additionally, spectral datasets offer a distinct category of a priori knowledge, facilitating enhanced quantitative precision in characterizing chromophore concentrations within multispectral DOT paradigms. A pivotal shift in the field is the emergence of reconstruction algorithms leveraged by artificial intelligence (AI) and high-performance computing. Recent progress in deep learning has fundamentally transformed both diffuse optical reconstruction [15–19] and the computational modeling of radiative transport [20]. These cutting-edge developments in biomedical optics now permit the high-fidelity visualization of deeply embedded vascular structures [21, 22]. Significant advancements in DOT are expected to soon enable a qualitative breakthrough in characterizing sub-surface chromophore distributions. Such progress underscores the potential of DOT as a cornerstone modality for early cancer detection and the assessment of cortical hemodynamics. *Despite these prospects, the inverse problem in infrared DOT remains complicated by multiple scattering events. These phenomena continue to pose significant hurdles, often limiting achievable spatial resolution and necessitating more robust algorithms to ensure quantitative accuracy.*

1.1.1.2 Fluorescence Molecular Tomography Methods and Systems

Optical molecular imaging methodologies are extensively utilized for disease diagnostics, the assessment of therapeutic efficacy, and pharmaceutical development. These techniques analyze the interaction of optical radiation with biological media to quantify the concentration of fluorescent or bioluminescent probes within target specimens. Among these, fluorescence molecular tomography (FMT) stands as a highly informative diagnostic tool, enabling the three-dimensional reconstruction of fluorescence distributions derived from surface photon flux density measurements [1, 2]. Quantitative analysis through FMT facilitates the identification of oncological biomarkers, typically employing short laser pulses for fluorescence excitation. Furthermore, time-resolved recording of fluorescence fields ensures high temporal resolution, permitting the extraction of critical preclinical data regarding the morphological and biochemical optical properties of the examined subjects. Time-resolved imaging facilitates the direct characterization of the intrinsic fluorescence lifetimes of molecular probes under pulsed laser excitation. While the fluorescence quantum yield – defined as the efficiency ratio of emitted to absorbed photons – enables the visualization of fluorophore topography and the spatiotemporal dynamics of molecular interactions, the fluorescence lifetime offers a more nuanced diagnostic parameter. Modulated by microenvironmental quenching factors such as local pH, ionic concentrations, and oxygen saturation, this lifetime is uniquely sensitive to pathological transitions. Furthermore, it demonstrates remarkable robustness against experimental fluctuations, remaining largely invariant to changes in fluorophore concentration and excitation power [3, 4]. Consequently, fluorescence lifetime imaging (FLIM) offers distinct analytical advantages over intensity-based approaches. Surface fluorescence intensity is inherently confounded by a complex interplay of fluorophore concentration, intrinsic optical properties, and underlying tissue anatomy. While conventional micro-CT provides essential anatomical frameworks for small-animal models, the recent maturation of photon-counting micro-CT (PCMCT) has revolutionized this integration. PCMCT facilitates sophisticated material decomposition and contrast enhancement, thereby refining the accuracy of structural *a priori* assessments. In preclinical oncology, PCMCT enables the precise delineation of diverse tumor morphologies in vivo, providing the necessary spatial constraints to stabilize FMT reconstructions [5]. The modeling of photon migration within biological media typically relies on the radiative transfer equation (RTE) [6] or Monte Carlo simulations [7]. The fidelity of tomographic reconstruction is fundamentally dependent on the precision of the underlying optical parameters; inaccuracies in these values can severely distort image data, compromising both the localization and quantitative characterization of molecular probes [8]. Addressing this challenge, a novel methodology leveraging PCMCT has been developed to derive high-fidelity optical parameters for heterogeneous biological tissues [9]. A salient feature of this methodology is its capacity for multienergy image acquisition, which facilitates the segmentation of various organs and histological components. By integrating the specific optical characteristics of these segmented regions, it becomes possible to analytically derive a comprehensive set of absorption and scattering coefficients [9]. Nevertheless, the high degree of infrared scattering within voluminous biological media significantly complicates the reconstruction of fluorescence yield and lifetime via FMT [10, 11]. To address these challenges, gradient-based optimization frameworks incorporating Tikhonov regularization are frequently employed [12, 13]. However, the efficacy of such analytical approaches is often constrained by the existence of multiple local minima, which are sensitive to initial conditions, model approximations, measurement noise, and specific algorithmic architectures. To identify the global optimum within this ill-posed inverse problem, a range of sophisticated analytical regularization techniques has been proposed [11, 14]. A prominent methodology in this field is Tikhonov regularization, which leverages structural *a priori* constraints on the initial fluorescence distribution; specifically, the L1 norm and its derivatives are utilized to enforce sparsity [15, 16]. While subsequent minimization of total variation (TV) promotes piecewise smoothness [17], nonconvex Lq $(0 < q < 1)$ and logarithmic regularization schemes frequently yield superior tomographic reconstruction quality compared to standard L2-norm approaches [18–20]. Despite these advancements, the framework is fundamentally challenged by the severe ill-posedness inherent in fluorescence inverse problems and by physical measurement constraints. Furthermore, the fidelity of these reconstructions remains highly contingent upon the empirical selection of regularization parameters and iteration counts, often introducing significant uncertainty and interpretive ambiguity into the resulting datasets. The preclinical PCMCT workflow comprises a sequence of rigorous analytical stages. Initially, *in vivo* scanning of the specimen (e.g., a murine model) is conducted to acquire a series of projection datasets across multiple energy thresholds. Subsequently, these projections are utilized to reconstruct multienergy PCMCT images, followed by sophisticated image segmentation to delineate internal anatomical features [5]. This process yields an extensive repository of data regarding both anatomical topography and the biochemical composition of tissues. Such granular information facilitates the derivation of optical absorption and scattering coefficients with significantly higher precision than that achievable via conventional current-integrated micro-CT [9]. Furthermore, the application of contrast-enhanced PCMCT markedly increases diagnostic sensitivity for tumor detection [21]. The administration of exogenous contrast agents enhances the visibility of organs and neoplastic tissues [22]. Within this framework, fluorescent probes are delivered to the subject to establish optical contrast; following a specific incubation period, these biomarkers accumulate within target tissues at concentrations sufficient for oncological assessment. Consequently, the signal-to-noise ratio is optimized as the concentration of the fluorescent agent in the localized pathology significantly exceeds that of the surrounding background parenchyma. These accumulated biomarkers are subsequently excited using pulsed laser radiation for time-resolved analysis [23]. Incident laser pulses propagate through the biological volume to reach the target site, where they induce the excitation of embedded fluorophores. This ensemble of excited molecules acts as a secondary emission source, subsequently emitting fluorescence photons at a redshifted wavelength. These emitted photons undergo further scattering and absorption as they migrate through the heterogeneous tissue media. A fraction of this diffuse radiation exits the boundary and is captured by time-resolved optical detectors equipped with appropriate spectral filters. While these datasets provide the necessary input for fluorescence image reconstruction, resolving the inverse source problem remains a formidable challenge. The severe ill-posedness, driven by the high degree of multiple scattering from morphological inhomogeneities, often leads to nonunique solutions. Consequently, the presence of diffuse scattering noise and signal ambiguity frequently results in uncertain reconstructions, hindering the clinical and preclinical translation of this imaging modality.

1.1.1.3 Bioluminescence Imaging Methods and Systems

Bioluminescence imaging (BLI) methodologies utilize a diverse range of light-emitting biological systems [1]. The phenomenon of bioluminescence involves the enzymatic oxidation of a low-molecular-weight substrate, luciferin, catalyzed by the enzyme luciferase within living organisms [1].

Currently, technological advancements in BLI are progressing across three interconnected domains [2–19]:

1. The identification of novel luciferases and the chemical synthesis of innovative luciferin analogs;
2. The application of protein engineering techniques to develop genetically encoded probes incorporating luciferases;
3. The integration of these components and detection systems into comprehensive imaging platforms for monitoring molecular events *in vivo*.

The phenomenon of bioluminescence is extensively utilized across diverse bioassays and molecular imaging modalities. As detailed in the comprehensive review [20], the primary advantages of this optical imaging technique stem from a unique combination of the following factors [21–51]:

- Exceptional Sensitivity: BLI facilitates the detection of molecular events even at trace concentrations of luciferase and its cognate substrate, luciferin.

- Minimal Background Interference and Superior Signal-to-noise Ratio (SNR): Unlike fluorescence-based methods, BLI systems exhibit negligible intrinsic background luminescence in mammalian tissues. This absence of autoluminescence ensures a high SNR by effectively eliminating presignal noise.
- High Quantum Yield (QY): The luciferase–luciferin reaction typically demonstrates a superior QY compared to alternative chemiluminescent processes, resulting in high photon emission per chemical reaction.
- Extended Dynamic Range: BLI platforms generally offer a broader dynamic range of signal detection than other optical methods, such as fluorescence imaging (FLI).
- Independence from External Illumination: By eliminating the need for an external excitation source, BLI avoids phototoxic effects on biological specimens and simplifies the instrumentation by removing the requirement for complex optical filtering systems.
- Analytical Versatility: BLI serves as a highly adaptable optical readout across various bioassay formats and model organisms, ranging from prokaryotes to complex vertebrates.
- Rapid Temporal Response: BLI allows for near-instantaneous signal acquisition without extensive sample pretreatment. While fluorescent proteins require a maturation period for fluorophore oxidation, luciferase reporters become functional immediately upon expression.
- Cost-effectiveness: BLI represents a significantly more economical imaging modality compared to high-end diagnostic techniques such as PET, SPECT, or MRI.
- Deep Tissue Penetration: The technique is suitable for noninvasive imaging of internal structures at depths ranging from several millimeters to centimeters.
- Biocompatibility and Safety: Derived from natural biological systems, BLI components are inherently biocompatible and nontoxic, offering a nonradioactive alternative for longitudinal studies *in vivo*.

The limitations and technical constraints of BLI can be categorized as follows:

- Inferior Optical Intensity: Compared to FLI, BLI typically exhibits lower photon emission flux, which can limit the absolute signal strength.
- Obligatory Substrate and Cofactor Requirements: The generation of a bioluminescent signal is strictly dependent on the availability of a specific substrate (luciferin) and, in many cases, essential cofactors such as Mg^{2+} and ATP.
- Substrate-dependent Optical Properties: The spectral profile and emission intensity are inherently dictated by the chemical structure and properties of the substrate, which serves as the primary luminophore.
- Sensitivity to Microenvironmental Variables: As an enzyme-mediated process, the kinetics of the bioluminescent reaction are susceptible to fluctuations in pH, ionic strength, temperature, and buffer composition. Such dependencies can potentially compromise the precision of quantitative longitudinal measurements.
- Limited Spatial Resolution: BLI generally offers lower spatial resolution relative to FLI. Whereas fluorescence techniques provide high-contrast, sharp cellular images, BLI signals often appear more diffuse and lack fine structural detail.

A comprehensive analysis of the mechanisms, methodologies, and practical applications of bioluminescence is available in the literature cited in the following sections.

1.1.1.4 Methods and Systems of Diffuse Photoacoustic Tomography

The investigation of inhomogeneities within continuous media holds significant implications across scientific, industrial, ecological, and medical sectors, necessitating the systematic categorization of empirical data [1–3]. Visualization systems operating in nonvisible spectral regions enable the integration of human cognitive visual processing into the interpretation of complex measurement datasets. By utilizing integral information channels, researchers can evaluate radiative power, spatial dimensions, the relative positioning of objects, and temporal dynamics. Historically, the detection and identification of such subjects have been facilitated by integral and differential infrared (IR) technologies [4].

This study aims to delineate contemporary advancements in photoacoustic imaging (PAI) toward clinical translation, identify persisting technical and regulatory barriers, and explore prospective trajectories for establishing PAI as a standard clinically relevant modality.

Photoacoustics represents an emergent biomedical imaging paradigm that characterizes optical absorbers within biological tissues via acoustic detection – essentially converting optical excitation into acoustic output. This modality offers substantial clinical potential due to its high spatial resolution, significant penetration depth, and compatibility with a diverse array of endogenous and exogenous contrast agents, all while remaining inherently nonionizing.

In recent years, substantial advancements in instrumentation and methodological frameworks have expanded the clinical utility of imaging technologies. These developments have facilitated innovative applications in functional neuroimaging, oncology – specifically for breast cancer screening, sentinel lymph node metastasis detection, and urological malignancies – as well as in dermatological diagnostics, surgical navigation, and reproductive therapy [1].

Concurrently, the rapid evolution of molecular imaging has led to its integration into standardized clinical practice, enhancing the characterization and longitudinal monitoring of various pathologies. While magnetic resonance imaging (MRI) and radionuclide-based modalities have driven the discovery of novel contrast agents, their broader adoption is often constrained by high operational costs and logistical requirements, such as magnetic shielding and radiation safety protocols [4–6]. Conversely, the increasing portability of ultrasound systems, coupled with the emergence of molecular techniques like elastography and microbubble-enhanced imaging, has established ultrasonography as an essential tool for both anatomical and molecular assessment [7].

PAI, also referred to as optoacoustic imaging, represents a burgeoning technology capable of augmenting ultrasound with high-contrast optical data. This modality offers a portable, cost-effective solution for the localized imaging of vasculature and other optically active agents, functioning either as a supplementary tool or a standalone diagnostic platform [8–16].

The principal advantages of PAI encompass high potential for excellent spatial and temporal resolution, clinically relevant imaging depths, the capacity to visualize both endogenous and exogenous chromophores, and the inherent safety of nonionizing radiation. Key endogenous chromophores include water (in both free and bound states), oxyhemoglobin, deoxyhemoglobin, melanin, and various lipids. Exogenous agents commonly comprise low-molecular-weight dyes like indocyanine green and methylene blue, nanoparticles, and reporter gene agents. Unlike conventional microbubble-based ultrasound contrast agents, PAI facilitates the imaging of small molecules that readily extravasate, target cell membrane components, or even penetrate intracellular compartments, thereby providing valuable molecular data for clinical assessment.

The underlying photoacoustic effect, characterized by the conversion of absorbed optical energy into acoustic pressure waves, was initially described by Alexander Graham Bell over a century ago. However, significant technological advancement remained stagnant until the advent of the laser, which dramatically enhanced signal generation efficiency. Subsequently, photoacoustics gained traction, initially in gas spectroscopy and later within the realm of biomedical applications.

Over the past two decades, PAI has rapidly transitioned from a theoretical concept to a sophisticated tomographic modality utilized in both small animal research and clinical diagnostic devices. The effect relies on the generation of transient pressure waves following the absorption of pulsed or modulated light. These induced acoustic waves experience significantly less scattering and absorption in biological tissue compared to visible or NIR light. The detection of these pressure transients at the tissue surface via acoustic sensors enables the precise reconstruction of the original optical absorption sites.

Owing to the inherent scalability of the PAI principle, the technology can be optimized across a broad range of depths and resolutions. Acoustic-resolution photoacoustic tomography can achieve submillimeter resolution at depths extending up to several centimeters, while optical-resolution photoacoustic microscopy offers submicron resolution, albeit limited to imaging depths of a few hundred microns. The typical depth-to-resolution ratio remains consistently around 200, and the field of view generally scales proportionally with the imaging depth and acquisition duration.

In recent years, advancements in light source technologies have facilitated the development of more portable, tunable, and cost-effective compact systems. Furthermore, innovative reconstruction algorithms now enable real-time mapping of initial pressure distributions with high fidelity and precision, even under suboptimal conditions, such as restricted fields of view, sparse detector coverage, or limited sensor bandwidth. By employing multiwavelength excitation, it is theoretically possible to quantify the concentrations of diverse chromophores simultaneously, thereby providing molecularly specific contrast [17].

Despite these advancements, quantitative accuracy and spectral resolution remain persistent challenges in PAI, primarily due to the phenomenon of spectral coloring [18]. A critical factor governing the clinical utility of PAI is its penetration depth. PAI is capable of resolving structures at depths of several centimeters, significantly outperforming purely optical modalities, such as confocal microscopy, which rely on the rejection of out-of-focus light [15].

Moreover, at comparable imaging depths, PAI offers superior spatial resolution relative to techniques based on diffuse photon migration, such as DOT. However, achieving reliable and clinically actionable *in vivo* imaging at depths exceeding 2–3 cm remains a formidable challenge [19]. This limitation is exacerbated by the fact that, unlike B-mode ultrasonography, PAI is fundamentally designed to be a quantitative or semi-quantitative modality. Consequently, while image intensity is expected to correlate positively with local optical absorption, this linear relationship becomes increasingly tenuous at greater depths due to the nonlinear attenuation of light.

Firstly, the optical fluence undergoes significant attenuation and its spatial distribution increases in complexity by at least an order of magnitude for every centimeter of tissue depth. In the absence of a robust fluence compensation scheme, a reconstructed photoacoustic image reflects the product of the local optical flux and the absorption coefficient, rather than the intrinsic optical absorption alone. Consequently, regions may appear hypo-intense due to insufficient light penetration rather than low absorption, and vice versa [20]. While optical modeling can partially mitigate these effects by predicting light attenuation [21], the accuracy of such compensation diminishes substantially as imaging depth increases. This challenge is further compounded in multispectral applications, where light fluence is nonuniformly perturbed across different wavelengths.

Secondly, many clinical implementations of PAI are constrained by a limited detection aperture, as it is often impossible to surround the volume of interest with sensors. Reconstructing the initial pressure distribution from such incomplete datasets leads to artifacts and errors that increase exponentially as the viewing angle narrows – a situation that worsens with increasing depth relative to the aperture size [22]. Furthermore, even under full-view conditions, the SNR deteriorates significantly at greater depths due to both optical and acoustic attenuation and dispersion. Consequently, the fidelity of the reconstruction regarding local fluence and absorption is compromised as imaging depth increases [23].

Thirdly, the acoustic properties of biological media – specifically the speed of sound and the Grüneisen parameter, which dictates the efficiency of thermoelastic conversion – are heterogeneous across various tissue types, exhibiting variations of several tens of percent. Such variability introduces further complexities in ensuring the quantitative accuracy of the imaging modality.

Specifically, the acoustic properties of tissue are temperature-dependent, which introduces variability as thermal profiles shift with increasing depth [24]. This temperature sensitivity may account for the notable discrepancy between the penetration depths achieved in phantoms (exceeding 6 cm) and those observed in *in vivo* clinical settings (typically restricted to 2–3 cm).

Another substantial barrier to the clinical adoption of PAI pertains to image contrast. Biological tissues comprise a complex milieu of endogenous chromophores, such as water, various hemoglobin states, and pigments, possessing overlapping absorption spectra [25]. While multispectral PAI can theoretically differentiate these components to monitor pathological shifts (e.g., lipid deposition in atherosclerosis or neoangiogenesis in malignancies), the diagnostic utility of visualizing vascular architecture and oxygenation alone remains limited for many conditions [26]. Such vascular alterations are often insufficiently pronounced for early-stage detection or lack the specificity required to differentiate between benign and malignant lesions. Furthermore, the inherent difficulties in achieving precise quantification at depth mean that the diagnostic accuracy of PAI-derived parameters remains a subject of ongoing investigation [27].

Technical constraints also include the requirement for "wet" acoustic coupling between the transducer and the tissue. Although alternative "dry" or noncontact coupling methods have been proposed [28], they typically suffer from reduced sensitivity and increased system complexity, hindering their practical implementation.

Many of these limitations could be mitigated through the use of targeted exogenous molecular agents [29]. These contrast agents offer superior absorption and distinct spectral signatures, facilitating their separation from background signals and providing critical data on biomarker expression or enzymatic activity [30]. However, despite extensive preclinical validation, such agents have yet to receive FDA approval. Currently, the clinical application of PAI is further hampered by depth constraints, quantification uncertainties, and a lack of standardized clinical systems. The future integration of PAI into routine medical practice likely depends on the development of robust hardware and the demonstration of clinical benefit that outweighs the current operational complexities [32–41].

1.1.1.5 Methods and Systems for Polarization Mapping and Biomedical Imaging

This section provides a concise overview of methodologies and systems for laser polarization mapping and biomedical imaging. A more exhaustive treatment of this subject is available in the authoritative review [18], which offers the most comprehensive synthesis of the field and its extensive literature to date [1–286].

The integration of polarimetric modalities with established optical techniques facilitates the acquisition of unique vector-based data, enhancing existing technologies such as microscopy and endoscopy [16, 24, 32, 33, 92, 180]. A significant methodological advantage of biomedical polarimetry is its capacity for label-free imaging, precluding the need for exogenous dyes [16, 22, 24]. Furthermore, polarimetry serves as a powerful tool for characterizing the vector properties of fluorescent markers, as the dipole orientation of a fluorophore is intrinsically encoded within the polarization state of the emitted radiation [154, 155]. Research has demonstrated that the polarization azimuth (PA) and the degree of linear polarization (SOP) are critical parameters for advancing biomedical super-resolution microscopy [153, 181, 182].

To date, a diverse array of biomedical polarimetry methods has been developed and successfully implemented, including:

- Polarization wide-field and confocal microscopy [16, 24, 183, 197];
- Spatial-frequency domain imaging in polarized light [184];
- Polarization-based endoscopy [185–190];
- Spectral polarimetry of scattered light and polarization fluorescence spectroscopy [18, 82, 191–196];
- Polarization Raman spectroscopy [198, 199];
- Polarization-sensitive optical coherence tomography (PS-OCT) [200–218];
- Polarization-resolved nonlinear microscopy and polarization speckle imaging [213, 227].

While certain polarimetric modalities have been extended to three-dimensional (3D) imaging through signal integration or volumetric segmentation, the majority of contemporary methods are primarily focused on two-dimensional (2D) analysis. The theoretical framework for modeling the interaction between polarized photons and biological media is predominantly based on Monte Carlo (MC) simulations. Within these computational models, the microstructural architecture of biological tissues is typically represented by fundamental geometric elements, such as spherical or cylindrical scatterers, birefringent domains, and multilayered geometries. MC simulations have proven highly effective in reproducing the essential polarimetric characteristics observed in biomedical specimens.

As previously established, biomedical polarimetry and imaging are categorized into label-free and label-based methodologies:

- Label-based Polarimetry: This approach is widely utilized in biomedical microscopy to extract vector-based information from dipole emitters, which is encoded in the polarization state of the detected light. Polarimetric detection of dipole orientation and fluorescence intensity is a cornerstone of fluorescence polarization microscopy (FPM). This technique has been instrumental in characterizing cytoskeletal components, such as actin, myosin, kinesin, microtubules, and septin, and in investigating dynamic molecular processes, including ATP/ADP binding kinetics and the heterogeneous behavior of subcellular lipid membranes.

The second avenue of research, label-free biomedical polarimetry, has demonstrated significant clinical utility in the detection and characterization of various malignancies [22–24]. Extensive studies have confirmed its efficacy in identifying pathologies of the skin [242], cervix [243–246], colon [166, 247–250], liver [163, 251], breast, and gastrointestinal tract [93–95, 252]. Bioinformatic processing of polarimetric datasets permits the quantitative evaluation of fibrosis across different oncological stages [94, 163]. Beyond assessing the degree of fibrotic progression, polarization data facilitates the pathological diagnosis of morphological distributions within fibrotic regions [164, 165, 171]. Furthermore, recent applications have extended polarimetric analysis to the detection of nonmalignant conditions, including Alzheimer's disease and bladder outlet obstruction [24, 253, 254].

Polarization-based techniques enhance the contrast of superficial tissue imaging by effectively filtering out multiply scattered photons originating from deeper structural layers [20–24]. Given that more than 85% of cancers develop within the superficial epithelium, these methods hold substantial promise for early cancer screening and diagnostic identification, particularly within the context of minimally invasive surgery (MIS) [24, 255].

A salient feature of polarimetric measurement is the capacity to characterize the optical anisotropy inherent in biological tissues of varying physiological and morphological states. For instance, structural anisotropy and linear birefringence are prominent in collagen-rich tissues, such as tendons, skin, and the bladder [156, 256–259], as well as in skeletal and myocardial muscle fibers and elastin [20–23, 164, 260–262]. The magnitude of this optical anisotropy is intrinsically linked to the orientation of the fast axes of linear birefringence [16, 165, 171]. Consequently, the extraction of such anisotropic vector data via polarimetry provides a robust framework for the differentiation and identification of diverse fibrous architectures within complex biological media [164, 165, 167]. To date, however, the majority of *ex vivo* investigations have focused primarily on fundamental research applications [24, 152, 171].

Typical polarimetric configurations in backscattering mode encompass a variety of modalities designed for *in vivo* clinical diagnostics, including polarization endoscopy [24], reflectance Mueller matrix (MM) microscopy [16, 266], MM colposcopy [180, 267], wide-field handheld polarimetry [16], and PS-OCT [268]. As a promising label-free diagnostic instrument, polarization endoscopic imaging has been utilized in murine abdominal studies, effectively distinguishing the small intestine, stomach, liver, and adipose tissues based on their unique polarimetric signatures [92].

Recent advancements have further introduced several specialized MM endoscopic systems [188–190, 256], some of which have been expanded into the multispectral domain using discrete fixed wavelengths [92]. Furthermore, volumetric analysis of complex biological structures, such as human pulmonary malignancies [159] and cutaneous tissues [269], underscores the significant potential of these techniques for future clinical diagnostic integration [16, 22–24].

Notwithstanding the substantial progress achieved in polarization mapping and biomedical imaging, several persistent limitations and technical challenges must be addressed to facilitate further maturation and clinical translation of the field.

The following are key limitations hindering the advancement and clinical implementation of laser polarimetry [287]:

1. Lack of a Unified Phenomenological Model: There is currently no standardized analytical model capable of comprehensively describing the transformation of laser radiation polarization parameters as light propagates through biological tissues characterized by diverse morphological structures and physiological states.
2. Absence of Standardized Analytical Frameworks: A unified system of analytical relationships is needed to link the measured distributions of scattered radiation polarization parameters directly to the specific optical characteristics, such as linear and circular birefringence, and dichroism, of the polycrystalline architecture found in biological tissues.
3. Paucity of Data on Biological Fluid Films: There is a notable absence of systematic data regarding the applicability and capabilities of laser polarimetry for analyzing the optical parameters and structures of human biological fluid films, an extremely critical class of biological objects.
4. Influence of Multiple Scattering Artifacts: Experimental results are frequently compromised by the significant informational distortions introduced by multiply scattered, depolarized background radiation, which obscures the primary signal.
5. Deficiency in Quantitative Evaluation Systems: A unified system for the multifunctional, quantitative evaluation of the statistical, correlational, scale-selective, and fractal properties within coordinate distributions or maps of polarization parameters is currently lacking.

1.1.1.6 Optical Coherence Tomography Methods and Systems

Optical coherence tomography (OCT) has established itself as a sophisticated modality for noninvasive optical imaging, offering substantial potential for biological investigation and *in situ* medical biopsy, particularly within 3D *in vivo* contexts. Typically utilizing broadband NIR laser sources, OCT achieves significantly greater penetration depths, higher resolution, and superior SNR compared to conventional microscopy. This spectral advantage minimizes blurring associated with multiple scattering, thereby facilitating the high-fidelity visualization of internal morphological architectures beyond superficial layers. The integration of advanced probing sources, such as multiplexed superluminescent diodes and ultra-broadband supercontinuum lasers, has further expanded clinical utility by enabling submicron resolution and high-speed data acquisition.

Initially conceptualized by David Huang in 1991, OCT represents a cost-effective, high-resolution imaging paradigm that reconstructs transverse cross-sections of biological tissues *in situ* by measuring backscattered light signals [1]. Over the subsequent three decades, the transition to ultra-high-resolution OCT, powered by broadband femtosecond laser sources, has realized axial and transverse resolutions of 1 and 3 μm, respectively, facilitating subcellular imaging [2]. Furthermore, the implementation of Fourier-domain OCT (FD-OCT) has catalyzed a qualitative advancement in acquisition speeds, solidifying the role of this 3D biomedical imaging technology in contemporary clinical and research settings.

Owing to advancements in acquisition speed and image fidelity, OCT has been extensively adopted across engineering and biomedical disciplines. In forensic science, for instance, OCT facilitates the reconstruction and analysis of latent evidence, including subsurface fingerprints [3], layered handwriting [4], and multilayered automotive coatings [5–7]. Furthermore, the increasing accessibility of high-power mid-infrared sources has expanded the utility of mid-infrared OCT in nondestructive testing, ceramic manufacturing, and the diagnostic conservation of historical artwork [8, 9]. Despite these diverse industrial applications, the primary impact of OCT remains in the biomedical sector, where it has become an indispensable standard for ophthalmic examinations.

The evolution of these imaging technologies has significantly enhanced the visualization and diagnosis of various pathologies [10–13]. The capacity to resolve finer morphological details within lesions supports earlier clinical intervention and improved therapeutic outcomes. Notably, OCT offers distinct advantages in spatial resolution and cost-effectiveness compared to established modalities such as MRI and CT.

Beyond structural visualization, various functional extensions of OCT have been developed to assess physiological parameters [14–21]. Doppler OCT, for example, enables the visualization of microvasculature and the estimation of hemodynamic velocity by differentiating between signals originating from flowing erythrocytes and stationary anatomical structures [22–25]. Additionally, specialized techniques such as polarization-sensitive OCT (PS-OCT) and phase-sensitive OCT provide enhanced structural contrast. PS-OCT, in particular, leverages the polarization state and tissue birefringence to resolve intricate structural data with high spatial resolution [26–30]. Consequently, modern commercial OCT systems, characterized by micron-scale resolution and millimeter-scale penetration depth, have become integral to fields such as ophthalmology, endoscopy, cardiology, and otology [14, 31–33].

A primary challenge associated with the aforementioned OCT methodologies is the prevalence of significant speckle noise within the laser-scanned cross-sectional images of the internal architecture of diffusely scattering biological tissues.

Current OCT techniques predominantly rely on diverse modeling approaches to characterize the structural and optical properties of reconstructed tomograms, including intensity distributions, polarization maps, and Mueller or Jones matrix representations. However, the quantitative analysis of these datasets remains neither systematically organized nor standardized across the field [34–58].

Furthermore, there is a notable deficiency in OCT-based data regarding the polycrystalline structure of dehydrated biological fluid films. These specimens represent a critical diagnostic resource that is arguably as informative as tissue biopsies while being significantly more accessible for clinical evaluation.

1.1.2 Unsolved Problems and Proposed Solutions via 3D Jones–Mueller Digital Holographic Mapping

This section provides a systematic overview of the critical unresolved challenges (italicized in Section 1.1) inherent to the current landscape of optical biomedical imaging modalities:

1. Diffuse Optical Tomography
   - Propagation Constraints: Significant technical barriers persist regarding multiple scattering and photon absorption during light propagation through biological media.
   - Reconstruction Inefficiency: Unlike X-ray CT, the application of back-projection for infrared image reconstruction is ineffective due to the high background of scattered radiation.
   - Analytical Limitations: DOT exhibits suboptimal spatial resolution and a lack of precision in the quantitative evaluation of optically inhomogeneous diffuse tissues.
   - Parametric Deficiencies: These systems are currently unable to extract data regarding the optical anisotropy of biological structures.
2. Fluorescence Molecular Tomography
   - Computational Complexity: Extensive multiple scattering of infrared radiation within biological volumes complicates the accurate reconstruction of fluorescence yield.
   - Algorithmic Instability: Analytical optimization is constrained by the existence of multiple local solutions, which are highly sensitive to initial conditions, model approximations, stochastic noise, and algorithmic architecture.
   - Methodological Ambiguity: The modality is characterized by ill-defined measurement limitations and a high degree of uncertainty in the derived informational content.
   - Regularization Sensitivity: Reconstruction fidelity is heavily dependent on the subjective selection of regularization parameters and iteration counts.
   - Structural Information Gap: FMT lacks the capability to characterize the structural anisotropy and optical activity of biological specimens.
3. Bioluminescence Imaging
   - Signal Flux Constraints: BLI typically exhibits lower optical intensity relative to fluorescence-based modalities.
   - Biochemical Dependencies: The technique requires specific exogenous substrates, with the resulting optical signatures (spectral profile and intensity) being strictly dictated by the chemical properties of these substrates and the localized reaction conditions.
   - Resolution and Fidelity Issues: BLI is characterized by low spatial resolution, often yielding diffuse and low-contrast cellular images.
   - Vectorial Limitations: Similar to the aforementioned modalities, BLI cannot currently retrieve information pertaining to structural anisotropy or the optical activity of the target tissue.
4. Diffuse Photoacoustic Tomography/PAI
   - Spectral Interference: Quantitative accuracy and spectral unmixing remain compromised by the persistent challenge of "spectral coloring," which distorts the spectral signatures of deep-seated chromophores.
   - Clinical Depth Limitations: There is significant ambiguity regarding the clinical efficacy of this modality at imaging depths exceeding 2–3 cm *in vivo*.

  - Fluence Dependency: Photoacoustic signals do not directly represent intrinsic optical absorption; rather, they reflect the product of local optical fluence and the absorption coefficient, leading to potential misinterpretation of hypo-intense regions.
  - Detection Aperture Constraints: Clinical implementation is frequently hindered by "limited-view" architectures, where the inability to surround the target volume with sensors results in reconstruction artifacts.
  - Heterogeneity of Acoustic Parameters: Essential parameters, such as the speed of sound and the Grüneisen coefficient, are nonuniform across different tissue types and exhibit significant fluctuations (up to several tens of percent).
  - Polarimetric Information Gap: The modality is currently unable to preserve or retrieve structural polarization data.
5. Laser Polarization Mapping and Biomedical Imaging
  - Theoretical Deficiencies: There is an absence of a unified phenomenological model to analytically describe the transformation of polarization parameters as laser radiation propagates through diverse morphological and physiological tissue states.
  - Analytical Disconnection: A standardized system of relationships between measured polarization distributions and the specific optical properties (linear/circular birefringence and dichroism) of polycrystalline biological architectures has yet to be established.
  - Paucity of Biofluid Data: Systematic empirical data regarding the polarimetric characterization of human biological fluid films-a critically informative class of specimens-is virtually nonexistent.
6. Methods and Systems for Polarization Mapping and Biomedical Imaging
  - Multiple Scattering Interference: The integrity of experimental results is frequently undermined by the distorting effects of a multiply scattered, depolarized background.
  - Quantitative Standardization: There is a lack of a unified framework for the multifunctional quantitative assessment of the statistical, correlational, scale-selective, and fractal dimensions of polarization coordinate maps.
7. OCT Methods and Systems
  - Speckle Noise Interference: A primary technical challenge is the high level of speckle background noise present in layered laser images of the internal architecture of diffusely scattering biological tissues.
  - Modeling Variability: Current approaches rely on diverse and often disparate model considerations for the formation of optical characteristics in reconstructed images, including intensity maps, polarization maps, and Mueller/Jones matrix representations.
  - Lack of Standardization: The quantitative analysis of OCT data is currently neither systematized nor standardized across research and clinical practice.
  - Data Deficit in Biofluids: There is a notable absence of systematic OCT data pertaining to the polycrystalline structure of dehydrated biological fluid films.

Comparative Analysis and Future Directions

A comparative analysis of existing unresolved issues across the spectrum of biomedical optical imaging modalities highlights promising avenues for qualitative enhancement and future development:

- Development of Integrated Systems: The creation of a new generation of "interdisciplinary" hardware and software information and diagnostic systems is proposed, synthesizing principles from polarimetry, interferometry, and tomography.
- Background Elimination and Sensitivity Enhancement: New diagnostic systems aim to mitigate the influence of depolarized background noise through polarization-interference recording of biospeckles. This is followed by digital holographic reconstruction and phase scanning to isolate the single-scattered component within the object field.
- Standardized Layer-by-layer Reconstruction: The objective is to develop unified principles for the layer-by-layer phase reconstruction of maps detailing the linear/circular birefringence and dichroism of biological tissues and fluid samples, utilizing the rigorous Mueller–Jones matrix formalism.
- Objective Multiparametric Analysis: Future work will focus on the objective, multifunctional quantitative evaluation of medical Mueller–Jones matrix images and optical anisotropy maps, integrating statistical, correlational, fractal, singular, and other advanced analytical approaches.
- Biomarker Development: The final direction involves the development and clinical validation of a complex suite of digital markers for the early diagnosis and differentiation of pathological and necrotic changes in the polycrystalline architecture of biological specimens.

References

Section 1.1.1.1

1 Hebden JC, Arridge SR, Delpy DT. Optical imaging in medicine: I. Experimental techniques. Phys Med Biol. 1997;42:825–840.
2 Arridge SR, Hebden JC. Optical imaging in medicine: II. Modelling and reconstruction. Phys Med Biol. 1997;42:841–853.
3 Boas DA, Brooks DH, Miller EL, DiMarzio CA, Kilmer M, Gaudette RJ, et al. Imaging the body with diffuse optical tomography. IEEE Sig Process Mag. 2001;18:57–75.
4 Hielscher AH, Bluestone AY, Abdoulaev GS, Klose AD, Lasker J, Stewart M, et al. Near-infrared diffuse optical tomography. Dis Markers. 2002;18:313–337.
5 Gibson AP, Hebden JC, Arridge SR. Recent advances in diffuse optical imaging. Phys Med Biol. 2005;50:R1–R43.
6 Yamada Y, Okawa S. Diffuse optical tomography: present status and its future. Opt Rev. 2014;21:185–205.
7 Hoshi Y, Yamada Y. Overview of diffuse optical tomography and its clinical applications. J Biomed Opt. 2016;21:091312.
8 Hawrysz DJ, Sevick-Muraca EM. Developments toward diagnostic breast cancer imaging using near-infrared optical measurements and fluorescent contrast agents. Neoplasia. 2000;2:388–417.
9 Tromberg BJ, Pogue BW, Paulsen KD, Yodh AG, Boas DA, Cerussi AE. Assessing the future of diffuse optical imaging technologies for breast cancer management. Med Phys. 2008;35:2443–2451.
10 Leff DR, Warren OJ, Enfield LC, Gibson A, Athanasiou T, Patten DK, et al. Diffuse optical imaging of the healthy and diseased breast: a systematic review. Breast Cancer Res Treat. 2008;108:9–22.
11 Taroni P. Diffuse optical imaging and spectroscopy of the breast: a brief outline of history and perspectives. Photochem Photobiol Sci. 2012;11:241–250.
12 Grosenick D, Rinneberg H, Cubeddu R, Taroni P. Review of optical breast imaging and spectroscopy. J Biomed Opt. 2016;21:091311.
13 Boas DA, Dale AM, Franceschini MA. Diffuse optical imaging of brain activation: approaches to optimizing image sensitivity, resolution, and accuracy. NeuroImage. 2004;23:S275–S288.
14 Lee CW, Cooper RJ, Austin T. Diffuse optical tomography to investigate the newborn brain. Pediatr Res. 2017;82:376–386.
15 Yoo J, Sabir S, Heo D, Kim KH, Wahab A, Choi Y, et al. Deep learning diffuse optical tomography. IEEE Trans Med Imaging. 2020;39:877–887.
16 Smith JT, Ochoa M, Faulkner D, Haskins G, Intes X. Deep learning in macroscopic diffuse optical imaging. J Biomed Opt. 2022;27:020901.
17 Balasubramaniam GM, Wiesel B, Biton N, Kumar R, Kupferman J, Arnon S. Tutorial on the use of deep learning in diffuse optical tomography. Electronics. 2022;11:305.
18 Hauptmann A, Cox BT. Deep learning in photoacoustic tomography: current approaches and future directions. J Biomed Opt. 2020;25:112903.
19 Deng H, Qiao H, Dai Q, Ma C. Deep learning in photoacoustic imaging: a review. J Biomed Opt. 2021;25:040901.
20 Yajima H, Abe M, Umemura M, Takamizu Y, Hoshi Y. TRINITY: a three-dimensional time-dependent radiative transfer code for in-vivo near-infrared imaging. J Quant Spectrosc Radiat Transf. 2022;277:107948.
21 Ntziachristos V. Going deeper than microscopy: the optical imaging frontier in biology. Nat Methods. 2010;7:603–614.
22 Wang LV, Fu S. Photoacoustic tomography: in vivo imaging from organelles to organs. Science. 2012;335:1458–1462.

Section 1.1.1.2

1 Rudin M, Rausch M, Stoeckli M. Molecular imaging in drug discovery and development: potential and limitations of nonnuclear methods. Mol Imaging Biol. 2005;7(1):5–13.
2 Serkova NJ, Glunde K, Haney CR, Farhoud M, De Lille A, Redente EF, et al. Preclinical applications of multi-platform imaging in animal models of cancer. Cancer Res. 2021;81(5):1189–1200.
3 Bloch S, Lesage F, McIntosh L, Gandjbakhche A, Liang K, Achilefu S. Whole-body fluorescence lifetime imaging of a tumor-targeted near-infrared molecular probe in mice. J Biomed Opt. 2005;10(5):054003.
4 Kumar AT, Raymond SB, Boverman G, Boas DA, Bacskai BJ. Time resolved fluorescence tomography of turbid media based on lifetime contrast. Opt Express. 2006;14(25):12255–12270.
5 Allphin AJ, Mowery YM, Lafata KJ, Clark DP, Bassil AM, Castillo R, et al. Photon counting CT and radiomic analysis enables differentiation of tumors based on lymphocyte burden. Tomography. 2022;8(2):740–753.

6 Liemert A, Kienle A. Infinite space Green's function of the time-dependent radiative transfer equation. Biomed Opt Express. 2012;3(3):543–551.
7 Shen H, Wang G. A study on tetrahedron-based inhomogeneous Monte Carlo optical simulation. Biomed Opt Express. 2010;2(1):44–57.
8 Alexandrakis G, Rannou FR, Chatziioannou AF. Effect of optical property estimation accuracy on tomographic bioluminescence imaging: simulation of a combined optical-PET (OPET) system. Phys Med Biol. 2006;51(8):2045–2053.
9 Cong W, Li M, Guo X, Wang G. Estimating optical parameters of biological tissues with photoncounting micro-CT. J Opt Soc Am A. 2022;39(5):841–846.
10 Cong A, Wang G. A finite-element-based reconstruction method for 3D fluorescence tomography. Opt Express. 2005;13(24):9847–9857.
11 Ntziachristos V. Fluorescence molecular imaging. Annu Rev Biomed Eng. 2006;8:1–33.
12 Intes X, Maloux C, Guven M, Yazici B, Chance B. Diffuse optical tomography with physiological and spatial a priori constraints. Phys Med Biol. 2004;49(12):N155–N163.
13 Fang Q, Selb J, Carp SA, Boverman G, Miller EL, Brooks DH, et al. Combined optical and X-ray tomosynthesis breast imaging. Radiology. 2011;258(1):89–97.
14 Davis SC, Dehghani H, Wang J, Jiang S, Pogue BW, Paulsen KD. Image-guided diffuse optical fluorescence tomography implemented with Laplacian-type regularization. Opt Express. 2007;15(7):4066–4082.
15 Shi J, Zhang B, Liu F, Luo J, Bai J. Efficient L1 regularization-based reconstruction for fluorescent molecular tomography using restarted nonlinear conjugate gradient. Opt Lett. 2013;38(18):3696–3699.
16 Han D, Tian J, Zhu S, Feng J, Qin C, Zhang B, et al. A fast reconstruction algorithm for fluorescence molecular tomography with sparsity regularization. Opt Express. 2010;18(8):8630–8646.
17 Behrooz A, Zhou HM, Eftekhar AA, Adibi A. Total variation regularization for 3D reconstruction in fluorescence tomography: experimental phantom studies. Appl Opt. 2012;51(34):8216–8227.
18 Zhu D, Li C. Nonconvex regularizations in fluorescence molecular tomography for sparsity enhancement. Phys Med Biol. 2014;59(12):2901–2912.
19 Okawa S, Hoshi Y, Yamada Y. Improvement of image quality of time-domain diffuse optical tomography with l sparsity regularization. Biomed Opt Express. 2011;2(12):3334–3348.
20 Zhao L, Yang H, Cong W, Wang G, Intes X. Lp regularization for early gate fluorescence molecular tomography. Opt Lett. 2014;39(14):4156–4159.
21 Cassol F, Portal L, Richelme S, Dupont M, Boursier Y, Arechederra M, et al. Tracking dynamics of spontaneous tumors in mice using photon-counting computed tomography. iScience. 2019;21:68–83.
22 Tan MJ, Fernandes N, Williams KC, Ford NL. In vivo micro-computed tomography imaging in liver tumor study of mice using Fenestra VC and Fenestra HDVC. Sci Rep. 2022;12(1):22399.
23 Deliolanis NC, Wurdinger T, Pike L, Tannous BA, Breakefield XO, Weissleder R, et al. In vivo tomographic imaging of red-shifted fluorescent proteins. Biomed Opt Express. 2011;2(4):887–900.

Section 1.1.1.3

1 Abe M, Nishihara R, Ikeda Y, Nakajima T, Sato M, Iwasawa N, et al. Near-infrared bioluminescence imaging with a through-bond energy transfer cassette. Chembiochem. 2019;20(15):1919–1923. https://doi.org/10.1002/cbic.201900149.
2 Adams ST, Miller SC. Enzymatic promiscuity and the evolution of bioluminescence. FEBS J. 2020;287(7):1369–1380. https://doi.org/10.1111/febs.15176.
3 Ando Y, Niwa K, Yamada N, Enomoto T, Irie T, Kubota H, et al. Firefly bioluminescence quantum yield and colour change by pH-sensitive green emission. Nat Photon. 2008;2(1):44–47. https://doi.org/10.1038/nphoton.2007.251.
4 Aron AT, Heffern MC, Lonergan ZR, Vander Wal MN, Blank BR, Spangler B, et al. In vivo bioluminescence imaging of labile iron accumulation in a murine model of Acinetobacter baumannii infection. Proc Natl Acad Sci U S A. 2017;114(48):12669–12674. https://doi.org/10.1073/pnas.1708747114.
5 Ataei F, Torkzadeh-Mahani M, Hosseinkhani S. A novel luminescent biosensor for rapid monitoring of IP3 by split-luciferase complementary assay. Biosens Bioelectron. 2013;41:642–648. https://doi.org/10.1016/j.bios.2012.09.037.
6 Auld DS, Inglese J. Interferences with luciferase reporter enzymes. Bethesda (MD): Eli Lilly & Company; 2016. Azad T, Janse van Rensburg HJ, Morgan J, Rezaei R, Crupi MJF, Chen R, et al. Luciferase-based biosensors in the era of the COVID-19 pandemic. ACS Nanosci Au. 2021;1(1):15–37. https://doi.org/10.1021/acsnanoscienceau.1c00009.

7 Bae Kim S, Fujii R, Natarajan A, Massoud TF, Paulmurugan R. Ligand-activated BRET9 imaging for measuring protein-protein interactions in living mice. Chem Commun. 2020;56(2):281–284. https://doi.org/10.1039/c9cc07634d.

8 Bakhtiarova A, Taslimi P, Elliman SJ, Kosinski PA, Hubbard B, Kavana M, et al. Resveratrol inhibits firefly luciferase. Biochem Biophysical Res Commun. 2006;351(2):481–484. https://doi.org/10.1016/j.bbrc.2006.10.057.

9 Berger F, Paulmurugan R, Bhaumik S, Gambhir SS. Uptake kinetics and biodistribution of 14C-D-luciferin – a radiolabeled substrate for the firefly luciferase catalyzed bioluminescence reaction: impact on bioluminescence based reporter gene imaging. Eur J Nucl Med Mol Imaging. 2008;35(12):2275–2285. https://doi.org/10.1007/s00259-008-0870-6.

10 Bevilaqua VR, Matsuhashi T, Oliveira G, Oliveira PSL, Hirano T, Viviani VR. Phrixotrix luciferase and 6′-aminoluciferins reveal a larger luciferin phenolate binding site and provide novel far-red combinations for bioimaging purposes. Sci Rep. 2019;9(1):8998. https://doi.org/10.1038/s41598-019-44534-3.

11 Branchini BR, Ablamsky DM, Davis AL, Southworth TL, Butler B, Fan F, et al. Red-emitting luciferases for bioluminescence reporter and imaging applications. Anal Biochem. 2010;396(2):290–297. https://doi.org/10.1016/j.ab.2009.09.009.

12 Branchini BR, Southworth TL, Fontaine DM, Kohrt D, Welcome FS, Florentine CM, et al. Red-emitting chimeric firefly luciferase for in vivo imaging in low ATP cellular environments. Anal Biochem. 2017;534:36–39. https://doi.org/10.1016/j.ab.2017.07.001.

13 Chan CT, Reeves RE, Geller R, Yaghoubi SS, Hoehne A, Solow-Cordero DE, et al. Discovery and validation of small-molecule heat-shock protein 90 inhibitors through multimodality molecular imaging in living subjects. Proc Natl Acad Sci U S A. 2012;109(37):E2476–E2485. https://doi.org/10.1073/pnas.1205459109.

14 Chang M, Anttonen KP, Cirillo SLG, Francis KP, Cirillo JD. Real-time bioluminescence imaging of mixed mycobacterial infections. PLoS One. 2014;9(9):e108341. https://doi.org/10.1371/journal.pone.0108341

15 Chu J, Oh Y, Sens A, Ataie N, Dana H, Macklin JJ, et al. A bright cyan excitable orange fluorescent protein facilitates dual-emission microscopy and enhances bioluminescence imaging in vivo. Nat Biotechnol. 2016;34(7):760–767. https://doi.org/10.1038/nbt.3550.

16 Dixon AS, Schwinn MK, Hall MP, Zimmerman K, Otto P, Lubben TH, et al. NanoLuc complementation reporter optimized for accurate measurement of protein interactions in cells. ACS Chem Biol. 2016;11(2):400–408. https://doi.org/10.1021/acschembio.5b00753.

17 Endo M, Miyasaki M, Li QJ, Kawamura G, Ozawa T. A detection method for GLUT4 exocytosis based on spontaneous split luciferase complementation. Anal Sci. 2019;35(8):835–838. https://doi.org/10.2116/analsci.19c003.

18 Fleiss A, Sarkisyan KS. A brief review of bioluminescent systems (2019). Curr Genet. 2019;65(4):877–882. https://doi.org/10.1007/s00294-019-00951-5.

19 Gregor C, Gwosch KC, Sahl SJ, Hell SW. Strongly enhanced bacterial bioluminescence with the ilux operon for single-cell imaging. Proc Natl Acad Sci U S A. 2018;115(5):962–967. https://doi.org/10.1073/pnas.1715946115.

20 Hall MP, Unch J, Binkowski BF, Valley MP, Butler BL, Wood MG, et al. Engineered luciferase reporter from a deep sea shrimp utilizing a novel imidazopyrazinone substrate. ACS Chem Biol. 2012;7(11):1848–1857. https://doi.org/10.1021/cb3002478.

21 Hall MP, Woodroofe CC, Wood MG, Que I, van't Root M, Ridwan Y, et al. Click beetle luciferase mutant and near infrared naphthyl-luciferins for improved bioluminescence imaging. Nat Commun. 2018;9:132. https://doi.org/10.1038/s41467-017-02542-9.

22 Ho PI, Yue K, Pandey P, Breault L, Harbinski F, McBride AJ, et al. Reporter enzyme inhibitor study to aid assembly of orthogonal reporter gene assays. ACS Chem Biol. 2013;8(5):1009–1017. https://doi.org/10.1021/cb3007264.

23 Homaei AA, Mymandi AB, Sariri R, Kamrani E, Stevanato R, Etezad SM, et al. Purification and characterization of a novel thermostable luciferase from Benthosema pterotum. J Photochem Photobiol B Biol. 2013;125:131–136. https://doi.org/10.1016/j. jphotobiol.2013.05.015.

24 Ikeda Y, Nomoto T, Hiruta Y, Nishiyama N, Citterio D. Ring-fused firefly luciferins: expanded palette of near-infrared emitting bioluminescent substrates. Anal Chem. 2020;92(6):4235–4243. https://doi.org/10.1021/acs.analchem.9b04562.

25 Inouye S, Sato J-I, Sahara-Miura Y, Tomabechi Y, Sumida Y, Sekine S-I, et al. Reverse mutants of the catalytic 19 kDa mutant protein (nanoKAZ/nanoLuc) from Oplophorus luciferase with coelenterazine as preferred substrate. PLoS One. 2022;17(9): e0272992. https://doi.org/10.1371/journal.pone.0272992.

26 Kamiya G, Kitada N, Furuta T, Hirano T, Maki SA, Kim SB. S-series coelenterazine-driven combinatorial bioluminescence imaging systems for mammalian cells. Int J Mol Sci. 2023;24(2):1420. https://doi.org/10.3390/ijms24021420.

27 Kim SB, Fujii R, Nishihara R, Bose RJC, Citterio D, Suzuki K, et al. Molecular imaging of retinoic acids in live cells using single-chain bioluminescence probes. ACS Comb Sci. 2019;21(6):473–481. https://doi.org/10.1021/acscombsci.9b00035.

28 Kim SB, Furuta T, Kamiya G, Kitada N, Paulmurugan R, Maki SA. Bright molecular strain probe templates for reporting protein-protein interactions. Sensors. 202323(7):3498. https://doi.org/10.3390/s23073498.

29 Kim SB, Furuta T, Kitada N, Maki SA. Creation of artificial luciferase 60s from sequential insights and their applications to bioassays. Sensors. 202323(14):6376. https://doi.org/10.3390/s23146376. Kim SB, Furuta T, Ohmuro-Matsuyama Y, Kitada N, Nishihara R, Maki SA. Bioluminescent imaging systems boosting near-infrared signals in mammalian cells. Photochem Photobiol Sci. 202322(6):1267–1278. https://doi.org/10.1007/s43630-023-00367-8.

30 Klein MA, Lazarev S, Gervasi C, Cowan C, Machleidt T, Friedman Ohana R. Luciferase calibrants enable absolute quantitation of bioluminescence power. ACS Meas Sci Au. 2023;3(6):496–503. https://doi.org/10.1021/acsmeasuresciau.3c00036.

31 Kojima R, Takakura H, Kamiya M, Kobayashi E, Komatsu T, Ueno T, et al. Development of a sensitive bioluminogenic probe for imaging highly reactive oxygen species in living rats. Angew Chem Int Ed. 2015;54(49):14768–14771. https://doi.org/10.1002/anie.201507530.

32 Li XY, Nakajima Y, Niwa K, Viviani VR, Ohmiya Y. Enhanced redemitting railroad worm luciferase for bioassays and bioimaging. Protein Sci. 2010;19(1):26–33. https://doi.org/10.1002/pro.279. Loening AM, Dragulescu-Andrasi A, Gambhir SS. A red-shifted Renilla luciferase for transient reporter-gene expression. Nat Methods. 2010; 7(1):5–6. https://doi.org/10.1038/nmeth0110-05.

33 Love AC, Caldwell DR, Kolbaba-Kartchner B, Townsend KM, Halbers LP, Yao Z, et al. Red-shifted coumarin luciferins for improved bioluminescence imaging. J Am Chem Soc. 2023;145(6):3335–3345. https://doi.org/10.1021/jacs.2c07220.

34 Misawa N, Kafi AKM, Hattori M, Miura K, Masuda K, Ozawa T. Rapid and high-sensitivity cell-based assays of protein-protein interactions using split click beetle luciferase complementation: an approach to the study of G-protein-coupled receptors. Anal Chem. 2010;82(6):2552–2560. https://doi.org/10.1021/ac100104q.

35 Mujawar A, Phadte P, Palkina KA, Markina NM, Mohammad A, Thakur BL, et al. Triple reporter assay: a non-overlapping luciferase assay for the measurement of complex macromolecular regulation in cancer cells using a new mushroom luciferase-luciferin pair. Sensors. 2023;23(17):7313. https://doi.org/10.3390/s23177313.

36 Neefjes M, Housmans BAC, van den Akker GGH, van Rhijn LW, Welting TJM, van der Kraan PM. Reporter gene comparison demonstrates interference of complex body fluids with secreted luciferase activity. Sci Rep. 2021;11(1):1359. https://doi.org/10.1038/s41598-020-80451-6. Nemergut M, Pluskal D, Horackova J, Sustrova T, Tulis J, Barta T, et al. Illuminating the mechanism and allosteric behavior of NanoLuc luciferase. Nat Commun. 2023; 14(1):7864. https://doi.org/10.1038/s41467-023-43403-y.

37 Nishihara R, Abe M, Nishiyama S, Citterio D, Suzuki K, Kim SB. Luciferase-specific coelenterazine analogues for optical contamination-free bioassays. Sci Rep. 2017;7:908. https://doi.org/10.1038/s41598-017-00955-6. Nishihara R, Dokainish HM, Kihara Y, Ashiba H, Sugita Y, Kurita R. Pseudo-luciferase activity of the SARS-CoV-2 spike protein for Cypridina luciferin luciferin. ACS Central Sci. 2024;10(2):283–290. https://doi.org/10.1021/acscentsci.3c00887.

38 Nishihara R, Niwa K, Tomita T, Kurita R. Coelenterazine analogue with human serum albumin-specific bioluminescence. Bioconjugate Chem. 2020;31(12):2679–2684. https://doi.org/10.1021/acs.bioconjchem.0c00536.

39 O'Sullivan JJ, Medici V, Heffern MC. A caged imidazopyrazinone for selective bioluminescence detection of labile extracellular copper(ii). Chem Sci. 2022;13(15):4352–4363. https://doi.org/10.1039/d1sc07177g.

40 Ozawa T, Yoshimura H, Kim SB. Advances in fluorescence bioluminescence imaging. Anal Chem. 2013;85(2):590–609. https://doi.org/10.1021/ac3031724.

41 Porterfield WB, Jones KA, McCutcheon DC, Prescher JA. A "caged" luciferin for imaging cell-cell contacts. J Am Chem. Soc. 2015;137(27):8656–8659. https://doi.org/10.1021/jacs.5b02774. Schenkmayerova A, Toul M, Pluskal D, Baatallah R, Gagnot G, Pinto GP, et al. Catalytic mechanism for Renilla-type luciferases. Nat Catal. 2023;6(1):23–38. https://doi.org/10.1038/s41929-022-00895-z.

42 Takenaka Y, Yamaguchi A, Tsuruoka N, Torimura M, Gojobori T, Shigeri Y. Evolution of bioluminescence in marine planktonic copepods. Mol Biol Evol. 2012;29(6):1669–1681. https://doi.org/10.1093/molbev/mss009.

43 Tamaki S, Kitada N, Kiyama M, Fujii R, Hirano T, Kim SB, et al. Color-tunable bioluminescence imaging portfolio for cell imaging. Sci Rep. 2021;11(1):2219. https://doi.org/10.1038/s41598-021-81430-1.

44 Weihs F, Dacres H. Red-shifted bioluminescence resonance energy transfer: improved tools and materials for analytical in vivo approaches. TrAC Trends Anal Chem. 2019;116:61–73. https://doi.org/10.1016/j.trac.2019.04.011.

45 Weihs F, Peh A, Dacres H. A red-shifted Bioluminescence resonance energy transfer (BRET) biosensing system for rapid measurement of plasmin activity in human plasma. Anal Chim Acta. 2020;1102:99–108. https://doi.org/10.1016/j.aca.2019.12.044.

46 Welsh JP, Patel KG, Manthiram K, Swartz JR. Multiply mutated Gaussia luciferases provide prolonged and intense bioluminescence. Biochem Biophysical Res Commun. 2009;389(4):563–568. https://doi.org/10.1016/j.bbrc.2009.09.006.

47 Wu N, Kobayashi N, Tsuda K, Unzai S, Saotome T, Kuroda Y, et al. Solution structure of Gaussia Luciferase with five disulfide bonds and identification of a putative coelenterazine binding cavity by heteronuclear NMR. Sci Rep. 2020;10(1):20069. https://doi.org/10.1038/s41598-020-76486-4.

48 Xu TT, Ripp S, Sayler GS, Close DM. Expression of a humanized viral 2A-mediated lux operon efficiently generates autonomous bioluminescence in human cells. PLoS One. 2014;9(5):e96347. https://doi.org/10.1371/journal.pone.0096347.

49 Yeh AH, Norn C, Kipnis Y, Tischer D, Pellock SJ, Evans D, et al. De novo design of luciferases using deep learning. Nature. 2023;614(7949):774–780. https://doi.org/10.1038/ s41586-023-05696-3.

50 Yeh HW, Karmach O, Ji A, Carter D, Martins-Green MM, Ai HW. Red-shifted luciferase-luciferin pairs for enhanced bioluminescence imaging. Nat Methods. 2017;14(10):971–974. https://doi.org/10.1038/Nmeth.4400.

51 Zhang CC, Cheng L, Dong GP, Han GX, Yang XY, Tang CC, et al. Novel photoactivatable substrates for Renilla luciferase imaging in vitro and in vivo. Org Biomol Chem. 2018;16(26):4789–4792. https://doi.org/10.1039/c8ob01192c.

Section 1.1.1.4

1 Jakovljevic M, Lamnisos D, Westerman R, Chattu VK, Cerda A. Future health spending forecast in leading emerging BRICS markets in 2030: health policy implications. Health Res Policy Syst. 2022;20(1):23. https://doi.org/10.1186/s12961–022–00822–5. Erratum in: Health Res Policy Syst. 2022;20(1):30. https://doi.org/10.1186/s12961–022–00836–z.

2 Cen W, Lang T, Wang J, Xiao M. High–Q Fano terahertz resonance based on bound states in the continuum in all–dielectric metasurface. Appl Surf Sci. 2022;575(6):e151723. https://doi.org/10.1016/j.apsusc.2021.151723.

3 Vafapour Z. Slowing down light using terahertz semiconductor metamaterial for dual–band thermally tunable modulator applications. Appl Opt. 2018;57(4):722–729. https://doi.org/10.1364/AO.57.000722.

4 Fu X, Liu Y, Chen Q, Fu Y, Cui TJ. Applications of terahertz spectroscopy in the detection and recognition of substances. Frontiers Phys. 2022;10:869537. https://doi.org/10.3389/fphy.2022.869537.

5 Nourinovin S, Navarro–Cia M, Rahman MM, Philpott MP, Abbasi QH, Alomainy A. Terahertz metastructures for noninvasive biomedical sensing and characterization in future health care (bioelectromagnetics). IEEE Antennas Propag Mag. 2022;64(2):60–70. https://doi.org/10.1109/MAP.2022.3145712.

6 Banerjee A, Vajandar S, Basu T. Chapter 13 – Prospects in medical applications of terahertz waves. In: Banerjee A, Chakraborty B, Inokawa H, Roy JN, editors. Terahertz Biomedical and Healthcare Technologies: Materials to Devices. Elsevier; 2020. pp. 225–239. ISBN: 978–0128185568.

7 Mukherjee M, Chatterjee S. Chapter 6 – THz medical imaging: current status and future outlook. In: Banerjee A, Chakraborty B, Inokawa H, Roy JN, editors. Terahertz Biomedical and Healthcare Technologies: Materials to Devices. Elsevier; 2020. pp. 113–125. ISBN: 978–0128185568.

8 Banerjee A, Chakraborty C, Rathi M. Chapter 8 – Medical imaging, artificial intelligence, Internet of Things, wearable devices in terahertz healthcare technologies. In: Banerjee A, Chakraborty B, Inokawa H, Roy JN, editors. Terahertz Biomedical and Healthcare Technologies: Materials to Devices. Elsevier; 2020. pp. 145–165. ISBN: 978–0128185568.

9 Taday PF, Pepper M, Arnone DD. Selected applications of terahertz pulses in medicine and industry. Appl Sci. 2022;12(12):6169. https://doi.org/10.3390/app12126169.

10 D'Arco A, Di Fabrizio M, Dolci V, Petrarca M, Lupi S. THz pulsed imaging in biomedical applications. Condens Matter. 2020;5(2):25. https://doi.org/10.3390/condmat5020025.

11 Vafapour Z, Keshavarz A, Ghahraloud H. The potential of terahertz sensing for cancer diagnosis. Heliyon. 2020;6(12):e05623. https://doi.org/10.1016/j.heliyon.2020.e05623.

12 Nikitkina AI, Bikmulina PY, Gafarova ER, Kosheleva NV, Efremov YM, Bezrukov EA, et al. Terahertz radiation and the skin: a review. J Biomed Opt. 2021;26(4):043005. https://doi.org/10.1117/1.JBO.26.4.043005.

13 Gong A, Qiu Y, Chen X, Zhao Z, Xia L, Shao Y. Biomedical applications of terahertz technology. Appl Spectrosc Rev. 2019;55(5):418–438. https://doi.org/10.1080/05704928.2019.1670202.

14 Zhang J, Li S, Le W. Advances of terahertz technology in neuroscience: current status and a future perspective. iScience. 2021;24(12):103548. https://doi.org/10.1016/j.isci.2021.103548.

15 Zhang Y, Wang C, Huai B, Wang S, Zhang Y, Wang D, et al. Continuous–wave THz imaging for biomedical samples. Appl Sci. 2021;11(1):71. https://doi.org/10.3390/app11010071.

16 Wang L. Terahertz imaging for breast cancer detection. Sensors. 2021;21(19):6465. https://doi.org/10.3390/s21196465.

17 Peng Y, Shi C, Wu X, Zhu Y, Zhuang S. Terahertz imaging and spectroscopy in cancer diagnostics: a technical review. BME Front. 2020; 2020:2547609. https://doi.org/10.34133/2020/2547609.

18 Yu C, Fan S, Sun Y, Pickwell–Macpherson E. The potential of terahertz imaging for cancer diagnosis: a review of investigations to date. Quant Imaging Med Surg. 2012;2(1):33–45. https://doi.org/10.3978/j.issn.2223–4292.2012.01.04.

19 Castro–Camus E, Koch M, Mittleman DM. Recent advances in terahertz imaging: 1999 to 2021. Appl Phys B: Lasers Opt. 2022;128(1):12. https://doi.org/10.1007/s00340–021–07732–4.

20 Cheon H, Yang HJ, Son JH. Toward clinical cancer imaging using terahertz spectroscopy. IEEE J Sel Top Quantum Electron. 2017;23(4):7929316. https://doi.org/10.1109/JSTQE.2017.2704905.

21 Son JH. Terahertz molecular imaging and its biomedical applications. In: Laser and Tera–Hertz Science and Technology: Int Photon Opto-electron Meetings. 2012. ISBN: 978–1–55752–960–2. https://doi.org/10.1364/LTST.2012.SF1A.1.

22 Lindley–Hatcher H, Stantchev RI, Chen X, Hernandez–Serrano AI, Hardwicke J, Pickwell–MacPherson E. Real time THz imaging–opportunities and challenges for skin cancer detection. Appl Phys Lett. 2021;11(23):230501. https://doi.org/10.1063/5.0055259.

23 Yang S, Du C, Feng L, Zhang C, Wu J, Jin B, et al. Real–time near– field terahertz spectroscopy imaging. In: 46th International Conference on Infrared, Millimeter and Terahertz Waves (IRMMW–THz). Chengdu, China. 2021; 1–1. https://doi.org/10.1109/IRMMW-THz50926.2021.9567448.

24 Didi SE, Halkhams I, Fattah M, Younes B, Mazer S, Bekkali M, et al. Study and design of the microstrip patch antenna operating at 120 GHz. In: Ghzaoui M, Das S, Lenka TR, Biswas A., editors. Terahertz Wireless Communication Components and System Technologies. Singapore: Springer; 2022. pp. 175–190. ISBN: 978–981–16–9181–2. https://doi.org/10.1007/978– 981–16–9182–9_12.

25 Burford NM, El–Shenawee MO. Review of terahertz photoconductive antenna technology. Opt Eng. 2017;56(1):010901. https://doi.org/10.1117/1.OE.56.1.010901.

26 Malhotra I, Jha KR, Singh G. Design of highly directive lens–less photoconductive dipole antenna array with frequency selective surface for terahertz imaging applications. Optik. 2018;173:206–219. https://doi.org/10.1016/J.IJLEO.2018.08.010.

27 Shur M, Liu X. Biomedical applications of terahertz technology. Proc. SPIE. 2022;11975:1–6.

28 Cristian CR, Thomas S, Vasile D, Stanciu GD, Alexa–Stratulat T, Luca A, et al. Research on functionalized gadolinium oxide nanoparticles for MRI and THz imaging. In: Proc Int Conf Expo Electr Power Eng (EPE). 2018; 646–9. https://doi.org/10.1109/ICEPE.2018.8559855.

29 Chen X, Lindley–Hatcher H, Stantchev RI, Wang J, Li K, Serrano AH, et al. Terahertz (THz) biophotonics technology: instrumentation, techniques, and biomedical applications. Chem Phys Rev. 2022;3(1):e011311. https://doi.org/10.1063/5.0068979.

30 Gezimati M, Singh G. Advances in terahertz technology for cancer detection applications. Opt Quantum Electron. 2023;55(2):151. https://doi.org/10.1007/s11082–022–04340–0.

31 Wang Q, Xie L, Ying Y. Overview of imaging methods based on terahertz time–domain spectroscopy. Appl Spectrosc Rev. 2022;57(3):249–264. https://doi.org/10.1080/05704928.2021.1875480.

32 Huang Y, Shen Y, Wang J. From terahertz imaging to terahertz wireless communications. Engineering. 2022;22(3):106–124. https://doi.org/10.1016/j.eng.2022.06.023.

33 Yin XX, Baghai–Wadji A, Zhang Y. A biomedical perspective in terahertz nano–communications – a review. IEEE Sens J. 2022;22(10): 9215–9227. https://doi.org/10.1109/JSEN.2022.3161013.

34 Liu J, Khan ZU, Sarjoghian S. Layered THz waveguides for SPPs, filter and sensor applications. J Opt. 2019;48(4):567–581. https://doi.org/10.1007/s12596–019–00569–3.

35 Poola PK, Krasnikov I, Seteikin A. Waveguides for terahertz endoscopy. In: Banerjee A, Chakraborty B, Inokawa H, Roy JN, editors. Terahertz Biomedical and Healthcare Technologies. Singapore: Elsevier; 2020. pp. 215–224. ISBN: 9780128185568. https://doi.org/10.1016/B978–0–12–818556–8.00012–4.

36 Fan S, Qian Z, Wallace VP. Collagen and muscle fibrous tissue as a contrast mechanism in the THz region. In: 2018 43rd Int Conf Infr, Millim, THz Waves (IRMMW–THz). Nagoya, Japan. 2018; e8509855. IBSN:978–1–5386–3810–1. https://doi.org/10.1109/IRMMW–THz.2018.850985.

37 Zhang P, Zhong S, Zhang J, Ding J, Liu Z, Huang Y, et al. Application of terahertz spectroscopy and imaging in the diagnosis of prostate cancer. Current Opt Photon Korea. 2020;4(1):31–43. https://doi.org/10.3807/ COPP.2020.4.1.031.

38 Banduryan B, Sokolenko V, Vinokurov E, Sibilyova R. Infrared–optical frequency method for determining iodomethane concentration in the air when simulating the flow composition in Npps ventilation system. Available at SSRN: https://ssrn.com/abstract=4818909. https://doi.org/10.2139/ssrn.4818909.

39 Didovets Y, Koloskov V, Bandurian B, Koloskova H. Method of Investigation of Soil Contamination with Heavy Metals at the Sites of Explosions. Key Eng Mater. 2024; 988:107–16. https://doi.org/10.4028/ p–n9oeAe. Available at: https://www.scientific.net/KEM.988.107.
40 Bandurian B, Kovalevskyj V, Tsvaigov D. Formalization of safety assessment and management. Technogenic and ecological safety. 2021;9(1), 26–30. Ukrainian. https://doi.org/10.52363/2522–1892.2021.1.4.
41 Boiko VV, editor. Radiophysical bases and experience of application of microwave electromagnetic fields in surgery. Kharkiv: Novoe slovo; 2009. 166 c. Russian. ISBN: 978–617–568–001–8.

Section 1.1.1.5

1 Ronchi V. Barocas, V. The Nature of Light: An Historical Survey. Harvard University Press; 1970.
2 Huard S. Polarization of Light. Wiley; 1997.
3 Goldstein D. Polarized Light. 2nd ed. Marcel Dekker; 2003.
4 Chipman RA, Lam WST, Young G. Polarized Light and Optical Systems. CRC Press; 2018.
5 Pérez JJG, Ossikovski R. Polarized Light and the Mueller Matrix Approach. CRC Press; 2017.
6 Zhan QW. Cylindrical vector beams: from mathematical concepts to applications. Adv Opt Photonics 2009;1:1–57. Rosales-Guzmán C, Ndagano B, Forbes A. A review of complex vector light fields and their applications. J Opt. 2018;20:123001.
7 Forbes A, De Oliveira M, Dennis MR. Structured light. Nat Photon. 2021;15:253–262.
8 Wang JW, Castellucci F, Franke-Arnold S. Vectorial light–matter interaction: exploring spatially structured complex light fields. AVS Quantum Sci. 2020;2:031702.
9 Slussarenko S, Alberucci A, Jisha CP, Piccirillo B, Santamato E, Assanto G, et al. Guiding light via geometric phases. Nat Photon. 2016;10:571–575.
10 Cloude S. Polarisation: Applications in Remote Sensing. OUP Oxford; 2009.
11 Ndagano B, Perez-Garcia B, Roux FS, McLaren M, Rosales-Guzman C, Zhang Y, et al. Characterizing quantum channels with non-separable states of classical light. Nat Phys. 2017;13:397–402.
12 Marrucci L, Manzo C, Paparo D. Optical spin-to-orbital angular momentum conversion in inhomogeneous anisotropic media. Phys Rev Lett. 2006;96:163905.
13 Bliokh KY, Rodríguez-Fortuño FJ, Nori F, Zayats AV. Spin-orbit interactions of light. Nat Photon. 2015;9:796–808.
14 Schulz M, Zablocki J, Abdullaeva OS, Brück S, Balzer F, Lützen A, et al. Giant intrinsic circular dichroism of prolinol-derived squaraine thin films. Nat Commun. 2018;9:2413.
15 He H, Liao R, Zeng N, Li P, Chen Z, Liu X, et al. Mueller matrix polarimetry – an emerging new tool for characterizing the microstructural feature of complex biological specimen. J Lightwave Technol. 2019;37:2534–2548.
16 Oldenbourg R. A new view on polarization microscopy. Nature. 1996;381:811–812.
17 Gurjar RS, Backman V, Perelman LT, Georgakoudi I, Badizadegan K, Itzkan I, Dasari RR, et al. Imaging human epithelial properties with polarized light-scattering spectroscopy. Nat Med. 2001;7:1245–1248.
18 Qiu L, Pleskow DK, Chuttani R, Vitkin E, Leyden J, Ozden N, et al. Multispectral scanning during endoscopy guides biopsy of dysplasia in Barrett's esophagus. Nat Med. 2010;16:603–606. He H, Chang J, Chen B, Ma H, Booth MJ. Polarisation optics for biomedical and clinical applications: a review. Light Sci Appl. 2021;22;10(1):194. https://doi.org/10.1038/s41377-021-00639-x. Page 15 of 20.
19 Ghosh N, Vitkin IA. Tissue polarimetry: concepts, challenges, applications, and outlook. J Biomed Opt. 2011;16:110801.
20 Novikova T, Meglinski I, Ramella-Roman JC, Tuchin VV. Special section guest editorial: polarized light for biomedical applications. J Biomed Opt. 2016;21:071001.
21 Tuchin VV. Polarized light interaction with tissues. J Biomed Opt. 2016;21:071114.
22 Ramella-Roman JC, Saytashev I, Piccini M. A review of polarization-based imaging technologies for clinical and preclinical applications. J Opt. 2020;22:123001.
23 Qi J, Elson DS. Mueller polarimetric imaging for surgical and diagnostic applications: a review. J Biophotonics. 2017;10:950–982.
24 Fujiwara H. Spectroscopic Ellipsometry: Principles and Applications. John Wiley & Sons; 2007.
25 Azzam RMA, Bashara NM, Ballard SS. Ellipsometry and polarized light. Phys Today. 1978;31:72.
26 Mendoza-Galván A, Muñoz-Pineda E, Ribeiro SJ, Santos MV, Järrendahl K, Arwin H. Mueller matrix spectroscopic ellipsometry study of chiral nanocrystalline cellulose films. J Opt. 2018;20:024001.

27 Azzam RMA. Stokes-vector and Mueller-matrix polarimetry. J Opt Soc Am A. 2016;33:1396–1408.
28 Azzam RMA. Mueller-matrix ellipsometry: a review. Proc SPIE. 1997;3121:369–405.
29 Song B, B, Gu H, Zhu S, Jiang H, Chen X, Zhang C, et al. Broadband optical properties of graphene and HOPG investigated by spectroscopic Mueller matrix ellipsometry. Appl Surf Sci. 2018;439:1079–1087.
30 Jiang H, Ma Z, Gu H, Chen X, Liu S. Characterization of volume gratings based on distributed dielectric constant model using Mueller matrix ellipsometry. Appl Sci. 2019;9:698.
31 Chang J, He H, Wang Y, Huang Y, Li X, He C, et al. Division of focal plane polarimeter-based 3×4 Mueller matrix microscope: a potential tool for quick diagnosis of human carcinoma tissues. J Biomed Opt. 2016;21:056002.
32 Fu Y, Huang Z, He H, Ma H, Wu J. Flexible 3×3 Mueller matrix endoscope prototype for cancer detection. IEEE Trans Instrum Meas. 2018;67:1700–1712.
33 De Boer JF, Hitzenberger CK, Yasuno Y. Polarization sensitive optical coherence tomography – a review. Biomed Opt Express. 2017;8:1838–1873.
34 Ghosh N, Patel HS, Gupta PK. Depolarization of light in tissue phantoms – effect of a distribution in the size of scatterers. Opt Express. 2003;11:2198–2205.
35 Kienle A, Hibst R. Light guiding in biological tissue due to scattering. Phys Rev Lett. 2006;97:018104.
36 Baravian C, Dillet J, Decruppe JP. Birefringence determination in turbid media. Phys Rev E. 2007;75:032501.
37 Wang XD, Wang LV. Propagation of polarized light in birefringent turbid media: a Monte Carlo study. J Biomed Opt. 2002;7:279–290.
38 Wang XD, Wang LV. Propagation of polarized light in birefringent turbid media: time-resolved simulations. Opt Express. 2001;9:254–259.
39 Du, E. He H, Zeng N, Guo Y, Liao R, He Y, et al. Two-dimensional backscattering Mueller matrix of spherecylinder birefringence media. J Biomed Opt. 2012;17:126016.
40 He, H, Zeng N, Liao R, Yun T, Li W, He Y, et al. Application of sphere-cylinder scattering model to skeletal muscle. Opt Express. 2010;18:15104–15112. Chen D, Zeng N, Xie Q, He H, Tuchin VV, Ma H. Mueller matrix polarimetry for characterizing microstructural variation of nude mouse skin during tissue optical clearing. Biomed Opt Express. 2017;8:3559–3570.
41 Donner C, Jensen HW. Light diffusion in multi-layered translucent materials. ACM Trans Graph. 2005;24:1032–1039.
42 Van De Hulst HC. Light Scattering by Small Particles. Dover Publications; 1981.
43 Wang LH, Jacques SL, Zheng LQ. MCML – Monte Carlo modeling of light transport in multi-layered tissues. Comput Methods Prog Biomed. 1995;47:131–146.
44 Yun, T, Zeng N, Li W, Li D, Jiang X, Ma H. Monte Carlo simulation of polarized photon scattering in anisotropic media. Opt Express. 2009;17:16590–16602.
45 Brosseau C, Fundamentals of Polarized Light: A Statistical Optics Approach, Wiley-Interscience; 1998.
46 Born M, Wolf E. Principles of Optics: Electromagnetic Theory of Propagation, Interference and Diffraction of Light. Elsevier; 2013.
47 Ling X, Zhou X, Huang K, Liu Y, Qiu CW, Luo H, et al. Recent advances in the spin Hall effect of light. Rep Prog Phys. 2017;80:066401.
48 Rubinsztein-Dunlop H, Forbes A, Berry MV, Dennis MR, Andrews DL, Mansuripur M, et al. Roadmap on structured light. J Opt. 2017;19:013001.
49 Bass M, DeCusatis C, Enoch J, Lakshminarayanan V, Li G, MacDonald C, et al. Handbook of Optics, Volume IV: Optical Properties of Materials. 3rd ed. McGraw-Hill Education; 2009. Lu SY, Chipman RA. Interpretation of Mueller matrices based on polar decomposition. J Opt Soc Am A. 1996;13:1106–1113.
50 Arteaga O, Canillas A. Pseudopolar decomposition of the Jones and Mueller–Jones exponential polarization matrices. J Opt Soc Am A. 2009;26:783–793.
51 Ossikovski R, De Martino A, Guyot S. Forward and reverse product decompositions of depolarizing Mueller matrices. Opt Lett. 2007;32:689–691.
52 Ortega-Quijano N, Arce-Diego JL. Mueller matrix differential decomposition. Opt Lett. 2011;36:1942–1944.
53 Arteaga O, Garcia-Caurel E, Ossikovski R. Anisotropy coefficients of a Mueller matrix. J Opt Soc Am A. 2011;28:548–553.
54 He H, Zeng N, Du E, Guo Y, Li D, Liao R, et al. A possible quantitative Mueller matrix transformation technique for anisotropic scattering media/Eine mögliche quantitative MüllerMatrix-Transformations-Technik für anisotrope streuende Medien. Photonics Lasers Med. 2013;2:129–137.

55 Ghosh N, Wood MFG, Vitkin IA. Mueller matrix decomposition for extraction of individual polarization parameters from complex turbid media exhibiting multiple scattering, optical activity, and linear birefringence. J Biomed Opt. 2008;13:044036.

56 Gil JJ. Characteristic properties of Mueller matrices. J Opt Soc Am A. 2000;17:328–334.

57 Ossikovski R. Analysis of depolarizing Mueller matrices through a symmetric decomposition. J Opt Soc Am A. 2009;26:1109–1118.

58 Vizet J, Ossikovski R. Symmetric decomposition of experimental depolarizing Mueller matrices in the degenerate case. Appl Opt. 2018;57:1159–1167.

59 Cloude SR. Group theory and polarisation algebra. Optik. 1985;75:26–36.

60 Pezzaniti JL, Chipman RA. Mueller matrix imaging polarimetry. Opt Eng. 1995;34:1558–1568.

61 Azzam RMA. Photopolarimetric measurement of the Mueller matrix by Fourier analysis of a single detected signal. Opt Lett. 1978;2:148–150.

62 Goldstein DH. Mueller matrix dual-rotating retarder polarimeter. Appl Opt. 1992;31:6676–6683.

63 Smith MH. Optimization of a dual-rotating-retarder Mueller matrix polarimeter. Appl Opt. 2002;41:2488–2493.

64 Dubreuil M, Rivet S, Le Jeune B, Cariou J. Snapshot Mueller matrix polarimeter by wavelength polarization coding. Opt Express. 2007;15:13660–13668.

65 He, C, Lin J, Chang J, Antonello J, Dai B, Wang J, et al. Full Poincare mapping for ultra-sensitive polarimetry. Preprint at https://arxiv.org/abs/2101.09372.2021. Sabatke DS, Descour MR, Dereniak EL, Sweatt WC, Kemme SA, Phipps GS. Optimization of retardance for a complete Stokes polarimeter. Opt Lett. 2000;25:802–804.

66 Tyo JS. Design of optimal polarimeters: maximization of signal-to-noise ratio and minimization of systematic error. Appl Opt. 2002;41:619–630.

67 He C, Chang J, Wang Y, Liao R, He H, Zeng N, et al. Linear polarization optimized Stokes polarimeter based on fourquadrant detector. Appl Opt. 2015;54:4458–4463.

68 He C, Chang J, Hu Q, Wang J, Antonello J, He H, et al. Complex vectorial optics through gradient index lens cascades. Nat Commun. 2019;10:4264.

69 Li X, Li H, Lin Y, Guo J, Yang J, Yue H, et al. Learning-based denoising for polarimetric images. Opt Express. 2020;28:16309–16321.

70 Abubakar A, Zhao X, Takruri M, Bastaki E, Bermak A. A hybrid denoising algorithm of BM3D and KSVD for Gaussian noise in DoFP polarization images. IEEE Access. 2020;8:57451–57459.

71 Bueno JM. Polarimetry using liquid-crystal variable retarders: theory and calibration. J Opt A: Pure Appl Opt. 2000;2:216–222. Skumanich A, Lites BW, Pillet VM, Seagraves P. The calibration of the advanced Stokes polarimeter. Astrophys J Suppl Ser. 1997;110:357–380.

72 Arteaga O, Freudenthal J, Wang B, Kahr B. Mueller matrix polarimetry with four photoelastic modulators: theory and calibration. Appl Opt. 2012;51:6805–6817.

73 Smith MH, Miller MA, Blumer RV, Stevens MA, Teale DM, Howe JD. Infrared Stokes polarimeter calibration. Proc SPIE. 2000;4133:55–64.

74 Tyo JS, Goldstein DL, Chenault DB, Shaw JA. Review of passive imaging polarimetry for remote sensing applications. Appl Opt. 2006;45:5453–5469. Jacques SL, Ramella-Roman JC, Lee K. Imaging skin pathology with polarized light. J Biomed Opt. 2002;7:329–340.

75 Jacques SL, Roman JR, Lee K. Imaging superficial tissues with polarized light. Lasers Surg Med. 2000;26:119–129.

76 Demos SG, Radousky HB, Alfano RR. Deep subsurface imaging in tissues using spectral and polarization filtering. Opt Express. 2000;7:23–28.

77 Demos SG, Alfano RR. Optical polarization imaging. Appl Opt. 1997;36:150–155. He et al. Light: Sci Appl. 2021;10:194, Page 16 of 20.

78 Groner W, Winkelman JW, Harris AG, Ince C, Bouma GJ, Messmer K, et al. Orthogonal polarization spectral imaging: a new method for study of the microcirculation. Nat Med. 1999;5:1209–1212.

79 Bargo PR, Kollias N. Measurement of skin texture through polarization imaging. Br J Dermatol. 2010;162:724–731.

80 Sridhar S, Da Silva A. Enhanced contrast and depth resolution in polarization imaging using elliptically polarized light. J Biomed Opt. 2016;21:071107.

81 Collett E. Measurement of the four Stokes polarization parameters with a single circular polarizer. Opt Commun. 1984;52:77–80.

82 Laude-Boulesteix B, De Martino A, Drévillon B, Schwartz L. Mueller polarimetric imaging system with liquid crystals. Appl Opt. 2004;43:2824–2832. Sornsin EA, Chipman RA. Mueller matrix polarimetry of electro-optic PLZT spatial light modulators. Proc SPIE. 1996;2873:196–201.

83 Peinado A, Lizana A, Campos J. Optimization and tolerance analysis of a polarimeter with ferroelectric liquid crystals. Appl Opt. 2013;52:5748–5757.

84 Alali S, Gribble A, Vitkin IA. Rapid wide-field Mueller matrix polarimetry imaging based on four photoelastic modulators with no moving parts. Opt Lett. 2016;41:1038–1041.

85 Qi J, Ye M, Singh M, Clancy NT, Elson DS. Narrow band 3 × 3 Mueller polarimetric endoscopy. Biomed Opt Express. 2013;4:2433–2449.

86 Dong Y, Liu S, Shen Y, He H, Ma H. Probing variations of fibrous structures during the development of breast ductal carcinoma tissues via Mueller matrix imaging. Biomed Opt Express. 2020;11:4960–4975.

87 Dong Y, Qi J, He H, He C, Liu S, Wu J, Quantitatively characterizing the microstructural features of breast ductal carcinoma tissues in different progression stages by Mueller matrix microscope. Biomed Opt Express. 2017;8:3643–3655.

88 Dong, Y, Wan J, Si L, Meng Y, Dong Y, Liu S, et al. Deriving polarimetry feature parameters to characterize microstructural features in histological sections of breast tissues. IEEE Trans Biomed Eng. 2021;68:881–892.

89 Hagen N, Oka K, Dereniak EL. Snapshot Mueller matrix spectropolarimeter. Opt Lett. 2007;32:2100–2102.

90 Azzam RMA. Division-of-amplitude photopolarimeter (DOAP) for the simultaneous measurement of all four Stokes parameters of light. Opt Acta: Int J Opt. 1982;29:685–689.

91 Jellison Jr. GE. Four-channel polarimeter for time-resolved ellipsometry. Opt Lett. 1987;12:766–768.

92 Compain E, Drevillon B. Broadband division-of-amplitude polarimeter based on uncoated prisms. Appl Opt. 1998;37:5938–5944.

93 Peinado A, Turpin A, Lizana A, Fernández E, Mompart J, Campos J. Conical refraction as a tool for polarization metrology. Opt Lett. 2013;38:4100–4103.

94 Haigh JA, Kinebas Y, Ramsay AJ. Inverse conoscopy: a method to measure polarization using patterns generated by a single birefringent crystal. Appl Opt. 2014;53:184–188.

95 Chang J, Zeng N, He H, He Y, Ma H. Single-shot spatially modulated Stokes polarimeter based on a GRIN lens. Opt Lett. 2014;39:2656–2659.

96 Bhandari P, Voss KJ, Logan L. An instrument to measure the downwelling polarized radiance distribution in the ocean. Opt Express. 2011;19:17609–17620.

97 Pezzaniti JL, Chenault DB. A division of aperture MWIR imaging polarimeter. Proc SPIE.2005;5888,239–250. Zimmerman BG, Ramkhalawon R, Alonso M, Brown TG. Pinhole array implementation of star test polarimetry. Proc SPIE. 2014; 8949: 164–173.

98 Chun CSL, Fleming DL, Torok EJ. Polarization-sensitive thermal imaging. Proc SPIE. 1994;2234:275–286. Nordin GP, Meier JT, Deguzman PC, Jones MW. Micropolarizer array for infrared imaging polarimetry. J Opt Soc Am A. 1999;16:1168–1174. Andreou AG, Kalayjian, ZK. Polarization imaging: principles and integrated polarimeters. IEEE Sens J. 2002;2:566–576.

99 Chen ZY, Wang X, Liang RG. Calibration method of microgrid polarimeters with image interpolation. Appl Opt. 2015;54:995–1001.

100 Gao SK, Gruev V. Bilinear and bicubic interpolation methods for division of focal plane polarimeters. Opt Express. 2011;19:26161–26173.

101 Gao SK, Gruev V. Gradient-based interpolation method for division-offocal-plane polarimeters. Opt Express. 2013;21:1137–1151.

102 Gruev V, Perkins R, York T. CCD polarization imaging sensor with aluminum nanowire optical filters. Opt Express. 2010;18:19087–19094.

103 Hsu, W. L. et al. Polarization microscope using a near infrared full-Stokes imaging polarimeter. Opt Express. 2015;23:4357–4368.

104 Liu, Y, York T, Akers W, Sudlow G, Gruev V, Achilefu S. Complementary fluorescence-polarization microscopy using division-of-focal-plane polarization imaging sensor. J Biomed Opt. 2012;17:116001. Millerd J, Brock N, Hayes J, North-Morris M, Kimbrough B, Wyant J. Pixelated phase-mask dynamic interferometers. In Osten, W. editor. Fringe 2005. Springer; 2006. pp. 640–647.

105 Ratliff BM, LaCasse CF, Tyo JS. Interpolation strategies for reducing IFOV artifacts in microgrid polarimeter imagery. Opt Express. 2009;17:9112–9125.

106 Ratliff BM, Tyo JS, Boger JK, Black WT, Bowers DL, Fetrow MP. Dead pixel replacement in LWIR microgrid polarimeters. Opt Express. 2007;15:7596–7609.

107 Tyo JS, LaCasse CF, Ratliff BM. Total elimination of sampling errors in polarization imagery obtained with integrated microgrid polarimeters. Opt Lett. 2009;34:3187–3189.

108 York T, Gruev V. Calibration method for division of focal plane polarimeters in the optical and near-infrared regime. Proc SPIE. 2011;8012:80120H.

109 York T, Powell SB, Gao S, Kahan L, Charanya T, Saha D, et al. Bioinspired polarization imaging sensors: from circuits and optics to signal processing algorithms and biomedical applications. Proc IEEE. 2014;102:1450–1469.

110 Zhang Z, Dong F, Cheng T, Qiu K, Zhang Q, Chu W, et al. Nano-fabricated pixelated micropolarizer array for visible imaging polarimetry. Rev Sci Instrum. 2014;85:105002.

111 Zhao X, Pan X, Fan X, Xu P, Bermak A, Chigrinov VG. Patterned dual-layer achromatic micro-quarter-wave-retarder array for active polarization imaging. Opt Express. 2014;22:8024–8034.

112 Oka K, Saito N. Snapshot complete imaging polarimeter using Savart plates. Proc SPIE. 2006;6295:629508.

113 Suárez-Bermejo JC, de Sande JC, Santarsiero M, Piquero G. Mueller matrix polarimetry using full Poincaré beams. Opt Lasers Eng. 2019;122:134–141. Goldstein DH, Chipman RA. Error analysis of a Mueller matrix polarimeter. J Opt Soc Am A. 1990;7:693–700. Ahmad JE, Takakura Y. Error analysis for rotating active Stokes–Mueller imaging polarimeters. Opt Lett. 2006;31:2858–2860.

114 Dai H, Yan CX. Measurement errors resulted from misalignment errors of the retarder in a rotating-retarder complete Stokes polarimeter. Opt Express. 2014;22:11869–11883.

115 Mu T, Zhang C, Li Q, Liang R. Error analysis of single-snapshot full-Stokes division-of-aperture imaging polarimeters. Opt Express. 2015;23:10822–10835.

116 Macias-Romero C, Török P. Eigenvalue calibration methods for polarimetry. J Eur Opt Soc Rapid Publ. 2012;7:12004.

117 Compain E, Poirier S, Drevillon B. General and self-consistent method for the calibration of polarization modulators, polarimeters, and Mueller-matrix ellipsometers. Appl Opt. 1999;38:3490–3502.

118 De Martino A, Garcia-Caurel E, Laude B, Drévillon B. General methods for optimized design and calibration of Mueller polarimeters. Thin Solid Films. 2004;455–456:112–119.

119 Marenko V, Molebnaya T. Optimization of stokes polarimeters employing a measurement of 4 intensities. Sov J Opt Technol. 1990;57:452–455.

120 Ambirajan A, Look DC, Jr. Optimum angles for a Mueller matrix polarimeter. Proc SPIE. 1994;2265:314–326. Ambirajan A, Jr. Look DC, Jr. Optimum angles for a polarimeter: part I. Opt Eng. 1995;34:1651–1655.

121 Tyo JS. Optimum linear combination strategy for an N-channel polarization-sensitive imaging or vision system. J Opt Soc Am A. 1998;15:359–366.

122 Tyo JS. Relation between system optimization and systematic errors in Stokes vector polarimeters. Proc SPIE. 2002;4481:22–30.

123 Tyo JS. Noise equalization in Stokes parameter images obtained by use of variable-retardance polarimeters. Opt Lett. 2000;25:1198–1200.

124 Azzam RMA, Elminyawi IM, El-Saba AM. General analysis and optimization of the four-detector photopolarimeter. J Opt Soc Am A. 1988;5:681–689.

125 Tyo JS. Considerations in polarimeter design. Proc SPIE. 2000;4133:65–74.

126 Peinado A, Lizana A, Vidal J, Iemmi C, Campos J. Optimization and performance criteria of a Stokes polarimeter based on two variable retarders. Opt Express. 2010;18:9815–9830.

127 Foreman MR, Favaro A, Aiello A. Optimal frames for polarization state reconstruction. Phys Rev Lett. 2015;115:263901.

128 Foreman MR, Goudail F. On the equivalence of optimization metrics in Stokes polarimetry. Opt Eng. 2019;58:082410.

129 Twietmeyer KM, Chipman RA. Optimization of Mueller matrix polarimeters in the presence of error sources. Opt Express. 2008;16:11589–11603.

130 Beckley AM, Brown TG, Alonso MA. Full Poincaré beams. Opt Express. 2010;18:10777–10785.

131 Beckley AM. Polarimetry and Beam Apodization Using Stress-engineered Optical Elements. PhD thesis. New York: University of Rochester; 2012.

132 Dewage AAG, Brown T. Interferometric polarimetry using full-Poincaré beams. Proc SPIE. 2021;117010N-1-9. Vella A, Alonso MA. Optimal birefringence distributions for imaging polarimetry. Opt Express. 2019;27:36799–36814. He et al. Light: Sci Appl. 2021;10:194, Page 17 of 20.

133 Vella A, Alonso MA. Chapter 7 – Maximum likelihood estimation in the context of an optical measurement. Prog Opt. 2020;65:231–311.

134 Vella AJ. Description and Applications of Space-variant Polarization States and Elements. PhD thesis. New York: University of Rochester; 2018.

135 Ramkhalawon RD, Brown TG, Alonso MA. Imaging the polarization of a light field. Opt Express. 2013;21:4106–4115.

136 Zimmerman BG, Brown TG. Star test image-sampling polarimeter. Opt Express. 2016;24:23154–23161.

137 Chue-Sang J, Gonzalez M, Pierre A, Laughrey M, Saytashev I, Novikova T, et al. Optical phantoms for biomedical polarimetry: a review. J Biomed Opt. 2019;24:030901.

138 Zhanghao K, Chen L, Yang XS, Wang MY, Jing ZL, Han HB, et al. Super-resolution dipole orientation mapping via polarization demodulation. Light: Sci Appl. 2016;5:e16166.

139 Zhanghao K, Gao J, Jin D, Zhang X, Xi P. Super-resolution fluorescence polarization microscopy. J Innovative Opt Health Sci. 2018;11:1730002.

140 Chen L, Chen X, Yang X, He C, Wang M, Xi P, et al. Advances of super-resolution fluorescence polarization microscopy and its applications in life sciences. Comput Struct Biotechnol J. 2020;18:2209–2216.

141 Wu PJ, Walsh Jr. JT. Stokes polarimetry imaging of rat-tail tissue in a turbid medium using incident circularly polarized light. Lasers Surg Med. 2005;37:396–406.

142 Macdonald C. Backscattering of circular polarized light from a disperse random medium influenced by optical clearing. Laser Phys Lett. 2011;8:324–328.

143 Qi J, He H, Lin J, Dong Y, Chen D, Ma H, et al. Assessment of tissue polarimetric properties using Stokes polarimetric imaging with circularly polarized illumination. J Biophotonics. 2018;11:e201700139.

144 Kunnen B, Macdonald C, Doronin A, Jacques S, Eccles M, Meglinski I. Application of circularly polarized light for non-invasive diagnosis of cancerous tissues and turbid tissue-like scattering media. J Biophotonics. 2015;8:317–323.

145 Xu M, Alfano RR. Circular polarization memory of light. Phys Rev E. 2005;72:065601.

146 Macdonald CM, Jacques SLV. Circular polarization memory in polydisperse scattering media. Phys Rev E. 2015;91:033204.

147 Wood MF, Ghosh N, Moriyama EH, Wilson BC, Vitkin IA. Proof-of-principle demonstration of a Mueller matrix decomposition method for polarized light tissue characterization in vivo. J Biomed Opt. 2009;14:014029.

148 Wang Y, He H, Chang J, He C, Liu S, Li M, et al. Mueller matrix microscope: a quantitative tool to facilitate detections and fibrosis scorings of liver cirrhosis and cancer tissues. J Biomed Opt. 2016;21:071112.

149 Liu T, Lu M, Chen B, Zhong Q, Li J, He H, et al. Distinguishing structural features between Crohn's disease and gastrointestinal luminal tuberculosis using Mueller matrix derived parameters. J Biophotonics. 2019;12:e201900151.

150 Shen YX, Huang R, He H, Liu S, Dong Y, Wu J, et al. Comparative study of the influence of imaging resolution on linear retardance parameters derived from the Mueller matrix. Biomed Opt Express. 2021;12:211–225.

151 Pierangelo A, Manhas S, Benali A, Fallet C, Totobenazara JL, Antonelli MR, et al. Multispectral Mueller polarimetric imaging detecting residual cancer and cancer regression after neoadjuvant treatment for colorectal carcinomas. J Biomed Opt. 2013;18:046014.

152 Sun M, He H, Zeng N, Du E, Guo Y, Liu S, et al. Characterizing the microstructures of biological tissues using Mueller matrix and transformed polarization parameters. Biomed Opt Express. 2014;5:4223–4234.

153 Li P, Lv D, He H, Ma H. Separating azimuthal orientation dependence in polarization measurements of anisotropic media. Opt Express. 2018;26:3791–3800.

154 Gil JJ. Invariant quantities of a Mueller matrix under rotation and retarder transformations. J Opt Soc Am A. 2016;33:52–58.

155 Iqbal M, Ahmad I, Khaliq A, Khan S. Comparative study of Mueller matrix transformation and polar decomposition for optical characterization of turbid media. Optik. 2020;224:165508.

156 Sun T, Liu T, He H, Wu J, Ma H. Distinguishing anisotropy orientations originated from scattering and birefringence of turbid media using Mueller matrix derived parameters. Opt Lett. 2018;43:4092–4095.

157 Khaliq A, Ashraf S, Gul B, Ahmad I. Comparative study of 3×3 Mueller matrix transformation and polar decomposition. Opt Commun. 2021;485:126756.

158 Tariq A, Li P, Chen D, Lv D, Ma H. Physically realizable space for the purity-depolarization plane for polarized light scattering media. Phys Rev Lett. 2017;119:033202.

159 Swami MK, Manhas S, Buddhiwant P, Ghosh N, Uppal A, Gupta PK. Polar decomposition of 3×3 Mueller matrix: a tool for quantitative tissue polarimetry. Opt Express. 2006;14:9324–9337.

160 Wang Y, Guo Y, Zeng N, Chen D, He H, Ma H. Study on the validity of 3×3 Mueller matrix decomposition. J Biomed Opt. 2015;20:065003.

161 Morio J, Goudail F. Influence of the order of diattenuator, retarder, and polarizer in polar decomposition of Mueller matrices. Opt Lett. 2004;29:2234–2236.

162 Ghosh N, Wood MFG, Vitkin IA. Influence of the order of the constituent basis matrices on the Mueller matrix decomposition-derived polarization parameters in complex turbid media such as biological tissues. Opt Commun. 2010;283:1200–1208.

163 Li P, Lee HR, Chandel S, Lotz C, Groeber-Becker FK, Dembski S, et al. Analysis of tissue microstructure with Mueller microscopy: logarithmic decomposition and Monte Carlo modeling. J Biomed Opt. 2020;25:015002.

164 Li P, Tariq A, He H, Ma H. Characteristic Mueller matrices for direct assessment of the breaking of symmetries. Opt Lett. 2020;45:706–709.

165 Vizet J, Rehbinder J, Deby S, Roussel S, Nazac A, Soufan R, et al. In vivo imaging of uterine cervix with a Mueller polarimetric colposcope. Sci Rep. 2017;7:2471.

166 Zhanghao K, Liu W, Li M, Wu Z, Wang X, Chen X, et al. High-dimensional super-resolution imaging reveals heterogeneity and dynamics of subcellular lipid membranes. Nat Commun. 2020;11:5890.

167 Zhanghao K, Chen X, Liu W, Li M, Liu Y, Wang Y, et al. Super-resolution imaging of fluorescent dipoles via polarized structured illumination microscopy. Nat Commun 2019;10:4694.

168 Spandana KU, Mahato KK, Mazumder N. Polarization-resolved Stokes–Mueller imaging: a review of technology and applications. Lasers Med Sci. 2019;34:1283–1293.

169 Yang B, Lesicko J, Sharma M, Hill M, Sacks MS, Tunnell JW. Polarized light spatial frequency domain imaging for nondestructive quantification of soft tissue fibrous structures. Biomed Opt Express. 2015;6:1520–1533.

170 Clancy NT, Arya S, Qi J, Stoyanov D, Hanna GB, Elson DS. Polarised stereo endoscope and narrowband detection for minimal access surgery. Biomed Opt Express. 2014;5:4108–4117.

171 Manhas S, Vizet J, Deby S, Vanel JC, Boito P, Verdier M, et al. Demonstration of full 4×4 Mueller polarimetry through an optical fiber for endoscopic applications. Opt Express. 2015;23:3047–3054.

172 Wood TC, Elson DS. Polarization response measurement and simulation of rigid endoscopes. Biomed Opt Express. 2010;1:463–470.

173 Vizet J, Manhas S, Tran J, Validire P, Benali A, Garcia-Caurel E, et al. Optical fiber-based full Mueller polarimeter for endoscopic imaging using a two-wavelength simultaneous measurement method. J Biomed Opt. 2016;21:071106.

174 Rivet S, Bradu A, Podoleanu A. 70 kHz full 4×4 Mueller polarimeter and simultaneous fiber calibration for endoscopic applications. Opt Express. 2015;23:23768–23786.

175 Forward S, Gribble A, Alali S, Lindenmaier AA, Vitkin IA. Flexible polarimetric probe for 3×3 Mueller matrix measurements of biological tissue. Sci Rep. 2017;7:11958.

176 Backman V, Gurjar R, Badizadegan K, Itzkan I, Dasari RR, Perelman LT, et al. Polarized light scattering spectroscopy for quantitative measurement of epithelial cellular structures in situ. IEEE J Sel Top Quantum Electron. 1999;5:1019–1026.

177 Chan D, Schulz B, Gloystein K, Müller HH, Rübhausen M. In vivo spectroscopic ellipsometry measurements on human skin. J Biomed Opt. 2007;12:014023. Banerjee P, Soni J, Purwar H, Ghosh N, Sengupta TK. Probing the fractal pattern and organization of Bacillus thuringiensis bacteria colonies growing under different conditions using quantitative spectral light scattering polarimetry. J Biomed Opt. 2013;18:035003.

178 Soni J, Purwar H, Lakhotia H, Chandel S, Banerjee C, Kumar U, et al. Quantitative fluorescence and elastic scattering tissue polarimetry using an Eigenvalue calibrated spectroscopic Mueller matrix system. Opt Express. 2013;21:15475–15489.

179 Jagtap J, Chandel S, Das N, Soni J, Chatterjee S, Pradhan A, et al. Quantitative Mueller matrix fluorescence spectroscopy for precancer detection. Opt Lett. 2014;39:243–246.

180 Satapathi S, Soni J, Ghosh N. Fluorescent Mueller matrix analysis of a highly scattering turbid media. Appl Phys Lett. 2014;104:131902.

181 Huse N, Schöenle A, Hell SW. Z-polarized confocal microscopy. J Biomed Opt. 2001;6:273–276.

182 Lim NSJ, Hamed Z, Yeow CH, Chan C, Huang Z. Early detection of biomolecular changes in disrupted porcine cartilage using polarized Raman spectroscopy. J Biomed Opt. 2011;16:017003.

183 Ahlawat S, Chowdhury A, Kumar N, Uppal A, Verma RS, Gupta PK. Polarized Raman spectroscopic investigations on hemoglobin ordering in red blood cells. J Biomed Opt. 2014;19:087002.

184 Chan KH, Chan AC, Fried WA, Simon JC, Darling CL, Fried D. Use of 2D images of depth and integrated reflectivity to represent the severity of demineralization in cross-polarization optical coherence tomography. J Biophotonics. 2015;8:36–45.

185 De Boer JF, Milner TE. Review of polarization sensitive optical coherence tomography and Stokes vector determination. J Biomed Opt. 2002;7:359–371.

186 Fan CM, Yao G. Imaging myocardial fiber orientation using polarization sensitive optical coherence tomography. Biomed Opt Express. 2013;4:460–465. He et al. Light: Sci Appl. 2021;10:194, Page 18 of 20.

187 Gladkova N, Kiseleva E, Robakidze N, Balalaeva I, Karabut M, Gubarkova E, et al. Evaluation of oral mucosa collagen condition with crosspolarization optical coherence tomography. J Biophotonics. 2013;6:321–329.

188 Hitzenberger CK, Götzinger E, Sticker M, Pircher M, Fercher AF. Measurement and imaging of birefringence and optic axis orientation by phase resolved polarization sensitive optical coherence tomography. Opt Express. 2001;9:780–790.

189 Hong YJ, Makita S, Sugiyama S, Yasuno Y. Optically buffered Jones-matrix-based multifunctional optical coherence tomography with polarization mode dispersion correction. Biomed Opt Express. 2015;6:225–243.

190 Jiao SL, Yao G, Wang LV. Depth-resolved two-dimensional Stokes vectors of backscattered light and Mueller matrices of biological tissue measured with optical coherence tomography. Appl Opt. 2000;39:6318–6324.

191 Lee RC, Kang H, Darling CL, Fried D. Automated assessment of the remineralization of artificial enamel lesions with polarization-sensitive optical coherence tomography. Biomed Opt Express. 2014;5:2950–2962.

192 Popescu DP, Sowa MG, Hewko MD, Choo-Smith LP. Assessment of early demineralization in teeth using the signal attenuation in optical coherence tomography images. J Biomed Opt. 2008;13:054053.

193 Sugita M, Pircher M, Zotter S, Baumann B, Roberts P, Makihira T, et al. Retinal nerve fiber bundle tracing and analysis in human eye by polarization sensitive OCT. Biomed Opt Express. 2015;6:1030–1054.

194 Vitkin A, Ghosh N, De Martino A. Tissue polarimetry. Photonics: Sci Found Technol Appl. 2015;4:239–321. Tuchin VV, Wang LV, Zimnyakov DA. Optical Polarization in Biomedical Applications. Springer; 2006.

195 Wang LV, Coté GL, Jacques SL. Special section guest editorial: tissue polarimetry. J Biomed Opt. 2002;7:278. Wang LV, Wu HI. Biomedical Optics: Principles and Imaging. John Wiley & Sons; 2009.

196 Wang S, Larin KV. Optical coherence elastography for tissue characterization: a review. J Biophotonics. 2015;8:279–302.

197 Yamanari M, Nagase S, Fukuda S, Ishii K, Tanaka R, Yasui T, et al. Scleral birefringence as measured by polarization-sensitive optical coherence tomography and ocular biometric parameters of human eyes in vivo. Biomed Opt Express. 2014;5:1391–1402.

198 Yao G, Wang LV. Two-dimensional depth-resolved Mueller matrix characterization of biological tissue by optical coherence tomography. Opt Lett. 1999;24:537–539.

199 Milione G, Dudley A, Nguyen TA, Chakraborty O, Karimi E, Forbes A, et al. Measuring the self-healing of the spatially inhomogeneous states of polarization of vector Bessel beams. J Opt. 2015;17:035617.

200 Mansfield JC, Winlove CP, Moger J, Matcher SJ. Collagen fiber arrangement in normal and diseased cartilage studied by polarization sensitive nonlinear microscopy. J Biomed Opt. 2008;13:044020.

201 Tanaka Y, Hase E, Fukushima S, Ogura Y, Yamashita T, Hirao T, et al. Motion-artifact-robust, polarization-resolved second-harmonic-generation microscopy based on rapid polarization switching with electro-optic Pockells cell and its application to in vivo visualization of collagen fiber orientation in human facial skin. Biomed Opt Express. 2014;5:1099–1113.

202 DeWalt EL, Sullivan SZ, Schmitt PD, Muir RD, Simpson GJ. Polarization-modulated second harmonic generation ellipsometric microscopy at video rate. Anal Chem. 2014;86:8448–8456.

203 Pavone FS, Campagnola PJ. Second Harmonic Generation Imaging. CRC Press; 2013.

204 Brasselet S. Polarization-resolved nonlinear microscopy: application to structural molecular and biological imaging. Adv Opt Photonics. 2011;3:205–271.

205 Kapsokalyvas D, Cicchi R, Bruscino N, Alfieri D, Prignano F, Massi D, et al. In-vivo imaging of psoriatic lesions with polarization multispectral dermoscopy and multiphoton microscopy. Biomed Opt Express. 2014;5:2405–2419.

206 Golaraei A, Cisek R, Krouglov S, Navab R, Niu C, Sakashita S, et al. Characterization of collagen in non-small cell lung carcinoma with second harmonic polarization microscopy. Biomed Opt Express. 2014;5:3562–3567.

207 Daly SM, Leahy MJ. 'Go with the flow': a review of methods and advancements in blood flow imaging. J Biophotonics 2013;6:217–255.

208 Wolf E. Introduction to the Theory of Coherence and Polarization of Light. Cambridge University Press; 2007.

209 Abrahamsson S, McQuilken M, Mehta SB, Verma A, Larsch J, Ilic R, et al. MultiFocus polarization microscope (MF-PolScope) for 3D polarization imaging of up to 25 focal planes simultaneously. Opt Express. 2015;23:7734–7754.

210 Chen DS, Zeng N, Wang Y, He H, Tuchin VV, Ma H. Study of optical clearing in polarization measurements by Monte Carlo simulations with anisotropic tissue-mimicking models. J Biomed Opt. 2016;21:081209.

211 Wang Y, He H, Chang J, Zeng N, Liu S, Li M, et al. Differentiating characteristic microstructural features of cancerous tissues using Mueller matrix microscope. Micron. 2015;79:8–15.

212 Hafi N, Grunwald M, Van Den Heuvel LS, Aspelmeier T, Chen JH, Zagrebelsky M, et al. Fluorescence nanoscopy by polarization modulation and polarization angle narrowing. Nat Methods. 2014;11:579–584.

213 Cruz CAV, Shaban HA, Kress A, Bertaux N, Monneret S, Mavrakis M, et al. Quantitative nanoscale imaging of orientational order in biological filaments by polarized superresolution microscopy. Proc Natl Acad Sci U S A. 2016;113:E820–E828.

214 Kampmann M, Atkinson CE, Mattheyses AL, Simon SM. Mapping the orientation of nuclear pore proteins in living cells with polarized fluorescence microscopy. Nat Struct Mol Biol. 2011;18:643–649.

215 Sase I, Miyata H, Ishiwata SI, Kinosita Jr. K. Axial rotation of sliding actin filaments revealed by singlefluorophore imaging. Proc Natl Acad Sci U S A. 1997;94:5646–5650.

216 Forkey JN, Quinlan ME, Alexander Shaw M, Corrie JE, Goldman YE. Three-dimensional structural dynamics of myosin V by single-molecule fluorescence polarization. Nature. 2003;422:399–404.

217 Sosa H, Peterman EJ, Moerner WE, Goldstein LS. ADP-induced rocking of the kinesin motor domain revealed by single-molecule fluorescence polarization microscopy. Nat Struct Biol. 2001;8:540–544.

218 DeMay BS, Bai X, Howard L, Occhipinti P, Meseroll RA, Spiliotis ET, et al. Septin filaments exhibit a dynamic, paired organization that is conserved from yeast to mammals. J Cell Biol. 2011;193:1065–1081.

219 DeMay BS, Noda N, Gladfelter AS, Oldenbourg R. Rapid and quantitative imaging of excitation polarized fluorescence reveals ordered septin dynamics in live yeast. Biophys J. 2011;101:985–994.

220 Axelrod D. Carbocyanine dye orientation in red cell membrane studied by microscopic fluorescence polarization. Biophys J. 1979;26:557–573.

221 Schütz GJ, Schindler H, Schmidt T. Imaging single-molecule dichroism. Opt Lett. 1997;22:651–653.

222 Du E, He H, Zeng N, Sun M, Guo Y, Wu J, Liu S, et al. Mueller matrix polarimetry for differentiating characteristic features of cancerous tissues. J Biomed Opt. 2014;19:076013.

223 Pierangelo A, Nazac A, Benali A, Validire P, Cohen H, Novikova T, et al. Polarimetric imaging of uterine cervix: a case study. Opt Express. 2013;21:14120–14130.

224 Rehbinder J, Haddad H, Deby S, Teig B, Nazac A, Novikova T, et al. Ex vivo Mueller polarimetric imaging of the uterine cervix: a first statistical evaluation. J Biomed Opt. 2016;21:071113.

225 Shukla P, Pradhan A. Mueller decomposition images for cervical tissue: potential for discriminating normal and dysplastic states. Opt Express. 2009;17:1600–1609.

226 Chue-Sang J, Bai Y, Stoff S, Straton D, Ramaswamy S, Ramella-Roman JC. Use of combined polarization-sensitive optical coherence tomography and Mueller matrix imaging for the polarimetric characterization of excised biological tissue. J Biomed Opt. 2016;21:071109.

227 Ahmad I, Ahmad M, Khan K, Ashraf S, Ahmad S, Ikram M. Ex vivo characterization of normal and adenocarcinoma colon samples by Mueller matrix polarimetry. J Biomed Opt. 2015;20:056012.

228 Pierangelo A, Benali A, Antonelli MR, Novikova T, Validire P, Gayet B, et al. Ex-vivo characterization of human colon cancer by Mueller polarimetric imaging. Opt Express. 2011;19:1582–1593.

229 Pierangelo A, Manhas S, Benali A, Fallet C, Antonelli MR, Novikova T, et al. Ex vivo photometric and polarimetric multilayer characterization of human healthy colon by multispectral Mueller imaging. J Biomed Opt. 2012;17:066009.

230 Dubreuil M, Babilotte P, Martin L, Sevrain D, Rivet S, Le Grand Y, et al. Mueller matrix polarimetry for improved liver fibrosis diagnosis. Opt Lett. 2012;37:1061–1063. Wang W, Lim LG, Srivastava S, Yan JS, Shabbir A, Liu Q. Roles of linear and circular polarization properties and effect of wavelength choice on differentiation between ex vivo normal and cancerous gastric samples. J Biomed Opt. 2014;19:046020.

231 Alali S, Aitken KJ, Schröder A, Gribble A, Bagli DJ, Vitkin IA. Assessment of local structural disorders of the bladder wall in partial bladder outlet obstruction using polarized light imaging. Biomed Opt Express. 2014;5:621–629.

232 Backman V, Wallace MB, Perelman LT, Arendt JT, Gurjar R, Müller MG, et al. Detection of preinvasive cancer cells. Nature. 2000;406:35–36.

233 Qi J, Elson DS. A high definition Mueller polarimetric endoscope for tissue characterisation. Sci Rep. 2016;6:25953.

234 Gan Y, Fleming CP. Extracting three-dimensional orientation and tractography of myofibers using optical coherence tomography. Biomed Opt Express. 2013;4:2150–2165. He et al. Light: Sci Appl. 2021;10:194, Page 19 of 20.

235 Pham HTT, Nguyen AL, Vo TV, Huynh KC, Phan QH. Optical parameters of human blood plasma, collagen, and calfskin based on the Stokes–Mueller technique. Appl Opt. 2018;57:4353–4359.

236 Lu R, Zhang Q, Zhi Y, Yao X. A polarization-sensitive light field imager for multi-channel angular spectroscopy of light scattering in biological tissues. Quant. Imaging Med Surg. 2015;5:1–8.

237 Ghosh N, Wood MF, Li SH, Weisel RD, Wilson BC, Li RK, et al. Mueller matrix decomposition for polarized light assessment of biological tissues. J Biophotonics. 2009;2:145–156.

238 Wood MF, Ghosh N, Wallenburg MA, Li SH, Weisel RD, Wilson BC, et al. Polarization birefringence measurements for characterizing the myocardium, including healthy, infarcted, and stem-cellregenerated tissues. J Biomed Opt. 2010;15:047009.

239 Ahmad I, Gribble A, Ikram M, Pop M, Vitkin A. Polarimetric assessment of healthy and radiofrequency ablated porcine myocardial tissue. J Biophotonics, 2016;9:750–759.

240 He H, He C, Chang J, Lv D, Wu J, Duan C, Zhou Q, et al. Monitoring microstructural variations of fresh skeletal muscle tissues by Mueller matrix imaging. J Biophotonics. 2017;10:664–673.

241 Sugita S, Matsumoto T. Quantitative measurement of the distribution and alignment of collagen fibers in unfixed aortic tissues. J Biomech. 2013;46:1403–1407.

242 Chen ZH, Meng R, Zhu Y, Ma H. A collinear reflection Mueller matrix microscope for backscattering Mueller matrix imaging. Opt Lasers Eng. 2020;129:106055.

243 Gonzalez M, Montejo KA, Krupp K, Srinivas V, DeHoog E, Madhivanan P, et al. Design and implementation of a portable colposcope Mueller matrix polarimeter. J Biomed Opt. 2020;25:116006.

244 Zotter S, Pircher M, Götzinger E, Torzicky T, Yoshida H, Hirose F, et al. Measuring retinal nerve fiber layer birefringence, retardation, and thickness using wide-field, high-speed polarization sensitive spectral domain OCT. Invest. Ophthalmol. Vis. Sci. 2013;54:72–84.

245 Dong Y, He H, Sheng W, Wu J, Ma H. A quantitative and non-contact technique to characterise microstructural variations of skin tissues during photo-damaging process based on Mueller matrix polarimetry. Sci Rep. 2017;7:14702.

246 Liu XY, Beaudette K, Wang X, Liu L, Bouma BE, Villiger M. Tissue-like phantoms for quantitative birefringence imaging. Biomed Opt Express. 2017;8:4454–4465. Swami MK, Patel HS, Geethu P, Uppal A, Kushwaha PK, Gupta PK. Effect of gold nanoparticles on depolarization characteristics of Intralipid tissue phantom. Opt Lett. 2013;38:2855–2857.

247 Guo YH, Liu C, Zeng N, He H, Du E, He Y, et al. Study on retardance due to well-ordered birefringent cylinders in anisotropic scattering media. J Biomed Opt. 2014;19:065001.

248 Li X, Liao R, Ma H, Leung PT, Yan M. Polarimetric learning: a Siamese approach to learning distance metrics of algal Mueller matrix images. Appl Opt. 2018;57:3829–3837.

249 Heinrich C, Rehbinder J, Nazac A, Teig B, Pierangelo A, Zallat J. Mueller polarimetric imaging of biological tissues: classification in a decision-theoretic framework. J Opt Soc Am A. 2018;35:2046–2057.

250 Wetzstein G, Ozcan A, Gigan S, Fan S, Englund D, Soljačić M, et al. Inference in artificial intelligence with deep optics and photonics. Nature. 2020;588:39–47.

251 Rivenson Y, Göröcs Z, Günaydin H, Zhang Y, Wang H, Ozcan A. Deep learning microscopy. Optica. 2017;4:1437–1443.

252 He C, Booth MJ. Vectorial adaptive optics: correction of polarization and phase. In Bifano TG, Gigan S, Ji N, editors. Adaptive Optics and Wavefront Control for Biological Systems VI. Vol. 11248 International Society for Optics and Photonics; 2020.

253 Hu Q, Dai Y, He C, Booth MJ. Arbitrary vectorial state conversion using liquid crystal spatial light modulators. Opt Commun. 2020;459:125028.

254 Hu Q, He C, Booth MJ. Arbitrary complex retarders using a sequence of spatial light modulators as the basis for adaptive polarisation compensation. J Opt. 2021;23:065602.

255 Dai YY, He C, Wang J, Turcotte R, Fish L, Wincott M, et al. Active compensation of extrinsic polarization errors using adaptive optics. Opt Express. 2019;27:35797–35810.

256 Rubin NA, D'Aversa G, Chevalier P, Shi Z, Chen WT, Capasso F. Matrix Fourier optics enables a compact full-Stokes polarization camera. Science. 2019;365:eaax1839.

257 Dorrah AH, Rubin NA, Zaidi A, Tamagnone M, Capasso F. Metasurface optics for on-demand polarization transformations along the optical path. Nat Photonics. 2021;15:287–296.

258 Pan T, Lu D, Xin H, Li B. Biophotonic probes for bio-detection and imaging. Light: Sci Appl. 2021;10:124.

259 Samim M, Krouglov S, Barzda V. Nonlinear Stokes-Mueller polarimetry. Phys Rev A. 2016;93:013847.

260 Kontenis L. Mirin, R. P., Gerrits, T., Nam, S. W., & Marsili, F. (2016). High-efficiency W-Si superconducting nanowire single-photon detectors for the 1.5 μm wavelength. In CLEO: Science and Innovations (pp. STu1R-3). Optical Society of America.

261 Okoro C. Second-harmonic Generation-based Mueller Matrix Polarization Analysis Of Collagen-rich Tissues. PhD thesis. University of Illinois, UrbanaChampaign, Urbana; 2018.

262 Azzam RMA. Arrangement of four photodetectors for measuring the state of polarization of light. Opt Lett. 1985;10:309–311.

263 Pishak V, Ushenko A, Gryhoryshyn P, Dubolazov O, Angelsky O, Zabolotna N, et al. Study of polarization structure of biospeckle fields in crosslinked tissues: I. Vector structure of skin biospeckles. Proc SPIE. 1997;3317:418–424.

264 Angelsky PO, Ushenko AG, Dubolazov AV, Sidor MI, Bodnar GB, Koval G, et al. The singular approach for processing polarization-inhomogeneous laser images of blood plasma layers. J Opt. 2013;15:044030.

265 Ushenko AG, Dubolazov AV, Ushenko VA, Novakovskaya OY. Statistical analysis of polarization-inhomogeneous Fourier spectra of laser radiation scattered by human skin in differentiation of benign and malignant formations. J Biomed Opt. 2016;21:071110.

266 Garazdyuk MS, Bachinskyi VT, Vanchulyak OYa, Ushenko AG, Dubolazov OV, Gorsky MP. Polarization-phase images of liquor polycrystalline films in determining time of death. Appl Opt. 2016;55:B67–B71.

267 Angelskii OV, Ushenko OG, Burkovets DN, Arkhelyuk OD, Ushenko YA. Polarization-correlation studies of multifractal structures in biotissues and diagnostics of their pathologic changes. Laser Phys. 2000;10:1136–1142.

268 Yermolenko S, Ushenko A, Ivashko P, Goudail F, Gruia I, Gavrilă C, et al. Spectropolarimetry of cancer change of biotissues. Proc SPIE. 2009;7388:404–410.

269 Jóźwicki R, Patorski K, Angelsky OV, Ushenko AG, Burkovets DN, Ushenko YA. Automatic polarimetric system for early medical diagnosis by biotissue testing. Opt Appl. 2002;32:603–612.

270 Ushenko AG, Dubolazov OV, Ushenko VA, Novakovskaya OY, Olar OV. Fourier polarimetry of human skin in differentiation of benign and malignant formations. Appl Opt. 2016;55:B56–B60.

271 Ushenko AG, Pishak VP. Coherent-domain optical methods: biomedical diagnostics, environmental and material science. In: Laser polarimetry of biological tissue: principles and applications. Kyiv; 2004.

272 Ushenko AG, Fediv AI, Marchuk YF. Correlation and fractal structure of Jones matrices of human bile secret. Proc SPIE. 2009;7368:73681Q.

273 Ushenko AG. Depolarization of a laser emission field as the correlation development of its polarization structure. Proc SPIE. 1997;3317:331–339.

274 Bachinskyi V, Boychuk T, Ushenko A. Laser Polarimetry of Biological Tissues and Fluids: Methods of Spatial-frequency, Singular and Correlation Analysis of Biological Layers Object Fields. LAP Lambert Academic Publishing; 2018.

275 Ushenko AG, Burkovets DN, Yermolenko SB, Arkhelyuk AD, Pishak VP, Yuzko AM, et al. Stokes polarimetry of biotissues. Proc SPIE. 1999;3904:527–533.

276 Ushenko VA, Dubolazov AV, Pidkamin LY, Sakchnovsky MY, Bodnar AB, Ushenko YA, et al. Mapping of polycrystalline films of biological fluids utilizing the Jones-matrix formalism. Laser Phys. 2018;28:025602.

277 Ushenko AG, Angelsky PO, Sidor M, Marchuk YF, Andreychuk DR, Pashkovskaya NV. Spatial-frequency selection of complex degree of coherence of laser images of blood plasma in diagnostics and differentiation of pathological states of human organism. Appl Opt. 2014;53:B172–B180.

278 Olar EI, Ushenko AG, Ushenko YA. Polarization correlation measurements of the phase tomograms of optically anisotropic biofractals. Laser Phys. 2004;14:1115–1121.

279 Guminetskiy SH, Ushenko OG, Polyanskiy IP, Motrych AV, Grynchuk FV. Optical method for investigation of the peritonitis progressing process. Proc SPIE. 2008;7008:626–632.

280 Ushenko AG, Fediv AI, Marchuk YF. Singular structure of polarization images of bile secret in diagnostics of human physiological state. In: European Conference on Biomedical Optics. Optica Publishing Group; 2009.

281 Ushenko Y, Ushenko A, Dubolazov A, Soltys I, Litvinenko O, Wanchuliak O, et al. Mueller-matrix interferometric multifractal scaling of optically anisotropic architectonics of diffuse blood facies: fundamental and applied aspects. J Biophotonics. 2025;18:e202400412. https://doi.org/10.1002/jbio.202400412.

282 Ushenko A, Soltys I, Dubolazov A, Ushenko Y, Bilookyi V, Bilookyi O, et al. 3D Jones-matrix thesiography of biological fluid facies. J Innov Opt Health Sci. 2025;18:2443002. https://doi.org/10.1142/S1793545824430028.

283 Ushenko A, Pavlyukovich N, Khukhlina O, Pavlyukovich O, Gorsky M, Soltys I, et al. Layer-by-layer multifractal scanning of optically anisotropic architectonics of blood plasma films: fundamental and applied aspects. Photonics. 2025;12:215. https://doi.org/10.3390/photonics12030215.

284 Ushenko A, Pavlyukovich N, Khukhlina O, Pavlyukovich O, Soltys I, Dubolazov A, et al. Blood plasma film multifractal scanning in COVID-19 consequences diagnostics. J Biophotonics. 2024;17:e202400356. https://doi.org/10.1002/jbio.202400356.

285 Ushenko A, Dubolazov A, Zheng J, Litvinenko A, Gorsky M, Ushenko Y, et al. 3D polarization-interference holographic histology for wavelet-based differentiation of biological tissues with different necrotic states. J Biomed Opt. 2024;29:052920. https://doi.org/10.1117/1.JBO.29.5.052920.

286 Ushenko O, Bilookyi O, Zheng J, Dubolazov A, Olar O, Ushenko Y, et al. 3D digital holographic polarimetry of laser speckle fields formed by polycrystalline blood films: a tool for differential diagnosis of thyroid pathology. Front Phys. 2024;12:1426469. https://doi.org/10.3389/fphy.2024.1426469.

287 Wójcik W, Hu Z, Ushenko Y, Smolarz A, Soltys I, Dubolazov O, et al. Optical sensor system for 3D Jones matrix reconstruction of optical anisotropy maps of self-assembled polycrystalline soft matter films. Sensors. 2024;24:1589. https://doi.org/10.3390/s24051589.

Section 1.1.1.6

1 Huang D, Swanson E, Lin C, Schuman J, Stinson W, Chang W, et al. Optical coherence tomography. Science. 1991;254(5035):1178–1181. https://doi.org/10.1126/science.1957169.

2 Drexler W, Morgner U, Kärtner FX, Pitris C, Boppart SA, Li XD, et al. In vivo ultrahigh-resolution optical coherence tomography. Opt Lett. 1999;24(17):1221–1223. https://doi.org/10.1364/OL.24.001221.

3 Zhang N, Wang C, Sun Z, Li Z, Xie L, Yan Y, et al. Detection of latent fingerprint hidden beneath adhesive tape by optical coherence tomography. Forensic Sci Int. 2018;287:81–87. https://doi.org/10.1016/j.forsciint.2018.03.030.

4 de Freitas AZ, Ribeiro LAD, Neto ON, Sarkis JES, Siqueira AN. Opticalcoherence-tomography-based algorithm for handwriting forensic analysis. In: Ai and Optical Data Sciences. Proceedings of SPIE. 2020;11299:11299P. https://doi.org/10.1117/12.2543356

5 Wang C, Zhang N, Sun Z, Li Z, Li Z, Xu X. Recovering hidden sublayers of repainted automotive paint by 3D optical coherence tomography. Aust J Forensic Sci. 2019;51(3):331–339. https://doi.org/10.1080/00450618.2017.1367418.

6 Anna T, Shakher C, Singh Mehta D. Simultaneous tomography and topography of silicon integrated circuits using full-field swept-source optical coherence tomography. J Opt A: Pure Appl Opt. 2009;11(4):045501. https://doi.org/10.1088/1464-4258/11/4/045501.

7 Song G, Harding K. OCT for industrial applications. In: Optical Metrology and Inspection for Industrial Applications Ii. Proceedings of SPIE. 2012:8563:8563. https://doi.org/10.1117/12.2000462.

8 Zorin I, Gattinger P, Brandstetter M, Heise B. Dual-band infrared optical coherence tomography using a single supercontinuum source. Opt Express. 2020;28(6):7858–7874. https://doi.org/10.1364/oe.386398.

9 Zorin I, Su R, Prylepa A, Kilgus J, Brandstetter M, Heise B. Mid-infrared Fourier-domain optical coherence tomography with a pyroelectric linear array. Opt Express. 2018;26(25):33428–33439. https://doi.org/10.1364/oe.26.033428.

10 Drexler W. Ultrahigh-resolution optical coherence tomography. J Biomed Opt. 2004;9(1):47–74. https://doi.org/10.1117/1.1629679.

11 Runge VM, Richter JK, Heverhagen JT. Speed in clinical magnetic resonance. Invest Radiol. 2017;52(1):1–17. https://doi.org/10.1097/rli.0000000000000330.

12 Holdsworth S, Bammer R. Magnetic resonance imaging techniques: fMRI, DWI, and PWI. Semin Neurol. 2008;28(4):395–406. https://doi.org/10.1055/s0028-1083697.

13 Muller NL. Computed tomography and magnetic resonance imaging: past, present and future. Eur Respir J Suppl. 2002;35:3s–12s. https://doi.org/10.1183/09031936.02.00248202.

14 Fercher AF, Drexler W, Hitzenberger CK, Lasser T. Optical coherence tomography – principles and applications. Rep Prog Phys. 2003;66(2):239–303. https://doi.org/10.1088/0034-4885/66/2/204.

15 Wang C, Kim J, Jin CT, Leong PHW, McEwan A. Near infrared spectroscopy in optical coherence tomography. J near Infrared Spectrosc. 2012;20(1):237–247. https://doi.org/10.1255/jnirs.975.

16 Drexler W, Liu M, Kumar A, Kamali T, Unterhuber A, Leitgeb RA. Optical coherence tomography today: speed, contrast, and multimodality. J Biomed Opt. 2014;19(7):071412. https://doi.org/10.1117/1.Jbo.19.7.071412.

17 Scholler J, Groux K, Goureau O, Sahel J-A, Fink M, Reichman S, et al. Dynamic full-field optical coherence tomography: 3D live-imaging of retinal organoids. Light Sci Appl. 2020;9(1):140. https://doi.org/10.1038/s41377-020-00375-8.

18 Stremplewski P, Auksorius E, Wnuk P, Kozoń Ł, Garstecki P, Wojtkowski M. In vivo volumetric imaging by crosstalk-free full-field OCT. Optica. 2019;6(5):608–617. https://doi.org/10.1364/optica.6.000608.

19 Ogien J, Dubois A. A compact high-speed full-field optical coherence microscope for high-resolution in vivo skin imaging. J Biophotonics. 2019;12(2):e201800208. https://doi.org/10.1002/jbio.201800208.

20 Ju MJ, Lee SJ, Min EJ, Kim Y, Kim HY, Lee BH. Evaluating and identifying pearls and their nuclei by using optical coherence tomography. Opt Express. 2010;18(13):13468–13477. https://doi.org/10.1364/OE.18.013468.

21 Choi WJ, Jeon DI, Ahn S-G, Yoon J-H, Kim S, Lee BH. Full-field optical coherence microscopy for identifying live cancer cells by quantitative measurement of refractive index distribution. Opt Express. 2010;18(22):23285–23295. https://doi.org/10.1364/oe.18.023285.

22 Leitgeb RA, Werkmeister RM, Blatter C, Schmetterer L. Doppler optical coherence tomography. Prog Retin Eye Res. 2014;41:26–43. https://doi.org/10.1016/j.preteyeres.2014.03.004.

23 Elahi S, Gu S, Thrane L, Rollins AM, Jenkins MW. Complex regression doppler optical coherence tomography. J Biomed Opt. 2018;23(4):1. https://doi.org/10.1117/1.Jbo.23.4.046009.

24 Lauri J, Myllylä R. Doppler OCT imaging of cytoplasm shuttle flow in *Physarum polycephalum*. J Biophoton. 2009;2(8-9):540–547. https://doi.org/10.1002/jbio.200910057.

25 White BR, Pierce MC, Nassif N, Cense B, Park BH, Tearney GJ, et al. In vivo dynamic human retinal blood flow imaging using ultra-high-speed spectral domain optical doppler tomography. Opt Express. 2003;11(25):3490–3497. https://doi.org/10.1364/oe.11.003490.

26 Beer F, Wartak A, Haindl R, Gröschl M, Baumann B, Pircher M, et al. Conical scan pattern for enhanced visualization of the human cornea using polarization-sensitive OCT. Biomed Opt Express. 2017;8(6):2906–2923. https://doi.org/10.1364/boe.8.002906.

27 de Boer JF, Milner TE. Review of polarization sensitive optical coherence tomography and Stokes vector determination. J Biomed Opt. 2002;7(3):359–371. https://doi.org/10.1117/1.1483879.

28 de Boer JF, Milner TE, van Gemert MJC, Nelson JS. Two-dimensional birefringence imaging in biological tissue by polarization-sensitive optical coherence tomography. Opt Lett. 1997;22(12):934–936. https://doi.org/10.1364/ol.22.000934.

29 Zhang Y, Dong B, Bai Y, Ye S, Lei Z, Zhou Y. Measurement of depth resolved thermal deformation distribution using phase-contrast spectral optical coherence tomography. Opt Express. 2015;23(21):28067–28075. https://doi.org/10.1364/oe.23.028067.

30 Yan Y, Ding Z, Wang L, Wang C, Shen Y. High-sensitive quantitative phase imaging with averaged spectral domain phase microscopy. Opt Commun. 2013;303:21–24. https://doi.org/10.1016/j.optcom.2013.04.015.

31 Wojtkowski M. High-speed optical coherence tomography: basics and applications. Appl Opt. 2010;49(16):D30–D61. https://doi.org/10.1364/ao.49.000d30.

32 Ramamoorthy S, Zhang Y, Petrie T, Fridberger A, Ren T, Wang R, et al. Minimally invasive surgical method to detect sound processing in the cochlear apex by optical coherence tomography. J Biomed Opt. 2016;21(2):025003. https://doi.org/10.1117/1.Jbo.21.2.025003.

33 Auksorius E, Borycki D, Stremplewski P, Liżewski K, Tomczewski S, Niedźwiedziuk P, et al. In vivo imaging of the human cornea with high-speed and high-resolution fourier-domain full-field optical coherence tomography. Biomed Opt Express. 2020;11(5):2849–65. https://doi.org/10.1364/boe.393801.

34 Pishak V, Ushenko A, Gryhoryshyn P, Dubolazov O, Angelsky O, Zabolotna N, et al. Study of polarization structure of biospeckle fields in crosslinked tissues: I. Vector structure of skin biospeckles. Proc SPIE. 1997;3317:418–424.

35 Angelsky PO, Ushenko AG, Dubolazov AV, Sidor MI, Bodnar GB, Koval G, et al. The singular approach for processing polarization-inhomogeneous laser images of blood plasma layers. J Opt. 2013;15:044030.

36 Ushenko AG, Dubolazov AV, Ushenko VA, Novakovskaya OY. Statistical analysis of polarization-inhomogeneous Fourier spectra of laser radiation scattered by human skin in differentiation of benign and malignant formations. J Biomed Opt. 2016;21:071110.

37 Garazdyuk MS, Bachinskyi VT, Vanchulyak OYa, Ushenko AG, Dubolazov OV, Gorsky MP. Polarization-phase images of liquor polycrystalline films in determining time of death. Appl Opt. 2016;55:B67–B71.

38 Angelskii OV, Ushenko OG, Burkovets DN, Arkhelyuk OD, Ushenko YA. Polarization-correlation studies of multifractal structures in biotissues and diagnostics of their pathologic changes. Laser Phys. 2000;10:1136–1142.

39 Yermolenko S, Ushenko A, Ivashko P, Goudail F, Gruia I, Gavrilă C, et al. Spectropolarimetry of cancer change of biotissues. Proc SPIE. 2009;7388:404–410.

40 Jóźwicki R, Patorski K, Angelsky OV, Ushenko AG, Burkovets DN, Ushenko YA. Automatic polarimetric system for early medical diagnosis by biotissue testing. Opt Appl. 2002;32:603–612.

41 Ushenko AG, Dubolazov OV, Ushenko VA, Novakovskaya OY, Olar OV. Fourier polarimetry of human skin in differentiation of benign and malignant formations. Appl Opt. 2016;55:B56–B60.

42 Ushenko AG, Pishak VP. Coherent-domain optical methods: biomedical diagnostics, environmental and material science. In: Laser polarimetry of biological tissue: principles and applications. Kyiv; 2004.

43 Ushenko AG, Fediv AI, Marchuk YF. Correlation and fractal structure of Jones matrices of human bile secret. Proc SPIE. 2009;7368:73681Q.

44 Ushenko AG. Depolarization of a laser emission field as the correlation development of its polarization structure. Proc SPIE. 1997;3317:331–339.

45 Bachinskyi V, Boychuk T, Ushenko A. Laser Polarimetry of Biological Tissues and Fluids: Methods of Spatial-Frequency, Singular and Correlation Analysis of Biological Layers Object Fields. LAP Lambert Academic Publishing; 2018.

46 Ushenko AG, Burkovets DN, Yermolenko SB, Arkhelyuk AD, Pishak VP, Yuzko AM, et al. Stokes polarimetry of biotissues. Proc SPIE. 1999;3904:527–533.

47 Ushenko VA, Dubolazov AV, Pidkamin LY, Sakchnovsky MY, Bodnar AB, Ushenko YA, et al. Mapping of polycrystalline films of biological fluids utilizing the Jones-matrix formalism. Laser Phys. 2018;28:025602.

48 Ushenko AG, Angelsky PO, Sidor M, Marchuk YF, Andreychuk DR, Pashkovskaya NV. Spatial-frequency selection of complex degree of coherence of laser images of blood plasma in diagnostics and differentiation of pathological states of human organism. Appl Opt. 2014;53:B172–B180.

49 Olar EI, Ushenko AG, Ushenko YA. Polarization correlation measurements of the phase tomograms of optically anisotropic biofractals. Laser Phys. 2004;14:1115–1121.

50 Guminetskiy SH, Ushenko OG, Polyanskiy IP, Motrych AV, Grynchuk FV. Optical method for investigation of the peritonitis progressing process. Proc SPIE. 2008;7008:626–632.

51 Ushenko AG, Fediv AI, Marchuk YF. Singular structure of polarization images of bile secret in diagnostics of human physiological state. In: European Conference on Biomedical Optics. Optica Publishing Group; 2009.

52 Ushenko Y, Ushenko A, Dubolazov A, Soltys I, Litvinenko O, Wanchuliak O, et al. Mueller-matrix interferometric multifractal scaling of optically anisotropic architectonics of diffuse blood facies: fundamental and applied aspects. J Biophotonics. 2025;18:e202400412. https://doi.org/10.1002/jbio.202400412.

53 Ushenko A, Soltys I, Dubolazov A, Ushenko Y, Bilookyi V, Bilookyi O, et al. 3D Jones-matrix thesiography of biological fluid facies. J Innov Opt Health Sci. 2025;18:2443002. https://doi.org/10.1142/S1793545824430028.

54 Ushenko A, Pavlyukovich N, Khukhlina O, Pavlyukovich O, Gorsky M, Soltys I, et al. Layer-by-layer multifractal scanning of optically anisotropic architectonics of blood plasma films: fundamental and applied aspects. Photonics. 2025;12:215. https://doi.org/10.3390/photonics12030215.

55 Ushenko A, Pavlyukovich N, Khukhlina O, Pavlyukovich O, Soltys I, Dubolazov A, et al. Blood plasma film multifractal scanning in COVID-19 consequences diagnostics. J Biophotonics. 2024;17:e202400356. https://doi.org/10.1002/jbio.202400356.

56 Ushenko A, Dubolazov A, Zheng J, Litvinenko A, Gorsky M, Ushenko Y, et al. 3D polarization-interference holographic histology for wavelet-based differentiation of biological tissues with different necrotic states. J Biomed Opt. 2024;29:052920. https://doi.org/10.1117/1.JBO.29.5.052920.

57 Ushenko O, Bilookyi O, Zheng J, Dubolazov A, Olar O, Ushenko Y, et al. 3D digital holographic polarimetry of laser speckle fields formed by polycrystalline blood films: a tool for differential diagnosis of thyroid pathology. Front Phys. 2024;12:1426469. https://doi.org/10.3389/fphy.2024.1426469.

58 Wójcik W, Hu Z, Ushenko Y, Smolarz A, Soltys I, Dubolazov O, et al. Optical sensor system for 3D Jones matrix reconstruction of optical anisotropy maps of self-assembled polycrystalline soft matter films. Sensors. 2024;24:1589. https://doi.org/10.3390/s24051589.

2

Physical Principles, Algorithms, and Methods of 3D Jones–Mueller Digital Holographic Mapping

Yurii Ushenko[1,2], *Weidong Dou*[1], *Oleksandr Ushenko*[2,3], *Oleksandr Dubolazov*[2], *Iryna Soltys*[2], *Viktor Bachinskiy*[4], *Marta Garazdyuk*[4], *and Oleksandr Olar*[2]

[1] *Shaoxing University, Shaoxing, China*
[2] *Yurii Fedkovych Chernivtsi National University, Chernivtsi, Ukraine*
[3] *Taizhou Institute of Zhejiang University, Taizhou, China*
[4] *Bukovinian State Medical University, Chernivtsi, Ukraine*

2.1 Physical Principles of 3D Jones–Mueller Digital Holographic Mapping

The core principles of this chapter are based on the objective fact that physiological or pathological processes in living organisms primarily involve specific changes in the structure of proteins and other organic molecules. These changes are highly informative for studying homeostasis at the molecular level and can form the basis for diagnosing various diseases at their earliest stages.

Modern laboratory and clinical diagnostics are mainly focused on two areas:

- determination of numerous quantitative indicators contained in cellular elements of biological fluids, products of the body's vital activity.
- analysis and processing of medical images – analog or digital distributions of the magnitude of attenuation of electromagnetic or acoustic radiation.

Historically, the morphological "imaging" analysis or introscopy of the polycrystalline structure in biological tissues and fluids, which serve as fundamental life support systems, has been underdeveloped in clinical practice. This situation defines the focus, subject, and methodology of the monograph, which aims to analyze and substantiate the latest hardware and software systems for laser polarization introscopy. These systems utilize intelligent processing of multiparametric Jones–Mueller matrix medical images of optically anisotropic, polycrystalline architectures in diffuse biological tissues, and fluids of human organs. The goal is to enable early diagnosis and differentiation of stages in common diseases such as thyroid cancer, prostate and uterine abnormalities, preinfarction conditions, myocardial infarction, type 2 diabetes, COVID-19, and others.

The objects of study are:

- Native histological sections of biological tissues with various structures and physiological states, free of staining and fixation, allow direct (without chemical distortions) polarization–interference methods. These methods enable layer-by-layer polarization, polarization-correlation maps, and Mueller-matrix images of the tissue's optically anisotropic components in the studied organ.
- Dehydrated films of biological fluids have a supramolecular optically anisotropic structure that contains information about the composition and ratio of dissolved substances. This allows for the potential to assess the organism's condition through analysis of its digitally holographically reconstructed layer-by-layer macroscopic Jones-matrix images and optical anisotropy maps.

The monograph's overall approach involves systematizing, analyzing, and predicting the evolution of innovative biomedical polarization Jones–Mueller matrix digital holographic imaging technologies. These are used for screening and early detection of pathological and necrotic changes in human organs [1–61].

2.2 Phenomenological Models of Optically Anisotropic Tissue and Fluid Samples

2.2.1 Partial Mueller Matrices of Optically Anisotropic Architecture of Biological Tissues

This investigation's theory is based on modeling the phase (circular and linear birefringence) and amplitude (circular and linear dichroism) anisotropy of the polycrystalline structure of biological layers.

Circular Birefringence

Mueller-matrix circular birefringence or optical activity of biochemical molecules has the form

$$\{\Omega\} = \begin{Vmatrix} 1 & 0 & 0 & 0 \\ 0 & \omega_{22} & \omega_{23} & 0 \\ 0 & \omega_{32} & \omega_{33} & 0 \\ 0 & 0 & 0 & 1 \end{Vmatrix}, \quad \omega_{ik} = \begin{cases} \omega_{22} = \omega_{33} = \cos 2\theta \\ \omega_{23} = -\omega_{32} = \sin 2\theta \end{cases}, \tag{2.1}$$

where θ is the rotation angle of the polarization plane of the light beam.

The Linear Birefringence

Linear birefringence of molecular chains is characterized by the Mueller matrix $\{D\}$ of the following form

$$\{D\} = \begin{Vmatrix} 1 & 0 & 0 & 0 \\ 0 & d_{22} & d_{23} & d_{24} \\ 0 & d_{32} & d_{33} & d_{34} \\ 0 & d_{42} & d_{43} & d_{44} \end{Vmatrix}, \quad d_{ik} = \begin{cases} d_{22} = \cos^2 2\rho + \sin^2 2\rho \cos \delta; \\ d_{23} = d_{32} = \cos 2\rho \sin 2\rho(1 - \cos \delta); \\ d_{33} = \sin^2 2\rho + \cos^2 2\rho \cos \delta; \\ d_{42} = -d_{24} = \sin 2\rho \sin \delta; \\ d_{34} = -d_{43} = \cos 2\rho \sin \delta; \\ d_{44} = \cos \delta. \end{cases} \tag{2.2}$$

Here ρ is the direction of optical axes; $\delta = \frac{2\pi}{\lambda}\delta nl$ is the phase shift between linearly polarized orthogonal components of the light beam amplitude; λ is the wavelength; Δn is the birefringence; and l is the geometrical thickness.

Circular Dichroism

The Mueller matrix for circular dichroism is recorded as follows:

$$\{\Phi\} = \begin{Vmatrix} 1 & 0 & 0 & \varphi_{14} \\ 0 & \varphi_{22} & 0 & 0 \\ 0 & 0 & \varphi_{33} & 0 \\ \varphi_{41} & 0 & 0 & 1 \end{Vmatrix}, \quad \varphi_{ik} = \begin{cases} \varphi_{22} = \varphi_{33} = \dfrac{1 - C^2}{1 + C^2} \\ \varphi_{14} = \varphi_{41} = \dfrac{\pm 2C}{1 + C^2} \end{cases}. \tag{2.3}$$

Here, $= \frac{g_{\otimes} - g_{\oplus}}{g_{\otimes} + g_{\oplus}}$, $g_{\otimes}$, $g_{\oplus}$ are the absorption coefficients of left- ($\otimes$) and right-hand ($\oplus$) circularly polarized components of the light beam amplitude.

Linear Dichroism

Linear dichroism of the Mueller matrix is measured in the following way:

$$\{\Psi\} = \begin{Vmatrix} 1 & \varphi_{12} & \varphi_{13} & 0 \\ \varphi_{21} & \varphi_{22} & \varphi_{23} & 0 \\ \varphi_{31} & \varphi_{32} & \varphi_{33} & 0 \\ 0 & 0 & 0 & \varphi_{44} \end{Vmatrix}, \quad \varphi_{ik} = \begin{cases} \varphi_{12} = \varphi_{21} = (1-\Delta\tau)\cos 2\rho; \\ \varphi_{13} = \varphi_{31} = (1-\Delta\tau)\sin 2\rho; \\ \varphi_{22} = (1-\Delta\tau)\cos^2 2\rho + 2\sqrt{\Delta\tau}\sin^2 2\rho; \\ \varphi_{13} = \varphi_{31} = (1-\Delta\tau)\sin 2\rho; \\ \varphi_{22} = (1-\Delta\tau)\sin^2 2\rho + 2\sqrt{\Delta\tau}\cos^2 2\rho; \\ \varphi_{44} = 2\sqrt{\Delta\tau}. \end{cases} \tag{2.4}$$

Here, $\Delta\tau = \frac{\tau_x}{\tau_y}$, $\begin{cases} \tau_x = \tau\cos\rho; \\ \tau_y = \tau\sin\rho. \end{cases}$, τ_x, τ_y are the absorption coefficients of linearly polarized orthogonal components of the light beam amplitude.

The resulting Mueller matrix of the generalized optical anisotropy is written in the following form:

$$\{F\} = \{\Phi\}\{\Psi\}\{D\}\{\Omega\}.$$

Here,

$$\{F\} = \prod_{i=1}^{4}\{F_i\} = f_{11}^{-1} \times \begin{Vmatrix} 1 & f_{12} & f_{13} & f_{14} \\ f_{21} & f_{22} & f_{23} & f_{24} \\ f_{31} & f_{32} & f_{33} & f_{34} \\ f_{41} & f_{42} & f_{43} & f_{44} \end{Vmatrix}. \tag{2.5}$$

Here, $\{F_i\}$ is the Mueller matrix that characterizes the phase anisotropy ($\{\Omega\}$, $\{D\}$) and also amplitude anisotropy ($\{\Phi\}$, $\{\Psi\}$) molecules and their aggregates.

2.2.1.1 Examples of Mueller-matrix Images of Native Histological Sections of Biological Tissues

This section presents examples of a set of Mueller-matrix images of native histological sections of myocardial fibrillar tissue (Figure 2.1) and liver parenchymal tissue (Figure 2.2).

2.2.1.2 A Brief Stokes-polarimetric Theory of the Formation of Polarization-inhomogeneous Object Fields of Biospeckles

Optical techniques for diagnosing various polycrystalline objects, including layers of biological tissues, and visualizing their structure are prominent because of their high informativeness and ability to enable multifunctional monitoring of the investigated environment, such as photometric, spectral, polarization, and correlation monitoring. However, there is currently no unified methodological approach for diagnosing these structures.

Polarimetric research has become a specialized area within optical analysis of biological tissues. Studying the polarization features of scattered light provides novel insights into the morphological and physiological conditions of tissues, such as cataracts in the lens, glucose levels in diabetic tissues, and malignant transformations.

A major breakthrough in biomedical polarimetric diagnostics is the creation of Mueller-matrix methods, which have given rise to Mueller-matrix microscopy as a distinct discipline.

This method for optical diagnostics integrates instrumental Mueller-matrix imaging polarimetry with diverse theoretical models and analysis techniques, such as the regression model of optical anisotropy, logarithmic Mueller-matrix decomposition, Monte Carlo simulations of polarized radiation conversion, and statistical analysis of Mueller-matrix images and optical anisotropy maps.

The core findings from MMM of biological tissues show promise for both clinical and preclinical research on the polycrystalline structure of biological tissues. Notably, it has been demonstrated that quantitative optical metrics can be derived to track the progression of gastric tissue from health to inflammation and then to cancer, using Mueller microscopy of gastric biopsies, along with a regression model of optical anisotropy and statistical analysis of the images.

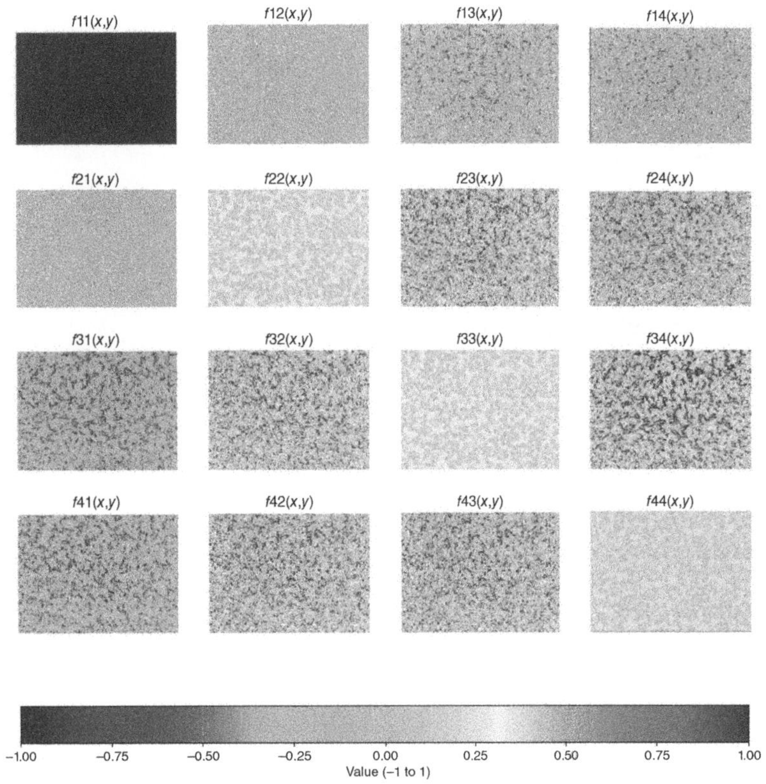

Figure 2.1 Mueller-matrix images of a native histological section of myocardial fibrillar tissue.

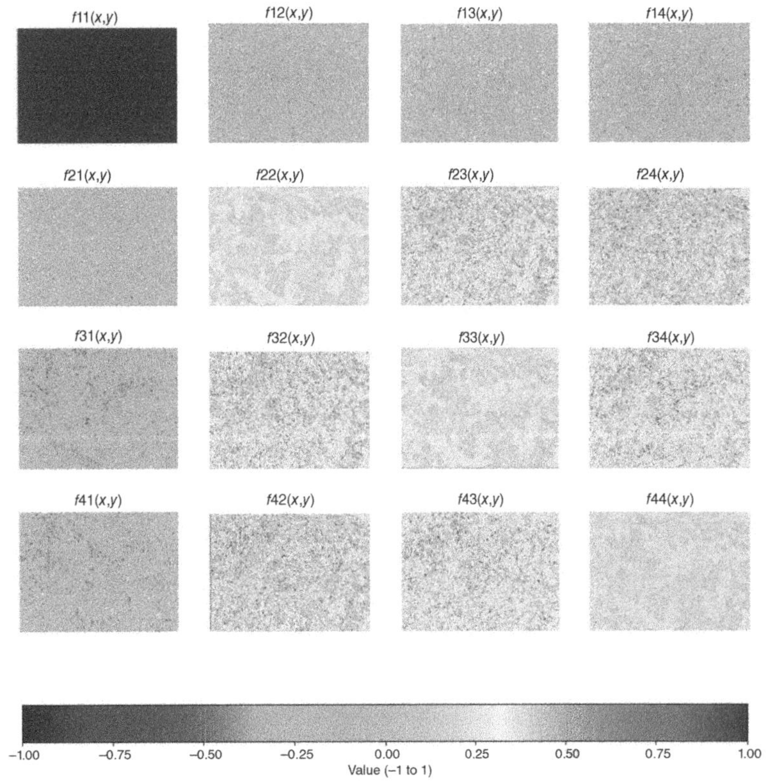

Figure 2.2 Mueller-matrix images of a native histological section of liver parenchymal tissue.

At the same time, methods and systems for laser polarimetry of biological tissue and liquid samples have been extensively developed, employing various analytical algorithms like statistical, correlation, fractal, wavelet, and Fourier analyses for quantitative data processing.

A key challenge in polarimetric mapping is its reliance on the depolarization degree of laser radiation. This dependence not only averages out polarization distributions but also hinders clear correlations between polarization-inhomogeneous object fields and the properties of polycrystalline soft matter layers. Consequently, developing new, reproducible, and more precise laser polarimetry techniques is of significant importance.

A major breakthrough in this area has been the successful combination of polarimetric and interferometric mapping techniques, which has paved the way for three-dimensional (3D) polarimetry. Currently, only preliminary yet promising research exists in this field. These studies have shown new potential for 3D polarimetric differential diagnosis in biomedical applications, particularly for distinguishing between benign and malignant tumors in human organs.

Nonetheless, the results remain somewhat empirical at this stage. There is currently no data on how this technique performs with tissue samples that vary in optical thickness, morphology, and pathology. From a physical perspective, the main challenge is to establish scientific relevance by linking the structure of polycrystalline networks in biological tissues to the layer-by-layer distributions of polarization parameters, such as azimuth and ellipticities.

A key future goal in polarimetric biomedical optics is to create a universal description of how optical fields are formed in anisotropic polycrystalline biological layers. This should include a broader scenario – detecting and selecting the polarimetric components of the object laser field, even when light scattering varies, through digital holographic reconstruction. Combining polarimetric and interferometric techniques is expected to yield new insights into the morphological and optical anisotropic structures of diffuse layers in biological tissues, which contain diverse architectural anisotropic polycrystalline elements.

2.2.1.2.1 A Brief Theory

In this section, we will briefly consider (without compromising the completeness of the analysis) the main theoretical relationships within the framework of the phase anisotropy approximation (linear LB and circular CB birefringence). These relationships illustrate how polarization structures form in both singly and multiply scattered components of the object laser field within the diffuse layer of biological tissue.

The diffuse layer of biological tissue is illuminated with a linearly polarized beam with an azimuth angle α_0 and the Stokes vector of the beam

$$S^0(\alpha_0) = \begin{pmatrix} 1 \\ \cos 2\alpha_0 \\ \sin 2\alpha_0 \\ 0 \end{pmatrix}. \tag{2.6}$$

To each optical anisotropic (Δn; $\Delta\mu$ is the linear and circular birefringence indices) inhomogeneity with a geometric size d and spatial orientation of the optical axis ρ in the volume of a biological tissue sample, we associate a 4×4 Mueller-matrix operator

$$\{F\}_i = \begin{pmatrix} f_{11}(\rho;\delta;\xi) & \cdots & f_{14}(\rho;\delta;\xi) \\ \vdots & \ddots & \vdots \\ f_{41}(\rho;\delta;\xi) & \cdots & f_{44}(\rho;\delta;\xi) \end{pmatrix}_i, \tag{2.7}$$

where f_{ik} is the set of matrix elements, $\delta = (2\pi/\lambda)\Delta n d$ and $\xi = (2\pi/\lambda)\Delta\mu d$ is the phase shift of linear LB and circular CB birefringence.

The process of local (ith) single transformation of the polarization structure of the probing beam $S^0(\alpha_0)$ is described by the following matrix equation

$$S_i^*(\alpha_i, \beta_i) = \{F\}_i S^0(\alpha_0). \tag{2.8}$$

As a result, a partial laser wave is formed with the following azimuth α_i and ellipticity β_i of polarization

$$\alpha_i = 0.5 \arctan\left(\frac{S_3^*}{S_2^*}\right); \tag{2.9}$$

$$\beta_i = 0.5\,\text{arcin}\left(\frac{S_4^*}{S_1^*}\right). \tag{2.10}$$

Here, $S_{i=1;2;3;4}^*$ is the parameters of the Stokes vector of a singly scattered laser beam $S_i^*(\alpha_i, \beta_i)$.

Thus, the singly scattered component of the object field of a diffuse biological layer is a polarization distribution of azimuths $A(\alpha_i)$ and ellipticities $B(\beta_i)$.

For multiple acts of interaction with the linearly polarized laser probe $S^0(\alpha_0)$, the matrix equation (2.9) takes the following form

$$S_i^n(\alpha_i^n; \beta_i^n) = \{F\}_n\{F\}_{n-1} \dots \{F\}_2\{F\}_1 S^0(\alpha_0). \tag{2.11}$$

This optical scenario leads to the formation of the following random values of azimuth α_i^n and ellipticity of polarization β_i^n at each point of the n times scattered object field

$$\alpha_i^n = 0.5\,\arctan\left(\frac{\sum_{i=1}^{n} S_{3i}^*}{\sum_{i=1}^{n} S_{2i}^*}\right); \tag{2.12}$$

$$\beta_i^n = 0.5\,\arcsin\left(\frac{\sum_{i=1}^{n} S_{4i}^*}{\sum_{i=1}^{n} S_{1i}^*}\right). \tag{2.13}$$

As a result, a polarimetrically inhomogeneous component of the diffuse field is formed with a different distribution of azimuth values $A^n(\alpha_i^n)$ and ellipticity values $B^n(\beta_i^n)$ of polarization.

Therefore, the polarization pattern of the multiply scattered laser light by a diffuse biological object can be viewed as a superposition of different components.

$$A^*(\alpha_i^*) = A^1(\alpha_i^1) + A^2(\alpha_i^2) + \dots + A^{n-1}(\alpha_i^{n-1}) + A^n(\alpha_i^n) = \sum_{i=1}^{n} A^i(\alpha_i^i); \tag{2.14}$$

$$B^*(\beta_i^*) = B^1(\beta_i^1) + B^2(\beta_i^2) + \dots + B^{n-1}(\beta_i^{n-1}) + B^n(\beta_i^n) = \sum_{i=1}^{n} B^i(\beta_i^i). \tag{2.15}$$

The Equations (2.7)–(2.15) are commonly used in standard polarimetry, also known as Stokes polarimetry, to analyze the intensity of the object field in biological samples.

Conversely, the coherence of laser radiation facilitates an alternative (amplitude-based) analytical description of the processes that create the polarization structure of the object field within a diffuse biological layer.

2.2.1.2.2 Amplitude Considerations

For coherent laser fields, there is a direct relationship between the parameters of the Stokes vector and the orthogonal components (E_{xi} and E_{yi}) of complex amplitudes.

Building on this, the earlier Equations (2.10) and (2.11) for the polarization parameters can be reformulated as follows:

$$\alpha_i = 0.5\,\arctan\left(\frac{\left(E_{xi}E_{yi}^* + E_{xi}^*E_{yi}\right)}{\left(E_{xi}E_{xi}^* - E_{yi}E_{yi}^*\right)}\right); \tag{2.16}$$

$$\beta_i = 0.5\,\text{arcin}\left(\frac{j\left(E_{xi}E_{yi}^* + E_{xi}^*E_{yi}\right)}{\left(E_{xi}E_{xi}^* - E_{yi}E_{yi}^*\right)}\right). \tag{2.17}$$

At the same time, another process occurs – the interference of partially coherent waves with different polarization – leading to the formation of an additional inhomogeneous component in the diffuse field of the object, influenced by polarization.

2.2.1.2.3 Interference Interaction

For the orthogonal components E_{xi} and E_{yi} of the complex amplitudes E_1 and E_2 of two partial singly scattered coherent waves at a local point of the object field, the following interference equations can be written

$$E_{xi} = (E_{x1} + E_{x2})_i = \left(|E_{x1}| + |E_{x2}| + 2\sqrt{|E_{x1}||E_{x2}|}\cos\varphi_{x12}\right)_i; \tag{2.18}$$

$$E_{yi} = (E_{y1} + E_{y2})_i = \left(|E_{y1}| + |E_{y2}| + 2\sqrt{|E_{y1}||E_{y2}|}\cos\varphi_{y12}\right)_i. \tag{2.19}$$

Here, $|E_{xi}|; |E_{yi}|$ is the modules of complex amplitudes; φ_{x12} and φ_{y12} is the phase shift between $(E_{x1}; E_{x2})$ and $(E_{y1}; E_{y2})$.

For the process of forming the orthogonal components of amplitudes E_x and E_y through n-fold interaction of the laser probe with optical inhomogeneities, the following expressions can be written

$$E_x = \sum_{i=1}^{n} E_{xi};\ E_y = \sum_{i=1}^{n} E_{yi};\ \varphi_{xy} = \sum_{i=1}^{n} \varphi_{xi} - \sum_{i=1}^{n} \varphi_{yi}. \tag{2.20}$$

The interference addition of two phase-shifted by φ_{xy} orthogonal components E_x and E_y results in the formation of an elliptically polarized wave

$$\frac{X^2}{E_x^2} + \frac{Y^2}{E_y^2} - \frac{2XY}{E_x E_y}\cos\varphi_{xy} = \sin^2\varphi_{xy}. \tag{2.21}$$

with the following "interference" (I) values of azimuth $\alpha(I)$ and ellipticity $\beta(I)$

$$\alpha(I) = 0.5\arcsin\left(\frac{\sin 2\left(\frac{E_y}{E_x}\right)}{\sqrt{1 + \tan^2\delta_{xy}\cos^2 2\left(\frac{E_y}{E_x}\right)}}\right); \tag{2.22}$$

$$\beta(I) = 0.5\arctan\left(\frac{\tan\delta_{xy}\sin 2\left(\frac{E_y}{E_x}\right)}{\sqrt{1 + \tan^2\delta_{xy}\cos^2 2\left(\frac{E_y}{E_x}\right)}}\right). \tag{2.23}$$

As shown, the secondary interference of differently polarized coherent partial waves also causes the formation of a polarimetrically inhomogeneous component (shown in Equations (2.22) and (2.23)) of the object field in the diffuse biological layer – $\mathrm{AI}^n(\alpha_i^n)$ and $\mathrm{BI}^n(\beta_i^n)$.

2.2.1.2.4 Examples of Polarization Maps

This section shows examples of polarization maps of azimuth and ellipticity in digital microscopic images of native histological sections, including the myocardium (Figure 2.3), brain (Figure 2.4), and liver (Figure 2.5), as well as dehydrated polycrystalline films of blood plasma (Figure 2.6) and synovial fluid (Figure 2.7).

2.2.2 Jones-matrix Formalism for Describing Optical Anisotropy of Polycrystalline Films of Dehydrated Biological Fluids

A key system maintaining the body's vital activity is composed of biological fluids involved in metabolic processes. In these fluids, there are rapid changes in molecular composition and the interactions among components during physiological, extreme, and pathological states.

Any physiological and pathological processes in a living organism essentially involve specific alterations in the structure of proteins and other organic molecules. These changes provide valuable insights into homeostasis at the molecular level and can form the basis for early diagnosis of various diseases.

Modern laboratory and clinical diagnostics mainly focus on quantifying various indicators in blood serum that result from essential bodily processes.

Morphological research in biology and medicine primarily concentrates on cellular components, whereas morphological analysis of BL – the fundamental life support system – has only recently become accessible for clinical use.

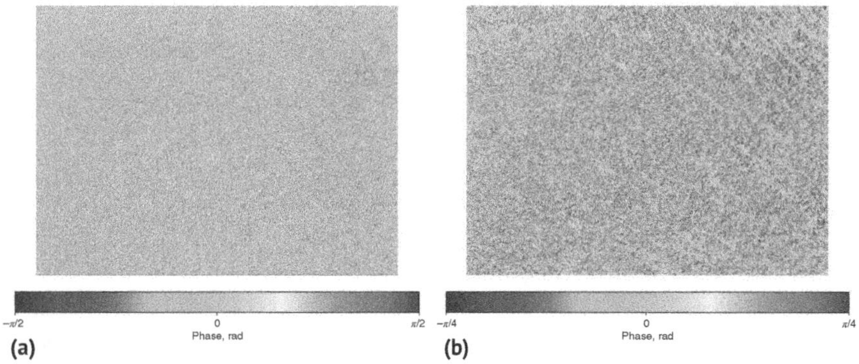

Figure 2.3 Polarization maps of the coordinate distributions of azimuth (a) and ellipticity (b) in a digital microscopic image (×4) of a native histological section of the myocardium.

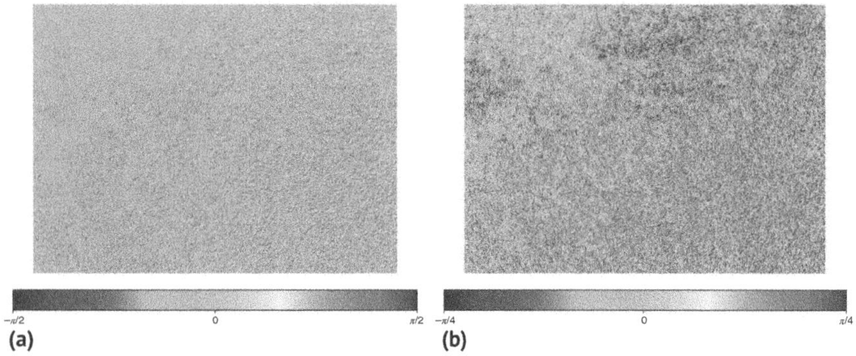

Figure 2.4 Polarization maps of the coordinate distributions of azimuth (a) and ellipticity (b) in a digital microscopic image (×4) of a native histological section of the brain.

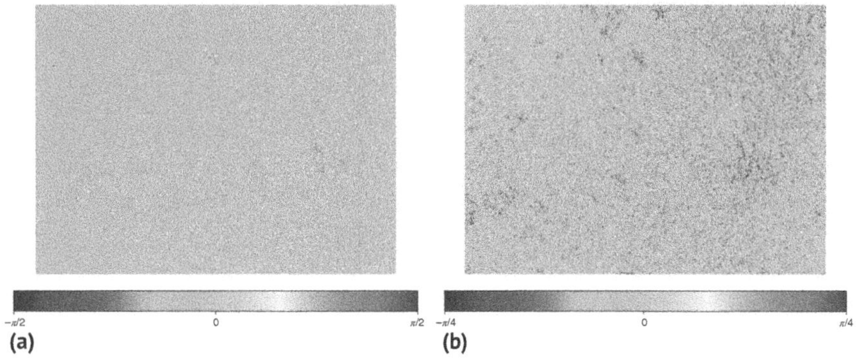

Figure 2.5 Polarization maps of the coordinate distributions of azimuth (a) and ellipticity (b) of a digital microscopic image (×4) of a native histological section of the liver.

A BL drop resting on a horizontal surface serves as a useful model for self-organizing systems, allowing the investigation of physical–chemical processes influenced by the dissolved substances in a liquid, external dehydration conditions, and the substrate material. During dehydration, the BL drop forms a dry film called a "facia," whose structure encodes information about the composition and interactions of the dissolved substances. By analyzing this system at a macroscopic level of self-organization, researchers can infer details about the system's molecular behavior.

So far, visual analysis – including microscopy at different magnifications, illumination with white or polarized light, and dark field techniques – has primarily been used to examine images of biological fluid facies.

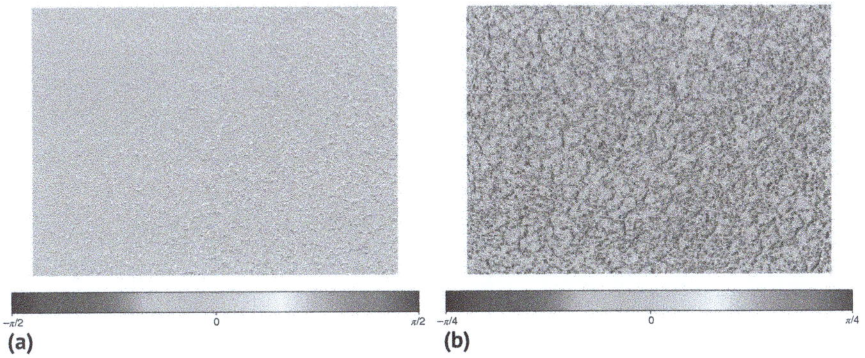

Figure 2.6 Polarization maps of the coordinate distributions of azimuth (a) and ellipticity (b) of digital microscopic images (×4) of a dehydrated polycrystalline film of blood plasma.

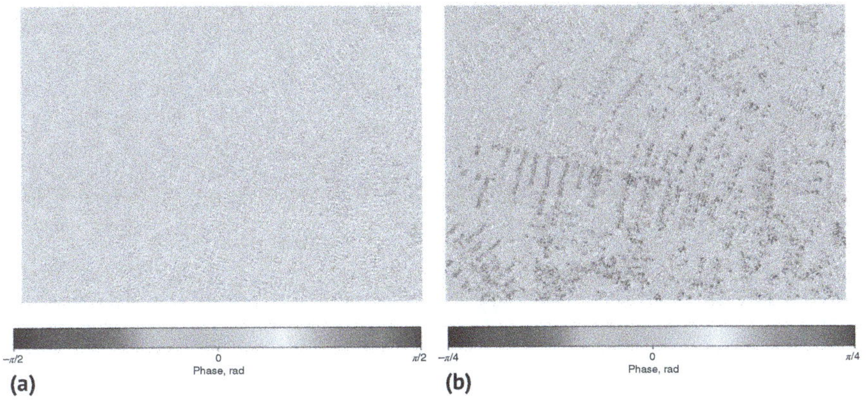

Figure 2.7 Polarization maps of the coordinate distributions of azimuth (a) and ellipticity (b) of a digital microscopic image (×4) of a dehydrated polycrystalline film of synovial fluid.

Current microscopic methods for analyzing the geometric structure of BL images rely on general algorithms that do not account for the specific features of the images being processed. Additionally, valuable information regarding the architectonics of supramolecular self-organized networks in BL molecular crystals, including their optical anisotropic properties accessible through polarization introscopy, remains completely unknown or inaccessible.

Currently, only initial steps in this area are known. Methods for 2D polarization mapping of microscopic images of blood plasma, blood, urine, and other biological fluids have been developed and tested experimentally, utilizing various data processing algorithms such as statistical, correlation, wavelet, and Fourier analysis.

Consequently, the initial objective criteria (markers) were identified, enabling the differential diagnosis of various pathologies in urology, gynecology, and oncology.

However, several factors have constrained the sensitivity and progress of these methods:

- distorting scattered background influence of the BL facies microscopic images polarization map structure;
- lack of algorithms for mapping the polycrystalline networks – digital reconstruction of BL facies optical anisotropy maps;
- coordinate BL facies are heterogeneous, geometric, and optical polycrystalline 3D structures.

These issues can be addressed using innovative laser techniques for 3D interference polarization mapping, which use the Jones-matrix formalism and digital layer-by-layer holographic reconstruction of optical anisotropy maps.

2.2.2.1 Partial Jones Matrices of Optically Anisotropic Architecture of Polycrystalline Dehydrated Films of Biological Fluids

A polycrystalline film of any biological fluid can be described as a series of partial layers featuring six distinct types of optical anisotropy, including linear and circular birefringence $\left(LB_{(0,90);(45,135)}, CB_{\otimes,\oplus}\right)$ and dichroism $\left(LD_{(0,90);(45,135)}, CD_{\otimes,\oplus}\right)$. These $LD_{0,90}$,$LD_{45,135}$, and $LB_{0,90}$,$LB_{45,135}$ – "linear dichroism – birefringence"; $CD_{\otimes;\oplus}$ and

$CB_{\otimes;\oplus}$ – "circular dichroism – birefringence" of biological layer optically anisotropic component for linearly (0° ÷ 90° and 45° ÷ 135°) and circularly right- (⊗) and left- (⊕) polarized orthes.

$$LB_{0;90} = \frac{2\pi}{\lambda}(n_0 - n_{90})h; \tag{2.24}$$

$$LB_{45;135} = \frac{2\pi}{\lambda}(n_{45} - n_{135})h; \tag{2.25}$$

$$CB_{\otimes;\oplus} = \frac{2\pi}{\lambda}(n_{\otimes} - n_{\oplus})h; \tag{2.26}$$

$$LD_{0;90} = \frac{2\pi}{\lambda}(\chi_0 - \chi_{90})h; \tag{2.27}$$

$$LD_{45;135} = \frac{2\pi}{\lambda}(\chi_{45} - \chi_{135})h; \tag{2.28}$$

$$CD_{\otimes;\oplus} = \frac{2\pi}{\lambda}(\chi_{\otimes} - \chi_{\oplus})h, \tag{2.29}$$

where $(n_0, n_{90}, n_{45}, n_{135})$ and $(\chi_0, \chi_{90}, \chi_{45}, \chi_{135})$ is the refractive indices – absorption for orthogonal amplitude components 0°–90° and 45°–135°; λ is the laser wavelength; h is the geometrical layer thickness.

- The mechanisms of linear birefringence and dichroism (structural anisotropy $LB_{(0,90);(45,135)}$, $LD_{(0,90);(45,135)}$) are inherent in the polycrystalline networks. Molecular complexes have optical activity $(CB_{\otimes,\oplus})$ and circular dichroism $(CD_{\otimes,\oplus})$.
- We detail how each optical anisotropy mechanism manifests in polarization using Jones partial matrix operators of this form

$$J(LB_{0;90}) = \left\|\begin{matrix} \exp(-0.5iLB_{0;90}) & 0 \\ 0 & \exp(0.5iLB_{0;90}) \end{matrix}\right\|; \tag{2.30}$$

$$J(LB_{45;135}) = \left\|\begin{matrix} \cos 0.5LB_{45;135} & -i\sin 0.5LB_{45;135} \\ -i\sin 0.5LB_{45;135} & \cos 0.5LB_{45;135} \end{matrix}\right\|; \tag{2.31}$$

$$J(CB_{\otimes;\oplus}) = \left\|\begin{matrix} \cos CB_{\otimes;\oplus} & \sin CB_{\otimes;\oplus} \\ -\sin CB_{\otimes;\oplus} & \cos CB_{\otimes;\oplus} \end{matrix}\right\|; \tag{2.32}$$

$$J(LD_{0;90}) = \left\|\begin{matrix} \exp(-0.5LD_{0;90}) & 0 \\ 0 & \exp(0.5LD_{0;90}) \end{matrix}\right\|; \tag{2.33}$$

$$J(LD_{45;135}) = \left\|\begin{matrix} \cosh 0.5LD_{45;135} & -i\sinh 0.5LD_{45;135} \\ -i\sinh 0.5LD_{45;135} & \cosh 0.5LD_{45;135} \end{matrix}\right\|; \tag{2.34}$$

$$J(CD_{\otimes;\oplus}) = \left\|\begin{matrix} \cosh CD_{\otimes;\oplus} & \sinh CD_{\otimes;\oplus} \\ -\sinh CD_{\otimes;\oplus} & \cosh CD_{\otimes;\oplus} \end{matrix}\right\|. \tag{2.35}$$

An analytical form of the generalized Jones matrix for a polycrystalline medium with complex anisotropy is as follows

$$\{J\} = \begin{Vmatrix} j_{11} & j_{12} \\ j_{21} & j_{22} \end{Vmatrix} = \begin{Vmatrix} \cos 0.5|V| - i\frac{L}{|V|}\sin 0.5|V| & \frac{(C-iL')}{|V|}\sin 0.5|V| \\ -\frac{(C+iL')}{|V|}\sin 0.5|V| & \cos 0.5|V| + i\frac{L}{|V|}\sin 0.5|V| \end{Vmatrix}. \tag{2.36}$$

Here, V is the generalized anisotropy vector.

$$V = V(L_{0;90}, L_{45;135}; -C_{\otimes;\oplus}), \tag{2.37}$$

where

$$\begin{cases} L_{0;90} = \mathrm{LB}_{0;90} - i\mathrm{LD}_{0;90}; \\ L_{45;135} = \mathrm{LB}_{45;135} - i\mathrm{LD}_{45;135}; \\ C_{\otimes;\oplus} = \mathrm{CB}_{\otimes;\oplus} - i\mathrm{CD}_{\otimes;\oplus}. \end{cases} \tag{2.38}$$

From Equations (2.37) and (2.38), an explicit form of the vector modulus V was obtained

$$\begin{aligned} |V| &= \left(|L_{0;90}| + |L_{45;135}| + |C_{\otimes;\oplus}|\right)^{0.5} \\ &= \left((\mathrm{LD}^2_{0;90} + \mathrm{LB}^2_{0;90}) + (\mathrm{LD}^2_{45;135} + \mathrm{LB}^2_{45;135}) + (\mathrm{CD}^2_{\otimes;\oplus} + \mathrm{CB}^2_{\otimes;\oplus})\right)^{0.5}. \end{aligned} \tag{2.39}$$

2.2.2.2 Examples of Real and Imaginary Components of Jones Matrix Images of Dehydrated Polycrystalline Films of Biological Fluids

This section provides examples of the real and imaginary parts of Jones-matrix images for dehydrated polycrystalline films of blood plasma (Figures 2.8 and 2.9) and gallbladder bile (Figures 2.10 and 2.11).

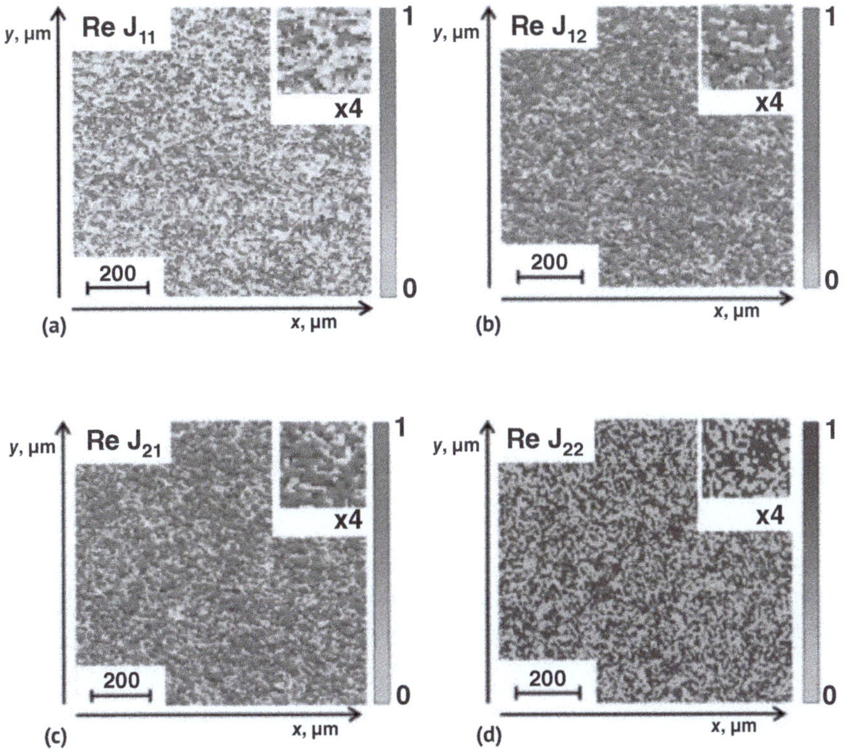

Figure 2.8 Real part of Jones-matrix images of a dehydrated polycrystalline film of blood plasma.

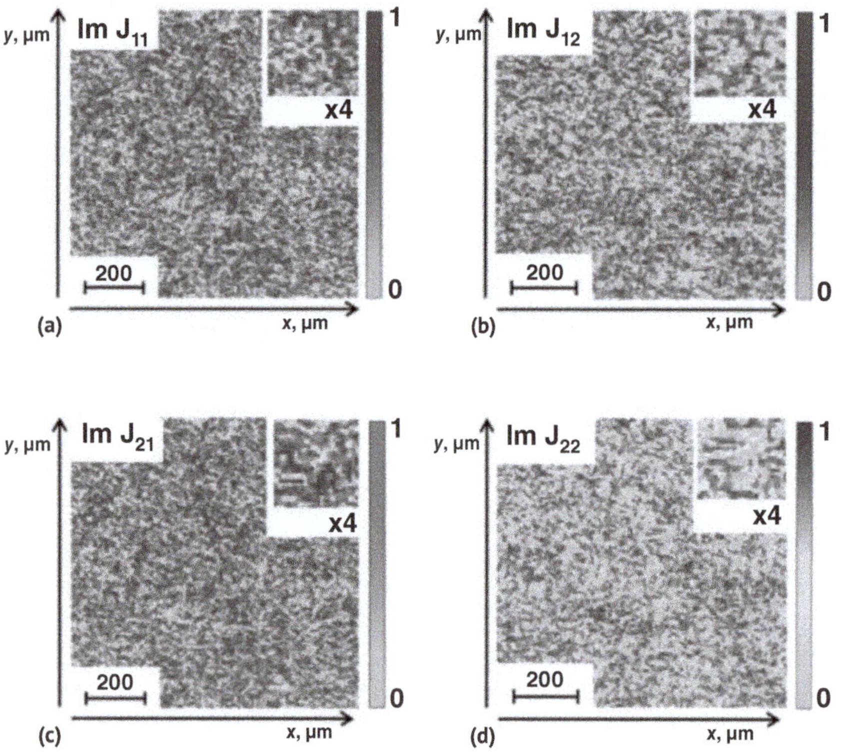

Figure 2.9 Imaginary part of Jones-matrix images of a dehydrated polycrystalline film of blood plasma.

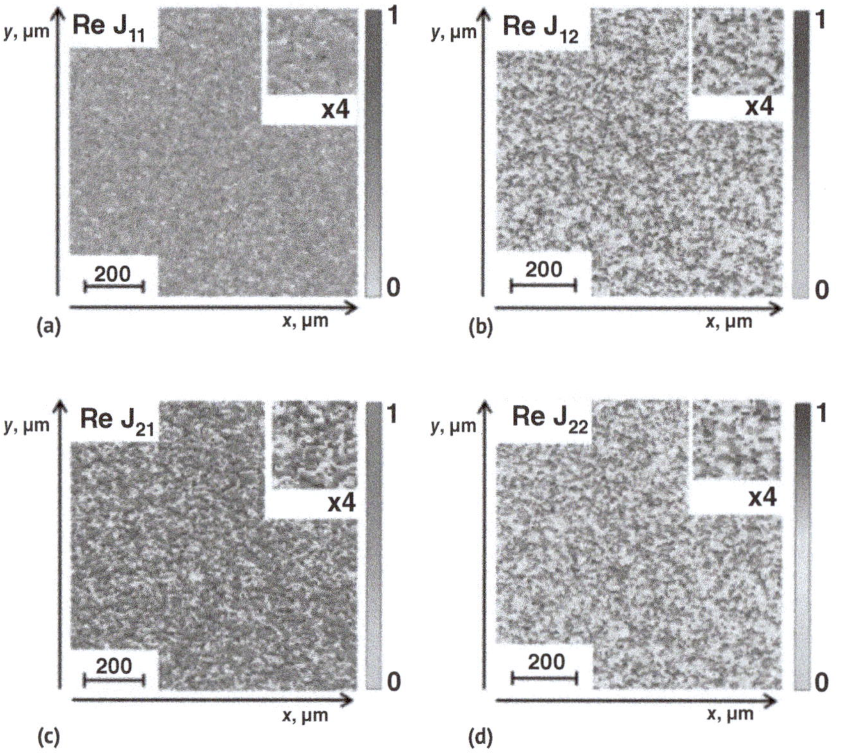

Figure 2.10 Real part of Jones-matrix images of a dehydrated polycrystalline film of gallbladder bile.

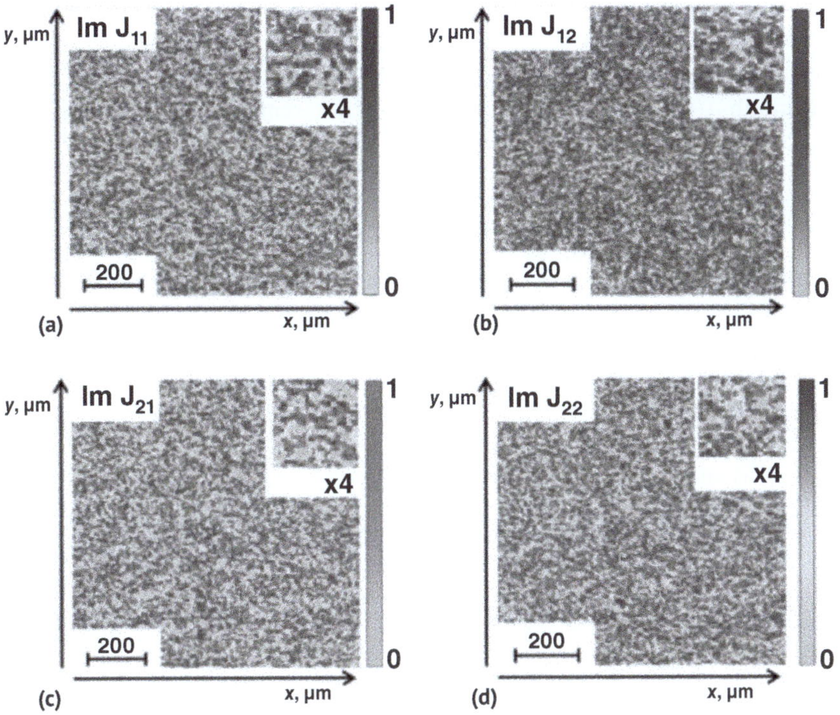

Figure 2.11 Imaginary component of Jones-matrix images of a dehydrated polycrystalline film of gallbladder bile.

2.2.2.3 A Brief Jones-polarimetric Theory of the Formation of the Polarization-correlation Structure of Biospeckle Object Fields

Our phenomenological approach relies on the correlation Stokes parametric formalism to describe polarization-inhomogeneous biospeckle fields. In this context, the expressions for the Stokes vector parameters are given by

$$SC^{(w)}_{l=1;2;3;4}(r_1;r_2) \equiv \begin{pmatrix} SC_1 \\ SC_2 \\ SC_3 \\ SC_4 \end{pmatrix}^{(w)}(r_1;r_2) = \begin{pmatrix} U^*_{x1}U_{x2} + U^*_{y1}U_{y2} \\ U^*_{x1}U_{x2} - U^*_{y1}U_{y2} \\ U^*_{x1}U_{y2} + U^*_{y1}U_{x2} \\ i\left(U^*_{y1}U_{x2} - U^*_{x1}U_{y2}\right) \end{pmatrix}^{(w)}(r_1;r_2). \tag{2.40}$$

Here, w is the polarization state of the illuminating beam.

In their expanded form, Equations (2.41) are expressed as follows

$$SC^{(w)}_{l=1;2;3;4}(r_1;r_2) \equiv \begin{cases} SC^{(w)}_1 = \left(|U_{x1}||U_{x2}| + |U_{y1}||U_{y2}|\cos(\delta_2-\delta_1) + i|U_{y1}||U_{y2}|\sin(\delta_2-\delta_1)\right)^{(w)} \\ \quad \equiv a^{(w)}_1 + ib^{(w)}_1; \\ SC^{(w)}_2 = \left(|U_{x1}||U_{x2}| + |U_{y1}||U_{y2}|\cos(\delta_2-\delta_1) - i|U_{y1}||U_{y2}|\sin(\delta_2-\delta_1)\right)^{(w)} \\ \quad \equiv a^{(w)}_2 + ib^{(w)}_2; \\ SC^{(w)}_3 = \left(|U_{x1}||U_{y2}|\cos\delta_2 + |U_{x2}||U_{y1}|\cos\delta_1 + i(|U_{x1}||U_{y2}|\sin\delta_2 - |U_{x2}||U_{y1}|\sin\delta_1)\right)^{(w)} \\ \quad \equiv a^{(w)}_3 + ib^{(w)}_3; \\ SC^{(w)}_4 = \left(|U_{x1}||U_{y2}|\sin\delta_2 + |U_{x2}||U_{y1}|\sin\delta_1 + i(|U_{x1}||U_{y2}|\cos\delta_2 - |U_{x2}||U_{y1}|\cos\delta_1)\right)^{(w)} \\ \quad \equiv a^{(w)}_4 + ib^{(w)}_4. \end{cases} \tag{2.41}$$

Here, | | is the absolute value of the orthogonal components (x, y) of complex amplitudes $U_{x1}; U_{x2}$ and $U_{y1}; U_{y2}$ in points $(r_1; r_2)$; δ_1 and δ_2 are phase shifts between $U_{x1}; U_{y1}$ and $U_{x2}; U_{y2}$; $a^{(w)}_{l=1;2;3;4} = Re\left(SC^{(w)}_{l=1;2;3;4}\right)$; $b^{(w)}_{l=1;2;3;4} = \mathrm{Im}\left(SC^{(w)}_{l=1;2;3;4}\right)$.

Considering the equations for the parameters of the "two-point" Stokes vector, the relationships for the complex values of the polarization-correlation parameters of azimuth $A(r_1; r_2)$ and ellipticity $B(r_1; r_2)$ of different points in the biospeckle field take the following form

$$\begin{cases} A(r_1;r_2) = 0.5\arctan\left(\dfrac{SC_3}{SC_2}\right)(r_1;r_2) = 0.5\arctan\left(\dfrac{E^*_{x1}E_{y2} + E^*_{y1}E_{x2}}{E^*_{x1}E_{x2} - E^*_{y1}E_{y2}}\right)(r_1;r_2) \equiv X_A + iY_A; \\ B(r_1;r_2) = 0.5\arcsin\left(\dfrac{SC_4}{SC_1}\right)(r_1;r_2) = 0.5\arctan\left(\dfrac{i(E^*_{y1}E_{x2} - E^*_{x1}E_{y2}}{E^*_{x1}E_{x2} + E^*_{y1}E_{y2}}\right)(r_1;r_2) \equiv X_B + iY_B. \end{cases} \tag{2.42}$$

Here, $X_A = Re(A(r_1; r_2))$; $Y_A = \mathrm{Im}(A(r_1; r_2))$; $X_B = Re(B(r_1; r_2))$; $Y_B = Re(B(r_1; r_2))$.

In the future, the distributions of the module (|⊣|) and argument (Arg) of the "two-point" azimuth A and ellipticity B will be called polarization-correlation maps (PCM) of the biospeckle field

$$\begin{cases} |A(r_1;r_2)| = \sqrt{(X_A)^2 + (Y_A)^2}; \\ \mathrm{Arg}(A(r_1;r_2)) = \arctan\left(\dfrac{Y_A}{X_A}\right); \end{cases} \tag{2.43}$$

$$\begin{cases} |B(r_1;r_2)| = \sqrt{(X_B)^2 + (Y_B)^2}; \\ \mathrm{Arg}(B(r_1;r_2)) = \arctan\left(\dfrac{Y_B}{X_B}\right). \end{cases} \tag{2.44}$$

Considering the Equations (2.34)–(2.37), the analytical representations of the PCC (2.33) are expressed in the following Jones-matrix form.

$$\begin{cases} A(r_1;r_2) = 0.5\arctan\left(\dfrac{(q_{11}(r_1))^*(q_{21}(r_2)) + (q_{21}(r_1))^*(q_{11}(r_2))}{(q_{11}(r_1))^*(q_{11}(r_2)) - (q_{21}(r_1))^*(q_{21}(r_2))}\right); \\ |A(r_1;r_2)| = \sqrt{Re\left(\dfrac{(q_{11}(r_1))^*(q_{21}(r_2)) + (q_{21}(r_1))^*(q_{11}(r_2))}{(q_{11}(r_1))^*(q_{11}(r_2)) - (q_{21}(r_1))^*(q_{21}(r_2))}\right)^2 + \mathrm{Im}\left(\dfrac{(q_{11}(r_1))^*(q_{21}(r_2)) + (q_{21}(r_1))^*(q_{11}(r_2))}{(q_{11}(r_1))^*(q_{11}(r_2)) - (q_{21}(r_1))^*(q_{21}(r_2))}\right)^2}; \\ \mathrm{Arg}(A(r_1;r_2)) = \arctan\left(\dfrac{Im\left(\frac{(q_{11}(r_1))^*(q_{21}(r_2))+(q_{21}(r_1))^*(q_{11}(r_2))}{(q_{11}(r_1))^*(q_{11}(r_2))-(q_{21}(r_1))^*(q_{21}(r_2))}\right)}{Re\left(\frac{(q_{11}(r_1))^*(q_{21}(r_2))+(q_{21}(r_1))^*(q_{11}(r_2))}{(q_{11}(r_1))^*(q_{11}(r_2))-(q_{21}(r_1))^*(q_{21}(r_2))}\right)}\right); \end{cases} \tag{2.45}$$

$$\begin{cases} B(r_1;r_2) = 0.5\arcsin\left(\dfrac{(q_{21}(r_1))^*(q_{11}(r_2)) + (q_{11}(r_1))^*(q_{21}(r_2))}{(q_{11}(r_1))^*(q_{11}(r_2)) + (q_{21}(r_1))^*(q_{21}(r_2))}\right); \\ |B(r_1;r_2)| = \sqrt{Re\left(\dfrac{(q_{21}(r_1))^*(q_{11}(r_2)) + (q_{11}(r_1))^*(q_{21}(r_2))}{(q_{11}(r_1))^*(q_{11}(r_2)) + (q_{21}(r_1))^*(q_{21}(r_2))}\right)^2 + \mathrm{Im}\left(\dfrac{(q_{21}(r_1))^*(q_{11}(r_2)) + (q_{11}(r_1))^*(q_{21}(r_2))}{(q_{11}(r_1))^*(q_{11}(r_2)) + (q_{21}(r_1))^*(q_{21}(r_2))}\right)^2}; \\ \mathrm{Arg}(\mathrm{B}(r_1;r_2)) = \arctan\left(\dfrac{Im\left(\frac{(q_{21}(r_1))^*(q_{11}(r_2))+(q_{11}(r_1))^*(q_{21}(r_2))}{(q_{11}(r_1))^*(q_{11}(r_2))+(q_{21}(r_1))^*(q_{21}(r_2))}\right)}{Re\left(\frac{(q_{21}(r_1))^*(q_{11}(r_2))+(q_{11}(r_1))^*(q_{21}(r_2))}{(q_{11}(r_1))^*(q_{11}(r_2))+(q_{21}(r_1))^*(q_{21}(r_2))}\right)}\right). \end{cases} \tag{2.46}$$

Under the assumption of weak phase fluctuations, the "two-point" Stokes vector (Equations (2.41) and (2.42)) and the PCC expressions (Equations (2.45) and (2.46)) can be reformulated into the following approximate analytical expressions

$$\begin{pmatrix} \mathrm{SC}_1 \\ \mathrm{SC}_2 \\ \mathrm{SC}_3 \\ \mathrm{SC}_4 \end{pmatrix}^{(w=0^0)} (r_1;r_2) \cong \begin{pmatrix} 1 - 0.5i(\mathrm{LB}_2\cos 2\rho_2 - \mathrm{LB}_1\cos 2\rho_1); \\ 1 - 0.5i(\mathrm{LB}_2\cos 2\rho_2 - \mathrm{LB}_1\cos 2\rho_1); \\ 0.5((\mathrm{CB}_1 + \mathrm{CB}_2) + i(\mathrm{LB}_1\sin 2\rho_1 + \mathrm{LB}_2\sin 2\rho_2)); \\ 0.5((\mathrm{LB}_1\sin 2\rho_1 + \mathrm{LB}_2\sin 2\rho_2) + i(\mathrm{CB}_1 + \mathrm{CB}_2)) \end{pmatrix}. \tag{2.47}$$

Here, $\mathrm{LB}_1 \equiv \mathrm{LB}(r_1)$; $\mathrm{LB}_2 \equiv \mathrm{LB}(r_2)$; $\rho_1 \equiv \rho(r_1)$; $\rho_2 \equiv \rho(r_2)$; $\mathrm{CB}_1 \equiv \mathrm{CB}(r_1)$; $\mathrm{CB}_2 \equiv \mathrm{CB}(r_2)$.

$$\begin{cases} A(r_1;r_2) \cong 0.5\arctan(0.5(\mathrm{CB}_1 + \mathrm{CB}_2) + i(\mathrm{LB}_1\sin 2\rho_1 + \mathrm{LB}_2\sin 2\rho_2)); \\ |A(r_1;r_2)| \cong 0.5\sin(0.25(\mathrm{CB}_1 + \mathrm{CB}_2))^2 + (\mathrm{LB}_1\sin 2\rho_1 + \mathrm{LB}_2\sin 2\rho_2)^2; \\ \mathrm{Arg}(A(r_1;r_2)) \cong 0.5\arctan\left(\dfrac{\mathrm{LB}_1\sin 2\rho_1 + \mathrm{LB}_2\sin 2\rho_2}{\mathrm{CB}_1 + \mathrm{CB}_2}\right); \end{cases} \tag{2.48}$$

$$\begin{cases} B(r_1;r_2) \cong 0.5\arcsin((\mathrm{LB}_1\sin 2\rho_1 + \mathrm{LB}_2\sin 2\rho_2) + 0.5i(\mathrm{CB}_1 + \mathrm{CB}_2)); \\ |B(r_1;r_2)| \cong 0.5\arcsin\left((\mathrm{LB}_1\sin 2\rho_1 + \mathrm{LB}_2\sin 2\rho_2)^2 + 0.25(\mathrm{CB}_1 + \mathrm{CB}_2)\right)^2; \\ \mathrm{Arg}(B(r_1;r_2)) \cong 0.5\arctan\left(\dfrac{\mathrm{CB}_1 + \mathrm{CB}_2}{\mathrm{LB}_1\sin 2\rho_1 + \mathrm{LB}_2\sin 2\rho_2}\right). \end{cases} \tag{2.49}$$

The distributions (2.42)–(2.46) and their interpretations (2.47)–(2.49) form the foundation of our proposed PCC method. This method aims to analyze "textural" information regarding the correlation between the polarization states of the biospeckle field and the anisotropy parameters of the biological layer at different coordinate points $(r_1; r_2)$.

2.2.2.4 Examples of Polarization-correlation Maps of Digital Microscopic Images of Dehydrated Films of Biological Fluids

This section shows examples of polarization-correlation maps of the modulus and argument of azimuth (Figure 2.12) and ellipticity (Figure 2.13) for a digital microscopic image of a dehydrated polycrystalline blood plasma film [1–64].

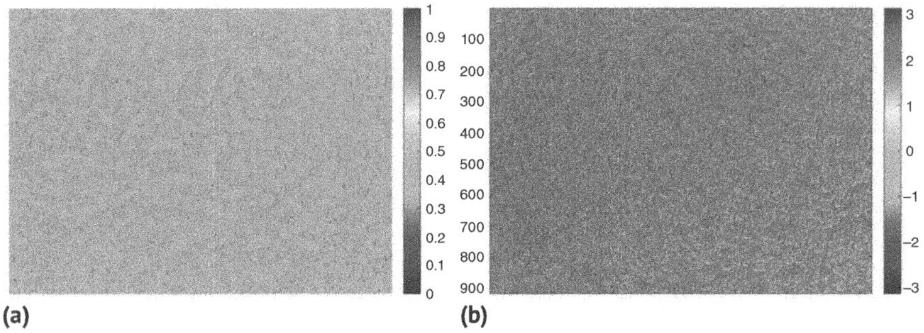

Figure 2.12 The modulus (a) and argument (b) of the polarization-correlation map of the azimuth of a digital microscopic image of a dehydrated polycrystalline blood plasma film.

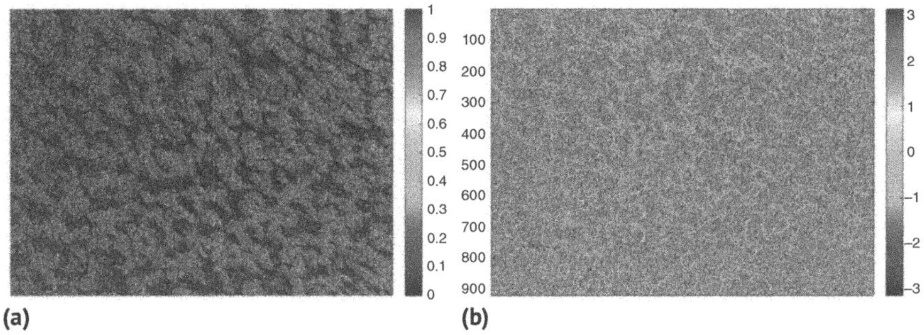

Figure 2.13 The modulus (a) and argument (b) of the polarization-correlation map of the ellipticity of a digital microscopic image of a dehydrated polycrystalline blood plasma film.

2.3 Reconstruction of Optical Anisotropy Parameters (Linear and Circular Birefringence and Dichroism)

2.3.1 Basic Algorithms for Mueller-matrix Polarization Reconstruction of Optical Anisotropy Parameter Maps of Biological Tissues

2.3.1.1 Differential Decomposition of the Mueller Matrix

Azam developed this layer-by-layer analysis method for the optical properties of nondepolarizing, phase-inhomogeneous objects within the Jones-matrix framework and extended it to the Mueller matrix framework.

The following algorithm describes the process of transforming the polarization structure of optical radiation through partial optically anisotropic layers, expressed in terms of the Stokes vector parameters $S = \{S_1;\ \ S_2;\ \ S_3;\ \ S_4\}$

$$\frac{dS}{dr} = \{M\}(r)S(r). \tag{2.50}$$

Here, $\{M\}(r)$ is the Mueller differential matrix; $S(r)$ is the Stokes vector of optical radiation at a distance r in a medium. Equation (2.50) expresses the optical properties of an anisotropic medium within the Mueller-matrix formalism.

$$\frac{d\{F\}(r)}{dr} = \{F\}(r)\{M\}(r), \tag{2.51}$$

where $\{F\}(r)$ – Mueller matrix of a phase-inhomogeneous object in the plane of a partial layer located at a distance r.

For single scattering (nondepolarizing) layers that transform the polarization of the probing radiation, the symmetry of the differential matrix $\{M\}(r)$ is a set of six elementary polarization properties. These parameters fully and exhaustively characterize the amplitude (linear $LD_{0,90}$, $LD_{45,135}$ and circular CD dichroism) and phase (linear $LB_{0,90}$, $LB_{45,135}$ and circular

CB birefringence) anisotropy of the biological layer.

$$\{M\} = \begin{Vmatrix} 0 & m_{12} & m_{13} & m_{14} \\ m_{21} & 0 & m_{23} & -m_{24} \\ m_{31} & -m_{32} & 0 & m_{34} \\ m_{41} & m_{42} & -m_{43} & 0 \end{Vmatrix} = \begin{Vmatrix} 0 & \mathrm{LD}_{0,90} & \mathrm{LD}_{45,135} & \mathrm{CD} \\ \mathrm{LD}_{0,90} & 0 & \mathrm{CB} & -\mathrm{LB}_{45,135} \\ \mathrm{LD}_{45,135} & -\mathrm{CB} & 0 & \mathrm{LB}_{0,90} \\ \mathrm{CD} & \mathrm{LB}_{45,135} & -\mathrm{LB}_{0,90} & 0 \end{Vmatrix}. \tag{2.52}$$

Here, $\mathrm{LD}_{0,90}$ and $\mathrm{LB}_{0,90}$ are linear dichroism and birefringence for the direction of the optical axis $\rho = 0°$; $\mathrm{LD}_{45,135}$ and $\mathrm{LB}_{45,135}$ are linear dichroism and birefringence for the direction of the optical axis $\rho = 45°$; CD and CB are circular dichroism and birefringence.

Elementary polarization properties, indicated by the elements of the differential matrix $\{M\}(r)$, are determined by the following relationships in accordance with the parameters of the phase and amplitude anisotropy of the optical anisotropic partial layer of the phase-inhomogeneous medium

- phase anisotropy

$$m_{34} = -m_{43} \Rightarrow \mathrm{LB}_{0,90} = \frac{2\pi}{\lambda}\Delta n_{0,90} r; \tag{2.53}$$

$$m_{42} = -m_{24} \Rightarrow \mathrm{LB}_{45,135} = \frac{2\pi}{\lambda}\Delta n_{45,135} r; \tag{2.54}$$

$$m_{23} = -m_{32} \Rightarrow \mathrm{CB} = \frac{2\pi}{\lambda}\Delta n_{\mathrm{CB}} r; \tag{2.55}$$

Here, λ is the wavelength; $\Delta n_{0,90}$; $\Delta n_{45,135}$ is the birefringence value for linearly polarized orthogonal components of the laser radiation amplitude with azimuths $0° \div 90°$ and $45° \div 135°$; Δn_{CB} – birefringence value for circularly right- ($\otimes$) and left- ($\oplus$) orthogonally polarized components of the laser radiation amplitude.

- amplitude anisotropy

$$m_{12} = m_{21} \Rightarrow \mathrm{LD}_{0,90} = \frac{2\pi}{\lambda}\Delta \mu_{0;90} r; \tag{2.56}$$

$$m_{13} = m_{31} \Rightarrow \mathrm{LD}_{45,135} = \frac{2\pi}{\lambda}\Delta \mu_{45;135} r; \tag{2.57}$$

$$m_{14} = m_{41} = \mathrm{CD} = \frac{2\pi}{\lambda}\Delta \mu_{\otimes;\oplus} r. \tag{2.58}$$

Here, $\Delta\mu_{0;90}$; $\Delta\mu_{45;135}$; $\Delta\mu_{\otimes;\oplus}$ is the difference in absorption coefficients μ_0; μ_{90}; μ_{45}; μ_{135}; $\mu_{\otimes}$; $\mu_{\oplus}$ of linearly polarized with azimuths $0° \div 90°$ and $45° \div 135°$ and circularly right- ($\otimes$) and left- ($\oplus$) orthogonally polarized components of the laser radiation amplitude.

In the future, to evaluate the extent of optical anisotropy in the polycrystalline structure of biological layers, we plan to use the generalized parameters of linear birefringence (LB) and dichroism (LD)

$$\mathrm{LB} = \sqrt{\mathrm{LB}^2_{0;90} + \mathrm{LB}^2_{45;135}}; \tag{2.59}$$

$$\mathrm{LD} = \sqrt{\mathrm{LD}^2_{0;90} + \mathrm{LD}^2_{45;135}}. \tag{2.60}$$

Another important theoretical foundation for developing the method of differential Mueller-matrix mapping of partially depolarizing biological layers is the feedback between the differential matrix $\{M\}$ and the Mueller matrix $\{F\}$ in the following form

$$\{F\}(r) = \exp(\{M\}(r)). \tag{2.61}$$

Analyzing Equations (2.51) and (2.52) together enables us to derive a logarithmic expression for the superposition of the fully polarized $P(r)$ and depolarized $T(r)$ components of the matrix algorithm (2.61)

$$\Lambda(r) = \ln\{F(r)\} = P(r) + T(r), \tag{2.62}$$

In (2.62), the components $P(r)$ and $T(r)$ are related to the mean ($\langle M\rangle$) and the variance of fluctuations ($\langle\widetilde{M}\rangle$) of the values of the differential matrix $\{M\}$ and are determined by the following Equations

$$P(r) = \langle\{M\}\rangle r; \tag{2.63}$$

$$T(r) = 0,5\langle\{\widetilde{M}\}\rangle r^2, \tag{2.64}$$

where

$$\begin{cases} P(r) = 0.5\left(\Lambda(r) - N\Lambda^T(r)N\right); \\ T(r) = 0.5\left(\Lambda(r) - N\Lambda^T(r)N\right); \\ N = \mathrm{diag}(1, -1, -1, -1). \end{cases} \tag{2.65}$$

Here, T is the transposition operation; N is the Minkowski metric matrix.

2.3.1.2 Differential Matrix of the First Order

In this section, the primary "information" object for our study of the optical anisotropy of partially depolarizing biological layers will be the coordinate distributions of the magnitude of the averaged elements of the average ($\langle f_{ik}\rangle$) differential matrix $\{F\}$.

We will develop an algorithm to identify the elements of this matrix through Stokes-polarimetric mapping of the sample's polycrystalline network, followed by capturing a series of Mueller-matrix images $f_{ik}(x, y)$.

Taking into account the relations (2.63)–(2.65), the polarization component $P(r)$ of the logarithmic matrix algorithm (2.21) $\Lambda(r)$ can be represented in the following form

$$\langle\{M\}\rangle = r^{-1}\left\|\begin{matrix} 0 & (p_{12}+p_{21}) & (p_{13}+p_{31}) & (p_{14}+p_{41}) \\ (p_{21}+p_{12}) & 0 & (p_{23}-p_{32}) & (p_{24}-p_{42}) \\ (p_{31}+p_{13}) & (p_{32}-p_{23}) & 0 & (p_{34}-p_{43}) \\ (p_{41}+p_{14}) & (p_{42}-p_{24}) & (p_{43}-p_{34}) & 0 \end{matrix}\right\|, \tag{2.66}$$

where

$$p_{ik} = \ln f_{ik}; \tag{2.67}$$

$$p_{ik} + p_{ki} = \ln(f_{ik} \times f_{ki}); \tag{2.68}$$

$$p_{ik} - p_{ki} = \ln\left(\frac{f_{ik}}{f_{ki}}\right). \tag{2.69}$$

Equation (2.66), considering Equations (2.67)–(2.69), can be reformulated into the following analytical expression that describes the relationship between the polarization component $\langle\{M\}\rangle$ of the differential matrix $\{M\}$ and the Mueller matrix $\{F\}$

$$\langle\{M\}\rangle = r^{-1}\begin{Vmatrix} 0 & \ln(f_{12}f_{21}) & \ln(f_{13}f_{31}) & \ln(f_{14}f_{41}) \\ \ln(f_{12}f_{21}) & 0 & \ln\left(\frac{f_{23}}{f_{32}}\right) & \ln\left(\frac{f_{24}}{f_{42}}\right) \\ \ln(f_{13}f_{31}) & \ln\left(\frac{f_{32}}{f_{23}}\right) & 0 & \ln\left(\frac{f_{34}}{f_{43}}\right) \\ \ln(f_{14}f_{41}) & \ln\left(\frac{f_{42}}{f_{24}}\right) & \ln\left(\frac{f_{43}}{f_{34}}\right) & 0 \end{Vmatrix}. \tag{2.70}$$

Therefore, by direct Mueller-matrix mapping $(f_{ik}(x, y))$ of an optically anisotropic partially depolarizing biological layer, it is possible to obtain information about the coordinate distributions $\langle\{m_{ik}(x, y)\}\rangle$, characterizing the averaged parameters of phase and amplitude anisotropy.

This information serves as the basis for reproducing the polarization of the distributions of average linear and circular birefringence and dichroism in the polycrystalline structure of the biological layer.

2.3.1.3 Algorithms for Reconstructing Distributions of Average Values of Optical Anisotropy Parameters of Biological Layers

By means of a joint analysis of the relationships between the distributions $\langle\{m_{ik}(x, y)\}\rangle$ and $f_{ik}(x, y)$ based on Equations (2.56)–(2.60) and (2.70), we obtained algorithms for the polarization reproduction of the average values of the parameters of the phase $(\langle\Delta n_{0;90}\rangle;\ \langle\Delta n_{45;135}\rangle;\ \langle\Delta n_{\otimes;\oplus}\rangle)$ and amplitude $(\langle\Delta\mu_{0;90}\rangle;\ \langle\Delta\mu_{45;135}\rangle;\ \langle\Delta\mu_{\otimes;\oplus}\rangle)$ anisotropy of the partially depolarizing biological layer

$$\langle\Delta n_{0,90}\rangle = \frac{\lambda}{2\pi r}\ln\left(\frac{f_{34}}{f_{43}}\right); \tag{2.71}$$

$$\langle\Delta n_{45;135}\rangle = \frac{\lambda}{2\pi r}\ln\left(\frac{f_{24}}{f_{42}}\right); \tag{2.72}$$

$$\langle\Delta n_{\otimes;\oplus}\rangle = \frac{\lambda}{2\pi r}\ln\left(\frac{f_{23}}{f_{32}}\right); \tag{2.73}$$

$$\langle\Delta\mu_{0,90}\rangle = \frac{\lambda}{2\pi r}\ln(f_{12}f_{21}); \tag{2.74}$$

$$\langle\Delta\mu_{45;135}\rangle = \frac{\lambda}{2\pi r}\ln(f_{13}f_{31}); \tag{2.75}$$

$$\langle\Delta\mu_{\otimes;\oplus}\rangle = \frac{\lambda}{2\pi r}\ln(f_{14}f_{41}). \tag{2.76}$$

The differential method for analyzing data from the direct Mueller-matrix mapping enabled the development of a set of algorithms (Equations (2.71)–(2.76)) to replicate the polarization distributions of the average linear and circular parameters of birefringence $(\langle\Delta n_{0;90}\rangle; \langle\Delta n_{45;135}\rangle; \langle\Delta n_{\otimes;\oplus}\rangle)$ and dichroism $(\langle\Delta\mu_{0;90}\rangle; \langle\Delta\mu_{45;135}\rangle; \langle\Delta\mu_{\otimes;\oplus}\rangle)$.

2.3.1.4 Examples of Optical Anisotropy Maps of the Polycrystalline Component of Biological Tissues

This section showcases examples of optical anisotropy maps derived from native histological sections of myocardium (see Figure 2.14).

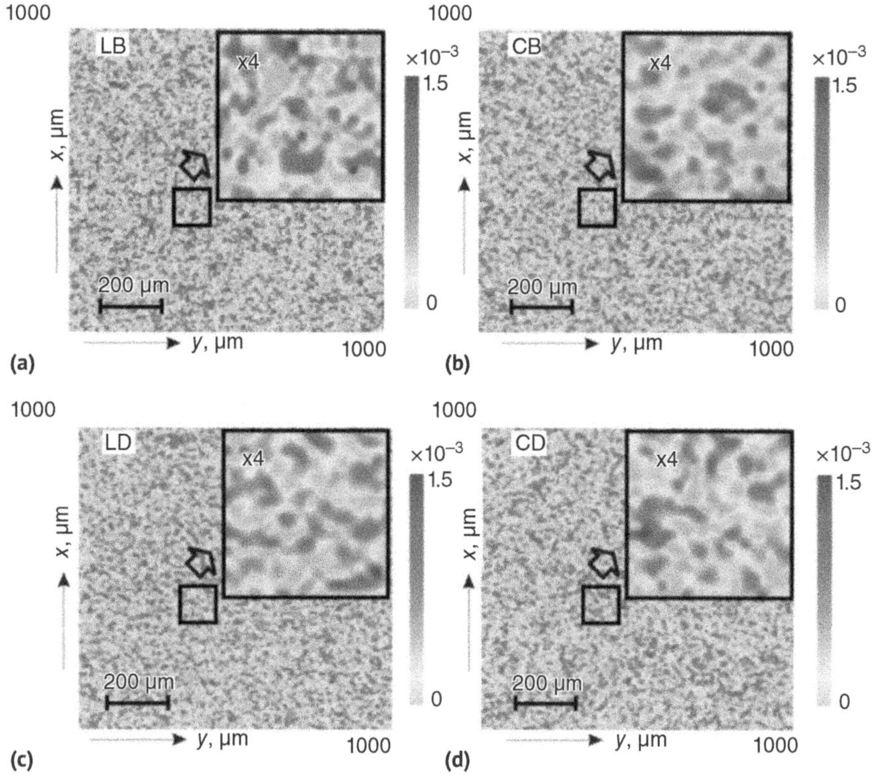

Figure 2.14 Optical anisotropy maps of the polycrystalline architecture of a native histological section of myocardium.

2.3.1.5 Differential Matrix of the Second Order

It can be shown that the symmetric component of $T(r)$ (Equation (2.67)) is determined by a matrix operator of the following symmetry.

$$\{\widetilde{M}\} = 0.5r^{-2}\left\|\begin{matrix} t_{11} & (t_{12}-t_{21}) & (t_{13}-t_{31}) & (t_{14}-t_{41}) \\ (t_{21}-t_{12}) & t_{22} & (t_{23}+t_{32}) & (t_{24}+t_{42}) \\ (t_{31}-t_{13}) & (t_{32}+t_{23}) & t_{33} & (t_{34}+t_{43}) \\ (t_{41}-t_{14}) & (t_{42}+t_{24}) & (t_{43}+t_{34}) & t_{44} \end{matrix}\right\|. \quad (2.77)$$

The matrix operator (2.77) is expressed in expanded form as follows

$$\{\widetilde{M}\} = 0.5r^{-2}\left\|\begin{matrix} \ln f_{11} & \ln\left(\frac{f_{12}}{f_{21}}\right) & \ln\left(\frac{f_{13}}{f_{31}}\right) & \ln\left(\frac{f_{14}}{f_{41}}\right) \\ \ln\left(\frac{f_{21}}{f_{12}}\right) & \ln f_{22} & \ln(f_{23}f_{32}) & \ln(f_{24}f_{42}) \\ \ln\left(\frac{f_{31}}{f_{13}}\right) & \ln(f_{32}f_{23}) & \ln f_{33} & \ln(f_{34}f_{43}) \\ \ln\left(\frac{f_{41}}{f_{14}}\right) & \ln(f_{42}f_{24}) & \ln(f_{43}f_{34}) & \ln f_{44} \end{matrix}\right\|. \quad (2.78)$$

The theoretical analysis of the MMI depolarization component $\{m\}$ is grounded in the generalized theory. Here, the value of each of the elementary polarization properties $\xi_{i=1-6}$ (Equation (2.65)) of the scattering medium is considered in the form of the average μ_i and fluctuating σ_i components

$$\xi_i = \mu_i + \sigma_i; \quad (2.79)$$

$$\overline{\xi}_i = \mu_i; \overline{\sigma}_i = 0. \quad (2.80)$$

here

$$\mu_i = \begin{pmatrix} \mu_1 = \mathrm{LD} \\ \mu_2 = \mathrm{CD} \\ \mu_3 = \mathrm{LD}' \\ \mu_4 = \mathrm{LB} \\ \mu_5 = \mathrm{CB} \\ \mu_6 = \mathrm{LB}' \end{pmatrix}; \sigma_i = \begin{pmatrix} \sigma_1 = \sqrt{D_{\mathrm{LD}}} \\ \sigma_2 = \sqrt{D_{\mathrm{CD}}} \\ \sigma_3 = \sqrt{D_{\mathrm{LD}'}} \\ \sigma_4 = \sqrt{D_{\mathrm{LB}}} \\ \sigma_5 = \sqrt{D_{\mathrm{CB}}} \\ \sigma_6 = \sqrt{D_{\mathrm{LB}'}} \end{pmatrix}, \tag{2.81}$$

where $\sqrt{D_i}$ is the standard deviation of fluctuations in polarization properties $\xi_{i=1-6}$.

In the approximation (2.79)–(2.81), the following equation is derived for the second-order differential matrix.

$$\{\widetilde{M}\} = = \left\| \begin{matrix} \left(\overline{\sigma}_4^2 + \overline{\sigma}_5^2 + \overline{\sigma}_6^2\right)_{11} & -0.5\left(\overline{\sigma}_2\overline{\sigma}_6 - \overline{\sigma}_3\overline{\sigma}_5\right)_{12} & -0.5\left(\overline{\sigma}_3\overline{\sigma}_4 - \overline{\sigma}_1\overline{\sigma}_6\right)_{13} & -0.5\left(\overline{\sigma}_1\overline{\sigma}_5 - \overline{\sigma}_2\overline{\sigma}_4\right)_{14} \\ 0.5\left(\overline{\sigma}_2\overline{\sigma}_6 - \overline{\sigma}_3\overline{\sigma}_5\right)_{21} & \left(\overline{\sigma}_4^2 - \overline{\sigma}_2^2 - \overline{\sigma}_3^2\right)_{22} & 0.5\left(\overline{\sigma}_1\overline{\sigma}_2 + \overline{\sigma}_4\overline{\sigma}_5\right)_{23} & 0.5\left(\overline{\sigma}_1\overline{\sigma}_3 + \overline{\sigma}_4\overline{\sigma}_5\right)_{24} \\ 0.5\left(\overline{\sigma}_3\overline{\sigma}_4 - \overline{\sigma}_1\overline{\sigma}_6\right)_{31} & 0.5\left(\overline{\sigma}_1\overline{\sigma}_2 + \overline{\sigma}_4\overline{\sigma}_5\right)_{32} & \left(\overline{\sigma}_5^2 - \overline{\sigma}_1^2 - \overline{\sigma}_3^2\right)_{33} & 0.5\left(\overline{\sigma}_2\overline{\sigma}_3 + \overline{\sigma}_5\overline{\sigma}_6\right)_{34} \\ 0.5\left(\overline{\sigma}_1\overline{\sigma}_5 - \overline{\sigma}_2\overline{\sigma}_4\right)_{41} & 0.5\left(\overline{\sigma}_1\overline{\sigma}_3 + \overline{\sigma}_4\overline{\sigma}_6\right)_{42} & 0.5\left(\overline{\sigma}_2\overline{\sigma}_3 + \overline{\sigma}_5\overline{\sigma}_6\right)_{43} & \left(\overline{\sigma}_6^2 - \overline{\sigma}_1^2 - \overline{\sigma}_3^2\right)_{44} \end{matrix} \right\|. \tag{2.82}$$

Analysis of Equation (2.82) uncovers the physical significance of the partial elements in the depolarization part of the Mueller matrix for an optically thick biological layer, which exhibits fluctuations in linear and circular birefringence and dichroism parameters.

- diagonal elements $\{\widetilde{m}\}_{ii}$ are determined by combinations of the magnitude of the dispersion D_i of the fluctuations of the parameters of various mechanisms of phase and amplitude anisotropy;
- the off-diagonal elements $\{\widetilde{m}\}_{ik}$ represent the degree of correlation between fluctuations of various polarization properties of linear and circular birefringence $(\overline{\sigma}_{4;5;6}))$ and dichroism $(\overline{\sigma}_{1;2;3})$.

It should be noted that as the multiplicity of light scattering within the biological layer increases, the correlation between the parameters of different mechanisms of phase and amplitude anisotropy diminishes and approaches zero ($\widetilde{m}_{ik} \to 0$). As a result, the second-order differential matrix operator is expressed as follows

$$\{\widetilde{m}\} = \left\| \begin{matrix} \left(\overline{\sigma}_4^2 + \overline{\sigma}_5^2 + \overline{\sigma}_6^2\right)_{11} & 0 & 0 & 0 \\ 0 & \left(\overline{\sigma}_4^2 - \overline{\sigma}_2^2 - \overline{\sigma}_3^2\right)_{22} & 0 & 0 \\ 0 & 0 & \left(\overline{\sigma}_5^2 - \overline{\sigma}_1^2 - \overline{\sigma}_3^2\right)_{33} & 0 \\ 0 & 0 & 0 & \left(\overline{\sigma}_6^2 - \overline{\sigma}_1^2 - \overline{\sigma}_3^2\right)_{44} \end{matrix} \right\|. \tag{2.83}$$

Thus, the off-diagonal elements $\{\widetilde{m}\}_{ik}$ are diagnostically sensitive to changes in the polycrystalline structure of biological layers under transient conditions of partial depolarization. Under boundary conditions of strong depolarization, it is important to study the diagonal matrix elements $\{\widetilde{m}\}_{ii}$, the value of which is correlated with the dispersion σ_i of fluctuations of linear and circular birefringence and dichroism.

2.3.1.6 Algorithms for Assessing the Polycrystalline Structure of Diffuse Biological Layers

For biological layers that strongly depolarize (diffuse), the degree of depolarization serves as the primary diagnostic parameter, since it results from the superposition of nonzero values of the Mueller matrix's diagonal elements matrix $f_{ii} \neq 0$

$$\Lambda = 1 - \frac{1}{3}\{f_{22} + f_{33} + f_{44}\}. \tag{2.84}$$

In the second stage, Mueller-matrix algorithms for diffusion tomography will be developed to reproduce the distributions of fluctuation magnitudes for linear and circular birefringence, as well as dichroism, in an optically thick biological layer.

To achieve this, we utilize the "two-wave" method, which was previously introduced for distinguishing the birefringence and dichroism effects in optically thin polycrystalline films of biological fluids.

Let's examine two scenarios. The first occurs when the mechanisms behind optically anisotropic absorption are negligible.

$$\mu_{1;2;3} = 0, \sigma_{1;2;3} = 0. \tag{2.85}$$

Such a condition can be created experimentally by choosing the correct wavelength of laser radiation used to probe the biological layer. It is well known that the absorption maxima of most protein molecules in biological tissues and amino acids in biological fluids occur in the ultraviolet region. Therefore, experimentally, the analytical conditions (2.73) are realized in the "red – λ_1" section of the spectral range.

Using this assumption, it can be shown that the dispersion of fluctuations of circular $(D_\phi(\lambda_1))$ and linear $(D_\delta(\lambda_1))$ birefringence of the polycrystalline component of the biological layer is determined by the following equations

$$D_\phi(\lambda_1) = \frac{1}{r^2} \ln\left(\frac{f_{44}(\lambda_1)}{f_{22}(\lambda_1) + f_{33}(\lambda_1)}\right); \tag{2.86}$$

$$D_\delta(\lambda_1) = \frac{1}{r^2} \ln\left(\frac{1}{f_{44}(\lambda_1)}\right), \tag{2.87}$$

and the dispersion of fluctuations in the magnitude of linear $(D_\tau(\lambda_1))$ and circular $(D_\chi(\lambda_1))$ dichroism of optically anisotropic absorption of polycrystalline structures of the biological layer is minimal

$$D_\tau(\lambda_1) \to 0; D_\chi(\lambda_1) \to 0. \tag{2.88}$$

In the short wave, "blue" (λ_2) region of the irradiation spectrum of histological sections (polycrystalline films) of biological samples, the equations describing the dispersion of the phase and amplitude anisotropy parameters are as follows

$$D_\phi(\lambda_2) = \frac{\lambda_1}{\lambda_2} \frac{1}{r^2} \ln\left(\frac{f_{44}(\lambda_2)}{f_{22}(\lambda_1) + f_{33}(\lambda_2)}\right); \tag{2.89}$$

$$D_\delta(\lambda_2) = \frac{\lambda_1}{\lambda_2} \frac{1}{r^2} \ln\left(\frac{1}{f_{44}(\lambda_2)}\right); \tag{2.90}$$

$$D_\tau(\lambda_2) = \frac{1}{r^2}\left(\ln\left(\frac{1}{f_{44}(\lambda_2)}\right) - \frac{\lambda_1}{\lambda_2} \ln\left(\frac{1}{f_{44}(\lambda_1)}\right)\right); \tag{2.91}$$

$$D_\chi = \frac{1}{r^2}\left(2\ln\left(\frac{1}{f_{22}(\lambda_2)}\right) - \ln\left(\frac{1}{f_{44}(\lambda_2)}\right) - \ln\left(\frac{1}{f_{44}(\lambda_1)}\right)\right). \tag{2.92}$$

Within the framework of the differential Mueller-matrix method for mapping the optically anisotropic structure of biological layers, we developed a set of algorithms (Equations (2.86)–(2.92)) to reproduce the distribution of fluctuation magnitudes in linear and circular birefringence $(D_\delta(\lambda_1), D_\phi(\lambda_1))$ and dichroism $(D_\tau(\lambda_2), D_\chi(\lambda_2))$ parameters.

In what follows, the coordinate distributions $\left\{\begin{matrix} D_\delta(x, y, \lambda_1) \\ D_\phi(x, y, \lambda_1) \\ D_\tau(x, y, \lambda_2) \\ D_\chi(x, y, \lambda_2) \end{matrix}\right\}$ will be called diffusion tomograms DT [1–62].

References

Section 2.1

1 Ronchi V, Barocas V. The Nature of Light: An Historical Survey. Harvard University Press; 1970.
2 Huard S. Polarization of Light. Wiley; 1997.
3 Goldstein D. Polarized Light. 2nd ed. Marcel Dekker; 2003.
4 Chipman RA, Lam WST, Young G. Polarized Light and Optical Systems. CRC Press; 2018.
5 Pérez JJG, Ossikovski R. Polarized Light and the Mueller Matrix Approach. CRC Press; 2017.
6 Zhan QW. Cylindrical vector beams: from mathematical concepts to applications. Adv Opt Photon. 2009;1:1–57.
7 Rosales-Guzmán C, Ndagano B, Forbes A. A review of complex vector light fields and their applications. J Opt. 2018;20:123001.
8 Forbes A, De Oliveira M, Dennis MR. Structured light. Nat Photon. 2021;15:253–262.
9 Wang JW, Castellucci F, Franke-Arnold S. Vectorial light–matter interaction: exploring spatially structured complex light fields. AVS Quantum Sci. 2020;2:031702.
10 Slussarenko S, Alberucci A, Jisha CP, Piccirillo B, Santamato E, Assanto G, et al. Guiding light via geometric phases. Nat Photon. 2016;10:571–575. Cloude S. Polarisation: Applications in Remote Sensing. OUP Oxford; 2009.
11 Ndagano B, Perez-Garcia B, Roux FS, McLaren M, Rosales-Guzman C, Zhang Y, et al. Characterizing quantum channels with non-separable states of classical light. Nat Phys. 2017;13:397–402.
12 Marrucci L, Manzo C, Paparo D. Optical spin-to-orbital angular momentum conversion in inhomogeneous anisotropic media. Phys Rev Lett. 2006;96:163905.
13 Bliokh KY, Rodríguez-Fortuño FJ, Nori F, Zayats AV. Spin-orbit interactions of light. Nat Photon. 2015;9:796–808.
14 Schulz M, Zablocki J, Abdullaeva OS, Brück S, Balzer F, Lützen A, et al. Giant intrinsic circular dichroism of prolinol-derived squaraine thin films. Nat Commun. 2018;9:2413.
15 Ghosh N, Vitkin IA. Tissue polarimetry: concepts, challenges, applications, and outlook. J Biomed Opt. 2011;16:110801.
16 Novikova T, Meglinski I, Ramella-Roman JC, Tuchin VV. Special section guest editorial: polarized light for biomedical applications. J Biomed Opt. 2016;21:071001.
17 Azzam RMA, Bashara NM, Ballard SS. Ellipsometry and polarized light. Phys Today. 1978;31:72.
18 Brosseau C. Fundamentals of Polarized Light: A Statistical Optics Approach. Wiley-Interscience; 1998.
19 Pishak V, Ushenko A, Gryhoryshyn P, Yermolenko SB, Rudeychuk VM, Pishak OV. Study of polarization structure of biospeckle fields in crosslinked tissues of human organism: 1. Vector structure of skin biospeckles. In Proceedings of SPIE-The International Society for Optical Engineering. 1997;418–424.
20 Ushenko AG. Depolarization of a laser emission field as the correlation development of its polarization structure. In International Conference on Correlation Optics. Proc SPIE. 1997;3317:331–339.
21 Ushenko AG, Burkovets DN, Yermolenko SB, Arkhelyuk AD, Pishak VP, Yuzko AM, et al. Stokes polarimetry of biotissues. In Fourth International Conference on Correlation Optics. Proc SPIE. 1999;3904:527–533.
22 Jóźwicki R, Patorski K, Angelsky OV, Ushenko AG, Burkovets DN, Ushenko YA. Automatic polarimetric system for early medical diagnosis by biotissue testing. Opt. Appl. 2002;32(4):603–612.
23 Ushenko AG, Pishak VP. Coherent-Domain Optical Methods: Biomedical Diagnostics, Environmental and Material Science. Laser Polarimetry of Biological Tissue. Principles and Applications. Boston: Kluwer Academic Publishers; 2004. pp. 67–93.
24 Olar EI, Ushenko AG, Ushenko YA. Correlation microstructure of the Jones matrices for multifractal networks of biotissues. Laser Phys. 2004;14(7):1012–1018.
25 Yermolenko S, Ushenko A, Ivashko P, Goudail F, Gruia I, Gavrilă C, et al. Spectropolarimetry of cancer change of biotissues. In Ninth International Conference on Correlation Optics. Proc SPIE. 2009;7388:404–410.
26 Ushenko AG, Fediv AI, Marchuk YF. Correlation and fractal structure of Jones matrices of human bile secret. In European Conference on Biomedical Optics. Proc SPIE. 2009;73681Q-1-6.
27 Angelsky OV, Ushenko AG, Ushenko YA, Pishak VP, Peresunko AP. Handbook of Photonics for Biomedical Science. CRC Press; 2010; pp. 22–67.
28 Angelsky OV, Ushenko AG, Zenkova CY, Felde CV, Bogatyryova HV, Mokhun II, et al. Optical Measurements: Polarization and Coherence of Light Fields. INTECH Open Access Publisher; 2012.
29 Garazdyuk MS, Bachinskyi VT, Vanchulyak OY, Ushenko AG, Dubolazov OV, Gorsky MP. Polarization-phase images of liquor polycrystalline films in determining time of death. Appl Opt. 2016;55(12):B67–B71.

30 Ushenko AG, Dubolazov OV, Ushenko VA, Novakovskaya OY, Olar OV. Fourier polarimetry of human skin in the tasks of differentiation of benign and malignant formations. Appl Opt. 2016;55(12):B56–B60.
31 Huang D, Swanson EA, Lin CP, Schuman JS, Stinson WG, Chang W, et al. Optical coherence tomography. Science. 1991;254(5035):1178–1181.
32 Hee MR, Huang D, Swanson EA, Fujimoto JG. Polarization-sensitive low-coherence reflectometer for birefringence characterization and ranging. J Opt Soc Am B. 1992;9(6):903–908.
33 Park BH, de Boer JF. Polarization sensitive optical coherence tomography. In Drexler W, Fujimoto JG, editors. Optical Coherence Tomography – Technology and Applications. Cham: Springer; 2015 pp. 1055–1101.
34 Hitzenberger CK, Pircher M. MUW approach of PS OCT. In Drexler W, Fujimoto JG, editors. Optical Coherence Tomography – Technology and Applications. Cham: Springer; 2015 pp. 1103–1136.
35 Yasuno Y, Ju MJ, Hong YJ, Makita S, Lim Y, Yamanari M. Jones matrix based polarization sensitive optical coherence tomography. In Drexler W, Fujimoto JG, editors. Optical Coherence Tomography – Technology and Applications. Cham: Springer; 2015. pp. 1137–1162.
36 de Boer JF, and Milner TE. Review of polarization sensitive optical coherence tomography and Stokes vector determination. J Biomed Opt. 2002;7(3):359–371.
37 Pircher M, Hitzenberger CK, Schmidt-Erfurth, U. Polarization sensitive optical coherence tomography in the human eye. Prog Retin Eye Res. 2011;30(6):431–451.
38 Stifter D. Beyond biomedicine: a review of alternative applications and developments for optical coherence tomography. Appl Phys B. 2007;88(3):337–357.
39 Angelsky PO, Ushenko AG, Dubolazov AV, Sidor MI, Bodnar GB, Koval G. et al. The singular approach for processing polarization-inhomogeneous laser images of blood plasma layers. J Opt. 2013;15(4):044030. IOP Publishing.
40 Ushenko AG, Dubolazov AV, Ushenko VA, Novakovskaya OY. Statistical analysis of polarization-inhomogeneous Fourier spectra of laser radiation scattered by human skin in the tasks of differentiation of benign and malignant formations. J Biomed Opt. 2016;21(7):071110.
41 Angelskii OV, Ushenko OG, Burkovets DN, Arkhelyuk OD, Ushenko YA. Polarization-correlation studies of multifractal structures in biotissues and diagnostics of their pathologic changes. Laser Phys. 2000;10(5):1136–1142.
42 Yermolenko S, Ushenko A, Ivashko P, Goudail F, Gruia I, Gavrilă C, et al. Spectropolarimetry of cancer change of biotissues. In Ninth International Conference on Correlation Optics. SPIE. 2009;7388:404–410.
43 Ushenko AG, Pishak VP. Laser polarimetry of biological tissue: Principles and applications. In Coherent-domain Optical Methods: Biomedical Diagnostics, Environmental and Material Science. Springer Science & Business Media; 2004.
44 Ushenko AG, Fediv AI, Marchuk YF. Correlation and fractal structure of Jones matrices of human bile secret. In European Conference on Biomedical Optics. 2009;7368:73681Q.
45 Bachinskyi V, Boychuk T, Ushenko A. Laser Polarimetry of Biological Tissues and Fluids: Chapter 4. Methods of Spatial-frequency, Singular and Correlation Analysis of Biological Layers Object Fields. LAP Lambert Academic Publishing; 2018.
46 Ushenko VA, Dubolazov AV, Pidkamin LY, Sakchnovsky MY, Bodnar AB, Ushenko YA, et al. Mapping of polycrystalline films of biological fluids utilizing the Jones-matrix formalism. Laser Phys. 2018;28(2):025602.
47 Ushenko AG, Angelsky PO, Sidor M, Marchuk YF, Andreychuk DR, Pashkovskaya NV. Spatial-frequency selection of complex degree of coherence of laser images of blood plasma in diagnostics and differentiation of pathological states of human organism of various nosology. Appl Opt. 2014;53(10):B172–B180.
48 Olar EI, Ushenko AG, Ushenko YA. Polarization correlation measurements of the phase tomograms of optically anisotropic biofractals. Laser Phys. 2004;14(8):1115–1121.
49 Guminetskiy SH, Ushenko OG, Polyanskiy IP, Motrych AV, Grynchuk FV. The optical method for investigation of the peritonitis progressing process. In Eighth International Conference on Correlation Optics. 2008;7008:626–632.
50 Ushenko AG, Fediv AI, Marchuk YuF. Singular structure of polarization images of bile secret in diagnostics of human physiological state. In: European Conference on Biomedical Optics. Optica Publishing Group; 2009.
51 Ushenko Y, Ushenko A, Dubolazov A, Soltys I, Litvinenko O, Wanchuliak O, et al. Mueller-matrix interferometric multifractal scaling of optically anisotropic architectonics of diffuse blood facies: fundamental and applied aspects. J Biophotonics. 2025;18(3):e202400412. https://doi.org/10.1002/jbio.202400412.
52 Ushenko A, Soltys I, Dubolazov A, Ushenko Y, Bilookyi V, Bilookyi O, et al. 3D Jones-matrix thesiography of biological fluid facies. J Innov Opt Health Sci. 2025;18(2):2443002. https://doi.org/10.1142/S1793545824430028.
53 Ushenko A, Pavlyukovich N, Khukhlina O, Pavlyukovich O, Gorsky M, Soltys I, et al. Layer-by-layer multifractal scanning of optically anisotropic architectonics of blood plasma films: fundamental and applied aspects. Photonics. 2025;12:215. https://doi.org/10.3390/photonics12030215.

54 Ushenko A, Pavlyukovich N, Khukhlina O, Pavlyukovich O, Soltys I, Dubolazov A, et al. Blood plasma film multifractal scanning in COVID-19 consequences diagnostics. J Biophotonics. 2024;17(11):e202400356. https://doi.org/10.1002/jbio.202400356.

55 Ushenko A, Dubolazov A, Zheng J, Litvinenko A, Gorsky M, Ushenko Y, et al. 3D polarization-interference holographic histology for wavelet-based differentiation of biological tissues with different necrotic states. J Biomed Opt. 2024;29(5): 052920. https://doi.org/10.1117/1.JBO.29.5.052920.

56 Ushenko O, Bilookyi O, Zheng J, Dubolazov A, Olar O, Ushenko Y, et al. 3D digital holographic polarimetry of laser speckle fields formed by polycrystalline blood films: a tool for differential diagnosis of thyroid pathology. Front Phys. 2024;12:1426469. https://doi.org/10.3389/fphy.2024.1426469.

57 Wójcik W, Hu Z, Ushenko Y, Smolarz A, Soltys I, Dubolazov O, et al. Optical sensor system for 3D Jones matrix reconstruction of optical anisotropy maps of self-assembled polycrystalline soft matter films. Sensors. 2024;24(5):1589. https://doi.org/10.3390/s24051589.

58 Ushenko A, Zheng J, Gorsky M, Dubolazov A, Ushenko Y, Soltys I, et al. 3D digital holographic polarimetry of diffuse optically anisotropic biological tissue object fields. Front Phys. 2023;11:1288935. https://doi.org/10.3389/fphy.2023.1288935.

59 Dubolazov A, Ushenko V, Trifonyuk L, Stashkevich A, Soltys I, Ushenko Y, et al. Polarization-singular approach to imaging Mueller-matrix polarimetry in differential diagnosis of histological sections of biopsy of tumors of the uterus and prostate. Front Phys. 2021;9:711212. https://doi.org/10.3389/fphy.2021.711212.

60 Ushenko VO, Trifonyuk L, Ushenko YA, Dubolazov OV, Gorsky MP, Ushenko AG. Polarization singularity analysis of Mueller-matrix invariants of optical anisotropy of biological tissue samples in cancer diagnostics. J Opt. 2021;23(6):064004. https://doi.org/10.1088/2040-8986/abf97a.

61 Ushenko YA, Koval GD, Ushenko AG, Dubolazov OV, Ushenko VA, Novakovskaia OY. Mueller-matrix of laser-induced autofluorescence of polycrystalline films of dried peritoneal fluid in diagnostics of endometriosis. J Biomed Opt. 2016;21(7):071116. https://doi.org/10.1117/1.JBO.21.7.071116.

Section 2.2

1 Hunter R. Foundations of Colloid Science. New York: Oxford University Press; 2004. p. 806.

2 Chen X, Chen PG, Ouazzani J, Liu Q. Numerical simulations of sessile droplet evaporating on heated substrate. Eur Phys J: Spec Top. 2017;226(6): 1325–1335.

3 Dash S, Garimella SV. Droplet evaporation on heated hydrophobic and superhydrophobic surfaces. Phys Rev E. 2014;89(4):042402.

4 Gleason K, Putnam SA. Microdroplet evaporation with a forced pinned contact line. Langmuir. 2014;30(34):10548–10555.

5 Hu H, Larson RG. Evaporation of a sessile droplet on a substrate. J. Phys Chem B. 2020;106:1334–1344.

6 Serra J. Image Analysis and Mathematical Morphology. New York: Academic Press; 1988. p. 411.

7 Shirvaikar MV, Trivedi, MM. Developing texture-based image clutter measures for object detection. Opt Eng. 1992;31(12): 2628–2639.

8 Thiele U. Patterned deposition at moving contact lines. Adv Colloid Interface Sci. 2014;206:399–413.

9 Makita S, Yamanari M, Yasuno Y. Generalized Jones matrix optical coherence tomography: performance and local birefringence imaging. Opt Express. 2010;18(2):854–876.

10 Chipman RA. Polarization analysis of optical systems. Opt Eng. 1989;28(2):90–99.

11 Lu SY, Chipman RA. Interpretation of Mueller matrices based on polar decomposition. J Opt Soc Am A. 1996;13(5):1106–1113.

12 Jiao S, Yao G, Wang LV. Depth-resolved two-dimensional stokes vectors of backscattered light and Mueller matrices of biological tissue measured with optical coherence tomography. Appl Opt. 2000;39(34):6318–6324.

13 Adie SG, Hillman TR, Sampson DD. Detection of multiple scattering in optical coherence tomography using the spatial distribution of Stokes vectors. Opt Express. 2007;15(26):18033–18049.

14 Lee SW, Kang JH, Yoo JY, Kang MS, Oh JT, Kim BM. Quantification of scattering changes using polarization-sensitive optical coherence tomography. J Biomed Opt. 2008;13(5):054032.

15 Götzinger E, Baumann B, Pircher M, Hitzenberger CK. Polarization maintaining fiber based ultra-high resolution spectral domain polarization sensitive optical coherence tomography. Opt Express. 2009;17(25):22704–22717.

16 Götzinger E, Pircher M, Baumann B, Schmoll T, Sattmann H, Leitgeb RA, et al. Speckle noise reduction in high speed polarization sensitive spectral domain optical coherence tomography. Opt Express. 2011;19(15):14568–14585.

17 Bonesi M, Sattmann H, Torzicky T, Zotter S, Baumann B, Pircher M, et al. High-speed polarization sensitive optical coherence tomography scan engine based on Fourier domain mode locked laser. Biomed Opt Express. 2012;3(11):2987–3000.
18 Sugita M, Pircher M, Zotter S, Baumann B, Saito K, Makihira T, et al. Analysis of optimum conditions of depolarization imaging by polarization-sensitive optical coherence tomography in the human retina. J Biomed Opt. 2015;20(1):016011.
19 Makita S, Hong YJ, Miura M, Yasuno Y. Degree of polarization uniformity with high noise immunity using polarization-sensitive optical coherence tomography. Opt Lett. 2014;39(24):6783–6786.
20 Jones CR. New calculus for the treatment of optical systems. VII. Properties of the N-matrices. J Opt Soc Am. 1948;38:671–685.
21 Azzam RMA. Propagation of partially depolarized light through anisotropic media with or without depolarization. A differential 4x4 matrix calculus. J Opt Soc Am. 1978;68:1756–1767.
22 Arteaga O, Canillas A. Analytic inversion of the Mueller-Jones polarization matrices for homogeneous media. Opt Lett. 2010;35:559–561.
23 Barakat R. Exponential versions of the Jones and Mueller–Jones polarization matrices. J Opt Soc Am A. 1996;13:158–163.
24 Lee HR, Lotz C, Groeber-Becker FK, Dembski S, Novikova T. Digital histology with Mueller polarimetry and FastDBSCAN. Appl Opt. 2022;61(32): 9616–9624.
25 Pishak V, Ushenko A, Gryhoryshyn P, et al. Study of polarization structure of biospeckle fields in crosslinked tissues of human organism: 1. Vector structure of skin biospeckles. Proceedings of SPIE-The International Society for Optical Engineering, 1997;418–424.
26 Ushenko AG. Depolarization of a laser emission field as the correlation development of its polarization structure. International Conference on Correlation Optics. 1997;3317:331–339.
27 Ushenko AG, Burkovets DN, Yermolenko SB, Arkhelyuk AD, Pishak VP, Yuzko AM, et al. Stokes polarimetry of biotissues. Fourth International Conference on Correlation Optics 1999;3904:527–533.
28 Jóźwicki R, Patorski K, Angelsky OV, Ushenko AG, Burkovets DN, Ushenko YA. Automatic polarimetric system for early medical diagnosis by biotissue testing. Opt Appl. 2002;32(4):603–612.
29 Ushenko AG, Pishak VP. Coherent-domain Optical Methods: Biomedical Diagnostics, Environmental and Material Science. Laser Polarimetry of Biological Tissue. Principles and Applications. Boston: Kluwer Academic Publishers; 2004. pp. 67–93.
30 Olar EI, Ushenko AG, Ushenko YA. Correlation microstructure of the Jones matrices for multifractal networks of biotissues. Laser Phys. 2004; 14(7): 1012–1018.
31 Yermolenko S, Ushenko A, Ivashko P, Goudail F, Gruia I, Gavrilă C, et al. Spectropolarimetry of cancer change of biotissues. Ninth International Conference on Correlation Optics, 2009;7388:404–410.
32 Ushenko AG, Fediv AI, Marchuk YF. Correlation and fractal structure of Jones matrices of human bile secret. In European Conference on Biomedical Optics. 2009;7368_1Q.
33 Angelsky OV, Ushenko AG, Ushenko YA, Pishak VP, Peresunko AP. Handbook of Photonics for Biomedical Science. CRC press; 2010. pp. 22–67.
34 Angelsky OV, Ushenko AG, Zenkova CY, et al. Optical Measurements: Polarization and Coherence of Light Fields. INTECH Open Access Publisher; 2012.
35 Garazdyuk MS, Bachinskyi VT, Vanchulyak OY, Ushenko AG, Dubolazov OV, Gorsky MP. Polarization-phase images of liquor polycrystalline films in determining time of death. Appl Opt. 2016;55(12):B67–B71.
36 Ushenko AG, Dubolazov OV, Ushenko VA, Novakovskaya OY, Olar OV. Fourier polarimetry of human skin in the tasks of differentiation of benign and malignant formations. Appl Opt. 2016;55(12):B56–B60.
37 Huang D, Swanson EA, Lin CP, Schuman JS, Stinson WG, Chang W, et al. Optical coherence tomography. Science. 1991;254(5035):1178–1181.
38 Hee MR, Huang D, Swanson EA, Fujimoto JG. Polarization-sensitive low-coherence reflectometer for birefringence characterization and ranging. J Opt Soc Am B. 1992;9(6):903–908.
39 Park BH, de Boer JF. Polarization sensitive optical coherence tomography. In Drexler W, Fujimoto JG, editors. Optical Coherence Tomography – Technology and Applications. Cham: Springer; 2015. pp. 1055–1101.
40 Hitzenberger CK, Pircher M. MUW approach of PS OCT. In Drexler W, Fujimoto JG, editors. Optical Coherence Tomography – Technology and Applications. Cham: Springer; 2015. pp. 1103–1136.
41 Yasuno Y, Ju MJ, Hong YJ, Makita S, Lim Y, Yamanari M. Jones matrix based polarization sensitive optical coherence tomography. In Drexler W, Fujimoto JG, editors. Optical Coherence Tomography - Technology and Applications. Cham: Springer; 2015. pp. 1137–1162.
42 de Boer JF, Milner TE. Review of polarization sensitive optical coherence tomography and Stokes vector determination. J Biomed Opt. 2002;7(3):359–371.

43 Pircher M, Hitzenberger CK, Schmidt-Erfurth, U. Polarization sensitive optical coherence tomography in the human eye. Prog Retin Eye Res. 2011;30(6):431–451.

44 Stifter D. Beyond biomedicine: a review of alternative applications and developments for optical coherence tomography. Appl Phys B. 2007;88(3):337–357.

45 Angelsky PO, Ushenko AG, Dubolazov AV, Sidor MI, Bodnar GB, Koval G, et al. The singular approach for processing polarization-inhomogeneous laser images of blood plasma layers. J Opt. 2013;15(4):044030.

46 Ushenko AG, Dubolazov AV, Ushenko VA, Novakovskaya OY. Statistical analysis of polarization-inhomogeneous Fourier spectra of laser radiation scattered by human skin in the tasks of differentiation of benign and malignant formations. J Biomed Opt. 2016;21(7):071110.

47 Angelskii OV, Ushenko OG, Burkovets DN, Arkhelyuk OD, Ushenko YA. Polarization-correlation studies of multifractal structures in biotissues and diagnostics of their pathologic changes. Laser Phys. 2000;10(5):1136–1142.

48 Bachinskyi V, Boychuk T, Ushenko A. (2018). Laser Polarimetry of Biological Tissues and Fluids: Chapter 4. Methods of Spatial-frequency, Singular and Correlation Analysis of Biological Layers Object Fields. LAP Lambert Academic Publishing; 2018.

49 Ushenko VA, Dubolazov AV, Pidkamin LY, Sakchnovsky MY, Bodnar AB, Ushenko YA, et al. Mapping of polycrystalline films of biological fluids utilizing the Jones-matrix formalism. Laser Phys. 2018; 28(2): 025602.

50 Ushenko AG, Angelsky PO, Sidor M, Marchuk YF, Andreychuk DR, Pashkovskaya NV. Spatial-frequency selection of complex degree of coherence of laser images of blood plasma in diagnostics and differentiation of pathological states of human organism of various nosology. Appl Opt. 2014;53(10):B172–B180.

51 Olar EI, Ushenko AG, Ushenko YA. Polarization correlation measurements of the phase tomograms of optically anisotropic biofractals. Laser Phys. 2004;14(8):1115–1121.

52 Guminetskiy SH, Ushenko OG, Polyanskiy IP, Motrych AV, Grynchuk FV. The optical method for investigation of the peritonitis progressing process. In Eighth International Conference on Correlation Optics. 2008;7008:626–632.

53 Ushenko AG, Fediv AI, Marchuk YuF. Singular structure of polarization images of bile secret in diagnostics of human physiological state. In: European Conference on Biomedical Optics. Optica Publishing Group; 2009.

54 Ushenko Y, Ushenko A, Dubolazov A, Soltys I, Litvinenko O, Wanchuliak O, et al. Mueller-matrix interferometric multifractal scaling of optically anisotropic architectonics of diffuse blood facies: fundamental and applied aspects. J Biophotonics. 2025; 18(3): e202400412. https://doi.org/10.1002/jbio.202400412.

55 Ushenko A, Soltys I, Dubolazov A, Ushenko Y, Bilookyi V, Bilookyi O, et al. 3D Jones-matrix thesiography of biological fluid facies. J Innov Opt Health Sci. 2025;18(2):2443002. https://doi.org/10.1142/S1793545824430028.

56 Ushenko A, Pavlyukovich N, Khukhlina O, Pavlyukovich O, Gorsky M, Soltys I, et al. Layer-by-layer multifractal scanning of optically anisotropic architectonics of blood plasma films: fundamental and applied aspects. Photonics. 2025;12:215. https://doi.org/10.3390/photonics12030215.

57 Ushenko A, Pavlyukovich N, Khukhlina O, Pavlyukovich O, Soltys I, Dubolazov A, et al. Blood plasma film multifractal scanning in COVID-19 consequences diagnostics. J Biophotonics. 2024;17(11):e202400356. https://doi.org/10.1002/jbio.202400356.

58 Ushenko A, Dubolazov A, Zheng J, Litvinenko A, Gorsky M, Ushenko Y, et al. 3D polarization-interference holographic histology for wavelet-based differentiation of biological tissues with different necrotic states. J Biomed Opt. 2024;29(5): 052920. https://doi.org/10.1117/1.JBO.29.5.052920.

59 Ushenko O, Bilookyi O, Zheng J, Dubolazov A, Olar O, Ushenko Y, et al. 3D digital holographic polarimetry of laser speckle fields formed by polycrystalline blood films: a tool for differential diagnosis of thyroid pathology. Front Phys. 2024;12:1426469. https://doi.org/10.3389/fphy.2024.1426469.

60 Wójcik W, Hu Z, Ushenko Y, Smolarz A, Soltys I, Dubolazov O, et al. Optical sensor system for 3D Jones matrix reconstruction of optical anisotropy maps of self-assembled polycrystalline soft matter films. Sensors. 2024;24(5):1589. https://doi.org/10.3390/s24051589.

61 Ushenko A, Zheng J, Gorsky M, Dubolazov A, Ushenko Y, Soltys I, et al. 3D digital holographic polarimetry of diffuse optically anisotropic biological tissue object fields. Front Phys. 2023;11:1288935. https://doi.org/10.3389/fphy.2023.1288935.

62 Dubolazov A, Ushenko V, Trifonyuk L, Stashkevich A, Soltys I, Ushenko Y, et al. Polarization-singular approach to imaging Mueller-matrix polarimetry in differential diagnosis of histological sections of biopsy of tumors of the uterus and prostate. Front Phys. 2021;9:711212. https://doi.org/10.3389/fphy.2021.711212.

63 Ushenko VO, Trifonyuk L, Ushenko YA, Dubolazov OV, Gorsky MP, Ushenko AG. Polarization singularity analysis of Mueller-matrix invariants of optical anisotropy of biological tissue samples in cancer diagnostics. J Opt. 2021;23(6):064004. https://doi.org/10.1088/2040-8986/abf97a.

64 Ushenko YA, Koval GD, Ushenko AG, Dubolazov OV, Ushenko VA, Novakovskaia OY. Mueller-matrix of laser-induced autofluorescence of polycrystalline films of dried peritoneal fluid in diagnostics of endometriosis. J Biomed Opt. 2016;21(7):071116. https://doi.org/10.1117/1.JBO.21.7.071116.

Section 2.3

1 Lu SY, Chipman RA. Interpretation of Mueller matrices based on polar decomposition. J Opt Soc Am A. 1996;13:1106–1113.
2 Arteaga O, Canillas A. Pseudopolar decomposition of the Jones and Mueller-Jones exponential polarization matrices. J Opt Soc Am A. 2009;26:783–793.
3 Ossikovski R, De Martino A, Guyot S. Forward and reverse product decompositions of depolarizing Mueller matrices. Opt Lett. 2007;32:689–691.
4 Ortega-Quijano N, Arce-Diego JL. Mueller matrix differential decomposition. Opt Lett. 2011;36:1942–1944.
5 Wu PJ, Walsh Jr. JT. Stokes polarimetry imaging of rat-tail tissue in a turbid medium using incident circularly polarized light. Lasers Surg Med. 2005;37:396–406.
6 Macdonald C, Meglinski I. Backscattering of circular polarized light from a disperse random medium influenced by optical clearing. Laser Phys Lett. 2011;8:324–328.
7 Qi J, He H, Lin J, Dong Y, Chen D, Ma H, et al. Assessment of tissue polarimetric properties using Stokes polarimetric imaging with circularly polarized illumination. J Biophotonics. 2018;11:e201700139.
8 Kunnen, B. Macdonald C, Doronin A, Jacques S, Eccles M, Meglinski I. Application of circularly polarized light for non-invasive diagnosis of cancerous tissues and turbid tissue-like scattering media. J Biophotonics. 2015;8:317–323. Xu M, Alfano RR. Circular polarization memory of light. Phys Rev E. 2005;72:065601; Macdonald CM, Jacques SL, Meglinski IV. Circular polarization memory in polydisperse scattering media. Phys Rev E. 2015;91:033204.
9 Wood MF, Ghosh N, Moriyama EH, Wilson BC, Vitkin IA. Proof-of-principle demonstration of a Mueller matrix decomposition method for polarized light tissue characterization in vivo. J Biomed Opt. 2009;14:014029; Wang Y, He H, Chang J, He C, Liu S, Li M, et al. Mueller matrix microscope: a quantitative tool to facilitate detections and fibrosis scorings of liver cirrhosis and cancer tissues. J Biomed Opt. 2016;21:071112.
10 Liu T, Lu M, Chen B, Zhong Q, Li J, He H, et al. Distinguishing structural features between Crohn's disease and gastrointestinal luminal tuberculosis using Mueller matrix derived parameters. J Biophotonics. 2019;12:e201900151.
11 Shen YX, Huang R, He H, Liu S, Dong Y, Wu J, et al. Comparative study of the influence of imaging resolution on linear retardance parameters derived from the Mueller matrix. Biomed Opt Express. 2021;12:211–225.
12 Pierangelo A, Manhas S, Benali A, Fallet C, Totobenazara JL, Antonelli MR, et al. Multispectral Mueller polarimetric imaging detecting residual cancer and cancer regression after neoadjuvant treatment for colorectal carcinomas. J Biomed Opt. 2013;18:046014.
13 Sun M, He H, Zeng N, Du E, Guo Y, Liu S, et al. Characterizing the microstructures of biological tissues using Mueller matrix and transformed polarization parameters. Biomed Opt Express. 2014;5:4223–4234; Li P, Lv D, He H, Ma H. Separating azimuthal orientation dependence in polarization measurements of anisotropic media. Opt Express. 2018;26:3791–3800.
14 Gil JJ. Invariant quantities of a Mueller matrix under rotation and retarder transformations. J Opt Soc Am A. 2016;33:52–58.
15 Iqbal M, Ahmad I, Khaliq A, Khan S. Comparative study of Mueller matrix transformation and polar decomposition for optical characterization of turbid media. Optik. 2020;224:165508.
16 Sun T, Liu T, He H, Wu J, Ma H. Distinguishing anisotropy orientations originated from scattering and birefringence of turbid media using Mueller matrix derived parameters. Opt Lett. 2018;43:4092–4095.
17 Khaliq A, Ashraf S, Gul B, Ahmad I. Comparative study of 3 × 3 Mueller matrix transformation and polar decomposition. Opt Commun. 2021;485:126756.
18 Tariq A, Li P, Chen D, Lv D, Ma H. Physically realizable space for the purity-depolarization plane for polarized light scattering media. Phys Rev Lett. 2017;119:033202.
19 Swami MK, Manhas S, Buddhiwant P, Ghosh N, Uppal A, Gupta PK. Polar decomposition of 3 × 3 Mueller matrix: a tool for quantitative tissue polarimetry. Opt Express. 2006;14:9324–9337.
20 Wang Y, Guo Y, Zeng N, Chen D, He H, Ma H. Study on the validity of 3 × 3 Mueller matrix decomposition. J Biomed Opt. 2015;20:065003.
21 Morio J, Goudail F. Influence of the order of diattenuator, retarder, and polarizer in polar decomposition of Mueller matrices. Opt Lett. 2004;29:2234–2236.
22 Ghosh N, Wood MFG, Vitkin IA. Influence of the order of the constituent basis matrices on the Mueller matrix decomposition-derived polarization parameters in complex turbid media such as biological tissues. Opt Commun. 2010;283:1200–1208.

23 He H, Zeng N, Du E, Guo Y, Li D, Liao R, et al. A possible quantitative Mueller matrix transformation technique for anisotropic scattering media/Eine mögliche quantitative Mueller Matrix-Transformations-Technik für anisotrope streuende Medien. Photon Lasers Med. 2013;2:129–137.

24 Ghosh N, Wood MFG, Vitkin IA. Mueller matrix decomposition for extraction of individual polarization parameters from complex turbid media exhibiting multiple scattering, optical activity, and linear birefringence. J Biomed Opt. 2008;13:044036.

25 Gil JJ. Characteristic properties of Mueller matrices. J Opt Soc Am A. 2000;17:328–334.

26 Ossikovski R. Analysis of depolarizing Mueller matrices through a symmetric decomposition. J Opt Soc Am A. 2009;26:1109–1118.

27 Vizet J, Ossikovski R. Symmetric decomposition of experimental depolarizing Mueller matrices in the degenerate case. Appl Opt. 2018;57:1159–1167.

28 Pishak V, Ushenko A, Gryhoryshyn P, Yermolenko SB, Rudeychuk VM, Pishak OV. Study of polarization structure of biospeckle fields in crosslinked tissues of human organism: 1. Vector structure of skin biospeckles. In Proceedings of SPIE-The International Society for Optical Engineering. 1997;418–424.

29 Ushenko AG. Depolarization of a laser emission field as the correlation development of its polarization structure. In International Conference on Correlation Optics. 1997;3317:331–339.

30 Ushenko AG, Burkovets DN, Yermolenko SB, Arkhelyuk AD, Pishak VP, Yuzko AM, et al. Stokes polarimetry of biotissues. In Fourth International Conference on Correlation Optics. 1999;3904:527–533.

31 Jóźwicki R, Patorski K, Angelsky OV, Ushenko AG, Burkovets DN, Ushenko YA. Automatic polarimetric system for early medical diagnosis by biotissue testing. Opt Appl. 2002;32(4):603–612.

32 Ushenko AG, Pishak VP. Coherent-domain Optical Methods: Biomedical Diagnostics, Environmental and Material Science. Laser Polarimetry of Biological Tissue. Principles and Applications. Boston: Kluwer Academic Publishers; 2004. p. 67–93.

33 Olar EI, Ushenko AG, Ushenko YA. Correlation microstructure of the Jones matrices for multifractal networks of biotissues. Laser Phys. 2004;14(7):1012–1018.

34 Yermolenko S, Ushenko A, Ivashko P, Goudail F, Gruia I, Gavrilă C, et al. Spectropolarimetry of cancer change of biotissues. In Ninth International Conference on Correlation Optics. 2009;7388:404–410.

35 Ushenko AG, Fediv AI, Marchuk YF. Correlation and fractal structure of Jones matrices of human bile secret. In European Conference on Biomedical Optics. 2009;7368_1Q.

36 Angelsky OV, Ushenko AG, Ushenko YA, Pishak VP, Peresunko AP. Handbook of Photonics for Biomedical Science. CRC press; 2010; p. 22–67.

37 Angelsky OV, Ushenko AG, Zenkova CY, Felde CV, Bogatyryova HV, Mokhun II, et al. Optical measurements: polarization and coherence of light fields. INTECH Open Access Publisher; 2012.

38 Garazdyuk MS, Bachinskyi VT, Vanchulyak OY, Ushenko AG, Dubolazov OV, Gorsky MP. Polarization-phase images of liquor polycrystalline films in determining time of death. Appl Opt. 2016;55(12):B67–B71.

39 Ushenko AG, Dubolazov OV, Ushenko VA, Novakovskaya OY, Olar OV. Fourier polarimetry of human skin in the tasks of differentiation of benign and malignant formations. Appl Opt. 2016;55(12):B56–B60.

40 Huang D, Swanson EA, Lin CP, Schuman JS, Stinson WG, Chang W, et al. Optical coherence tomography. Science. 1991;254(5035):1178–1181.

41 Hee MR, Huang D, Swanson EA, Fujimoto JG. Polarization-sensitive low-coherence reflectometer for birefringence characterization and ranging. J Opt Soc Am B. 1992;9(6):903–908.

42 Park BH, de Boer JF. Polarization sensitive optical coherence tomography. In Drexler W, Fujimoto JG, editors. Optical Coherence Tomography – Technology and Applications. Cham: Springer; 2015. pp. 1055–1101.

43 Hitzenberger CK, Pircher M. MUW approach of PS OCT. In Drexler W, Fujimoto JG, editors. Optical Coherence Tomography – Technology and Applications. Cham: Springer; 2015. pp. 1103–1136.

44 Angelskii OV, Ushenko OG, Burkovets DN, Arkhelyuk OD, Ushenko YA. Polarization-correlation studies of multifractal structures in biotissues and diagnostics of their pathologic changes. Laser Phys. 2000;10(5):1136–1142.

45 Ushenko AG, Pishak VP. Laser polarimetry of biological tissue: Principles and applications. In Coherent-domain optical methods: Biomedical diagnostics, environmental and material science. 2004.

46 Bachinskyi V, Boychuk T, Ushenko A. Laser Polarimetry of Biological Tissues and Fluids: Chapter 4. Methods of Spatial-frequency, Singular and Correlation Analysis of Biological Layers Object Fields. LAP Lambert Academic Publishing; 2018.

47 Ushenko VA, Dubolazov AV, Pidkamin LY, Sakchnovsky MY, Bodnar AB, Ushenko YA, et al. Mapping of polycrystalline films of biological fluids utilizing the Jones-matrix formalism. Laser Phys. 2018;28(2):025602.

48 Ushenko AG, Angelsky PO, Sidor M, Marchuk YF, Andreychuk DR, Pashkovskaya NV. Spatial-frequency selection of complex degree of coherence of laser images of blood plasma in diagnostics and differentiation of pathological states of human organism of various nosology. Appl Opt. 2014;53(10):B172–B180.
49 Olar EI, Ushenko AG, Ushenko YA. Polarization correlation measurements of the phase tomograms of optically anisotropic biofractals. Laser Phys. 2004;14(8):1115–1121.
50 Guminetskiy SH, Ushenko OG, Polyanskiy IP, Motrych AV, Grynchuk FV. The optical method for investigation of the peritonitis progressing process. In Eighth International Conference on Correlation Optics. 2008;7008:626–632.
51 Ushenko AG, Fediv AI, Marchuk YuF. Singular structure of polarization images of bile secret in diagnostics of human physiological state. In: European Conference on Biomedical Optics. Optica Publishing Group; 2009.
52 Ushenko Y, Ushenko A, Dubolazov A, Soltys I, Litvinenko O, Wanchuliak O, et al. Mueller-matrix interferometric multifractal scaling of optically anisotropic architectonics of diffuse blood facies: fundamental and applied aspects. J Biophotonics. 2025;18(3):e202400412. https://doi.org/10.1002/jbio.202400412.
53 Ushenko A, Soltys I, Dubolazov A, Ushenko Y, Bilookyi V, Bilookyi O, et al. 3D Jones-matrix thesiography of biological fluid facies. J Innov Opt Health Sci. 2025;18(2):2443002. https://doi.org/10.1142/S1793545824430028.
54 Ushenko A, Pavlyukovich N, Khukhlina O, Pavlyukovich O, Gorsky M, Soltys I, et al. Layer-by-layer multifractal scanning of optically anisotropic architectonics of blood plasma films: fundamental and applied aspects. Photonics. 2025;12:215. https://doi.org/10.3390/photonics12030215.
55 Ushenko A, Pavlyukovich N, Khukhlina O, Pavlyukovich O, Soltys I, Dubolazov A, et al. Blood plasma film multifractal scanning in COVID-19 consequences diagnostics. J Biophotonics. 2024;17(11):e202400356. https://doi.org/10.1002/jbio.202400356.
56 Ushenko A, Dubolazov A, Zheng J, Litvinenko A, Gorsky M, Ushenko Y, et al. 3D polarization-interference holographic histology for wavelet-based differentiation of biological tissues with different necrotic states. J Biomed Opt. 2024;29(5):052920. https://doi.org/10.1117/1.JBO.29.5.052920.
57 Ushenko O, Bilookyi O, Zheng J, Dubolazov A, Olar O, Ushenko Y, et al. 3D digital holographic polarimetry of laser speckle fields formed by polycrystalline blood films: a tool for differential diagnosis of thyroid pathology. Front Phys. 2024;12:1426469. https://doi.org/10.3389/fphy.2024.1426469.
58 Wójcik W, Hu Z, Ushenko Y, Smolarz A, Soltys I, Dubolazov O, et al. Optical sensor system for 3D Jones matrix reconstruction of optical anisotropy maps of self-assembled polycrystalline soft matter films. Sensors. 2024;24(5):1589. https://doi.org/10.3390/s24051589.
59 Ushenko A, Zheng J, Gorsky M, Dubolazov A, Ushenko Y, Soltys I, et al. 3D digital holographic polarimetry of diffuse optically anisotropic biological tissue object fields. Front Phys. 2023;11:1288935. https://doi.org/10.3389/fphy.2023.1288935.
60 Dubolazov A, Ushenko V, Trifonyuk L, Stashkevich A, Soltys I, Ushenko Y, et al. Polarization-singular approach to imaging Mueller-matrix polarimetry in differential diagnosis of histological sections of biopsy of tumors of the uterus and prostate. Front Phys. 2021;9:711212. https://doi.org/10.3389/fphy.2021.711212.
61 Ushenko VO, Trifonyuk L, Ushenko YA, Dubolazov OV, Gorsky MP, Ushenko AG. Polarization singularity analysis of Mueller-matrix invariants of optical anisotropy of biological tissue samples in cancer diagnostics. J Opt. 2021;23(6):064004. https://doi.org/10.1088/2040-8986/abf97a.
62 Ushenko YA, Koval GD, Ushenko AG, Dubolazov OV, Ushenko VA, Novakovskaia OY. Mueller-matrix of laser-induced autofluorescence of polycrystalline films of dried peritoneal fluid in diagnostics of endometriosis. J Biomed Opt. 2016;21(7):071116. https://doi.org/10.1117/1.JBO.21.7.071116.

3

Algorithms and Methods of 3D Jones–Mueller Digital Holographic Mapping

Yurii Ushenko[1,2], Weidong Dou[1], Oleksandr Ushenko[2,3], Oleksandr Dubolazov[2], Iryna Soltys[2], Mykhailo Gorsky[2], Mykhailo Gavryliak[2], Viktor Bachinskiy[4], Oleh Vanchuliak[4], Nataliia Pavlyukovich[4], Oleksandr Pavlyukovich[4], and Vitaliy Rozhko[4]

[1] *Shaoxing University, Shaoxing, People's Republic of China*
[2] *Yurii Fedkovych Chernivtsi National University, Chernivtsi, Ukraine*
[3] *Taizhou Institute of Zhejiang University, Taizhou, People's Republic of China*
[4] *Bukovinian State Medical University, Chernivtsi, Ukraine*

3.1 Digital Holographic Reconstruction and Phase Scanning

3.1.1 Polarization Method of Mueller Matrix Experimental Measurements

The coordinate distributions of Mueller-matrix elements were measured in the setup (Figure 3.1) [1–10].

Illumination of samples was performed by parallel ($\varnothing = 2 \times 10^3$ μm) weakly intensive (W = 5 mW) beam 1 of He—Ne laser ($\lambda = 0.6328$ μm).

Using a collimated illuminating beam provides the same conditions of transformation of the polarization states at different points of the illuminated area of the object. In this case, this scheme provides the resolution of 4.65 μm × 4.65 μm. This scale is sufficient to evaluate the optical properties of the average (5–20 μm) crystalline structural elements of the biological samples.

Polarization light source consisted of quarter-wave plate 3 and polarizer 4. The image of sample 6 was projected in the plane of light-sensitive plane of the CCD camera 10 (The Imaging Source DMK 41AU02.AS, monochrome 1/2" CCD, Sony ICX205AL (progressive scan); resolution – 1280 × 960; size of light-sensitive plate – 5952 μm × 4464 μm; sensitivity – 0.05 lx; dynamic range – 8 bit; SNR – 9 bit, nonlinearity does not exceed 3–5%) by means of optical system 7.

In this experimental arrangement, seven designates an image-forming apparatus, which consists of strain-free objective (Nikon CFI Achromat P, working distance, 30 mm; focal distance, 50 mm; NA, – 0.1; magnification, 4×) and tube lens (focal distance 200 mm). Polarization analysis of the sample images was performed by means of quarter-wave plate 8 and polarizer–analyzer 9.

3.1.1.1 Stokes Polarimetry

The object of the study is a microscopic image of a native histological cut of myocardium [11–15] (Figures 3.2 and 3.3). In Figure 3.2, the polarization-produced images $I_{0;90;45;135;\otimes;\oplus}(x, y)$ of the myocardial sample for a line-polarized laser beam $P = 0°$ are given.

To determine the totality of the coordinate distributions of the parameters of the Stokes vector of the microscopic image of the biological drug, the ratio traditional for polarimetry is used [16–19]:

$$\begin{cases} S_1^{0;90;45;\otimes} = I_0^{0;90;45;\otimes} + I_{90}^{0;90;45;\otimes}; \\ S_2^{0;90;45;\otimes} = I_0^{0;90;45;\otimes} - I_{90}^{0;90;45;\otimes}; \\ S_3^{0;90;45;\otimes} = I_{45}^{0;90;45;\otimes} - I_{135}^{0;90;45;\otimes}; \\ S_4^{0;90;45;\otimes} = I_{\otimes}^{0;90;45;\otimes} - I_{\oplus}^{0;90;45;\otimes}; \end{cases} \tag{3.1}$$

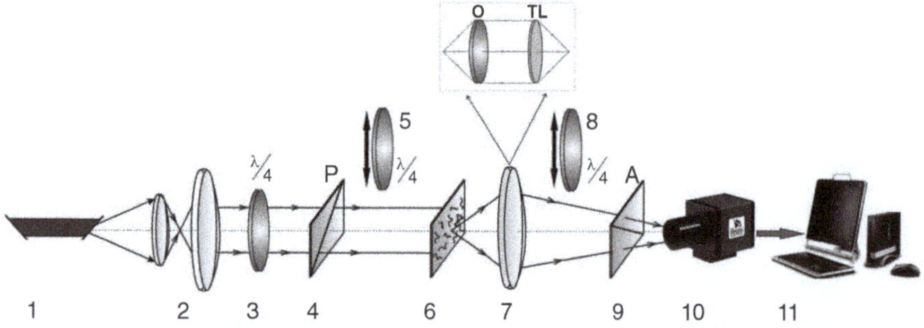

Figure 3.1 Optical scheme of the experimental setup. Here, 1 – He–Ne laser; 2 – collimator; 3 – stationary quarter-wave plate; 5, 8 – mechanically movable quarter-wave plates; 4, 9 – polarizer and analyzer, respectively; 6 – object of investigation; 7 – optical system; 10 – CCD camera; 11 – PC.

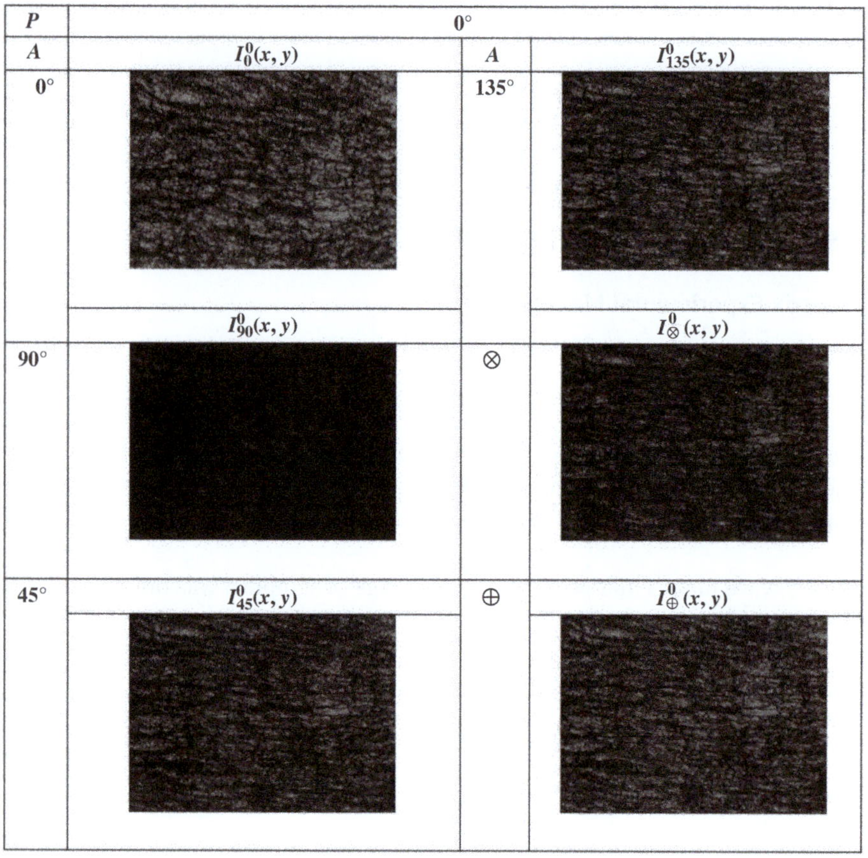

Figure 3.2 Coordinate distributions of intensity $I_{0;90;45;135;\otimes;\oplus}(x, y)$ of polarization images of the native histological section of the myocardium.

Here, $S^{0;90;45;\otimes}_{1;2;3;4}$ – Stokes vector parameters for the line (0°; 90°; 45°) and write circular polarized (⊗) laser beams.

3.1.1.2 Multichannel Stokes Polarimetry

To measure the coordinate distributions of the 16 elements of the Mueller matrix (we took, for example, another type of tissue – the nervous tissue of the human brain), use the following methodology [17–24]:

1. They illuminate a sample of the native histological cut of the brain with a series of four differently polarized laser bundles: 0°; 90°; 45°; ⊗.

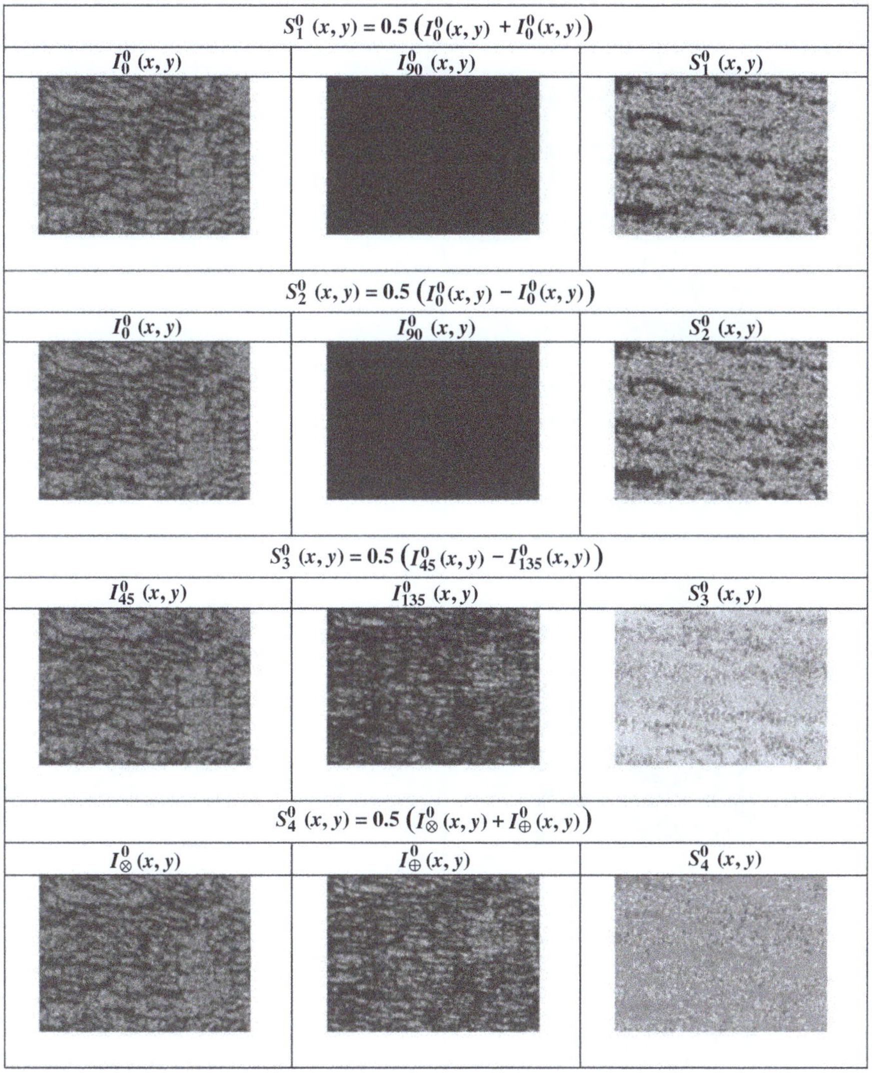

Figure 3.3 The scheme of the image Stokes polarimetry of the microscopic image of the histological cut of the myocardium.

2. For each of the illuminating bundles using the polarization filter, six polarization measurements (0°; 90°; 45°; 135°; ⊗; ⊕) of coordinate distributions $\boldsymbol{I}^{0;90;45;\otimes}_{0;90;45;135;\otimes;\oplus}(\boldsymbol{x},\boldsymbol{y})$ are made (Figures 3.4 and 3.5).

3. By the totality of experimentally measured data, $I^{0;90;45;\otimes}_{0;90;45;135;\otimes;\oplus}(x,y)$ is calculated for each polarization state (0°; 90°; 45°; ⊗) an probe laser beam set of the parameters of the Stokes vector (equation (2.1)) (Figures 3.6–3.9).

3.1.1.3 Mueller Matrix Polarimetry

The calculation within each pixel of the digital camera of the set of elements of the Mueller matrix $f_{ik}(x,y)$ of brain sample was carried out in accordance with the algorithm [22–27]

$$f_{i=1;k=1;2;3;4} = \begin{cases} f_{11} = 0.5(S_1^0 + S_1^{90}); \\ f_{12} = 0.5(S_1^0 - S_1^{90}); \\ f_{13} = S_1^{45} - f_{11}; \\ f_{14} = S_1^{\otimes} - f_{11}. \end{cases} \tag{3.2}$$

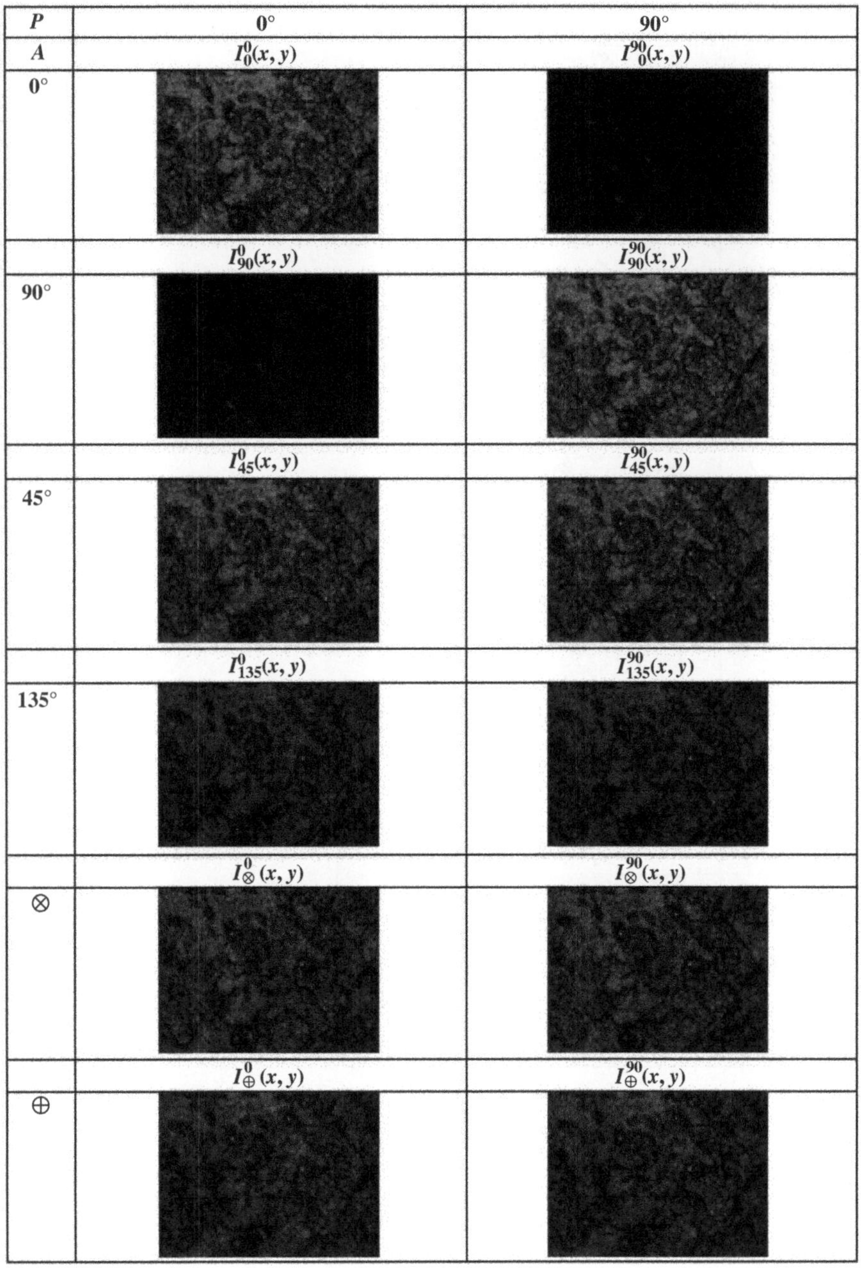

Figure 3.4 Polarization-produced microscopic images of the native histological cut of the brain $I_{0;90;45;135;\otimes;\oplus}^{0;90}(x, y)$.

Here, $S_{i=2;3;4}^{0;45;90;\otimes}$ – Stokes vector parameters for linear $(0°; 45°; 90°)$ and right circular $(\otimes)$ polarized laser beams (Figure 3.10).

The calculation within each pixel of the digital camera of the set of elements of the Mueller matrix $f_{ik}(x, y)$ of brain sample was carried out in accordance with the algorithm (Figures 3.11–3.13)

$$f_{i=2;k=1;2;3;4} = \begin{cases} f_{21} = 0.5(S_2^0 + S_2^{90}); \\ f_{22} = 0.5(S_2^0 - S_2^{90}); \\ f_{23} = S_2^{45} - f_{21}; \\ f_{24} = S_2^{\otimes} - f_{21}. \end{cases} \tag{3.3}$$

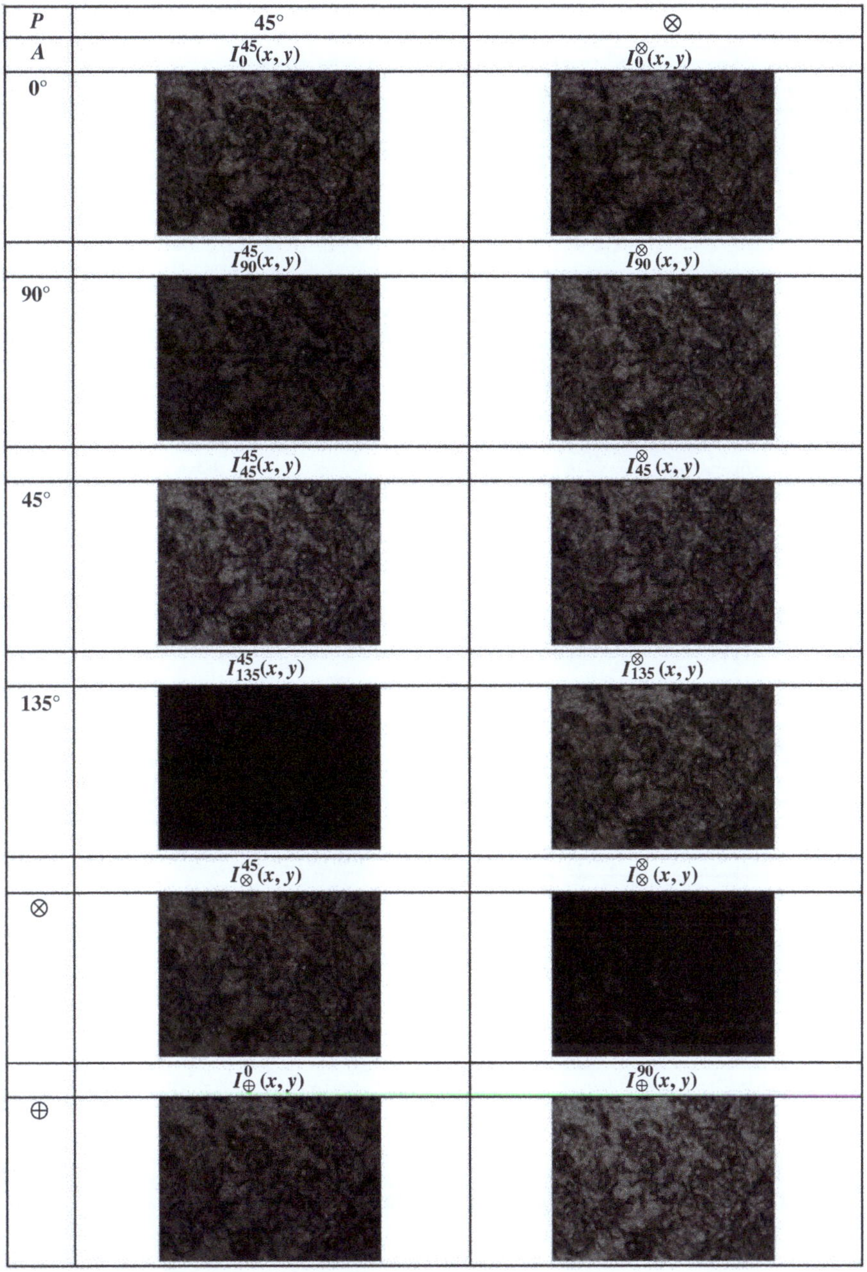

Figure 3.5 Polarization-produced microscopic images of the native histological cut of the brain $I^{45;\otimes}_{0;90;45;135;\otimes;\oplus}(x, y)$.

The calculation within each pixel of the digital camera of the set of elements of the Mueller matrix $f_{ik}(x, y)$ of brain sample was carried out in accordance with the algorithm

$$f_{i=3;k=1;2;3;4} = \begin{cases} f_{31} = 0.5\left(S_3^0 + S_3^{90}\right); \\ f_{32} = 0.5\left(S_3^0 - S_3^{90}\right); \\ f_{33} = S_3^{45} - f_{31}; \\ f_{34} = S_{32}^{\otimes} - f_{31}. \end{cases} \tag{3.4}$$

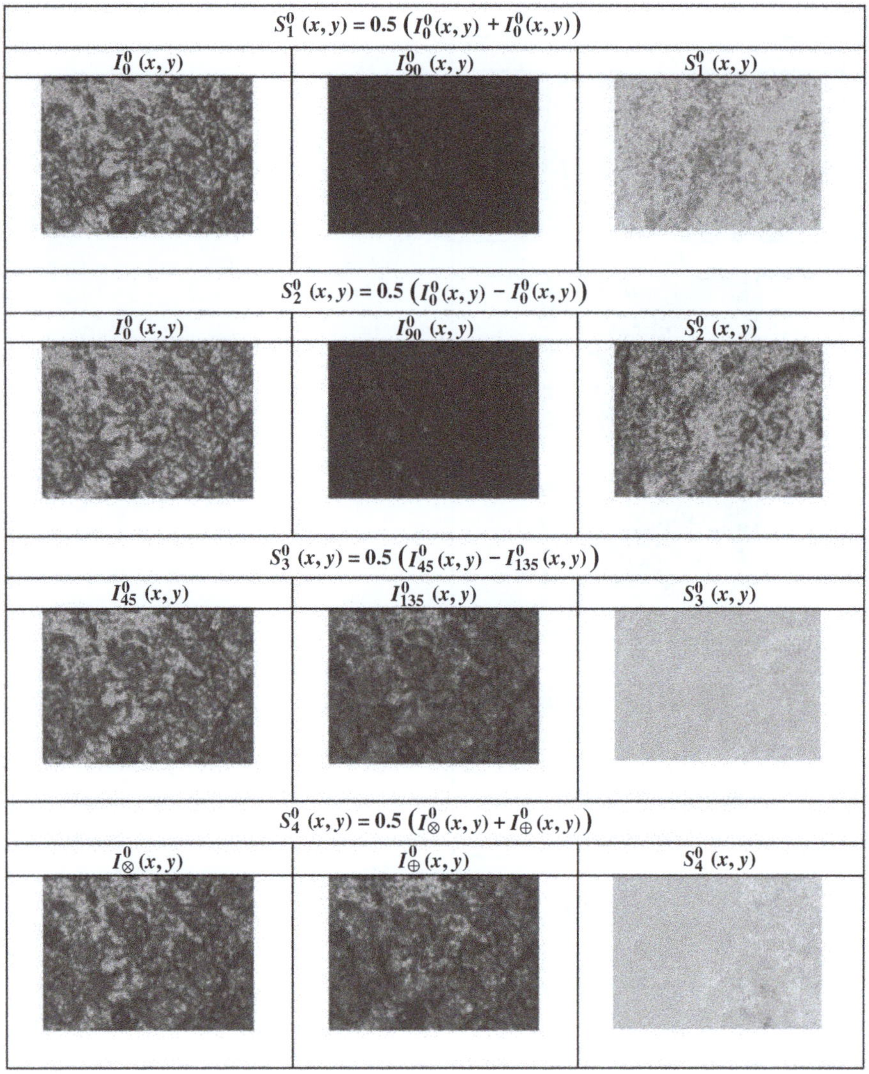

Figure 3.6 The totality of the coordinate distributions of the parameters of the Stokes vector $S^0_{1;2;3;4}(x, y)$ microscopic image of the native histological cut of the brain.

The calculation within each pixel of the digital camera of the set of elements of the Mueller matrix $f_{ik}(x, y)$ of brain sample was carried out in accordance with the algorithm

$$f_{i=4;k=1;2;3;4} = \begin{cases} f_{41} = 0.5(S_4^0 + S_4^{90}); \\ f_{42} = 0.5(S_4^0 - S_4^{90}); \\ f_{43} = S_4^{45} - f_{41}; \\ f_{44} = S_4^{\otimes} - f_{41}. \end{cases} \tag{3.5}$$

3.1.2 Polarization-Interference Method of Mueller Matrix Experimental Measurements

The experimental three-dimensional (3D) studies were carried out in the arrangement of the Mach-Zehnder polarization interferometer (Figure 3.14) [28–34]. We used this interferometer in a number of studies of layered maps of azimuth and ellipticity of polarization, Mueller-matrix images of the optically anisotropic component of biological tissues of human organs. These works present a detailed characteristic of the optical arrangement and a description of its optical-electronic components. Therefore, to facilitate the perception of our new experimental material, we present its optical scheme and a detailed description of the technique for layered determination of Jones-matrix images of biological fluid facies.

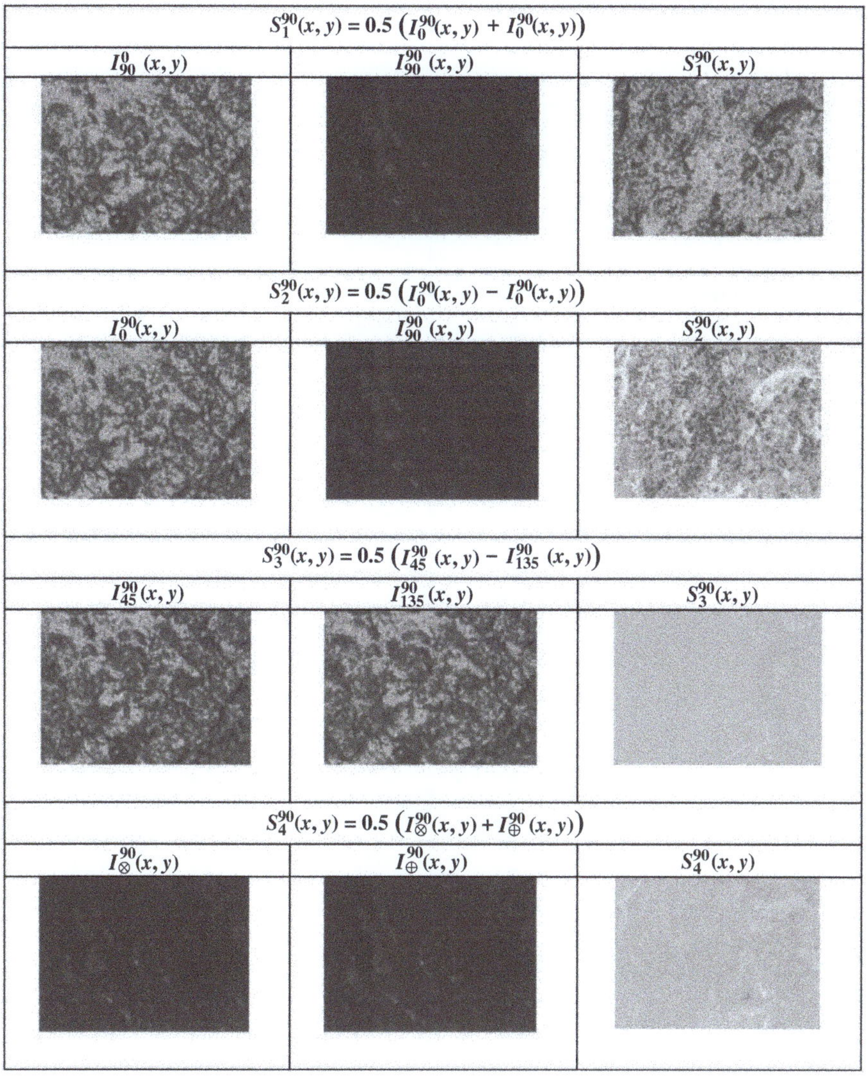

Figure 3.7 The totality of the coordinate distributions of the parameters of the Stokes vector $S^{90}_{1;2;3;4}(x, y)$ microscopic image of the native histological cut of the brain.

The polarization interferometer includes the following main blocks:

A. Polarizing illuminator, which generates a circularly polarized laser wave:
- He–Ne ($\lambda = 0.6328\ \mu\text{m}$) laser 1;
- collimator 2;
- polarizing filter (polarizer 3 and quarter-wave plate 4, the angle between the polarizer 3 transmission axes and the plate 4 highest speed axis is 45°).

B. Reference channel:
- beam splitter 5;
- polarization intensity attenuator 6;
- rotating mirror 7;
- universal polarization filter (quarter-wave plates 12, 14, and polarizer 13) can form arbitrary polarization states.

C. Object channel:
- beam splitter 5;
- rotating mirror 8;
- universal polarization filter (quarter-wave plates 9, 11, and polarizer 10), which generates arbitrary polarization states;
- biological liquid facie sample 15.

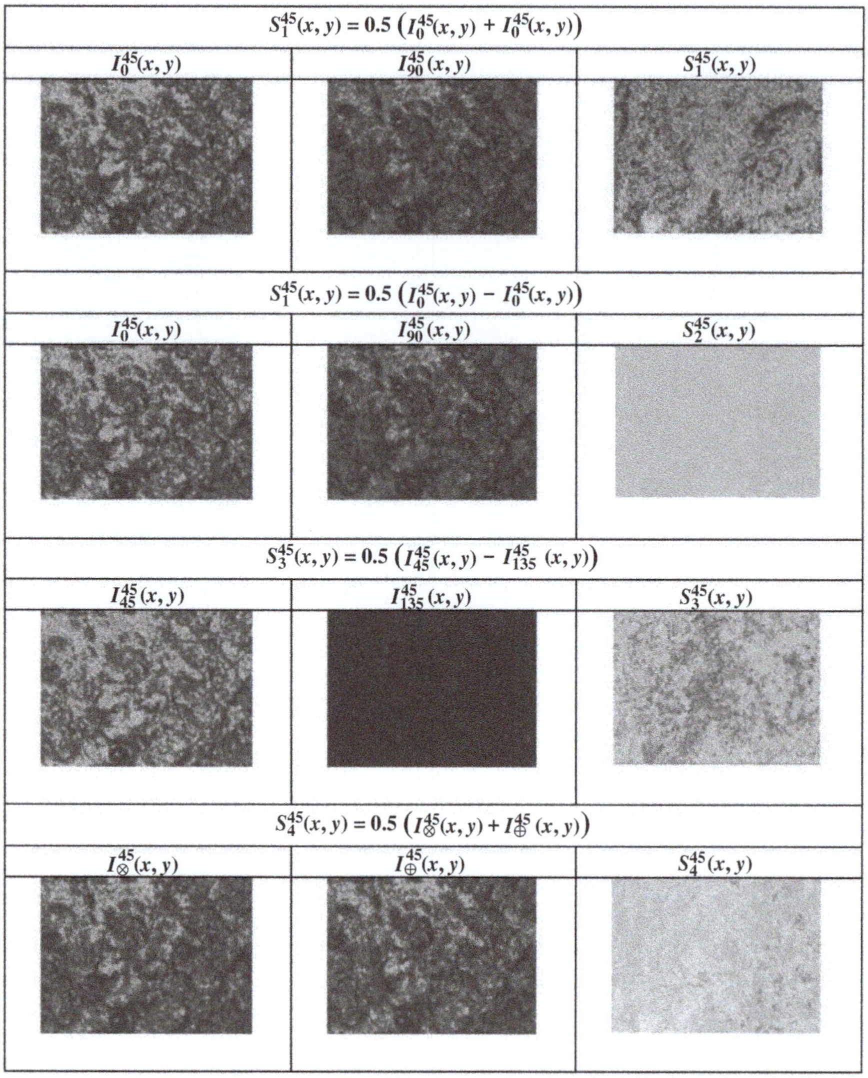

Figure 3.8 The totality of the coordinate distributions of the parameters of the Stokes vector $S_{1;2;3;4}^{45}(x, y)$ microscopic image of the native histological cut of the brain.

D. Polarization filtering channel of the object field interference distribution:
- optical mixer 16;
- polarizing microlens 17;
- polarizer–analyzer 18 with an orientation of the transmission plane at angles $\Omega = 0°$; $\Omega = 90°$.

E. Digital recording and computer analysis unit
- digital camera 19;
- PC 20.

Parallel ($\varnothing = 2 \times 10^3 \mu m$) He–Ne ($\lambda = 0.6328\ \mu m$) laser 1 beam, formed by spatial-frequency filter 2, with 50% beam splitter 3 is divided into "object" and "reference" ones.

The "object" beam with the rotating mirror 5 is directed through the polarizing filters 6–7 (Achromatic True Zero-Order Waveplate and manufacturer – B + W Kaesemann XS-Pro Polarizer MRC Nano) in the biological liquid facie 8 direction. The biological liquid facie polarization-inhomogeneous image is projected by the strain-free objective 12 (manufacturer – Nikon CFI Achromat P, focal length, 30 mm, numerical aperture – 0.1, magnification - 4×) into the digital camera 14 (The Imaging Source DMK 41AU02. AS, monochrome 1/2 "CCD, Sony ICX205AL (progressive scan) resolution – 1280 × 960, size of the photosensitive area – 7600 μm × 6200 μm; sensitivity – 0.05 lx; dynamic range – 8 bit, SNR – 9 bit), the photosensitive area, which contains $m \times n = 1280 \times 960$ pixels) plane.

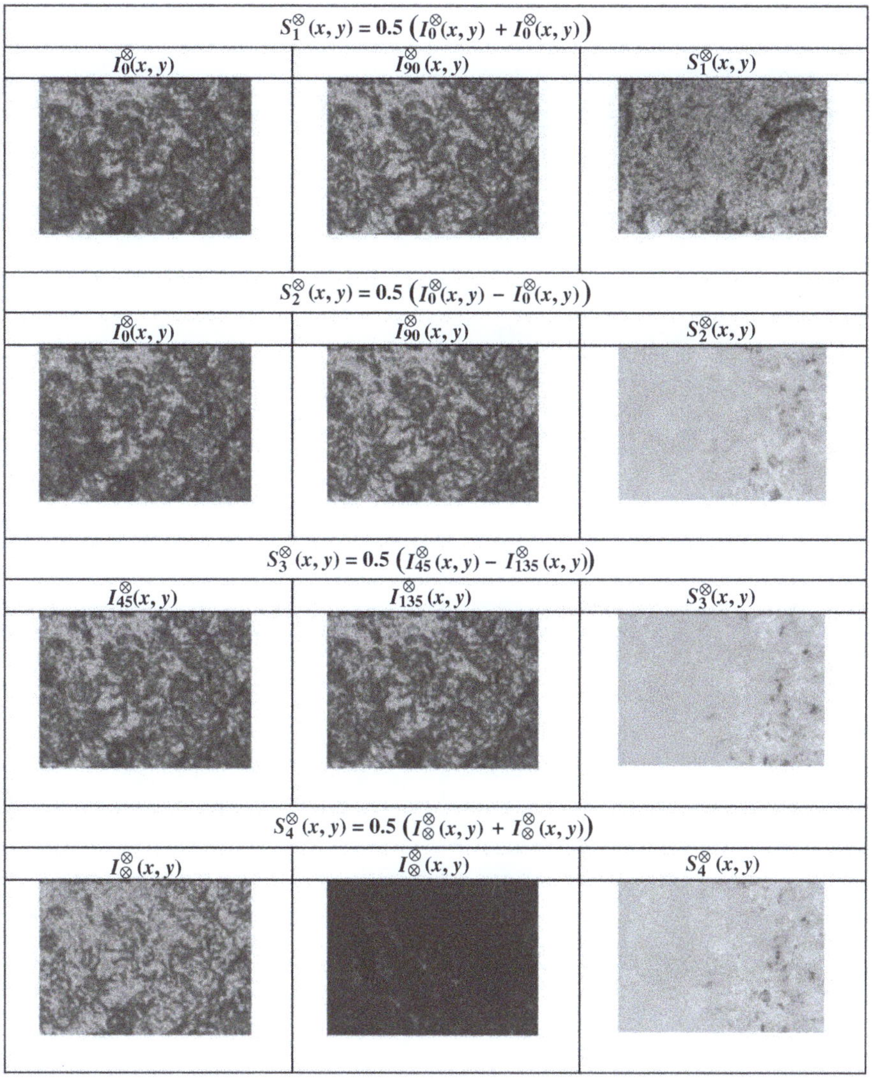

Figure 3.9 The coordinate distributions of the parameters of the Stokes vector $S_{1;2;3;4}^{\otimes}(x, y)$ microscopic image of the native histological cut of the brain.

The "reference" beam is directed by mirror 4 through the polarization filters 9–10 (Achromatic True Zero-Order Waveplate and manufacturer: B + W Kaesemann XS-Pro Polarizer MRC Nano) into the biological liquid facie polarization image plane.

As a result, an interference pattern is formed, the coordinate intensity distribution of which is recorded by a digital camera 14 through a polarizer 13.

3.2 Multichannel Polarization-Interference Measurement and Reconstruction of Anisotropy Maps

3.2.1 Mueller-matrix Polarization-Interference Reconstruction of Anisotropy Maps of Biological Tissues

The use of a reference wave of laser radiation, which in the scheme of optical interferometer is superimposed on a polarization-inhomogeneous image of a biological layer is fundamental. The resulting interference pattern is recorded using a digital camera. With the use of diffraction integrals, the operation of digital holographic reproduction of distributions of complex amplitudes $\{E_x(x, y);\quad E_y(x, y)\}$ of the objective field of a biological layer [35–42].

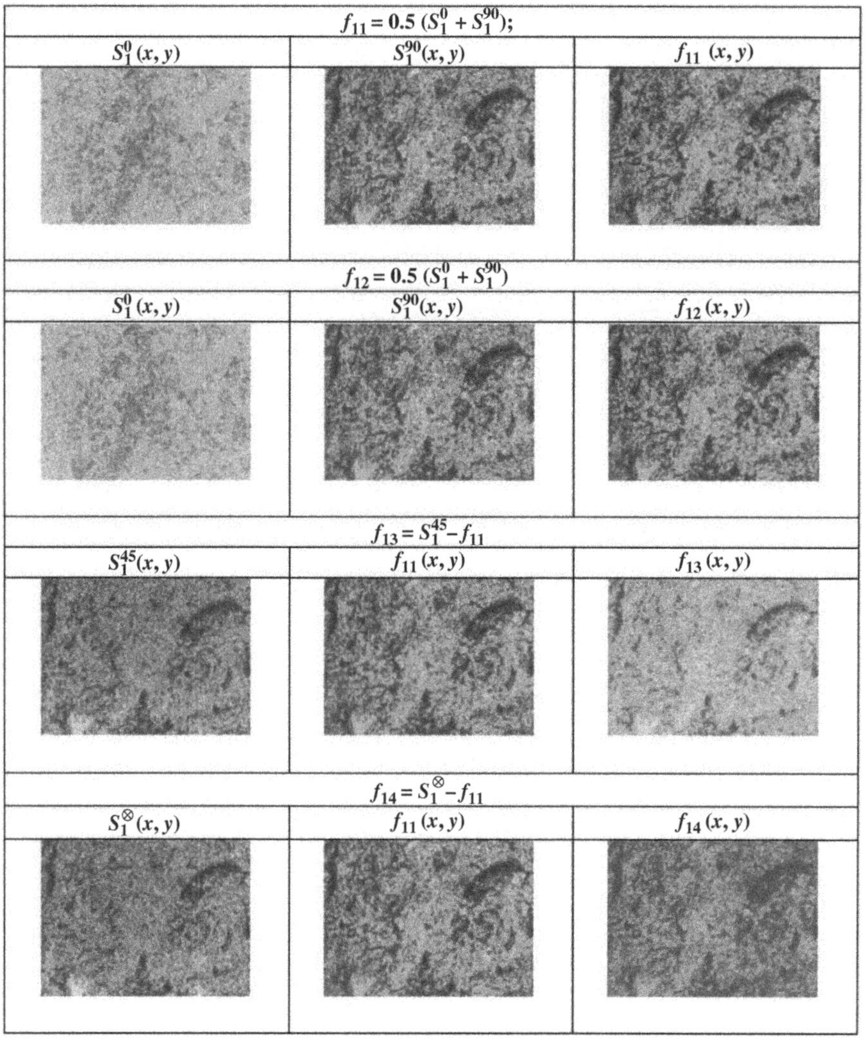

Figure 3.10 Coordinate distributions of Mueller matrix elements – Mueller matrix images $f_{i=1;k=1;2;3;4}(x, y)$.

For each state of the irradiating beam, the reconstructed distributions of the Stokes vector parameters of the object field of a biological layer are calculated according to the reproduced distributions of complex amplitudes $\{E_x(x, y);\;\; E_y(x, y)\}$

$$\begin{aligned} S_1(0°, 90°, 45°, \otimes) &= |E_x|^2 + \left|E_y\right|^2; \\ S_2(0°, 90°, 45°, \otimes) &= |E_x|^2 - \left|E_y\right|^2; \\ S_3(0°, 90°, 45°, \otimes) &= 2\mathrm{Re}\left|E_x E_y^*\right|; \\ S_4(0°, 90°, 45°, \otimes) &= 2\mathrm{Im}\left|E_x E_y^*\right|. \end{aligned} \tag{3.6}$$

Here 0°, 90°, and 45° are polarization azimuths of linearly polarized irradiating beams, ⊗ is the the right-circularly polarized beam.

Furthermore, on the basis of the set of Mueller matrix elements is calculated by the following Stokes-polarimetric equations:

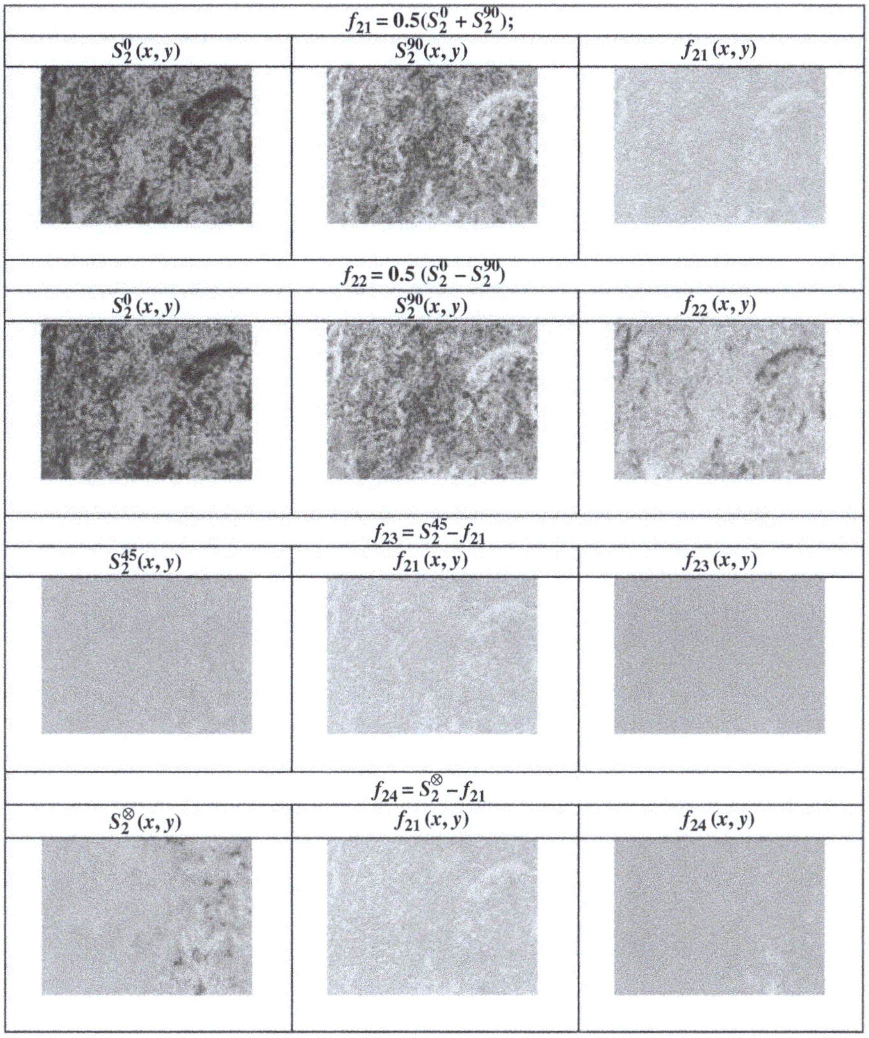

Figure 3.11 Mueller matrix images $f_{i=2;k=1;2;3;4}(x, y)$.

- for Stokes vectors of linearly polarized probing beams $S^0(0°)$; $S^0(90°)$

$$\left.\begin{Bmatrix} \left[S^0(0^\circ) = \{F\} \begin{pmatrix} 1 \\ 1 \\ 0 \\ 0 \end{pmatrix} \to S(0^\circ) = \begin{pmatrix} f_{11} + f_{12} \\ f_{21} + f_{22} \\ f_{31} + f_{32} \\ f_{41} + f_{42} \end{pmatrix} \right]; \\ \left[S^0(90^\circ) = \{F\} \begin{pmatrix} 1 \\ -1 \\ 0 \\ 0 \end{pmatrix} \to S(90^\circ) = \begin{pmatrix} f_{11} - f_{12} \\ f_{21} - f_{22} \\ f_{31} - f_{32} \\ f_{41} - f_{42} \end{pmatrix} \right] \end{Bmatrix} \right\} \Rightarrow f_{ik} = \begin{Vmatrix} f_{11} & f_{12} \\ f_{21} & f_{22} \\ f_{31} & f_{32} \\ f_{41} & f_{42} \end{Vmatrix}; \tag{3.7}$$

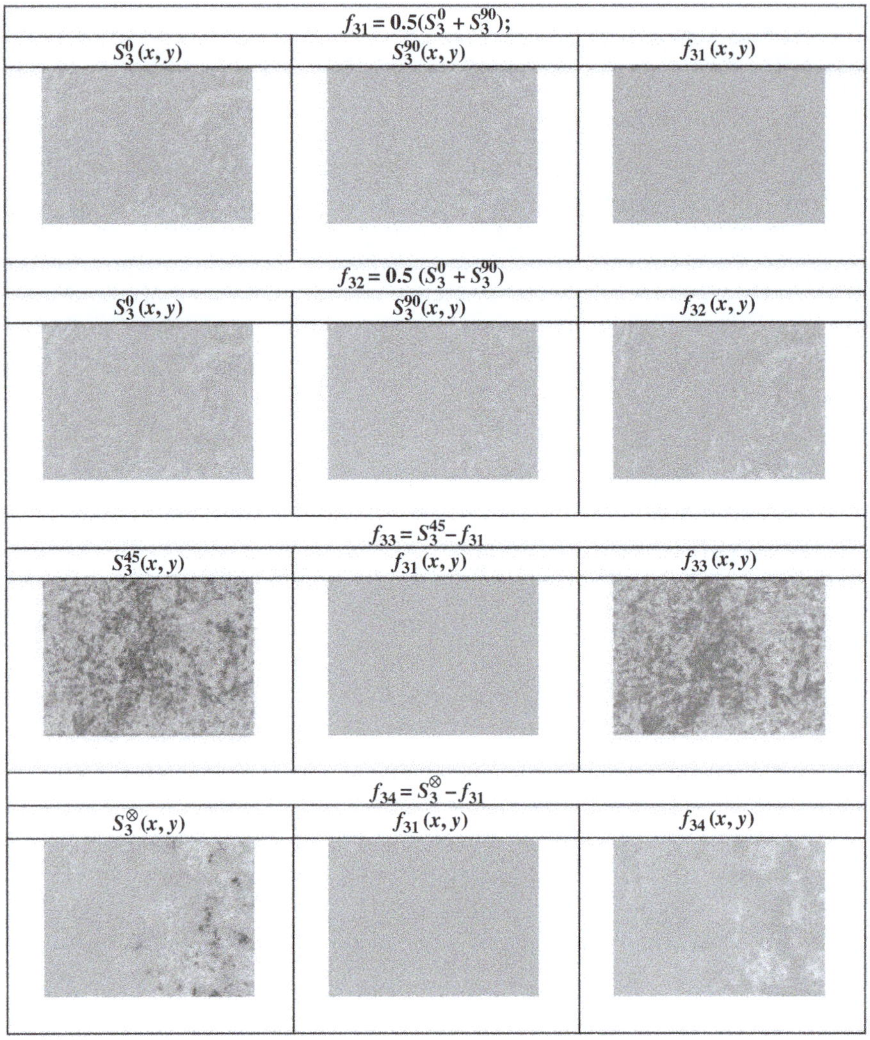

Figure 3.12 Mueller matrix images $f_{i=3;k=1;2;3;4}(x, y)$.

- for Stokes vectors of linearly polarized probing beams $S^0(45°)$; $S^0(135°)$

$$\left\{\begin{matrix} \left[S^0(45°) = \{F\}\begin{pmatrix} 1 \\ 0 \\ 1 \\ 0 \end{pmatrix} \rightarrow S(45°) = \begin{pmatrix} f_{11} + f_{13} \\ f_{21} + f_{23} \\ f_{31} + f_{33} \\ f_{41} + f_{43} \end{pmatrix} \right]; \\ \left[S^0(135°) = \{F\}\begin{pmatrix} 1 \\ 0 \\ -1 \\ 0 \end{pmatrix} \rightarrow S(135°) = \begin{pmatrix} f_{11} - f_{13} \\ f_{21} - f_{23} \\ f_{31} - f_{33} \\ f_{41} - f_{43} \end{pmatrix} \right] \end{matrix}\right\} \Rightarrow f_{ik} = \begin{Vmatrix} f_{11} & f_{13} \\ f_{21} & f_{23} \\ f_{31} & f_{33} \\ f_{41} & f_{43} \end{Vmatrix}; \tag{3.8}$$

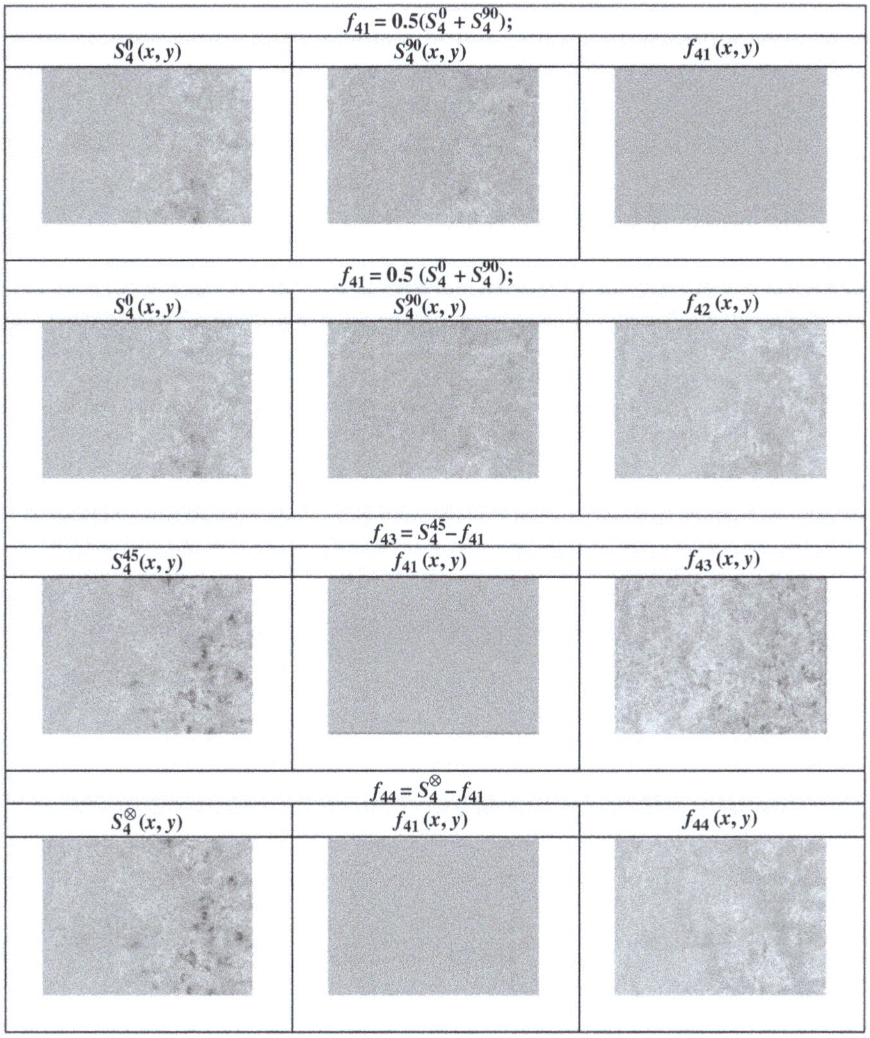

Figure 3.13 Mueller matrix images $f_{i=4;k=1;2;3;4}(x, y)$.

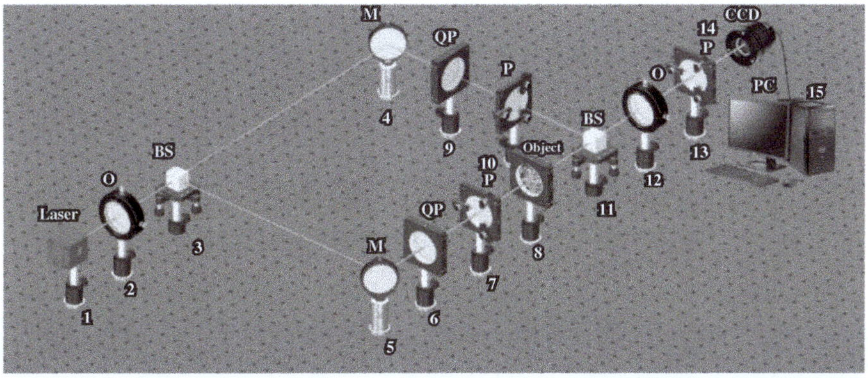

Figure 3.14 Optical scheme for the Jones matrix elements polarization-interference mapping. 1 – He–Ne laser; 2 – collimator – "O"; 3,11 – beam splitters – "BS"; 4,5 – mirrors – "M"; 7,10,13 – polarizer's "P"; 6,9 – quarter-wave plates – "QP"; 8 – object; 12 – polarization objective – "O"; 14 – digital camera – "CCD"; 15 – personal computer – "PC."

- For Stokes vectors of right- and left-circularly polarized probing beams $S^0(\otimes)$; $S^0(\oplus)$

$$\left.\begin{cases}\left[S^0(\otimes) = \{F\}\begin{pmatrix}1\\0\\0\\1\end{pmatrix} \rightarrow S(\otimes) = \begin{pmatrix}f_{11}+f_{14}\\f_{21}+f_{24}\\f_{31}+f_{34}\\f_{41}+f_{44}\end{pmatrix}\right]; \\ \left[S^0(\oplus) = \{F\}\begin{pmatrix}1\\0\\0\\-1\end{pmatrix} \rightarrow S(\oplus) = \begin{pmatrix}f_{11}-f_{14}\\f_{21}-f_{24}\\f_{31}-f_{34}\\f_{41}-f_{44}\end{pmatrix}\right]\end{cases}\right\} \Rightarrow f_{ik} = \begin{Vmatrix}f_{11} & f_{14}\\f_{21} & f_{24}\\f_{31} & f_{34}\\f_{41} & f_{44}\end{Vmatrix}; \tag{3.9}$$

$$\{M\} = 0,5\begin{Vmatrix}(S_1^0+S_1^{90}) & (S_1^0-S_1^{90}) & (S_1^{45}-S_1^{135}) & (S_1^{\otimes}-S_1^{\oplus})\\(S_2^0+S_2^{90}) & (S_2^0-S_2^{90}) & (S_2^{45}-S_2^{135}) & (S_2^{\otimes}-S_2^{\oplus})\\(S_3^0+S_3^{90}) & (S_3^0-S_3^{90}) & (S_3^{45}-S_3^{135}) & (S_3^{\otimes}-S_3^{\oplus})\\(S_4^0+S_4^{90}) & (S_4^0-S_4^{90}) & (S_4^{45}-S_4^{135}) & (S_4^{\otimes}-S_4^{\oplus})\end{Vmatrix}. \tag{3.10}$$

Traditionally, in the right side of equation (3.10), the value of the integral across the entire thickness l of the biological layer of phase shift $\delta(z = l)$ "appears"

$$S(0°, 90°, 45°, \otimes) = \begin{pmatrix}S_1(0°, 90°, 45°, \otimes)\\S_2(0°, 90°, 45°, \otimes)\\S_3(0°, 90°, 45°, \otimes)\\S_4(0°, 90°, 45°, \otimes)\end{pmatrix}\begin{matrix}=\\=\\=\\=\end{matrix}\begin{pmatrix}|E_x|^2 + |E_y|^2;\\|E_x|^2 - |E_y|^2;\\2|E_x||E_y|\cos\delta(z=l);\\2|E_x||E_y|\sin\delta(z=l).\end{pmatrix} = S(E_x, E_y, \delta). \tag{3.11}$$

Therefore, the direct Mueller-matrix mapping results in two-dimensional distributions of the values of matrix elements $f_{ik} = q_j(S_{z=1;2;3;4}(x, y, \delta))$, averaging $(\delta(z = l))$ over the entire thickness l of the biological layer [43–48].

$$f_{jk} = \begin{cases}f_{11} = \left((E_x^*E_x + E_y^*E_y)^0 + (E_x^*E_x + E_y^*E_y)^{90}\right);\ f_{12} = \left((E_x^*E_x + E_y^*E_y)^0 - (E_x^*E_x + E_y^*E_y)^{90}\right);\\ f_{13} = \left((E_x^*E_x + E_y^*E_y)^{45} - (E_x^*E_x + E_y^*E_y)^{135}\right);\ f_{14} = \left((E_x^*E_x + E_y^*E_y)^{RC} + (E_x^*E_x + E_y^*E_y)^{LC}\right);\\ f_{21} = \left((E_x^*E_x - E_y^*E_y)^0 + (E_x^*E_x - E_y^*E_y)^{90}\right);\ f_{22} = \left((E_x^*E_x - E_y^*E_y)^0 - (E_x^*E_x - E_y^*E_y)^{90}\right);\\ f_{23} = \left((E_x^*E_x - E_y^*E_y)^{45} - (E_x^*E_x - E_y^*E_y)^{135}\right);\ f_{24} = \left((E_x^*E_x - E_y^*E_y)^{RC} + (E_x^*E_x - E_y^*E_y)^{LC}\right);\\ f_{31} = \left((E_x^*E_y + E_y^*E_x)^0 + (E_x^*E_y + E_y^*E_x)^{90}\right);\ f_{32} = \left((E_x^*E_y + E_y^*E_x)^0 - (E_x^*E_y + E_y^*E_x)^{90}\right);\\ f_{32} = \left((E_x^*E_y + E_y^*E_x)^{45} - (E_x^*E_y + E_y^*E_x)^{135}\right);\ f_{24} = \left((E_x^*E_y + E_y^*E_x)^{RC} + (E_x^*E_y + E_y^*E_x)^{LC}\right);\\ f_{41} = \left((i(E_y^*E_x - E_x^*E_y))^0 + (i(E_y^*E_x - E_x^*E_y))^{90}\right);\ f_{42} = \left((i(E_y^*E_x - E_x^*E_y))^0 - (i(E_y^*E_x - E_x^*E_y))^{90}\right);\\ f_{43} = \left((i(E_y^*E_x - E_x^*E_y))^{45} - (i(E_y^*E_x - E_x^*E_y))^{135}\right);\ f_{42} = \left((i(E_y^*E_x - E_x^*E_y))^{RC} + (i(E_y^*E_x - E_x^*E_y))^{LC}\right);\end{cases} \tag{3.12}$$

1. Beam splitter "BS" 3 forms a collimated (diameter $\oslash = 2000\ \mu m$) laser beam 1–2 into "illuminating" Ir and "reference" Ref.
2. Using polarizing filters "QP-P" 6–7 and 9–10, collinear polarization states are sequentially formed. $(w \equiv 0°; 90°; 45°; RC)$ $Ir(w)$ and $Ref(w)$.

3. Beam splitter "BS" 11 superimposes a series of reference waves Ref (w) onto polarization images "Im (w)" of a biological tissue sample.
4. Interference patterns "$IntIm(w)$" formed by a polarizing microlens (×4) are projected onto the plane of the photosensitive area ($m \times n$; $M = 1120$; $N = 960$) of digital camera CCD 14 through polarizer P 13.
5. Digital camera CCD 14 records the coordinate distributions of the intensity of the series of interference patterns $I^{w}_{x;y}(m \times n)$ for two orientations of the transmission plane of polarizer-analyzer P 13. ($\theta = 0° \longleftrightarrow (I^{w}_{x}(m \times n)$; $\theta = 90° \longleftrightarrow (I^{w}_{y}(m \times n))$.
6. For each partial interference distribution $I^{w}_{x;y}(m \times n)$ a 2D digital discrete Fourier transform is applied $\Phi_{x,y}(\upsilon, \nu)$.

$$\Phi_{x,y} = \frac{1}{m \times n} \sum_{m=0}^{M-1} \sum_{n=0}^{N-1} I_{x,y}(m \times n) \exp\left[-i2\pi\left(\frac{m \times \upsilon}{M} + \frac{n \times \nu}{N}\right)\right] \tag{3.13}$$

where $(\upsilon \times \nu)$ are the spatial frequencies; $(m \times n)$ – CCD pixels; $I_{x,y}$ is "polarization projections" of $I(m \times n)$ value for two rotations ($\theta = 0°$ and $\theta = 90°$) of polarizer–analyzer.

7. Algorithm (3.13) provides the possibility of digital reconstruction of the distribution of orthogonal components $E_{x;y}(m \times n)$ of complex amplitudes $E(m \times n)$ of the object field

$$\begin{cases} \Phi_x \to (|E_x(\theta = 0°)|)(m \times n); \\ \Phi_y \to \left(\left|E_y(\theta = 90°)\right| \exp(i\delta_{x;y})\right)(m \times n); \\ \delta_{xy} = \left(\delta_y^{(\theta=90)} - \delta_x^{(\theta=0)}\right)(m \times n). \end{cases} \tag{3.14}$$

8. In each partial (v) phase plane $\left(\delta^{(w)}_{xy}(\upsilon)\right)$ the distributions $\left(\left|E^{(w)}_x(\theta = 0°)\right|\right)(m, n)$ and $\left(\left|E^{(w)}_y(\theta = 90°)\right| \exp\left(i\delta^{(w)}_{xy}\right)\right)(m, n)$ are scanned line by line, and the set of Mueller-matrix images of the polycrystalline architecture of the biological layer is calculated (equation (3.12)).

3.2.2 Jones-matrix Polarization–Interference Reconstruction of Anisotropy Maps of Polycrystalline Dehydrated Films of Biological Fluids

The algorithm for calculating the generalized Jones matrix $||J||$ elements include [49–55]:

- Consistent irradiation of a facie sample by linearly polarized beams with azimuths $0°\left(\text{Jones vector } H^0 = \begin{pmatrix}1\\0\end{pmatrix}\right)$ and $90°\left(\text{Jones vector } H^{90} = \begin{pmatrix}0\\1\end{pmatrix}\right)$;
- For each probing laser beam (H^0 and H^{90}), two polarization projections $\left((U^0_x; U^0_y)\right.$ and $\left.(U^{90}_x; U^{90}_y)\right)$ of the object field of complex amplitudes are determined by orienting the transmission plane of the linear analyzer at angles $\Omega = 0°$ and $\Omega = 90°$, respectively

$$U^0 = ||J|| \times H^0 \Rightarrow \begin{pmatrix} U^0_x \\ U^0_y \end{pmatrix} = \begin{Vmatrix} j_{11} & j_{12} \\ j_{21} & j_{22} \end{Vmatrix} \times \begin{pmatrix} 1 \\ 0 \end{pmatrix} \Rightarrow \begin{pmatrix} j_{11} \\ j_{21} \end{pmatrix}; \tag{3.15}$$

$$\begin{cases} U^0_x = ||P(\Omega = 0°)|| \times U^0 = \begin{Vmatrix} 1 & 0 \\ 0 & 0 \end{Vmatrix} \times \begin{pmatrix} j_{11} \\ j_{12} \end{pmatrix} = j_{11}; \\ U^0_y = ||P(\Omega = 90°)|| \times U^0 = \begin{Vmatrix} 0 & 0 \\ 0 & 1 \end{Vmatrix} \times \begin{pmatrix} j_{11} \\ j_{12} \end{pmatrix} = j_{21}; \end{cases} \tag{3.16}$$

$$U^{90} = ||J|| \times H^{90} \Rightarrow \begin{pmatrix} U^{90}_x \\ U^{90}_y \end{pmatrix} = \begin{Vmatrix} j_{11} & j_{12} \\ j_{21} & j_{22} \end{Vmatrix} \begin{pmatrix} 0 \\ 1 \end{pmatrix} \Rightarrow \begin{pmatrix} j_{12} \\ j_{22} \end{pmatrix}; \tag{3.17}$$

$$\begin{cases} U_x^{90} = \|P(\Omega = 0°)\| \times U^{90} = \begin{Vmatrix} 1 & 0 \\ 0 & 0 \end{Vmatrix} \times \begin{pmatrix} j_{12} \\ j_{22} \end{pmatrix} = j_{21}; \\ U_y^{90} = \|P(\Omega = 90°)\| \times U^{90} = \begin{Vmatrix} 0 & 0 \\ 0 & 1 \end{Vmatrix} \times \begin{pmatrix} j_{12} \\ j_{22} \end{pmatrix} = j_{22}. \end{cases} \tag{3.18}$$

It is possible to experimentally implement equations (3.15)–(3.18) using polarization–interference mapping and subsequent application of the algorithm of 2D digital Fourier reconstruction of complex amplitudes of the object field of the biological fluid facies.

Let us consider this method in more detail. To measure the coordinate ($m \times n$ is the number of pixels of the digital camera) distributions of the elements of the Jones matrix $j_{ik}(m, n)$, it is necessary to perform the following steps:

1. Two states of linear polarization (with next azimuth 0° and 90°) are sequentially formed in the "irradiating" (Ir) and "reference" (Re) parallel laser beams – $Ir(H^0) - \mathrm{Re}(H^0)$ and $Ir(H^{90}) - \mathrm{Re}(H^{90})$.
2. In the process of superposition of reference laser waves $\mathrm{Re}(H^{0;90})$ and object fields $U^{0;90}(m, n)$ of the biological fluid facies, interference patterns $E^{0;90}(m, n)$ are formed with the following distribution of complex amplitudes
3.

$$E^0(m, n) = (U^0 + H^0)(m, n) = \left(\left|U^0\right| + \left|H^0\right| + 2\sqrt{|U^0||H^0|}\right)(m, n); \tag{3.19}$$

$$E^{90}(m, n) = (U^{90} + H^{90})(m, n) = \left(\left|U^{90}\right| + \left|H^{90}\right| + 2\sqrt{|U^{90}||H^{90}|}\cos\,\Delta\varphi^{0;90}\right)(m, n); \tag{3.20}$$

$$\Delta\varphi_{0;90} = \varphi_{90} - \varphi_0. \tag{3.21}$$

Here $\left|U^{0;90}\right|$; $\left|H^{0;90}\right|$ is the modules of complex amplitudes; $\varphi^{0;90}$ is the phase shifts between $U^{0;90}$ and $H^{0;90}$.

4. By rotating the plane of transmission of the analyzer by angles 0° and $\Omega = 90°$, two "polarization projections" of the complex amplitudes of the object field of the biological fluid facies are extracted from the interference distributions (3.19), (3.20)

$$\Omega = 0° \Rightarrow \begin{cases} E_0^0(m, n) = \left(\left|U_0^0\right| + \left|H_0^0\right| + 2\sqrt{\left|U_0^0\right|\left|H_0^0\right|}\right)(m, n); \\ E_{90}^0(m, n) = \left(\left|U_{90}^0\right| + \left|H_{90}^0\right| + 2\sqrt{\left|U_{90}^0\right|\left|H_{90}^0\right|}\cos\,\Delta\varphi_{0;90}^0\right)(m, n); \\ \Delta\varphi_{0;90}^0 = \varphi_{90}^0 - \varphi_0^0; \end{cases} \tag{3.22}$$

$$\Omega = 90° \Rightarrow \begin{cases} E_0^{90}(m, n) = \left(\left|U_0^{90}\right| + \left|H_0^{90}\right| + 2\sqrt{\left|U_0^{90}\right|\left|H_0^{90}\right|}\right)(m, n); \\ E_{90}^{90}(m, n) = \left(\left|U_{90}^{90}\right| + \left|H_{90}^{90}\right| + 2\sqrt{\left|U_{90}^{90}\right|\left|H_{90}^{90}\right|}\cos\,\Delta\varphi_{0;90}^{90}\right)(m, n); \\ \Delta\varphi_{0;90}^{90} = \varphi_{90}^{90} - \varphi_0^{90}. \end{cases} \tag{3.23}$$

5. For each partial interference distributions (3.22) and (3.23), we perform a two-dimensional discrete Fourier transform $\Phi(\upsilon, \nu)$

$$\begin{cases} j_{11} \equiv \Phi_0^0 = \dfrac{1}{M \times N} \sum\limits_{m=0}^{M-1} \sum\limits_{n=0}^{N-1} E_0^0(m,n)(E_x^0)^*(m,n) \exp\left[-i2\pi\left(\dfrac{m \times \upsilon}{M} + \dfrac{n \times \nu}{N}\right)\right]; \\ j_{21} \equiv \Phi_{90}^0 = \dfrac{1}{M \times N} \sum\limits_{m=0}^{M-1} \sum\limits_{n=0}^{N-1} E_{90}^0(m,n)(E_y^0)^*(m,n) \exp\left[-i2\pi\left(\dfrac{m \times \upsilon}{M} + \dfrac{n \times \nu}{N}\right)\right]; \\ j_{21} \equiv \Phi_0^{90} = \dfrac{1}{M \times N} \sum\limits_{m=0}^{M-1} \sum\limits_{n=0}^{N-1} E_0^{90}(m,n)(E_x^{90})^*(m,n) \exp\left[-i2\pi\left(\dfrac{m \times \upsilon}{M} + \dfrac{n \times \nu}{N}\right)\right]; \\ j_{22} \equiv \Phi_{90}^{90} = \dfrac{1}{M \times N} \sum\limits_{m=0}^{M-1} \sum\limits_{n=0}^{N-1} E_{90}^{90}(m,n)(E_y^{90})^*(m,n) \exp\left[-i2\pi\left(\dfrac{m \times \upsilon}{M} + \dfrac{n \times \nu}{N}\right)\right]. \end{cases} \tag{3.24}$$

The obtained equations (3.22)–(3.24), taking into account the found analytical equations (3.12)–(3.13), form the basis for the implementation of the technique of polarization–interference layer-by-layer Jones matrix theziography of supramolecular networks of biological fluid facies. For this purpose, the algorithm is rewritten as

$$Re\begin{cases} j_{11} \equiv \Phi_0^0 = \dfrac{1}{M \times N} \sum\limits_{m=0}^{M-1} \sum\limits_{n=0}^{N-1} E_0^0(m,n)(E_0^0)^*(m,n) \cos\left(\dfrac{m \times \upsilon}{M} + \dfrac{n \times \nu}{N}\right); \\ j_{21} \equiv \Phi_{90}^0 = \dfrac{1}{M \times N} \sum\limits_{m=0}^{M-1} \sum\limits_{n=0}^{N-1} E_{90}^0(m,n)(E_{90}^0)^*(m,n) \cos\left(\dfrac{m \times \upsilon}{M} + \dfrac{n \times \nu}{N}\right); \\ j_{21} \equiv \Phi_0^{90} = \dfrac{1}{M \times N} \sum\limits_{m=0}^{M-1} \sum\limits_{n=0}^{N-1} E_{90}^{90}(m,n)(E_0^{90})^*(m,n) \cos\left(\dfrac{m \times \upsilon}{M} + \dfrac{n \times \nu}{N}\right); \\ j_{22} \equiv \Phi_{90}^{90} = \dfrac{1}{M \times N} \sum\limits_{m=0}^{M-1} \sum\limits_{n=0}^{N-1} E_{90}^{90}(m,n)(E_{90}^{90})^*(m,n) \cos\left(\dfrac{m \times \upsilon}{M} + \dfrac{n \times \nu}{N}\right) \end{cases} \tag{3.25}$$

$$Im\begin{cases} j_{11} \equiv \Phi_0^0 = \dfrac{1}{M \times N} \sum\limits_{m=0}^{M-1} \sum\limits_{n=0}^{N-1} E_0^0(m,n)(E_0^0)^*(m,n) \cos\left(\dfrac{m \times \upsilon}{M} + \dfrac{n \times \nu}{N}\right); \\ j_{21} \equiv \Phi_{90}^0 = \dfrac{1}{M \times N} \sum\limits_{m=0}^{M-1} \sum\limits_{n=0}^{N-1} E_{90}^0(m,n)(E_{90}^0)^*(m,n) \cos\left(\dfrac{m \times \upsilon}{M} + \dfrac{n \times \nu}{N}\right); \\ j_{21} \equiv \Phi_0^{90} = \dfrac{1}{M \times N} \sum\limits_{m=0}^{M-1} \sum\limits_{n=0}^{N-1} E_{90}^{90}(m,n)(E_0^{90})^*(m,n) \cos\left(\dfrac{m \times \upsilon}{M} + \dfrac{n \times \nu}{N}\right); \\ j_{22} \equiv \Phi_{90}^{90} = \dfrac{1}{M \times N} \sum\limits_{m=0}^{M-1} \sum\limits_{n=0}^{N-1} E_{90}^{90}(m,n)(E_{90}^{90})^*(m,n) \cos\left(\dfrac{m \times \upsilon}{M} + \dfrac{n \times \nu}{N}\right) \end{cases} \tag{3.26}$$

and determined ((3.12)–(3.13)) the set of Jones matrix images.

6. The results of the digital Fourier transforms (equations (3.25) and (3.26)) are used to obtain layered distributions $JT(\Theta_k, m, n)$ by discrete phase scanning with step Θ of the set of values $\Delta\varphi_{0;90}^0$ and $\Delta\varphi_{0;90}^{90}$ in the field complex amplitudes according to the following algorithms

$$\begin{cases} E_0^0 \Rightarrow \left|E_0^0(\Omega = 0°)\right|; \\ E_{90}^0(\Theta_k) \Rightarrow \left|E_{90}^0(\Omega = 90°)\right| \exp(i(\varphi_{90}^0 - \varphi_0^0)); \end{cases} \tag{3.27}$$

$$\begin{cases} E_0^{90} \Rightarrow \left|E_0^0(\Omega = 0°)\right|; \\ E_{90}^{90}(\Theta_k) \Rightarrow \left|E_{90}^0(\Omega = 90°)\right| \exp(i(\varphi_{90}^0 - \varphi_0^0)); \end{cases} \tag{3.28}$$

$$\begin{cases} \Delta\varphi_{0;90}^0 \Rightarrow 0; \Theta; 2\Theta; \dots k\Theta; \dots 2\pi; \\ \Delta\varphi_{0;90}^{90} \Rightarrow 0; \Theta; 2\Theta; \dots k\Theta; \dots 2\pi. \end{cases} \tag{3.29}$$

3.2.3 Objects of Investigations

Examples of digital polarization microscopic images (×4) of native histological sections of fibrillar and parenchymatous biological tissues (Figures 3.15–3.24) and dehydrated polycrystalline films of biological fluids (Figures 3.25–3.28) are presented here [52–66].

Each figure consists of four segments:

- Topographic structure of a biomedical specimen obtained in coaxial (a) and crossed polarizer–analyzers (c).
- Coordinate distributions of intensity values (b, d) of various polarization projections of microscopic images.

3.2.3.1 Native Histological Sections of Human Biological Tissues

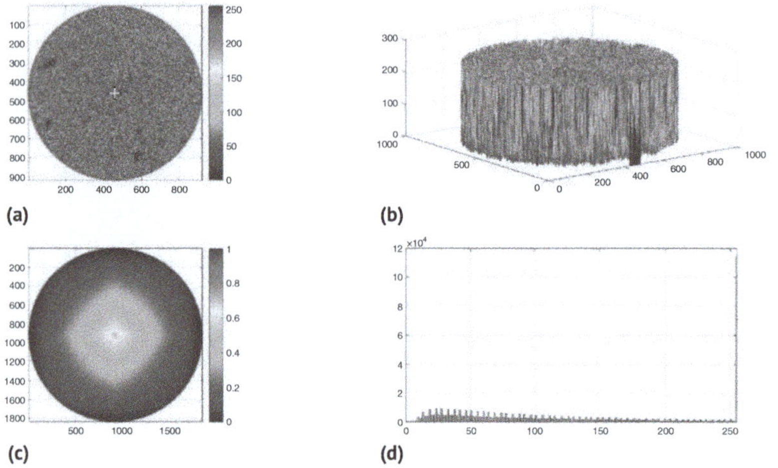

Figure 3.15 Intensity maps (a, b), autocorrelation function (c) and intensity distribution histogram (d) in a digital polarization image of a native histological brain section in coaxial transmission planes of the polarizer and analyzer.

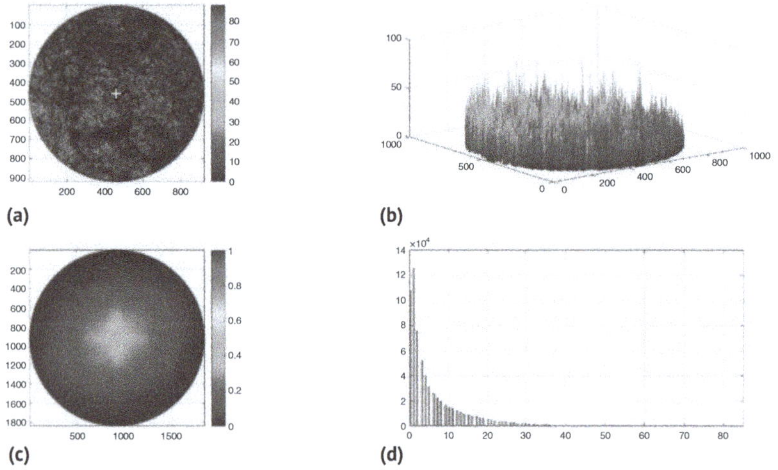

Figure 3.16 Intensity maps (a, b), autocorrelation function (c) and intensity distribution histogram (d) in a digital polarization image of a native histological brain section in crossed transmission planes of the polarizer and analyzer.

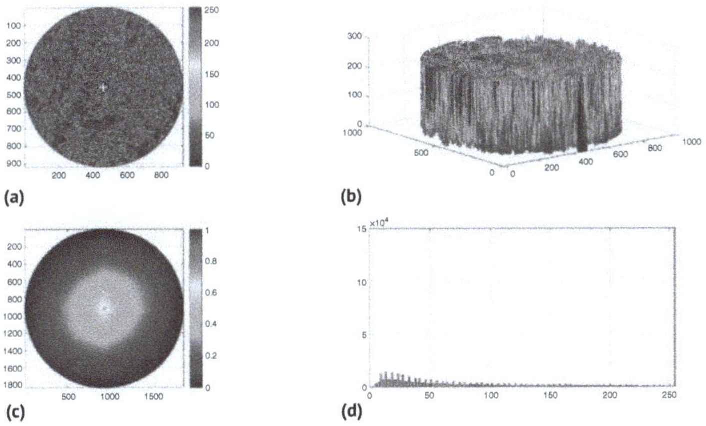

Figure 3.17 Intensity maps (a, b), autocorrelation function (c) and intensity distribution histogram (d) in a digital polarization image of a native histological section of the prostate in coaxial transmission planes of the polarizer and analyzer.

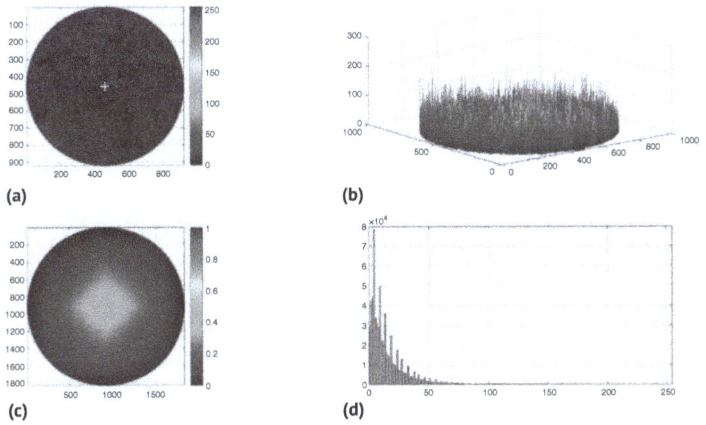

Figure 3.18 Intensity maps (a, b), autocorrelation function (c) and intensity distribution histogram (d) in a digital polarization image of a native histological section of the prostate in crossed transmission planes of the polarizer and analyzer.

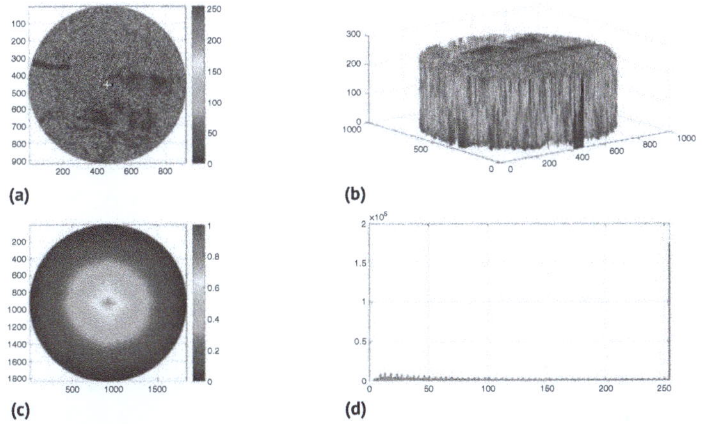

Figure 3.19 Intensity maps (a, b), autocorrelation function (c) and intensity distribution histogram (d) in a digital polarization image of a native histological section of the cervix in coaxial transmission planes of the polarizer and analyzer.

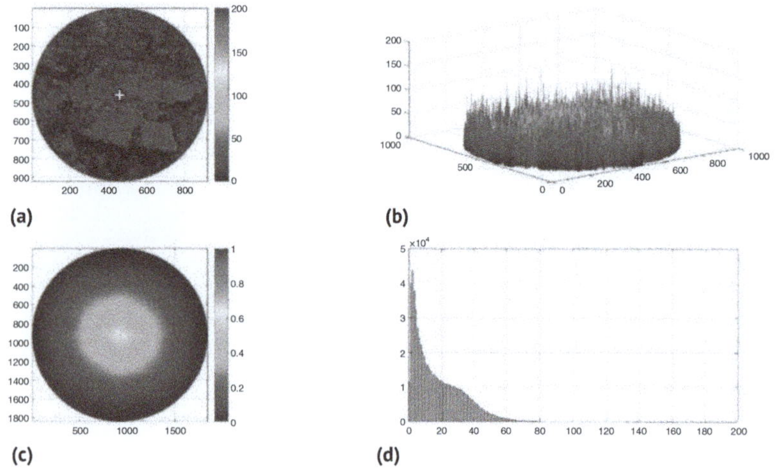

Figure 3.20 Intensity maps (a, b), autocorrelation function (c) and intensity distribution histogram (d) in a digital polarization image of a native histological section of the cervix in crossed transmission planes of the polarizer and analyzer.

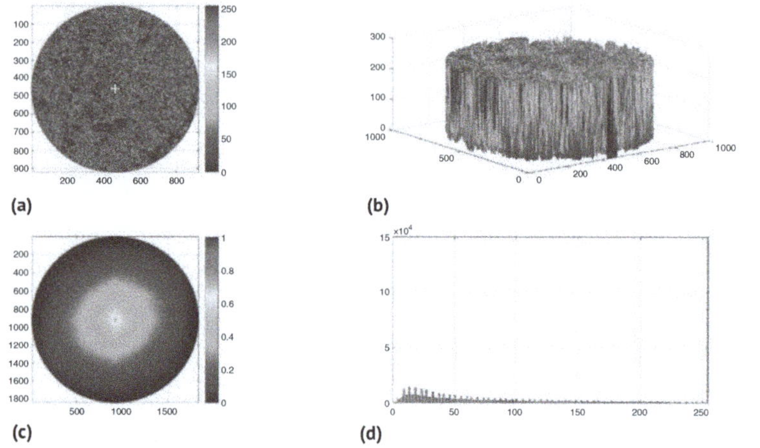

Figure 3.21 Intensity maps (a, b), autocorrelation function (c) and intensity distribution histogram (d) in a digital polarization image of a native histological section of the endometrium in coaxial transmission planes of the polarizer and analyzer.

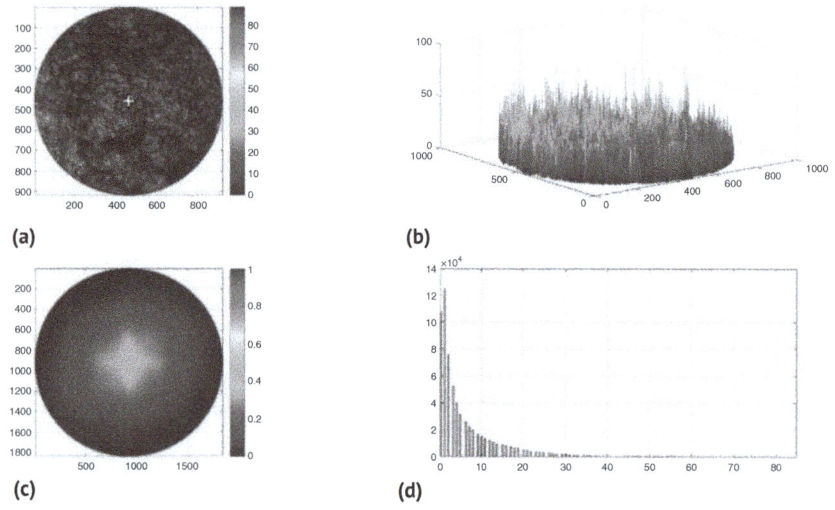

Figure 3.22 Intensity maps (a, b), autocorrelation function (c) and intensity distribution histogram (d) in a digital polarization image of a native histological section of the endometrium in crossed transmission planes of the polarizer and analyzer.

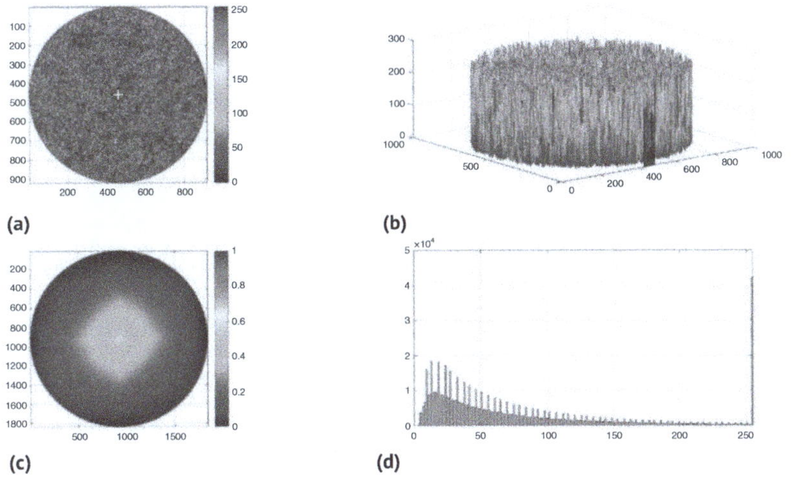

Figure 3.23 Intensity maps (a, b), autocorrelation function (c), and intensity distribution histogram (d) in a digital polarization image of a native liver histological section in coaxial transmission planes of the polarizer and analyzer.

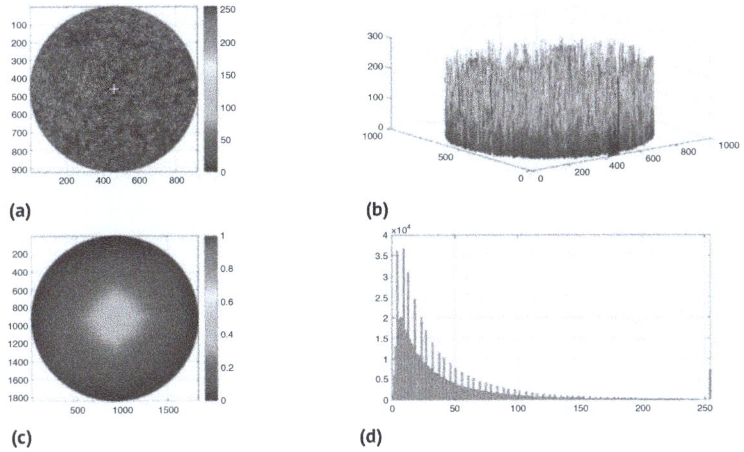

Figure 3.24 Intensity maps (a, b), autocorrelation function (c) and intensity distribution histogram (d) in a digital polarization image of a native liver histological section in crossed transmission planes of the polarizer and analyzer.

3.2.3.2 Dehydrated Polycrystalline Films of Biological Fluids from Human Organs

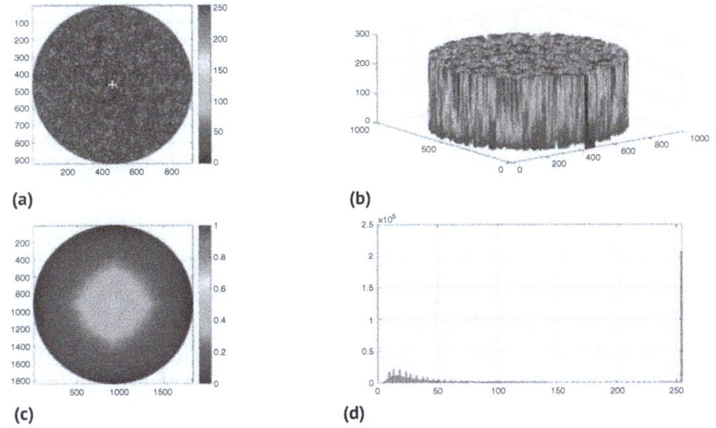

Figure 3.25 Intensity maps (a, b), autocorrelation function (c), and intensity distribution histogram (d) in a digital polarization image of a dehydrated blood plasma film in coaxial transmission planes of the polarizer and analyzer.

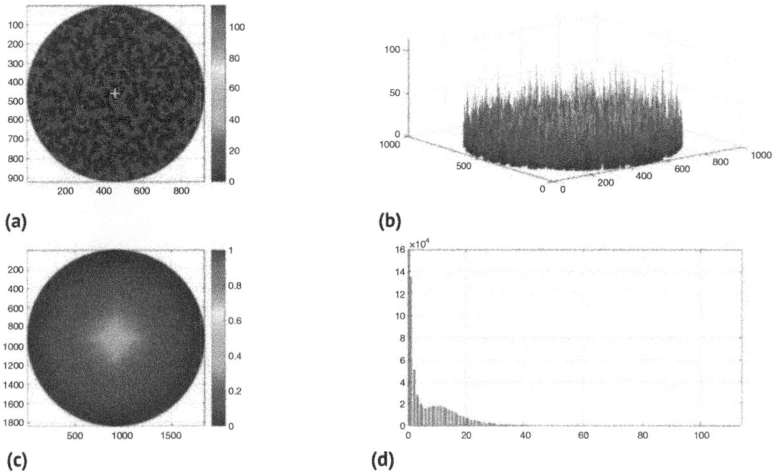

Figure 3.26 Intensity maps (a, b), autocorrelation function (c), and intensity distribution histogram (d) in a digital polarization image of a dehydrated blood plasma film in crossed transmission planes of the polarizer and analyzer.

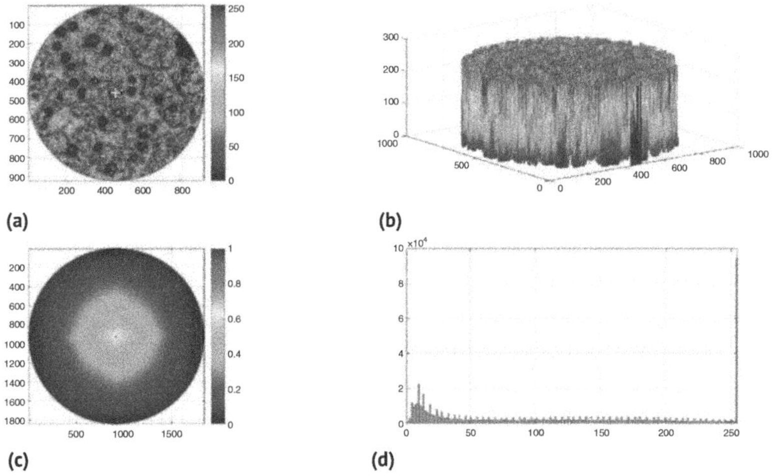

Figure 3.27 Intensity maps (a, b), autocorrelation function (c), and intensity distribution histogram (d) in a digital polarization image of a dehydrated bile film in coaxial transmission planes of the polarizer and analyzer.

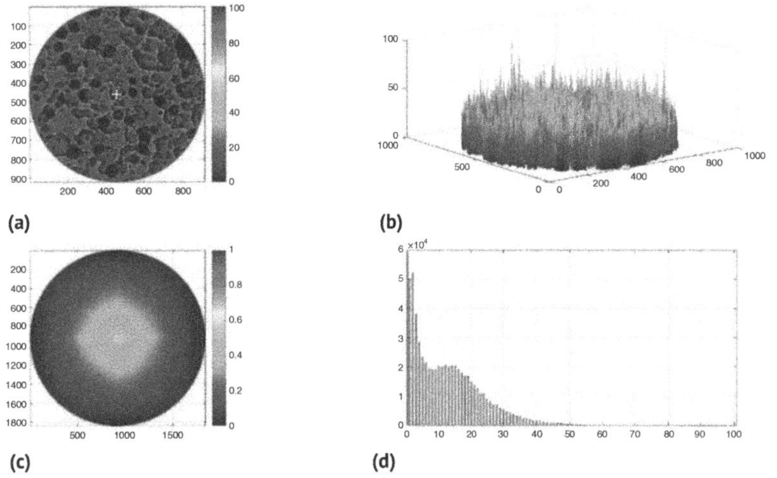

Figure 3.28 Intensity maps (a, b), autocorrelation function (c), and intensity distribution histogram (d) in a digital polarization image of a dehydrated bile film in coaxial transmission planes of the polarizer and analyzer.

References

1 He HH. Liao R, Zeng N, Li P, Chen Z, Liu X, et al. Mueller matrix polarimetry – an emerging new tool for characterizing the microstructural feature of complex biological specimen. J Lightwave Technol. 2019;37:2534–2548.
2 Oldenbourg R. A new view on polarization microscopy. Nature. 1996;381:811–812.
3 Gurjar RS. Backman V, Perelman LT, Georgakoudi I, Badizadegan K, Itzkan I, et al. Imaging human epithelial properties with polarized light-scattering spectroscopy. Nat Med. 2001;7:1245–1248.
4 Ramella-Roman JC, Saytashev I, Piccini M. A review of polarization-based imaging technologies for clinical and preclinical applications. J Opt. 2020;22:123001.
5 Qi J, Elson DS. Mueller polarimetric imaging for surgical and diagnostic applications: a review. J Biophotonics 2017;10:950–982.
6 Fujiwara H. Spectroscopic Ellipsometry: Principles and Applications. John Wiley & Sons; 2007.
7 Mendoza-Galván, A. et al. Mueller matrix spectroscopic ellipsometry study of chiral nanocrystalline cellulose films. J Opt. 2018;20:024001.
8 Azzam RMA. Stokes-vector and Mueller-matrix polarimetry. J Optical Soc Am A. 2016;33:1396–1408.
9 Azzam RMA. Mueller-matrix ellipsometry: a review. Proc SPIE. 1997;3121:369–405.
10 Song BK, Gu H, Zhu S, Jiang H, Chen X, Zhang C, et al. Broadband optical properties of graphene and HOPG investigated by spectroscopic Mueller matrix ellipsometry. Appl Surf Sci. 2018;439:1079–1087.
11 Jiang H, Ma Z, Gu H, Chen X, Liu S. Characterization of volume gratings based on distributed dielectric constant model using Mueller matrix ellipsometry. Appl Sci. 2019;9:698.
12 Chang J, He H, Wang Y, Huang Y, Li X, He C, et al. Division of focal plane polarimeter-based 3×4 Mueller matrix microscope: a potential tool for quick diagnosis of human carcinoma tissues. J Biomed Opt. 2016;21:056002.
13 Fu Y, Huang Z, He H, Ma H, Wu J. Flexible 3×3 Mueller matrix endoscope prototype for cancer detection. IEEE Trans Instrum Meas. 2018;67:1700–1712.
14 de Boer JF, Milner, TE, van Gemert, MJC, Nelson, JS. Two-dimensional birefringence imaging in biological tissue by polarization-sensitive optical coherence tomography. Opt Lett. 1997;22(12):934–936.
15 De Boer, JF, Srinivas SM, Malekafzali A, Chen ZP, Nelson JS. Imaging thermally damaged tissue by polarization sensitive optical coherence tomography. Opt Express 1998;3(6):212–218.
16 de Boer JF, Milner TE, Nelson JS. Determination of the depth-resolved Stokes parameters of light backscattered from turbid media by use of polarization-sensitive optical coherence tomography. Opt Lett. 1999;24(5):300–302.
17 Hitzenberger CK, Goetzinger E, Sticker M, Pircher M, Fercher AF. Measurement and imaging of birefringence and optic axis orientation by phase resolved polarization sensitive optical coherence tomography. Opt Express 2001;9(13):780–790.
18 Schoenenberger K, Colston BW, Maitland DJ, Da Silva, LB Everett MJ. Mapping of birefringence and thermal damage in tissue by use of polarization-sensitive optical coherence tomography. Appl Opt. 1998;37(25):6026–6036.
19 Yao G, Wang LV. Two-dimensional depth-resolved Mueller matrix characterization of biological tissue by optical coherence tomography. Opt Lett. 1999;24(8):537–539.
20 Jiao S, Wang LV. Two-dimensional depth-resolved Mueller matrix of biological tissue measured with double-beam polarization-sensitive optical coherence tomography. Opt Lett. 2002;27(2):101–103.
21 Jiao S, Yu W, Stoica G, Wang LV. Contrast mechanisms in polarization-sensitive Mueller-matrix optical coherence tomography and application in burn imaging. Appl Opt. 2003;42(25):5191–5197.
22 Jiao S, Yu W, Stoica G, Wang LV. Optical-fiber-based Mueller optical coherence tomography. Opt Lett. 2003;28(14):1206–1208.
23 de Boer JF, Milner TE, Nelson JS. Two dimensional birefringence imaging in biological tissue using phase and polarization sensitive optical coherence tomography. In Fujimoto JG, Patterson MS, editor. Trends in Optics and Photonics (TOPS): Advances in Optical Imaging and Photon Migration. Washington, DC: Optical Society of America; 1998. pp. 321–324.
24 Al-Qaisi MK, Akkin T. Polarization-sensitive optical coherence tomography based on polarization-maintaining fibers and frequency multiplexing. Opt Express. 2008;16(17):13032–13041.
25 Trasischker W, Zotter S, Torzicky T, Baumann B, Haindl R, Pircher M, Hitzenberger CK. Single input state polarization sensitive swept source optical coherence tomography based on an all single mode fiber interferometer. Biomed Opt Express. 2014;5(8):2798–2809.
26 Lippok N, Villiger M, Jun C, Bouma BE. Single input state, single-mode fiber-based polarization sensitive optical frequency domain imaging by eigenpolarization referencing. Opt Lett. 2015;40(9):2025–2028.

27 Yamanari M, Makita S, Yasuno Y. Polarization-sensitive swept-source optical coherence tomography with continuous source polarization modulation. Opt Express. 2008;16(8):5892–5906.

28 Oh WY, Yun SH, Vakoc BJ, Shishkov M, Desjardins AE, Park BH, et al. High-speed polarization sensitive optical frequency domain imaging with frequency multiplexing. Opt Express. 2008;16(2):1096–1103.

29 Kim KH, Park BH, Tu Y, Hasan T, Lee B, Li J, et al. Polarization-sensitive optical frequency domain imaging based on unpolarized light. Opt Express. 2011;19(2):552–561.

30 Pishak V, Ushenko A, Gryhoryshyn P, Yermolenko SB, Rudeychuk VM, Pishak OV. Study of polarization structure of biospeckle fields in crosslinked tissues of human organism: 1. Vector structure of skin biospeckles. In Proceedings of SPIE-The International Society for Optical Engineering. 1997;3317:418–424.

31 Ushenko AG. Depolarization of a laser emission field as the correlation development of its polarization structure. In International Conference on Correlation Optics. 1997;3317:331–339.

32 Ushenko AG, Burkovets DN, Yermolenko SB, Arkhelyuk AD, Pishak VP, Yuzko AM, et al. Stokes polarimetry of biotissues. In Fourth International Conference on Correlation Optics. 1999;3904:527–533.

33 Jóźwicki R, Patorski K, Angelsky OV, Ushenko AG, Burkovets DN, Ushenko YA. Automatic polarimetric system for early medical diagnosis by biotissue testing. Opt Appl. 2002;32(4):603–612.

34 Ushenko AG, Pishak VP. Coherent-domain Optical Methods: Biomedical Diagnostics, Environmental and Material Science. Laser Polarimetry of Biological Tissue. Principles and Applications. Boston: Kluwer Academic Publishers; 2004. pp. 67–93.

35 Olar EI, Ushenko AG, Ushenko YA. Correlation microstructure of the Jones matrices for multifractal networks of biotissues. Laser Phys. 2004;14(7):1012–1018.

36 Yermolenko S, Ushenko A, Ivashko P, Goudail F, Gruia I, Gavrilă C, et al. Spectropolarimetry of cancer change of biotissues. In Ninth International Conference on Correlation Optics. 2009;7388:404–410.

37 Ushenko AG, Fediv AI, Marchuk YF. Correlation and fractal structure of Jones matrices of human bile secret. In European Conference on Biomedical Optics. 2009;7368:7368_1Q.

38 Angelsky OV, Ushenko AG, Ushenko YA, Pishak VP, Peresunko AP. Handbook of Photonics for Biomedical Science. CRC press; 2010. p. 22–67.

39 Angelsky OV, Ushenko AG, Zenkova CY, Felde CV, Bogatyryova HV, Mokhun II, et al. Optical Measurements: Polarization and Coherence of Light Fields. INTECH Open Access Publisher; 2012.

40 Garazdyuk MS, Bachinskyi VT, Vanchulyak OY, Ushenko AG, Dubolazov OV, Gorsky MP. Polarization-phase images of liquor polycrystalline films in determining time of death. Appl Opt. 2016;55(12):B67–B71.

41 Ushenko AG, Dubolazov OV, Ushenko VA, Novakovskaya OY, Olar OV. Fourier polarimetry of human skin in the tasks of differentiation of benign and malignant formations. Appl Opt. 2016;55(12):B56–B60.

42 Huang D, Swanson EA, Lin CP, Schuman JS, Stinson WG, Chang W, et al. Optical coherence tomography. Science. 1991;254(5035):1178–1181.

43 Hee MR, Huang D, Swanson EA, Fujimoto JG. Polarization-sensitive low-coherence reflectometer for birefringence characterization and ranging. J Opt Soc Am B. 1992;9(6):903–908.

44 Park BH, de Boer JF. Polarization sensitive optical coherence tomography. In Drexler W, Fujimoto JG, editors. Optical Coherence Tomography – Technology and Applications. Cham: Springer; 2015. pp. 1055–1101.

45 Hitzenberger CK, Pircher M. MUW approach of PS OCT. In Drexler W, Fujimoto JG, editors. Optical Coherence Tomography – Technology and Applications. Cham: Springer; 2015. pp. 1103–1136.

46 Yasuno Y, Ju MJ, Hong YJ, Makita S, Lim Y, Yamanari M. Jones matrix based polarization sensitive optical coherence tomography. In Drexler W, Fujimoto JG, editors. Optical Coherence Tomography – Technology and Applications. Cham: Springer; 2015. pp. 1137–1162.

47 de Boer JF, Milner TE. Review of polarization sensitive optical coherence tomography and Stokes vector determination. J Biomed Opt. 2002;7(3):359–371.

48 Pircher M, Hitzenberger CK, Schmidt-Erfurth U. Polarization sensitive optical coherence tomography in the human eye. Prog Retin Eye Res. 2011;30(6):431–451.

49 Stifter D. Beyond biomedicine: a review of alternative applications and developments for optical coherence tomography. Appl Phys B. 2007;88(3):337–357.

50 Angelsky PO, Ushenko AG, Dubolazov AV, Sidor MI, Bodnar GB, Koval G, et al. The singular approach for processing polarization-inhomogeneous laser images of blood plasma layers. J Opt. 2013;15(4):044030.

51 Ushenko AG, Dubolazov AV, Ushenko VA, Novakovskaya OY. Statistical analysis of polarization-inhomogeneous Fourier spectra of laser radiation scattered by human skin in the tasks of differentiation of benign and malignant formations. J Biomed Opt. 2016;21(7):071110.

52 Angelskii OV, Ushenko OG, Burkovets DN, Arkhelyuk OD, Ushenko YA. Polarization-correlation studies of multifractal structures in biotissues and diagnostics of their pathologic changes. Laser Phys. 2000;10(5):1136–1142.

53 Bachinskyi V, Boychuk T, Ushenko A. Laser Polarimetry of Biological Tissues and Fluids: Chapter 4. Methods of Spatial-frequency, Singular and Correlation Analysis of Biological Layers Object Fields. LAP Lambert Academic Publishing; 2018.

54 Ushenko VA, Dubolazov AV, Pidkamin LY, Sakchnovsky MY, Bodnar AB, Ushenko YA, et al. Mapping of polycrystalline films of biological fluids utilizing the Jones-matrix formalism. Laser Phys. 2018;28(2):025602.

55 Ushenko AG, Angelsky PO, Sidor M, Marchuk YF, Andreychuk DR, Pashkovskaya NV. Spatial-frequency selection of complex degree of coherence of laser images of blood plasma in diagnostics and differentiation of pathological states of human organism of various nosology. Appl Opt. 2014;53(10):B172–B180.

56 Olar EI, Ushenko AG, Ushenko YA. Polarization correlation measurements of the phase tomograms of optically anisotropic biofractals. Laser Phys. 2004;14(8):1115–1121.

57 Guminetskiy SH, Ushenko OG, Polyanskiy IP, Motrych AV, Grynchuk FV. The optical method for investigation of the peritonitis progressing process. In Eighth International Conference on Correlation Optics. 2008;7008:626–632.

58 Ushenko AG, Fediv AI, Marchuk YuF. Singular structure of polarization images of bile secret in diagnostics of human physiological state. In: European Conference on Biomedical Optics. Optica Publishing Group; 2009.

59 Ushenko Y, Ushenko A, Dubolazov A, Soltys I, Litvinenko O, Wanchuliak O, et al. Mueller-matrix interferometric multifractal scaling of optically anisotropic architectonics of diffuse blood facies: fundamental and applied aspects. J Biophotonics. 2025;18(3):e202400412. https://doi.org/10.1002/jbio.202400412.

60 Ushenko A, Soltys I, Dubolazov A, Ushenko Y, Bilookyi V, Bilookyi O, et al. 3D Jones-matrix thesiography of biological fluid facies. J Innov Opt Health Sci. 2025;18(2):2443002. https://doi.org/10.1142/S1793545824430028.

61 Ushenko A, Pavlyukovich N, Khukhlina O, Pavlyukovich O, Gorsky M, Soltys I, et al. Layer-by-layer multifractal scanning of optically anisotropic architectonics of blood plasma films: fundamental and applied aspects. Photonics. 2025;12:215. https://doi.org/10.3390/photonics12030215.

62 Ushenko A, Pavlyukovich N, Khukhlina O, Pavlyukovich O, Soltys I, Dubolazov A, et al. Blood plasma film multifractal scanning in COVID-19 consequences diagnostics. J Biophotonics. 2024;17(11):e202400356. https://doi.org/10.1002/jbio.202400356.

63 Ushenko A, Dubolazov A, Zheng J, Litvinenko A, Gorsky M, Ushenko Y, et al. 3D polarization-interference holographic histology for wavelet-based differentiation of biological tissues with different necrotic states. J Biomed Opt. 2024;29(5):052920. https://doi.org/10.1117/1.JBO.29.5.052920.

64 Ushenko O, Bilookyi O, Zheng J, Dubolazov A, Olar O, Ushenko Y, et al. 3D digital holographic polarimetry of laser speckle fields formed by polycrystalline blood films: a tool for differential diagnosis of thyroid pathology. Front Phys. 2024;12:1426469. https://doi.org/10.3389/fphy.2024.1426469.

65 Wójcik W, Hu Z, Ushenko Y, Smolarz A, Soltys I, Dubolazov O, et al. Optical sensor system for 3D Jones matrix reconstruction of optical anisotropy maps of self-assembled polycrystalline soft matter films. Sensors. 2024;24(5):1589. https://doi.org/10.3390/s24051589.

66 Fercher AF, Hitzenberger CK. Optical coherence tomography. Prog Opt. 2002;44:215–302.

4

Basic Algorithms for Processing 3D Jones–Mueller Digital Holographic Mapping Data

Yurii Ushenko[1,2], Weidong Dou[1], Oleksandr Ushenko[2,3], Oleksandr Dubolazov[2], Iryna Soltys[2], Yuriy Tomka[2], Mykhailo Gorsky[2], Mykhailo Gavryliak[2], Viktor Bachinskiy[4], Oleh Vanchuliak[4], Marta Garazdyuk[4], Nataliia Pavlyukovich[4], Oleksandr Pavlyukovich[4], and Vitaliy Rozhko[4]

[1] *Shaoxing University, Shaoxing, People's Republic of China*
[2] *Yurii Fedkovych Chernivtsi National University, Chernivtsi, Ukraine*
[3] *Taizhou Institute of Zhejiang University, Taizhou, People's Republic of China*
[4] *Bukovinian State Medical University, Chernivtsi, Ukraine*

4.1 Statistical Analysis

To assess the distributions of random values q, the set of statistical moments of the first and fourth orders $Z_{i=1;2;3;4}$ is calculated [1–8].

$$Z_1 = \frac{1}{N}\sum_{i=1}^{N}\left|(q)_i\right|;\ Z_2 = \sqrt{\frac{1}{N}\sum_{i=1}^{N}(q)_i^2};\ Z_3 = \frac{1}{Z_2^3}\frac{1}{N}\sum_{i=1}^{N}(q)_i^3;\ Z_4 = \frac{1}{Z_2^4}\frac{1}{N}\sum_{i=1}^{N}(q)_i^4. \tag{4.1}$$

Here, N is the number of (1280×960) pixels of CCD camera.

4.2 Correlation Analysis

To calculate the autocorrelation function $\overline{K}(\Delta x)$ with a scanning step $\Delta x = 1$ pix of q polarization parameters of values distribution, average within all n lines $(1, \ldots, m)$ of photosensitive plate of digital camera, the following classical equation is used [9–14]

$$K_{j=1\div n}(\Delta x) = \lim_{x\to 0}\frac{1}{m}\int_0^x \left[q_j^*(x = 1 \div m)\right][q^*(x - \Delta x)]dx. \tag{4.2}$$

4.3 Wavelet Analysis

If a prototype function is taken as a specific wavelet function possessing a finite base in both coordinate and frequency spaces, then one can expand into a series the one-dimensional distribution of polarization parameters q [15–22]

$$q(x) = \sum_{a,b=-\infty}^{\infty} C_{ab}\Psi_{ab}(x), \tag{4.3}$$

where $\Psi_{ab}(x) = \Psi(ax - b)$ is a base function formed from the function-prototype by shifting $\boldsymbol{b}$ and scaling $\boldsymbol{a}$, while the coefficients of this expansion are determined as follows

$$C_{ab} = \int q(x)\Psi_{ab}(x)dx. \tag{4.4}$$

The result of this wavelet transformation for the one-dimensional distribution of polarization parameters is a two-dimensional array of the coefficients $W_\beta(a, b)$, which are defined by the following equation

$$W(a, b) = \frac{1}{\sqrt{|a|}} \int_{-\infty}^{\infty} f(q)\Psi\left(\frac{x - b}{a}\right)dx. \tag{4.5}$$

where a is a scale parameter, b is a spatial coordinate, and ω is a soliton-like function (wavelet) constructed on the basis of derivatives of the Gaussian function [20–26].

In our work, the second derivative ($m = 2$) or MHAT wavelet is used

$$\omega^{(m)} = (-1)^m \frac{\partial m}{\partial x^m}\left[\exp\left(\frac{x^2}{2}\right)\right] \Longrightarrow \omega^{(2)} = \frac{\partial^2}{\partial x^2}\left[\exp\left(\frac{x^2}{2}\right)\right]. \tag{4.6}$$

4.4 Fractal and Multifractal Analysis

4.4.1 Fractal Analysis

Fractal analysis was performed by calculating the logarithmic dependences $\log J(q) - \log(v)$ of power spectra of random values q distributions. Here, $\nu = d^{-1}$ – spatial frequencies defined by the physical sizes (d) of structural elements of maps of polarization inhomogeneous images of a biological layer. Then dependences are approximated by least squares method in curves $\Phi(\eta)$. With the appearance of curves, the classification of coordinate distributions q is carried out [27–33]:

- q distributions – fractal dependences $\Phi(\eta)$ at the constant value of slope angle η = const. for 2–3 decades of size changes d;
- q distributions – multifractal at the condition of availability of several constant slope angles $\Phi(\eta)$;
- q distributions – random at the condition of absence of several constant slope angles $\Phi(\eta)$ throughout the whole range of size changes.

4.4.2 Multifractal Analysis

We have already noted that fractality is inherent not only in geometric structures, but also in various processes that can form scale-like distributions (fractal measures) – in our case, maps of polarization parameters $\begin{pmatrix} q_{11}^{\odot} & \cdots & q_{1t}^{\odot} \\ \vdots & \ddots & \vdots \\ q_{u1}^{\odot} & \cdots & q_{ut}^{\odot} \end{pmatrix}$ of microscopic images of biological layers [34–38].

It is known from the fractal theory that such measures $(q^{\odot}(a, b))$ can be of two types – fractal $(q^{\odot})^h$ or multifractal $(q^{\odot})^{F(h)}$.

For fractal distributions $(q^{\odot})^h$, the concepts of one fractal dimension $h_i \equiv h_0$ are introduced.

Multifractal distributions $(q^{\odot})^{F(h_i)}$ are characterized by a set of fractal dimensions h_i or a singularities spectrum $F(h_i)$.

To calculate the fractal dimensions of h_i, a special approach is used, which is clearly based on the introduction of partial functions or generalized statistical sums

$$Z(k, d) = \sum_{q=1}^{P(d)} (q^{\odot})_q^k(d). \tag{4.7}$$

Here, $P(d)$ is the number of multifractal distribution coating elements $(q^{\odot})$; $k \in K$.

As a rule, the dependence (4.7) has a stepwise character

$$Z(k, d) \sim d^{(k-1)h_r}, \tag{4.8}$$

where h_r is the generalized fractal dimension, which for the case of partial functions is written as scaling exponents

$$\tau(k) = (k-1)h_k. \tag{4.9}$$

Here, the exponent $(k-1)$ is introduced into the exponent in order to automatically fulfill the equality $Z(1, d) = 1$, which means the normalization conditions for the distribution $\left(q^{\odot}\right)_q^k(d)$.

In particular, for monofractals $\tau(r) = \text{const}$.

For more complex scale-like distributions, dependence (4.9) is transformed into a specific multifractal spectrum $F(h_r)$.

For analytical determination of the dependence $F(h_r)$, the method of wavelet transforms modules maxima (MWTM) is used.

4.4.3 Method of Wavelet Transform Modules Maxima

The MWTM method includes two steps [37–40]:

- Wavelet transformation $W(a, b)$ of the $\left(\beta^{\odot}(u, t)\right)$ map set and the skeleton $\sup|W(a^*, x_i(a^*))|$ determination – the maximum modulus local surface extremes values wavelet coefficients $W(a, b)$ lines for each scale a of scanning soliton-like function $\Psi(b)$;
- Construction of wavelet coefficient modules $Z(k, d, \sup|W(a^*, x_i(a^*))|)$ local maxima partial functions and fractal dimension spectrum $F(h_k)$ determination.

The maps $\left(\beta^{\odot}(u, t)\right)$ wavelet analysis is based on an analytical transformation consisting in the decomposition of a distribution over a basis constructed from a soliton-like function (wavelet) by means of large-scale changes and transfers.

The continuous wavelet transforms of the function $f(x)$ is defined by the following equation

$$W(a, b) = \frac{1}{\sqrt{a}} \int_{-\infty}^{\infty} f(x) M\left(\frac{x-b}{a}\right) dx, \tag{4.10}$$

where a is a scale parameter, b is a spatial coordinate, and M is a soliton-like function (wavelet) constructed on the basis of the Gaussian function derivatives.

In our work, the second derivative $(m = 2)$ or MHAT wavelet is used

$$M^{(m)} = (-1)^m \frac{\partial^m}{\partial x^m}\left[\exp\left(\frac{x^2}{2}\right)\right] \Longrightarrow M^{(2)} = \frac{\partial^2}{\partial x^2}\left[\exp\left(\frac{x^2}{2}\right)\right]. \tag{4.11}$$

The wavelet equation (4.11) for ellipticity maps can be written as the following equation

$$W^q(a, b) = \frac{1}{\sqrt{a}} \int_{-\infty}^{\infty} q^{\odot}(a, x) M\left(\frac{x-b}{a}\right) dx. \tag{4.12}$$

The equation (4.12) analysis shows that the wavelet decomposition of biological fluids facies is a superposition parameters multiscale distributions (fractal or multifractal measures $\left(q^{\odot}\right)_q^k(d)$ of the supramolecular polycrystalline networks with linear and circular birefringence).

In the following, we will focus on the search for the possibility's scaling exponents $\tau(k)$ determination that characterizes $\left(q^{\odot}\right)_q^k(d)$, based on the use of skeletons $\sup|W(a^*, x_i(a^*))|$.

This analytical procedure is based on the construction of partial functions using the following equation

$$Z(k, d) = \sum_{l \in L(a)} \left(\sup|W(a^*, x_i(a^*))|\right)^k. \tag{4.13}$$

The dependence $Z(k, d) \sim d^{\tau(k)}$ holds, where the fractal dimension $\tau(r)$ for some r is determined by setting the scaling exponent $\ln Z(k, d)/\ln d$ dependence slope.

Variation of r in construction $\sum_{l\in L(a)}(\sup|W(a^*, x_i(a^*))|)^r$ allows you to get:

- for fractal distributions $\left(q^{\odot}\right)_q^k(d)$ line dependences $\tau(k)$, $-\left(h(k) = \frac{d\tau}{dk} = \text{const.}\right)$;
- for multifractal distributions $\left(q^{\odot}\right)_q^k(d)$ none line dependencies $\tau(k) = kh - D(h)$ with big number of fractal dimensions $h(k) = {}^{d\tau}/_{dk}$, which characterized the multifractal spectrum $F(h_k)$.

In our case, we will study the multifractal spectra of $\begin{pmatrix} q_{11}^{\odot} & \cdots & q_{1t}^{\odot} \\ \vdots & \ddots & \vdots \\ q_{u1}^{\odot} & \cdots & q_{ut}^{\odot} \end{pmatrix}$ maps based on the partial functions $Z(r, d, \sup|W(a^*, x_i(a^*))|)$ wavelet maximal modules amplitudes skeletons $\sup|W(a^*, x_i(a^*))|$.

4.5 Singular Analysis

A new approach to the description of the biological tissue laser images, based on the analysis of coordinate distributions of polarization singularities, was developed using the above-mentioned statistical methods. Linearly (S-points) and circularly (C-points) polarized states of light oscillations belong to them. For S-points, the direction of the electric-intensity vector's rotation is indefinite (singular). For a C-point, the polarization azimuth of the electric-intensity vector is indefinite. The use of the fourth parameter of the Stokes vector appears to be a suitable and widely applied means of such singularity's representation $V_4 = \begin{cases} 0 \leftrightarrow S; \\ \pm 1 \leftrightarrow C. \end{cases}$ Investigation of laser images of the connective tissue layers revealed a well-developed network of polarization singularities, which was quantitatively estimated in the form of distribution of the amount of S- and C-points (Table 4.1).

Table 4.1 Interconnection between characteristic values of Mueller-matrix f_{ik} elements of biological tissues and polarization singularities described by the fourth Stokes parameter S_4.

f_{ik}		"+S"-point ($\delta = 0$)	"−S"-point ($\delta = \pi$)	"+C"-point ($\delta = +\pi/2$)	"−C"-point ($\delta = -\pi/2$)	S_4
	0	–	–	$+(\rho = +\pi/4)$	$+(\rho = -\pi/4)$	±1
f_{22}	1	$+(\delta = 0)$	–	–	–	0
	−1	–	$+(\delta = \pi)$	–	–	0
	0	$+(\rho = +\pi/4)$	$+(\rho = -\pi/4)$	–	–	0
$f_{23} = f_{32}$	1	–	–	$+\rho = +\pi/4$	$+\rho = -\pi/4$	+1
	−1	–	–	$+(\rho = -\pi/4)$	$+(\rho = +\pi/4)$	−1
	0	$+(\rho = 0)$	$+(\rho = \pi/2)$	–	–	0
$f_{24} = -f_{42}$	1	–	–	$+(\rho = +\pi/4)$	$+(\rho = -\pi/4)$	+1
	−1	–	–	$+(\rho = -\pi/4)$	$+(\rho = +\pi/4)$	−1
	0	–	–	$+(\rho = 0)$	$+(\rho = 0)$	±1
f_{33}	1	$+(\delta = 0)$	–	–	–	0
	−1	–	$+(\delta = \pi)$	–	–	0
	0	$+(\rho = +\pi/4)$	$+(\rho = -\pi/4)$	–	–	0
$f_{34} = -f_{43}$	1	–	–	$+(\rho = 0)$	$+(\rho = +\pi/2)$	+1
	−1	–	–	$+(\rho = +\pi/2)$	$+(\rho = 0)$	−1
	0	–	–	$+(\rho = 0 \div \pi)$	$+(\rho = 0 \div \pi)$	±1
f_{44}	1	$+(\delta = 0)$	–	–	–	0
	−1	–	$+(\delta = \pi)$	–	–	0

Table 4.2 Threshold levels of balanced accuracy.

Diagnostic accuracy assessment	Accuracy, Ac, %
Unsatisfactory	≤80
Satisfactory	81–85
Good	86–90
Very good	91–95
Excellent	>95

Thus, the development of laser polarimetry techniques based on determination of singular interconnections "object – field" to find new methods of diagnostics of transformation of the BT extracellular matrix orientation–phase structure connected with precancerous changes of their physiological state is very important. To solve such a problem, we should revert to the analysis of optical properties of biological crystals' nets, comprehensively described by the polarization matrix, though within a singular approach.

4.6 Information Analysis

Information analysis uses the operational characteristics of evidence-based medicine:

- Sensitivity (Se) is the proportion of correct positive results (A) of the diagnostic method among all samples from experimental group 2, group 4(N) – $\text{Se} = \left(\frac{A}{N}\right)100\%$.
- Specificity (Sp) is the proportion of correct negative results (B) of the method among control group 1, group 3(H) – $\text{Sp} = (B/H)100\%$.
- Accuracy (Ac) is the proportion of correct test results ($A + B$) among all samples ($N + H$) – $\text{Ac} = [(A+B)/(N+H)]100\%$.

If $(A + B) = (N + H)$, then Ac is called balanced accuracy. The following scale for evaluating the diagnostic accuracy is used (Table 4.2).

4.7 Conclusions

In this chapter, we presented a systematic presentation of the fundamental principles of a model for the optical anisotropy of the polycrystalline architecture of a biological layer.

The analytical approach is based on the Jones–Mueller mathematical matrix formalism. This approach provided a comprehensive, informative solution to the direct problem – determining the parameters of polarization (Stokes vector and Mueller matrix) and polarization-correlation (complex amplitudes and Jones matrix) maps of digital microscopic images of native histological sections of biological tissues and dehydrated polycrystalline films of biological fluids. In parallel, algorithms were developed for solving the inverse problem – the Jones–Mueller matrix reconstruction of the linear and circular birefringence-dichroism distributions of the polycrystalline architecture of biomedical samples.

A description of experimental setups and methods for 2D and 3D layer-by-layer Jones–Mueller matrix mapping of biological tissue and fluid samples is presented.

Algorithms for phase scanning of biospeckle fields and methods for polarization-interference (Stokes polarimetric and Jones–Mueller matrix mapping) with digital holographic reconstruction of layer-by-layer polarization and matrix images, as well as optical anisotropy maps of native histological sections of biological tissues and dehydrated films of biological fluids, are described. Examples of algorithmic reconstruction of polarization maps, polarization-correlation maps, Mueller-matrix images, the real and imaginary parts of Jones-matrix images, and maps of optical anisotropy parameters of the polycrystalline architecture of biological specimens are provided. Presents the basic algorithms for multifunctional (statistical, correlation, wavelet, multifractal, and singular) digital analysis of polarization-interference Stokes polarimetric and Jones-Mueller matrix mapping data.

References

1 Pishak V, Ushenko A, Gryhoryshyn P, Yermolenko SB, Rudeychuk VM, Pishak OV. Study of polarization structure of biospeckle fields in crosslinked tissues of human organism: 1. Vector structure of skin biospeckles. In Proceedings of SPIE-The International Society for Optical Engineering. 1997;3317:418–424.
2 Ushenko AG. Depolarization of a laser emission field as the correlation development of its polarization structure. In International Conference on Correlation Optics. 1997;3317:331–339.
3 Ushenko AG, Burkovets DN, Yermolenko SB, Arkhelyuk AD, Pishak VP, Yuzko AM, et al. Stokes polarimetry of biotissues. In Fourth International Conference on Correlation Optics. 1999;3904:527–533.
4 Jóźwicki R, Patorski K, Angelsky OV, Ushenko AG, Burkovets DN, Ushenko YA. Automatic polarimetric system for early medical diagnosis by biotissue testing. Opt Appl. 2002;32(4):603–612.
5 Ushenko AG, Pishak VP. Coherent-Domain Optical Methods: Biomedical Diagnostics, Environmental and Material Science. Laser Polarimetry of Biological Tissue. Principles and Applications. Boston: Kluwer Academic Publishers; 2004. pp. 67–93.
6 Olar EI, Ushenko AG, Ushenko YA. Correlation microstructure of the Jones matrices for multifractal networks of biotissues. Laser Phys. 2004;14(7):1012–1018.
7 Yermolenko S, Ushenko A, Ivashko P, Goudail F, Gruia I, Gavrilă C, et al. Spectropolarimetry of cancer change of biotissues. In Ninth International Conference on Correlation Optics. 2009;7388:404–410.
8 Ushenko AG, Fediv AI, Marchuk YF. Correlation and fractal structure of Jones matrices of human bile secret. In European Conference on Biomedical Optics. 2009;7368:7368_1Q.
9 Angelsky OV, Ushenko AG, Ushenko YA, Pishak VP, Peresunko AP. Handbook of Photonics for Biomedical Science. CRC Press; 2010, pp. 22–67.
10 Angelsky OV, Ushenko AG, Zenkova CY, Felde CV, Bogatyryova HV, Mokhun II, et al. Optical Measurements: Polarization and Coherence of Light Fields. INTECH Open Access Publisher; 2012.
11 Garazdyuk MS, Bachinskyi VT, Vanchulyak OY, Ushenko AG, Dubolazov OV, Gorsky MP. Polarization-phase images of liquor polycrystalline films in determining time of death. Appl Opt. 2016;55(12):B67–B71.
12 Ushenko AG, Dubolazov OV, Ushenko VA, Novakovskaya OY, Olar OV. Fourier polarimetry of human skin in the tasks of differentiation of benign and malignant formations. Appl Opt. 2016;55(12):B56–B60.
13 Huang D, Swanson EA, Lin CP, Schuman JS, Stinson WG, Chang W, et al. Optical coherence tomography. Science. 1991;254(5035):1178–1181.
14 Hee MR, Huang D, Swanson EA, Fujimoto JG. Polarization-sensitive low-coherence reflectometer for birefringence characterization and ranging. J Opt Soc Am B. 1992;9(6):903–908.
15 Park BH, de Boer JF. Polarization sensitive optical coherence tomography. In Drexler W, Fujimoto JG, editors. Optical Coherence Tomography – Technology and Applications. Cham: Springer; 2015. pp. 1055–1101.
16 Hitzenberger CK, Pircher M. MUW approach of PS OCT. In Drexler W, Fujimoto JG, editors. Optical Coherence Tomography – Technology and Applications. Cham: Springer; 2015. pp. 1103–1136.
17 Yasuno Y, Ju MJ, Hong YJ, Makita S, Lim Y, Yamanari M. Jones matrix based polarization sensitive optical coherence tomography. In Drexler W, Fujimoto JG, editors. Optical Coherence Tomography – Technology and Applications. Cham: Springer; 2015. pp. 1137–1162.
18 de Boer JF, Milner TE. Review of polarization sensitive optical coherence tomography and Stokes vector determination. J Biomed Opt. 2002;7(3):359–371.
19 Pircher M, Hitzenberger CK, Schmidt-Erfurth U. Polarization sensitive optical coherence tomography in the human eye. Prog Retin Eye Res. 2011;30(6):431–451.
20 Stifter D. Beyond biomedicine: a review of alternative applications and developments for optical coherence tomography. Appl Phys B. 2007;88(3):337–357.
21 Angelsky PO, Ushenko AG, Dubolazov AV, Sidor MI, Bodnar GB, Koval G, et al. The singular approach for processing polarization-inhomogeneous laser images of blood plasma layers. J Opt. 2013;15(4):044030.
22 Ushenko AG, Dubolazov AV, Ushenko VA, Novakovskaya OY. Statistical analysis of polarization-inhomogeneous Fourier spectra of laser radiation scattered by human skin in the tasks of differentiation of benign and malignant formations. J Biomed Opt. 2016;21(7):071110.
23 Angelskii OV, Ushenko OG, Burkovets DN, Arkhelyuk OD, Ushenko YA. Polarization-correlation studies of multifractal structures in biotissues and diagnostics of their pathologic changes. Laser Phys. 2000;10(5):1136–1142.

24 Bachinskyi V, Boychuk T, Ushenko A. Laser Polarimetry of Biological Tissues and Fluids: Chapter 4. Methods of Spatial-frequency, Singular and Correlation Analysis of Biological Layers Object Fields. LAP Lambert Academic Publishing; 2018.

25 Ushenko VA, Dubolazov AV, Pidkamin LY, Sakchnovsky MY, Bodnar AB, Ushenko YA, et al. Mapping of polycrystalline films of biological fluids utilizing the Jones-matrix formalism. Laser Phys. 2018;28(2):025602.

26 Ushenko AG, Angelsky PO, Sidor M, Marchuk YF, Andreychuk DR, Pashkovskaya NV. Spatial-frequency selection of complex degree of coherence of laser images of blood plasma in diagnostics and differentiation of pathological states of human organism of various nosology. Appl Opt. 2014:53(10):B172–B180.

27 Olar EI, Ushenko AG, Ushenko YA. Polarization correlation measurements of the phase tomograms of optically anisotropic biofractals. Laser Phys. 2004;14(8):1115–1121.

28 Guminetskiy SH, Ushenko OG, Polyanskiy IP, Motrych AV, Grynchuk FV. The optical method for investigation of the peritonitis progressing process. In Eighth International Conference on Correlation Optics. 2008;7008:626–632.

29 Ushenko AG, Fediv AI, Marchuk YuF. Singular structure of polarization images of bile secret in diagnostics of human physiological state. In: European Conference on Biomedical Optics. Optica Publishing Group; 2009.

30 Ushenko Y, Ushenko A, Dubolazov A, Soltys I, Litvinenko O, Wanchuliak O, et al. Mueller-matrix interferometric multifractal scaling of optically anisotropic architectonics of diffuse blood facies: fundamental and applied aspects. J Biophotonics. 2025;18(3):e202400412. https://doi.org/10.1002/jbio.202400412.

31 Ushenko A, Soltys I, Dubolazov A, Ushenko Y, Bilookyi V, Bilookyi O, et al. 3D Jones-matrix thesiography of biological fluid facies. J Innov Opt Health Sci. 2025;18(2):2443002. https://doi.org/10.1142/S1793545824430028.

32 Ushenko A, Pavlyukovich N, Khukhlina O, Pavlyukovich O, Gorsky M, Soltys I, et al. Layer-by-layer multifractal scanning of optically anisotropic architectonics of blood plasma films: fundamental and applied aspects. Photonics. 2025;12:215. https://doi.org/10.3390/photonics12030215.

33 Ushenko A, Pavlyukovich N, Khukhlina O, Pavlyukovich O, Soltys I, Dubolazov A, et al. Blood plasma film multifractal scanning in COVID-19 consequences diagnostics. J Biophotonics. 2024;17(11):e202400356. https://doi.org/10.1002/jbio.202400356.

34 Ushenko A, Dubolazov A, Zheng J, Litvinenko A, Gorsky M, Ushenko Y, et al. 3D polarization-interference holographic histology for wavelet-based differentiation of biological tissues with different necrotic states. J Biomed Opt. 2024;29(5):052920. https://doi.org/10.1117/1.JBO.29.5.052920.

35 Ushenko O, Bilookyi O, Zheng J, Dubolazov A, Olar O, Ushenko Y, et al. 3D digital holographic polarimetry of laser speckle fields formed by polycrystalline blood films: a tool for differential diagnosis of thyroid pathology. Front Phys. 2024;12:1426469. https://doi.org/10.3389/fphy.2024.1426469.

36 Wójcik W, Hu Z, Ushenko Y, Smolarz A, Soltys I, Dubolazov O, et al. Optical sensor system for 3D Jones matrix reconstruction of optical anisotropy maps of self-assembled polycrystalline soft matter films. Sensors. 2024;24(5):1589. https://doi.org/10.3390/s24051589.

37 Ushenko A, Zheng J, Gorsky M, Dubolazov A, Ushenko Y, Soltys I, et al. 3D digital holographic polarimetry of diffuse optically anisotropic biological tissue object fields. Front Phys. 2023;11:1288935. https://doi.org/10.3389/fphy.2023.1288935.

38 Dubolazov A, Ushenko V, Trifonyuk L, Stashkevich A, Soltys I, Ushenko Y, et al. Polarization-singular approach to imaging Mueller-matrix polarimetry in differential diagnosis of histological sections of biopsy of tumors of the uterus and prostate. Front Phys. 2021;9:711212. https://doi.org/10.3389/fphy.2021.711212.

39 Ushenko VO, Trifonyuk L, Ushenko YA, Dubolazov OV, Gorsky MP, Ushenko AG. Polarization singularity analysis of Mueller-matrix invariants of optical anisotropy of biological tissue samples in cancer diagnostics. J Opt. 2021;23(6):064004. https://doi.org/10.1088/2040-8986/abf97a.

40 Ushenko YA, Koval GD, Ushenko AG, Dubolazov OV, Ushenko VA, Novakovskaia OY. Mueller-matrix of laser-induced autofluorescence of polycrystalline films of dried peritoneal fluid in diagnostics of endometriosis. J Biomed Opt. 2016;21(7):071116. https://doi.org/10.1117/1.JBO.21.7.071116.

5

Diagnostic Efficiency of 3D Mueller Digital Holographic Mapping

Yurii Ushenko[1,2], Weidong Dou[1], Oleksandr Ushenko[2,3], Oleksandr Dubolazov[2], Iryna Soltys[2], Viktor Bachinskiy[4], Oleh Vanchuliak[4], Marta Garazdyuk[4], Nataliia Pavlyukovich[4], Oleksandr Pavlyukovich[4], Vitaliy Rozhko[4], and Oleksandr Olar[2]

[1] *Shaoxing University, Shaoxing, People's Republic of China*
[2] *Yurii Fedkovych Chernivtsi National University, Chernivtsi, Ukraine*
[3] *Taizhou Institute of Zhejiang University, Taizhou, People's Republic of China*
[4] *Bukovinian State Medical University, Chernivtsi, Ukraine*

5.1 Phase Waves of Local Depolarization in Biological Tissues

5.1.1 Introduction

Optical methods for diagnosing polycrystalline media – including biological tissue layers – occupy a central position in modern biomedical optics due to their high informational content and the ability to perform multifunctional monitoring of the examined medium (photometric, spectral, polarization, and correlation monitoring). Despite this progress, a unified methodological framework for diagnosing such structurally complex media is still lacking [1–7].

Polarimetric analysis has emerged as an independent and powerful direction in the optical characterization of biological tissues. Examination of the polarization properties of scattered radiation provides qualitatively new insights into the morphological and physiological state of tissues, enabling the detection of cataract formation in the lens, glucose concentration in diabetic tissue, and malignant transformations.

A major advance in biomedical polarimetric diagnostics has been the development of Mueller-matrix–based techniques, which has led to the formation of Mueller-matrix microscopy (MMM) as a distinct research field. This direction represents a successful integration of instrumental Mueller-matrix imaging polarimetry with a wide range of theoretical models and analytical approaches, including the regression model of optical anisotropy, logarithmic Mueller-matrix decomposition, Monte Carlo simulations of polarized radiation propagation, and statistical analysis of Mueller-matrix images combined with optical anisotropy mapping [8–13].

The fundamental results obtained within MMM are highly promising for both clinical and preclinical applications in the study of the polycrystalline architecture of biological tissues. In particular, MMM of gastric biopsy samples – combined with a regression model of optical anisotropy and statistical analysis of the resulting images – has demonstrated the feasibility of extracting quantitative optical metrics capable of characterizing the progression of gastric tissue from a healthy state through inflammation to cancer.

In parallel, laser polarimetry methods and systems for analyzing biological tissue samples have undergone rapid development, incorporating a broad spectrum of analytical algorithms – statistical, correlation-based, fractal, wavelet, Fourier, and others – for quantitative data processing.

At present, a major and unresolved challenge in biomedical laser polarimetric diagnostics is the strong dependence of experimental results on the degree of depolarization of laser radiation. One approach to addressing this issue has involved direct studies of the formation mechanisms of integral depolarization in object fields of diffuse biological tissue layers. This phenomenon has become a key subject of investigation in structural diagnostics of diffuse biological media. A relationship between the degree of integral depolarization and the morphological structure of biological tissues was demonstrated previously. These findings were further generalized using Monte Carlo simulations for anisotropic media and extended by the introduction of polarization Mueller-matrix decomposition principles, enabling the separation of polarimetric contributions into retardance, depolarization, and attenuation components. The resulting integral model of the polarization properties of diffuse biological tissue layers has subsequently been applied to cancer diagnostics.

A significant advancement in this area has been the successful integration of polarimetric and interferometric mapping techniques, leading to the development of several promising branches of three-dimensional (3D) polarimetry for biological objects and media [14–18].

Among these, polarization-sensitive optical coherence tomography (PSOCT) has emerged as one of the most powerful and rapidly evolving approaches. The fundamental principles and major achievements of PSOCT in biological tissue research over the past 25 years have been comprehensively summarized in a series of review articles. This laser-based polarization–interference technique enables the acquisition of depth-resolved images from shallow tissue layers (up to ~2 mm) and provides access to coordinate-resolved distributions of Jones and Mueller-matrix elements, as well as maps of optical anisotropy. These capabilities have been demonstrated most effectively in studies of transparent ocular tissues.

Studies of layered polarization and matrix images of scattering (depolarizing) biological tissues with varying morphological structures and physiological states have extended the functionality of PSOCT. Specifically, these approaches enabled accurate diagnosis of fibrosis and differentiation of tumor sites with low fibrosis. However, the sensitivity of such systems is limited by the distorting effects of high levels of depolarized laser speckle background, which reduces the contrast of polarimetric images of tissue layers [19–25].

Despite these advances, the results remain largely empirical. Currently, there is limited information on the capabilities of this method for tissue samples with varying optical thicknesses, morphological structures, and pathological conditions. From a physical standpoint, the primary objective is to establish the scientific significance of the approach: to determine the algorithmic relationships between the structure of polycrystalline fibrillar networks in biological tissues and the spatial distribution of depolarization parameters in the scattered object field [23–29].

In this context, depolarization of optical radiation is understood as the integral (phase-nonstationary) averaging of scattered waves with varying polarization states ($\alpha(r)$, $\beta(r)$) at each point in the object field, where stationary-phase shifts ($\delta(r)$) between the waves are absent. Integral depolarization results in a reduction of the degree of polarization of the object field, formed by scattering events of low multiplicity. This component of the object field is highly informative and strongly correlated with the morphological structure of the biological sample. Consequently, the presence of a high level of depolarized background remains a central challenge in identifying diagnostic relationships between the parameters of polarization distributions and the volumetric architecture of biological tissues.

A promising solution involves treating the degree of depolarization of multiply scattered laser radiation itself as an informative parameter.

In this work, we focus on the theoretical investigation and applied utilization of birefringent fibrillar networks forming biological tissues and their contribution to local depolarization in laser speckle fields. We propose a novel polarization–interference method for phase scanning of diffuse speckle fields, incorporating digital holographic reconstruction of complex amplitude distributions. This approach enables the study of the physical mechanisms governing local depolarization distributions and their phase dynamics. Compared to traditional polarization and Mueller-matrix mapping techniques, as well as PSOCT, this method significantly extends functional capabilities by allowing differential diagnostics not only at the boundary zone (image plane) of the speckle field but also across the full geometric range of field propagation, including the Fresnel and Fraunhofer diffraction zones.

The applied objective of this study is the identification of new diagnostic markers for differentiating various necrotic changes in diffuse histological samples of the myocardium under conditions of a highly depolarized background, specifically in deceased individuals with ischemic heart disease (IHD) and acute coronary insufficiency (ACI).

5.1.2 Materials and Methods

5.1.2.1 Brief Theory and Fundamental Analytical Equations

This section provides a concise theoretical overview of the mechanisms underlying the formation of local depolarization in phase-stationary laser object fields generated by birefringent biological tissues [30–36].

Traditionally, the degree of depolarization in a diffusely scattered object field is quantified using the following equation

$$\langle T\rangle = 1 - \langle P\rangle = 0.5\left[\frac{2\langle I_{\alpha\leftrightarrow 0.5\pi+\alpha}\rangle}{\langle I_{\alpha\leftrightarrow\alpha}\rangle + \langle I_{\alpha\leftrightarrow 0.5\pi+\alpha}\rangle}\right]. \tag{5.1}$$

Here, the signals integrally averaged over the photodetector area are considered for coaxial ($\alpha \leftrightarrow \alpha$) and crossed ($\alpha \leftrightarrow 0.5\pi + \alpha$) configurations of the polarizer and analyzer.

For a partially elliptically polarized wave at a point of the object field with coordinate (r), equation (5.1) can be rewritten in the form

$$T(r, \alpha_i, \beta_i) = \frac{\cos\alpha_i(r) + \tan\beta_i(r)\sin\alpha_i(r)}{(\cos\alpha_i(r) + \sin\alpha_i(r))(1 + \tan\beta(r)_i)}. \tag{5.2}$$

For a set $(n \in N)$ of such waves, for which no phase-stationary shifts are present ($\delta_n = \text{const.}$), equation (5.2) takes the following ensemble-averaged ($\langle\rangle$) form

$$\langle T(r, \alpha_i, \beta_i)\rangle_{n\in N} = \frac{\cos\langle\alpha_i(r)\rangle_{n\in N} + \tan\langle\beta_i(r)\rangle_{n\in N}\sin\langle\alpha_i(r)\rangle_{n\in N}}{\left(\cos\langle\alpha_i(r)\rangle_{n\in N} + \sin\langle\alpha_i(r)\rangle_{n\in N}\right)\left(1 + \tan\left\langle\langle\alpha_i(r)\rangle_{n\in N_i}(r)\right\rangle_{n\in N}\right)}. \tag{5.3}$$

Finally, for all points $(r \in R)$ of the object field projected onto the photodetector plane, equation (5.3) is transformed into the following form

$$T_{\text{integr}} = \left\langle\langle T(r, \alpha_i, \beta_i)\rangle_{n\in N}\right\rangle_{r\in R}. \tag{5.4}$$

Equation (5.4) represents the "integral," or measurable, depolarization, which increases monotonically with rising orders of light-scattering multiplicity ($i \gg 1$).

A fundamentally different physical scenario of depolarization formation occurs in phase-stationary scattered laser object fields of biological samples. At each point (r) of such a coherent field, the amplitudes of partially polarized waves with different polarization states interfere. Consequently, an ensemble of spatially distributed, fully polarized speckles with locally varying polarization parameters ($\alpha(r), \beta(r)$) is formed. The depolarization arising in this polarization-inhomogeneous laser field will hereafter be referred to as *local* or *true* depolarization, denoted as (T).

5.1.2.2 Formation of Local Depolarization

Local depolarization arises through the generation of an orthogonally polarized component within each speckle, relative to the polarization state of the probing laser beam interacting with the biological layer [37–41].

To elucidate the mechanisms of local depolarization formation in a phase-stationary laser object field, let us consider birefringent fibrillar biological tissue as a representative example.

At each point (r) within the fibrillar network, the polarization response associated with linear birefringence is described by the partial Mueller matrix $\{\nabla(r)\}$

$$\{\nabla(r)\} = \left\|\begin{matrix} 1 & 0 & 0 & 0 \\ 0 & \nabla_{22} & \nabla_{23} & \nabla_{24} \\ 0 & \nabla_{32} & \nabla_{33} & \nabla_{34} \\ 0 & \nabla_{42} & \nabla_{43} & \nabla_{44} \end{matrix}\right\|(r), \tag{5.5}$$

where

$$\nabla_{ik}(r) = \begin{pmatrix} \nabla_{22} \\ \nabla_{23} = \nabla_{32} \\ \nabla_{33} \\ \nabla_{42} = -\nabla_{24} \\ \nabla_{34} = -\nabla_{43} \\ \nabla_{44} \end{pmatrix}(r) = \begin{pmatrix} \cos^2 2\gamma + \sin^2 2\gamma\cos\delta \\ \cos 2\gamma\sin 2\gamma(1 - \cos\delta) \\ \sin^2 2\gamma + \cos^2 2\gamma\cos\delta \\ \sin 2\gamma\sin\delta \\ \cos 2\gamma\sin\delta \\ \cos\delta \end{pmatrix}(r). \tag{5.6}$$

Here, $\gamma(r)$ denotes the local orientation of the optical axis at point (r) determined by the direction of fibrillation. The quantity $\delta(r) = \frac{2\pi}{\lambda}\Delta nl$ represents the stationary-phase retardation between the orthogonal components of the laser wave at point (r); here λ is the wavelength, Δn is the birefringence, and l is the geometric thickness of the biological layer.

In the general case, the polarization state of the probing laser beam is described by the Stokes vector of the form

$$S^{(\text{zond})}(\alpha_0, \beta_0) = \begin{pmatrix} 1 \\ \cos 2\alpha_0 \cos 2\beta_0 \\ \sin 2\alpha_0 \cos 2\beta_0 \\ \sin 2\beta_0 \end{pmatrix}. \quad (5.7)$$

The process of local polarization transformation at point (r) by a birefringent fibrillar element is described by the following matrix equation

$$S^{(\text{object})}(r, \alpha^*, \beta^*) = \|\nabla(r)\| S^{(\text{zond})}(\alpha_0, \beta_0). \quad (5.8)$$

Solving equation (5.8) yields the polarization parameters of the object field at point (r)

$$\alpha^*(r, \alpha_0, \beta_0, \gamma, \delta) = 0.5 \arctan\left(\frac{\nabla_{32} \cos 2\alpha_0 \cos 2\beta_0 + \nabla_{33} \sin 2\alpha_0 \cos 2\beta_0 + \nabla_{34} \sin 2\beta_0}{\nabla_{22} \cos 2\alpha_0 \cos 2\beta_0 + \nabla_{23} \sin 2\alpha_0 \cos 2\beta_0 + \nabla_{24} - \sin 2\beta_0} \right); \quad (5.9)$$

$$\beta^*(r, \alpha_0, \beta_0, \gamma, \delta) = 0.5 \arcsin(\nabla_{24} \cos 2\alpha_0 \cos 2\beta_0 + \nabla_{34} \sin 2\alpha_0 \cos 2\beta_0 + \nabla_{44} \sin 2\beta_0). \quad (5.10)$$

Careful analysis of equations (5.9) and (5.10) indicates that the full polarization of speckles in the object field is determined by the combined influence of the birefringence parameters (α^*, β^*) of the fibrillar network and the polarization state (α_0, β_0) of the incident probing beam

$$\langle \alpha^*(\alpha_0, \beta_0) \rangle_{r \in (R = m \times n)} = \begin{pmatrix} \alpha^*_{11} & \cdots & \alpha^*_{1n} \\ \vdots & \ddots & \vdots \\ \alpha^*_{m1} & \cdots & \alpha^*_{mn} \end{pmatrix} (\alpha_0, \beta_0); \quad (5.11)$$

$$\langle \beta^*(\alpha_0, \beta_0) \rangle_{r \in (R = m \times n)} = \begin{pmatrix} \beta^*_{11} & \cdots & \beta^*_{1n} \\ \vdots & \ddots & \vdots \\ \beta^*_{m1} & \cdots & \beta^*_{mn} \end{pmatrix} (\alpha_0, \beta_0). \quad (5.12)$$

Without compromising the completeness of the analysis, we consider two limiting scenarios for the formation of local depolarization in phase-stationary speckle fields. The first scenario involves the illumination of a birefringent layer of biological tissue with a linearly polarized laser beam, where the polarization azimuth is set to $\alpha_0 = \frac{\pi}{4}$

$$S^{(\text{zond})}\left(\alpha_0 = \frac{\pi}{4}\right) = \begin{pmatrix} 1 \\ 0 \\ 1 \\ 0 \end{pmatrix}. \quad (5.13)$$

The second scenario involves the illumination of the biological layer with a circularly polarized laser beam

$$S^{(\text{zond})}\left(\beta_0 = \frac{\pi}{4}\right) = \begin{pmatrix} 1 \\ 0 \\ 0 \\ 1 \end{pmatrix}. \quad (5.14)$$

For $S^{(\text{zond})}\left(\alpha_0 = \frac{\pi}{4}\right)$, the polarization parameters at the point (r) of the object field are expressed as follows

$$\alpha^*\left(\alpha_0 = \frac{\pi}{4}, r, \gamma, \delta\right) = \left(0.5 \arctan\left(\frac{\sin^2 2\gamma + \cos^2 2\gamma \cos \delta}{\cos 2\gamma \sin 2\gamma (1 - \cos \delta)} \right)\right)(r); \quad (5.15)$$

$$\beta^*\left(\alpha_0 = \frac{\pi}{4}, r, \gamma, \delta\right) = (0.5 \arcsin(\cos 2\gamma \sin \delta))(r). \quad (5.16)$$

For $S^{(zond)}\left(\beta_0 = \frac{\pi}{4}\right)$ the polarization parameters at the point (r) of the object field assume a different form

$$\alpha^{**}\left(\beta_0 = \frac{\pi}{4}, r, \gamma\right) = \gamma(r); \tag{5.17}$$

$$\beta^{**}\left(\beta_0 = \frac{\pi}{4}, r, \gamma, \delta\right) = (0.5 \arcsin(\cos \delta))(r). \tag{5.18}$$

Analysis of equations (5.15–5.18) shows that the magnitudes of the polarization parameters of speckles in each stationary-phase region (δ_k) of the laser field are unique and determined by the random orientations (γ) of the optical axes of the fibrillar crystals.

5.1.2.3 Postulates of Probability Theory

To quantify the degree of local depolarization $(T(\delta_l))$ within a series of stationary-phase regions (δ_l) of the laser object field in a birefringent fibrillar layer of biological tissue, we applied established postulates of probability theory.

The specific analytical form of the probability distribution (F) in each phase region was assumed to follow a normal distribution, reflecting the random orientations of the optical axes of the birefringent fibrils

$$F^{\gamma} = \left(\frac{1}{\sqrt{2\pi\sigma}}\right) \exp\left\{-\left[\frac{\gamma^2}{2\sigma^2}\right]\right\}. \tag{5.19}$$

Here, σ denotes the standard deviation of the random orientations.

By defining the inverse function $\gamma = \Pi_{\delta_l}^{-1}(T)$, we obtain an analytical expression for the distribution function of the random values of the local depolarization degree across different stationary-phase regions (δ_l) of the speckle field

$$F_{\delta_l}^{T} = \left|\frac{d\left(\Pi_{\delta_l}^{-1}(T)\right)}{dT}\right| \exp\left\{-\left[\frac{\left(\Pi_{\delta_l}^{-1}(T)\right)^2}{2\sigma^2}\right]\right\}. \tag{5.20}$$

To derive the explicit form of equation (5.20), the probabilistic distributions of the polarization azimuth and ellipticity were determined

$$F_{\delta_l}^{\alpha^*} = \Omega^{\delta_l}(\gamma); \tag{5.21}$$

$$F_{\delta_l}^{\beta^*} = \Theta^{\delta_l}(\gamma). \tag{5.22}$$

Since α and β are independent random variables, their joint probability distribution function can be expressed as

$$\Omega^{\delta_l}(\gamma)\Theta^{\delta_l}(\gamma) \equiv \Pi^{\delta_l}(\gamma). \tag{5.23}$$

Thus, the phase-dependent fluctuations in the magnitude of the "true" depolarization of the laser speckle field can be determined using the following equation

$$T(\delta_l) = \int_0^1 T \times F^T(\delta_l) dT. \tag{5.24}$$

The results of our analytical review (equations 5.13–5.24) are systematically summarized in Tables 5.1 $\left(\alpha_0 = \frac{\pi}{4}\right)$ and 5.2 $\left(\beta_0 = \frac{\pi}{4}\right)$, where the expression $T^{\beta_0} = (1 - T^*)$ was used as a parameter of local depolarization.

From the analysis of the theoretical data presented in equations (5.13–5.24) and summarized in Tables 5.1 and 5.2, the following conclusions can be drawn:

1. Regardless of variations in the polarization states of the laser beams probing the birefringent biological tissue, ensembles of fully polarized speckles are consistently formed.
2. At different phase-stationary levels of the object laser speckle field, the probability distributions of the random variables representing the azimuth and ellipticity of polarization are individual and unique.
3. Variations in the birefringence of the biological layer give rise to phase waves – harmonic oscillations in the magnitude – of the "true" degree of local depolarization of the object laser field, denoted as $T^{\alpha_0}(\delta_k)$ and $T^{\beta_0}(\delta_k)$.

Table 5.1 Probabilistic structure of a polarization-inhomogeneous object field in a birefringent biological layer for $\alpha_0 = \frac{\pi}{4}$.

δ	F^{α^*}	F^{β^*}	F^T	T^{α_0}
0	$\alpha^* = \frac{\pi}{4}$ F^{α^*}	$\beta^* = 0$ F^{β^*}	$\alpha^* = \frac{\pi}{4}$	0
$\frac{\pi}{2}$	$\alpha^* = \gamma$ $F^{\alpha^*} \leftrightarrow F^{\gamma}(\gamma)$	$\beta^* = 0.5\arcsin(\cos 2\gamma)$ $F^{\beta^*} \leftrightarrow W^{\gamma}(0.5\arcsin(\cos 2\gamma))$	$F^T \leftrightarrow$ $F^{\gamma}(\gamma) \times W^{\gamma}(\gamma)$	$0 \leq T^{\alpha_0} \leq 0.5$
π	$\alpha^* = -\arctan(\text{cotan}\, 2\gamma)$ $F^{\alpha^*} \leftrightarrow Q^{\gamma}(\arctan(\text{cotan}\, 2\gamma))$	$\beta^* = 0$	$F^T \leftrightarrow Q^{\gamma}(\gamma)$	1
$\frac{3\pi}{2}$	$\alpha^* = \gamma$ $F^{\alpha^*} \leftrightarrow F^{\gamma}(\gamma)$	$\beta^* = -0.5\arcsin(\cos 2\gamma)$ $W^{\gamma}(0.5\arcsin(\cos 2\gamma))$	$F^T \leftrightarrow$ $F^{\gamma}(\gamma) \times W^{\gamma}(\gamma)$	$0 \leq T^{\alpha_0} \leq 0.5$
2π	$\alpha^* = \frac{\pi}{4}$	$\beta^* = 0$	$\alpha^* = \frac{\pi}{4}$	0

Table 5.2 Probabilistic structure of the polarization-inhomogeneous object field of a birefringent biological layer for $\beta_0 = \frac{\pi}{4}$.

δ	$F^{\alpha^{**}}$	$F^{\beta^{**}}$	G^T	$T^{\beta_0} = (1 - T^*)$
0	$\alpha^{**} = \gamma$	$\beta^{**} = \frac{\pi}{4}$	$G^T = 1$	0
$\frac{\pi}{2}$	$\alpha^{**} = \gamma$ $F^{\alpha^{**}} \leftrightarrow F^{\gamma}(\gamma)$	$\beta^{**} = 0$ $F^{\beta^{**}} \leftrightarrow 1$	$G^T = F^{\alpha^{**}} \times F^{\beta^{**}}$ $\rightarrow F^{\gamma}(\gamma)$	$0 \leq T^{\beta_0} \leq 0.5$
π	$\alpha^{**} = \gamma$	$\beta^{**} = -\frac{\pi}{4}$	$G^T = 1$	1
$\frac{3\pi}{2}$	$\alpha^{**} = \gamma$ $F^{\alpha^{**}} \leftrightarrow F^{\gamma}(\gamma)$	$\beta^{**} = 0$ $F^{\beta^{**}} \leftrightarrow 1$	$G^T = F^{\alpha^{**}} \times F^{\beta^{**}}$ $\rightarrow F^{\gamma}(\gamma)$	$0 \leq T^{\beta_0} \leq 0.5$
2π	$\alpha^{**} = \gamma$	$\beta^{**} = \frac{\pi}{4}$	$G^T = 1$	0

From a physical standpoint, the results of the analytical consideration of the formation processes of the polarization structure of object laser fields can be interpreted as follows.

First, as a polarized laser wave propagates through a birefringent biological layer, its amplitude is decomposed into two orthogonal, linearly polarized components. The relative amplitudes of these components at each point of the fibrillar network are determined by the orientation of the optical axis. A phase shift is induced between the resulting orthogonal components.

The magnitude of this phase shift is proportional to the birefringence of the fibrils and the geometric path length of the laser wave through the layer. Upon exiting the birefringent medium, the orthogonal amplitude components interfere, producing an elliptically polarized wave

$$\left(\frac{x^2}{(\cos\gamma(\gamma(r)))^2}\right) + \left(\frac{y^2}{(\sin\gamma(r))^2}\right) - \left(\frac{2xy}{\sin\gamma(r)\cos\gamma(r)}\right)\cos\delta_k = \sin^2\delta_k. \tag{5.25}$$

Second, a sequential increase in the phase shift results in a corresponding and periodic (2π) variation of the polarization states of the laser beam (Table 5.3).

Third, harmonic phase variations of the polarization states ($\alpha^*(r), \beta^*(r)$) induce analogous fluctuations in the magnitude of local depolarization. As the phase retardation increases within the range ($\delta_k \rightarrow 0 \div \pi$), the degree of local depolarization ($T(\delta_k)$) rises from 0 to 1. Beyond this range ($\delta_k \rightarrow \pi \div 2\pi$), the process reverses, reflecting an inverted scenario.

Fourth, our analytical review (equations (5.5)–(5.24), Tables 5.1 and 5.2) demonstrates the emergence of phase waves of local depolarization within a birefringent biological layer.

The modulation depth of these waves for different polarization states of the probing laser beam is defined by the extreme values: $\left(T(\widetilde{\alpha}_0, r) = \begin{cases} 0 \leftrightarrow \delta = 2k\pi \\ 1 \leftrightarrow \delta = (2k+1)\pi \end{cases} \text{ and } T(\widetilde{\beta}_0, r) = \begin{cases} 1 \leftrightarrow \delta = 2k\pi \\ 0 \leftrightarrow \delta = (2k+1)\pi \end{cases}\right)$

$$K(\alpha_0, r) = \frac{(T(\alpha_0, r)_{\max} - T(\alpha_0, r)_{\min})}{(T(\alpha_0, r)_{\max} + T(\alpha_0, r)_{\min})}; \tag{5.26}$$

Table 5.3 Phase transformations of the polarization state of the object field in a birefringent biological layer $\left(\left(\alpha_0 = \frac{\pi}{4}\right) \text{ and } \left(\beta_0 = \frac{\pi}{4}\right)\right)$.

δ	Equations of polarization	α^*	β^*	$T(\alpha_0, r)$	$T(\beta_0, r)$
0	$\frac{x}{\cos\gamma(r)} - \frac{y}{\sin\gamma(r)} = 0; \gamma(r) = \frac{\pi}{4}$	$\alpha^*(r) = \gamma(r) = \frac{\pi}{4} = \alpha_0$	$\beta^*(r) = 0$	0	1
$\frac{\pi}{2}$	$\frac{x^2}{(\cos\gamma(r))^2} + \frac{y^2}{(\sin\gamma(r))^2} = 1$	$0 \leq \gamma(r) \leq \pi$	$\beta^*(r) = \frac{\pi}{4}$	0.5	0.5
π	$\frac{x}{\cos\gamma(r)} + \frac{y}{\sin\gamma(r)} = 0; \gamma(r) = -\frac{\pi}{4}$	$\alpha^*(r) = \gamma(r) = -\frac{\pi}{4} = -\alpha_0$	$\beta^*(r) = 0$	1	1
$\frac{3\pi}{2}$	$\frac{x^2}{(\cos\gamma(r))^2} + \frac{y^2}{\sin\gamma(r))^2} = 1$	$0 \leq \gamma(r) \leq \pi$	$\beta^*(r) = \frac{\pi}{4}$	0.5	0.5
2π	$\frac{x}{\cos\gamma(r)} - \frac{y}{\sin\gamma(r)} = 0; \gamma(r) = \frac{\pi}{4}$	$\alpha^*(r) = \gamma(r) = \frac{\pi}{4} = \alpha_0$	$\beta^*(r) = 0$	0	1

$$K(\beta_0, r) = \frac{\left(T(\beta_0, r)_{\max} - T(\beta_0, r)_{\min}\right)}{\left(T(\beta_0, r)_{\max} + T(\beta_0, r)_{\min}\right)}. \tag{5.27}$$

Fifth, the consideration presented in equations (5.26–5.27) and Table 5.3 refers to a single propagation of a polarized (α_0, β_0) laser wave through a biological layer with a given geometric thickness and birefringence. In most practical cases, multiple scattering occurs within the volume of the fibrillar network of biological tissue. As a result, sub-ensembles of speckles with different levels of local depolarization are formed in each phase plane (δ_k) of the object field. This additional set $T_q(\delta_k \pm \Delta\delta(k \gg 1))$ has its own probability distribution H^{T_q}. In other words, the experimentally measured value in the phase region δ_k ($T(\alpha_0, r)$ and $T(\beta_0, r)$) is increased by the integral average depolarization level $T\langle T_q(\delta_k \pm \Delta\delta(k \gg 1))\rangle = \int_0^1 T_q H^{T_q} dT$.

Taking this integral correction into account, the equation of modulation of phase waves of local depolarization takes the following form

$$K^*(\alpha_0, r) = \frac{\left[\left(T(\alpha_0, r)_{\max} - T(\alpha_0, r)_{\min}\right) - 2\langle T_q(\delta_k \pm \Delta\delta)\rangle\right]}{\left(T(\alpha_0, r)_{\max} + T(\alpha_0, r)_{\min}\right)}; \tag{5.28}$$

$$K^*(\beta_0, r) = \frac{\left[\left(T(\beta_0, r)_{\max} - T(\beta_0, r)_{\min}\right) - 2\langle T_q(\delta_k \pm \Delta\delta)\rangle\right]}{\left(T(\beta_0, r)_{\max} + T(\beta_0, r)_{\min}\right)}. \tag{5.29}$$

It follows from (5.28) and (5.29) that as the diversity of light scattering increases, the modulation depth of depolarization phase waves decreases. In the extreme case, it vanishes when the analytical condition is met

$$\left(T(\alpha_0, r)_{\max} - T(\alpha_0, r)_{\min}\right) = 2\langle T_q(\delta_k \pm \Delta\delta)\rangle. \tag{5.30}$$

Based on the theoretical analysis presented, two tasks were considered: fundamental and applied.

The fundamental task is to determine whether experimental manifestations of phase waves of local depolarization exist in the laser object fields of actual biological tissues.

The applied task is to explore how the presence of phase fluctuations in local depolarization can be utilized to detect pathological or necrotic changes in birefringent fibrillar networks.

Two principal prognostic scenarios for changes in the modulation depth of local depolarization phase waves in laser speckle fields of diffuse biological tissue samples can be identified.

The first scenario involves pathological changes, which are accompanied by an increase in the birefringence parameters of the polycrystalline architecture of biological tissues. As a result, the phase retardation between the orthogonal components of the complex amplitudes of the laser speckle field increase, leading to a corresponding increase in the modulation coefficient (5.28 and 5.29).

The opposite trend – a decrease in the phase modulation of local depolarization waves – occurs during necrotic changes in the morphology of biological tissue. These changes are accompanied by degradation or “destruction” of the polycrystalline architecture and a corresponding reduction in birefringence. These processes constitute the primary focus of the studies presented in this work.

5.1.2.4 Measurement Methodology

Experimental measurements were performed using a Mach–Zehnder polarization interferometer, see Chapter 3, Section 3.1.2.

To perform serial azimuthally invariant measurements that remain unaffected by the rotation of the sample relative to the probing beam, circularly polarized radiation was employed (see equation (5.14), Table 5.2). This choice ensures the consistency of group measurements across native myocardial histological sections. Using other polarization states of the illuminating laser beam on a single myocardial histological section would lead to the formation of different elliptical polarization states at the same level of birefringence. This effect arises from variations in the angle $\gamma(r)$ between the orthogonal amplitude components (see equation (5.25), Table 5.3) as the polarization azimuth α_0 of the probing beam changes. Consequently, the modulation coefficient also varies with α_0, which, within a myocardial histological sample, would undermine the reliability of the data or lead to "information chaos."

1. Polarizing filters 6–7 and 9–10 sequentially generate circular polarization ($\otimes$) in the "irradiating" (Ir) and "reference" (Re) parallel laser beams, producing $Ir(\otimes) - \mathrm{Re}(\otimes)$.
2. For this circular polarization ($\otimes$) state, two partial interference patterns are recorded through the polarizer-analyzer 13, with the transmission axis oriented at $\Omega = 0°$; $\Omega = 90°$.
3. A piezoceramic element enabled micro-movements of mirror 4, sequentially generating phase shifts $\delta_k = \frac{\pi}{8}$.
4. For each phase shift and for each pixel $(m \times n)$, the microscopic interference images of myocardial samples were analytically processed using a digital Fourier transform $\mathrm{FT}(\upsilon, \nu)$

$$\mathrm{FT}_{x;y}(\upsilon, \nu) = \frac{1}{M \times N} \sum_{m=0}^{M-1} \sum_{n=0}^{N-1} I_{x,y}(\Omega = 0°; 90°) \exp\left[-i2\pi\left(\frac{m \times \upsilon}{M} + \frac{n \times \nu}{N}\right)\right]. \tag{5.31}$$

where $\begin{cases} I_x^{\otimes}(\Omega = 0°)(m, n) = U_x^{\otimes}(\Omega = 0°)(U_x^{\otimes})^*(\Omega = 0°)(m, n); \\ I_y^{\otimes}(\Omega = 90°)(m, n) = U_y^{\otimes}(\Omega = 90°)(U_y^{\otimes})^*(\Omega = 90°)(m, n); \end{cases}$ $U_{x,y}^{\otimes}$ - orthogonal

components of the complex amplitude for the two orientations orientation $\Omega = 0°$; $\Omega = 90°$; $*$ denotes the complex conjugation operation (υ, ν) are the spatial frequencies; and $(m = 1120, n = 960)$ represent the number of pixels of the CCD camera.

1. The results of the digital Fourier transform (5.31) are then used to obtain the complex amplitude distributions according to the following algorithms

$$U_{0°}(m, n) \to \left|U_x^{\otimes}(\Omega = 0°; m, n)\right|; \tag{5.32}$$

$$U_{90°}(m, n) \to \left|U_y^{\otimes}(\Omega = 90°; m, n)\right| \exp(i(\delta_x^{\otimes} - \delta_y^{\otimes})); \tag{5.33}$$

$$\delta_k^{\otimes} = \delta_x^{\otimes} - \delta_y^{\otimes}. \tag{5.34}$$

2. The data obtained (5.32–5.34) were used to determine the set of coordinate distributions of the Stokes vector parameters

$$\begin{pmatrix} S_1(\delta_k^{\otimes}; m, n) = (U_x^*U_x + U_y^*U_y)(\delta_k^{\otimes}; m, n); \\ S_2(\delta_k^{\otimes}; m, n) = (U_x^*U_x - U_y^*U_y)(\delta_k^{\otimes}; m, n); \\ S_3(\delta_k^{\otimes}; m, n) = (U_y^*U_x + U_x^*U_y)(\delta_k^{\otimes}; m, n); \\ S_4(\delta_k^{\otimes}; m, n) = i(U_y^*U_x - U_x^*U_y)(\delta_k^{\otimes}; m, n) \end{pmatrix}. \tag{5.35}$$

3. Using equation (5.35), the coordinate distributions of the polarization parameters were determined for each phase section $(\delta_k^{\otimes})$

$$\alpha(\delta_k^{\otimes}; m, n) = \arctan\left[\frac{S_3(\delta_k^{\otimes}; m, n)}{S_2(\delta_k^{\otimes}; m, n)}\right]; \tag{5.36}$$

$$\beta(\delta_k^{\otimes}; m, n) = \arcsin\left[\frac{S_4(\delta_k^{\otimes}; m, n)}{S_1(\delta_k^{\otimes}; m, n)}\right]. \tag{5.37}$$

4. Based on equations (5.36) and (5.37), the phase dependences of the magnitude of local depolarization were determined at each point of the object laser speckle field

$$T(\delta_k^{\otimes}; m, n) = \frac{(\cos \alpha(\delta_k^{\otimes}; m, n) + \tan \beta(\delta_k^{\otimes}; m, n)\sin \alpha(\delta_k^{\otimes}; m, n))}{((\cos \alpha(\delta_k^{\otimes}; m, n) + \sin \alpha(\delta_k^{\otimes}; m, n))(1 + \tan \beta(\delta_k^{\otimes}; m, n)))}. \tag{5.38}$$

5. In this section, we focus on the regularities of formation and the phase fluctuations of the mean value of the distribution (5.38), $\langle T(\delta_k^{\otimes}; m, n)\rangle$

$$\langle T(\delta_k^{\otimes}; m, n)\rangle = \frac{\sum_{g=1}^{m\times n} (\langle T(\delta_k^{\otimes}; m, n)\rangle)_g}{m \times n}. \tag{5.39}$$

5.1.2.5 Objects of Research

Myocardial tissue was selected as the object of study. Morphologically, this biological tissue is composed of networks of myosin fibrils, whose spatial and angular configuration is both ordered and diverse.

From an optical perspective, these fibrillar myosin networks generate structural anisotropy, or optical birefringence. In other words, myocardial tissue can be regarded as a system of probabilistically distributed, optically uniaxial, birefringent biological crystals.

To experimentally identify and investigate the characteristics of local depolarization phase waves in object speckle fields, we used healthy myocardial tissue obtained from individuals deceased as a result of accidents.

For applied research, myocardial samples were obtained from individuals who died due to IHD and ACI.

Native histological sections of the myocardium were examined to eliminate any potential distortions caused by optically active formalin, diffuse paraffin, or staining agents on the experimental results.

Native histological sections of the myocardium were prepared using a microtome followed by rapid freezing according to standard procedures.

To investigate the influence of the magnitude of the phase shift due to linear birefringence $\left(\delta = \left(\frac{2\pi}{\lambda}\right)\Delta nl\right)$, five groups of histological sections (14 samples per group) were produced with varying geometric thicknesses. The thickness of each section was chosen to achieve a specific level of phase shift (δ) during a single passage of the laser wave.

Table 5.4 presents the optical and geometric parameters of these samples.

The extinction coefficient (τ, cm^{-1}) of the biological tissue samples was determined using the standard photometric method of beam attenuation, with measurements performed in an integrating sphere for diffuse light scattering. The integral degree of depolarization (Φ, %) of the tissue samples was measured using a standard Mueller-matrix polarimeter. Histological sections were prepared using the conventional rapid-freezing microtomy technique.

5.1.3 Results

At the first stage, we measured the coordinate distributions of random values of the integral depolarization of circularly polarized laser radiation (5.14) obtained from a series of myocardial histological sections (Table 5.4) using the conventional

Table 5.4 Optical-geometric parameters of native histological sections of the myocardium.

	$\Delta n = 10^{-3}; \lambda = 0.6328\,\mu\text{m}$				
Phase shift, $\delta(k = 1)$, rad	$\frac{\pi}{10}$	$\frac{\pi}{2}$	π	$\frac{3\pi}{2}$	2π
Geometrical thickness, l, μm	10 ± 0.5	50 ± 0.5	100 ± 0.5	150 ± 0.5	200 ± 0.5
Attenuation coefficient, τ	0.07 ± 0.001	0.18 ± 0.004	0.29 ± 0.007	0.34 ± 0.004	0.42 ± 0.005
Integral depolarization, Φ, %	8 ± 0.5	37 ± 0.85	59 ± 1.7	74 ± 2.1	87 ± 3.4

technique (5.1.1). A set of fragments in Figure 5.1 presents 2D and 3D maps of the distributions of the integral depolarization degree $\Delta(m, n)$ of the diffuse laser object field for native myocardial sections of various geometric thicknesses.

Analysis of the obtained data revealed the presence of an integral depolarized background in the scattered laser fields for all groups of native myocardial sections. The magnitude of this depolarized background varies depending on the geometric thickness of the samples. This is demonstrated by the individual coordinate structures (Figure 5.1a–e) and by the 3D distributions (Figure 5.1f–j) of the random values of $\Delta(m, n)$.

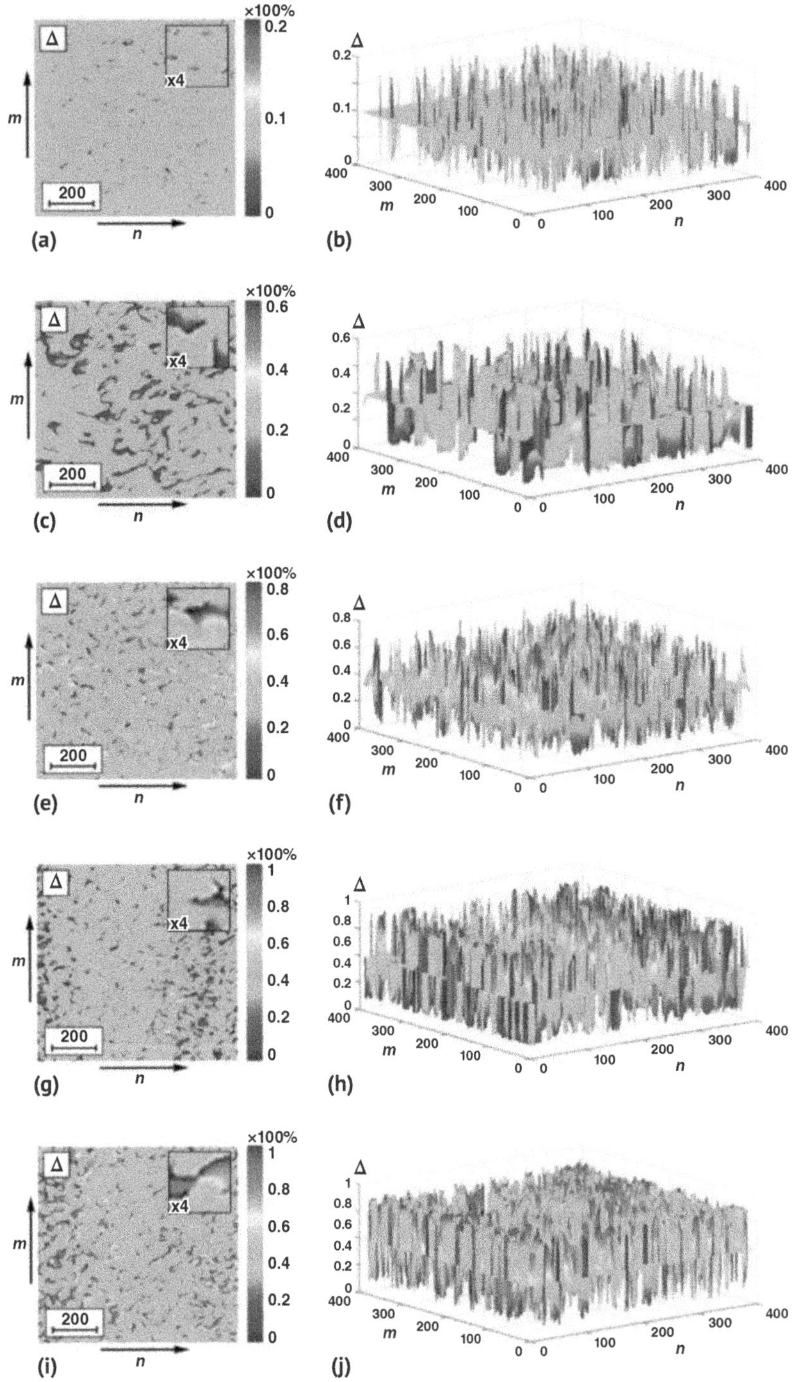

Figure 5.1 Two-dimensional (a)–(e) and 3D (f)–(j) maps of the integral degree of depolarization of the object speckle field for native myocardial histological sections of varying geometric thickness.

With increasing geometric thickness (and thus increasing multiplicity of light scattering), the average value of the $\Delta(m, n)$ distributions increases monotonically, remaining within the same order of magnitude (0.08–0.87).

The results correlate well with the traditional concepts of statistical averaging of the partial polarization states of optical radiation multiply scattered in biological tissue.

It should be noted that within the framework of this approach, the phase-stationary structure of the laser fields remains unaccounted for. Moreover, the obtained maps and histograms of the depolarization parameter distributions (Figure 5.1) do not provide any information on the phase manifestations of optical anisotropy (birefringence) in biological tissue. Such information can be retrieved using our proposed method (5.31–5.38) of polarization–interference analysis of the phase-stationary structure of object speckle fields in birefringent myocardial samples.

For this purpose, interference patterns were recorded for circularly polarized probing and reference beams within a Mach–Zehnder interferometer configuration. The interference distributions obtained by piezoceramic scanning (5.34) across different phase planes of the speckle field were then analyzed using polarizer 13 at two orthogonal orientations of its transmission axis ($\Omega = 0°$; $\Omega = 90°$).

As an experimental example, Figure 5.2 presents a series of interferograms corresponding to the phase levels of the laser object speckle fields of a native myocardial histological slice with a geometric thickness of $l = 100\,\mu\text{m}$.

Analysis of the obtained polarization–interference phase-scanning results revealed the following:

- interference distributions are present at all phase levels of the object laser speckle field (Figure 5.2a–j);
- the topographic structures of the interferograms are unique for each phase plane (Figure 5.2a–j);
- the interference fringes exhibit varying contrast across different regions of the interferogram plane (Figure 5.2a–j);
- with increasing phase-scan shift, the contrast of the interference patterns recorded at ($\Omega = 0°$) changes periodically – decreasing as $\delta_k \to \pi$ and then increasing as ($\delta_k \to 2\pi$) (Figure 5.2a, c, e, g, i);
- the opposite trend is observed for the orthogonal orientation of the analyzer's transmission axis ($\Omega = 90°$) the contrast of the polarization interferograms initially increases ($\delta_k \to \pi$), followed by a decrease within the phase interval ($\delta_k \to 2\pi$) (Figure 5.2b, d, f, h, j).

The observed results can be physically explained by the coordinate and polarization structure of each phase plane of the coherent object field in the birefringent myocardial layer. As previously shown (5.8–5.18), individual probability distributions of the azimuth and ellipticity of the polarization of laser oscillations in a speckle ensemble are formed within these planes. These arrays of random polarization parameters also exhibit complex coordinate (topographic) distributions. The structure of these topographic polarization maps is determined by the geometric and spatial-angular organization of the birefringent fibrillar myosin networks in the myocardium (5.20–5.24).

Consequently, at different points of a given phase section (δ_k) of such a polarization-inhomogeneous laser object field, the degree of correlation between the local polarization states of the speckles and the reference wave varies. This results in the formation of individual topographic distributions of the interference fringe system (Figure 5.2).

Moreover, for different polarization projections $(\Omega = 0°; 90° \leftrightarrow U_x; U_y)$ of the field, the phase dependences of the correlation degree are opposite. For the coaxial orientation of the analyzer ($\Omega = 0°$), only the vertical components (U_x) of the local polarization amplitudes are present in the "filtered" speckle field. These components first decrease ($\downarrow$) $U_x(\delta_k = 0 \div 0.5\pi)$ over the phase interval and then increase ($U_x \uparrow$) with further phase shift ($\delta_k = 0.5\pi \div \pi$). Accordingly, the contrast of the interference distributions of the polarization-filtered object speckle field changes as well (Figure 5.2a, c, e, g, i).

The opposite scenario is observed for the horizontal components (U_y), which determine the magnitude of local depolarization (Figure 5.2b, d, f, h, j).

Similar patterns of formation and phase transformation of interferograms of object speckle fields were observed for other native myocardial histological sections.

The obtained results were used as the basis for the algorithmic determination (5.38, 5.39) of the average values of the local depolarization degree, $\langle T(\delta_k^{\otimes}; m, n)\rangle$.

Figure 5.3 presents a series of fragments showing the phase-dependent fluctuations in the average values of local depolarization, $(\langle T(\delta_k^{\otimes}; m, n)\rangle)$, of the object speckle fields for a set of native myocardial histological sections.

Analysis of the data revealed a correlation between the theoretical model of the birefringence of fibrillar networks and the processes underlying the formation of local depolarization in laser speckle fields (5.5–5.24). We have experimentally confirmed the existence of phase waves of local depolarization in coherent laser speckle fields, which are generated by the birefringent networks of the myocardium.

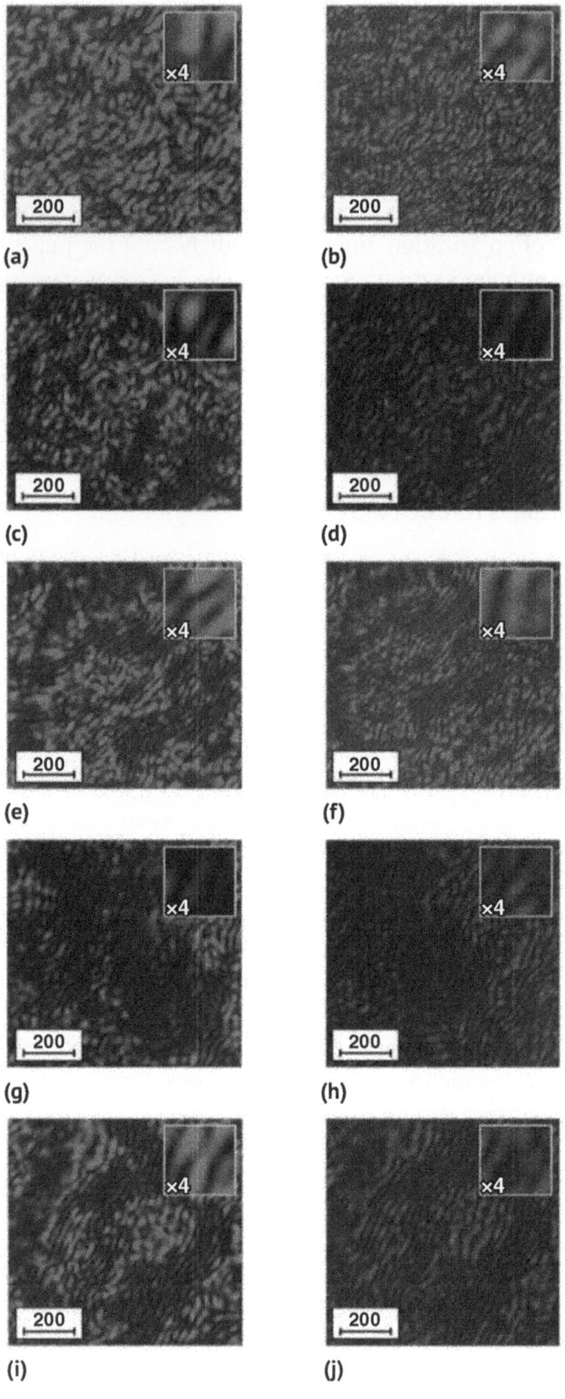

Figure 5.2 Phase interferograms of the speckle fields of native myocardial histological sections ($l = 100\,\mu\text{m}$) for orthogonal orientations of the polarizer–analyzer transmission axis: (a, c, e, g, i) $\Omega = 0°$; (b, d, f, h, j) correspond to $\Omega = 90°$.

For all types of native myocardial histological sections studied, the following observations were made:

- The local depolarization $\langle T(\delta_k^{\otimes}; m, n)\rangle$ demonstrates a phase-dependent fluctuating behavior with a full-cycle periodicity of (2π) (Figure 5.3a–e).
- The modulation depth $K^*(\beta_0, r)$ (5.29) attains its highest values within the first phase interval $(0 \div 2\pi)$ and then gradually diminishes over the subsequent intervals $(2\pi \div 4\pi)$ and $(4\pi \div 6\pi)$.
- For optically thin samples (Figure 5.3a, b), $K^*(\beta_0, r)$ ranges from 0.92 to 0.81.

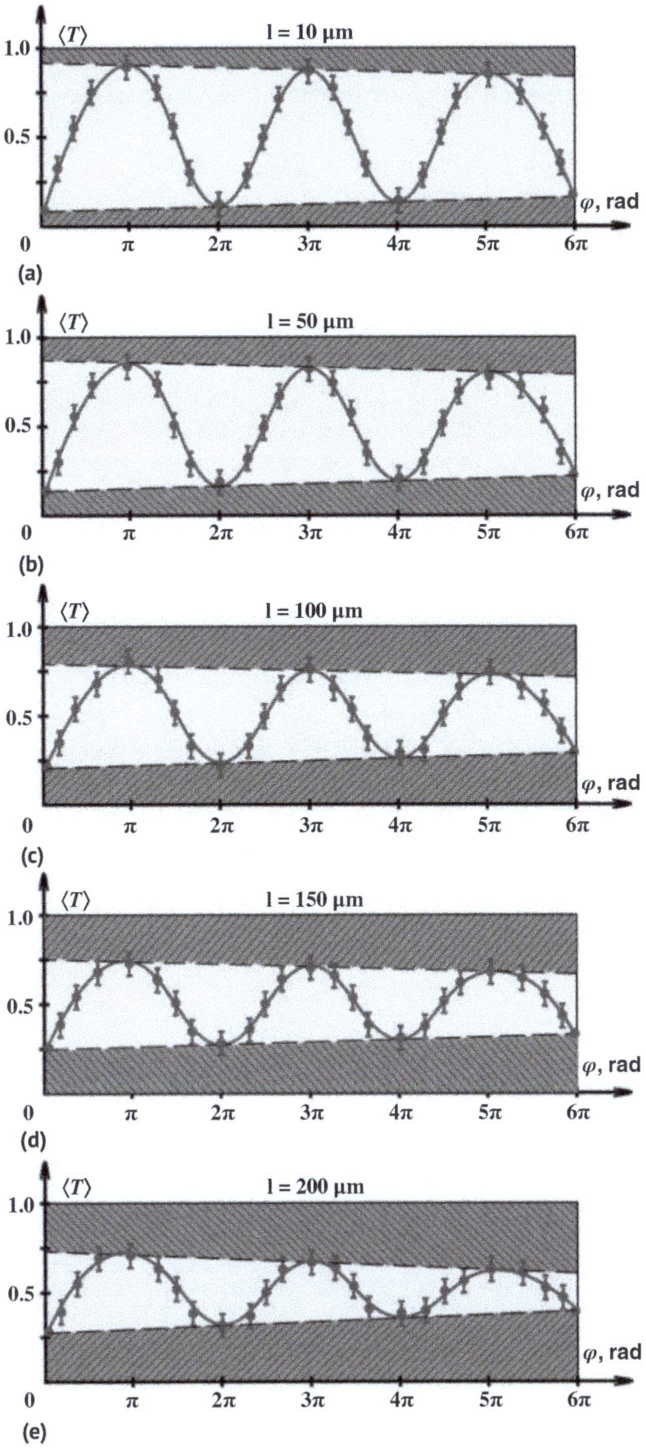

Figure 5.3 Phase waves of local depolarization in the object speckle fields of native myocardial histological sections: $l = 10\,\mu m$ (a); $l = 50\,\mu m$ (b); $l = 100\,\mu m$ (c) $l = 150\,\mu m$ (d); and $l = 200\,\mu m$ (e).

- As the samples become optically thicker, this range narrows progressively – down to 0.65–0.51 (Figure 5.3c) and further to 0.53–0.44 (Figure 5.3d).
- In diffuse samples, the modulation depth of the phase waves of local depolarization is reduced to 0.33–0.22, yet remains appreciably nonzero (Figure 5.3a, e).

The results obtained from polarization–interference phase scanning of object speckle fields (Figure 5.3) can be interpreted in terms of the following physical factors:

- Phase variations in the degree of local depolarization arise from the action of birefringent fibrils, which modify the polarization states of the laser radiation (5.5–5.18).
- At interferometric shifts (φ_k) within the corresponding phase planes (δ_k) of the object field, maximum correlation between the polarization states of the speckles and the probing layer of myocardial tissue is maintained (5.17, 5.18). In this regime, the degree of local depolarization is $\langle T(\delta_k^{\otimes}; m, n)\rangle = 0$.
- Increasing the phase shift ($\varphi_k \equiv \delta_k$) from 0 to π leads to the formation of orthogonally polarized speckle states, which produce the maximum degree of local depolarization, $\langle T(\delta_k^{\otimes}; m, n)\rangle$.
- Further growth of the phase-scan parameter ($\varphi_k \equiv \delta_k$) results in a periodic recurrence of the modulation of $\langle T(\delta_k^{\otimes}; m, n)\rangle$.
- For larger phase shifts (δ_k), integral averaging due to multiple scattering (($k \gg 1$)) strongly influences the oscillations of $\langle T(\delta_k^{\otimes}; m, n)\rangle$ because of the depolarized background. This effect is described by the integral depolarization correction $\langle T_q(\delta_k \pm \Delta\delta(k \gg 1))\rangle$ (5.30). As a result, the modulation depth $K^*(\beta_0, r)$ of the local depolarization waves decreases significantly, although it does not vanish completely (Figure 5.3d, e).

These results were subsequently applied to an important forensic task – differentiating the causes of myocardial death.

5.1.4 Applied Aspect

In this section, we examine the application of experimentally obtained parameters of phase depolarization waves for the differential diagnosis of myocardial tissue from individuals who died due to IHD and ACI.

Previous research in this area has primarily addressed polarization and Mueller-matrix mapping of optically thin (non-depolarizing) layers of biological tissues, including the myocardium. However, geometrically thin myocardial samples are often blood-filled and therefore exhibit strong depolarization.

To overcome this limitation, we performed a comparative study to evaluate the effectiveness of conventional polarization and Mueller-matrix mapping versus our polarization–interference phase scanning method under conditions of pronounced depolarized background. For this purpose, we analyzed groups of diffuse myocardial samples with different geometric thicknesses, corresponding to varying degrees of depolarization:

- Group A (thickness 50 μm) – integral degree of depolarization: 37.3%
  - Subgroup A1 – IHD (21 samples)
  - Subgroup A2 – ACI (21 samples)
- Group B (thickness 200 μm) — integral degree of depolarization: 82.7%
  - Subgroup B1 – IHD (21 samples)
  - Subgroup B2 – ACI (21 samples)

The following markers were used for differential diagnosis:

- Mean azimuthal and ellipticity distributions of the polarization parameters.
- Modulation depth of fluctuations in local depolarization.

Figures 5.4 and 5.5 present the phase dependences of the degree of local depolarization in the myocardial object speckle fields for Group A (Figure 5.4) and Group B (Figure 5.5). In these plots, lower intensity curves correspond to subgroups A1 and B1, whereas higher intensity curves represent subgroups A2 and B2.

Statistical analysis of the obtained data demonstrates that the phase waves of depolarization in the object laser speckle fields of native myocardial histological sections from Groups A and B differ markedly in the magnitude of the modulation coefficient T:

- Group A:
  - IHD – 0.53 ± 0.044
  - ACI – 0.79 ± 0.061
- Group B:
  - IHD – 0.21 ± 0.019
  - ACI – 0.34 ± 0.026

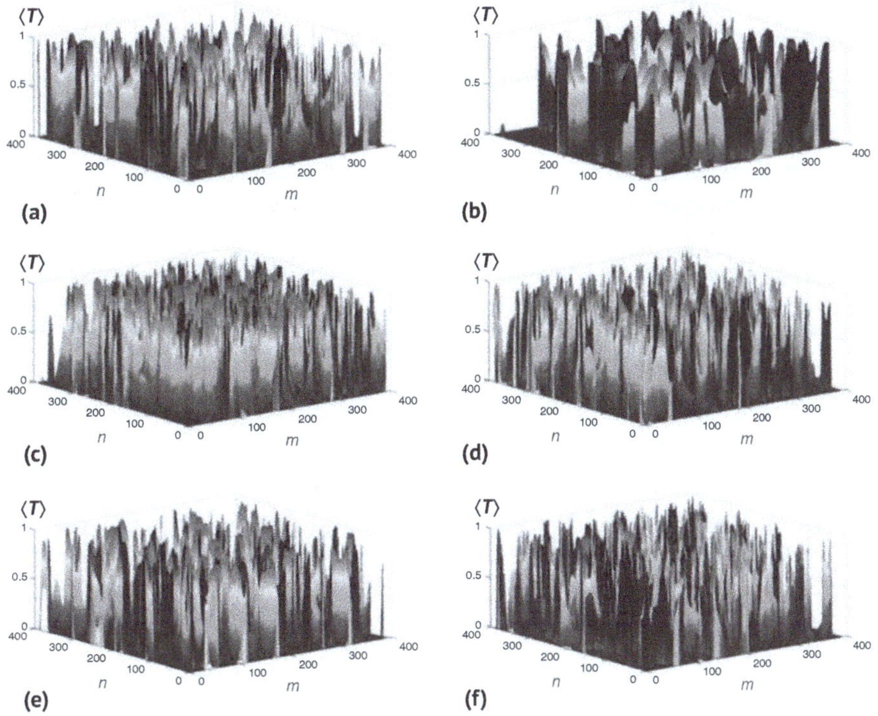

Figure 5.4 Phase dependences of the degree of local depolarization in the object speckle field of myocardial samples from Group A (geometric thickness: 50 μm). (a), (c), (e) Correspond to IHD, whereas (b), (d), (f) represent ACI.

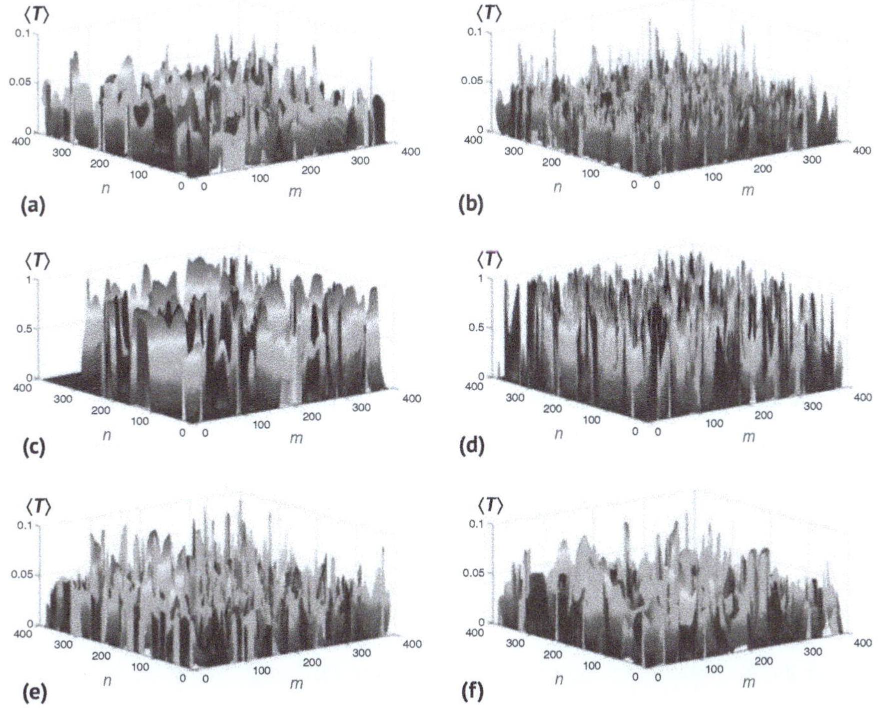

Figure 5.5 Phase dependences of the degree of local depolarization in the object speckle field of myocardial samples from Group B (geometric thickness: 200 μm). (a), (c), (e) Correspond to IHD, while (b), (d), (f) represent ACI.

From a physical perspective, the observed distinctions in the structure of phase waves of local depolarization can be attributed to the following considerations.

The maximal level of structural anisotropy – practically indistinguishable from that of healthy tissue – is characteristic of the fibrillar networks in the myocardium of individuals who died from ACI. Consequently, such tissue produces the strongest phase transformations of partial polarization states. This, in turn, leads to a relatively high level of local depolarization superimposed on the background of repeatedly scattered radiation.

In contrast, the fibrillar network in individuals who died from IHD undergoes necrotic alterations: the fibrils become thinned, partially degraded, and lose their orientational order. As a result, the overall structural anisotropy is substantially reduced. Consequently, the phase fluctuations of local depolarization are significantly diminished and approach the level of the integral depolarized background.

Tables 5.5 and 5.6 present the results of a comparative information analysis of polarization mapping, Mueller-matrix mapping, and polarization–interference phase scanning of object speckle fields in native myocardial histological sections – Group A (geometric thickness: 50 μm) and Group B (geometric thickness: 200 μm).

Within the framework of evidence-based medicine, the operational characteristics of each method's diagnostic performance were evaluated, including sensitivity, specificity, and accuracy for the differential diagnosis of IHD and ACI.

Analysis of the data presented in Table 5.5 (geometric thickness: 50 μm – average depolarization level) revealed the following:

- Polarization mapping provided low diagnostic accuracy for the differential diagnosis of depolarizing myocardial samples: satisfactory for IHD (76.18%) and unsatisfactory for ACI (61.9%).
- Mueller-matrix mapping improved the diagnostic accuracy to satisfactory (76.18%) for IHD and good (80.95%) for ACI.
- Polarization–interference phase scanning of object speckle fields achieved excellent accuracy (95.2%) for IHD and very good accuracy (90.45%) for ACI.

Analysis of the data presented in Table 5.6 (geometric thickness: 200 μm – high depolarization level) revealed the following:

- Polarization mapping showed low diagnostic accuracy for the differential diagnosis of diffuse myocardial samples: unsatisfactory (66.7% for IHD and 59.5% for ACI).
- Mueller-matrix mapping provided a minor improvement in diagnostic accuracy: satisfactory (80.4%) for IHD and unsatisfactory (69.05%) for ACI.
- Polarization–interference phase scanning of object speckle fields achieved very good diagnostic accuracy (90.45%) for both IHD and ACI.

Table 5.5 Operational characteristics of the diagnostic performance of polarimetry, Mueller-matrix mapping, and polarization–interference methods for Group A (geometric thickness: 50 μm).

Methods	**Polarization mapping**	
Parameters	**Group A1 (IHD – 21 samples)**	**Group A2 (ACI – 21 samples)**
Sensitivity, %	80.95; ($a = 17$; $b = 4$)	66.7; ($a = 14$; $b = 7$)
Specificity, %	71.4; ($c = 15$; $d = 6$)	57.1; ($c = 12$; $d = 9$)
Accuracy, %	76.18	61.9
Methods	**Mueller-matrix mapping**	
Parameters	**Group A1 (IHD – 21 samples)**	**Group A2 (ACI – 21 samples)**
Sensitivity, %	85.7; ($a = 18$; $b = 3$)	80.95; ($a = 17$; $b = 5$)
Specificity, %	76.2; ($c = 16$; $d = 5$)	71.4; ($c = 15$; $d = 7$)
Accuracy, %	80.95	76.18
Methods	**Phase scanning**	
Parameters	**Group A1 (IHD – 21 samples)**	**Group A2 (ACI – 21 samples)**
Sensitivity, %	95.2; ($a = 20$; $b = 1$)	95.2; ($a = 20$; $b = 1$)
Specificity, %	95.2; ($c = 20$; $d = 1$)	85.7; ($c = 18$; $d = 3$)
Accuracy, %	95.2	90.45

Table 5.6 Operational characteristics of the diagnostic performance of polarimetry, Mueller-matrix mapping, and polarization–interference methods for Group B (geometric thickness: 200 μm).

Methods	Polarization mapping	
Parameters	**Group A1 (IHD – 21 samples)**	**Group A2 (ACI – 21 samples)**
Sensitivity, %,	71.4; ($a = 15$; $b = 6$)	66.7; ($a = 14$; $b = 7$)
Specificity, %,	61.9; ($c = 13$; $d = 8$)	52.4; ($c = 11$; $d = 10$)
Accuracy, %	66.7	59.55
Methods	**Mueller-matrix mapping**	
Parameters	**Group A1 (IHD – 21 samples)**	**Group A2 (ACI – 21 samples)**
Sensitivity, %	76.2; ($a = 16$; $b = 5$)	71.4 ($a = 15$; $b = 6$)
Specificity, %	71.4; ($c = 15$; $d = 6$)	66.7; ($c = 14$; $d = 7$)
Accuracy, %	73.8	69.05
Methods	**Phase scanning**	
Parameters	**Group A1 (IHD – 21 samples)**	**Group A2 (ACI – 21 samples)**
Sensitivity, %	95.2; ($a = 20$; $b = 1$)	95.2; ($a = 19$; $b = 2$)
Specificity,%	85.7; ($c = 12$; $d = 2$)	85.7; ($c = 18$; $d = 3$)
Accuracy, %	90.45	90.45

Here, b and b denote the number of false-positive and false-negative results, respectively.

5.1.5 Conclusions

1. The theoretical foundations for the formation of local depolarization phase waves in diffuse laser speckle fields have been established, and the experimental confirmation has been achieved using a series of native birefringent histological sections of the myocardium with varying levels of integral depolarization.
2. A polarization–interference method for extracting and estimating the modulation depth of phase waves of local depolarization in diffusely scattered laser fields from histological sections of birefringent myocardial tissues has been developed and experimentally tested. Additionally, the parameters of phase waves of local depolarization in speckle fields of native myocardial histological sections, with depolarization capacities ranging from 8% to 82%, were experimentally investigated.
3. To evaluate the extent of necrotic changes in the optical anisotropy of diffuse myocardial layers, a new quantitative parameter has been introduced: the modulation depth of fluctuations in the amplitude of local depolarization waves.
4. A comparative study of the effectiveness of traditional techniques of polarization, Mueller-matrix mapping and polarization–interference phase scanning of scattered speckle fields in the differential diagnosis of IHD and ACI was conducted for myocardial samples with an average (group A – geometric thickness of 50 μm, integral degree of depolarization of 37%) and high (group B – geometric thickness of 200 μm, integral degree of depolarization of 83%) level of depolarized background.
5. Excellent (95.2%) and very good (90.45%) accuracy of differential diagnosis of IHD and ACI for myocardial samples from group A and group B, respectively, was demonstrated.

References

1 Durduran T, Choe R, Baker WB, Yodh AG. Diffuse optics for tissue monitoring and tomography. Rep Prog Phys. 2010;73(7):076701.

2 Ghosh N, Vitkin I. Tissue polarimetry: concepts, challenges, applications, and outlook. J Biomed Opt. 2011;16(11):110801.

3 Jacques S. Polarized light imaging of biological tissues. In Boas DA, Pitris C, Ramanujam N, editors. Handbook of Biomedical Optics. Boca Raton, Florida: CRC Press; 2011. pp. 649–669.

4 Ushenko AG, Pishak VP. Laser Polarimetry of Biological Tissues: Principles and Applications. In Tuchin VV, editor. Handbook of Coherent Domain Optical Methods. US: Springer; 2004. pp. 93–138.
5 Ghosh N, Wood M, Vitkin A. Polarized light assessment of complex turbid media such as biological tissues using Mueller matrix decomposition. In Tuchin VV, editor. Handbook of Photonics for Biomedical Science. Boca Raton: CRC Press; 2010. pp. 253–282.
6 Layden D, Ghosh N, Vitkin IA. Quantitative polarimetry for tissue characterization and diagnosis. In Wang RK, Tuchin VV, editors. Advanced Biophotonics: Tissue Optical Sectioning. Boca Raton: CRC Press; 2013. pp. 73–108.
7 Vitkin A, Ghosh N, de Martino A. Tissue polarimetry. In Andrews DL, editor. Photonics: Scientific Foundations, Technology and Applications. John Wiley & Sons, Ltd.; 2015. pp. 239–321.
8 de Boer JF, Milner TE. Review of polarization sensitive optical coherence tomography and Stokes vector determination. J Biomed Opt. 2002;7(3):359–371.
9 Ku S, Mahato KK, Mazumder N. Polarization-resolved Stokes-Mueller imaging: a review of technology and applications. Lasers Med Sci. 2019;34(7):1283–1293.
10 Willemse J, Gräfe MGO, Verbraak FD, de Boer JF. In vivo 3D determination of peripapillary scleral and retinal layer architecture using polarization-sensitive optical coherence tomography. Transl Vis Sci Technol. 2021;9(11):21.
11 Baumann B, Augustin M, Lichtenegger A, Harper D, Muck M, Eugui P, et al. Polarization-sensitive optical coherence tomography imaging of the anterior mouse eye. J Biomed Opt. 2018;23(8):1–12.
12 Hariri LP, Adams DC, Applegate MB, Miller AJ, Roop BW, Villiger M, et al. Distinguishing tumor from associated fibrosis to increase diagnostic biopsy yield with polarization-sensitive optical coherence tomography. Clin Cancer Res. 2019;25(17):5242–5249.
13 Tao K, Sun K, Ding Z, Ma Y, Kuang H, Zhao H, et al. Catheter-based polarization sensitive optical coherence tomography using similar Mueller matrix method. IEEE Trans Biomed Eng. 2020;67(1):60–68.
14 Angelsky PO, Ushenko AG, Dubolazov AV, Sidor MI, Bodnar GB, Koval G, et al. The singular approach for processing polarization-inhomogeneous laser images of blood plasma layers. J Opt. 2013;15(4):044030.
15 Ushenko AG, Dubolazov AV, Ushenko VA, Novakovskaya OY. Statistical analysis of polarization-inhomogeneous Fourier spectra of laser radiation scattered by human skin in the tasks of differentiation of benign and malignant formations. J Biomed Opt. 2016;21(7):071110.
16 Garazdyuk MS, Bachinskyi VT, Vanchulyak OY, Ushenko AG, Dubolazov OV, Gorsky MP. Polarization-phase images of liquor polycrystalline films in determining time of death. Appl Opt. 2016;55(12):B67–B71.
17 Angelskii OV, Ushenko OG, Burkovets DN, Arkhelyuk OD, Ushenko YA. Polarization-correlation studies of multifractal structures in biotissues and diagnostics of their pathologic changes. Laser Phys. 2000;10(5):1136–1142.
18 Yermolenko S, Ushenko A, Ivashko P, Goudail F, Gruia I, Gavrilă C, et al. Spectropolarimetry of cancer change of biotissues. In Ninth International Conference on Correlation Optics. SPIE. 2009;7388:404–410.
19 Jóźwicki R, Patorski K, Angelsky OV, Ushenko AG, Burkovets DN, Ushenko YA. Automatic polarimetric system for early medical diagnosis by biotissue testing. Opt Appl. 2002;32(4):603–612.
20 Ushenko AG, Dubolazov OV, Ushenko VA, Novakovskaya OY, Olar OV. Fourier polarimetry of human skin in the tasks of differentiation of benign and malignant formations. Appl Opt. 2016;55(12), B56–B60.
21 Ushenko AG, Pishak VP. Coherent-domain Optical Methods: Biomedical Diagnostics, Environmental and Material Science. Laser Polarimetry of Biological Tissue: Principles and Applications. Boston: Kluwer Academic Publishers; 2004. pp. 67–93.
22 Ushenko AG, Fediv AI, Marchuk YF. Correlation and fractal structure of Jones matrices of human bile secret. In European Conference on Biomedical Optics. 2009;7368:73681Q.
23 Ushenko AG. Depolarization of a laser emission field as the correlation development of its polarization structure. In International Conference on Correlation Optics. 1997;3317:331–339.
24 Bachinskyi V, Boychuk T, Ushenko A. Laser Polarimetry of Biological Tissues and Fluids: Chapter 4. Methods of Spatial-frequency, Singular and Correlation Analysis of Biological Layers Object Fields. LAP Lambert Academic Publishing; 2018.
25 Ushenko AG, Burkovets DN, Yermolenko SB, Arkhelyuk AD, Pishak VP, Yuzko AM, et al. Stokes polarimetry of biotissues. In Fourth International Conference on Correlation Optics. SPIE. 1999;3904:527–533.
26 Ushenko VA, Dubolazov AV, Pidkamin LY, Sakchnovsky MY, Bodnar AB, Ushenko YA, et al. Mapping of polycrystalline films of biological fluids utilizing the Jones-matrix formalism. Laser Phys. 2018;28(2):025602.
27 Ushenko AG, Angelsky PO, Sidor M. Marchuk YF, Andreychuk DR, Pashkovskaya NV. Spatial-frequency selection of complex degree of coherence of laser images of blood plasma in diagnostics and differentiation of pathological states of human organism of various nosology. Appl Opt. 2014;53(10):B172–B180.

28 Olar EI, Ushenko AG, Ushenko YA. Polarization correlation measurements of the phase tomograms of optically anisotropic biofractals. Laser Phys. 2004;14(8):1115–1121.

29 Guminetskiy SH, Ushenko OG, Polyanskiy IP, Motrych AV, Grynchuk FV. The optical method for investigation of the peritonitis progressing process. In Eighth International Conference on Correlation Optics. 2008;7008:626–632.

30 Ushenko AG, Fediv AI, Marchuk YF. Singular structure of polarization images of bile secret in diagnostics of human physiological state. In: European Conference on Biomedical Optics. Optica Publishing Group; 2009.

31 Ushenko Y, Ushenko A, Dubolazov A, Soltys I, Litvinenko O, Wanchuliak O, et al. Mueller-matrix interferometric multifractal scaling of optically anisotropic architectonics of diffuse blood facies: fundamental and applied aspects. J Biophotonics. 2025;18(3):e202400412. https://doi.org/10.1002/jbio.202400412.

32 Ushenko A, Soltys I, Dubolazov A, Ushenko Y, Bilookyi V, Bilookyi O, et al. 3D Jones-matrix thesiography of biological fluid facies. J Innov Opt Health Sci. 2025;18(2):2443002. https://doi.org/10.1142/S1793545824430028.

33 Ushenko A, Pavlyukovich N, Khukhlina O, Pavlyukovich O, Gorsky M, Soltys I, et al. Layer-by-layer multifractal scanning of optically anisotropic architectonics of blood plasma films: fundamental and applied aspects. Photonics. 2025;12:215. https://doi.org/10.3390/photonics12030215.

34 Ushenko A, Pavlyukovich N, Khukhlina O, Pavlyukovich O, Soltys I, Dubolazov A, et al. Blood plasma film multifractal scanning in COVID-19 consequences diagnostics. J Biophotonics. 2024;17(11):e202400356. https://doi.org/10.1002/jbio.202400356.

35 Ushenko A, Dubolazov A, Zheng J, Litvinenko A, Gorsky M, Ushenko Y, et al. 3D polarization-interference holographic histology for wavelet-based differentiation of biological tissues with different necrotic states. J Biomed Opt. 2024;29(5):052920. https://doi.org/10.1117/1.JBO.29.5.052920.

36 Ushenko O, Bilookyi O, Zheng J, Dubolazov A, Olar O, Ushenko Y, et al. 3D digital holographic polarimetry of laser speckle fields formed by polycrystalline blood films: a tool for differential diagnosis of thyroid pathology. Front Phys. 2024;12:1426469. https://doi.org/10.3389/fphy.2024.1426469.

37 Wójcik W, Hu Z, Ushenko Y, Smolarz A, Soltys I, Dubolazov O, et al. Optical sensor system for 3D Jones matrix reconstruction of optical anisotropy maps of self-assembled polycrystalline soft matter films. Sensors. 2024;24(5):1589. https://doi.org/10.3390/s24051589.

38 Ushenko A, Zheng J, Gorsky M, Dubolazov A, Ushenko Y, Soltys I, et al. 3D digital holographic polarimetry of diffuse optically anisotropic biological tissue object fields. Front Phys. 2023;11:1288935. https://doi.org/10.3389/fphy.2023.1288935.

39 Dubolazov A, Ushenko V, Trifonyuk L, Stashkevich A, Soltys I, Ushenko Y, et al. Polarization-singular approach to imaging Mueller-matrix polarimetry in differential diagnosis of histological sections of biopsy of tumors of the uterus and prostate. Front Phys. 2021;9:711212. https://doi.org/10.3389/fphy.2021.711212.

40 Ushenko VO, Trifonyuk L, Ushenko YA, Dubolazov OV, Gorsky MP, Ushenko AG. Polarization singularity analysis of Mueller-matrix invariants of optical anisotropy of biological tissue samples in cancer diagnostics. J Opt. 2021;23(6):064004. https://doi.org/10.1088/2040-8986/abf97a.

41 Ushenko YA, Koval GD, Ushenko AG, Dubolazov OV, Ushenko VA, Novakovskaia OY. Mueller-matrix of laser-induced autofluorescence of polycrystalline films of dried peritoneal fluid in diagnostics of endometriosis. J Biomed Opt. 2016;21(7):071116. https://doi.org/10.1117/1.JBO.21.7.071116.

6

Differential Mueller-matrix Mapping of Polycrystalline Tissue Components

Yurii Ushenko[1,2], Weidong Dou[1], Oleksandr Ushenko[2,3], Oleksandr Dubolazov[2], Iryna Soltys[2], Viktor Bachinskiy[4], Oleh Vanchuliak[4], Marta Garazdyuk[4], Nataliia Pavlyukovich[4], Oleksandr Pavlyukovich[4], Vitaliy Rozhko[4], and Oleksandr Olar[2]

[1] *Shaoxing University, Shaoxing, People's Republic of China*
[2] *Yurii Fedkovych Chernivtsi National University, Chernivtsi, Ukraine*
[3] *Taizhou Institute of Zhejiang University, Taizhou, People's Republic of China*
[4] *Bukovinian State Medical University, Chernivtsi, Ukraine*

6.1 Differential Mueller-matrix Imaging of Polycrystalline Networks in Partially Depolarizing Biological Tissues

A method of differential Mueller-matrix mapping for reconstructing the distributions of linear and circular birefringence and dichroism parameters in partially depolarizing layers of biological tissues with different morphological structures is proposed and substantiated [1–9]. Coordinate distributions of the first-order differential Mueller-matrix elements were obtained for histological sections of the brain tissue containing a spatially structured, optically anisotropic fibrillar network, as well as for parenchymatous tissue of the rectal wall characterized by an "islet" polycrystalline structure. Statistical analysis of polarization-reconstructed distributions of the averaged phase and amplitude-anisotropy parameters demonstrated a pronounced sensitivity of the third- and fourth-order statistical moments to changes in the polycrystalline structure of partially depolarizing biological layers [10–14].

6.1.1 Introduction

Among the various approaches to optical diagnostics of biological tissues, Mueller-matrix polarimetry (MMP) occupies an important position. This method provides comprehensive information about the polarization manifestations of the optical properties of biological tissues. Traditionally, MMP techniques and instrumentation have been developed within two limiting approximations [15–19].

The first approximation focuses on identifying relationships between the angular (indicatrix-based) and coordinate (Mueller-matrix images, MMI) distributions of the Mueller-matrix elements and the structure of optically thick, strongly depolarizing biological tissues, for which the depolarization degree approaches $\Lambda \to 100\%$ [20–25].

The second approximation is MMP of optically thin, nondepolarizing ($\Lambda \to 0\%$) biological layers, followed by reconstruction of phase and amplitude-anisotropy parameters [26–32].

However, most biological tissues are partially depolarizing, with depolarization degrees in the range ($0\% < \Lambda < 100\%$). Therefore, further development and generalization of MMP techniques applicable to polycrystalline biological layers with varying levels of multiple scattering and differing depolarization abilities are essential.

6.1.2 Brief Description of Research Facilities

The selection of the studied objects was based on both fundamental and applied considerations.

6.1.2.1 Fundamental Considerations

Across the diversity of biological tissues, the morphological structure consistently contains optically anisotropic components. Two limiting structural cases can be distinguished within optically anisotropic biological layers.

6.1.2.2 First Case: Spatially Structured Fibrillar Protein Networks

These networks consist of proteins, such as collagen, elastin, and myosin, characterized by a distribution of optical-axis orientations across individual fibrils. Due to the presence of long-range order, the mechanism of linear birefringence dominates and significantly exceeds the contribution of circular birefringence arising from the chirality of amino acids (LB, LB′ $\gg$ CB).

6.1.2.3 Second Case: Parenchymatous or "Islet" Optically Anisotropic Structures

In these formations, long-range fibrillar order is absent. Their morphological structure is predominantly defined by spatially disordered polypeptide chains formed by optically active molecules exhibiting circular birefringence. In this case, the contribution of linear birefringence is weaker or comparable to circular birefringence (LB, LB′ $\leq$ CB).

Figure 6.1 presents microscopic images (×4) of the biological-layer samples obtained using polarization visualization in a crossed polarizer–analyzer configuration. Shown are partially depolarizing histological sections of the brain tissue, featuring a spatially structured fibrillar network (Figure 6.1a), and sections of rectal-wall tissue, characterized by "islet-like" parenchymatous structures (Figure 6.1b).

The polarization images of histological sections of biological tissues shown in Figure 6.1 illustrate the diversity of topographic and geometric structures within optically anisotropic formations. This diversity enables the identification of the characteristic features and regularities of the optical manifestations of phase and amplitude anisotropy in partially depolarizing layers of biological tissues.

6.1.3 Discussion and Analysis of Experimental Data

6.1.3.1 First-order Differential Matrix of Spatially Structured Optically Anisotropic Fibrillar Networks

The results of study of the two-dimensional structure of the elements $((\langle f_{12}\rangle = \langle f_{21}\rangle)$; $(\langle f_{13}\rangle = \langle f_{31}\rangle)$; $(\langle f_{14}\rangle = \langle f_{41}\rangle)$; $(\langle f_{23}\rangle = -\langle f_{32}\rangle)$; $(\langle f_{24}\rangle = -\langle f_{42}\rangle)$; $(\langle f_{34}\rangle = -\langle f_{43}\rangle))$ of the first-order differential matrix of the histological section of brain tissue (geometrical thickness $z = 60\,\mu$m; attenuation coefficient $\tau = 0.21$; degree of depolarization $\Lambda = 43\%$) are presented as a series of distributions, maps and histograms m_{ik} values, shown in Figures 6.2 and 6.3

The obtained data show that all off-diagonal elements of the first-order differential Mueller matrix – excluding the diagonal components $\langle f_{11}\rangle, \langle f_{22}\rangle, \langle f_{33}\rangle, \langle f_{44}\rangle$ – exhibit nonzero values that characterize the phase and amplitude anisotropy of the brain-tissue layer (see equations (5.2) and (5.10)–(5.15)). A good correspondence was found between the theoretical prediction of symmetry (equation (5.2)) and the experimental results (Figure 6.2) for the first-order differential Mueller matrix of a partially depolarizing biological layer: $(\langle f_{12}\rangle = \langle f_{21}\rangle)$; $(\langle f_{13}\rangle = \langle f_{31}\rangle)$; $(\langle f_{14}\rangle = \langle f_{41}\rangle)$; $(\langle f_{23}\rangle = -\langle f_{32}\rangle)$; $(\langle f_{24}\rangle = -\langle f_{42}\rangle)$; $(\langle f_{34}\rangle = -\langle f_{43}\rangle))$.

In other words, the results of the experimental measurements of the six independent elements of this matrix operator make it possible to directly analyze the average parameters of optical anisotropy of a multiple-scattering biological layer: $\left(\text{LD} = \langle f_{12;21}\rangle;\ \text{CD} = \langle f_{13;31}\rangle;\ \text{LD}' = \langle f_{14;41}\rangle;\ \text{LB} = \langle f_{34;43}\rangle;\ \text{CB} = \langle f_{23;32}\rangle;\ \text{LB}' = \langle f_{24;42}\rangle\right)$.

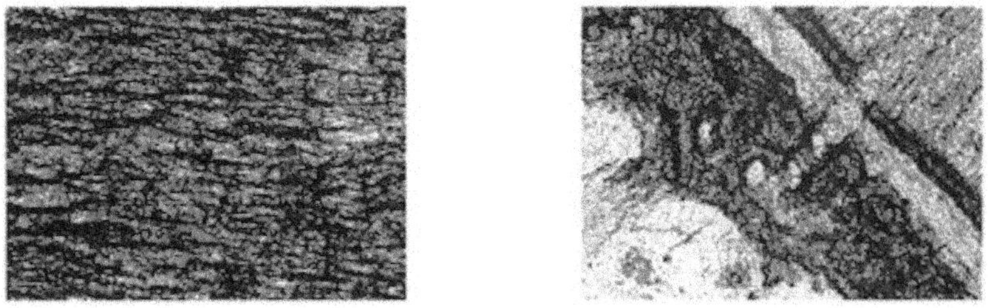

(a) (b)

Figure 6.1 Polarization-visualized images of optically anisotropic structures in histological sections of brain tissue (a) and rectal-wall tissue (b).

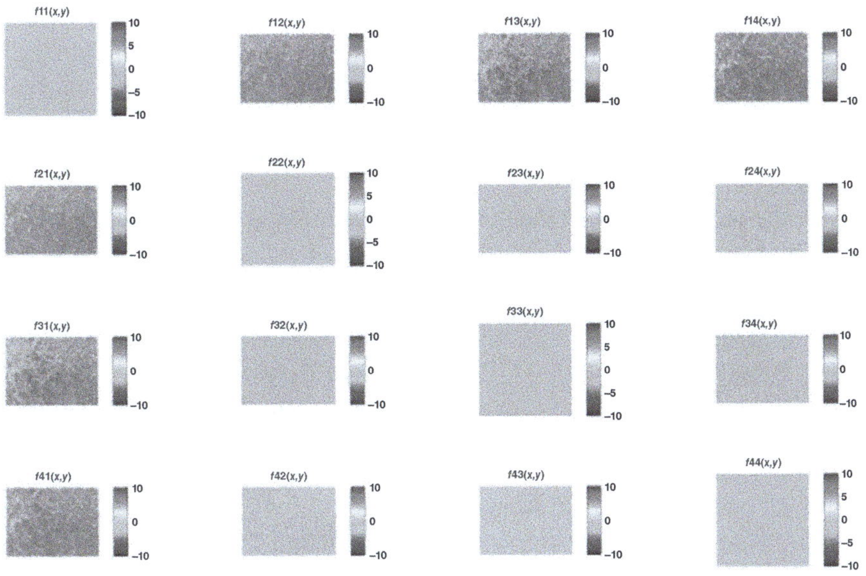

Figure 6.2 Maps of elements of the first-order differential Mueller matrix of a histological section of brain tissue ($z = 60\,\mu\text{m}$; $\tau = 0.21$; $\Lambda = 43\%$).

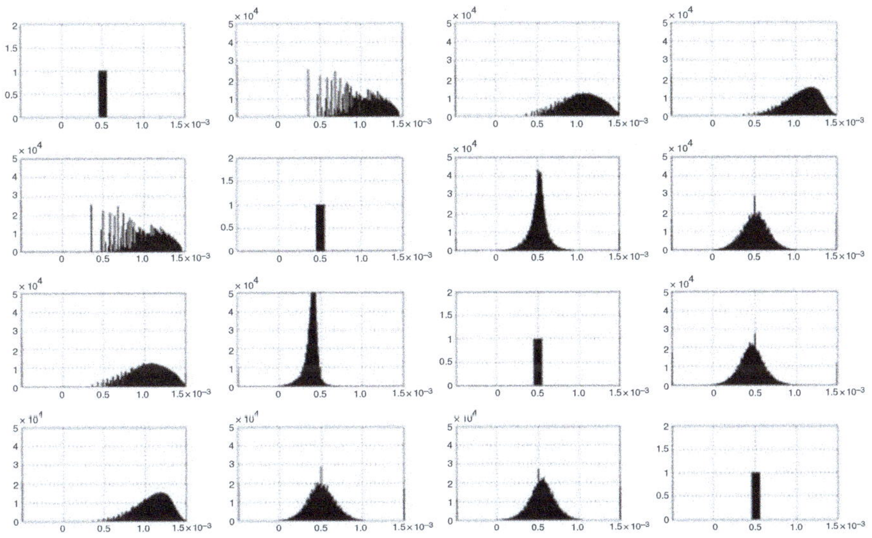

Figure 6.3 Histograms of distribution of values of the first-order differential matrix elements of a histological section of the brain ($z = 60\,\mu\text{m}$; $\tau = 0.21$; $\Lambda = 43\%$).

Figure 6.3 demonstrates that the histograms $N_{j=1-6}(\langle f_{ik}\rangle)$ are characterized by a wide spread in the magnitudes of all off-diagonal elements $\langle f_{i\neq k}\rangle$ of the first-order differential Mueller matrix. This indicates a complex polycrystalline structure of the biological tissue, with distinct manifestations of multiple birefringence and dichroism mechanisms.

A comparative analysis of these manifestations shows that the histograms of the linear and circular dichroism parameters exhibit asymmetric distributions. In contrast, the histograms for the linear and circular birefringence parameters exhibit symmetric, approximately bell-shaped distributions (Figure 6.3).

From a physical standpoint, this difference can be attributed to the unequal multiplicity of "non-absorbing" interactions (phase anisotropy, or "q-acts") and "absorbing" interactions (amplitude anisotropy, or "k-acts") of the laser radiation with optically anisotropic structures of brain tissue, where $q \gg k$. This inequality also reflects the fact that, in the red spectral range, the probability of absorption is much lower than the probability of Fresnel-type transformations of the laser field by birefringent refractive collagen fibrils formed by optically active protein molecules.

Considering these factors, and in accordance with the central limit theorem, the average values of the linear $(\langle f_{24;42}\rangle, \langle f_{34;43}\rangle)$ and circular $(\langle f_{23;32}\rangle)$ birefringence parameters appear to follow nearly normal distributions. In contrast, the values of the linear $(\langle f_{12;21}\rangle, \langle f_{14;41}\rangle)$ and circular $(\langle f_{13}, f_{31}\rangle)$ dichroism parameters exhibit asymmetric distributions, as shown in Figure 6.3.

The quantitative parameters characterizing the distributions of the first-order differential Mueller-matrix elements for the partially depolarizing brain-tissue layer are presented in Table 6.1.

Analysis of these statistical data indicates that the differing probabilities of absorption events and phase-shift formation between the orthogonal components of the laser amplitude within brain tissue lead to significant differences – up to an order of magnitude – between the first-through fourth-order statistical moments describing the distributions of the birefringence parameters LB; CB; LB′ and the dichroism parameters LD; CD; LD′, as summarized in Table 6.1.

6.1.3.2 First-order Differential Matrices of "Islet" Polycrystalline Structures in Parenchymatous Tissues

Figures 6.4 and 6.5 present the experimentally obtained two-dimensional maps and histograms of the distributions of the first-order differential Mueller-matrix elements for a partially depolarizing biological-tissue layer with an "islet" polycrystalline structure, represented by a histological section of the rectal wall.

The observed results can be attributed to the high multiplicity of light scattering by phase-shifting fibrillar networks and "islet-like" formations. Owing to this repeated averaging, the spatial–geometric structure of tissue samples of both types becomes substantially smoothed. This is evidenced by the close values of the first- and second-order statistical moments characterizing the histograms of the anisotropy parameters: $\mathrm{LD} = \langle f_{12;21}\rangle$; $\mathrm{CD} = \langle f_{13;31}\rangle$; $\mathrm{LD}' = \langle f_{14;41}\rangle$; $\mathrm{LB} = \langle f_{34;43}\rangle$; $\mathrm{CB} = \langle f_{23;32}\rangle$; $\mathrm{LB}' = \langle f_{24;42}\rangle$. However, for the third- and fourth-order statistical moments – which describe the distributions of the linear $(\langle f_{12;21}\rangle, \langle f_{13;31}\rangle)$ and circular $(\langle f_{14}, f_{41}\rangle)$ dichroism parameters – significant differences are observed, as summarized in Table 6.2.

Table 6.1 Statistical moments of the first to fourth orders characterizing the distributions of the elements of the first differential Mueller matrix for a histological section of brain tissue ($z = 60\,\mu\mathrm{m}$; $\tau = 0.21$; $\Lambda = 43\%$).

M_i	$\langle f_{12;21}\rangle$	$\langle f_{13;31}\rangle$	$\langle f_{14;41}\rangle$	$\langle f_{23;32}\rangle$	$\langle f_{24;42}\rangle$	$\langle f_{34;43}\rangle$
M_1	3.21	5.12	6.37	0.12	0.11	0.09
M_2	4.73	4.22	3.92	0.21	0.24	0.21
M_3	1.86	0.62	1.65	0.33	0.16	0.12
M_4	0.95	0.53	0.77	0.57	0.21	0.33

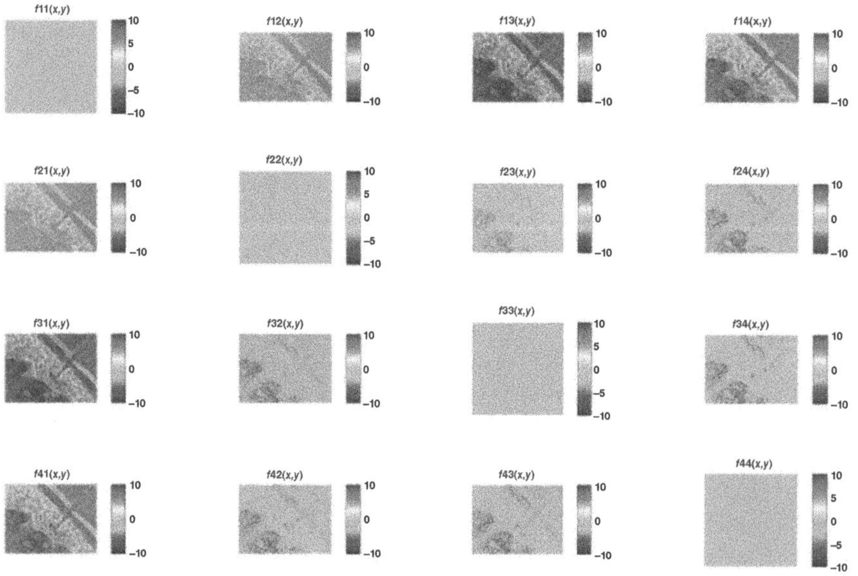

Figure 6.4 Maps of the first differential Mueller-matrix elements for a histological section of rectal-wall tissue ($z = 60\,\mu\mathrm{m}$; $\tau = 0.32$; $\Lambda = 54\%$).

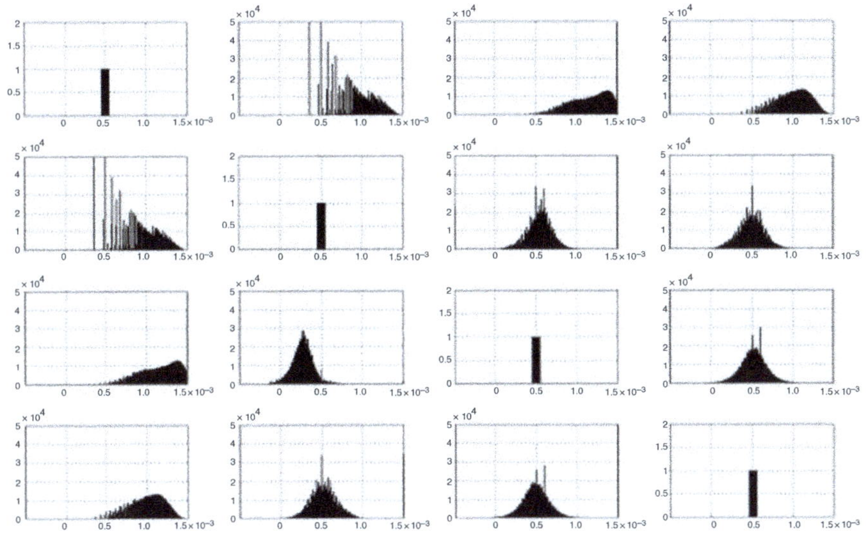

Figure 6.5 Histograms of the distributions of the first differential Mueller-matrix elements for a histological section of rectal-wall tissue ($z = 60\,\mu$m; $\tau = 0.32$; $\Lambda = 54\%$).

Table 6.2 Statistical moments of the first and fourth orders characterizing the distributions of the first-order differential Mueller-matrix elements for a histological section of rectal-wall tissue ($z = 60\,\mu$m; $\tau = 0.32$; $\Lambda = 54\%$) histological section.

M_i	$\langle f_{12;21}\rangle$	$\langle f_{13;31}\rangle$	$\langle f_{14;41}\rangle$	$\langle f_{23;32}\rangle$	$\langle f_{24;42}\rangle$	$\langle f_{34;43}\rangle$
M_1	1.98	6.19	5.59	0.14	0.13	0.14
M_2	3.28	3.94	2.15	0.28	0.21	0.26
M_3	4.63	2.92	3.52	0.21	0.29	0.17
M_4	3.37	3.14	2.63	0.38	0.38	0.21

A comparative analysis of the structures of the first-order differential Mueller matrices of partially depolarizing rectal-wall tissue (Figures 6.4 and 6.5) and brain tissue (Figures 6.2 and 6.3) revealed statistically similar manifestations of the optical-anisotropy mechanisms associated with fibrillar networks and the ensemble of spatially nonstructured "islet" protein chains. In particular, the distributions of the linear and circular dichroism parameters remain asymmetric. In contrast, the distributions of the linear and circular birefringence parameters exhibit symmetric, bell-shaped profiles (Figure 6.5).

For a sample of brain tissue, the probability of absorption is lower compared with rectal-wall tissue. Consequently, the histograms of the distributions of the elements $\langle f_{12;21;13;31;14;41}\rangle$ for this type of sample are more asymmetric and exhibit sharper peaks (Figure 6.6).

As a result, the differences between the higher-order statistical moments of the distributions of the differential Mueller-matrix elements for spatially structured versus parenchymatous tissues reach factors of three to four.

6.1.3.3 Conclusions

A differential Mueller-matrix mapping method for reconstructing spatial distributions of linear and circular birefringence and dichroism in partially depolarizing layers of biological tissues with different morphological structures has been proposed and substantiated.

Coordinate distributions of the first-order differential Mueller-matrix elements were obtained for histological sections of brain tissue ($z = 60\,\mu$m; $\tau = 0.21$; $\Lambda = 43\%$) with a spatially structured, optically anisotropic fibrillar network, as well as for parenchymatous rectal-wall tissue ($z = 60\,\mu$m; $\tau = 0.32$; $\Lambda = 54\%$) characterized by an "islet" polycrystalline structure.

Statistical analysis of the polarization-reconstructed distributions of the averaged phase and amplitude-anisotropy parameters demonstrated significant sensitivity of the third- and fourth-order statistical moments to changes in the polycrystalline architecture of partially depolarizing biological tissues.

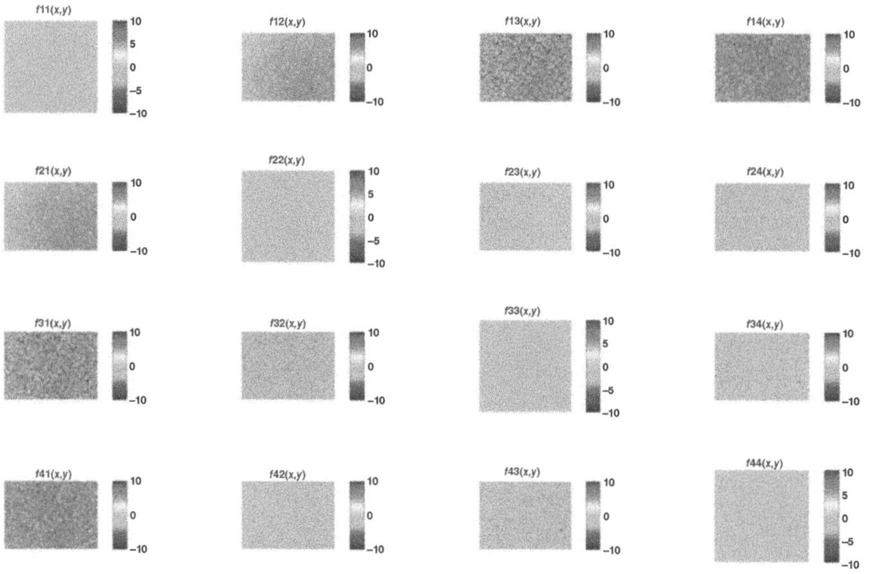

Figure 6.6 Maps of the first-order differential Mueller-matrix elements of a polycrystalline whole-blood film from a healthy donor ($\tau = 0.42$; $\Lambda = 34\%$).

6.2 Mueller-matrix Reconstruction of Optical-anisotropy Parameters of Partially Depolarizing of Polycrystalline Films of Blood in Cancer Diagnostics

6.2.1 Introduction

In recent years, a new conceptual basis has emerged in the field of biomedical optics for a well-established diagnostic modality – MMP of biological tissues. Traditionally, this approach has been applied primarily to multiply scattering, strongly depolarizing biological tissues. Diagnostic information is obtained by identifying relationships between the spatial distributions of Mueller-matrix elements (MMI) and the underlying optical properties of biological media, such as scattering and attenuation coefficients, as well as linear and circular birefringence and dichroism [33–40].

However, a substantial number of biological layers exhibit low depolarization while still powerfully transforming the polarization state of probing radiation. These include eye tissues, cellular monolayers, mucous membranes, and optically thin polycrystalline films of various biological fluids. Their characterization is typically performed using laser polarimetry methods formulated within the Mueller-matrix framework. At the same time, most biological objects in practical diagnostics are only partially depolarizing. This creates a clear need for further development and generalization of Mueller-matrix mapping techniques applicable to polycrystalline structures of such intermediate optical behavior.

This section explores the capabilities of Mueller-matrix reconstruction for partially depolarizing blood films, to identify objective criteria for differential diagnosis of breast cancer [41–46].

6.2.2 Discussion and Analysis of Experimental Data

The results presented in this section demonstrate the diagnostic potential for differentiating depolarizing, optically turbid polycrystalline films of whole blood obtained from healthy donors (36 samples) and patients with breast cancer (36 samples). The samples fall within the range of attenuation coefficients $\tau = 0.41 \div 0.48$ and depolarization degrees $\Lambda = 33\% \div 39\%$.

Although most contemporary studies focus on optically thin blood plasma films from patients with various pathologies, the practical significance of whole-blood samples remains substantial. Whole blood is an intrinsically more complex, optically anisotropic medium than plasma. In addition to the protein (albumin–globulin) fraction, it contains an ensemble of absorbing and circularly birefringent cellular elements, including erythrocytes, platelets, and monocytes. This contributes to a richer set of optical-anisotropy mechanisms and enhances the information content of Mueller-matrix measurements.

The two-dimensional coordinate distributions of the first-order differential Mueller-matrix elements for polycrystalline films of both groups of whole-blood samples are shown in Figures 6.6 and 6.7. These figures illustrate the spatial structure

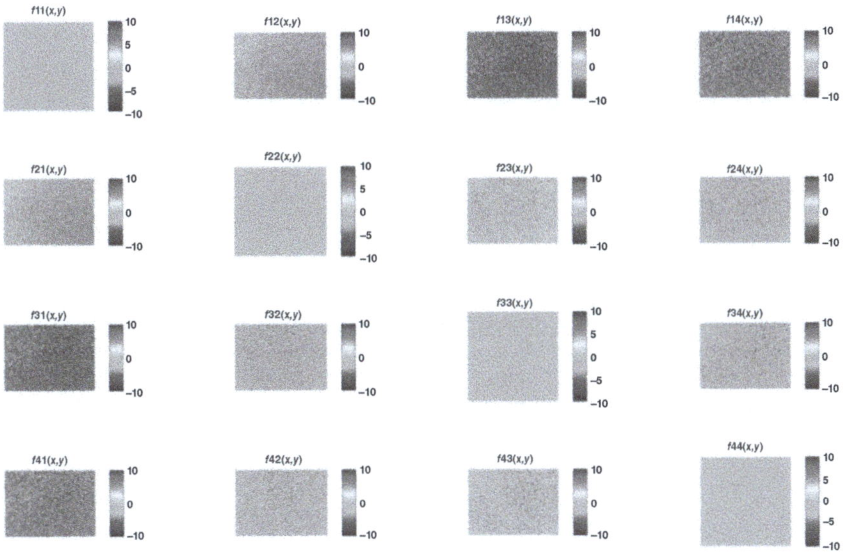

Figure 6.7 Maps of the first-order differential Mueller-matrix elements of a polycrystalline whole-blood film from a breast cancer patient ($\tau = 0.47$; $\Lambda = 39\%$).

and statistical behavior of the differential matrix elements that characterize the phase and amplitude anisotropy of partially depolarizing biological layers.

It follows from the obtained data that all nondiagonal elements of the first-order differential Mueller matrix (except $\langle f_{11;22;33;44}\rangle$), which characterize the phase and amplitude anisotropy of the polycrystalline blood films (equations (5.2), (5.10)–(5.15)), differ from zero for samples from both healthy donors and patients with breast cancer. A good correspondence is observed between the theoretical symmetry properties of the first-order differential matrix (Chapter 2, equations (2.50)–(2.76)) and the experimentally measured elements (Figures 6.6 and 6.7): $(\langle f_{12}\rangle = \langle f_{21}\rangle)$; $(\langle f_{13}\rangle = \langle f_{31}\rangle)$; $(\langle f_{14}\rangle = \langle f_{41}\rangle)$; $(\langle f_{23}\rangle = -\langle f_{32}\rangle)$; $(\langle f_{24}\rangle = -\langle f_{42}\rangle)$; $(\langle f_{34}\rangle = -\langle f_{43}\rangle)$.

Thus, the experimentally determined ensemble of six independent matrix elements enables direct evaluation of the average optical-anisotropy parameters of the samples: linear dichroism samples $LD = \langle f_{12;21}\rangle$, circular dichroism $CD = \langle f_{13;31}\rangle$, linear dichroism of the second type $LD' = \langle f_{14;41}\rangle$, linear birefringence $LB = \langle f_{34;43}\rangle$, circular birefringence $CB = \langle f_{23;32}\rangle$, and linear birefringence of the second type $LB' = \langle f_{24;42}\rangle$.

Figures 6.8 and 6.9 show the histograms of the distributions of birefringence and dichroism parameters for polycrystalline films of whole blood from healthy donors and breast cancer patients.

Tables 6.3 and 6.4 summarize the mean values, associated errors, and statistical moments of the first to fourth orders derived from these histograms for both groups. To assess the potential clinical applicability of differential Mueller-matrix mapping, the operational characteristics standard in evidence-based medicine were calculated for each group of whole-blood samples, quantifying the diagnostic performance of the method.

A comparative analysis of the obtained data revealed a pronounced increase in circular birefringence and a corresponding decrease in linear birefringence in whole-blood films from patients with breast cancer. This behavior can be attributed to an elevated concentration of circularly birefringent globulin molecules and an increased number of formed elements within the sample.

Quantitative assessment based on the first- to fourth-order statistical moments confirmed a substantial transformation of the polycrystalline structure of the samples in the breast cancer group toward enhanced circular birefringence. This is evidenced by a significant (2–10 times) increase in all statistical moments describing the distributions of the differential Mueller-matrix element $\langle\{f_{23;32}\}\rangle$, which reflects the magnitude of circular birefringence in optically anisotropic whole-blood films.

For the distributions of the other phase-anisotropy elements – $\langle\{f_{24;42}\}\rangle$ and $\langle\{f_{34;43}\}\rangle$, associated with linear birefringence – an opposite trend is observed: the statistical moments of the o first to fourth orders decrease by a factor of 2.5 to 11.

Such a pronounced intergroup contrast yields a highly informative diagnostic indicator, with the balanced accuracy of differentiating polycrystalline whole-blood films exceeding $Ac > 90\%$.

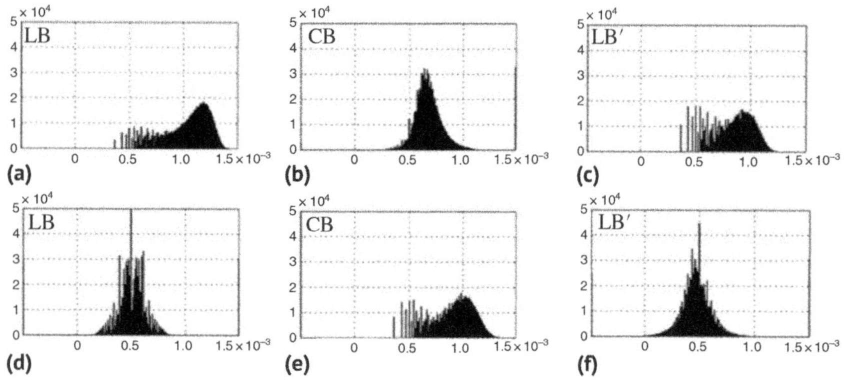

Figure 6.8 Histograms of the distributions of linear (a), (c), (d), (f) and circular (b), (e) birefringence values for polycrystalline films of whole blood from healthy donors ($\tau = 0.42$; $\Lambda = 34\%$; (a)–(c)) and breast cancer patients ($\tau = 0.47$; $\Lambda = 39\%$; (d)–(f)).

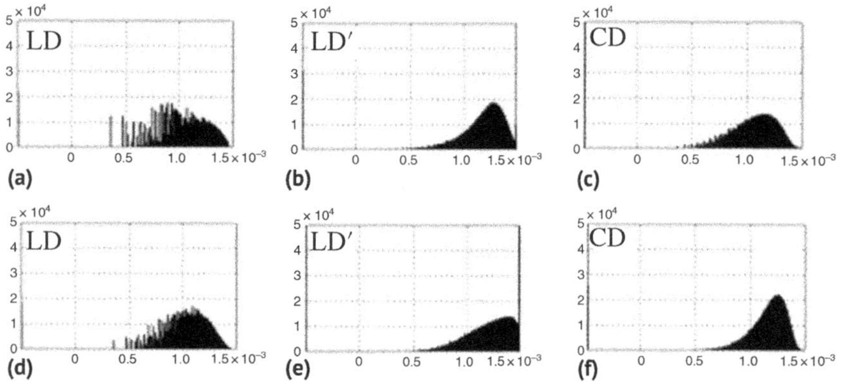

Figure 6.9 Histograms of the distributions of linear (a), (c), (d), (f), and circular (b), (e) dichroism values for polycrystalline films of whole blood from healthy donors ($\tau = 0.42$; $\Lambda = 34\%$, (a)–(c)) and patients with breast cancer ($\tau = 0.47$; $\Lambda = 39\%$, (d)–(f)).

Table 6.3 Statistical moments of the first to fourth orders characterizing the distributions of linear and circular birefringence values for polycrystalline whole-blood films from healthy donors and patients with breast cancer.

M_i	$\langle\{f_{23;32}\}\rangle$		$\langle\{f_{24;42}\}\rangle$		$\langle\{f_{34;43}\}\rangle$		Ac (%)
M_1	6.51 ± 0.44	0.53 ± 0.032	1.24 ± 0.077	6.11 ± 0.41	4.23 ± 0.29	0.87 ± 0.053	93–96
M_2	4.14 ± 0.28	1.87 ± 0.11	1.79 ± 0.11	3.86 ± 0.24	3.16 ± 0.16	1.56 ± 0.088	94–96
M_3	1.95 ± 0.13	0.24 ± 0.015	0.19 ± 0.011	1.84 ± 0.22	1.63 ± 0.09	0.25 ± 0.015	95–97
M_4	0.91 ± 0.053	0.18 ± 0.011	0.16 ± 0.09	0.79 ± 0.08	0.68 ± 0.04	0.14 ± 0.082	94–96

Table 6.4 Statistical moments of the first to fourth orders characterizing the distributions of linear and circular birefringence values for polycrystalline whole-blood films from healthy donors and breast cancer patients.

M_i	$\langle\{f_{12;21}\}\rangle$		$\langle\{f_{13;31}\}\rangle$		$\langle\{f_{14;41}\}\rangle$		*I* (%)
M_1	5.32 ± 0.33	5.81 ± 0.37	7.77 ± 0.47	8.13 ± 0.53	6.33 ± 0.45	7.41 ± 0.46	55–63
M_2	3.74 ± 0.28	3.32 ± 0.26	4.16 ± 0.32	2.47 ± 0.16	3.44 ± 0.21	2.29 ± 0.16	53–91
M_3	0.87 ± 0.055	0.69 ± 0.043	1.71 ± 0.098	1.22 ± 0.075	1.26 ± 0.078	0.78 ± 0.046	72–90
M_4	0.52 ± 0.029	0.31 ± 0.018	2.09 ± 0.12	1.14 ± 0.06	1.74 ± 0.099	0.92 ± 0.057	83–90

Table 6.5 Comparative diagnostic efficiency of direct and differential Mueller-matrix mapping applied to polycrystalline films of plasma and whole blood.

	Mueller-matrix mapping		
$M_{i=1;2;3;4}$	Blood plasma (%)	Whole blood (%)	Differential mapping (%)
M_1	81	77	95
M_2	82	71	90
M_3	84	72	92
M_4	83	74	94

6.2.3 Comparative Studies of the Diagnostic Efficiency of Direct and Differential Mueller-matrix Mapping of Polycrystalline Films of Plasma and Whole Blood

This section presents comparative results on the diagnostic efficiency of differentiating nondepolarizing polycrystalline plasma films and partially depolarizing whole-blood layers from healthy donors and patients with breast cancer using both direct and differential Mueller-matrix mapping [44–51]. The corresponding efficiency metrics are summarized in Table 6.5.

Balanced accuracy (Ac) for each method was determined separately for the statistical moments of the first to fourth orders $(M_{i=1;2;3;4})$ describing the distributions of linear and circular dichroism and birefringence parameters. The maximum value of $\mathrm{Ac}(M_{i=1;2;3;4})$ was selected as the primary diagnostic indicator.

6.2.4 Conclusions

1. A differential Mueller-matrix mapping method for reconstructing the spatial distributions of linear and circular birefringence and dichroism parameters in partially depolarizing polycrystalline blood films has been proposed and substantiated.
2. The coordinate distributions of the first-order differential matrix elements for polycrystalline films of whole blood from healthy donors ($\tau = 0.42$; $\Lambda = 34\%$) and patients with breast cancer ($\tau = 0.47$; $\Lambda = 39\%$) have been determined and analyzed within the statistical framework.
3. High diagnostic performance has been achieved: good quality $(\Delta Z_{2;3;4}(\mathrm{LD, CD, LD'}) \Rightarrow \mathrm{Ac} > 80\%)$ and excellent quality $(\Delta M_{1;2;3;4}(\mathrm{LB, CB, LB'}) \Rightarrow \mathrm{Ac} > 90\%)$ in the differentiation of breast cancer using Mueller-matrix mapping of partially depolarizing polycrystalline blood films.
4. A comparative assessment of the diagnostic efficiency of direct and differential Mueller-matrix mapping applied to polycrystalline films of plasma and whole blood has been performed, and the functional limitations of each approach have been identified.

6.3 Differential Mueller-matrix Diagnostics of Necrotic Changes in the Optical Anisotropy of Depolarizing Myocardial Tissues

6.3.1 Introduction

Postmortem diagnostics of acute coronary syndrome is currently a key direction in morphological and forensic investigations. The balanced accuracy of detecting ACI using conventional histochemical methods typically does not exceed 65–70% and requires a relatively long processing time. Therefore, the development of new, more accurate, and rapid diagnostic approaches based on modern optical–physical methodologies for the formation and analysis of microscopic images of myocardial histological sections is highly relevant [51–61].

This section addresses the potential of differential Mueller-matrix reconstruction for characterizing the polycrystalline structure of partially depolarizing histological sections of myocardium tissue from individuals who died of ischemic heart disease and acute coronary insufficiency (ACI).

6.3.2 Discussion and Analysis of Experimental Data

This study examines the diagnostic potential of differential Mueller-matrix mapping applied to the polarized polycrystalline component of optically thick myocardium layers (attenuation coefficient $\tau = 0.83 \div 0.91$) obtained from patients who died of coronary heart disease (CHD; 31 samples, group 1) and ACI (31 samples, group 2).

Histological sections of the myocardium were prepared using a standard cryosectioning technique with varying geometric thickness z.

From a morphological perspective, CHD is characterized by thinning and disorientation of the myosin fibrillar network. In contrast, ACI manifests as degradation of actin and myosin protein molecules while essentially preserving the spatial fibrillar scaffold. Optically, these conditions differ in the degree to which the mechanisms of linear (CHD) and circular (ACI) birefringence and dichroism are diminished.

The analysis of the two-dimensional distributions of the first-order differential Mueller-matrix elements for myocardium sections from CHD ($z = 60\,\mu$m; $\tau = 0.84$; $\Lambda = 44\%$) and ACI ($z = 60\,\mu$m; $\tau = 0.91$; $\Lambda = 52\%$) cases are presented in Figures 6.10 and 6.11.

The obtained data demonstrate that all nondiagonal elements $\langle\{f_{ik}\}\rangle$ of the first-order differential Mueller matrix – which characterize the amplitude and phase anisotropy of myocardium samples of both types – are nonzero. A good correspondence is observed between the theoretical predictions and the experimental results (Figures 6.10 and 6.11) regarding the symmetry equations of the first-order differential matrix for partially depolarizing myocardium sections from CHD and ACI cases: $(\langle f_{12}\rangle = \langle f_{21}\rangle)$; $(\langle f_{13}\rangle = \langle f_{31}\rangle)$; $(\langle f_{14}\rangle = \langle f_{41}\rangle)$; $(\langle f_{23}\rangle = -\langle f_{32}\rangle)$; $(\langle f_{24}\rangle = -\langle f_{42}\rangle)$; $(\langle \mathrm{f}_{34}\rangle = -\langle \mathrm{f}_{43}\rangle)$.

Thus, the experimentally measured ensemble of these six matrix elements enables direct analysis of the average optical-anisotropy parameters of multiply scattering myocardium tissue:

$$D = \langle f_{12;21}\rangle;\ \mathrm{CD} = \langle f_{13;31}\rangle;\ \mathrm{LD}' = \langle f_{14;41}\rangle;\ \mathrm{LB} = \langle f_{34;43}\rangle;\ \mathrm{CB} = \langle f_{23;32}\rangle;\ \mathrm{LB}' = \langle f_{24;42}\rangle.$$

Figures 6.12 and 6.13 show the series of histograms of distributions of the values of the averaged parameters of linear and circular birefringence (LB; CB; LB′ – Figure 6.12) and dichroism (LD; CD; LD′ – Figure 6.13) of myocardium histological sections of the patients who died of CHD and ACI.

A comparative analysis of the histograms of the distributions of phase-anisotropy parameters (Figure 6.12) and amplitude-anisotropy parameters (Figure 6.13) for myocardium layers of both groups confirms the previously stated conclusion: linear birefringence and linear dichroism decrease in CHD samples (Figures 6.12a, c, d, f and 6.13a, c, d, f), whereas circular dichroism and circular birefringence decrease in ACI samples (Figures 6.12a, b, e and 6.13a, b). These tendencies are reflected in the reduced range of variation and the diminished prominence of the main peaks in the corresponding histograms.

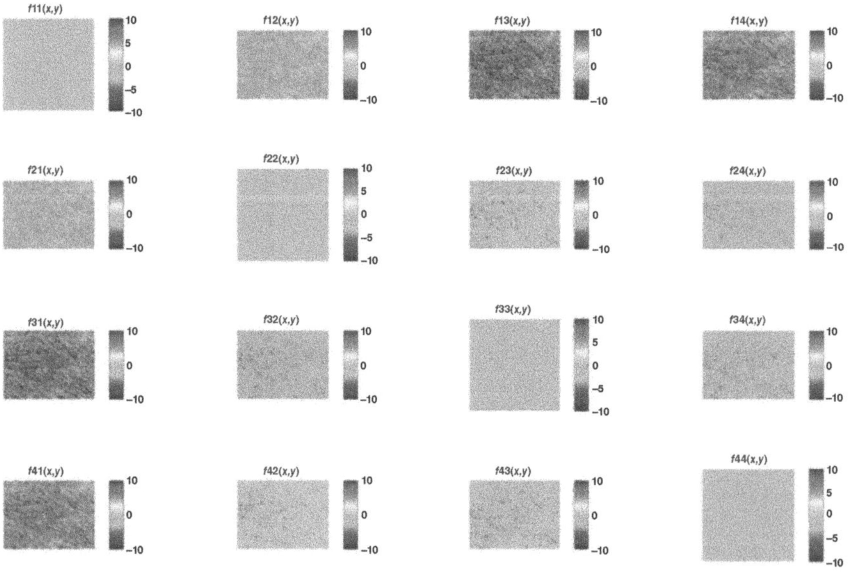

Figure 6.10 Maps of the first -order differential Mueller-matrix elements for a histological section of myocardium from a patient who died of CHD ($z = 60\,\mu$m; $\tau = 0.84$; $\Lambda = 44\%$).

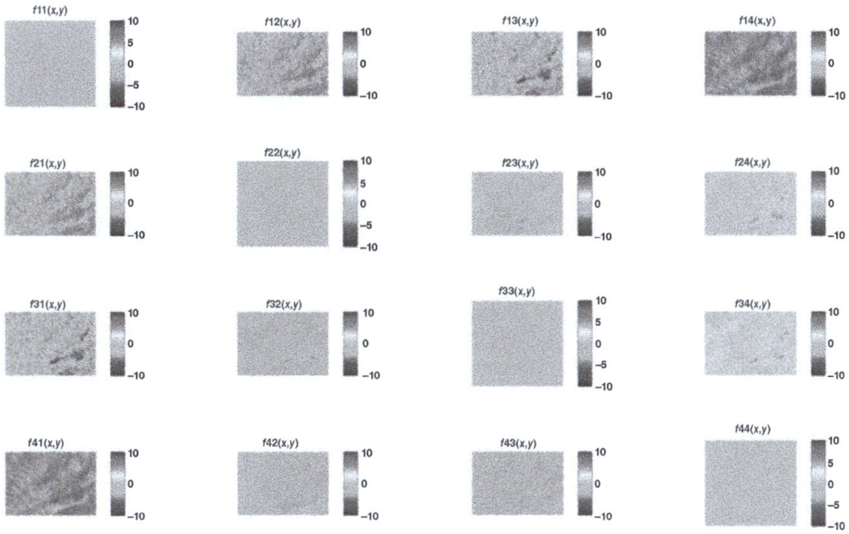

Figure 6.11 Maps of differential matrix elements of the first order of a histological section of myocardium of the patients who died of ACI ($z = 60\,\mu\text{m}$; $\tau = 0.91$; $\Lambda = 52\%$).

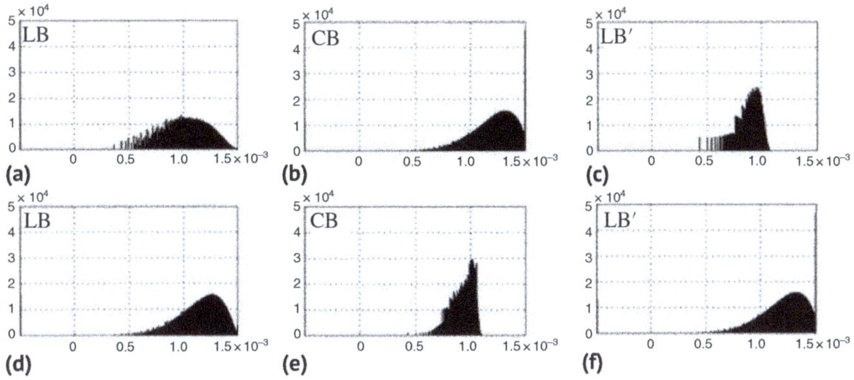

Figure 6.12 Histograms of value distribution of linear (a), (c), (d) and circular (b), (e) birefringence of myocardium histological sections of the patients who died of CHD (a)–(c) and ACI (d)–(f).

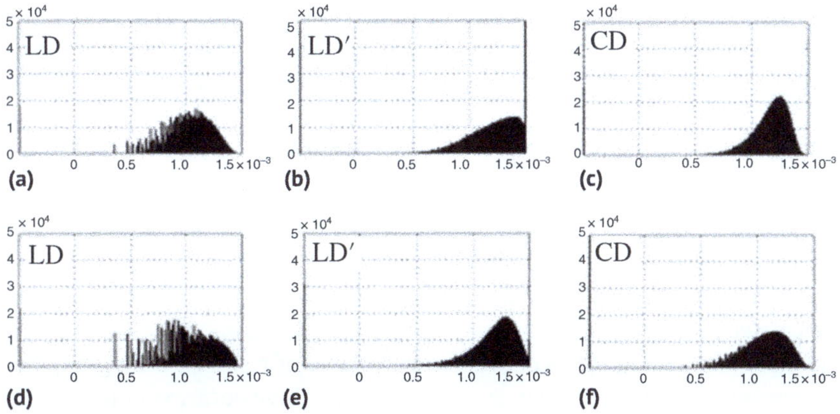

Figure 6.13 Histograms of value distribution of linear (a), (c), (d) and circular (b), (e) dichroism of myocardium histological sections of the patients who died of CHD (a)–(c) and ACI (d)–(f).

Table 6.6 Statistical moments of the first to fourth-orders characterizing distributions LB; CB; LB′ of myocardium histological sections of the patients who died of CHD and ACI.

M_i	$\langle\{f_{23;32}\}\rangle$		$\langle\{f_{24;42}\}\rangle$		$\langle\{f_{34;43}\}\rangle$	
M_1	5.75 ± 0.33	8.16 ± 0.57	3.27 ± 0.16	7.09 ± 0.47	3.57 ± 0.19	7.34 ± 0.49
M_2	4.23 ± 0.28	2.87 ± 0.16	1.93 ± 0.12	3.41 ± 0.18	1.84 ± 0.11	3.67 ± 0.21
M_3	0.62 ± 0.036	1.69 ± 0.11	2.12 ± 0.13	0.87 ± 0.052	1.92 ± 0.14	0.73 ± 0.044
M_4	0.41 ± 0.028	0.88 ± 0.054	1.96 ± 0.13	0.41 ± 0.025	1.71 ± 0.09	0.34 ± 0.019

Table 6.7 Statistical moments of the first to fourth orders characterizing distributions LD; CD; LD′ of myocardium histological sections of the patients who died of CHD and ACI.

M_i	$\langle\{f_{12;21}\}\rangle$		$\langle\{f_{13;31}\}\rangle$		$\langle\{f_{14;41}\}\rangle$	
M_1	4.12 ± 0.26	5.29 ± 0.33	7.86 ± 0.45	6.13 ± 0.39	6.21 ± 0.41	5.85 ± 0.37
M_2	6.22 ± 0.39	4.74 ± 0.27	4.15 ± 0.23	3.36 ± 0.18	3.41 ± 0.19	2.72 ± 0.15
M	0.85 ± 0.051	0.66 ± 0.038	1.96 ± 0.12	0.87 ± 0.052	1.14 ± 0.058	0.74 ± 0.043
M_4	0.74 ± 0.044	0.43 ± 0.027	0.82 ± 0.053	0.51 ± 0.031	0.72 ± 0.044	0.44 ± 0.029

Quantitatively, the structural transformations of optically thick myocardium layers are reflected in the statistical analysis of the two-dimensional distributions of the first-order differential matrix elements of the polarized component of the depolarizing polycrystalline networks.

Tables 6.6 and 6.7 present the mean values and standard deviation $(M_{i=1;2;3;4})$ of the statistical parameters characterizing the distributions of $\langle\{f_{ik}\}\rangle$ for myocardium sections from CHD and ACI groups.

For the potential clinical application of differential Mueller-matrix mapping, diagnostic performance was assessed using the operational characteristics standard in evidence-based medicine.

A comparative examination of the first- to fourth-order statistical moments identified a subset of parameters – highlighted in gray – that exhibit the highest sensitivity to changes in phase and amplitude anisotropy in myocardium layers under the depolarized background of multiply scattered radiation.

It is evident that the most sensitive and diagnostically accurate parameters – reaching excellent diagnostic quality Ac > 90% – are third- fourth-order statistical moments that characterize the distributions of linear and circular birefringence.

This effect can be attributed to the high probability of absorption of laser radiation by the formed elements of whole blood, saturating the myocardium tissue. As a consequence, the histograms of the differential matrix elements $\langle\{f_{12;21}\}\rangle$, $\langle\{f_{13;31}\}\rangle$, and $\langle\{f_{14;41}\}\rangle$, which describe linear and circular dichroism, acquire an approximately bell-shaped form.

Therefore, the differences in asymmetry and kurtosis $(\Delta Z_{3;4}(\mathrm{LD, CD, LD'}))$ of these distributions become less pronounced compared to those of the phase-anisotropy parameters $(\langle\{f_{23;32}\}\rangle, \langle\{f_{24;42}\}\rangle, \langle\{f_{34;43}\}\rangle)$. Nevertheless, even for these dichroism-related parameters, the balanced accuracy remains high, achieving a good diagnostic level (Ac > 80%).

6.3.3 Comparative Studies of the Diagnostic Efficiency of Polarization, Direct Mueller-matrix, and Differential Matrix Mapping

The dependence of the accuracy of the differential Mueller-matrix mapping method $(\langle f_{ik}\rangle)$ on the optical thickness (τ) or depolarization degree (Λ) of myocardium histological sections is a critical issue. It is also necessary to obtain comparative information for other widely used polarimetric techniques, including polarization $(\{\alpha, \beta\})$ mapping and direct Mueller-matrix $(\{f_{ik}\})$ mapping of polycrystalline biological networks.

For this purpose, a series of myocardium samples was prepared (31 samples for each value of τ and Λ) from groups 1 and 2, characterized by different geometric thicknesses ($z = 20$, 40, 50, 60, 80, and 100 μm). These correspond to optical thicknesses ($\tau = 0.25$, 0.45, 0.62, 0.91, 1.37, and 1.93) and depolarization degrees $\Lambda = 28\%$, 35%, 43%, 52%, 68%, and 89%.

Tables 6.8 and 6.9 present the results of the comparative analysis of the diagnostic potential of the $\{\alpha, \beta\}$, $\{M_{ik}\}$, $\langle\{f_{ik}\}\rangle$ methods for the statistical moments of the first to fourth orders describing the distributions of the averaged phase (Table 6.8) and amplitude (Table 6.9) anisotropy parameters of myocardium samples from patients who died of CHD and ACI.

Table 6.8 Comparative diagnostic efficiency of the methods of polarization ($\{\alpha, \beta\}$), direct ($\{M_{ik}\}$), and differential ($\langle\{f_{ik}\}\rangle$) matrix mapping of phase anisotropy.

τ		0.25			0.45			0.62	
l, μm		20			40			50	
Λ, %		28			35			43	
Ac($M_{i=1;2;3;4}$)	$\{\alpha, \beta\}$	$\{M_{ik}\}$	$\langle\{f_{ik}\}\rangle$	$\{\alpha, \beta\}$	$\{M_{ik}\}$	$\langle\{f_{ik}\}\rangle$	$\{\alpha, \beta\}$	$\{M_{ik}\}$	$\langle\{f_{ik}\}\rangle$
Ac(M_1)	68	72	90	65	68	87	61	63	85
Ac(M_2)	70	74	92	66	69	89	62	66	87
Ac(M_3)	73	80	94	69	73	95	64	69	93
Ac(M_4)	71	81	96	68	75	95	65	71	94
τ		0.91			1.37			1.93	
l, μm		60			80			100	
Λ, %		52			68			89	
Ac($M_{i=1;2;3;4}$)	$\{\alpha, \beta\}$	$\{M_{ik}\}$	$\langle\{f_{ik}\}\rangle$	$\{\alpha, \beta\}$	$\{M_{ik}\}$	$\langle\{f_{ik}\}\rangle$	$\{\alpha, \beta\}$	$\{M_{ik}\}$	$\langle\{f_{ik}\}\rangle$
Ac(M_1)	56	60	81	53	55	64	51	53	52
Ac(M_2)	55	61	84	52	54	66	52	56	54
Ac(M_3)	58	64	91	56	59	73	53	57	61
Ac(M_4)	57	66	94	55	62	75	52	60	62

Table 6.9 Comparative diagnostic efficiency of the methods of polarization $\{\alpha, \beta\}$, direct $\{M_{ik}\}$ and differential $\langle\{f_{ik}\}\rangle$ matrix mapping.

τ		0.25			0.45			0.62	
l, μm		20			40			50	
Λ, %		28			35			43	
Ac($M_{i=1;2;3;4}$)	$\{\alpha, \beta\}$	$\{M_{ik}\}$	$\langle\{f_{ik}\}\rangle$	$\{\alpha, \beta\}$	$\{M_{ik}\}$	$\langle\{f_{ik}\}\rangle$	$\{\alpha, \beta\}$	$\{M_{ik}\}$	$\langle\{f_{ik}\}\rangle$
Ac(M_1)	64	67	79	61	63	71	58	60	85
Ac(M_2)	66	69	82	63	66	73	60	63	87
Ac(M_3)	69	77	92	65	71	90	63	67	89
Ac(M_4)	70	78	90	67	72	87	65	70	85
τ		0.91			1.37			1.93	
l, μm		60			80			100	
Λ, %		52			68			89	
Ac($M_{i=1;2;3;4}$)	$\{\alpha, \beta\}$	$\{M_{ik}\}$	$\langle\{f_{ik}\}\rangle$	$\{\alpha, \beta\}$	$\{M_{ik}\}$	$\langle\{f_{ik}\}\rangle$	$\{\alpha, \beta\}$	$\{M_{ik}\}$	$\langle\{f_{ik}\}\rangle$
Ac(M_1)	54	57	71	52	54	62	51	52	52
Ac(M_2)	53	58	74	52	53	63	50	53	53
Ac(M_3)	56	60	87	54	57	72	52	55	58
Ac(M_4)	55	62	83	53	60	70	52	58	56

Comparative analysis of the polarimetric accuracy of diagnostic methods for detecting changes in the phase and amplitude anisotropy of optically thick myocardium layers demonstrated the following:

- Polarization mapping $\{\alpha, \beta\}$ provides satisfactory accuracy in differentiating necrotic changes in myocardium tissue ($Ac(\tau = 0.25) = 70\% \div 73\%$) only under conditions of minimal depolarization ($\Lambda = 28\%$). For greater optical thickness, the diagnostic performance of this method becomes insufficient ($Ac > 70\%$).
- The accuracy of differential diagnosis based on direct Mueller-matrix mapping $\{f_{ik}\}$ varies from good ($Ac(\tau = 0.25; \Lambda = 28\%) = 80\% \div 81\%$) to satisfactory ($Ac(\tau = 0.45; \Lambda = 35\%) = 71\% \div 72\%$) within the interval corresponding to a slight increase in depolarization ($0.25 < \tau < 0.62 \Leftrightarrow 28\% < \Lambda < 43\%$).
- Differential matrix mapping ($\langle\{f_{ik}\}\rangle$) remains diagnostically effective ($70\% < Ac < 96\%$) over a wide range of optical thickness values of multiply scattering myocardium samples ($0, 25 < \tau < 1, 37 \Leftrightarrow 28\% < \Lambda < 68\%$).

6.3.4 Conclusions

1. A method of differential Mueller-matrix mapping for reconstructing the distributions of linear and circular birefringence and dichroism parameters of partially depolarizing myocardium layers with different degrees of necrotic alteration is proposed and substantiated.
2. The coordinate distributions of the first-order differential matrix elements of myocardium histological sections from patients who died of CHD ($z = 60\,\mu\text{m}$; $\tau = 0.84$; $\Lambda = 44\%$) and ACI ($z = 60\,\mu\text{m}$; $\tau = 0.91$; $\Lambda = 52\%$) are determined and analyzed using a statistical approach.
3. The most sensitive objective parameters of necrotic modifications in the polycrystalline structure of partially depolarizing myocardium layers are identified. These parameters correspond to the third- and fourth-order statistical moments describing the polarimetrically reconstructed distributions of phase and amplitude anisotropy in fibrillar networks.
4. High diagnostic accuracy is achieved by differential Mueller-matrix mapping of partially depolarizing myocardium histological sections from patients who died of CHD and ACI: good quality $\left(\Delta M_{3;4}(\text{LD, CD, LD}') \Rightarrow \text{Ac} > 80\%\right)$ and excellent quality $\left(\Delta M_{3;4}(\text{LB, CB, LB}') \Rightarrow \text{Ac} > 90\%\right)$.
5. A comparative study of the diagnostic efficiency of polarization, direct Mueller-matrix, and differential mapping as a function of changes in the optical thickness of myocardium samples is performed, and the operational limits of each method are identified.

6.4 Three-dimensional Mueller-matrix Diffuse Tomography of Phase and Amplitude Anisotropy of Depolarizing Layers of Benign and Malignant Tumors

6.4.1 Introduction

Methods and technologies for polarimetric diagnostics of the optically anisotropic structure of diffuse biological tissues are actively evolving in biomedical optics. This development covers several original directions, including MMP, polar decomposition of Mueller matrices, and two-dimensional Mueller-matrix mapping within various model approximations.

Traditionally, experimental and analytical results obtained from measuring and analyzing Mueller matrices are presented as data integrally averaged over the entire sample volume, in the form of:

- one-dimensional angular dependences of the magnitude (indicatrix) of Mueller-matrix elements;
- two-dimensional coordinate distributions (MMI) of the values of matrix elements.

However, the morphological and, consequently, the polycrystalline structure of real biological tissues possesses a complex three-dimensional spatial and orientational organization. This motivates the development of MMP diagnostics capable of operating in a fully three-dimensional format for polycrystalline networks of diffuse biological tissues.

Such capability can be achieved by combining Stokes polarimetry with digital holographic reconstruction of the complex-amplitude distributions of the object field in phase-inhomogeneous layers.

This section is devoted to the analysis of a polarization-holographic method of diffuse tomography of biological tissues. The method enables Mueller-matrix reconstruction of three-dimensional distributions of linear and circular birefringence and dichroism fluctuations in depolarizing histological biopsy sections of benign (adenoma) and malignant (carcinoma) prostate tumors.

6.4.2 Analysis and Discussion of the Experimental Data of the 3D Diffuse Tomography Method of a Polycrystalline Structure of Optically Thick Biological Layers

The optical technology for differential diagnosis of prostate tumor samples consists of the following steps:

1. Determination of a series of phase layer-by-layer distributions $\text{DT}(x, y, \vartheta_k)$ for both groups of samples.
2. Calculation of the statistical moments of the first to fourth orders $M_{i=1;2;3;4}\{\text{DT}(x, y, \vartheta_k)\}$ for each phase section ϑ_k.
3. Construction of phase dependences of the statistical moments $M_{i=1;2;3;4}\{\text{DT}(\vartheta_1, \vartheta_2, \ldots\vartheta_k)\}$ and correlation moments ${K^k}_{i=2;4}\{\text{DT}(\vartheta_1, \vartheta_2, \ldots\vartheta_k)\}$ for histological sections of groups 1 and 2.
4. Identification of phase planes ϑ^* in which the maximum differences between the statistical $\left(\Delta M^*_{i=1;2;3;4} \equiv \Delta M_{i=1;2;3;4}(\vartheta^*) \rightarrow \max\right)$ and correlation $\left(\Delta\left(K^k_{2;4}\right)^* \equiv \Delta K^*_{2;4}(\vartheta^*) \rightarrow \max\right)$ moments are observed.

5. Determination of the mean values $\left(\Delta M^*_{i=1;2;3;4}, \Delta \bar{K}^{k*}_{i=2;4}\right)$ and the corresponding deviation $(\sigma(\Delta M^*_i), \sigma(\Delta K^{k*}_i))$ of the statistical and correlation moments of diffuse tomograms for histological sections from groups 1 and 2 within the phase plane ϑ^*.

A series of Figure 6.14 ($D(\text{LB}, x, y, \lambda_1)$), Figure 6.15 ($D(\text{LD}, x, y, \lambda_2)$), Figure 6.16 ($D(\text{CB}, x, y, \lambda_1)$), Figure 6.17 ($D(\text{CD}, x, y, \lambda_2)$) presents the phase sections ($\Delta M^*_{i=1;2;3;4} \equiv \Delta M_{i=1;2;3;4}(\vartheta^*) \to \max$ and $\Delta(K^k_{2;4})^* \equiv \Delta K^*_{2;4}(\vartheta^*) \to \max$) of diffuse tomograms DT of the polycrystalline structure of depolarizing layers of prostate tumors.

Figures 6.18–6.21 illustrate the cross-correlation functions $K(\Delta x)$, $K(\Delta y)$ obtained from the set of diffuse tomograms DT.

Analysis of the results obtained using the 3D Mueller-matrix two-wave polarization reconstruction of the distributions of fluctuations of optical-anisotropy parameters revealed the following:

- A phase section $\vartheta^* = 1.2$ rad was identified, in which the maximum differences between the statistical moments $\left(\Delta M^*_{i=1;2;3;4} \equiv \Delta M(\vartheta^*) \to \max\right)$ and correlation moments $\Delta\left(K^k_{2;4}\right)^* \equiv \Delta K^*_{2;4}(\vartheta^*) \to \max$ are formed. These parameters characterize the set of diffuse tomograms DT of tumor samples from groups 1 and 2.
- The most pronounced differences between the polarization manifestations of the polycrystalline structure of depolarizing histological biopsy sections of adenoma and carcinoma were observed for the statistical and correlation parameters of diffuse tomograms of linear birefringence $D(\text{LB}, \vartheta^*)$ and dichroism $D(\text{LD}, \vartheta^*)$.
- Similar, although less pronounced, trends were also observed in the comparative analysis of diffuse tomograms of circular birefringence $D(\text{CB}, x, y, \lambda_2)$ and dichroism $D(\text{CD}, x, y, \lambda_2)$.

Let us consider these results from a physical point of view.

The mechanisms of structural anisotropy – LB and LD – are predominant in benign prostate tumors due to the intensive formation and growth of newly developed fibrillar networks.

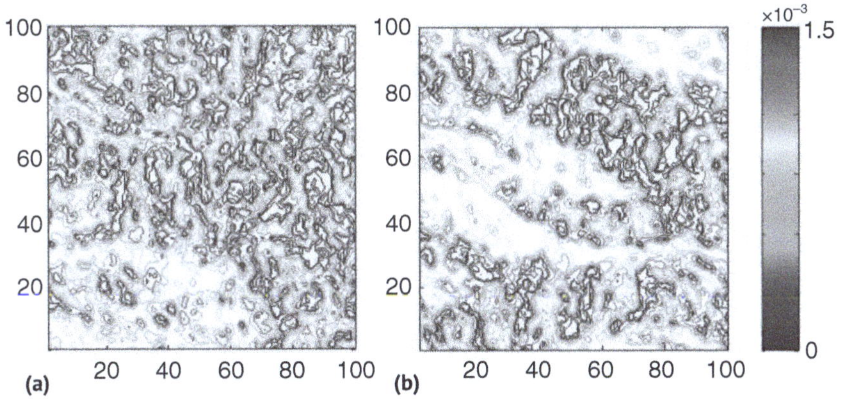

Figure 6.14 Coordinate distributions of fluctuations $D(\text{LB}, x, y, \lambda_1)$ in the magnitude of linear birefringence (LB) in the phase plane $\vartheta^* = 1.2$ rad of optically thick histological sections of adenoma biopsy (a) and carcinoma (b) of the prostate.

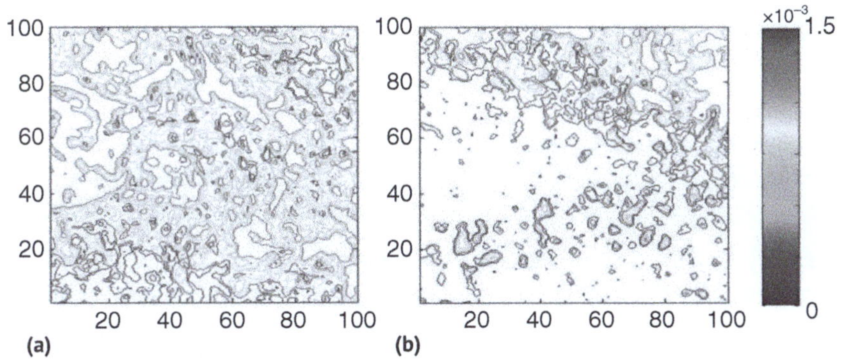

Figure 6.15 Coordinate distributions of fluctuations $D(\text{LD}, x, y, \lambda_2)$ of linear dichroism magnitude (LD) in the phase plane $\vartheta^* = 1.2$ rad of optically thick histological sections of adenoma biopsy (a) and carcinoma (b) of the prostate.

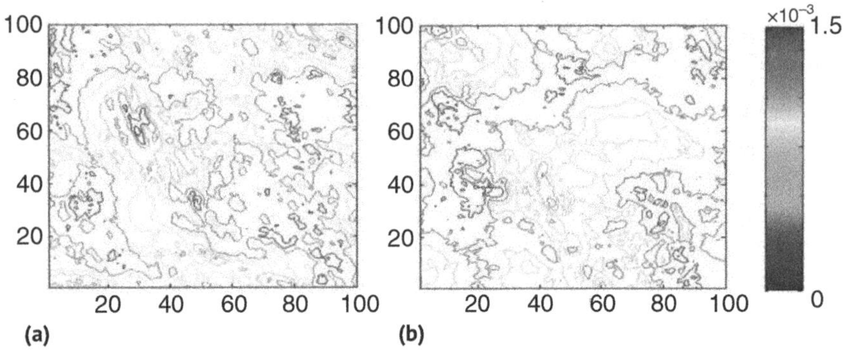

Figure 6.16 Coordinate distributions of fluctuations $D(\mathrm{CB}, x, y, \lambda_2)$ in the magnitude of CB in the phase plane $\vartheta^* = 1.2$ rad of optically thick histological sections of adenoma biopsy (a) and carcinoma (b) of the prostate.

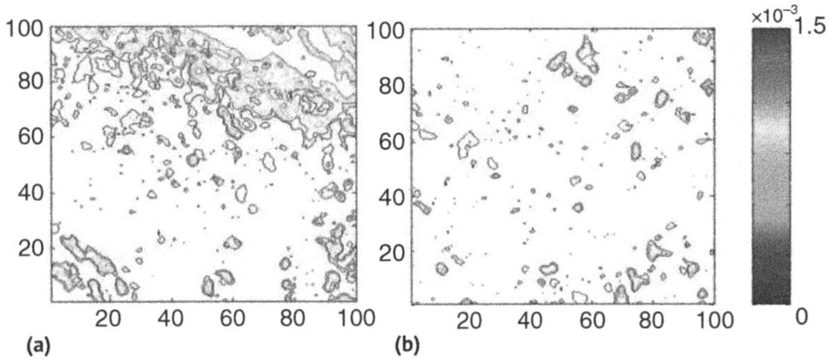

Figure 6.17 Coordinate distributions of fluctuations $D(\mathrm{CD}, x, y, \lambda_2)$ of CD fluctuations in the phase plane $\vartheta^* = 1.2$ rad of optically thick histological sections of adenoma biopsy (a) and carcinoma (b) of the prostate.

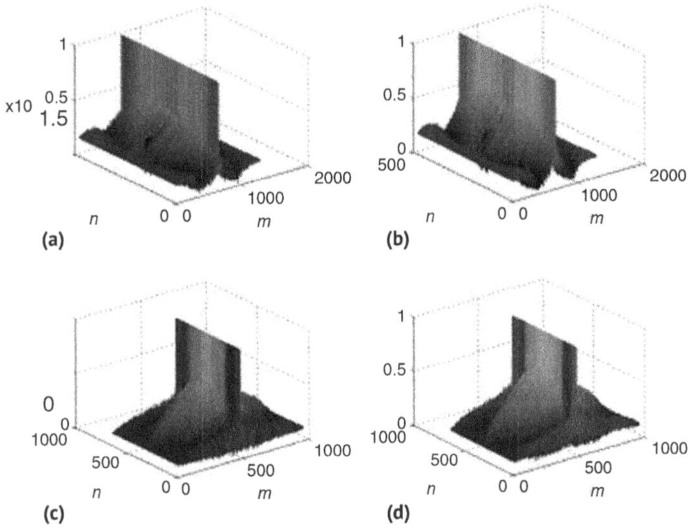

Figure 6.18 Cross-correlation functions (a)–(d) coordinate distributions of the variance of fluctuations $D(\mathrm{LB}, x, y, \lambda_1)$ of the magnitude of LB in the phase plane $\vartheta^* = 1.2$ rad of optically thick histological biopsy sections of adenoma biopsy (a), (c) and carcinomas (b), (d)) prostate.

This finding is confirmed by the comparative analysis of the corresponding diffuse tomograms $D(\mathrm{LB}, \vartheta^*)$ and $D(\mathrm{LD}, \vartheta^*)$, which characterize the distributions of the magnitude and dispersion of fluctuations of these polarization manifestations of structural anisotropy (Figures 6.14 and 6.16).

As shown, both the mean values and the dispersions of fluctuations of LB and LD in optically thick adenoma layers (Figures 6.14 and 6.16a) are significantly higher than the corresponding values obtained for the polycrystalline layer structure of carcinoma samples (Figures 6.14 and 6.16b).

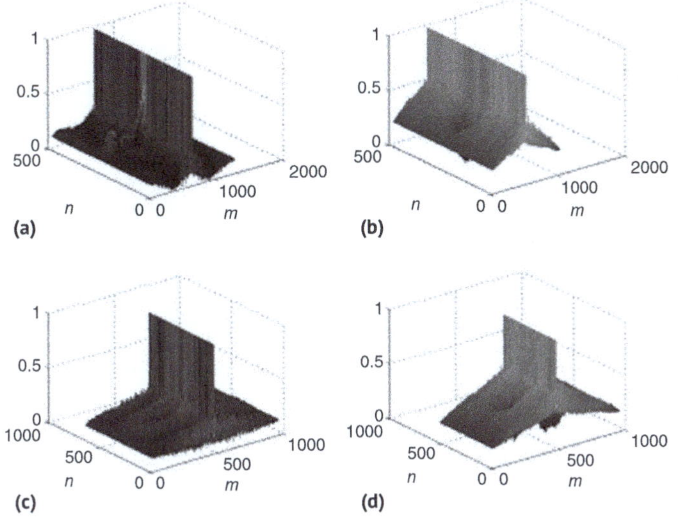

Figure 6.19 Cross-correlation functions (a–d) of coordinate distributions $D(\text{LD}, x, y, \lambda_2)$ of LD magnitude in the phase plane $\vartheta^* = 1.2$ rad of optically thick histological biopsies of adenomas (a), (c) and carcinomas (b), (d) prostate.

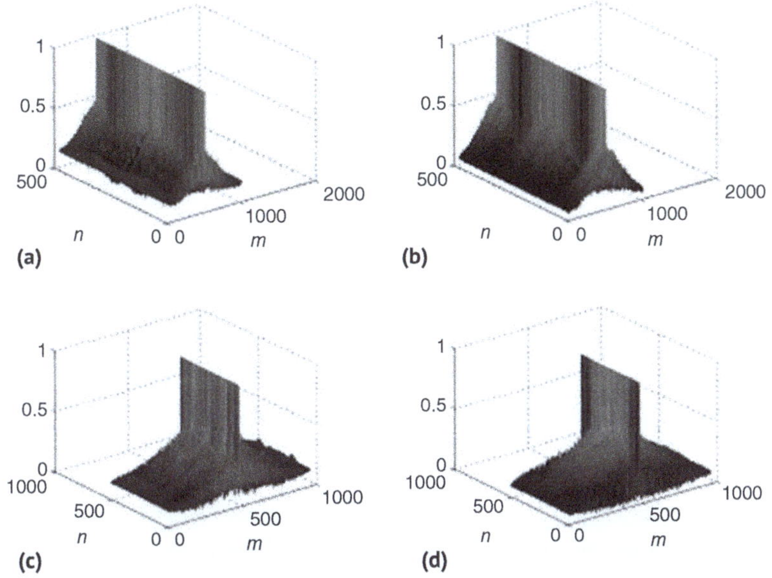

Figure 6.20 Cross-correlation functions (a)–(d) coordinate distributions $D(\text{CB}, x, y, \lambda_2)$ of the magnitude of CB in the phase plane $\vartheta^* = 1.2$ rad of optically thick histological adenoma biopsy sections (a), (c) and carcinomas (b), (d) prostate.

This difference is additionally associated with the fact that malignant transformation leads to degradation of fibrillar structures, resulting in a reduction of phase and amplitude anisotropy in spatially oriented protein networks.

Cross-correlation analysis of diffuse tomograms of linear birefringence $D(\text{LB}, \vartheta^*)$ and dichroism $D(\text{LD}, \vartheta^*)$ in depolarizing layers of prostate tumors of both types further confirms the statistically significant differences between the polarization manifestations of structural anisotropy in fibrillar networks of adenoma and carcinoma samples.

For benign tumors, the topographic structure of the diffuse tomograms $D(\text{LB}, x, y, \vartheta^*)$ and $D(\text{LD}, x, y, \vartheta^*)$ exhibits a small-scale domain ensemble in which regions with similar variances of fluctuations of LB and LD are formed (Figures 6.15 and 6.17a). In malignant tumors, the domain size increases, whereas the magnitudes of the fluctuations $D(\text{LB}, \vartheta^*)$ and $D(\text{LD}, \vartheta^*)$ decrease. As a consequence, the half-widths of the corresponding cross-correlation functions

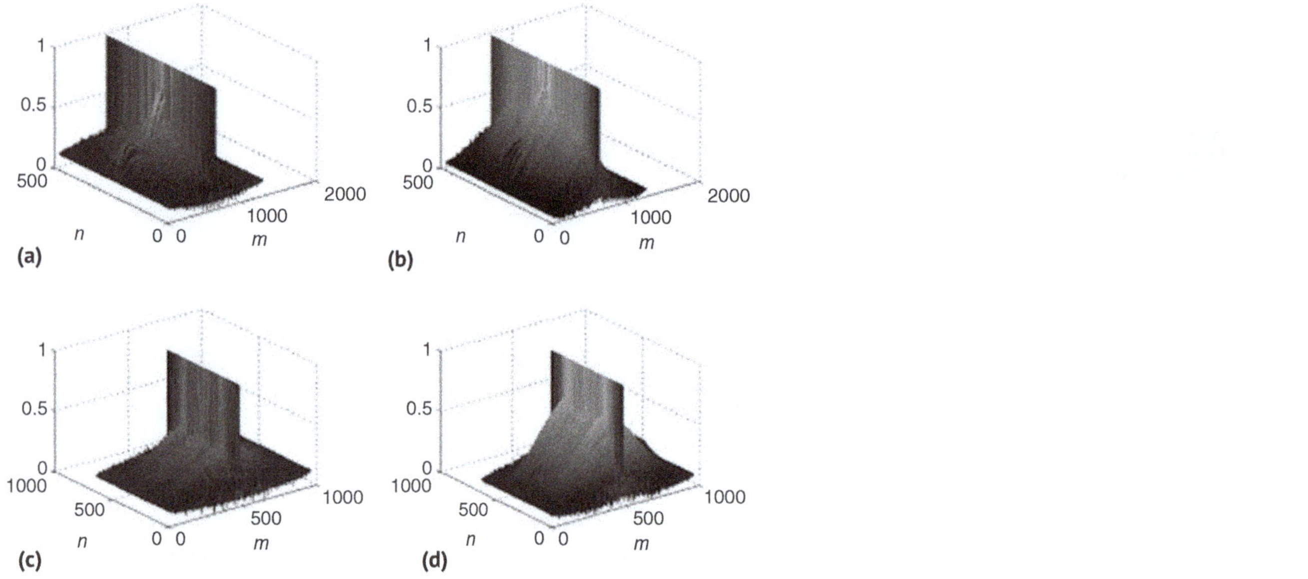

Figure 6.21 Cross-correlation functions (a)–(d) of coordinate distributions $D(\text{CD}, x, y, \lambda_2)$ of the magnitude of CD in the phase plane $\vartheta^* = 1.2$ rad of optically thick histological sections of adenoma biopsy (a), (c) and carcinomas (b), (d) prostate.

Table 6.10 Statistical and correlation criteria for the differential diagnosis of benign and malignant prostate tumors based on phase anisotropy.

Condition	Adenoma		Carcinoma		Accuracy, Ac, %	
Options	**LB**	**CB**	**LB**	**CB**	**LB**	**CB**
M_1	0.59 ± 0.037	0.29 ± 0.016	0.33 ± 0.019	0.21 ± 0.012	79	76
M_2	0.37 ± 0.019	0.12 ± 0.008	0.23 ± 0.014	0.08 ± 0.005	3	78
M_3	0.64 ± 0.042	1.02 ± 0.062	1.05 ± 0.069	1.29 ± 0.074	91	83
M_4	0.79 ± 0.043	1.26 ± 0.075	1.28 ± 0.074	1.49 ± 0.084	92	86
K_2^k	0.095 ± 0.005	0.055 ± 0.003	0.19 ± 0.009	0.07 ± 0.004	75	74
K_4^k	1.82 ± 0.091	2.08 ± 0.096	0.98 ± 0.058	1.64 ± 0.098	93	83

$K_i(\Delta x)$, $K_j(\Delta y)$, which characterize the degree of coordinate homogeneity of these distributions in two mutually perpendicular directions, increase $(K_2^k(\text{LB, LD}) \uparrow)$, while the kurtosis decreases $(K_4^k(\text{LB, LD}) \downarrow)$, as shown in Figures 6.14 and 6.16.

The degradation of structural anisotropy in fibrillar networks of carcinoma samples is accompanied by a corresponding reduction in the concentration of CB and absorbing protein complexes. As a result, the modulation depth of the variance of fluctuations of CB and CD is reduced. Consequently, the topographic structures of the diffuse tomograms $D(\text{CB}, x, y, \vartheta^*)$ and $D(\text{CD}, x, y, \vartheta^*)$ become more homogeneous due to a decrease in both the size and the number of domains (in Figures 6.18b and 6.20b).

Accordingly, as in the previous case of cross-correlation analysis of diffuse tomograms of structural anisotropy, the same pattern of changes in the correlation moments $K_2^k(\text{CB, CD}) \uparrow$ and $K_4^k(\text{CB, CD}) \downarrow$ is observed.

Tables 6.10 and 6.11 present the values of the statistical parameters $(\Delta \bar{M}^*_{i=1;2;3;4} \pm \sigma)$ and correlation $\left(\Delta K^k_{j=2;4} \pm \sigma\right)$ characterizing the distribution $\text{DT}(x, y, \vartheta^*)$ of the polycrystalline structure of depolarizing histological biopsy sections of adenoma and carcinoma of the prostate, along with the balanced accuracy Ac, %. of the diffuse tomography method.

An analysis of the aggregate of statistical $(M_{i=1;2;3;4})$ and correlation $(K^k_{i=2;3})$ moments characterizing the distribution of fluctuations in linear and circular birefringence (Table 6.10) and dichroism (Table 6.11) of prostate tumor samples was confirmed by this physical analysis.

Table 6.11 Statistical and correlation criteria for the differential diagnosis of benign and malignant prostate tumors based on amplitude anisotropy.

Condition	Adenoma		Carcinoma		Accuracy, Ac, %	
Options	LD	CD	LD	CD	LD	CD
M_1	0.12 ± 0.007	0.065 ± 0.0044	0.093 ± 0.052	0.072 ± 0.0047	73	72
M_2	0.14 ± 0.008	0.083 ± 0.0056	0.11 ± 0.006	0.094 ± 0.052	75	74
M_3	0.89 ± 0.051	1.49 ± 0.082	1.28 ± 0.071	1.16 ± 0.067	93	82
M_4	1.43 ± 0.087	1.73 ± 0.099	1.96 ± 0.113	1.29 ± 0.085	94	84
K_2^k	0.06 ± 0.045	0.056 ± 0.0044	0.08 ± 0.005	0.069 ± 0.056	74	72
K_4^k	2.42 ± 0.085	2.31 ± 0.14	1.68 ± 0.11	1.94 ± 0.086	95	87

According to the canons of evidence-based medicine, the most sensitive (grayed out) diagnostic parameters were identified:

- $\begin{cases} \mathrm{Ac}(M_{3;4}(\mathrm{CB}, \mathrm{CD})) = 82\text{–}86\%; \\ \mathrm{Ac}(M_4^k(\mathrm{CB}, \mathrm{CD})) = 83\text{–}87\% \end{cases}$ – good quality diagnostic test;
- $\begin{cases} \mathrm{Ac}(K_{3;4}(\mathrm{LB}, \mathrm{LD})) = 91\text{–}94\%; \\ \mathrm{Ac}(K_4^k(\mathrm{LB}, \mathrm{LD})) = 93\text{–}95\% \end{cases}$ – excellent diagnostic test quality;

6.5 Mueller-matrix Mapping of Fluctuations of Optical-anisotropy Parameters of Biological Tissues Diffusion Layers for Differentiating the Severity Degree of Endometriosis

6.5.1 Introduction

This part is devoted to the development and experimental validation of a polarization mapping method for reconstructing MMI of fluctuations of optical-anisotropy parameters (Chapter 2, equations (2.77)–(2.83)) in depolarizing layers of biological tissues, as well as to the determination of objective diagnostic criteria for differentiating diffuse endometrium layers corresponding to second- and third severity degrees of endometriosis.

6.5.2 Analysis and Discussion of Experimental Data

Two groups of histological sections obtained from surgically removed endometrium cysts corresponding to the second (group 1) and third (group 2) severity degrees of endometriosis were examined. The optical thickness of the biological tissue layers was within the range $2.11 \leq \tau \leq 2.39$, and the degree of depolarization was – $79\% \leq \Lambda \leq 88\%$.

Under these conditions, the symmetry of the matrix operator $\{\Delta f_{ik}^2\}$ becomes practically diagonal. Therefore, as diagnostic parameters, the statistical moments of the first to fourth orders were considered; these moments characterize the distributions of the values of the diagonal elements $\{\Delta f^2\}_{ii}$ of the second-order differential matrix for the examined diffuse endometrium layers.

Figures 6.22–6.24 show the maps and histograms of the distribution of the values of the diagonal elements $\{\Delta f_{22}^2\}$ (Figure 6.22), $\{\Delta f_{33}^2\}$ (Figure 6.23), $\{\Delta f_{44}^2\}$ (Figure 6.24) of histological sections of the endometrium.

From the data obtained, it is evident that endometriosis of the third degree of severity is accompanied by a significant increase in the value of linear birefringence (LB, LB′) on the background of linear (LD, LD′) and circular (CD) dichroism.

Quantitatively, this fact is illustrated by the histograms of distributions of the value of the elements of the second-order differential matrix $\{\Delta f_{22}^2\} = \sigma_4^2(\mathrm{LB}) - \sigma_1^2(\mathrm{LD}) - \sigma_2^2(\mathrm{CD})$ and $\{\Delta f_{44}^2\} = \sigma_6^2(\mathrm{LB}') - \sigma_1^2(\mathrm{LD}) - \sigma_3^2(\mathrm{LD}')$ of the histological sections of the endometrium biopsy with different severity degrees of the pathology (Figures 6.22b and 6.24d).

An increase in the average and range of changes in the value of $\{\Delta f_{22}^2\}$ and $\{\Delta f_{44}^2\}$ is found. Physically, this can be related to the trend $\sigma_4^2(\mathrm{LB}) \uparrow$ and $\sigma_4^2(\mathrm{LB}') \uparrow$ due to individual growth of linearly bipolar birefringent protein fibrils in different directions. Therefore, in the volume of such a multiply scattering pathologically altered fibrillar polycrystalline network the dispersion of the linear birefringence fluctuations increases.

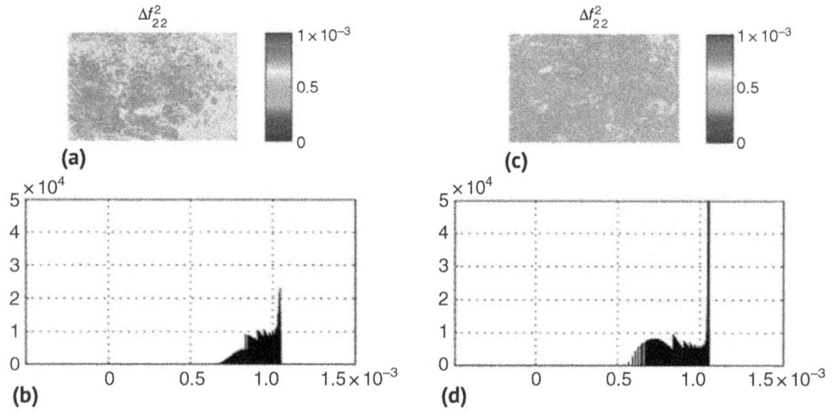

Figure 6.22 Maps (a), (c) and distribution histograms (b), (d) of diagonal element $\{\Delta f_{22}^2\}$ value of histological sections of the endometrium biopsy with different pathology.

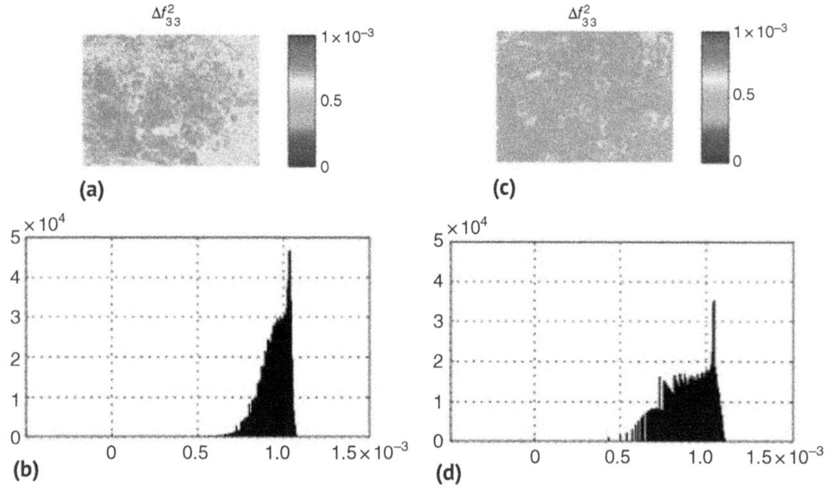

Figure 6.23 Maps (a), (c) and distribution histograms (b), (d) of diagonal element $\{\Delta f_{33}^2\}$ value of histological sections of the endometrium biopsy with different pathology.

Statistical distributions of CB fluctuations value, which characterize the histograms of metric elements $\{\Delta f_{33}^2\} = \sigma_5^2(\text{CB}) - \sigma_1^2(\text{LD}) - \sigma_3^2(\text{LD}')$ (Figure 6.23b, d) of the histological sections of both types of endometrium samples, undergo significantly smaller changes. This fact can be explained by the commensurate level of circular birefringence of samples from groups 1 and 2 due to the stable concentration of optically active protein molecules.

Table 6.12 shows the average values and errors of statistical $(M_{i=1;2;3;4})$ parameters that characterize value distributions of the diagonal elements of the second-order differential matrix of endometrium samples for both groups of pathology.

The comparative analysis of the set of statistical moments of the first and fourth showed the following ones, most sensitive to changes in the optical anisotropy of the endometrium layers on the depolarized background of multiply scattered radiation (highlighted in gray)

- $\begin{cases} \Delta M_2(\{\Delta f_{22}^2\}) \leftrightarrow \text{Ac} = 87\%; \\ \Delta M_3(\{\Delta f_{22}^2\}) \leftrightarrow \text{Ac} = 91\%; \\ \Delta M_4(\{\Delta f_{22}^2\}) \leftrightarrow \text{Ac} = 93\%; \end{cases}$
- $\{\Delta M_3(\{\Delta f_{33}^2\}) \leftrightarrow \text{Ac} = 79\%;$
- $\{\Delta M_4(\{\Delta f_{44}^2\}) \leftrightarrow \text{Ac} = 90\%.$

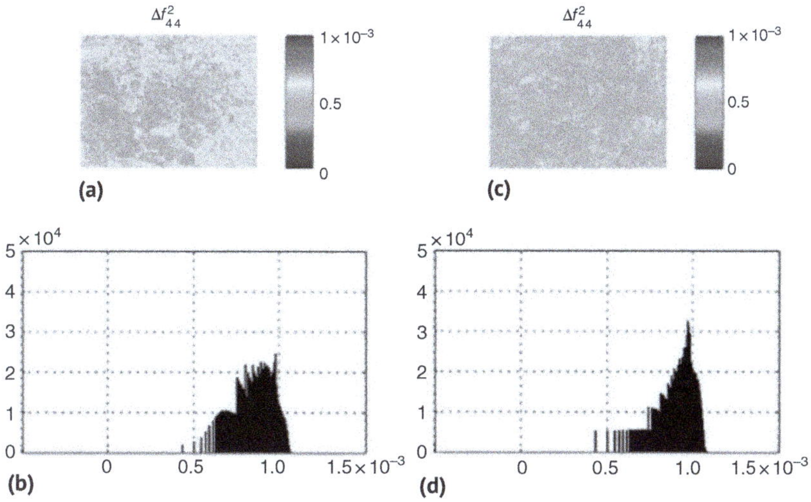

Figure 6.24 Maps (a), (c) and distribution histograms (b), (d) of diagonal element $\{\Delta f^2_{44}\}$ value of histological sections of the endometrium biopsy with different pathology.

Table 6.12 Statistical moments of the first and fourth orders that characterize the fluctuations distributions of optical anisotropy of histological sections of the endometrium biopsy of the second (group 1) and third degree (group 2) of severity of endometriosis.

	$\{\Delta f^2_{22}\}$		$\{\Delta f^2_{33}\}$		$\{\Delta f^2_{44}\}$	
M_i	**Group 1**	**Group 2**	**Group 1**	**Group 2**	**Group 1**	**Group 2**
M_1	2.61 ± 0.15	3.12 ± 0.17	2.81 ± 0.16	2.06 ± 0.11	3.18 ± 0.21	4.22 ± 0.31
M_2	2.09 ± 0.12	4.16 ± 0.27	2.61 ± 0.14	3.11 ± 0.18	2.41 ± 0.16	3.74 ± 0.21
M_3	0.92 ± 0.052	2.21 ± 0.13	1.19 ± 0.062	1.52 ± 0.082	0.61 ± 0.032	0.91 ± 0.062
M_4	1.33 ± 0.071	3.72 ± 0.19	1.83 ± 0.11	1.27 ± 0.091	1.48 ± 0.077	2.97 ± 0.17

Apparently, the statistical moments of the third and fourth orders characterizing the fluctuations distribution of the parameters of linear birefringence proved to be the most sensitive and diagnostically accurate (excellent quality Ac > 90%).

6.5.3 Conclusions

1. A set of layer-by-layer diffuse tomograms representing fluctuations of optical anisotropy in histological biopsy sections of benign and malignant prostate tumors was obtained, and the optimal phase section for their differential diagnosis was determined.
2. The statistical moments $(M_{3;4})$ and the correlation moment (K^k_4), which characterize the structure of diffuse tomograms of phase and amplitude anisotropy, were identified as the parameters most sensitive to changes in the polycrystalline structure of adenoma and carcinoma samples.
3. A good (Ac(CB, CD) = 82 − 87%) and excellent (Ac(LB, LD) = 91 − 95%) level of balanced accuracy of the diffuse tomography method in the differential diagnosis of benign and malignant conditions was achieved when using the third- and fourth-order statistical moments and the fourth-order correlation moment characterizing the distributions of circular (CB, CD), and linear (LB, LD) birefringence and dichroism fluctuations.
4. The coordinate distributions of the diagonal elements of the second-order differential matrix were determined for histological sections of endometrium biopsy samples corresponding to the second ($z = 100\,\mu$m; $\tau = 2.18$; $\Lambda = 84\%$) and third ($z = 100\,\mu$m; $\tau = 2.31$; $\Lambda = 87\%$) severity degrees of endometriosis using a statistical approach.
5. The most sensitive objective parameters for detecting pathological modifications in the polycrystalline structure of diffuse endometrium layers were found to be the third- and fourth-order statistical moments characterizing the polarimetrically reconstructed distributions $(\{\Delta f^2_{22;44}\})$ of fluctuations of optical-anisotropy parameters.
6. A comparative study of the diagnostic efficiency of polarization, direct Mueller-matrix, and differential Mueller-matrix mapping methods for diffuse endometrium layers with different severity degrees of endometriosis was performed.

6.6 Polarization-singular Approach to Imaging MMP of Optically Anisotropic Biological Tissues

6.6.1 Introduction

From an optical standpoint, biological tissues can be classified into two groups:

1. highly scattering (optically turbid) tissues such as skin, brain, vessel walls, blood, and sclera, whose optical properties can be described by a multiple-scattering model of scalar waves in a randomly inhomogeneous absorbing medium;
2. weakly scattering (transparent) tissues, including ocular tissues (cornea, lens) and thin histological sections of various biological structures, which can be adequately characterized within a single-scattering approximation.

However, for the analysis of real biological objects, scalar approximations are insufficient because they do not account for the vector nature of incident and scattered optical fields. Complementary to the scalar models is the vector approach, which enables the extraction of additional information through Mueller matrices that characterize both the structure of biological tissues and their polarization-inhomogeneous images.

In polarization-inhomogeneous fields, the existence of lines or surfaces is possible at which one of the polarization parameters becomes undefined (singular). These include:

1. points of circular polarization (C-points), where the polarization ellipse degenerates into a circle and the azimuth of its central axis becomes undefined; and
2. points of linear polarization, which are degenerate with respect to the sense of rotation of the electric field vector.

The principles of singular analysis of complex vector fields were established in the framework of optical polarization theory, and their extension to biological media has enabled the introduction of singular descriptors into biomedical diagnostics. For the first time in biomedical optics, analytical conditions were formulated for the formation of singly degenerate (linear) and doubly degenerate (circular) polarization singularities in images of linearly birefringent biological tissues. Experimental studies revealed that the third- and fourth-order statistical moments, which characterize the skewness and kurtosis of the distributions of singular points, are highly sensitive to variations in the optical-geometric organization of fibrillar networks in biological tissues of different morphological and physiological states.

Despite these advances, the application of singular-optics concepts to biomedical diagnostics remains limited. Further development is required to extend the singular field ideology to one of the most effective optical technologies – MMP of the polycrystalline component of biological tissues.

The objectives of this section are:

- to identify analytical relationships between the polarization-singular states of the object field formed by birefringent optically anisotropic layers of fibrillar biological tissues and the characteristic values of their MMI;
- to develop a new Mueller-matrix polarization-singular approach for differential diagnosis of changes in optical anisotropy associated with pathological transformations in biological tissues.

6.6.2 Investigation of the Relationship Between the Characteristic Values of MMI and Polarization Singularities of the Object Field of Fibrillar Biological Tissues

Traditionally, polarization-singular states are detected by measuring the characteristic values of the fourth parameter of the Stokes vector of the object field (Table 6.13).

Table 6.13 Relationship between polarization singularities of the biological tissue image and the characteristic values of the fourth parameters of the Stokes vector.

Polarization singularity type	Polarization state	Stokes vector parameter values
"+C" point	$\beta = +\pi/4$	$S_4 = +1$
"−C" point	$\beta = -\pi/4$	$S_4 = -1$
"S" point	$\beta = 0$	$S_4 = 0$

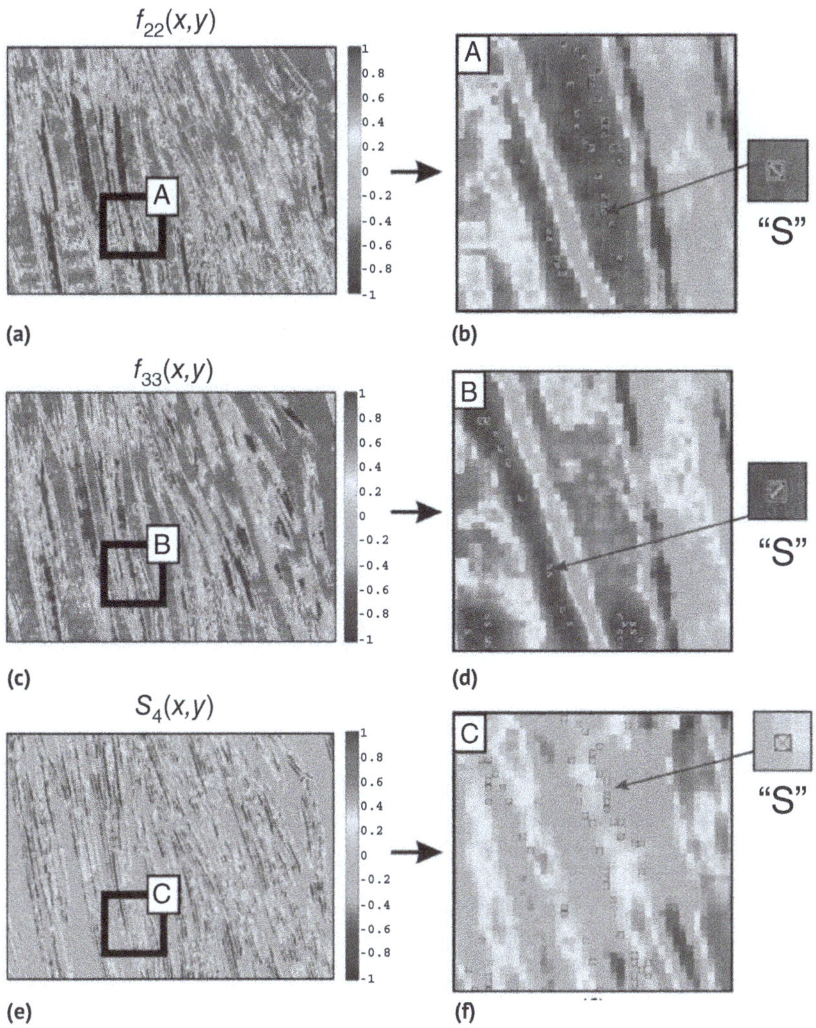

Figure 6.25 MMI f_{22} ("a") and f_{33} ("c") of muscle tissue and maps of the parameter S_4 ("e") of the Stokes vector of its polarization image. Enlarged views of regions A, B, and C are shown in (b), (d), and (f), respectively. A, B, and C indicate coordinate distributions of characteristic values f_{22} ("b") and f_{33} ("c") of the parameter S_4 ("e") of the Stokes vector, respectively.

Optically thin histological sections of skeletal muscle tissue formed by a birefringent network of myosin fibrils were used as the object of the study.

Figures 6.25–6.27 illustrate the correlation between the coordinate localization of the characteristic values of Mueller-matrix and Stokes-parametric images for S and $\pm C$ singular states.

Figure 6.27 shows the results of the Mueller-matrix differentiation S of the states of the polarization image of a histological section of skeletal muscle.

The obtained experimental confirmation of the coordinate relationship of the characteristic values of the elements $f_{ik}(x, y)$ of the Mueller matrix and the fourth parameter of the Stokes vector $S_4(x, y)$ in the image of a histological section of muscle tissue (Figures 6.25–6.27) was used as a basis for testing this approach for differential diagnosis of pathological conditions of biological tissues.

6.6.3 Mueller-matrix Polarization-singular Diagnostics of Pathological Alterations in the Fibrillar Networks of Female Reproductive Tissue

The objects of the experimental study were histological sections of myometrial tissue of two types:

1. biopsy samples of healthy tissue of the female reproductive system (type A);
2. biopsy samples of conditionally normal tissue taken from a region adjacent to benign fibroid formation (type B).

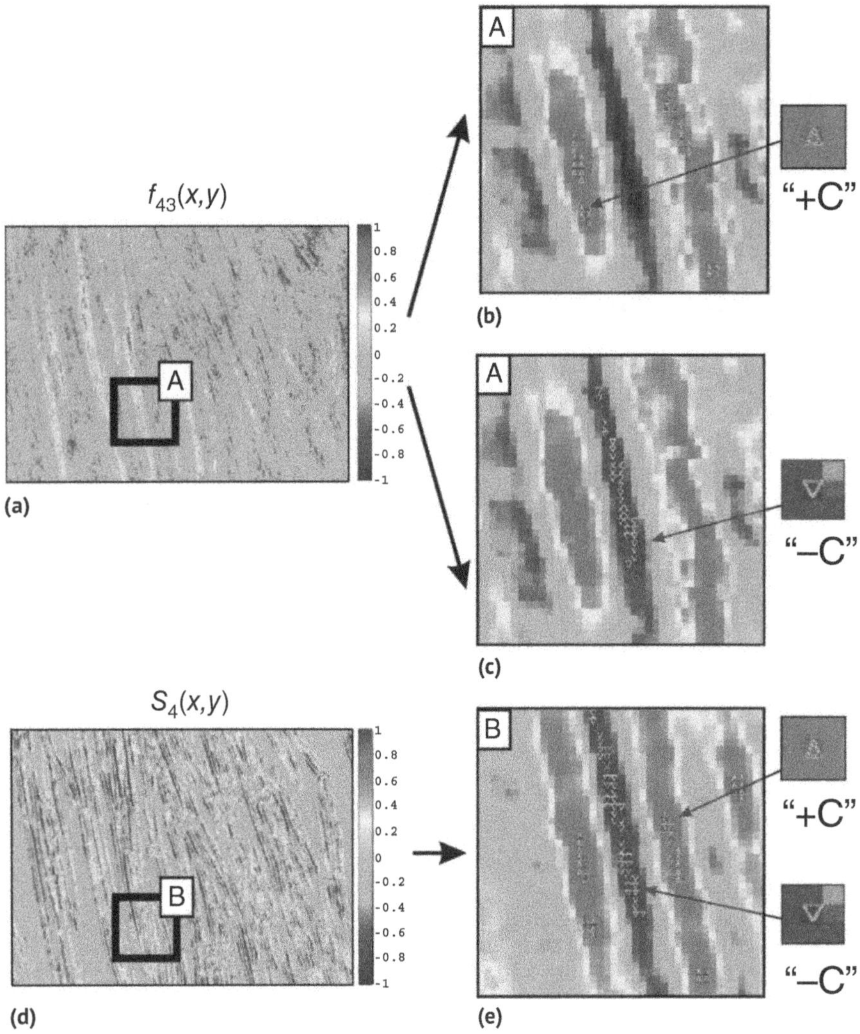

Figure 6.26 MMI f_{43} ("a") of muscle tissue and of the parameter S_4 ("d") of the Stokes vector of its polarization image. A and B indicates coordinate distributions of singular values of the matrix element f_{43} ("b," "c") and of the parameter S_4 ("e") of the Stokes vector, respectively.

Figure 6.28 presents the MMI of the element f_{44} for myometrium samples of types A and B.

The obtained two-dimensional distributions $f_{44}(x, y)$ characterize, from an optical perspective, the degree of phase anisotropy of the examined samples. It is evident that the magnitude of birefringence in myometrium samples of types A and B is low and practically identical. This is confirmed by the minimal number of characteristic values $f_{44}(x, y) = 0 \Leftrightarrow \pm C$ observed in both MMI.

At the same time, a noticeable ordering of the directions of the optical axes of anisotropic structures is observed in type B myometrium samples. Therefore, to establish objective criteria for Mueller-matrix polarization-singular S-differentiation of the optical properties of myometrium samples of types A and B:

- the MMI of the elements $f_{24,42}$ and $f_{34,43}$ were sequentially measured, the characteristic values (± 1 - S - states) of which are formed by fibrils with orthogonal orientations of the optical axes $(\rho = 0°, 90° \rightarrow f_{34,43} = \pm 1;\ \rho = +45°, -45° \rightarrow f_{24,42} = \pm 1$ (Table 6.14);
- coordinate grids of distribution of characteristic values were determined $f_{34,43}(x, y) = \pm 1$ and $f_{24,42}(x, y) = \pm 1$;
- a set of dependencies of the number of points of characteristic values of MMI was determined $N(\rho = 0°)$, $N(\rho = 90°)$, $N(\rho = 45°)$, $N(\rho = -45°)$;
- further dependencies $N(\rho_i)$ were processed according to the following algorithm

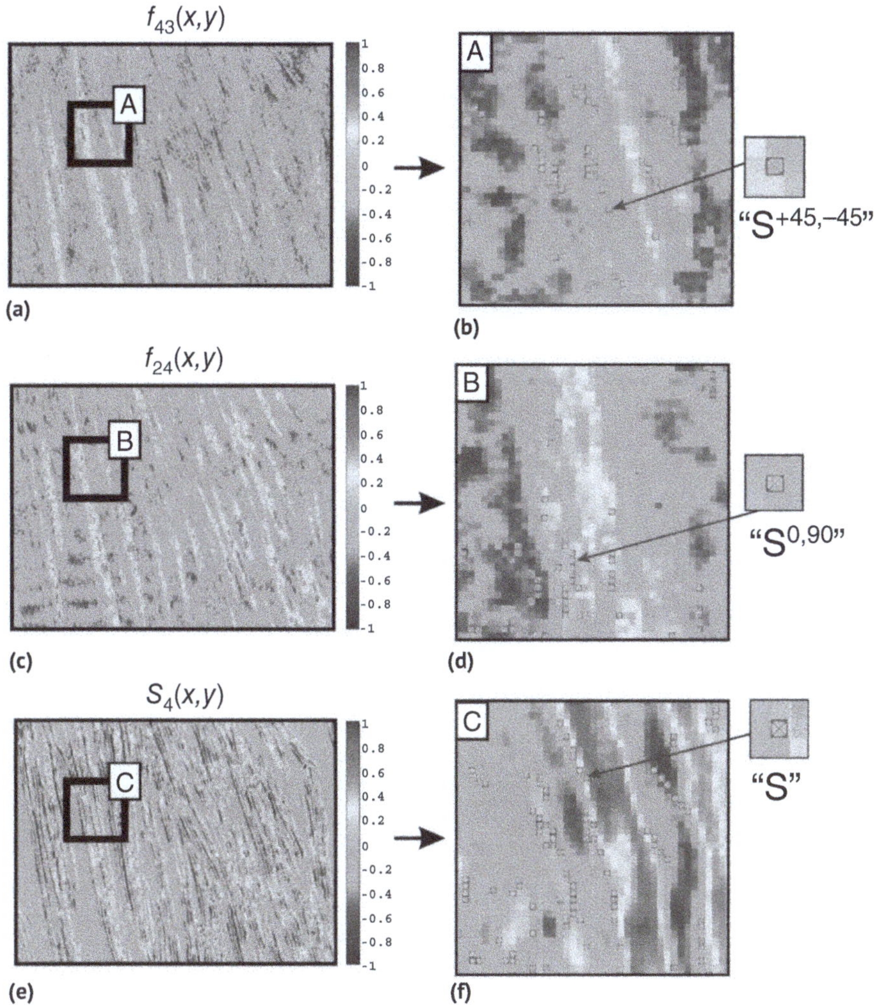

Figure 6.27 MMI f_{43} ("a") and f_{24} ("c") of muscle tissue and parameter S_4 ("e") of the Stokes vector of its polarization image. A, B and C indicate coordinate distributions of the characteristic values of the matrix element f_{43} ("b") and f_{24} ("d") and the parameter S_4 ("f").

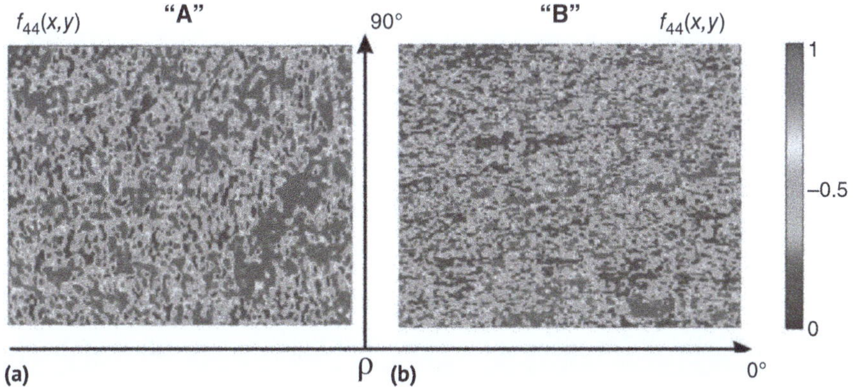

Figure 6.28 MMI of myometrial tissue element f_{44} "A" ("a") and "B" ("b") types.

Table 6.14 Asymmetry coefficient of the distributions of the number of MMI of singularities of myometrial tissue.

M	Myometrium (norm) (25 samples)	Myometrium (pathology) (23 samples)
M_A^{0-90}	0.03 ± 0.0014	0.45 ± 0.021
M_σ^{0-90}	0.02 ± 0.0009	0.12 ± 0.005
$M_A^{45,-45}$	0.025 ± 0.0011	0.28 ± 0.013
$M_\sigma^{45,-45}$	0.03 ± 0.0014	0.14 ± 0.006

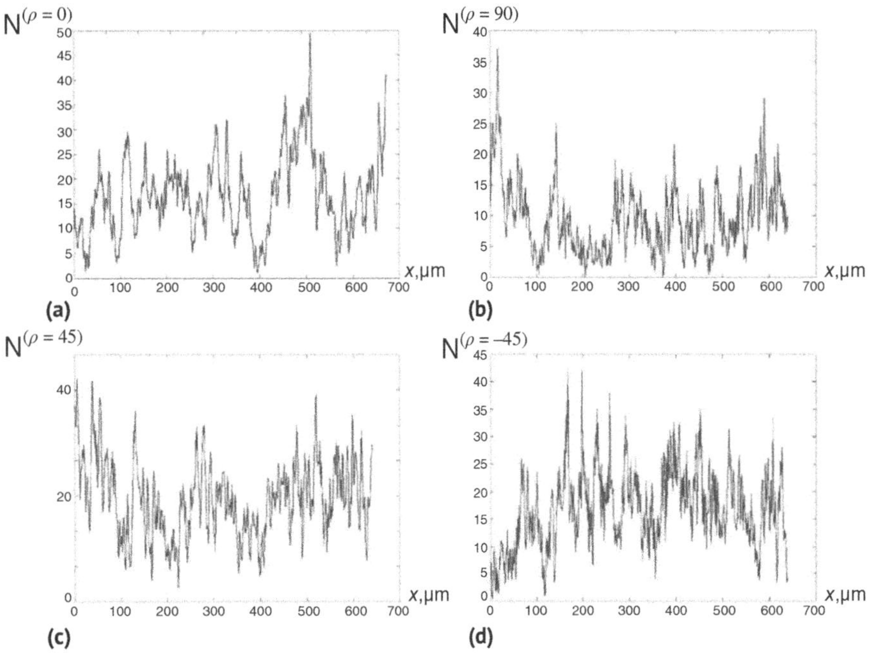

Figure 6.29 Dependencies of the number of singularities $N^{(j)}(x)$ in myometrium tissues of type "A."

$$\begin{cases} M_A^{(f_{34,43})} = \dfrac{M(\rho = 0°) - M(\rho = 90°)}{M(\rho = 0°) + M(\rho = 90°)}; \\ M_A^{(f_{24,42})} = \dfrac{M(\rho = 45°) - M(\rho = -45°)}{M(\rho = 45°) + M(\rho = -45°)}. \end{cases} \quad (6.1)$$

$$\begin{cases} M_\sigma^{(f_{34,43})} = \dfrac{\sigma(\rho = 0°) - \sigma(\rho = 90°)}{\sigma(\rho = 0°) + \sigma(\rho = 90°)}; \\ M_\sigma^{(f_{24,42})} = \dfrac{\sigma(\rho = 45°) - \sigma(\rho = -45°)}{\sigma(\rho = 45°) + \sigma(\rho = -45°)}. \end{cases} \quad (6.2)$$

Here, A and σ – average and variance characterizing the distributions of the number of characteristic values of the elements of the Mueller matrix.

In Figures 6.29 and 6.30, the distribution of the amount of singularity $N^{(j)}(x)$ of myometrium tissues of "A" and "B" types are shown.

Comparative analysis of the obtained data shows that the number of characteristic values of the MMI $f_{34,43}(x, y)$ and $f_{24,42}(x, y)$ for myometrium samples of type A is practically independent of the distribution of orientations of the optical axes of myosin fibrils.

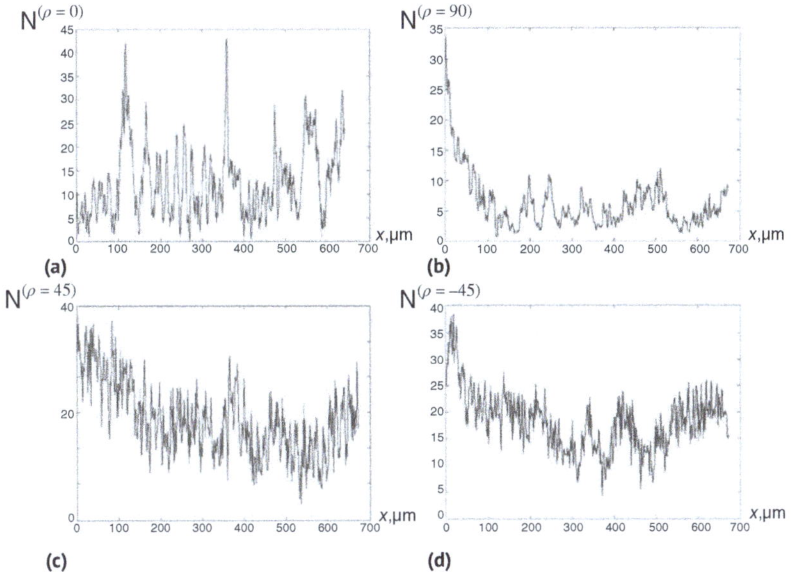

Figure 6.30 Dependencies of the number of singularities $N^{(j)}(x)$ in myometrium tissues of type "B."

From a physical standpoint, this behavior can be attributed to the azimuthal statistical symmetry of myosin fibril packing in myometrial tissue. These fibrillar orientations, described by the optical axis direction $\rho(x, y)$, determine the formation of characteristic values of the Mueller-matrix elements (Table 6.13).

For myometrium samples of type B exhibiting pathological alterations, a distinct asymmetry is observed between the ranges of the dependences $N(\rho = 0^\circ)$ (Figure 6.30) and $N(\rho = 90^\circ)$. This asymmetry can be explained by the fact that the orientational structure of the MMI of the element f_{44} in type B tissue contains characteristic points $f_{44}(x, y) = 1$ that are distributed asymmetrically along the direction $\varrho\rho = 90^\circ$.

Consequently, the maximum number of characteristic values of the element $f_{34}(x, y) = -1$ should be expected, compared with the characteristic values $f_{34,43}(x, y) = 1$ and $f_{24,42}(x, y) = \pm 1$.

The statistically established asymmetry in the distributions of the characteristic values of MMI for myometrium tissues of both types was quantified using the asymmetry coefficients defined in equations (6.1) and (6.2).

Table 6.14 shows the statistically averaged values of the coefficients M_A and M_σ within two groups of myometrial tissue samples "A" and "B."

From the data presented in Table 6.14, the following conclusions can be drawn:

- the first- and second-order statistical moments that characterize the asymmetry coefficients of the distributions of the number of characteristic values of MMI of healthy myometrium tissue are practically equal to zero;
- the first- and second-order statistical moments that characterize the asymmetry coefficients of the corresponding distributions for pathologically altered myometrium tissue increase by nearly an order of magnitude [62–65].

6.6.4 Conclusions

For the first time, a Mueller-matrix approach to describing the formation of polarization-singular states in microscopic images of birefringent fibrillar biological tissues is proposed. The relationship between the characteristic values of Mueller-matrix elements and the types of singularities in polarization-inhomogeneous images of birefringent biological tissues has been established.

It has been demonstrated that statistical analysis of the distributions of the number of characteristic values of Mueller-matrix elements is effective for differentiating the polarization properties of biological tissues associated with pathological modifications of the orientational structure of birefringent fibrillar networks.

References

1 Ghosh N, Vitkin IA. Tissue polarimetry: concepts, challenges, applications, and outlook. J Biomed Opt. 2011;16:110801.
2 Ku S, Mahato KK, Mazumder N. Polarization-resolved Stokes-Mueller imaging: a review of technology and applications. Lasers Med Sci. 2019;34(7):1283–1293. https://doi.org/10.1007/s10103-019-02752-1
3 Qi J, Elson DS. Mueller polarimetric imaging for surgical and diagnostic applications: a review. J Biophotonics. 2017;10(8):950–982. https://doi.org/10.1002/jbio.201600152.
4 Chue-Sang J, Gonzalez M, Pierre A, Laughrey M, Saytashev I, Novikova T, et al. Optical phantoms for biomedical polarimetry: a review. J Biomed Opt. 2019;24(3):1–12. https://doi.org/10.1117/1.JBO.24.3.030901.
5 He C, He H, Chang J, Chen B, Ma H, Booth MJ. Polarisation optics for biomedical and clinical applications: a review. Light Sci Appl. 2021;10(1):194. https://doi.org/10.1038/s41377-021-00639-x.
6 Jones CR. New calculus for the treatment of optical systems. VII. Properties of the N-matrices. J Opt Soc Am. 1948;38:671–685.
7 Azzam RMA. Propagation of partially depolarized light through anisotropic media with or without depolarization. A differential 4x4 matrix calculus. J Opt Soc Am. 1978;68:1756–1767.
8 Arteaga O, Canillas A. Analytic inversion of the Mueller-Jones polarization matrices for homogeneous media. Opt Lett. 2010;35:559–561.
9 Barakat R. Exponential versions of the Jones and Mueller–Jones polarization matrices. J Opt Soc Am A. 1996;13:158–163.
10 Ma H, He H, Ramella-Roman JC. Mueller matrix microscopy. In Ramella-Roman JC, Novikova T. editors. Polarized Light in Biomedical Imaging and Sensing. Cham: Springer; 2023.
11 Lu SY, Chipman RA. Interpretation of Mueller matrices based on polar decomposition. J Opt Soc Am A. 1996;13:1106–1113.
12 Arteaga O, Canillas A. Pseudopolar decomposition of the Jones and Mueller-Jones exponential polarization matrices. J Opt Soc Am A. 2009;26:783–793.
13 Ossikovski R, De Martino A, Guyot S. Forward and reverse product decompositions of depolarizing Mueller matrices. Opt Lett. 2007;32:689–691.
14 Ortega-Quijano N, Arce-Diego JL. Mueller matrix differential decomposition. Opt Lett. 2011;36:1942–1944.
15 Wu PJ, Walsh Jr. JT. Stokes polarimetry imaging of rat-tail tissue in a turbid medium using incident circularly polarized light. Lasers Surg Med. 2005;37:396–406.
16 Wood MF, Ghosh N, Moriyama EH, Wilson BC, Vitkin IA. Proof-of-principle demonstration of a Mueller matrix decomposition method for polarized light tissue characterization in vivo. J Biomed Opt. 2009;14:014029.
17 Wang Y, He H, Chang J, He C, Liu S, Li M, et al. Mueller matrix microscope: a quantitative tool to facilitate detections and fibrosis scorings of liver cirrhosis and cancer tissues. J Biomed Opt. 2016;21:071112.
18 Iqbal M, Ahmad I, Khaliq A, Khan S. Comparative study of Mueller matrix transformation and polar decomposition for optical characterization of turbid media. Optik. 2020;224:165508.
19 Khaliq A, Ashraf S, Gul B, Ahmad I. Comparative study of 3 × 3 Mueller matrix transformation and polar decomposition. Opt Commun. 2021;485:126756.
20 Swami MK, Manhas S, Buddhiwant P, Ghosh N, Uppal A, Gupta PK. Polar decomposition of 3 × 3 Mueller matrix: a tool for quantitative tissue polarimetry. Opt Express. 2006;14:9324–9337.
21 Wang Y, Guo Y, Zeng N, Chen D, He H, Ma H. Study on the validity of 3 × 3 Mueller matrix decomposition. J Biomed Opt. 2015;20:065003.
22 Freund I, Soskin MS, Mokhun AI. Elliptic critical points in paraxial optical fields. Opt Commun. 2002;207:223–253.
23 Soskin MS, Denisenko V, Freund I. Optical polarization singularities and elliptic stationary points. Opt Lett. 2003;28:1475–1477.
24 Dennis MR. Polarization singularities in paraxial vector fields: morphology and statistics. Opt Commun. 2002;213:201–221.
25 Hajnal JV. Singularities in the transverse fields of electromagnetic waves. I. Theory. Proc R Soc Lond A. Math Phys Sci. 1987;414:433–446.
26 Hajnal JV. Singularities in the transverse fields of electromagnetic waves II. Observations on the electric field. Proc R Soc Lond A. Math Phys Sci. 1987;414:447–468.
27 Nye JF. Lines of circular polarization in electromagnetic wave fields. Proc R Soc Lond A. Math Phys Sci. 1983;389:279–290.
28 Bliokh KY. Geometrical optics of beams with vortices: berry phase and orbital angular momentum Hall effect. Phys Rev Lett. 2006;97:043901.
29 Flossmann F, Schwarz UT, Maier M, Dennis MR. Polarization singularities from unfolding an optical vortex through a birefringent crystal. Phys Rev Lett. 2005;95:253901.

30 Flossmann F, Schwarz UT, Maier M, Dennis MR. Stokes parameters in the unfolding of an optical vortex through a birefringent crystal. Opt Express. 2006;14:11402–11411.

31 Angelsky OV, Besaha RN, Mokhun AI, Mokhun II, Sopin MO, Soskin MS, et al. Singularities in vectoral fields. In Fourth International Conference on Correlation Optics. SPIE Proc. 1999;3904:40–55.

32 Freund I, Mokhun AI, Soskin MS, Angelsky OV, Mokhun II. Stokes singularity relations. Opt Lett. 2002;27(7):545–547.

33 Angelsky O, Mokhun A, Mokhun I, Soskin MJ. The relationship between topological characteristics of component vortices and polarization. Opt Commun. 2002;207:56–57.

34 Nye JF. Line singularities in wave fields. Phil Trans R Soc Lond. 1997;355:2065–2069.

35 Berry MV, Dennis MR. Polarization singularities in isotropic random vector waves. Proc R Soc Lond. 2001;457:141–155.

36 Angelsky OV. Polarization singularities of the object field of skin surface. J Opt A: Pure Appl Opt. 2006;39:3547–3558.

37 Angelsky OV, Ushenko AG, Angelska AO, Ushenko YA. Correlation- and singular-optical approaches in diagnostics of polarization inhomogeneity of coherent optical fields from biological tissues. J Phys Opt. 2007;8(2):105–123.

38 Angelsky PO, Ushenko AG, Dubolazov AV, Sidor MI, Bodnar GB, Koval G, et al. The singular approach for processing polarization-inhomogeneous laser images of blood plasma layers. J Opt. 2013;15(4):044030.

39 Ushenko AG, Dubolazov AV, Ushenko VA, Novakovskaya OY. Statistical analysis of polarization-inhomogeneous Fourier spectra of laser radiation scattered by human skin in the tasks of differentiation of benign and malignant formations. J Biomed Opt. 2016;21(7):071110.

40 Garazdyuk MS, Bachinskyi VT, Vanchulyak OY, Ushenko AG, Dubolazov OV, Gorsky MP. Polarization-phase images of liquor polycrystalline films in determining time of death. Appl Opt. 2016;55(12):B67–B71.

41 Angelskii OV, Ushenko OG, Burkovets DN, Arkhelyuk OD, Ushenko YA. Polarization-correlation studies of multifractal structures in biotissues and diagnostics of their pathologic changes. Laser Phys. 2000;10(5):1136–1142.

42 Yermolenko S, Ushenko A, Ivashko P, Goudail F, Gruia I, Gavrilă C, et al. Spectropolarimetry of cancer change of biotissues. In Ninth International Conference on Correlation Optics. SPIE. 2009;7388:404–410.

43 Jóźwicki R, Patorski K, Angelsky OV, Ushenko AG, Burkovets DN, Ushenko YA. Automatic polarimetric system for early medical diagnosis by biotissue testing. Opt Appl. 2002;32(4):603–612.

44 Ushenko AG, Dubolazov OV, Ushenko VA, Novakovskaya OY, Olar OV. Fourier polarimetry of human skin in the tasks of differentiation of benign and malignant formations. Appl Opt. 2016;55(12):B56–B60.

45 Ushenko AG, Pishak VP. Coherent-domain optical methods: biomedical diagnostics, environmental and material science. In Laser polarimetry of biological tissue: Principles and applications. 2004.

46 Ushenko AG, Fediv AI, Marchuk YF. Correlation and fractal structure of Jones matrices of human bile secret. In European Conference on Biomedical Optics. 2009;7368:73681Q.

47 Ushenko AG. Depolarization of a laser emission field as the correlation development of its polarization structure. In International Conference on Correlation Optics. 1997;3317:331–339.

48 Bachinskyi V, Boychuk T, Ushenko A. Laser Polarimetry of Biological Tissues and Fluids: Chapter 4. Methods of Spatial-frequency, Singular and Correlation Analysis of Biological Layers Object Fields. LAP Lambert Academic Publishing; 2018.

49 Ushenko AG, Burkovets DN, Yermolenko SB, Arkhelyuk AD, Pishak VP, Yuzko AM, et al. Stokes polarimetry of biotissues. In Fourth International Conference on Correlation Optics. SPIE. 1999;3904:527–533.

50 Ushenko VA, Dubolazov AV, Pidkamin LY, Sakchnovsky MY, Bodnar AB, Ushenko YA, et al. Mapping of polycrystalline films of biological fluids utilizing the Jones-matrix formalism. Laser Phys. 2018;28(2):025602.

51 Ushenko AG, Angelsky PO, Sidor M, Marchuk YF, Andreychuk DR, Pashkovskaya NV. Spatial-frequency selection of complex degree of coherence of laser images of blood plasma in diagnostics and differentiation of pathological states of human organism of various nosology. Appl Opt. 2014;53(10):B172–B180.

52 Olar EI, Ushenko AG, Ushenko YA. Polarization correlation measurements of the phase tomograms of optically anisotropic biofractals. Laser Phys. 2004;14(8):1115–1121.

53 Guminetskiy SH, Ushenko OG, Polyanskiy IP, Motrych AV, Grynchuk FV. The optical method for investigation of the peritonitis progressing process. In Eighth International Conference on Correlation Optics. 2008;7008:626–632.

54 Ushenko AG, Fediv AI, Marchuk YF. Singular structure of polarization images of bile secret in diagnostics of human physiological state. In: European Conference on Biomedical Optics. Optica Publishing Group; 2009.

55 Ushenko Y, Ushenko A, Dubolazov A, Soltys I, Litvinenko O, Wanchuliak O, et al. Mueller-matrix interferometric multifractal scaling of optically anisotropic architectonics of diffuse blood facies: fundamental and applied aspects. J Biophotonics. 2025;18(3):e202400412. https://doi.org/10.1002/jbio.202400412.

56 Ushenko A, Soltys I, Dubolazov A, Ushenko Y, Bilookyi V, Bilookyi O, et al. 3D Jones-matrix thesiography of biological fluid facies. J Innov Opt Health Sci. 2025;18(2):2443002. https://doi.org/10.1142/S1793545824430028.

57 Ushenko A, Pavlyukovich N, Khukhlina O, Pavlyukovich O, Gorsky M, Soltys I, et al. Layer-by-layer multifractal scanning of optically anisotropic architectonics of blood plasma films: fundamental and applied aspects. Photonics. 2025;12:215. https://doi.org/10.3390/photonics12030215.

58 Ushenko A, Pavlyukovich N, Khukhlina O, Pavlyukovich O, Soltys I, Dubolazov A, et al. Blood plasma film multifractal scanning in COVID-19 consequences diagnostics. J Biophotonics. 2024;17(11):e202400356. https://doi.org/10.1002/jbio.202400356.

59 Ushenko A, Dubolazov A, Zheng J, Litvinenko A, Gorsky M, Ushenko Y, et al. 3D polarization-interference holographic histology for wavelet-based differentiation of biological tissues with different necrotic states. J Biomed Opt. 2024;29(5):052920. https://doi.org/10.1117/1.JBO.29.5.052920.

60 Ushenko O, Bilookyi O, Zheng J, Dubolazov A, Olar O, Ushenko Y, et al. 3D digital holographic polarimetry of laser speckle fields formed by polycrystalline blood films: a tool for differential diagnosis of thyroid pathology. Front Phys. 2024;12:1426469. https://doi.org/10.3389/fphy.2024.1426469.

61 Wójcik W, Hu Z, Ushenko Y, Smolarz A, Soltys I, Dubolazov O, et al. Optical sensor system for 3D Jones matrix reconstruction of optical anisotropy maps of self-assembled polycrystalline soft matter films. Sensors. 2024;24(5):1589. https://doi.org/10.3390/s24051589.

62 Ushenko A, Zheng J, Gorsky M, Dubolazov A, Ushenko Y, Soltys I, et al. 3D digital holographic polarimetry of diffuse optically anisotropic biological tissue object fields. Front Phys. 2023;11:1288935. https://doi.org/10.3389/fphy.2023.1288935.

63 Dubolazov A, Ushenko V, Trifonyuk L, Stashkevich A, Soltys I, Ushenko Y, et al. Polarization-singular approach to imaging Mueller-matrix polarimetry in differential diagnosis of histological sections of biopsy of tumors of the uterus and prostate. Front Phys. 2021;9:711212. https://doi.org/10.3389/fphy.2021.711212.

64 Ushenko VO, Trifonyuk L, Ushenko YA, Dubolazov OV, Gorsky MP, Ushenko AG. Polarization singularity analysis of Mueller-matrix invariants of optical anisotropy of biological tissue samples in cancer diagnostics. J Opt. 2021;23(6):064004. https://doi.org/10.1088/2040-8986/abf97a.

65 Ushenko YA, Koval GD, Ushenko AG, Dubolazov OV, Ushenko VA, Novakovskaia OY. Mueller-matrix of laser-induced autofluorescence of polycrystalline films of dried peritoneal fluid in diagnostics of endometriosis. J Biomed Opt. 2016;21(7):071116. https://doi.org/10.1117/1.JBO.21.7.071116.

7

Optical Interference-based Approach for Tumor Differentiation

Yurii Ushenko[1,2], Weidong Dou[1], Oleksandr Ushenko[2,3], Oleksandr Dubolazov[2], Iryna Soltys[2], Viktor Bachinskiy[4], Oleh Vanchuliak[4], Marta Garazdyuk[4], Nataliia Pavlyukovich[4], Oleksandr Pavlyukovich[4], Vitaliy Rozhko[4], and Oleksandr Olar[2]

[1] *Shaoxing University, Shaoxing, China*
[2] *Yurii Fedkovych Chernivtsi National University, Chernivtsi, Ukraine*
[3] *Taizhou Institute of Zhejiang University, Taizhou, China*
[4] *Bukovinian State Medical University, Chernivtsi, Ukraine*

7.1 Algorithms and Methods of Polarization-interference Mapping of Object Fields of Biological Layers

7.1.1 Brief Theory of the Polarization–Interference Mapping Method

The formation of interference patterns in the general case of laser fields with inhomogeneous polarization can be described as follows [1–12]

$$\begin{cases} U_x(r) + U_y(r)\exp(-i\delta); \\ E_{0x} + E_{0y}. \end{cases} \tag{7.1}$$

Here, U_x, U_y are the amplitudes of the orthogonal components of the polarization ellipse, δ is the phase shift between them; E_{0x}, E_{0y} are the orthogonal amplitude components of a linearly polarized reference wave with polarization azimuth Θ.

$$\begin{cases} E_{0x} = E_0 \cos\Theta; \\ E_{0y} = E_0 \sin\Theta. \end{cases} \tag{7.2}$$

The interference pattern resulting from adding two fields, as described in (7.1) and (7.2), is given by the following interference relations

$$\begin{cases} U_x^*(r) = E_{0x} + U_x(r) + 2\sqrt{E_{0x}U_x(r)}\cos\phi; \\ U_y^*(r) = E_{0y} + U_y(r) + 2\sqrt{E_{0y}U_y(r)}\cos(\phi - \delta(r)). \end{cases} \tag{7.3}$$

Here, the phase difference ϕ between the reference wave and the object wave depends on the coordinate (r) of the object field.

In terms of intensity, equation (7.3) can be represented in the form of interference (coplanar to the oscillation plane (Θ) of the reference wave amplitude E_0 and additive (orthogonal Θ to the reference wave amplitude $\tilde{I}(r)$) $\bar{I}(r)$ components

$$\begin{cases} \tilde{I}(r) = I_0 + I_x(r)\cos^2\Theta + I_y(r)\sin^2\Theta + 2\sqrt{I_x(r)I_y(r)}\sin\Theta\cos\Theta\cos\delta(r) \\ \qquad + 2\sqrt{I_0 I_x(r)}\cos\Theta\cos\phi + 2\sqrt{I_0 I_y(r)}\sin\Theta\cos(\phi - \delta(r)); \\ \bar{I}(r) = I_x(r)\sin^2\Theta + I_y(r)\cos^2\Theta. \end{cases} \tag{7.4}$$

To simplify the analysis, it can be assumed that $\phi \equiv 0$. In this case, equation (7.4) will be reformulated as follows

$$\begin{cases} \tilde{I}(r) = I_0 + I_x(r)\cos^2\Theta + I_y(r)\sin^2\Theta + 2\sqrt{I_0 I_x(r)}\cos\Theta \\ \qquad + 2\sqrt{I_y(r)}\sin\Theta\left(\sqrt{I_0} + \sqrt{I_x(r)}\cos\Theta\right)\cos\delta(r); \\ \bar{I}(r) = I_x(r)\sin^2\Theta + I_y(r)\cos^2\Theta. \end{cases} \tag{7.5}$$

The interference pattern's contrast, created by field components with different polarizations (7.3), is described by an equation of the form

$$K(r_{i+1}, r_i) = \frac{[\tilde{I}(r_{i+1}) + \bar{I}(r_{i+1})] - [\tilde{I}(r_i) + \bar{I}(r_i)]}{[\tilde{I}(r_{i+1}) + \bar{I}(r_{i+1})] + [\tilde{I}(r_i) + \bar{I}(r_i)]}. \tag{7.6}$$

Here, r_i, r_{i+1} are the coordinates of the nearest points, the vibration intensities of which take extreme values (I_{min}; I_{max}).

Analysis of equation (7.6), taking into account equations (7.3)–(7.5), reveals that the value $W(r_{i+1}, r_i)$ depends significantly on the state of polarization of points of the object field of biological tissue. This can be clearly illustrated on the basis of the relationship between the intrinsic parameters (azimuth $\alpha(r)$ and ellipticity $\beta(r)$ of light oscillations) of the state of elliptical polarization and the recording in the form of orthogonal components of its complex amplitude. In this interpretation, the intensity value $I(r)$ at each point of the object field will be written as follows

$$I(r) = U_x^2(r) + U_y^2(r) = a^2(r) + b^2(r) = a^2(r)\left[1 + tg^2\beta(r)\right]. \tag{7.7}$$

Considering (7.7), equation (7.5) is expressed as

$$\begin{cases} \tilde{I}(r) = I_0 + I(r)(1 + tg^2\beta(r))^{-1}\left[\cos^2(\alpha(r) - \Theta) + tg^2\beta(r)\sin^2(\alpha(r) - \Theta)\right] \\ \qquad + 2\sqrt{I_0 I(r)}tg\beta(r)(1 + tg^2\beta(r))^{-1}\sin(\alpha(r) - \Theta) \\ \qquad + 2(1 + tg^2\beta(r))^{-1}\left[I(r)tg\beta(r)\sqrt{0.5\sin^2(\alpha(r) - \Theta)} + \sqrt{I_0 I(r)}\cos(\alpha(r) - \Theta)\right]\cos\delta(r); \\ \bar{I}(r) = I(r)(1 + tg^2\beta(r))^{-1}\left[\sin^2(\alpha(r) - \Theta) + tg^2\beta(r)\cos^2(\alpha(r) - \Theta)\right], \end{cases} \tag{7.8}$$

the analytical equation for the contrast of the interference pattern is expressed as a function depending on the ratio of the intensity and polarization of the reference (I_0, Θ) and object (I, α, β) fields.

$$K = \frac{[\tilde{I}(r_{i+1}) + \bar{I}(r_{i+1})] - [\tilde{I}(r_i) + \bar{I}(r_i)]}{[\tilde{I}(r_{i+1}) + \bar{I}(r_{i+1})] + [\tilde{I}(r_i) + \bar{I}(r_i)]} = G[I_0, I, \Theta, \alpha, \beta]. \tag{7.9}$$

Suppose (without reducing the completeness of the analysis) that $(x_1) \approx I(x_2) \approx I_0$ and possibly write the interference equations for such points.

$$\begin{cases} I_1 = I_0 + I(x_1) + 2\sqrt{I_0 I(x_1)}\cos\delta_1; \\ I_2 = I_0 + I(x_2) + 2\sqrt{I_0 I(x_2)}\cos\delta_2. \end{cases} \tag{7.10}$$

With (7.10), we obtain an equation for the local contrast of the interference pattern at the points x_1; $x_2 = x_1 + 1$ pix.

$$W = \frac{I_2 - I_1}{I_2 + I_1} = \frac{(\cos\delta_2 - \cos\delta_1)}{2 + (\cos\delta_2 + \cos\delta_1)}. \tag{7.11}$$

7.1.2 Matrix Description of the Formation of Interference Distributions of the Object Field of Optically Anisotropic Biological Samples

This section discusses the formation of interference fringes in the plane of polarization-inhomogeneous images of an optically thin, nondepolarizing polycrystalline film of a biological fluid [13–18].

It is possible to show that during the superimposing of a plane reference wave E is the Jones vector) on the polarization-inhomogeneous image $\{D\}A$ ($\{D\}$ is the Jones matrix of the object, A is the Jones vector of the illuminating beam) at the point with coordinate r the following field will be formed

$$U(r) = \{D\}(r)A(r) + E(r). \tag{7.12}$$

The Jones vectors for the illuminating and reference beams are given by the following form

$$\begin{pmatrix} A_x \\ A_y \end{pmatrix} = \begin{pmatrix} |A_x| \exp(i\psi_x) \\ |A_y| \exp(i\psi_y) \end{pmatrix}; \tag{7.13}$$

$$\begin{pmatrix} E_x \\ E_y \end{pmatrix} = \begin{pmatrix} |E_x| \exp(i\varphi_x) \\ |E_y| \exp(i\varphi_y) \end{pmatrix} \exp(i\phi), \tag{7.14}$$

Here, $|A_x|, |A_y|; |E_x|, |E_y|$ are the moduli, and $|\psi_x|, |\psi_y|; |\varphi_x|, |\varphi_y|$ are the phases of orthogonal components of complex amplitudes A and E, ϕ is the phase shift between the reference (E) and object (U) waves in the point r of polarization-inhomogeneous image of the polycrystalline layer.

The matrix equation (7.11) can be expressed in extended form as follows

$$\begin{pmatrix} U_x \\ U_y \end{pmatrix}(r) = \begin{Vmatrix} d_{11}(r) & d_{12}(r) \\ d_{21}(r) & d_{22}(r) \end{Vmatrix} \begin{pmatrix} A_x \\ A_y \end{pmatrix}(r) + \begin{pmatrix} E_x \\ E_y \end{pmatrix}(r). \tag{7.15}$$

The magnitude of intensity $I(r)$ of the interference pattern is defined by the following equation

$$I(r) = |U_x(r)|^2 + |U_y(r)|^2, \tag{7.16}$$

which, according to (7.12) – (7.15), can be rewritten in the form of orthogonal components $U_x(r)$ and $U_y(r)$ of complex amplitude $U(r)$

$$\begin{cases} U_x(r) = (d_{11}A_x + d_{12}A_y + E_x)(r); \\ U_y(r) = (d_{21}A_x + d_{22}A_y + E_y)(r). \end{cases} \tag{7.17}$$

Here, d_{ik} are the elements of the Jones matrix of the birefringent polycrystalline layer in the point r with the direction of the optical axis $\rho(r)$ and phase shift $\delta(r)$

$$\begin{cases} d_{11}(r) = \cos^2\rho + \sin^2\rho \exp(-i\delta); \\ d_{12}(r) = d_{21}(r) = \cos\rho \sin\rho(1 - \exp(-i\delta)); \\ d_{22}(r) = \sin^2\rho + \cos^2\rho \exp(-i\delta). \end{cases} \tag{7.18}$$

Let's examine a specific case of azimuthally-invariant polarization-interference mapping, which does not reduce the analysis's completeness and remains independent of the sample's rotation angle to the illumination direction:

- The illuminating and reference waves are circularly polarized

$$\begin{cases} \psi_y - \psi_x = 0.5\pi; \\ \varphi_y - \varphi_x \equiv 0.5\pi; \end{cases} \begin{cases} |A_x| = |E_x| \equiv 1; \\ |A_y| = |E_y| \equiv 1; \end{cases} \tag{7.19}$$

$$A = E = \begin{pmatrix} 1 \\ i \end{pmatrix}; \tag{7.19}$$

- The phase fluctuations within the polycrystalline layer are minimal

$$\delta \to 0:\ \cos\delta \approx 1;\ \sin\delta \approx \delta. \tag{7.20}$$

It should be noted that the approximation of weak phase modulation ($\delta \leq 0, 12$ rad) is not artificial for optically thin polycrystalline films of synovial fluid. It can be shown that for laser radiation with the wavelength $\lambda = 0.63$ μm within the geometrical thickness $l = 15$ μm of a fully optically anisotropic layer ($\Delta n \approx 10^{-4} \div .5.3.5 \times 10^{-3}$ the maximal phase shift $\left(\delta_{12} \equiv \delta = \frac{2\pi}{\lambda}\Delta nl\right)$ is deviated within $0.015 \leq \delta \leq 0.25$. In accordance with this, (7.15)–(7.17) have the form

$$\begin{cases} d_{11} = 1 - i\delta \sin^2\rho; \\ d_{12} = d_{21} = i\delta \cos\rho \sin\rho; \\ d_{22} = 1 - i\cos^2\rho. \end{cases} \tag{7.21}$$

$$\begin{pmatrix} U_x \\ = \\ U_y \end{pmatrix} = \begin{Vmatrix} \left(1 - i\delta \sin^2\rho\right)_{11} & (i\delta \cos\rho \sin\rho)_{12} \\ (i\delta \cos\rho \sin\rho)_{21} & \left(1 - i\delta \cos^2\rho\right)_{22} \end{Vmatrix} \begin{pmatrix} 1 \\ i \end{pmatrix} + \begin{pmatrix} \cos\phi + i\sin\phi \\ -\sin\phi + i\cos\phi \end{pmatrix}; \tag{7.22}$$

$$I(r) = \delta^2(r) - 2\delta(r)\sin\phi(r) + 2\cos\phi(r). \tag{7.23}$$

The analysis of (7.23) reveals that, depending on the phase shift variations between the reference and object waves, the intensity in the microscopic image plane varies within these bounds

$$\begin{cases} \phi(r) = 0 \leftrightarrow I(r)^2(r)_{\min} \\ \phi(r) = \pi \leftrightarrow I(r)^2(r)_{\max} \end{cases} \tag{7.24}$$

Analysis of (7.14) and (7.24) indicates that variations in the interference pattern of the polarization-inhomogeneous image of the polycrystalline film are solely determined by phase modulation caused by birefringence $\left(\delta = \frac{2\pi}{\lambda}\Delta nl\right)$ of its substance.

Therefore, polarization–interference mapping of the microscopic image of the polycrystalline biological fluid film enables the extraction of coordinate distribution (within the $m \times n$ pixels of the digital camera) of the values of local contrast

$$W(m \times n) = W \begin{pmatrix} r_{11} & \dots & r_{1n} \\ \cdot & \dots & \cdot \\ r_{m1} & \dots & r_{mn} \end{pmatrix} \tag{7.25}$$

and assess the small variations in the birefringence of that layer

$$\begin{pmatrix} U_x \\ = \\ U_y \end{pmatrix} = \begin{Vmatrix} \left(1 - i\delta \sin^2\rho\right)_{11} & (i\delta \cos\rho \sin\rho)_{12} \\ (i\delta \cos\rho \sin\rho)_{21} & \left(1 - i\delta \cos^2\rho\right)_{22} \end{Vmatrix} \begin{pmatrix} 1 \\ 0 \end{pmatrix} + \begin{pmatrix} \cos\phi + i\sin\phi \\ 0 \end{pmatrix}; \tag{7.26}$$

$$I(r) = \delta(r)\sin^2\rho(r)(\delta(r) - 2\sin\phi(r)) + 2(1 + \cos\phi(r)); \tag{7.27}$$

$$\begin{cases} \phi(r) = 0; \\ I(r) = \delta^2(r)\sin^2\rho(r); \end{cases} \tag{7.28}$$

$$W(\delta, r_1, r_2) = \frac{\delta^2(r_1)\sin^2\rho(r_1) - \delta^2(r_2)\sin^2\rho(r_2)}{\delta^2(r_1)\sin^2\rho(r_1) + \delta^2(r_2)\sin^2\rho(r_2)}. \tag{7.29}$$

Here, r_1, r_2 are the coordinates of the closest points under the scanning with the step $\Delta r = 1$ pix $= 2$ μm.

7.1.3 Conclusions

A concise theory is provided for how interference distributions form in a polarization-inhomogeneous field when a reference coherent wave is used. The amplitude-phase approach is used to analytically explain how the coordinate distribution of complex amplitudes of the object field develops in an optically anisotropic biological layer. Additionally, an analytical expression for the local contrast of points within the polarization-inhomogeneous interference distribution has been derived.

A Jones-matrix formalism is introduced to establish analytical relationships between the optical anisotropy parameters of the polycrystalline structure in the biological layer and the local contrast within the interference distribution of the object laser speckle field.

7.2 Polarization-Interference Mapping Biological Layers in Differential Diagnosis of Benign and Malignant Tumors

7.2.1 Introduction

Prostate cancer ranks as the second most common cancer worldwide in men, and in certain countries, it has become the most frequently diagnosed cancer. Early detection, timely intervention, and proper management can greatly enhance patient outcomes. It is crucial to distinguish malignant (adenocarcinoma) prostate tissue types.

Optical tissue analysis provides clear benefits: it is nondestructive, demands minimal sample preparation, and is inexpensive and quick [19–24].

Analyzing the polarization properties of tissues, known as polarimetric diagnostics, shows significant potential. The field of polarimetric diagnostics for optically anisotropic biological tissues is quickly advancing within biomedical optics.

Several important areas of research have developed, such as scattering matrices, Mueller matrix polarimetry, polar decomposition of Mueller matrices, and 2D Mueller matrix mapping with different approximations.

Several Mueller-matrix diagnostic methods rely on directly detecting polarization to understand the phase modulation mechanisms of complex-field amplitudes.

Combining polarization and interference techniques for detecting the phase structure of object fields shows promise for enhancing Mueller-matrix diagnostics.

7.2.2 Biological Samples

Three main groups of tumor histology samples were categorized according to International Society of Urological Pathology (ISUP) grades:

- Group 1 consisted of highly differentiated (ISUP 1: 3 + 3 on Gleason's pattern scale) adenocarcinoma samples;
- Group 2 consisted of moderately differentiated (ISUP 2 − 3: 3 + 4; 4 + 3 on Gleason's pattern scale) adenocarcinoma samples;
- Group 3 consisted of poorly differentiated** (ISUP 4: 4 + 4; 3 + 5; 5 + 3 on Gleason's Pattern scale) adenocarcinoma samples.

An independent evaluation of stained histological samples was used to determine the type of prostate tumors (see Figure 7.1).

The histological analysis procedure involved these steps:

- fixation of prostate tissue using 40% formaldehyde aqueous solution (formalin);
- washing samples in running water for 24°;
- dehydration with alcohols at increasing concentrations (70–100%) within 48°;
- fix the material by immersing it in a xylene–paraffin mixture for one to two hours at 520–560 °C, then cut out a block that contains the sample;
- production of histological sections using a standard microtome;
- staining of histological sections with hematoxylin-eosin;
- microscopic examination of the images of the prepared samples, including grading their structure and identifying the sample's location within the prostate tumors according to the Gleason scale.

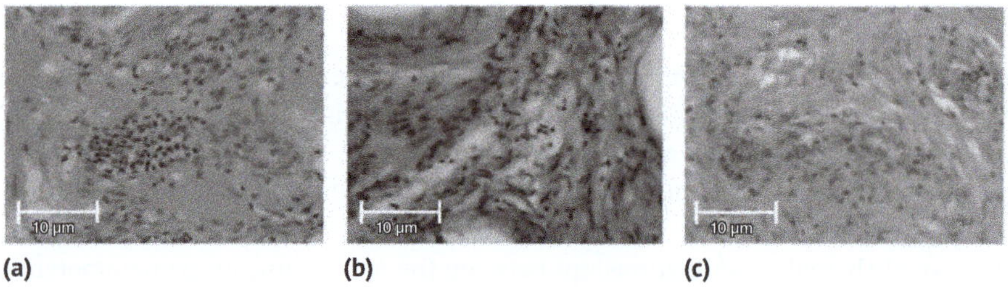

Figure 7.1 Representative microscopic images (magnification 50×) of: (a) a high differentiated (ISUP 1: 3 + 3 on Gleason's pattern scale) adenocarcinoma, (b) moderately differentiated (ISUP 2, 3: 3 + 4) adenocarcinoma, and (c) poorly differentiated (ISUP 4: 4 + 4) prostate adenocarcinoma. *Source*: Springer Nature / CC BY 4.0.

Table 7.1 Optical and geometric parameters of prostate tumor histological sections.

Parameter	Group 1	Group 2	Group 3
Geometric thickness h, µm	30 ± 0.45	30 ± 0.45	30 ± 0.45
Attenuation (extinction) coefficient τ, cm^{-1}	0.81 ± 0.041	0.79 ± 0.037	0.8 ± 0.039
Depolarization degree Λ, %	41 ± 0.38	39 ± 0.37	40 ± 0.37

Table 7.1 displays the optical and geometric parameters of native histological sections from prostate tumor biopsies across all groups.

The geometric thickness (h, µm) of histological prostate tissue sections was determined according to the standard scale values of the freezing microtome.

7.2.3 Experimental Results and Discussion

7.2.3.1 Polarization Maps of Prostate Tumor Samples

Initially, the polarization structure of microscopic images of adenocarcinoma samples was examined by analyzing the coordinate distributions of the polarization azimuth $\alpha(m, n)$ and ellipticity $\beta(m, n)$ using the traditional Stokes polarimetry technique.

A series of fragments in Figures 7.2 and 7.3 illustrate the results of polarization mapping on microscopic images of prostate tumor histological sections.

The results obtained (Figures 7.2 and 7.3) clearly show significant heterogeneity in the polarization in microscopic images of all prostate tumor tissue types. The range of random variations of the polarization parameters is $-0.5\pi \leq \alpha \leq 0.5\pi$ i $-0.25\pi \leq \beta \leq 0.25\pi$.

Comparative analysis of the $\alpha(m, n)$ and $\beta(m, n)$ distributions ((a), (c), (e)) revealed a high level of polarization homogeneity in microscopic images of histological sections from adenocarcinoma biopsies.

This conclusion is based objectively on the observation that the main extrema N_{max} of the histograms of azimuth and ellipticity distributions in malignant tumor images (Figures 7.2d, f and 7.3d, f) exceed the corresponding statistical parameters of polarization maps of highly differentiated tumors by a factor of 1.15–1.2 (Figures 7.2b and 7.3b).

Quantitatively, these polarization features are characterized by a set of statistical moments $\boldsymbol{M}_i$ describing the azimuth $\alpha(m, n)$ and ellipticity $\beta(m, n)$ polarization maps w (Table 7.2).

This may manifest physically as necrotic changes in the prostate tissue's structure, resulting in decreased optical anisotropy.

The data showed the greatest differences between groups in the third- and fourth-order statistical moments, which measure the skewness and kurtosis of the polarization parameter distributions w:

- Group 1–2 $\Delta M_3(\alpha) = 1.17$; $\Delta M_4(\alpha) = 1.31$ i $\Delta M_3(\beta) = 1.25$; $\Delta M_4(\beta) = 1.14$;
- Group 1–3 $\Delta M_3(\alpha) = 1.35$; $\Delta M_4(\alpha) = 1.44$ i $\Delta M_3(\beta) = 1.5$; $\Delta M_4(\beta) = 1.4$;
- Group 2–3 $\Delta M_3(\alpha) = 1.12$; $\Delta M_4(\alpha) = 1.24$ i $\Delta M_3(\beta) = 1.2$; $\Delta M_4(\beta) = 1.23$.

Table 7.3 presents the sensitivity Se, specificity Sp, and balanced accuracy Ac values for the method of polarization mapping of microscopic images of malignant prostate tumor samples.

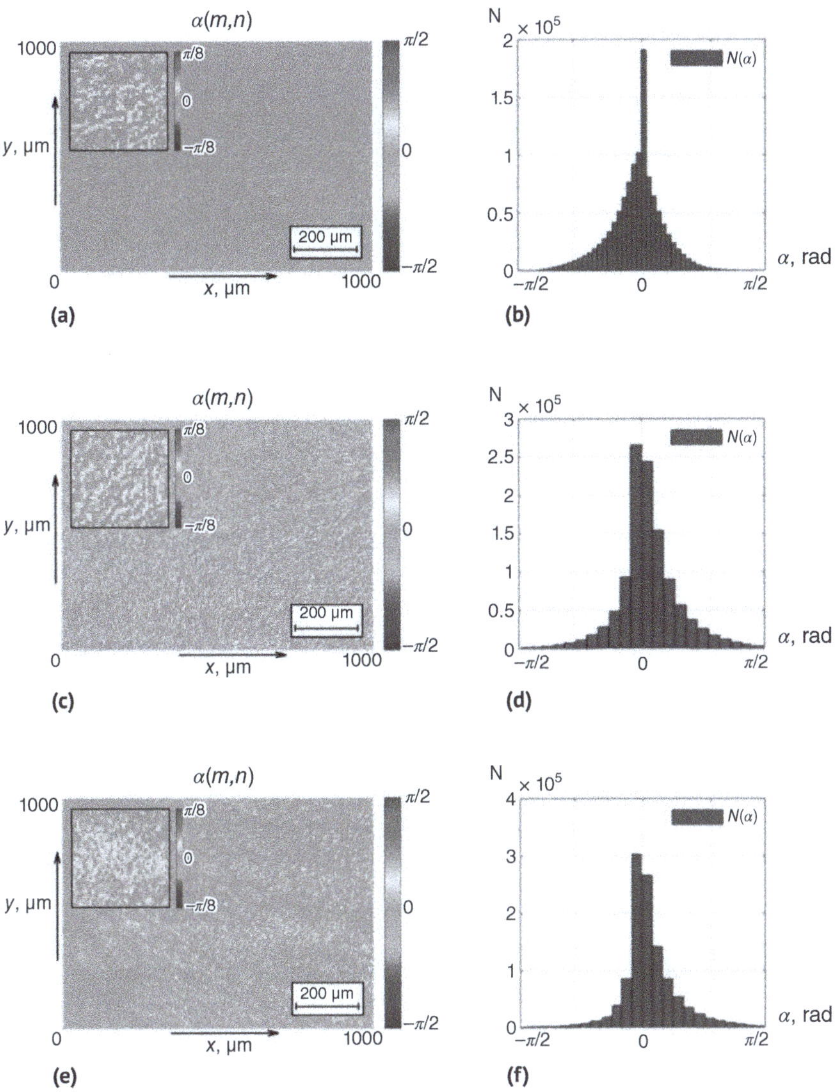

Figure 7.2 Maps of $\alpha(m, n)$ ((a), (c), (e)) and histograms of the polarization azimuth distributions ((b), (d), (f)) for microscopic images of histological sections from adenocarcinoma biopsies with differentiation grades 3 + 3 ((a), (b)), 3 + 4 ((c), (d)), and 4 + 4 ((e), (f)).

Analysis of the data in Table 7.3 indicates that the highest differential diagnosis accuracy for prostate adenocarcinomas (grades 3 + 3 to 4 + 4) using direct polarization mapping is 80.5%. However, the accuracy in distinguishing intermediate grades (3 + 4 vs. 4 + 4) remains unsatisfactory at 70.8–72.2%. This modest outcome is primarily due to strong background depolarization, which diminishes the contrast in polarization maps of prostate adenocarcinoma biopsy sections. To counteract the scattering effects that distort polarization signatures of pathological changes in the polycrystalline structure, preparing histological sections with smaller geometric thickness could be a potential solution. However, stable geometric parameters for native frozen prostate tissue sections can only be achieved at thicknesses $h \geq 30\ \mu m$, which correspond to a significant degree of depolarization of multiply scattered laser radiation (Table 7.1). One promising approach to enhance the balanced accuracy of rapid and differential diagnostics for malignant prostate tumors involves employing more sensitive interference-based methods to detect the structure of polarization-inhomogeneous fields.

7.2.3.2 Polarization–Interference Maps of Prostate Tumor Samples

Figure 7.4 displays a series of fragments showing polarization–interference maps of microscopic images from histological sections. These include highly differentiated adenocarcinoma (ISUP 1, 3 + 3; Figure 7.4a), moderately differentiated (ISUP 2–3, 3 + 4; Figure 7.4b), and poorly differentiated adenocarcinoma (ISUP 4, 4 + 4; Figure 7.4c) [25–29].

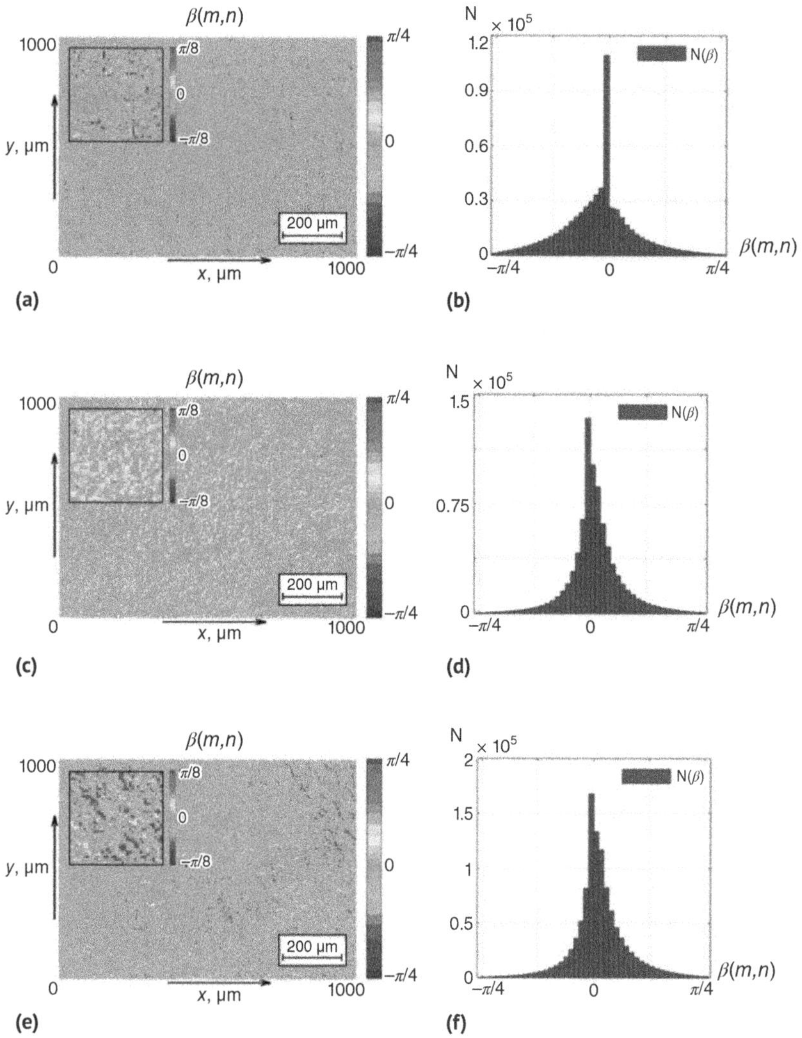

Figure 7.3 Maps of $\beta(m, n)$ (a), (c), (e) and histograms of the polarization ellipticity distributions (b), (d), (f) for microscopic images of histological sections from adenocarcinoma biopsies with differentiation grades 3 + 3 (a), (b), 3 + 4 (c), (d), and 4 + 4 (e), (f).

Table 7.2 Statistical parameters of polarization maps w of microscopic images of malignant prostate tumor histological sections.

w	$\alpha(m, n)$			$\beta(m, n)$		
M_i	Group 1	Group 2	Group 3	Group 1	Group 2	Group 3
M_1	0.044 ± 0.021	0.041 ± 0.0019	0.034 ± 0.0016	0.028 ± 0.016	0.027 ± 0.014	0.024 ± 0.012
M_2	0.25 ± 0.013	0.23 ± 0.012	0.19 ± 0.011	0.22 ± 0.011	0.19 ± 0.009	0.15 ± 0.008
M_3	1.08 ± 0.039	1.21 ± 0.044	1.42 ± 0.058	1.09 ± 0.043	1.31 ± 0.051	1.64 ± 0.068
M_4	1.21 ± 0.042	1.41 ± 0.052	1.75 ± 0.065	1.51 ± 0.053	1.86 ± 0.068	2.12 ± 0.089

The data show that the intensity distribution of interference fringes in prostate tissue images varies across different regions and has a wide range of values in all groups. In other words, the local contrast $W(m \times n)$ is also nonuniformly distributed across the microscopic image. This behavior can be explained using different physical principles.

The value of the local contrast $W(m \times n)$ is determined by the following factors:

- the ratio between the intensities of the object field $\tilde{I}(m, n)$ and the reference field $I_0 - \tilde{I} = I_0 \Leftrightarrow W(m, n) \rightarrow \max$;
- the match between the polarization state of the reference wave (Ω) and the local polarization state of the object wave (α, β) waves – $\alpha \rightarrow \Omega; \beta \rightarrow 0$;

Table 7.3 Operational characteristics of the diagnostic performance of the polarization mapping method on microscopic images of prostate tumor sections.

w	$\alpha(m, n)$			$\beta(m, n)$		
Groups	**Group 1–2**	**Group 1–3**	**Group 2–3**	**Group 1–2**	**Group 1–3**	**Group 2–3**
Se (%)	80.5	83.3	72.2	77.8	83.3	72.2
Sp (%)	77.8	80.5	69.4	72.2	80.5	72.2
Ac (%)	79.15	81.9	70.8	75	81.9	72.2

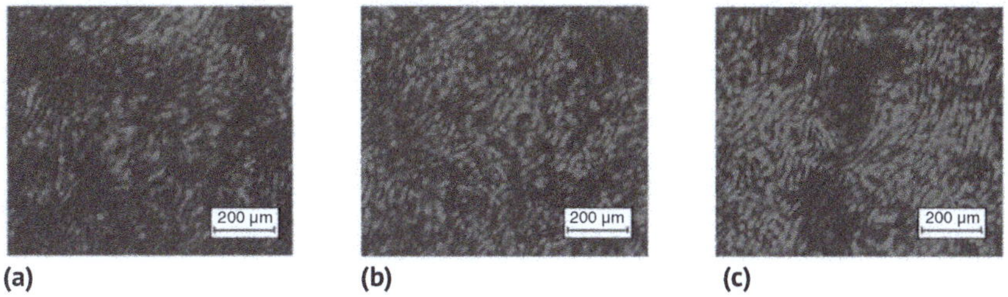

Figure 7.4 Polarization–interference maps of microscopic images of histological sections of malignant prostate tumors.

Variations in the orientation and geometric parameters of birefringent fibrils cause inhomogeneous distributions of optical axis directions $\rho(m, n)$ and phase shifts $\delta(m, n)$ arise between orthogonal linearly polarized components of the laser field amplitude U. As a laser beam travels through an anisotropic layer, it experiences complex topographic modulation of its amplitude $(\widetilde{U})$, phase $(\widetilde{\delta})$, and polarization $(\widetilde{\alpha}, \widetilde{\beta})$.

When a monopolarized laser reference wave with constant amplitude E_0 and $\phi \equiv 0$ interferes with an object wave exhibiting polarization $(\widetilde{\alpha}, \widetilde{\beta})$, amplitude $(\widetilde{U})$, and phase $(\widetilde{\delta})$ inhomogeneities, a multiparametric interference pattern is formed. The addition of amplitudes E_0 and $\widetilde{U}$ across different coordinates (m, n) produces an interference distribution containing regions of both high and low local contrast $W(m, n)$.

Coordinate-dependent and statistical variations in $W(m, n)$ observed in microscopic images of adenocarcinoma histological sections are presented in Figure 7.5.

Comparative analysis revealed that the ranges of local contrast values differ significantly between groups (Figure 7.5b, d, f) high differentiation $(3 + 3 - 0 \leq W \leq 0.6)$ moderate differentiation $(3 + 4 - 0 \leq W \leq 0.75)$ low differentiation $(4 + 4 - 0 \leq W \leq 0.85)$.

These differences result from the significant polarization heterogeneity seen in images of highly differentiated (3 + 3) prostate tumors (Figures 7.4a, b and 7.5a, b). As a result, the decorrelation between the polarization state of the reference wave (Ω) and the local polarization states (α, β) at coordinates (m, n) is greater than in samples of moderately (3 + 4) and poorly differentiated (4 + 4) adenocarcinomas. In all cases, the phase modulation of the multiply scattered component of the object field is similar. As a result, microscopic images of highly differentiated tumors generally show lower local contrast values $W(m, n)$.

Quantitatively, these processes are characterized by a set of statistical moments Z_i that describe the local contrast maps $W(m, n)$, Table 7.4.

The most significant differences between groups were seen in the second- and third- order statistical moments, which indicate the variance and skewness of local-contrast distributions values for W:

- Group 1–2 $\Delta M_2(W) = 1.32$; $\Delta M_3(W) = 1.46$;
- Group 1–3 $\Delta M_2(W) = 1.95$; $\Delta M_3(W) = 2.48$;
- Group 2–3 $\Delta M_2(W) = 1.28$; $\Delta M_3(W) = 1.69$.

The range of differences between groups in the magnitudes of the second- and third-order moments that characterize local-contrast distributions W increased by a factor of about 1.3–1.5 compared with the data from direct polarization mapping (m, n) and $\beta(m, n)$. Table 7.4 presents these values.

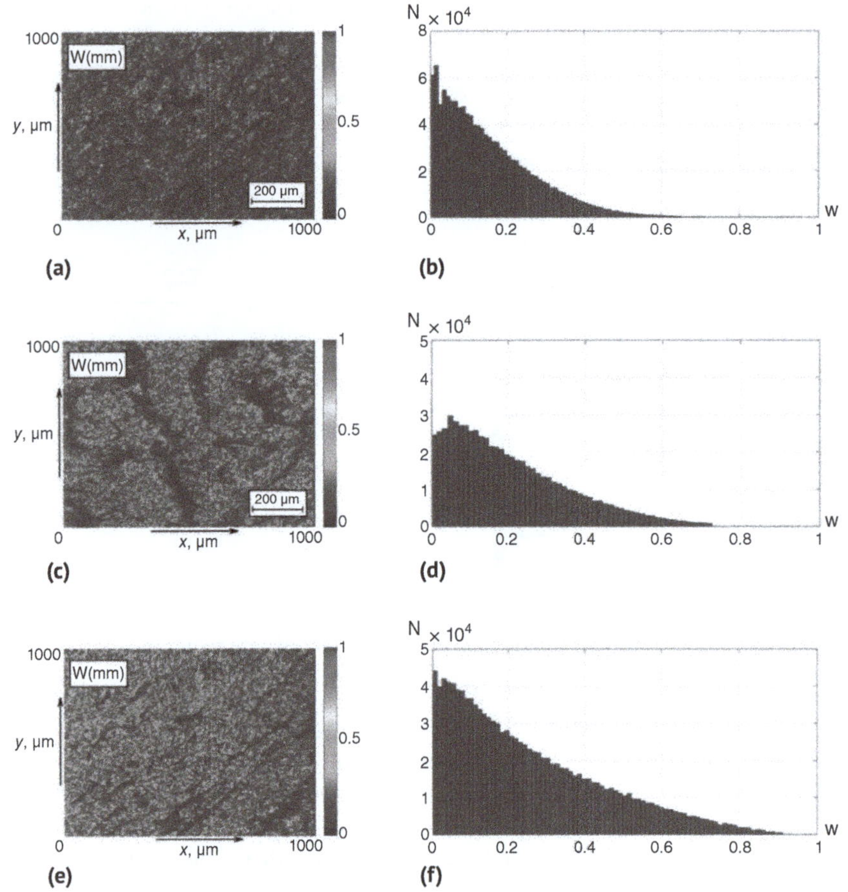

Figure 7.5 Polarization interference maps ((a), (c), (e)) and histograms of local contrast values $W(m, n)$ ((b), (d), (f)) in microscopic images of histological sections of adenocarcinomas with high (3 + 3; (a), (b)), moderate (3 + 4; (c), (d)), and poor (4 + 4; (e), (f)) differentiation.

Table 7.5 shows the sensitivity Se, specificity Sp, and balanced accuracy Ac of the polarization–interference mapping method.

The polarization–interference mapping of microscopic images of prostate tissue sections demonstrated excellent differentiation accuracy (3 + 3–4 + 4), ranging from 91.7% to 93.1%, for identifying malignant tumors. Additionally, a high balanced accuracy (90.3%) was achieved in determining the severity of the oncological process (3 + 4–4 + 4).

7.2.4 Conclusions

The polarization–interference mapping method of prostate tissue layers, using local-contrast maps, is supported by analytical evidence. Based on polarization interferometry, this technique was experimentally validated for quick (under 15 minutes) differential diagnosis of adenocarcinomas with different severity levels. A representative set of histological biopsy sections of malignant tumors was examined.

The most sensitive statistical parameters were identified as the variance and skewness of the coordinate distributions of local-contrast values within the interference pattern of a polarization-inhomogeneous object field.

The operational characteristics – sensitivity Se, specificity Sp, and balanced accuracy Ac – were determined, demonstrating the diagnostic power of polarization and polarization–interference mapping.

The use of statistical analysis of local-contrast maps provided excellent differentiation accuracy $\left(Ac_{12;13}(W) = 95.8\%\right)$ for distinguishing malignant conditions 3 + 3–4 + 4) in prostate tissue.

Furthermore, the method proved effective in accurately diagnosing histological sections of malignant tumors that have moderate (3 + 4) and poor (4 + 4) differentiation $Ac_{23}(W) = 90.3\%$ (W-based metrics).

Table 7.4 Statistical parameters of the local contrast map $W(m, n)$ for microscopic images of malignant prostate tumor histological sections.

w	$W(m,n)$		
M_i	Group 1	Group 2	Group 3
M_1	0.34 ± 0.015	0.42 ± 0.019	0.49 ± 0.023
M_2	0.21 ± 0.011	0.32 ± 0.014	0.41 ± 0.018
M_3	0.82 ± 0.039	0.56 ± 0.025	0.33 ± 0.014
M_4	0.45 ± 0.021	0.37 ± 0.016	0.32 ± 0.014

Table 7.5 Operational characteristics of the diagnostic power of polarization-interference.

w	$W(m,n)$		
Groups	Group 1-2	Group 1-3	Group 2-3
Se (%)	91.7	94.4	91.7
Sp (%)	91.7	91.7	88.9
Ac (%)	91.7	93.1	90.3

Table 7.6 Optical parameters of polycrystalline blood plasma films.

Parameter	Group 1	Group 2	Group 3
Attenuation (extinction) coefficient τ, cm^{-1}	0.31 ± 0.021	0.33 ± 0.023	0.32 ± 0.022
Depolarization degree Λ, %	18 ± 0.08	22 ± 0.11	20 ± 0.105

7.3 Polarization-Interference Mapping of Polycrystalline Blood Plasma Films in the Differential Diagnosis of Benign and Malignant Prostate Tumors

7.3.1 Biological Samples

Three representative groups of blood plasma smears were formed:

- Group 1 consisted of $k = 36$ polycrystalline film samples of blood and its plasma taken from patients with well-differentiated adenocarcinoma (3 + 3 on Gleason's Pattern scale - ISUP grade 1);
- Group 2 consisted of $k = 36$ polycrystalline film samples of blood and its plasma taken from patients with moderately differentiated adenocarcinoma (3 + 4; 4 + 3 on Gleason's Pattern scale - ISUP grade 2-3);
- Group 3 consisted of $n = 36$ poorly differentiated polycrystalline adenocarcinoma blood plasma samples (4 + 4; 5 + 3; 3 + 5 on Gleason's Pattern scale - ISUP grade 4).

Table 7.6 shows the optical parameters of polycrystalline blood plasma films obtained from patients with prostate tumors.

To differentiate malignant tumors, for each of the statistical moments $Z_{i=1;2;;3;4}$ (equation (7.18)), the sensitivity $\left(\mathrm{Se}_{12} = \frac{a_{12}}{a_{12}+b_{12}}100\%;\ \mathrm{Se}_{12} = \frac{a_{13}}{a_{13}+b_{13}}100\%\right)$, specificity $\left(\mathrm{Sp}_{12} = \frac{c_{12}}{c_{12}+d_{12}}100\%;\ \mathrm{Sp}_{13} = \frac{c_{13}}{c_{13}+d_{13}}100\%\right)$, and balanced accuracy $(\mathrm{Ac}_{12} = 0.5(\mathrm{Se}_{12} + \mathrm{Sp}_{12}); \mathrm{Ac}_{13} = 0.5(\mathrm{Se}_{13} + \mathrm{Sp}_{13})$ were calculated. Here, $a_{12}(a_{13})$ and $b_{12}(b_{13})$ are the numbers of correct and incorrect diagnoses within group 2 (group 3); and $c_{12}(c_{13})$ and $d_{12}(d_{13})$ are the same within group 1. Similarly, to differentiate the grade of cancer, the sensitivity $\left(\mathrm{Se}_{23} = \frac{a_{23}}{a_{23}+b_{23}}100\%\right)$, specificity $\left(\mathrm{Sp}_{23} = \frac{c_{23}}{c_{23}+d_{23}}100\%\right)$, and balanced accuracy $(\mathrm{Ac}_{23} = 0.5(\mathrm{Se}_{23} + \mathrm{Sp}_{23})$ were calculated for each of the statistical moments $Z_{i=1;2;3;4}$. Here, a_{23} and b_{23} are the numbers of correct and incorrect diagnoses within group 3, while c_{23} and d_{23} are the same within group 2 [30–33].

7.3.2 Experimental Results and Discussion

Figure 7.6 presents a series of fragments showing polarization–interference maps of microscopic images of blood plasma layers from patients with different levels of adenocarcinoma differentiation: high (3 + 3, Figure 7.6a), moderate (3 + 4, Figure 7.6b), and low (4 + 4, Figure 7.6c).

The data show that the distribution of interference fringe intensity in the image plane of polycrystalline plasma film samples across all patient groups is characterized by topographic inhomogeneity and a wide range of variation. In other words, the value of the local contrast $W(m \times n)$ is also inhomogeneously distributed in the plane of the microscopic image. This can be attributed to the following physical considerations. It has been shown that the value is determined by the phase consistency between the reference (phase shift ϕ) and object (phase shift $\delta(m, n)$) waves.

Due to the dispersion of phase shifts $\delta(m, n)$ introduced by birefringent crystals of albumin and globulin of fibrils, an interference distribution is formed with areas of maximum and minimum local contrast $W(m \times n)$.

Coordinate and statistical differences in the distributions $W(m, n)$ of images of polycrystalline blood plasma films from patients with adenocarcinoma with different degrees of differentiation are shown in Figure 7.7.

Comparative analysis of the obtained data revealed a different range of changes in the value of local contrast (Figure 7.7b, d, f) for polycrystalline films of blood plasma from patients with adenocarcinomas of high ($0 \le W \le 0.35$), moderate ($0 \le W \le 0.25$), and low ($0 \le W \le 0.15$) degrees of differentiation.

From a physical standpoint, the results obtained can be associated with changes in the albumin-to-globulin ratio in prostate tumors. It is known that a decrease in the concentration of birefringent needle albumin molecules accompanies malignant processes. Therefore, there is a decrease in phase modulation in the plane of the microscopic images of polycrystalline films of blood plasma from patients with malignant prostate tumors. On this basis, the local contrast $W(m, n)$ at the points of the corresponding interference distributions decreases.

Such processes are quantitatively characterized by a set of statistical moments M_i describing the local contrast maps $W(m, n)$, Table 7.7.

The largest between-group differences were observed for the third– and fourth-order statistical moments, which characterize the skewness and kurtosis of the distributions of local contrast values of W:

- Group 1–2 $\Delta M_3(W) = 1.5$; $\Delta M_4(W) = 1.63$;
- Group 1–3 $\Delta M_3(W) = 1.97$; $\Delta M_4(W) = 2.15$;
- Group 2–3 $\Delta M_3(W) = 1.31$; $\Delta M_4(W) = 1.32$.

Table 7.8 shows the values of the sensitivity Se, specificity Sp, and balanced accuracy Ac for the method of polarization–interference mapping of polycrystalline blood plasma films from patients with malignant prostate tumors.

Using the polarization–interference mapping method on microscopic images of polycrystalline blood plasma films, we achieved excellent differentiation accuracy of 93.1% to 95.8% between adenocarcinomas with high (3 + 3) and low (4 + 4) degrees of differentiation, as shown in Figure 7.8e, f. Additionally, a high balanced accuracy of 90.3% was obtained when distinguishing the severity of the oncological process, separating moderately differentiated adenocarcinomas (3 + 4) from low-differentiation tumors (4 + 4) in Figure 7.8c–f.

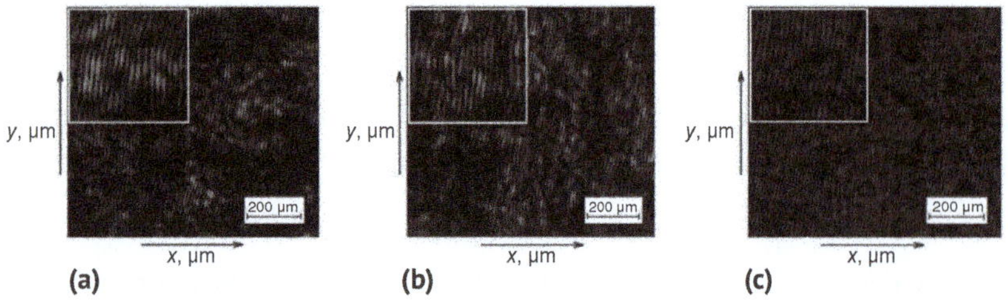

Figure 7.6 Polarization–interference maps of microscopic images of polycrystalline blood plasma films from patients with malignant prostate tumors.

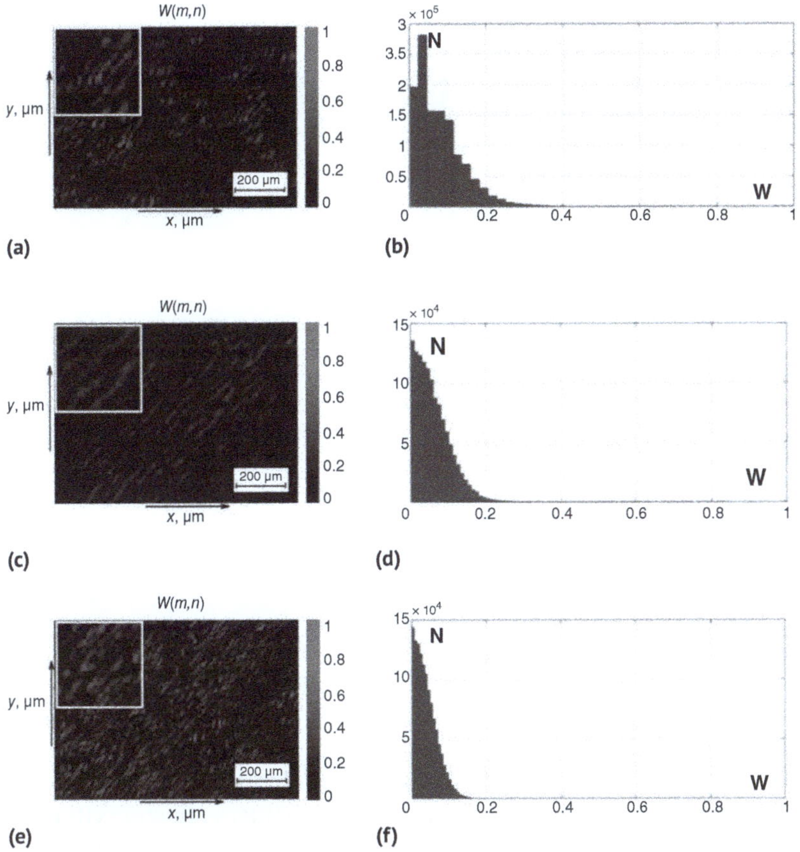

Figure 7.7 Polarization–interference maps (a, c, e) and histograms of distributions of local contrast values $W(m, n)$ (b, d, f) in microscopic images of polycrystalline blood plasma films from patients with adenocarcinomas of high (3 + 3, a, b), moderate (3 + 4, c, d), and low (4 + 4, e, f) degrees of differentiation.

Table 7.7 Statistical parameters of the local contrast map $W(m, n)$ for microscopic images of polycrystalline blood plasma films from patients with malignant prostate tumors.

w	$W(m, n)$		
M_i	**Group 1**	**Group 2**	**Group 3**
M_1	0.11 ± 0.005	0.08 ± 0.004	0.06 ± 0.003
M_2	0.09 ± 0.004	0.07 ± 0.003	0.04 ± 0.002
M_3	0.98 ± 0.039	1.47 ± 0.072	1.93 ± 0.084
M_4	1.45 ± 0.071	2.37 ± 0.096	3.12 ± 0.14

Table 7.8 Operational characteristics of the diagnostic power of the polarization-interferometry method.

w	$W(m, n)$		
Groups	**Group 1–2**	**Group 1–3**	**Group 2–3**
Se (%)	94.4	97.2	91.7
Sp (%)	91.7	94.4	88.9
Ac (%)	93.1	95.8	90.3

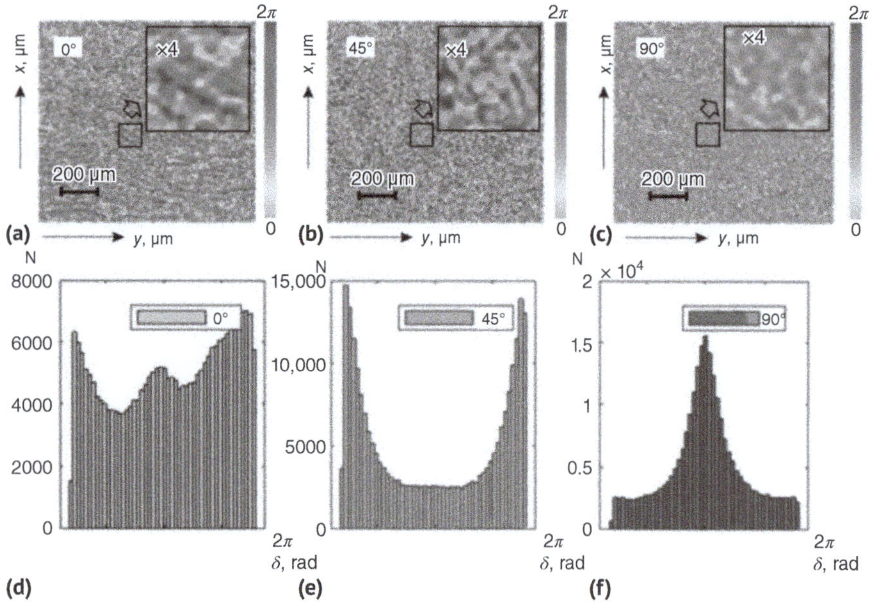

Figure 7.8 Coordinate ($x = 0 - 1000\ \mu m$; $y = 0 - 1000\ \mu m$) and the statistical structure of polarization azimuth maps obtained from microscopic images of the myocardium histological section.

7.3.3 Conclusions

The polarization–interference mapping method for polycrystalline blood plasma films from patients with prostate tumors, including the creation of local contrast maps, has been analytically validated. Using polarization interferometry as its basis, this method was experimentally evaluated for rapid (under 15 minutes) differential diagnosis of adenocarcinomas with different levels of differentiation.

A representative sample of blood plasma films from patients with malignant prostate tumors was examined. The most sensitive statistical measures were the skewness and kurtosis of the coordinate distributions of local contrast values within the interference pattern of the polarization-inhomogeneous object field.

The operational characteristics (sensitivity Se, specificity Sp, and balanced accuracy Ac) were determined, demonstrating the diagnostic power of polarization- and polarization–interference mapping.

The method demonstrated excellent differentiation accuracy, ranging from 93.1% to 95.8%, between high (3 + 3) and low (4 + 4) grade adenocarcinomas ((e), (f)). Additionally, it achieved a high balanced accuracy of 90.3% in distinguishing the severity of the oncological process, effectively separating moderately differentiated adenocarcinomas (3 + 4; 4 + 3) from low-differentiation tumors (4 + 4; 3 + 5; 5 + 3) ((c), (d), (e), (f)).

7.4 3D Digital Holographic Polarimetry of Diffuse Optically Anisotropic Biological Tissue Object Fields

This section discusses the development and experimental testing of techniques for 3D polarization–interference digital holographic polarimetry of multiply scattered object fields in biological tissues. The research examines two types of morphological boundaries: the organized fibrous networks in the myocardium and the disordered structures of liver parenchyma. The goal is to evaluate how layered phase selection can enhance holographic reconstruction of different scattered components in the polarization-inhomogeneous field within a diffuse biological layer [34–39].

7.4.1 Objects of Investigation

The choice of research subjects is grounded in the synthesis of analytical and applied results from polarimetric studies of biological tissues. These results encompass the following:

- Any biological tissue layer's optical properties can be modeled as a two-component "amorphous–polycrystalline" matrix.
- The polarization properties of biological tissue's polycrystalline architecture result from both structural anisotropy, such as dendritic or spatially organized fibrillary protein networks, and chiral anisotropy, which involves clusters of optically active molecular domains.

In actual biological tissues, the architectonics display both structural anisotropy, such as linear birefringence and dichroism, and chiral anisotropy, including circular birefringence and dichroism.

The Mueller-matrix formalism provides the most comprehensive and universal description of how optical radiation interacts with these biological crystal ensembles and the resulting sets of polarization maps of the object field.

The statistical parameters of azimuth and ellipticity in polarization maps relate directly to the features of structural and chiral anisotropy in biological tissues. Therefore, exploring these relationships analytically and experimentally via polarimetric analysis of "boundary structures" within the optically anisotropic architecture of biological crystals is a crucial area of both fundamental and applied research.

Fundamentally, among the diverse types of biological tissue structures, two "limiting" cases of how anisotropic polarization manifests can be distinguished:

- Structural anisotropy in fibrillar tissues refers to the spatially ordered arrangements of myocardial myosin fibrils.
- Chiral anisotropy in parenchymal tissues refers to clusters of optically active parenchymal molecules, such as those found in the liver.

The statistical properties of the polarization maps can also be utilized in diagnostic applications within the universal Mueller-matrix framework to identify pathological alterations in the optical anisotropy across a wide range of human biological tissues.

Morphologically, the myocardium is made up of well-structured networks of myosin fibers, while the liver consists of disorganized, island-like clusters known as Langerhans islets.

Optically, these tissues possess both shared and distinct features.

Common characteristics – each tissue layer studied represents a two-component isotropic–anisotropic matrix.

The anisotropic component induces phase modulation, δ_p and ξ_p, between the orthogonal components of the propagating laser radiation within the soft tissue.

Distinct characteristics – fibrous networks create structural anisotropy, which appears as linear birefringence (LB). This leads to the creation of ensembles of polarized waves, each characterized by specific azimuth and ellipticity values, as shown in relationships (7.4), (7.5), (7.7), and (7.8). In parenchymal structures, circular birefringence (CB) predominates, producing spatially distributed regions in the laser field with varying polarization azimuths.

Histological sections were created using a standard microtomy method with rapid freezing.

The optical and geometric parameters of the histological sections are presented in Table 7.9.

7.4.2 Myocardium Histological Sections Object Fields Polarization Azimuths Maps

Figure 7.8 presents a series of fragments illustrating integral maps of polarization azimuth (a)–(c) and corresponding histograms (d)–(f) that display the distribution of polarization azimuth values in microscopic images of the myocardial sample. Multichannel irradiation was utilized with linearly polarized laser beams $\alpha_0 = 0°$ – (a), (d); $\alpha_0 = 45°$ – (b), (e); $\alpha_0 = 90°$ – (c), (f).

Analysis of the results showed that the polarization azimuth maps of the microscopic images in the diffuse histological section of the myocardium exhibit both coordinate-dependent features and unique structural variations for each $\alpha(\alpha_0, m \times n)$, (a)–(c).

Table 7.9 Optical and geometric parameters of biological tissue samples.

Biological tissues	Myocardium	Liver
Geometric thickness, h, μm	35–40	35–40
Coefficient of attenuation, τ μm	0.11–0.14	0.12–0.15
Degree of depolarization, Δ, %	41–48	44–52

The experimentally measured histograms $G(\alpha, \alpha_0)$, representing the distributions of polarization azimuth values, are asymmetric and exhibit substantial ranges of azimuth variation – (d)–(f).

The quantitative statistical structure of the polarization maps $\alpha(\alpha_0, m \times n)$ is characterized by a set of statistical moments of the first through fourth orders, denoted as $M_{g=1,2,3,4}$). The corresponding values of these statistical moments are presented in Table 7.10.

A comparison of the results obtained from the statistical analysis of the integral polarization azimuth maps $\alpha(\alpha_0, m \times n)$ revealed that all coordinate polarization distributions deviate from the normal Gaussian – all statistical moments $M_{g=1,2,3,4} \neq 0$. Variations in the magnitudes of these statistical moments $M_{g=1,2,3,4}$ by factors of 2–7 were observed depending on changes in the polarization state of the incident laser beam.

Physically, these results can be explained by the combined effect of two factors:

1. *Object-related factor*: This arises from the complex orientation structure (ρ_p) of multiscale fibrillar myocardial networks (characterized by size (d)), which forms spatial phase distributions δ_p.
2. Because of single and multiple laser radiation interactions with this polycrystalline component, a polarization-inhomogeneous, "object-related" contribution to the laser field arises.
3. When the polarization azimuth α_0 of the incident beam changes, the distribution of angles $\alpha_0 \pm \rho_p(m, n)$, is transformed, leading to the formation of individual topographic maps $\alpha(\alpha_0, m \times n)$.
4. *Field-related factor*: This occurs due to secondary interference from partially polarized laser waves produced by optically anisotropic fibrous networks.
5. As a result, an extra "diffuse" inhomogeneous polarization component develops in the laser field of the object.

A key challenge in solving inverse problems for diagnosing polycrystalline structures in diffuse biological layers is to establish conditions that reduce the impact of multiple scattering on the vector structure of the object field.

A method designed for this purpose is the phase scanning technique, which is used on holographically reconstructed distributions of the complex amplitudes of laser radiation.

The results of phase scanning and the reconstruction of layered maps $\alpha(\theta, m \times n)$ for $\theta = \pi/6$ *i* $\theta = \pi/8$ are presented in Figures 7.9 and 7.10, respectively.

A comparative analysis of the layered polarization azimuth maps reconstructed through the phase scanning method with digital holographic reconstruction of layered distributions of complex amplitudes in partially depolarizing, optically thick myocardium histological sections showed that, across all examined phase planes θ_i, the maps $\alpha(\theta_i, m \times n)$ exhibit distinct spatial heterogeneity dependent on variations in the polarization state α_0 of the incident laser beam (see Figures 7.9 and 7.10a–c and d–f).

Table 7.11 shows the statistical analysis results for the layered polarization azimuth maps obtained in various phase planes of the microscopic images of the myocardium histological section.

It was found that, as the value of θ decreases, the dispersion $M_2(\theta; \alpha)$ – representing the spread of random polarization azimuth values in the phase planes of the object field-systematically decreases by a factor of 2–3.

In contrast, the higher-order statistical moments characterizing skewness $M_3(\theta; \alpha)$ and kurtosis $M_4(\theta; \alpha)$ of the distributions $\sum_{l=1}^{n} A_l\left(\alpha_l^{\otimes}\right) + \sum_{l=1}^{n} C_l(\alpha(I)_l)$ increase to comparable magnitudes (see Tables 7.10 and 7.11).

Further reduction of the parameter $< \pi/8$ for the phase cross-section of the object field practically does not lead to changes in the statistical structure of the polarization maps. This suggests that, in these phase planes, a single-scattering regime occurs within the myocardial diffuse layer. Therefore, using the phase-scanning method with digital holographic reconstruction of the microscopic image phase specifically, polarization-azimuth maps-enables experimental differentiation of various multiplicity components. This approach isolates the "object" component, which is practically single-scattered, from the broader "diffuse" background created by high-order interactions.

Table 7.10 Statistical parameters of polarization azimuth maps for microscopic images of the myocardium histological section.

α_0	0°	45°	90°
M_1	0.53	0.13	0.08
M_2	1.63	0.64	0.97
M_3	0.36	0.24	0.19
M_4	0.41	0.29	0.22

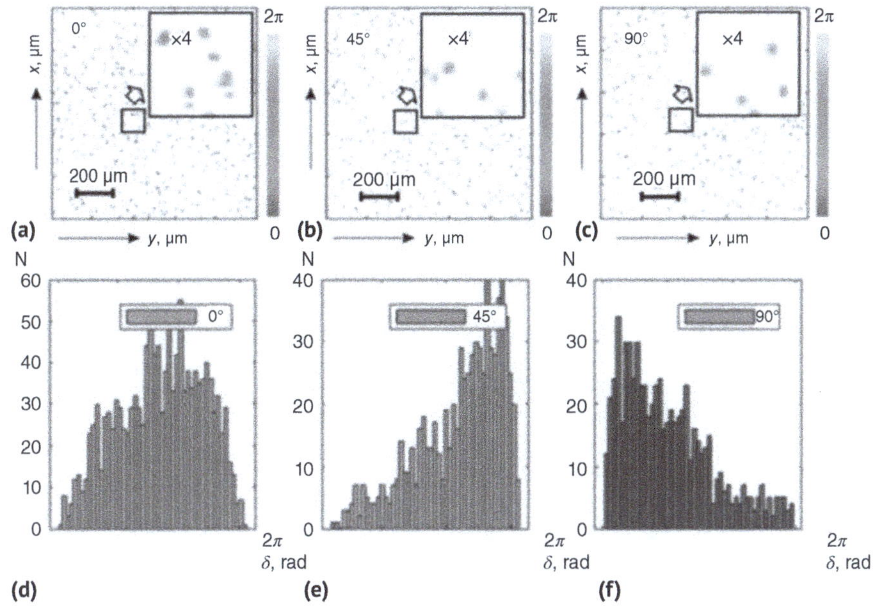

Figure 7.9 Coordinate ($x = 0 - 1000\ \mu\text{m}$; $y = 0 - 1000\ \mu\text{m}$) and the statistical structure of polarization azimuth maps of microscopic images of the myocardium histological section in the phase plane $\theta = \pi/6$.

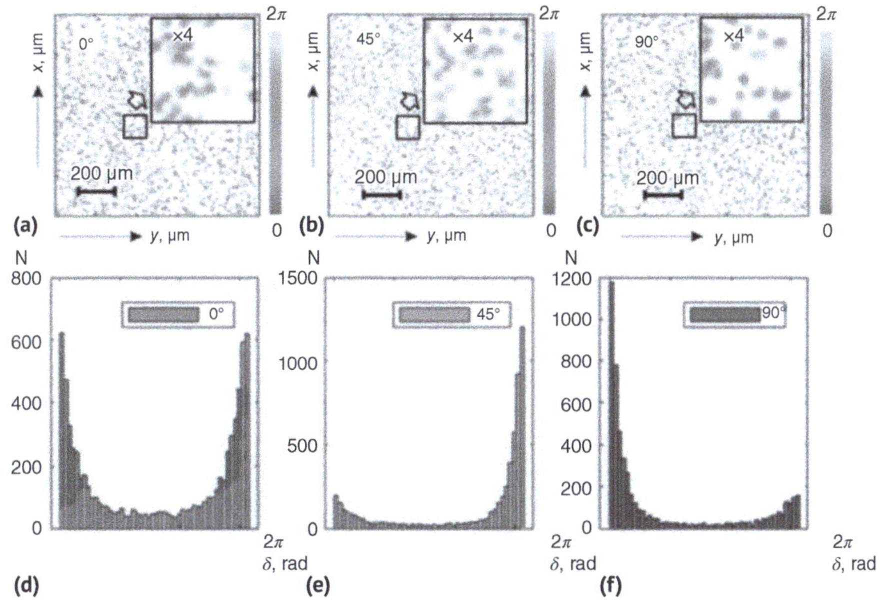

Figure 7.10 Coordinate ($x = 0 - 1000\ \mu\text{m}$; $y = 0 - 1000\ \mu\text{m}$) and the statistical structure of polarization azimuth maps of microscopic images of the myocardium histological section in the phase plane $\theta = \pi/8$.

7.4.3 Parenchymal Liver Histological Sections Object Fields Azimuths Polarization Maps

Figure 7.11 shows a series of experimentally measured integral azimuth polarization maps (a)–(c) and histograms (d)–(f) of the liver histological section microscopic images.

Both coordinate and individual heterogeneity of the experimentally measured azimuth polarization maps $\alpha(\alpha_0, m \times n)$ in the parenchymal liver tissues, microscopic images are observed, similar to the findings in the case of myocardium (Figure 7.11a–c). The histograms $G(\alpha, \alpha_0)$ of the azimuth values in the liver tissue exhibit significantly larger ranges of variation in random azimuth values compared to the fibrous myocardium (Figures 7.8d, f and 7.11d, f). On the other hand, the "polarization" differences between $G(\alpha, \alpha_0)$ are not as pronounced.

Table 7.11 Statistical parameters of polarization azimuth maps for microscopic images of the myocardium histological section.

θ		$\pi/6$			$\pi/8$	
α_0	**0°**	**45°**	**90°**	**0°**	**45°**	**90°**
M_1	0.39	0.096	0.064	0.83	0.97	0.32
M_2	0.63	0.36	0.49	0.46	0.23	0.34
M_3	0.66	0.54	0.41	1.41	2.04	2.16
M_4	0.94	0.89	0.77	1.95	2.09	2.27

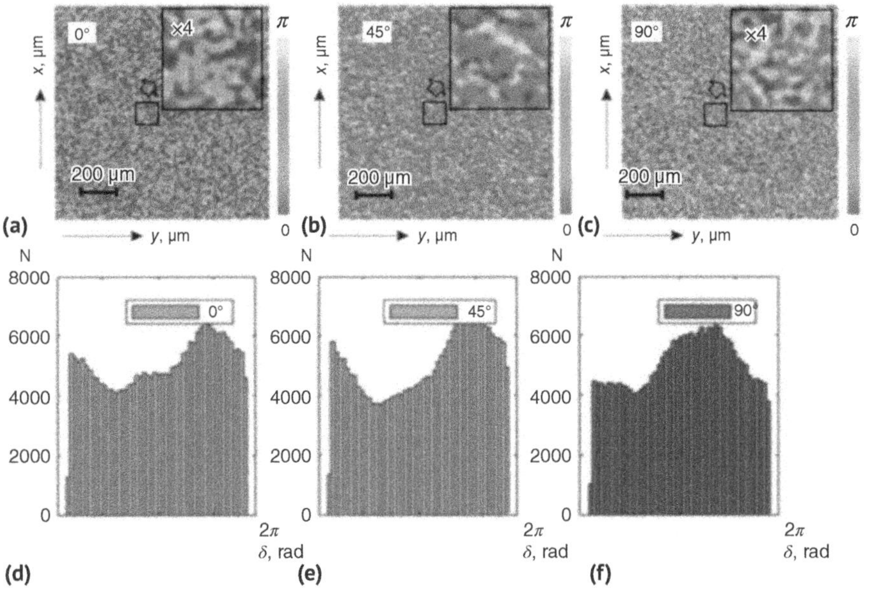

Figure 7.11 Coordinate ($x = 0 - 1000$ μm; $y = 0 - 1000$ μm) and statistical structure of the liver histological section microscopic images polarization azimuth maps.

Table 7.12 Statistical parameters of the liver histological section microscopic images polarization azimuth maps.

α_0	**0°**	**45°**	**90°**
M_1	0.74	0.13	0.072
M_2	1.33	0.98	1.18
M_3	0.105	0.12	0.13
M_4	0.13	0.16	0.11

Quantitatively, the coordinate statistical structure of the parenchymal liver layer images maps $\alpha(m \times n)$ is characterized by the set of statistical moments $M_{g=1,2,3,4}$, which is listed in Table 7.12.

Statistical analysis of microscopic images of a partially depolarizing hepatic parenchyma layer, using integral polarization azimuth maps, showed a significant decrease-up to threefold-in higher-order statistical moments $M_{3,4}$ compared to skewness and kurtosis, which characterize the coordinate distributions of myocardial microscopic images polarization azimuth maps. Conversely, the variance value M_2 of the random variable α is larger by 1.5 times (Tables 7.10 and 7.12).

From a physical perspective, the results mainly stem from the dominant effect of the object factor-circular birefringence *CB* of the hepatic parenchyma-along with the secondary interference of linearly polarized partial laser waves.

Consequently, a polarization-inhomogeneous "diffuse" component of the object laser radiation develops, marked by distributions of randomly varying azimuth angles, which are represented as the sum of individual elements components: $\sum_{l=1}^{n} A_l\left(\alpha_l^{\otimes}\right) + \sum_{l=1}^{n} C_l(\alpha(I)_l)$.

Furthermore, the spatial arrangement of optically anisotropic domains with azimuthal symmetry in the liver's morphological structure shows a relatively weak dependence in their distributions $\alpha(\alpha_0, m \times n)$ on variations in the polarization azimuth α_0 of the laser probe.

The results of phase scanning and reconstruction of layered maps $\alpha(\theta, m \times n)$ for $\theta = \pi/6$ *i* $\theta = \pi/8$ are presented in Figures 7.12 and 7.13, respectively.

Table 7.13 presents the data from the statistical analysis of the optically thick liver histological sections, microscopic images, and layered polarization azimuth maps.

A comparative analysis of polarization azimuth $\alpha(\theta, m \times n)$ layer-by-layer maps allows for the experimental identification of the 'object' single-scattered vector component of the laser field, similar to studies in other biological tissues like structured myocardium. This is performed within the context of the overall "diffuse" background generated by numerous interaction acts. This follows from the fact that when the phase parameter θ_k is reduced to the level of $\theta_k \leq \pi/8$, there is a successive decrease in the variance of the spread of the random values of the polarization azimuths $M_2(\alpha)$ by 3–4 times (Tables 7.12 and 7.13) with the subsequent "stabilization" of the values of the set of statistical moments $M_{g=1;2;3;4}(\alpha)$, which characterize the distributions $\sum_{l=1}^{n} A_l\left(\alpha_l^{\otimes}\right) + \sum_{l=1}^{n} C_l(\alpha(I)_l)$. In addition, the differences between the calculated values of $M_3(\alpha)$ and $M_4(\alpha)$ in the given phase plane of the object field of the myocardium (Table 7.12) and liver (Table 7.13) reach 2–2.5 times.

7.4.4 Myocardium Histological Sections Object Fields Polarization Maps Ellipticity

On a series of fragments of Figure 7.14 presents integral maps $\beta(\alpha_0; m, n)$ (a)–(c) and histograms (d)–(f) of myocardium histological sections microscopic images polarization ellipticity distributions [39–41].

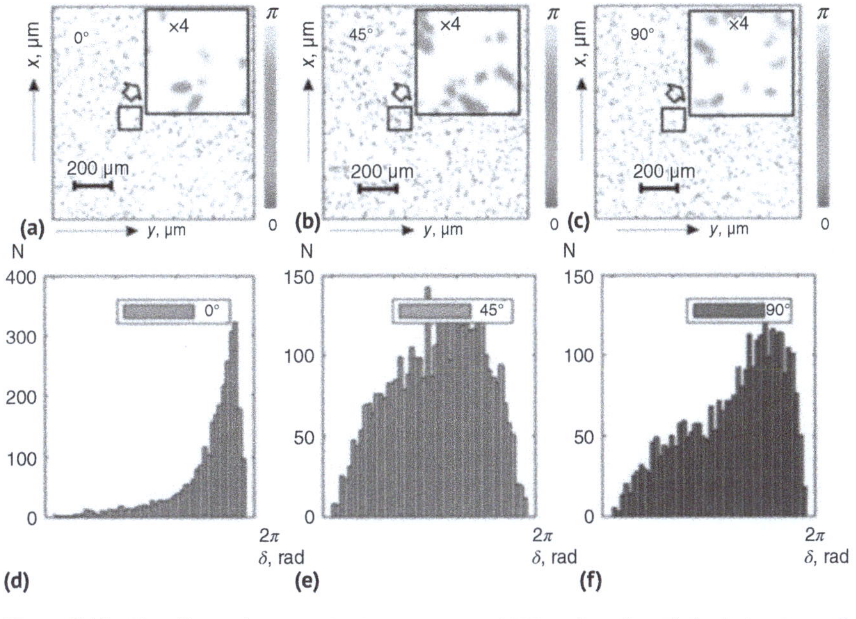

Figure 7.12 Coordinate ($x = 0 - 1000$ μm; $y = 0 - 1000$ μm) and statistical structure of the liver histological section microscopic images polarization azimuth maps in the phase cross-section $\theta = \pi/6$.

Table 7.13 Statistical parameters of the liver histological section microscopic images polarization azimuth maps.

θ	$\pi/6$			$\pi/8$		
α_0	0°	45°	90°	0°	45°	90°
M_1	0.65	0.11	0.057	0.39	0.087	0.032
M_2	0.43	0.62	0.51	0.23	0.43	0.37
M_3	0.55	0.48	0.63	0.69	0.57	0.44
M_4	0.63	0.55	0.72	0.44	0.86	0.73

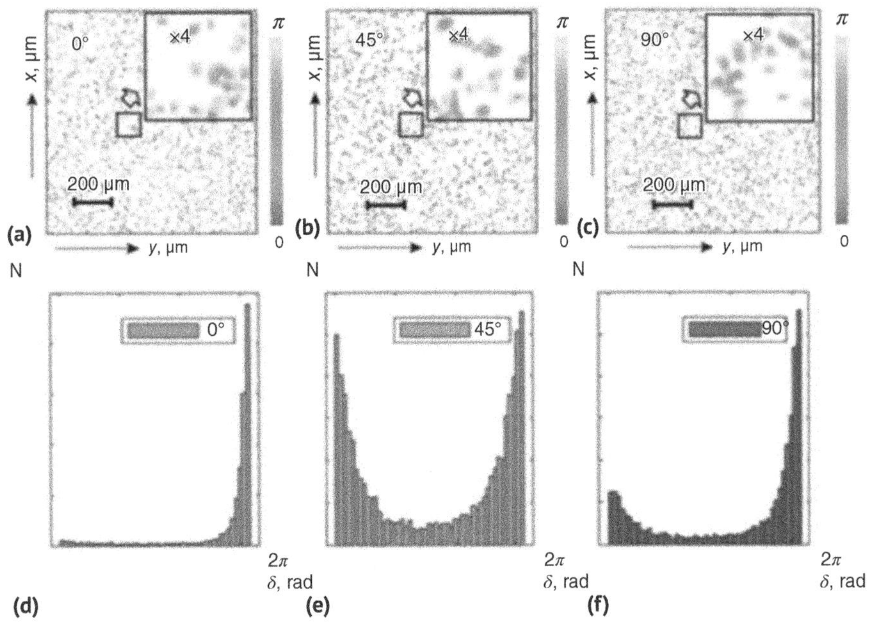

Figure 7.13 Coordinate ($x = 0 - 1000\ \mu m$; $y = 0 - 1000\ \mu m$) and statistical structure of the liver histological section microscopic images polarization azimuth maps in the phase section $\theta = \pi/8$.

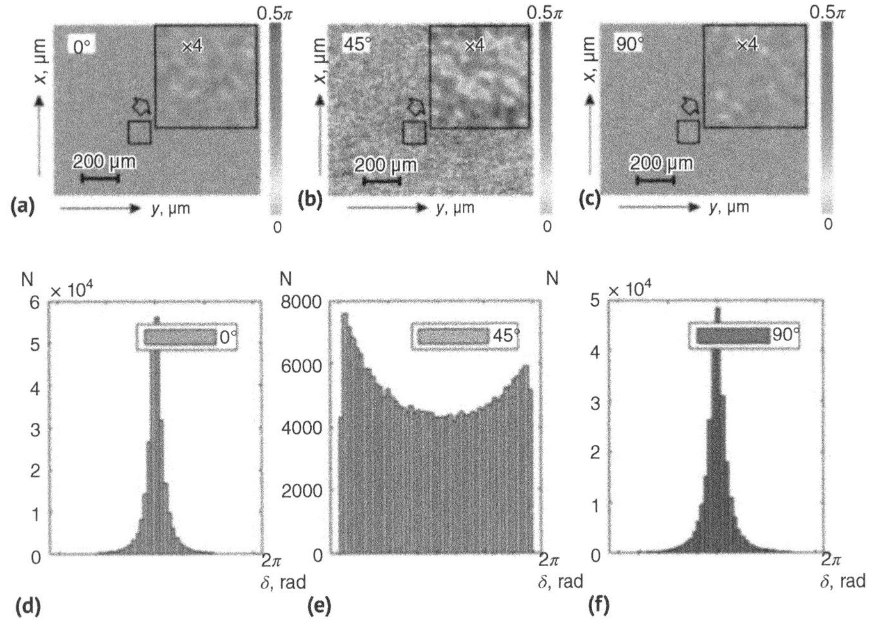

Figure 7.14 Coordinate ($x = 0 - 1000\ \mu m$; $y = 0 - 1000\ \mu m$) and statistical structure of the myocardium histological section, microscopic images polarization ellipticity maps.

The results showed the individual coordinate and statistical structure of the polarization maps parameter – polarization ellipticity $\beta(\alpha_0, m, n)$. Quantitatively, this is shown by the various intervals of "polarization" change, which remain within one order of magnitude, in the values of the statistical set moment $M_{g=1;2;3;4}$, which characterize the distributions $\sum_{l=1}^{n} B_l\left(\beta_l^{\otimes}\right) + \sum_{l=1}^{n} D_l(\beta(I)_l)$, Table 7.14.

From a physical point of view, the obtained results of polarimetry of $\beta(\alpha_0; m, n)$ ellipticity maps can be associated with the complex influence of the object (δ_l) and interference (φ_l) phase modulation between orthogonal components of complex amplitudes of partial coherent waves in volumes of the diffuse layer of the myocardium.

Table 7.14 Statistical parameters of the myocardium microscopic image polarization ellipticity maps.

α_0	0°	45°	90°
M_1	0.095	0.39	0.106
M_2	0.11	0.78	0.12
M_3	2.14	0.17	2.47
M_4	3.08	0.31	3.33

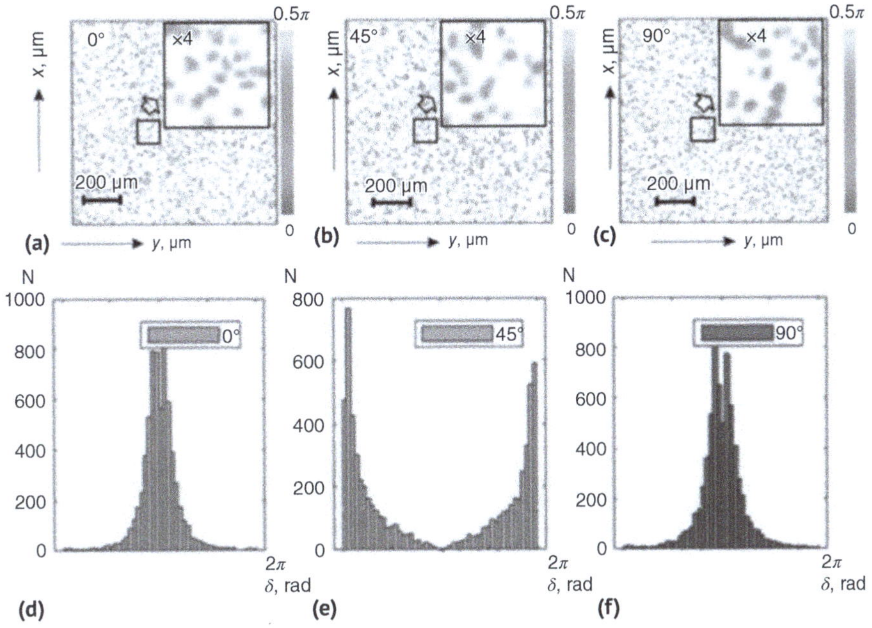

Figure 7.15 Coordinate ($x = 0 - 1000$ μm; $y = 0 - 1000$ μm) and statistical structure of the myocardium histological section microscopic images polarization ellipticity maps in the phase section $\theta = \pi/8$.

Table 7.15 Statistical parameters of the myocardium histological section microscopic images polarization ellipticity maps in the phase section $\theta = \pi/8$.

α_0	0°	45°	90°
M_1	0.109	0.24	0.14
M_2	0.14	0.65	0.25
M_3	1.09	0.62	1.16
M_4	1.41	1.15	1.52

Possibilities of phase selection of components of the object field with different scattering multiplicity are illustrated by layer-by-layer maps $\beta(\theta, m, n)$ for $\theta = \pi/8$ presented in Figure 7.15, as well as the results of their statistical analysis shown in Table 7.15.

The analysis of the layer-by-layer transformation in the statistical structure $\beta(\theta, m, n)$ revealed the possibility of selection by the algorithm, digital holographic reproduction and step-by-step phase scanning of the single-scattered component in the diffuse myocardial sample polarization-inhomogeneous object field – $M_{g=1;2;3;4}(\theta \leq \pi/8) \approx const.$

7.4.5 Parenchymal Liver Histological Sections Object Fields Polarization Ellipticity Maps

Figure 7.16 presents integral maps (a)–(c) and histograms (d)–(f) of a liver histological section microscopic image polarization ellipticity $\beta(\alpha_0; m, n)$.

Quantitative statistical characteristics of distributions $\beta(\alpha_0; m, n)$ are illustrated in Table 7.16.

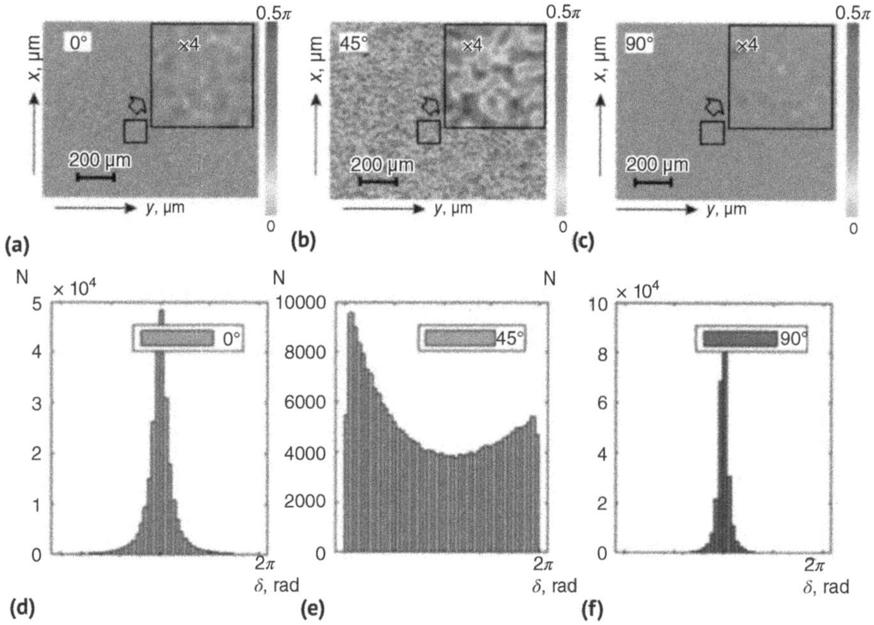

Figure 7.16 Coordinate ($x = 0 - 1000\ \mu m; y = 0 - 1000\ \mu m$) and statistical structure of the liver histological section microscopic images polarization ellipticity maps.

Table 7.16 Statistical parameters of the liver microscopic image polarization ellipticity maps.

α_0	0°	45°	90°
M_1	0.015	0.19	0.022
M_2	0.012	0.68	0.019
M_3	6.14	0.47	5.23
M_4	9.36	0.38	8.89

Statistical analysis of polarization ellipticity maps from integrated topographic microscopic images of the parenchymal liver layer showed significantly smaller (by 8–10 times) dispersion of random values β compared to maps of polarization distributions $\beta(\alpha_0, m, n)$ of the myocardium object field (Table 7.16). The well-established fact relates to the dominant role of liver parenchyma's circular birefringence CB, which defines the coordinate structure of the polarization azimuths in the object field. A more distinctly demonstrated characteristic of how linearly polarized states in the object field form is shown by the results of phase scanning and layer-by-layer holographic reproduction map $\beta(\theta, m \times n)$ for $\theta = \pi/8$ presented in Figure 7.17 and in Table 7.17.

Based on a comparative analysis of holographically reproduced microscopic images of liver diffuse histological sections and layer-by-layer elliptical polarization maps, it was found that $M_2(\beta) \to 0$. In accordance with this, the value of the statistical moments of higher orders, which characterize the skewness $M_3(\beta)$ and kurtosis $M_4(\beta)$ of the distributions $\beta(\theta, m \times n)$ of the object field in comparison with the fibrillar myocardium, increases in the parenchymal tissue layer sharply (up to 10 times) (Table 7.15). In addition, the differences between the calculated values of $M_3(\beta)$ and $M_4(\beta)$ in the given phase plane of the object field of the myocardium (Table 7.15) and liver (Table 7.17) reach three to four times.

7.4.6 An Example of Clinical Application

Two groups of myocardium histological samples were created:

- Coronary heart disease (CHD) – group 1 (12 samples).
- Ischemic heart disease (IHD) – group 2 (12 samples).

Table 7.18 presents the optical-geometric parameters of the samples.

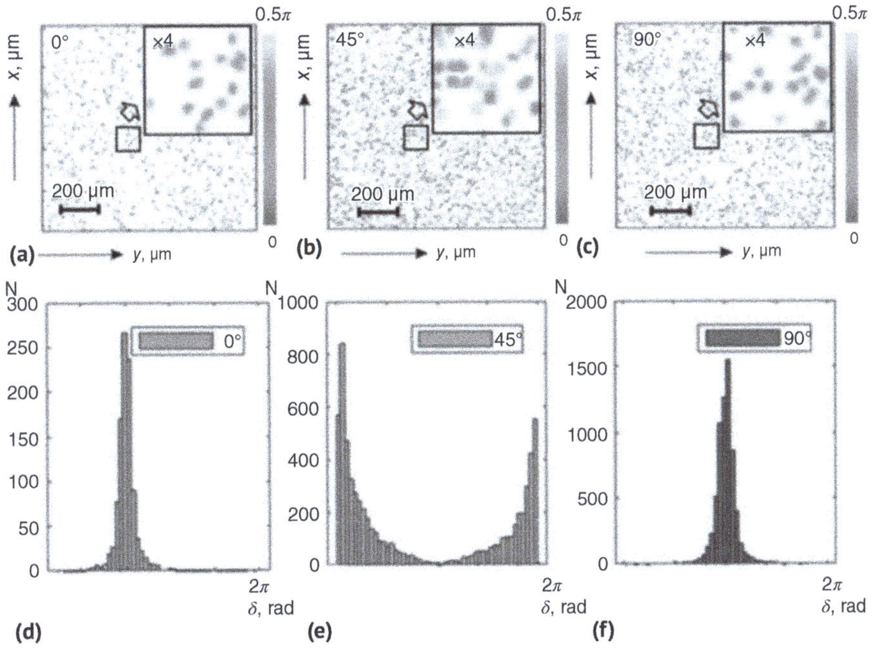

Figure 7.17 Coordinate ($x = 0 - 1000$ μm; $y = 0 - 1000$ μm) and statistical structure of the liver histological section microscopic images polarization ellipticity maps in the phase section $\theta = \pi/8$.

Table 7.17 Statistical parameters of the liver histological section microscopic images polarization ellipticity maps in the phase section $\theta = \pi/8$.

α_0	0°	45°	90°
M_1	0.031	0.17	0.043
M_2	0.014	0.37	0.044
M_3	5.28	0.68	3.19
M_4	9.06	1.04	4.69

Table 7.18 The optical-geometric parameters of the samples.

Parameters	**Optically thick**
Geometric thickness, h, μm	40–45
Coefficient of attenuation, τ μm	0.12–0.15
Degree of depolarization, Δ, %	44–49

From a morphological perspective, cases of CHD and IHD cause different necrotic alterations in myosin fibers and their organized networks.

CHD causes myosin fibrils to become thinner and reduces the spatial organization of fibrillar networks.

In IHD, some myosin fibrils are torn, but the overall spatial arrangement of the myosin network remains intact.

Figure 7.18 presents the results of the investigation into myocardium polarization inhomogeneous object fields for individuals who died from CHD and IHD.

- is integrated polarization ellipticity maps $\beta(m \times n)$ of the myocardium histological sections microscopic images from both groups ((a), (b));
- layer-by-layer myocardial histological section phase polarization ellipticity $\beta(\theta, m \times n)$ maps from both groups for the phase scanning steps $\theta = \pi/4$ ((e), (f)) and $\theta = \pi/8$ ((i), (j));
- histograms of the magnitude polarization ellipticity $G(\beta)$ distributions ((c), (d));
- histograms of layer-by-layer polarization ellipticity value $(\theta = \pi/4, \beta)$ distributions ((g), (h)) and $G(\theta_k = \pi/8, \beta)$ ((k), (l)).

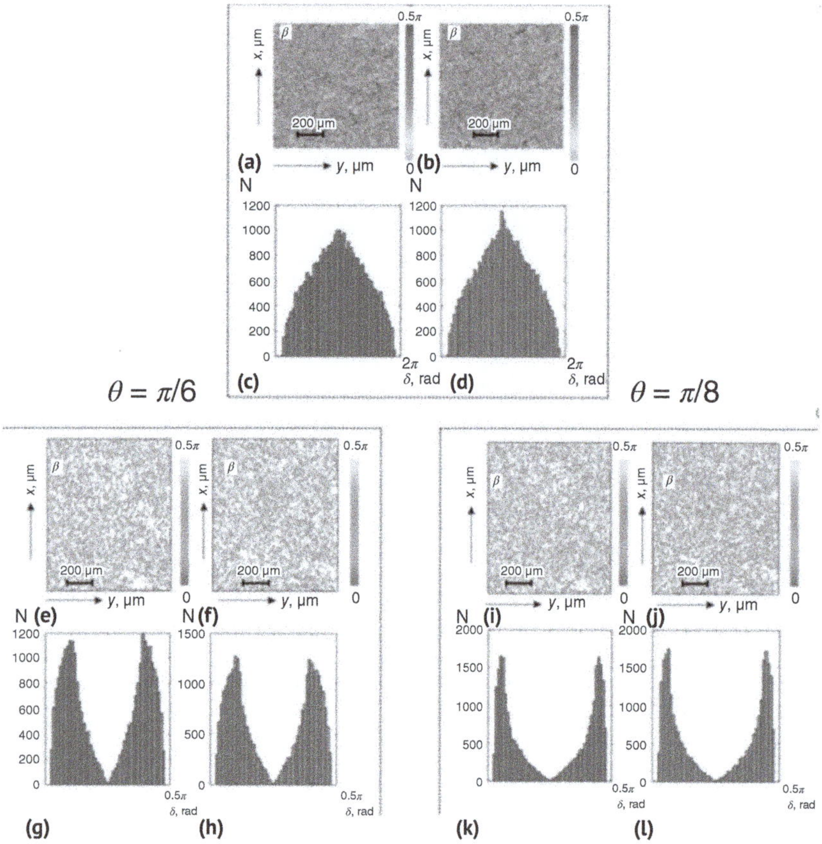

Figure 7.18 Coordinate ($x = 0 - 1000\ \mu m$; $y = 0 - 1000\ \mu m$) and statistical structure of layer-by-layer myocardium histological sections microscopic images polarization ellipticity maps of the patients who died of CHD (left column) and IHD (right column).

Table 7.19 illustrates the results of the experimentally determined polarization ellipticity maps $\beta(m \times n)$ and $\beta(\theta, m \times n)$ statistical analysis.

The results from integral and phase mapping can be linked to necrotic changes in birefringent fibrous networks.

For IHD, narrowing the range of variation in the orientation of optical axes $\Delta\rho(m \times n) \rightarrow \min$, and reduction of their phase-shifting capability, lead to a decrease in the mean $M_1(\beta, \rho, \delta) \downarrow$ and the dispersion $M_2(\beta, \rho, \delta) \downarrow$ of the polarization ellipticity random variables for the corresponding myocardial histological sections. In contrast, the third and fourth order moments, which characterize the skewness and kurtosis of the distributions $\beta(m \times n)$, increase $M_{3,4}(\beta, \rho, \delta) \uparrow$.

In CHD, acute coronary insufficiency causes rapid morphological changes in the myocardial fibrous networks, but these show only minor alterations in their orientation and phase structure $\Delta\rho(m \times n) \approx const.$ and $\Delta\delta(m \times n) \approx const.$

As a result, the following statistical relationships are established:

$$\begin{pmatrix} M_{i=1,2}(\text{IHD}, (\beta), (\theta)) < M_{i=1,2}(\text{CHD}, (\beta), (\theta)) \\ M_{i=3,4}(\text{IHD}, (\beta), (\theta)) > M_{i=3,4}(\text{CHD}, (\beta), (\theta)) \end{pmatrix}.$$

Birefringence in optically uniaxial biological crystals mainly produces elliptically polarized states, while the formation of linearly polarized states is almost negligible.

Because of the advanced statistics of elliptically polarized states in the myocardial fibrous layers' object field, the polarization ellipticity distributions are much more sensitive to changes in the myocardium's orientation and phase polycrystalline structure than the polarization azimuth maps from histological sections of samples in groups 1 and 2.

The informational analysis results show balanced accuracy (Ac, %) for the diagnostic method:

- integral maps of polarization ellipticity – good level with $\text{Ac}(M_{3,4}(\beta)) = 87.9\%$;
- layered polarization ellipticity phase maps – good level with $\text{Ac}(M_3(\beta, \theta = \pi/4)) = 87.5\%$, very good level with $\text{Ac}(M_4(\beta, \theta = \pi/8)) = 91.7\%$, and excellent level with $\text{Ac}(M_{3,4}(\beta, \theta = \pi/8)) = 95.8\%$.

Table 7.19 Statistical moments of the first and fourth orders, which characterize the layer-by-layer polarization ellipticity maps.

Parameters	Group 1	Group 2	AC, %
	Myocardium (Histological Section)		
M_1	0.07 ± 0.004	0.09 ± 0.005	83.3
M_2	0.38 ± 0.019	0.44 ± 0.023	83.3
M_3	0.12 ± 0.007	0.18 ± 0.0095	87.5
M_4	0.16 ± 0.009	0.21 ± 0.011	87.5
	Myocardium (phase section $\theta = \pi/4$)		
M_1	0.05 ± 0.003	0.07 ± 0.004	83.3
M_2	0.14 ± 0.008	0.19 ± 0.01	87.5
M_3	0.71 ± 0.036	1.03 ± 0.055	91.7
M_4	1.22 ± 0.069	1.68 ± 0.091	91.7
	Myocardium (phase section $\theta = \pi/8$)		
M_1	0.09 ± 0.005	0.07 ± 0.004	87.5
M_2	0.22 ± 0.012	0.15 ± 0.008	91.7
M_3	0.88 ± 0.049	1.29 ± 0.069	95.8
M_4	1.31 ± 0.071	1.94 ± 0.105	95.8

7.4.7 Conclusions

1. The proposed comprehensive Stokes polarimetric and interferometric analysis enables the formation of polarization-inhomogeneous components with varying scattering multiplicities within the object laser field of diffuse layers in biological tissues, which include different types of optically anisotropic fibrous and parenchymal structures.
2. Experimental validation has confirmed the methods of direct Stokes polarimetry and digital holographic 3D layered reconstruction of complex amplitudes across various phase planes of microscopic images of diffuse histological tissue sections. This process also includes the subsequent reconstruction of maps for polarization azimuth and ellipticity.
3. The comparison of the results of statistical analysis of integral maps of polarization azimuth $\alpha(\alpha_0, m, n)$ and ellipticity $\beta(\alpha_0, m, n)$ revealed the following:
   - differences of all coordinate polarization distributions from normal or Gaussian: all statistical moments $M_{g=1,2,3,4} \neq 0$.
   - variations in the magnitude (from two to seven times) of all statistical moments $M_{g=1,2,3,4}$ depending on changes in the polarization state α_0 of the incident laser beam.
4. The comparative analysis of layered maps showing polarization azimuth and ellipticity, reconstructed through the phase scanning method combined with digital holographic reconstruction of layered complex amplitude distributions, revealed that as the magnitude of θ_k there is a consecutive reduction (by two to three times) in the dispersion $M_2(\theta_k; \alpha)$ of the random values of polarization parameters in the phase cross-sectional planes of the object field.
5. It has been established that starting from a certain value of the phase parameter θ_k, the statistical structure of the polarization azimuth and ellipticity maps "stabilizes" – $M_{g=1;2;3;4}(\theta_k \leq {}^{\pi}/_{8}) \approx const$, which corresponds to a single-scattered object component of the laser field.
6. Biomedical validation of the method demonstrated a high level $((\mathrm{Ac}(M_{3,4}(\beta, \theta_k = \pi/8)) = 95.8\%)$ of diagnostic accuracy in distinguishing diffuse myocardium samples from individuals who died from ischemic heart disease and acute coronary insufficiency.

The study demonstrated that applying the 3D polarization–interference method, combined with phase-scanning and digital holographic reconstruction, to biological diffuse-tissue fields produces layered maps of polarization-azimuth and ellipticity. These maps reveal different architectural optical anisotropic components, facilitating the experimental identification of components with varying scattering multiplicities. This approach allows for isolating the "object"-essentially a singly scattered component-from the overall "diffuse" background created by high-multiplicity interactions.

The observed differences, which range from two to four times in the magnitudes of the statistical moments $M_{g=1;2;3;4}(\theta_k \leq {}^{\pi}/_{8})$, characterize the polarization maps of azimuth and ellipticity for the singly scattered component of the object field in tissues of both types. These findings suggest that the 3D polarization–interference method has high diagnostic sensitivity to changes in the optical anisotropic structure of diffuse tissue layers, making it promising for various clinical differential diagnosis applications.

7.5 Polarization Interferometry of Optically Anisotropic Architectonics Endometrium Object Fields

7.5.1 Object of Investigations

The study focused on native histological sections of the endometrium, prepared using a standard microtome with rapid freezing. For a comparative analysis of the diagnostic effectiveness of the Mueller matrix method under different scattering conditions, samples were categorized into optically thin (group 1) and optically thick (group 2). These included specimens with benign and precancerous alterations.

Table 7.20 shows the optical-geometric parameters of the samples.

The extinction coefficient of the uterine tissue samples was determined following standard procedures. Within groups 1 and 2, two representative subgroups of histological section samples were created:

1. Benign samples (myoma) include control group 1 with 14 samples and control group 2 with 14 samples.
2. Premalignant samples (endometriosis) include experimental group 1 with 14 samples and experimental group 2 with 14 samples.

7.5.2 Integral and Layered Polarization Azimuth Maps

Figure 7.19 shows: integral polarization azimuth $\alpha(m, n)$ maps; layered phase polarization azimuth $\alpha(\theta_k, m, n)$ maps for $\theta_k = \pi/4$ and $\theta_k = \pi/8$; histograms $G(\alpha)$; histograms $G(\theta_k = \pi/4, \alpha)$ and $G(\theta_k = \pi/8, \alpha)$.

The analysis of the obtained data revealed the following [39–45]:

Coordinate the inhomogeneity and different statistical patterns observed in the integral and layered phase azimuth polarization maps within myoma and endometrium histological sections.

Asymmetric histograms of integral $G(\alpha)$ (Figure 7.19) and layered $(G(\theta_k = \pi/4, \alpha), G(\theta_k = \pi/8, \alpha))$.

The statistical analysis results for the polarization azimuth maps $\alpha(m, n)$ and $\alpha(\theta_k, m, n)$ are presented in Table 7.21.

It has been shown that:

- All first to fourth order statistical moments that characterize the histograms of (α) and the layered phase maps $G(\theta_k = \pi/4, \alpha), G(\theta_k = \pi/8, \alpha)$ are nonzero: $M_{i=1,2,3,4}((\alpha), (\theta_k = \pi/4, \alpha), (\theta_k = \pi/8, \alpha)) \neq 0$.
- The higher-order moments, which characterize the skewness and kurtosis of the distributions $G(\alpha)$ and $G(\theta_k = \pi/4, \alpha), G(\theta_k = \pi/8, \alpha)$, are the most sensitive to changes in the polarization azimuth map.
- The scenarios of the first and fourth-order moments, describing the polarization azimuth maps of myoma (M) and endometrium (E) histological sections, show that the changes in moments satisfy:

$$\begin{pmatrix} M_{i=1,2}(M, (\alpha), (\theta_k)) < M_{i=1,2}(E, (\alpha), (\theta_k)) \\ M_{i=3,4}(M, (\alpha), (\theta_k)) > M_{i=3,4}(E, (\alpha), (\theta_k)) \end{pmatrix}.$$

From a physical perspective, the results can be linked to pathological changes in the connective tissue. These changes cause an increase in the structural anisotropy of collagen networks by promoting the geometric and orientational growth of

Table 7.20 Female reproductive system histological sections optical and geometric parameters.

Parameters	Optical thin	Optical thick
Geometric thickness, h, μm	20–25	40–45
Coefficient of attenuation, τ μm	0.093–0.012	0.14–0.18
Degree of depolarization, Δ, %	6–9	27–32

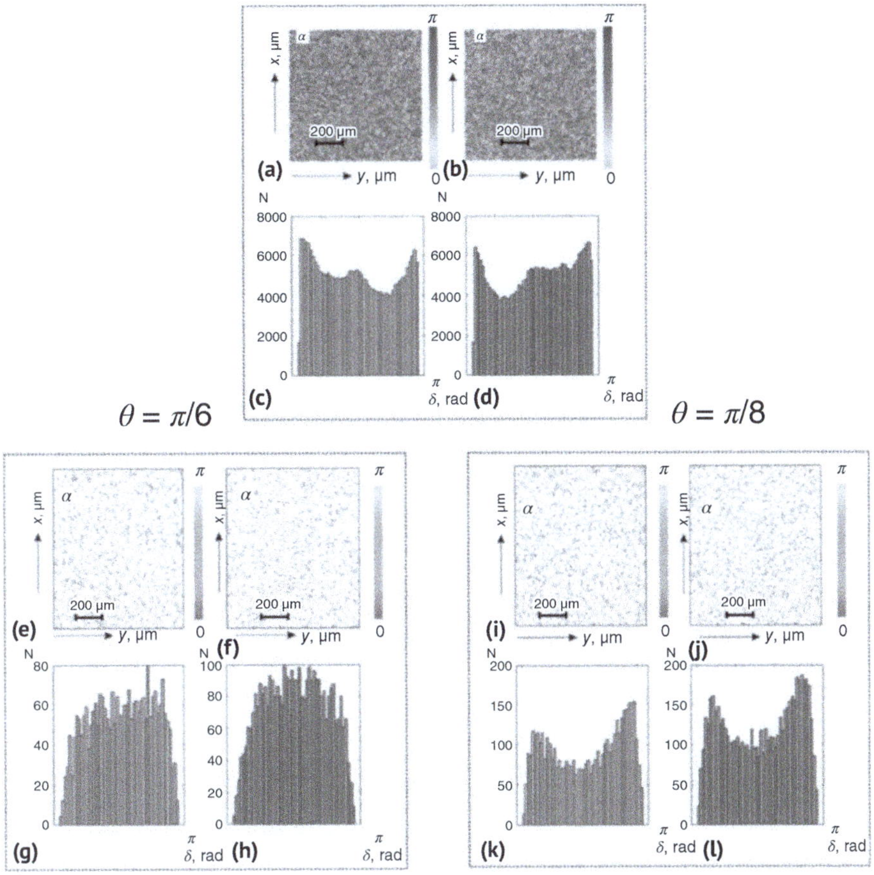

Figure 7.19 Coordinate and statistical structure of azimuth polarization maps for optically thin ($\tau \leq 0.01$) myoma and precancerous endometrium histological sections - left and right columns, respectively.

Table 7.21 Statistical moments of the first and fourth orders characterizing the layer-by-layer polarization azimuth maps

Parameters	Group 1	Group 2	AC, %
M_1	0.105 ± 0.006	0.12 ± 0.007	75
M_2	0.81 ± 0.04	0.72 ± 0.04	75
M_3	0.91 ± 0.05	0.84 ± 0.05	78.6
M_4	1.21 ± 0.07	0.97 ± 0.051	82.1
	Phase shift $\theta = \pi/4$		
M_1	0.93 ± 0.049	1.04 ± 0.05	78.6
M_2	0.69 ± 0.038	0.75 ± 0.04	78.6
M_3	0.76 ± 0.041	0.61 ± 0.03	85.7
M_4	0.88 ± 0.047	0.71 ± 0.04	85.7
	Phase shift $\theta = \pi/8$		
M_1	0.74 ± 0.04	0.86 ± 0.05	82.1
M_2	0.76 ± 0.04	0.91 ± 0.05	82.1
M_3	1.12 ± 0.06	0.88 ± 0.05	89.2
M_4	1.33 ± 0.07	1.04 ± 0.05	89.2

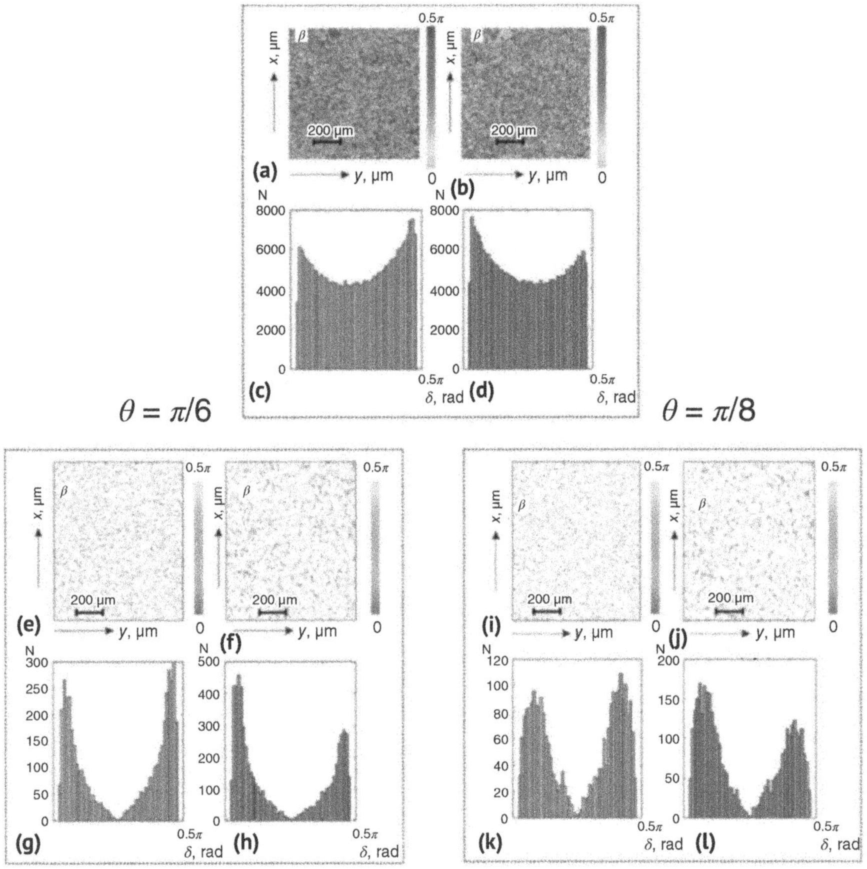

Figure 7.20 Coordinate and statistical structure of layered ellipticity polarization maps for optically thin ($\tau \leq 0.01$)) myoma and precancerous endometrium histological sections-the left and right columns, respectively.

protein fibers. As a consequence, there is an enlargement in the distribution of orientations of optical axes $\Delta\rho(m,n) \to \max$ of fibrillar crystals, as well as their birefringence $\delta(m,n) = \frac{2\pi}{\lambda}\Delta nd(m,n)$.

The precancerous pathological changes will lead to an increase in the mean $M_1(\alpha,\rho,\delta)\uparrow$ and the variance $M_2(\alpha,\rho,\delta)\uparrow$ of the polarization azimuth fluctuations in the corresponding carcinoma biopsy histological section microscopic image. Simultaneously, the size of the 3rd and 4th order statistical moments, which describe the skewness and kurtosis, respectively, of the distribution $\alpha(m,n)$ will decrease, i.e., $M_{3,4}(\alpha,\rho,\delta)\downarrow$.

The analysis results identify a group of diagnostic markers along with their accuracy levels in differentiating benign from precancerous conditions:

- integrated polarization azimuth maps - satisfactory level $\text{Ac}(M_4(\alpha)) = 82.1\%$.
- layered polarization azimuth maps - good level $\text{Ac}\big(M_{3,4}(\alpha, \theta_k = \pi/4, \pi/8)\big) = 85.7 - 89.2\%$.

7.5.3 Integral and Layered Polarization Ellipticity Maps

Figure 7.20 shows the results of the analysis of benign myoma and precancerous endometriosis object fields, with notations similar to those in Figure 7.19.

The analysis of the data showed a consistent yet diverse topographic and coordinate structure both in the integral data (Figure 7.20a, b) and in the layer-by-layer polarization ellipticity maps (Figure 7.20e, f, i, and j).

Table 7.22 presents the results of the statistical analysis of the $\beta(m,n)$ and $\beta(\theta_k,m,n)$ maps.

The following was found:

- The statistical distributions of the polarization parameters differ from normal and exhibit individual characteristics: $M_{i=1,2,3,4}((\beta), (\theta_k = \pi/4, \beta), (\theta_k = \pi/8, \beta)) \neq 0$.

- Diagnostic markers of benign and precancerous changes include higher-order statistical moments characterizing the skewness and kurtosis of the distributions $G(\beta)$ and $G(\theta_k = \pi/4, \beta)$, $G(\theta_k = \pi/8, \beta)$.
- Precancerous alterations in the polycrystalline component of endometrial samples (group 2) are accompanied by an increase in the mean $M_1(\beta, \rho, \delta) \uparrow$ and the variance $M_2(\beta, \rho, \delta) \uparrow$. At the same time, the third- and fourth-order statistical moments decrease: $M_{3,4}(\beta, \rho, \delta) \downarrow - \begin{pmatrix} M_{i=1,2}(M, (\beta), (\theta_k)) < M_{i=1,2}(E, (\beta), (\theta_k)); \\ M_{i=3,4}(M, (\beta), (\theta_k)) > M_{i=3,4}(E, (\beta), (\theta_k)) \end{pmatrix}$

The data analysis showed a steady improvement in differentiating benign from precancerous conditions of the female reproductive system:

- integral polarization ellipticity maps demonstrate a good accuracy level: $\text{Ac}(M_{3,4}(\beta)) = 85.7\%$;
- layer-wise phase polarization ellipticity maps show the following accuracy levels: good – $\text{Ac}(M_3(\beta, \theta_k = \pi/4)) = 89.2\%$, very good – $\text{Ac}(M_4(\beta, \theta_k = \pi/8)) = 91.7\%$, very good – $\text{Ac}(M_{3,4}(\beta, \theta_k = \pi/8)) = 92.8\%$;
- starting from $\theta_k \leq \pi/8$, the balanced accuracy remains practically unchanged: $\text{Ac}(M_{3,4}(\beta, \theta_k = \pi/8)) \approx const.$

7.5.4 Optically Thick Layer Polarization Azimuth Maps

Figure 7.21 shows maps and histograms that depict the distribution of polarization azimuth values $\alpha(m, n)$ in microscopic images of histological sections from both diffuse benign uterine fibroids and precancerous endometriosis tissues.

The analysis of the results showed, consistent with findings for optically thin layers, topographic and coordinate inhomogeneity in both the integral (Figure 7.21a, b and layer-wise phase (Figure 7.21e, f, i, and j) polarization azimuth maps of benign and precancerous uterine tissues in microscopic images of optically thick histological sections.

The evaluation of diagnostic sensitivity for integral and phase-polarization azimuth mapping in microscopic images of optically thick histological sections from both groups showed that secondary interference mechanisms mainly influenced the results. These mechanisms generate an elliptically polarized diffuse component of the object field in almost all phase planes (except for $\theta_k \leq \pi/8$). As a consequence, the diagnostic sensitivity of the integral polarization azimuth maps was found to be unsatisfactory (Table 7.21), with $\text{Ac}(M_{3,4}(\alpha)) < 80\%$. The layer-by-layer phase polarization azimuth maps, however, showed an improvement in diagnostic accuracy to a satisfactory level: $\text{Ac}(M_{3,4}(\alpha, \theta_k = \pi/8)) = 85.7\%$.

Table 7.22 First to fourth order statistical moments characterizing the layer-by-layer polarization azimuth maps for optically thick histological sections.

Parameters	Group 1	Group 2	AC, %
M_1	0.51 ± 0.03	0.59 ± 0.03	71.4
M_2	0.72 ± 0.04	0.79 ± 0.04	75
M_3	0.63 ± 0.03	0.56 ± 0.04	78.6
M_4	0.82 ± 0.04	0.76 ± 0.05	78.6
Phase shift $\theta = \pi/4$			
M_1	0.71 ± 0.04	0.78 ± 0.04	75
M_2	0.77 ± 0.04	0.86 ± 0.05	78.6
M_3	0.58 ± 0.03	0.47 ± 0.03	78.6
M_4	0.42 ± 0.02	0.34 ± 0.02	82.1
Phase shift $\theta = \pi/8$			
M_1	0.44 ± 0.02	0.49 ± 0.03	82.1
M_2	0.59 ± 0.03	0.67 ± 0.04	82.1
M_3	0.71 ± 0.04	0.82 ± 0.04	85.7
M_4	0.88 ± 0.05	0.97 ± 0.05	85.7

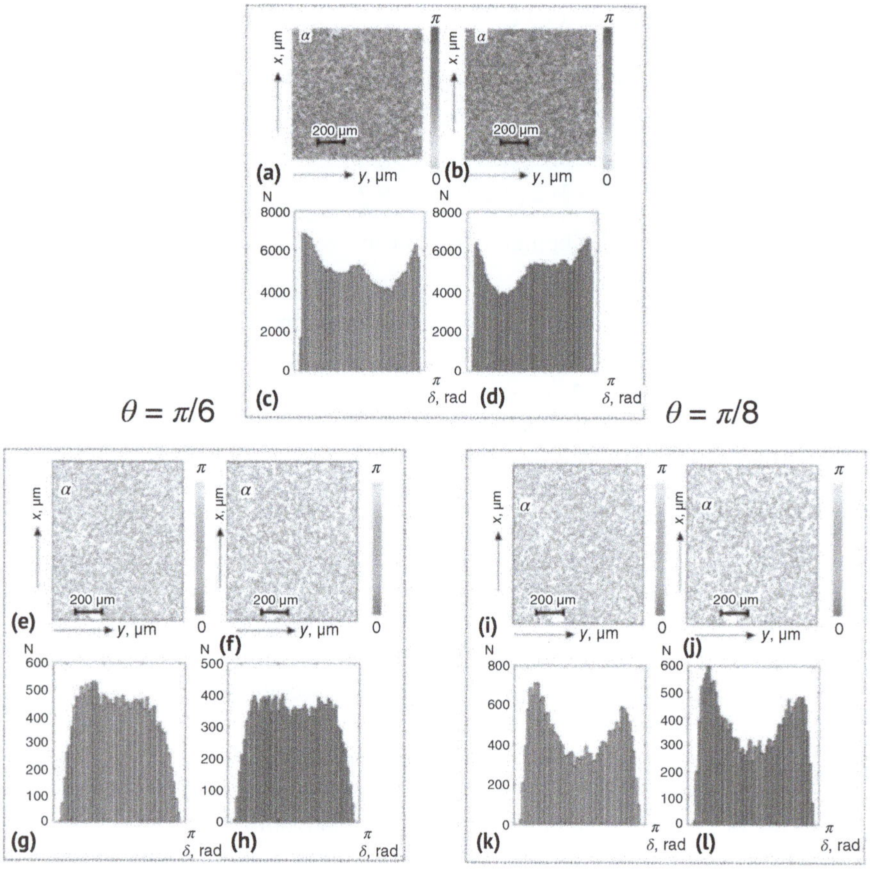

Figure 7.21 Coordinate and statistical structure of layer-by-layer polarization azimuth maps for optically thick ($\tau > 0.01$) myoma and precancerous endometrium histological sections – the left and right columns, respectively.

7.6 Optical Thick Layers Polarization Ellipticity Maps

Figure 7.22 presents the polarization maps $\beta(m, n)$ of histological sections of uterine tissue. Analysis of the results reveals a sufficiently similar, albeit complex, topographic and coordinate structure of the polarization-ellipticity maps (Figure 7.22a, b, e, f, i, and j).

Statistical analysis results for the polarization-ellipticity maps maps $\beta(m, n)$ and $\beta(\theta_i, m, n)$ (Table 7.23).

It has already been noted that pathological changes in the polycrystalline component of diffuse endometrial samples from group 2 are associated with opposite trends in the transformation of statistical marker values specifically, an increase in the average $M_1(\beta, \rho, \delta) \uparrow$ and variance $M_2(\beta, \rho, \delta) \uparrow$ and, conversely, a decrease of the third and fourth-orders statistical moments $M_{3;4}(\beta, \rho, \delta) \downarrow$.

A comparative analysis of layer-by-layer maps $\beta(\theta_i, m, n)$ revealed that as the value of θ_k decreases, there is a consistent dispersion $M_2(\theta_k; \beta)$ decrease (up to three times) (Table 7.23). At the same time, the value of $M_{3,4}(\beta, \rho, \delta)$ increases up to three to four times.

Further decreasing the object field phase parameter $\theta < \pi/8$ essentially does not alter the statistical structure of the polarization maps. This suggests that, for these phase planes, the most "diagnostically favorable" mode of single scattering within the diffuse layers of uterine tissue is achieved. Therefore, it is in the phase plane $\theta = \pi/8$ that there is the greatest difference (40–50%) between the statistical moments $M_{3;4}(\theta_k; \beta)$, which characterize the distributions of $\Psi(\beta_j; \delta_j)$ in object fields of samples of benign and precancerous uterine tissues.

As a result, the accuracy of balanced differential diagnosis in the phase plane $\theta = \pi/8$ of female reproductive diffuse tissue samples under precancerous conditions reaches the same level as that for optically thin samples (Table 7.22), $\text{Ac}(M_{3,4}(\beta, \theta_k = \pi/8)) = 92.8\%$.

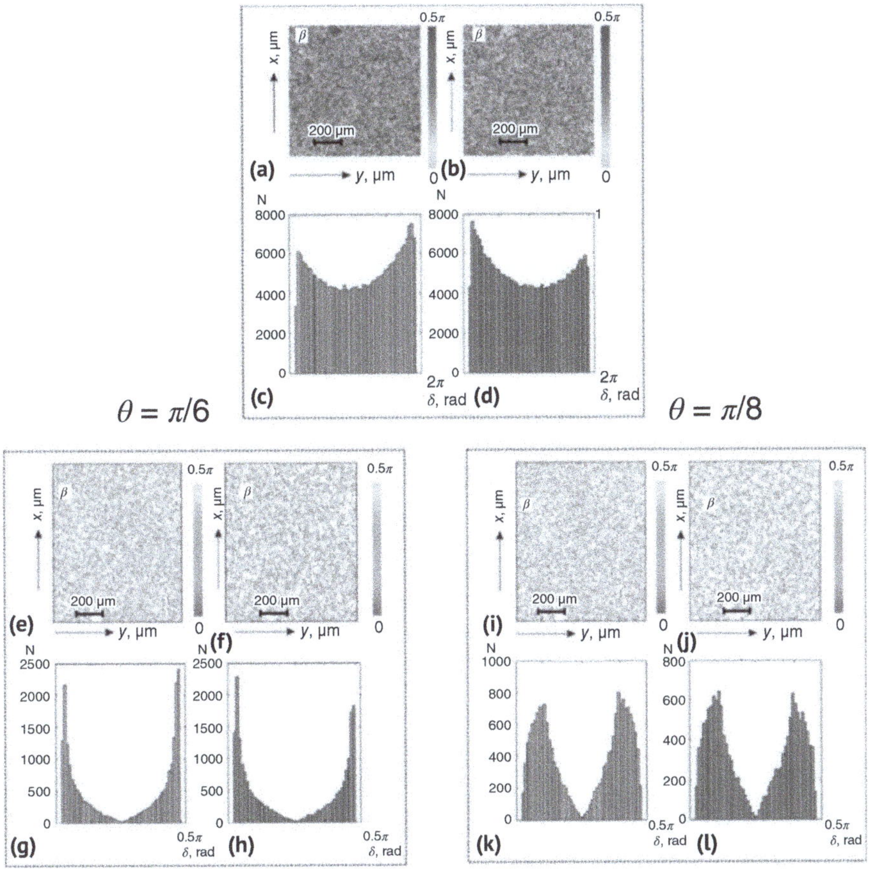

Figure 7.22 Coordinate and statistical structure of layer-by-layer polarization-ellipticity maps for optically thick ($\tau > 0.01$) myoma and precancerous endometrium histological sections (left and right columns, respectively).

In the context of the statistical method for analyzing experimental data, it was determined that:

- The polarization-ellipticity maps do not follow a normal distribution. Specifically, $M_{i=1,2,3,4}((\beta), (\theta_k = \pi/4, \beta), (\theta_k = \pi/8, \beta)) \neq 0$.
- Diagnostic markers are higher-order statistical moments that characterize the skewness and tails of the distributions $G(\beta)$ and $G(\theta_k = \pi/4, \beta)$, $G(\theta_k = \pi/8, \beta)$.

Pathological changes observed in the polycrystalline component of diffuse endometrial samples from group 2 are associated with opposite shifts in the statistical markers, specifically showing an increase in the mean $M_1(\beta, \rho, \delta) \uparrow$ and in the variance $M_2(\beta, \rho, \delta) \uparrow$, and, conversely, a decrease in the third and fourth order moments $M_{3;4}(\beta, \rho, \delta) \downarrow$. A comparative analysis of layer-by-layer maps $\beta(\theta_i, m, n)$ revealed that as θ_k decreases, there is a consistent decrease in $M_2(\theta_k; \beta)$ dispersion (up to about three times) (Table 7.23). At the same time, $M_{3,4}(\beta, \rho, \delta)$ increase by about three times.

Further lowering the parameter for the object-field phase section $\theta < \pi/8$ essentially does not alter the statistical structure of the polarization maps. This suggests that, in these phase planes, the most "diagnostically advantageous' mode of single scattering within the diffuse layers of uterine tissue volume is achieved.

Therefore, in the phase plane $\theta = \pi/8$, the greatest difference (40–50%) between the statistical moments $M_{3;4}(\theta_k; \beta)$, which characterize the distributions $\Psi(\beta_j; \delta_j)$ in the object fields of benign and precancerous uterine tissues, is observed.

As a result, the precision of balanced differential diagnosis within the phase plane $\theta = \pi/8$ for diffuse tissue samples of the female reproductive system in precancerous conditions reaches the same level as for optically thin samples (Table 7.27), $\mathrm{Ac}(M_{3,4}(\beta, \theta_k = \pi/8)) = 92.8\%$.

Table 7.23 Statistical moments (first and fourth order) characterizing layer-by-layer ellipticity polarization maps for optically thick histological sections.

Parameters	Group 1	Group 2	AC, %
M_1	0.27 ± 0.01	0.31 ± 0.02	67.9
M_2	0.33 ± 0.02	0.37 ± 0.02	71.4
M_3	0.42 ± 0.02	0.34 ± 0.03	75
M_4	0.55 ± 0.05	0.48 ± 0.04	78.6
	Phase shift $\theta = \pi/4$		
M_1	0.15 ± 0.008	0.18 ± 0.01	82.1
M_2	0.17 ± 0.009	0.22 ± 0.01	82.1
M_3	0.75 ± 0.04	0.66 ± 0.04	85.7
M_4	0.92 ± 0.05	0.81 ± 0.04	85.7
	Phase shift $\theta = \pi/8$		
M_1	0.11 ± 0.006	0.14 ± 0.008	82.1
M_2	0.15 ± 0.008	0.19 ± 0.01	85.7
M_3	1.38 ± 0.07	1.01 ± 0.05	85.7
M_4	2.21 ± 0.1	1.68 ± 0.09	92.8

7.6.1 Conclusions

1. The experimental results for diagnostic applications, obtained through vector-parametric polarization mapping and polarization–interference phase scanning of tissue fields from the female reproductive system, were presented and analyzed physically. This method includes layer-by-layer digital holographic reproduction of polarization maps and aids in differentiating benign (myoma) from precancerous (endometriosis) conditions.
2. For all pathology types, detailed statistical data for integral and layered polarization maps were created, covering all moments from the 1st to the 4th, which describe the histograms $G(\alpha, \beta)$ and $G(\theta_k = \pi/4, \alpha, \beta)$, $G(\theta_k = \pi/8, \alpha, \beta)$. The higher-order moments $M_{i=1,2,3,4}((\alpha, \beta), (\theta_k = \pi/4, \alpha, \beta), (\theta_k = \pi/8, \alpha, \beta)) \neq 0$ are not equal to zero.
3. The highest levels of differential-diagnostic accuracy for precancerous conditions, when using optically thin histological sections, were identified as follows:
   3.1 Integral polarization azimuth maps – satisfactory level $\text{Ac}(M_4(\alpha)) = 82.1\%$.
   3.2 Layer-by-layer phase polarization azimuth maps – good level $\text{Ac}\big(M_{3,4}(\alpha, \theta_k = \pi/4, \pi/8)\big) = 85.7 - 89.2\%$.
   3.3 Integral polarization ellipticity maps – good level $\text{Ac}\big(M_{3,4}(\beta)\big) = 85.7\%$.
   3.4 Layer-by-layer phase polarization ellipticity maps – good $\text{Ac}(M_3(\beta, \theta_k = \pi/4)) = 89.2\%$, very good $\text{Ac}(M_4(\beta, \theta_k = \pi/8)) = 91.7\%$, and very good $\text{Ac}\big(M_{3,4}(\beta, \theta_k = \pi/8)\big) = 92.8\%$.
4. The results for optically thick histological sections were as follows:
   4.1 Diagnostic sensitivity for integral polarization azimuth maps was unsatisfactory: $\text{Ac}\big(M_{3,4}(\alpha)\big) < 80\%$.
   4.2 Layer-wise phase polarization azimuth maps showed improvement to a satisfactory level: $\text{Ac}\big(M_{3,4}(\alpha, \theta_k = \pi/8)\big) = 85.7\%$.
5. The detection accuracy for precancerous states in diffuse tissue samples using polarization ellipticity maps in the phase plane $\theta = \pi/8$ reached a level similar to that for optically thin samples: $\text{Ac}(M_4(\beta, \theta_k = \pi/8)) = 92.8\%$.

7.7 3D Mueller-matrix Images of Polycrystalline Networks of Biological Tissues with Different Morphological Structures

7.7.1 Brief Description of Research Objects

The experimental samples consisted of histological sections of biological tissues from two types, prepared using the standard freezing microtome method.

Myocardium with coefficient of attenuation $\tau = 0.91$ and a corresponding depolarization level of the object field $\Lambda = 53\%$.

This sample represents biological layers with a morphological structure composed of spatially organized protein fibrils. Each fibril includes optically active myosin molecules. In essence, the myocardium layer is an optically anisotropic structure that exhibits all mechanisms of linear and circular birefringence and dichroism.

Rat liver parenchymal tissue with ($\tau = 0, 99$; $\Lambda = 61\%$). The optically anisotropic part of this sample is made up of tiny "islet" structures, which are formed by polypeptide chains of optically active protein molecules.

The primary physical features of the polycrystalline structure in parenchymal biological tissue include proportional aspects of linear and circular birefringence and dichroism.

A list of additional literature, consisting of 65 titles, is included at the end of this section.

7.7.2 Layer-by-layer Maps of Azimuthally Invariant Mueller Matrix Elements for Spatially Structured Fibrillar Networks

Figure 7.23 through Figure 7.26 show a series of layer-by-layer images of Mueller-matrix invariants for selected phase sections of the 3D distributions: $\phi = 0.3$ rad (fragment a); $\phi = 0.9$ rad (fragment b); $\phi = 1.5$ rad, (fragment c). Also shown are the dependencies of the first–fourth order statistical moments that characterize the 3D distributions of the Mueller matrix invariants (MMI) for the partially depolarizing myocardium section ($\tau = 0.91$).

The examination of 3D Mueller-matrix tomography in a partially depolarizing myocardium layer revealed:

1. Distinct topographic structures appear across various phase cross-sections of 3D distributions for all Mueller-matrix invariants (MMIs) (Figures 7.23–7.26). These MMIs characterize polarization manifestations of linear (($f_{44}(\phi, x, y)$, Figure 7.23) and circular ($\Delta f(\phi, x, y)$, Figure 7.24) birefringence, as well as linear ($f_{41}(\phi, x, y)$, Figure 7.25) and circular ($f_{14}(\phi, x, y)$, Figure 7.26) dichroism. This indicates different levels of light-scattering multiplicity within the myocardium volume across various phase cross-sections. Consequently, the topographic distribution of any MMI value is a result of convolution of polarization effects caused by phase and amplitude anisotropy of the fibrillar network in the optically thin layer ($\phi \rightarrow \min$) with the level of multiplicity of the interaction of laser radiation with that layer ($\phi \rightarrow \max$). Therefore, for different phase sections ϕ, an individual coordinate distribution of the MMI value is formed.

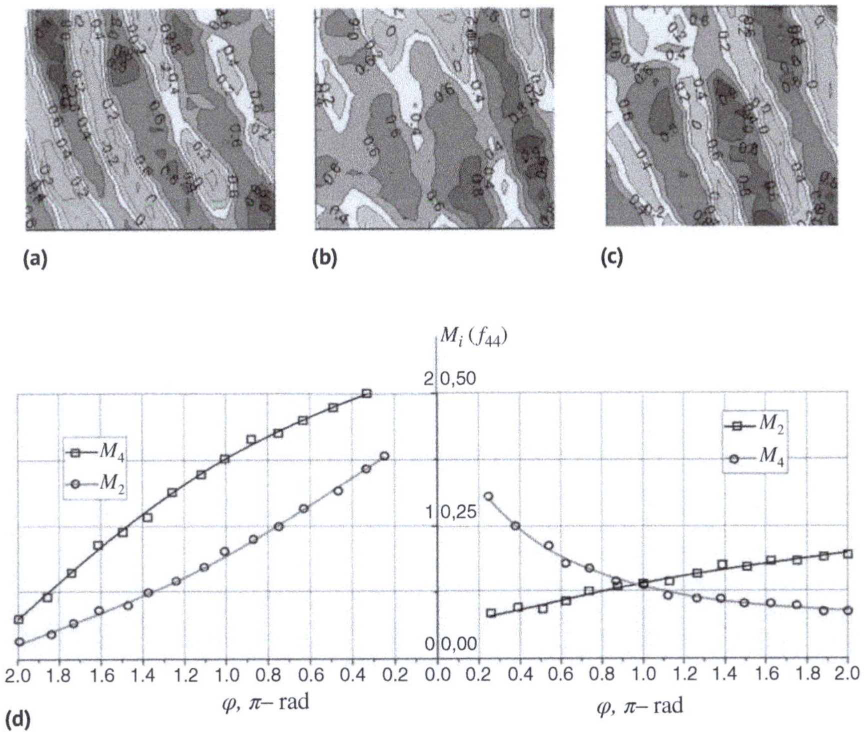

Figure 7.23 Layer-by-layer Mueller-matrix images and the dependencies of the first–fourth order statistical moments that characterize the 3D distributions of the azimuthally invariant element f_{44} for the partially depolarizing ($\tau = 0.91$) histological section of myocardium.

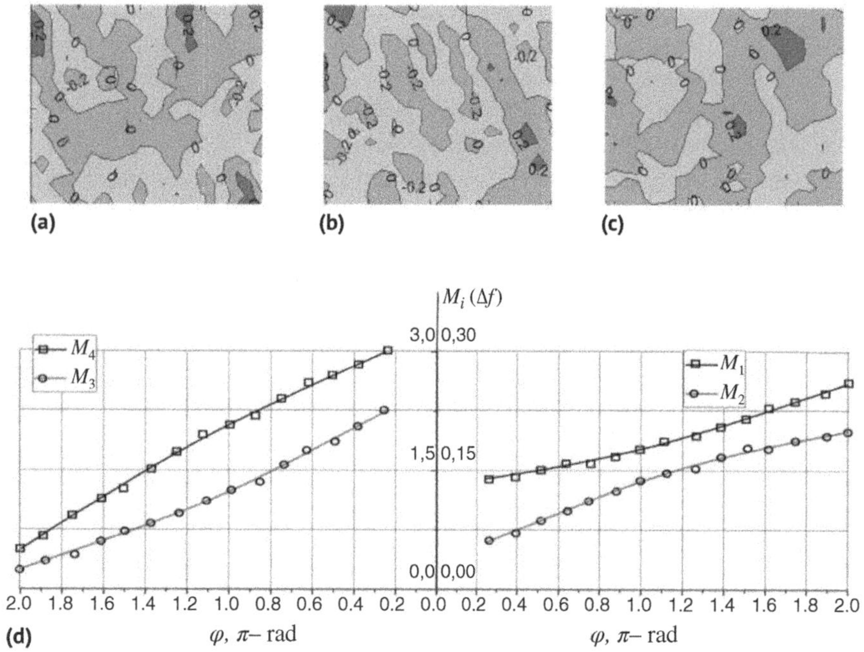

Figure 7.24 Layer-by-layer Mueller-matrix images and the dependencies of the first to fourth order statistical moments that characterize the 3D distributions of the azimuthally invariant element Δf for the histological section of myocardium.

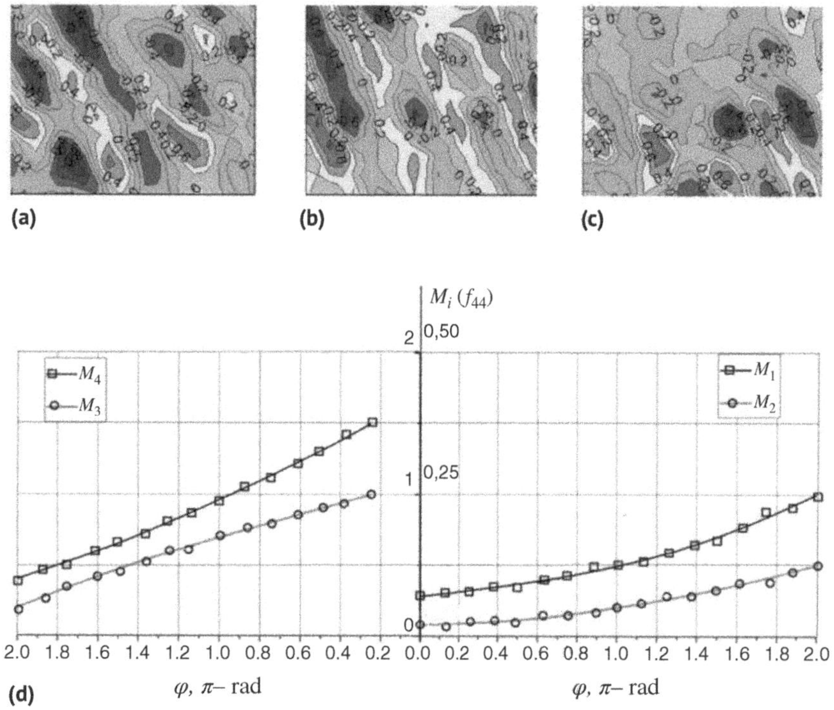

Figure 7.25 Layer-by-layer matrix images and the dependencies of the first to fourth order statistical moments that characterize the 3D distributions of the azimuthally invariant element f_{41} for the histological section of myocardium.

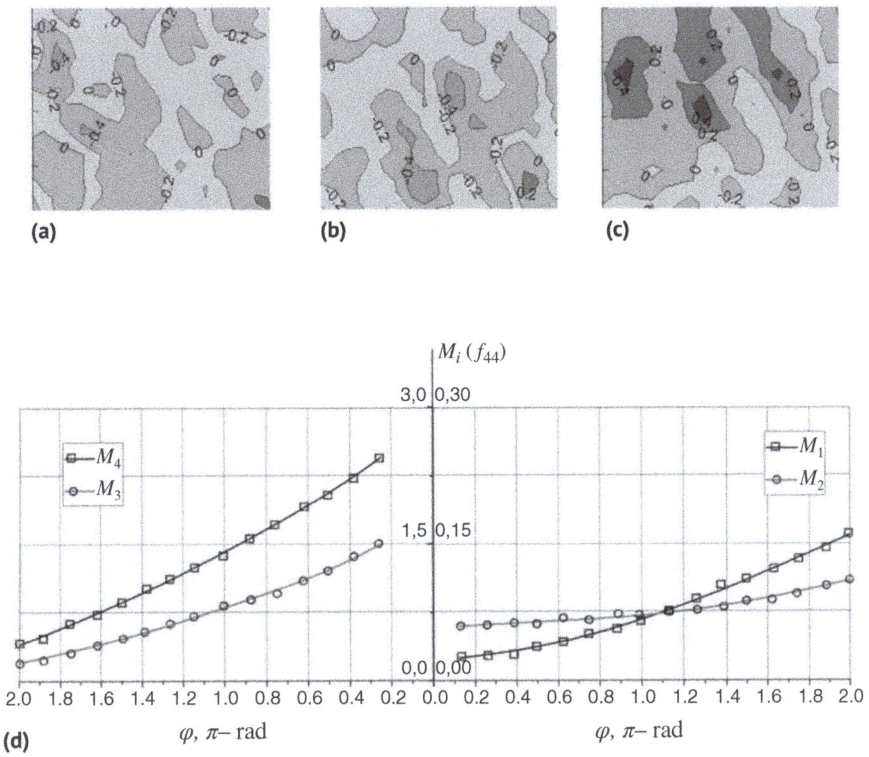

Figure 7.26 Layer-by-layer matrix images and the dependencies of the first to fourth order statistical moments that characterize the 3D distributions of the azimuthally invariant element f_{41} for the partially depolarizing ($\tau = 0.91$) histological section of myocardium.

2. The increase of the mean (M_1(MMI) ↑) and dispersion (M_2(MMI) ↑) of the spread of MMI values with the increase in the level of the "phase" section (ϕ ↑) of the 3D distributions (Figures 7.23d–7.26d) and, conversely, the decrease the skewness (M_3(MMI) ↓) and kurtosis (M_4(MMI) ↓) of such distributions – Figures 7.23d–7.26d).
3. The mechanisms of linear birefringence and dichroism in myocardium fibrillar networks have an advantage over the manifestations of circular birefringence ($f_{44}(\phi, x, y) > \Delta f(\phi, x, y)$) and dichroism ($f_{41}(\phi, x, y) > f_{41}(\phi, x, y)$) of optically active myosin molecules. This observation aligns with the tissue's morphological structure. The size of filamentous mesh structures greatly surpasses that of optically anisotropic myosin molecules; consequently, the degree of linear phase and amplitude anisotropy in the system of spatially aligned fibrils is greater than that of circular birefringence and dichroism.
4. The mean value (M_1(MMI) ↑) and the dispersion (M_2(MMI) ↑) with increasing phase section level (ϕ ↑) in the 3D distributions (Figures 7.23d–7.26d). Conversely, the skewness (M_3(MMI) ↓) and the kurtosis (M_4(MMI) ↓) (see Figures 7.23d–7.26d).

The observed scenario of change in the statistical structure $\left(\phi \uparrow \Leftrightarrow \begin{cases} M_{1;2}(\phi) \uparrow; \\ M_{3;4}(\phi) \downarrow \end{cases}\right)$ of the MMIs coordinate distributions, which characterize the optical anisotropy of the myocardium, can be attributed to the fact that, in the region of small phase values $\phi < 0.6$, predominantly a single interaction of the laser radiation with the fibrillar mesh of the myocardium occurs. As a result, the optogeometric structure of the network shows a maximum correlation with the topographical distribution of a specific MMI [46–49].

Quantitatively, this manifests as complex, asymmetric distributions of MMI values and simultaneously non-zero values of the first- to fourth-order statistical moments that characterize these distributions. As the phase cross-section level ϕ increases beyond $\phi > 0.6$, the effect of scattering multiplicity in the myocardium layer increases as well. Therefore, the distribution of MMI values, based on the central limit theorem, approaches a normal distribution. This is also accompanied by a notable reduction in the higher-order terms moment $(M_{3;4} \downarrow)$.

7.7.3 Layer-by-layer Maps of Mueller-matrix Invariants of "Islet" Polycrystalline Structures in Parenchymal Tissues

The series demonstrates how layered images of Mueller-matrix invariants associated with phase and amplitude anisotropy are combined in a partially depolarizing layer ($\tau = 0.99$; $\Lambda = 61\%$) of rat liver and the distributions of the statistical moments of the first to fourth orders (Figures 7.27d–7.30d).

Analysis of Mueller-matrix tomography on the partially depolarizing layer of rat liver tissue, similar to studies of the myocardium's volumetric polycrystalline structure, uncovered a unique topographic configuration. This structure varies across phase cross-sections due to different light-scattering multiplicities in the 3D distributions of all MMIs that depict the polarization features of phase ($f_{44}(\phi, x, y)$, Figure 7.27; $\Delta f(\phi, x, y)$, Figure 7.28) and amplitude ($f_{41}(\phi, x, y)$, Figure 7.29; $f_{14}(\phi, x, y)$, Figure 7.30) anisotropy.

A comparative analysis of MMI value fluctuations across various phase sections reveals the underlying polarization mechanisms that are consistent for both linear and circular birefringence ($f_{44}(\phi, x, y) \approx \Delta f(\phi, x, y)$) and dichroism ($f_{41}(\phi, x, y) \approx f_{41}(\phi, x, y)$) for islet structures with polypeptide chains of optically active protein molecules. This observation aligns with the morphological features of this parenchymal tissue, where the birefringence's geometric scale in twisted polypeptide chains is comparable to that of molecular protein complexes. The overall level of linear birefringence (f^*_{44}) and dichroism (f^*_{41}) in the parenchymal rat liver tissue is significantly lower than the corresponding properties ($f^*_{44} > f_{44}$ and $f^*_{41} < f_{41}$) in the spatially structured myosin network of the myocardium.

This can be attributed to two main factors. First, amorphous liver parenchyma has a much lower concentration of optically anisotropic structures compared to the myocardium, which mainly consists of polycrystalline tissue with circular birefringent myosin molecules. Second, the myocardial fibrils display notable linear birefringence and dichroism because of their larger size and organized structure.

Therefore, the appropriate distribution histogram values of Mueller-matrix images $f^*_{44}(\phi, x, y)$ and $f^*_{41}(\phi, x, y)$ for such mechanisms of anisotropy of parenchymatous tissue for small values of the phase section ($\phi < 0.6$) are characterized by lower values of the mean and dispersion, and, on the contrary, by greater skewness and peak sharpness. Because of this, among the values of the statistical moments that characterize the corresponding Mueller-matrix images of the biological layers of both types, the following relations are formed: $\begin{cases} f^*_{1;2} < f_{1;2}; \\ f^*_{3;4} > f_{3;4} \end{cases}$.

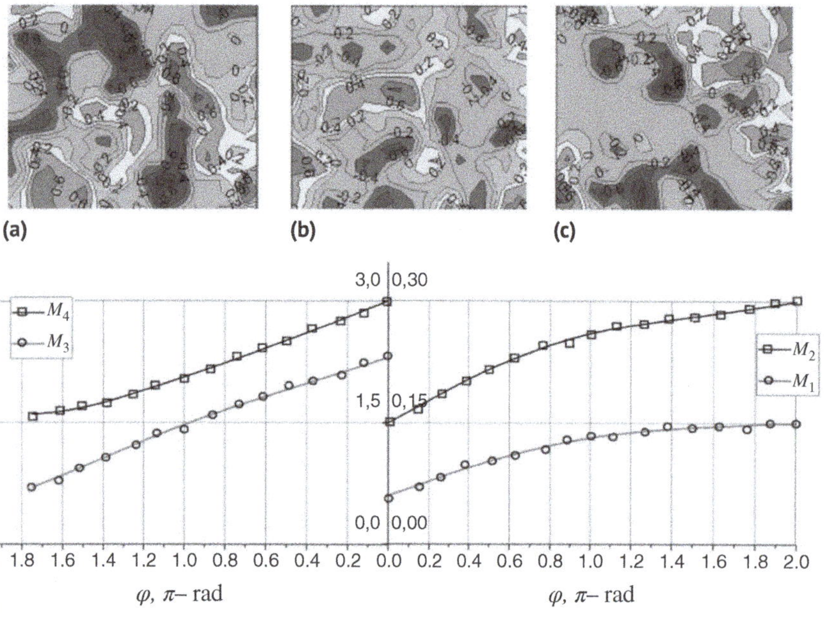

Figure 7.27 Layer-by-layer Mueller-matrix images and the dependencies of the first- to fourth-order statistical moments describing the 3D distributions of the azimuthally invariant element f_{44} for partially depolarizing ($\tau = 0.99$) rat liver histological sections.

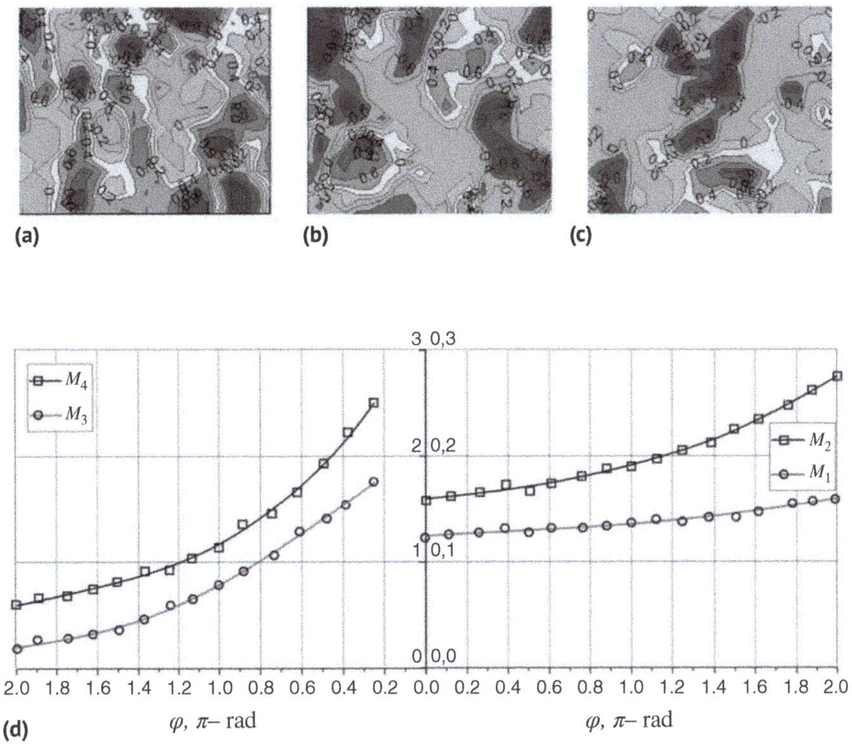

Figure 7.28 Layer-by-layer Mueller-matrix images and the dependencies of the first- to fourth-order statistical moments describing the 3D distributions of the azimuthally invariant element Δf for partially depolarizing ($\tau = 0.99$) rat liver histological sections.

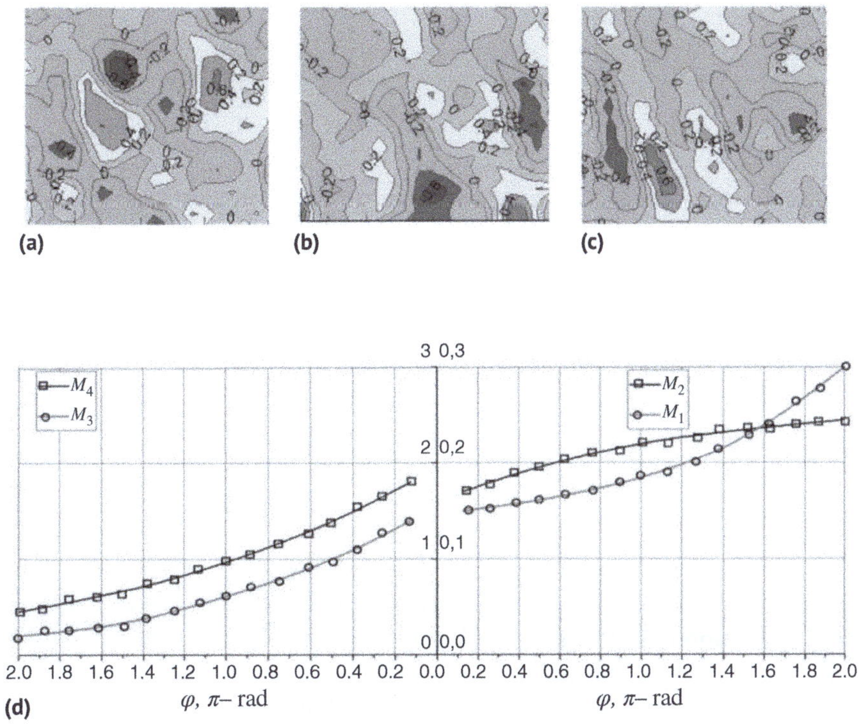

Figure 7.29 Layer-by-layer Mueller-matrix images and the dependencies of the first to fourth-order statistical moments describing the 3D distributions of the azimuthally invariant element M_{41} for partially depolarizing ($\tau = 0.99$) rat liver histological sections.

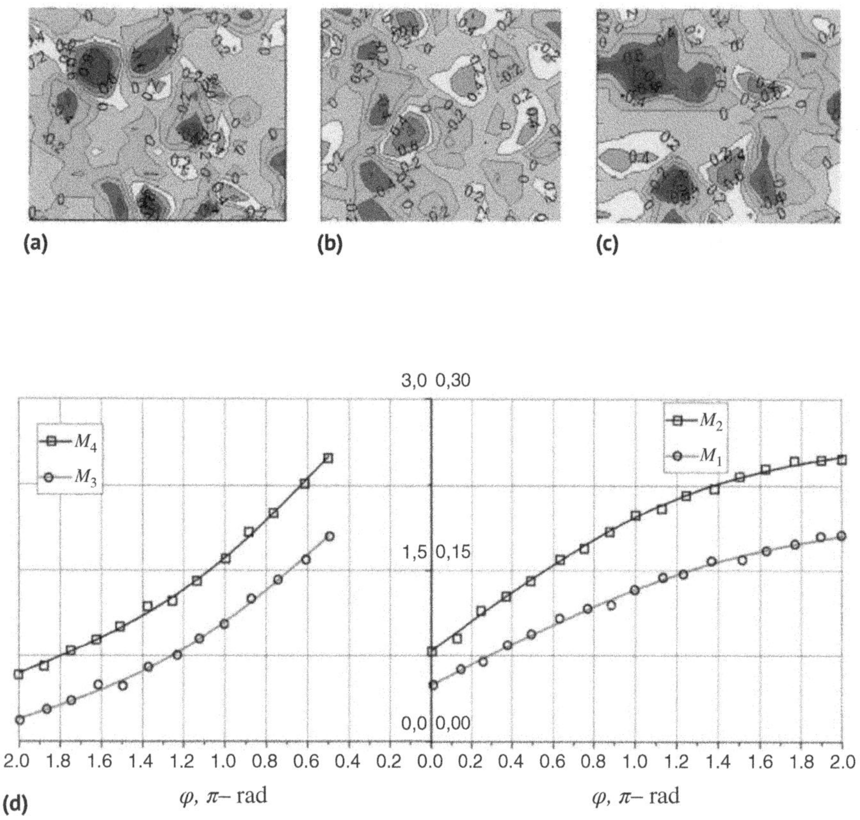

Figure 7.30 Layer-by-layer Mueller-matrix images and the dependencies of the first- to fourth-order statistical moments describing the 3D distributions of the azimuthally invariant element f_{14} for partially depolarizing ($\tau = 0.99$) rat liver histological sections.

With the increase in the level of the phase cross-section ($\phi \uparrow$) of the 3D distributions of such MMIs, as in the case of a partially depolarizing myocardium layer, there is an increase in the mean (M_1(MMI) $\uparrow$) and dispersion (M_2(MMI) $\uparrow$), and, conversely, a decrease of skewness M_3(MMI) $\downarrow$) and kurtosis (M_4(MMI) $\downarrow$) - Figures 7.27–7.30.

7.7.4 Comparative Characteristics of 2D and 3D Methods of Mueller-matrix Mapping of Biological Layers of Different Morphological Structures

This section shows the results of the statistical analysis of the MMI value distribution in histological sections of myocardium and rat liver tissue, using two different approximations:

- Statistical moments of the first to fourth orders $M^{2D}_{i=1;2;3;4}$, which characterize the MMI distributions averaged over the entire volume – the traditional 2D Mueller-matrix mapping;
- Statistical moments of the first to fourth orders $M^{3D}_{i=1;2;3;4}(\phi)$, which characterize the layers of "phase" distributions – 3D Mueller-matrix mapping.

Tables 7.24 and 7.25 show the differences $\Delta M_i(\phi) = \frac{M^{3D}_i(\phi)}{M^{2D}_i}$ between the statistical moments $M^{2D}_{i=1;2;3;4}$ and $M^{3D}_{i=1;2;3;4}(\phi)$.

A comparative analysis of experimental data from both polarization-mapping methods across tissues with different structures found that 3D Mueller-matrix tomography is notably more responsive to variations in the polycrystalline structure of biological layers. As indicated (in grey), the distribution of all MMI values in histological tissue sections varies most in internal layers that have low phase-section levels (0.3 rad $\leq \phi \leq$ 0.9 rad).

The most sensitive are the statistical moments of the third and fourth orders, which characterize the distributions of MMIs related to the polarization effects of linear birefringence. ($f_{44}(x, y, \phi)$) and dichroism ($f_{41}(x, y, \phi)$) of different types of optically anisotropic structures. As the multiplicity of light scattering ($\phi \uparrow$) increases, the diagnostic capabilities of polarization mapping in differentiating different types of polycrystalline structures significantly decrease – $\Delta M_i(\phi)\frac{M^{3D}_i(\phi)}{M^{2D}_i} \to 1$.

Table 7.24 Correlation between statistical moments $M^{2D}_{i=1;2;3;4}$ and $M^{3D}_{i=1;2;3;4}(\phi)$, which characterize the MMI of phase anisotropy f_{44} and Δf.

	f_{44}			Δf		
$\Delta M_i(\phi)$	$\phi = 0.3$	$\phi = 0.9$	$\phi = 1.5$	$\phi = 0.3$	$\phi = 0.9$	$\phi = 1.5$
$\Delta M_1(\phi)$	1.43	1.32	1.19	1.34	1.23	1.08
$\Delta M_2(\phi)$	1.39	1.28	1.16	1.23	1.12	1.06
$\Delta M_3(\phi)$	2.12	1.89	1.45	1.81	1.68	1.24
$\Delta M_4(\phi)$	2.46	2.08	1.52	1.94	1.75	1.35

Table 7.25 The relationship between statistical moments $M^{2D}_{i=1;2;3;4}$ and $M^{3D}_{i=1;2;3;4}(\phi)$, which characterize the MMI of amplitude anisotropy f_{41} and f_{14}.

	f_{41}			f_{14}		
$\Delta M_i(\phi)$	$\phi = 0.3$	$\phi = 0.9$	$\phi = 1.5$	$\phi = 0.3$	$\phi = 0.9$	$\phi = 1.5$
$\Delta M_1(\phi)$	1.54	1.39	1.22	1.31	1.19	1.07
$\Delta M_2(\phi)$	1.61	1.42	1.31	1.21	1.14	1.11
$\Delta M_3(\phi)$	2.21	1.98	1.48	1.74	1.55	1.29
$\Delta M_4(\phi)$	2.54	2.11	1.59	1.81	1.67	1.38

7.7.5 Conclusions

1. The azimuthally invariant 3D Mueller-matrix mapping method for analyzing the distribution of phase and amplitude anisotropy parameters in partially depolarizing layers of biological tissues with various morphological structures is proposed and validated.
2. In the volume of biological samples, the coordinate distributions of the values of the MMIs for myocardium tissue ($\tau = 0.91$; $\Lambda = 58\%$) histological sections with a spatially structured optically anisotropic fibrillar network, as well as for rat liver parenchymatous tissue ($\tau = 0.99$; $\Lambda = 61\%$) with an "islet" polycrystalline structure, are obtained.
3. The phase dependences of the first- to fourth-order statistical moments, which describe the distributions of MMIs of the polarization manifestations of linear and circular birefringence and dichroism parameters in the polycrystalline component of various biological tissues, are established.
4. Comparative studies of the potential of differentiating changes in optical anisotropy parameters using traditional 2D and 3D Mueller-matrix mapping methods are carried out.
5. The optimal conditions for differentiating polycrystalline structures of biological tissues-the range of phase cross sections ($0.3\text{ rad} \leq \phi \leq 0.9\text{ rad}$) – and the most sensitive parameters – the statistical moments of the third and fourth orders, which characterize the distributions of MMIs associated with the polarization manifestations of linear birefringence ($f_{44}(x, y, \phi)$) and dichroism ($f_{41}(x, y, \phi)$) of different types of optically anisotropic structures-are determined.

7.8 3D Mueller-matrix Differential Diagnostics of Polycrystalline Structure of Benign and Malignant Tumors

7.8.1 Principles of Differential Diagnosis of Benign and Malignant Prostate Tumors

To evaluate how effectively the 3D Mueller-matrix mapping method distinguishes between benign and malignant tumors in human organs, two groups of histological sections from biopsy samples of surgically removed prostates were prepared. These sections had varying degrees of depolarization $\Lambda \leq 50\%$:

- benign tumor (adenoma: 26 samples) – group 1 (attenuation coefficient $0.79 < \tau < 0.85$, $43\% < \Lambda < 48\%$);
- malignant tumor (carcinoma: 26 samples) – group 2 ($0.81 < \tau < 0.84$, $45\% < \Lambda < 47\%$).

The optical technology for differential diagnostics of such samples includes the following steps:

1. Definition of a series of phase-layered images of 3D MMI distributions $\{f_{44}; \Delta f; f_{41}; f_{14}\}(\phi_1 = 0, 3; 2\phi_1, ..., 6\phi_1)$ that characterize volumetric polarization manifestations of phase and amplitude anisotropy within the two groups of samples.
2. For each "phase" section of the 3D distributions of the MMI value, a set of statistical moments of the first- to fourth-order $M_{i=1;2;3;4}\{[f_{44}; \Delta f; f_{41}; f_{14}](\phi_k, x, y)\}$ is calculated.
3. "Phase" dependences $M_{i=1;2;3;4}\{[f_{44}; \Delta f; f_{41}; f_{14}](\phi_1, \phi_2, ...\phi_k)\}$ of the value of each statistical moment are constructed for histological sections of groups 1 and 2.
4. The "phase" planes (ϕ^*) are determined in the 3D distributions of MMIs, where the maximal differences between the values of statistical moments $\left(\Delta M^*_{i=1;2;3;4} \equiv \Delta M_{i=1;2;3;4}(\phi^*) \to \max\right)$ that characterize the distributions of the values of matrix elements $f_{44}; \Delta f; f_{41}; f_{14}$ in these planes are realized.
5. In the "phase" plane ϕ^*, the mean $\Delta \overline{M}^*_{i=1;2;3;4}$ and the error $\sigma(\Delta M^*_i)$ within the histological sections of groups 1 and 2 are determined.

7.8.2 Analysis and Discussion of Experimental Results

Figures 7.31 and 7.32 present the coordinate maps illustrating the 3D distributions of MMIs related to optical anisotropy in histological sections of prostate adenoma and carcinoma.

Examples of MMIs $f_{44}; \Delta f$, which characterize the linear and circular birefringence of optically thick samples $(\tau = 1.07 \div 1.12)$ obtained for the phase cross section $\phi^* = 0.45$ rad and corresponding to the conditions $\Delta M^*_{i=1;2;3;4} \equiv \Delta M_{i=1;2;3;4}(\phi^*) \to \max$, are provided.

Coordinate distributions of the MMI $f_{41}; f_{14}$, characterizing polarization manifestations of amplitude anisotropy, for histological sections of benign and malignant tumors are obtained for $\phi^* = 0.75$ rad.

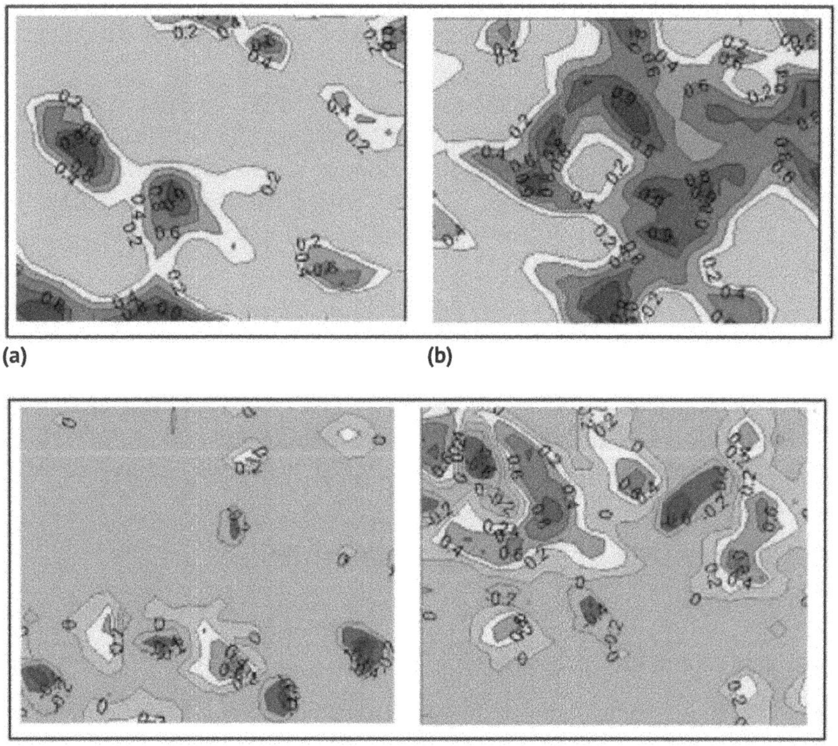

(a) (b) (c) (d)

Figure 7.31 Coordinate distributions of the MMI f_{44} at phase section $\phi^* = 0.45$ rad, characterizing the linear birefringence (a) and (b) and circular birefringence (c) and (d) of the prostate tissue layers: adenoma (a) and (c) and carcinoma (b) and (d).

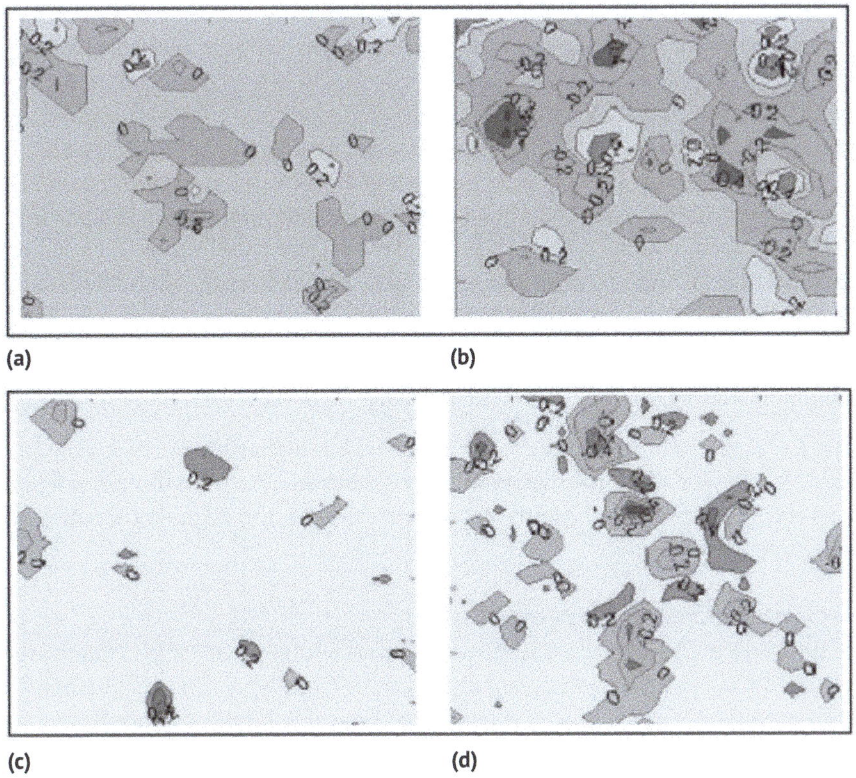

Figure 7.32 Coordinate distributions of Mueller-matrix invariants at phase section ($\phi^* = 0.75$ rad) that characterize the linear f_{41} (a) and (b) and circular f_{14} (c) and (d) dichroism of the prostate adenoma (a) and (c) and carcinoma (b) and (d) layers.

7.8.2.1 Statistical Analysis of MMI

The comparative analysis of distributions ($f_{ik}(\phi, x, y)$), which describe the phase and amplitude anisotropy in adenoma and carcinoma samples across various planes of the 3D distributions, uncovered a combined dependence of their structure. With increasing $\phi \uparrow$ (increasing the multiplicity of light scattering), the following scenario of changes occurs in the values of the statistical moments of the first to fourth orders, which characterize the histograms of $f_{ik}(\phi, x, y) - \begin{cases} M_{1;2}(f_{ik}(\phi, x, y)) \downarrow; \\ M_{3;4}(f_{ik}(\phi, x, y)) \uparrow \end{cases}$.

This transformation in the statistical structure of MMI across different phase planes is due to an increase in the scattering multiplicity of polarized laser radiation within the prostate tissue volume. In particular, for small $\phi < 0.9$ rad, the optical-geometric (morphological) structure of the polycrystalline component in the studied samples is the primary factor influencing the distribution of all types of MMIs.

At low scattering multiplicities, a nearly clear relationship exists between MMI values and the parameters of phase and amplitude anisotropy. As a result, within this range of phase sections in the 3D distributions $f_{ik}(\phi, x, y)$, there are the most distinct individual differences between the dependencies $M_{i=1;2;3;4}(f_{ik}(\phi, x, y))$.

With increasing scattering multiplicity ($\phi > 0.9$ rad) in the volume of the prostate layers, the averaging process occurs with a gradual reduction of skewness and a diminution of the peak of the distributions $f_{ik}(\phi, x, y)$ in the corresponding phase sections of the volumetric MMI $- \begin{cases} M_{1;2}(f_{ik}(\phi, x, y)) \uparrow; \\ M_{3;4}(f_{ik}(\phi, x, y)) \downarrow \end{cases}$.

According to the central limit theorem, with strong scattering these distributions tend to normal $- \begin{cases} M_{1;2}(f_{ik}(\phi, x, y)) \neq 0; \\ M_{3;4}(f_{ik}(\phi, x, y)) \to 0 \end{cases}$. As a result, the differences between the values of the statistical moments $M_{i=1;2;3;4}(f_{ik}(\phi, x, y))$ are significantly reduced – $\Delta M_i(f_{ik}) \downarrow$.

Furthermore, we interpret the results from a physical perspective. Regarding the linear and circular birefringence in both sample types (Figure 7.31), the differing pathology of prostate tissue leads to specific changes in the polycrystalline structure. The benign tumor shows a well-developed, newly formed linear birefringent network ($f_{44}(\phi, x, y)$) formed by the growth in concentration of optically active protein molecules ($\Delta f(\phi, x, y)$).

Quantitatively, this is shown by the relationships between the statistical moments from the first to fourth orders, which describe the distributions of MMI values for the linear and circular birefringence of this biological layer: $\begin{cases} M_{1;2}(f_{44}) > M_{1;2}(\Delta f); \\ M_{3;4}(f_{44}) < M_{3;4}(\Delta f) \end{cases}$.

Oncological processes involve the destruction of the fibrillar network, indicated by a decrease in linear birefringence in prostate tissue. Consequently, this affects the distribution histograms $f_{44}(\phi, x, y)$. It becomes narrower and more asymmetric with a decreasing mean value. A similar scenario (through reduction of concentration and decomposition of optically active protein molecules) occurs for the MMI $\Delta f(\phi, x, y)$, which characterizes circular birefringence (Figure 7.33). This process is most evident in the region of small phase cross sections ($0.3\text{ rad} < \phi < 0.6\text{ rad} \Rightarrow \phi^* = 0.45\text{ rad}$), where the differences between $M_3(f_{44}, \Delta f)$ and $M_4(f_{44}, \Delta f)$ reach maximum values (Figure 7.32).

Similar transformations happen in the distributions of optically anisotropic absorption for linear and circular dichroism of the partially depolarizing layers in adenoma and carcinoma: $\begin{cases} M_{1;2}(f_{41}) > M_{1;2}(f_{14}); \\ M_{3;4}(f_{41}) < M_{3;4}(f_{14}) \end{cases}$. However, the largest separation between the statistical moments $M_{i=1;2;3;4}(f_{41}; f_{14})$ is achieved in the range of somewhat higher phase cross sections ($0.6\text{ rad} < \phi < 0.9\text{ rad} \Rightarrow \phi^* = 0.75\text{ rad}$) (Figure 7.32). In our view, this peculiarity occurs because, as the number of light scattering events increases at a specific optical thickness range within prostate tumor volumes, the degree of amplitude anisotropy enhances differently for benign and malignant tumors.

7.8.2.2 Information Analysis of Statistical and Cross-correlation Structure of MMI

Tables 7.26 and 7.27 present the results of the statistical analysis of coordinate distributions within the MMI set, which are determined by linear and circular birefringence and dichroism in prostate tumor samples of both types. These samples are taken from the most diagnostically sensitive phase cross-sections. The MMIs related to phase and amplitude anisotropy for adenoma and carcinoma, which are most responsive to distribution differences, are the third and fourth order statistical moments, highlighted in gray. According to evidence-based medicine standards, these parameters demonstrate an excellent level of balanced accuracy when applying 3D Mueller-matrix mapping of partially depolarizing layers ($\Lambda \leq 50\%$) in the differential diagnosis of benign and malignant prostate tumors is achieved: $\mathrm{Ac}\left(M_{3;4}; \left(f_{44;41;14}, \Delta f\right)\right) \sim 91 - 94\%$.

7.8.3 Comparative Characteristics of Methods of Polarization, 2D, and 3D Mueller-matrix Mapping in the Differentiation of Benign and Malignant Tumors

This subsection shows comparative results on how well different methods distinguish benign from malignant prostate tumors based on depolarization. It compares polarization, 2D Mueller-matrix mapping, and 3D Mueller-matrix mapping. For the analysis, two groups of 25 histological section samples each-one from adenoma and one from carcinoma biopsies-were examined, with varying geometric (l, μm), optical (τ) and depolarization (Λ, %) characteristics were studied (Table 7.28).

The balanced accuracy (Ac) of the methods was determined for each of the statistical moments of the first to fourth orders $\left(M_{i=1;2;3;4}\right)$ that characterize the distributions of the polarization azimuth (α) and ellipticity (β), as well as the 2D and 3D sets of MMI $f_{14;41;44}$, Δf values that characterize the optical manifestations of the linear and circular birefringence and dichroism of the polycrystalline component of benign and malignant prostate tumors. The maximum level of balanced

Table 7.26 Statistical criteria for the differentiation of benign and malignant prostate tumors on the basis of phase anisotropy.

Parameters	Adenoma		Carcinoma		Accuracy, AC, %	
MMI	f_{44}	Δf	f_{44}	Δf	f_{44}	Δf
$M_1\phi^* = 0.45$ rad	0.29 ± 0.017	0.12 ± 0.007	0.44 ± 0.029	0.07 ± 0.004	85	82
$M_2(\phi^* = 0.45$ rad)	0.21 ± 0.012	0.15 ± 0.008	0.14 ± 0.006	0.11 ± 0.005	81	79
$M_3(\phi^* = 0.45$ rad)	0.46 ± 0.029	0.63 ± 0.041	0.69 ± 0.037	0.92 ± 0.055	91	89
$M_4(\phi^* = 0.45$ rad)	0.57 ± 0.033	0.88 ± 0.053	1.03 ± 0.059	1.39 ± 0.084	92	87

Table 7.27 Statistical and correlation criteria for differentiation of prostate tumors on the basis of amplitude anisotropy.

Parameters	Adenoma		Carcinoma		Accuracy, AC, %	
MMI	f_{44}	Δf	f_{44}	Δf	f_{44}	Δf
$M_1\phi^* = 0.45$ rad	0.21 ± 0.013	0.075 ± 0.0042	0.15 ± 0.091	0.093 ± 0.005	81	73
$M_2(\phi^* = 0.45$ rad)	0.14 ± 0.008	0.09 ± 0.005	0.095 ± 0.006	0.12 ± 0.071	79	77
$M_3(\phi^* = 0.45$ rad)	0.77 ± 0.043	0.98 ± 0.052	1.18 ± 0.063	0.61 ± 0.037	92	91
$M_4(\phi^* = 0.45$ rad)	0.96 ± 0.054	1.31 ± 0.079	1.65 ± 0.093	0.99 ± 0.056	94	93

Table 7.28 Comparative diagnostic efficiency of polarization methods ($\{\alpha, \beta\}$), 2D and 3D Mueller-matrix mapping for prostate tumor layers with different depolarizing ability.

τ		0.25			0.45			0.62	
l, μm		20			40			50	
Λ, %		28			35			43	
$Ac(M_{i=1;2;3;4})$	$\{\alpha, \beta\}$	2D	3D	$\{\alpha, \beta\}$	2D	3D	$\{\alpha, \beta\}$	2D	3D
$Ac(M_1)$	73	78	91	69	71	87	63	65	85
$Ac(M_2)$	76	79	93	70	74	89	65	68	87
$Ac(M_3)$	76	84	95	72	77	95	68	71	93
$Ac(M_4)$	77	83	96	73	79	95	70	75	94
τ		0.91			1.37			1.93	
l, μm		60			80			100	
Λ, %		52			68			89	
$Ac(M_{i=1;2;3;4})$	$\{\alpha, \beta\}$	2D	3D	$\{\alpha, \beta\}$	2D	3D	$\{\alpha, \beta\}$	2D	3D
$Ac(M_1)$	59	63	83	53	57	64	54	53	52
$Ac(M_2)$	58	65	85	55	56	66	52	56	54
$Ac(M_3)$	61	68	91	57	59	78	53	57	65
$Ac(M_4)$	63	69	94	56	63	79	55	60	68

accuracy $Ac(M_{i=1;2;3;4})$ was selected as the base value. The comparative analysis of polarimetry methods' accuracy in diagnosing changes in phase and amplitude anisotropy of depolarizing layers in benign and malignant prostate tumors revealed the following findings:

- Polarization mapping $\{\alpha, \beta\}$ provides satisfactory accuracy of differentiating prostate adenoma and carcinoma ($Ac(\tau = 0.25) = 76 - 77\%$) only at a minimal depolarization level ($\Lambda = 28\%$). This method is not suitable for higher optical thickness ($Ac < 70\%$) levels.
- The accuracy of differential diagnosis using the 2D Mueller-matrix mapping method ranges from good ($Ac(\tau = 0.25; \Lambda = 28\%) = 83 - 84\%$) to satisfactory ($Ac(\tau = 0.62; \Lambda = 43\%) = 75\%$) within a small increase in depolarization ($0.25 < \tau < 0.62 \Leftrightarrow 28\% < \Lambda < 43\%$).
- 3D Mueller-matrix mapping is diagnostically effective ($70\% < Ac < 96\%$) over a broad range of optical thickness changes ($0.25 < \tau < 1.37 \Leftrightarrow 28\% < \Lambda < 68\%$).

7.8.4 Conclusions

1. A method for 3D Mueller-matrix mapping that is azimuthally invariant has been proposed and validated. It analyzes distributions of phase and amplitude anisotropy in partially depolarizing layers of both benign and malignant prostate tumor biopsies.
2. Coordinate distributions of Mueller-matrix invariants are obtained layer-by-layer for histological sections of adenoma and carcinoma within the volume of the biological samples.

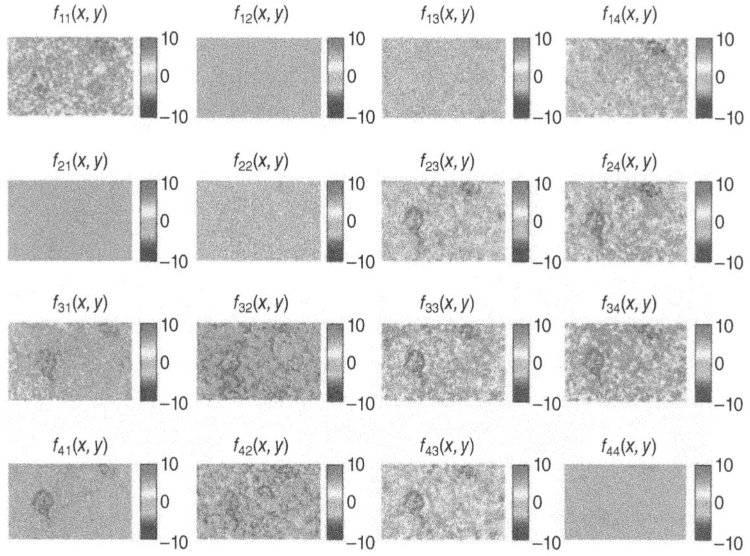

Figure 7.33 Maps showing the diagonal elements of the second-order differential Mueller matrix for rectum wall tissue.

Table 7.29 Statistical moments $M_{i=1;2;3;4}$ of the distributions of anisotropy parameters of the depolarizing layer of rectum wall tissue.

$M_{i=1;2;3;4}$	Δf^2_{11}	Δf^2_{22}	Δf^2_{33}	Δf^2_{44}
$M_{i=1}$	1.87	1.79	2.02	1.65
$M_{i=2}$	1.37	1.55	1.29	1.42
$M_{i=3}$	0.45	0.73	1.18	0.61
$M_{i=4}$	0.78	1.27	1.85	0.94

3. The phase-dependent statistical moments of the first to fourth order, which describe the distribution of MMI values for linear and circular birefringence and dichroism in the polycrystalline parts of various prostate tumor types, are calculated.
4. The optimal conditions for distinguishing the polycrystalline structures of benign and malignant tumors are specified, including the range of phase sections and the most sensitive parameters – the third- fourth-order moments. These moments characterize the MMI distributions linked to the polarization effects of linear birefringence and dichroism in adenoma and carcinoma.

7.9 Differential 3D Mueller-matrix Mapping of Optically Anisotropic Depolarizing Biological Layers

7.9.1 Experimental Results of the Differential Mapping Method

The experimental validation results for the differential mapping method, applied to the depolarizing component of the polycrystalline structure in a diffuse biological layer, are demonstrated using an optically thick rectal tissue layer ($z = 100\ \mu m$; $\tau = 1.36$; $\Lambda = 83\%$, z – geometrical thickness; τ – extinction coefficient; Λ – depolarization degree) (Figure 7.34). A series of maps and histograms of the distributions m_{ik} are presented in Figures 7.35 and 7.36.

Table 7.29 provides the statistical analysis (first- to fourth-order moments $M_{i=1;2;3;4}$) of the coordinate distributions of the diagonal elements of the differential Mueller matrix of the second order.

It was observed that the values $M_{i=3;4}$ show individual sensitivity to the specific characteristics of the coordinate distribution of fluctuations in optical anisotropy parameters for the partially depolarizing layer of rectum wall tissue.

This finding formed the basis for using statistical analysis of coordinate distributions of fluctuations in birefringence and dichroism values within the diffusive layers of rectal wall tissue for biomedical applications.

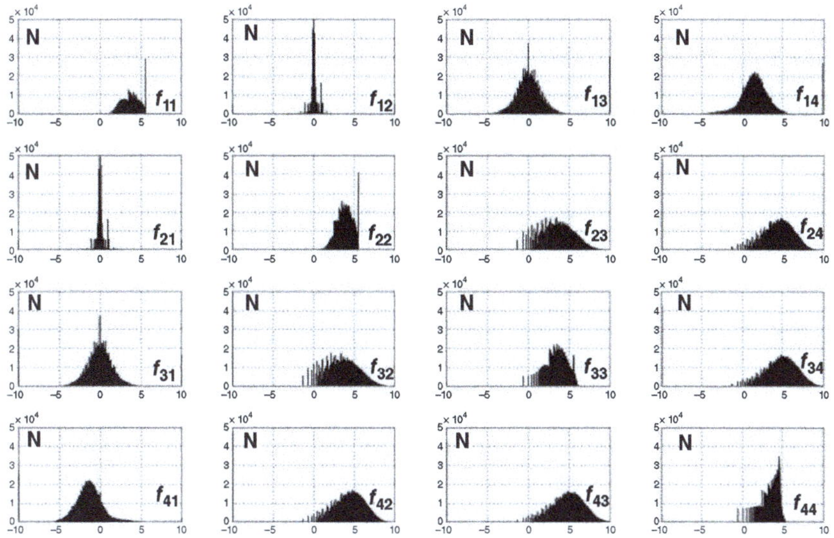

Figure 7.34 Histograms of the distributions of the second-order differential Mueller-matrix elements for a histological section of rectum wall tissue ($z = 100\ \mu m$; $\tau = 1.36$; $\Lambda = 83\%$).

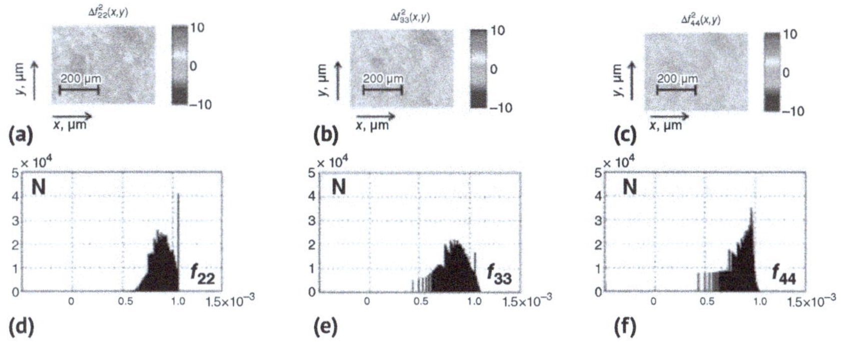

Figure 7.35 Maps (top row) and histograms (bottom row) of the distributions of the diagonal elements Δf_{ii}^2 of the second-order differential Mueller matrix for histological sections of vaginal wall biopsy from donors.

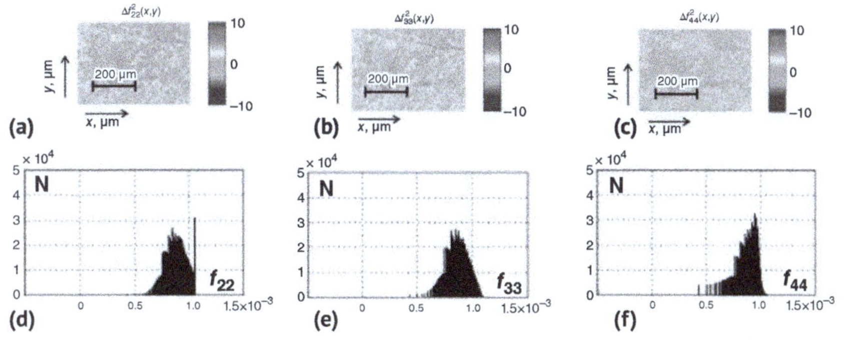

Figure 7.36 Maps (top row) and histograms (bottom row) of the distributions of the diagonal elements Δf_{ii}^2 of the second-order differential Mueller matrix for histological sections of vaginal wall biopsy from prolapse patients.

7.9.2 Clinical Application of Mueller-matrix Mapping of Biological Layers in Differential Diagnostics of Vaginal Wall Tissue with Prolapse

Two groups of vaginal wall biopsy samples were examined:

- Healthy donors – control group 1 (30 samples)
- Prolapse patients – investigated group 2 (30 samples)

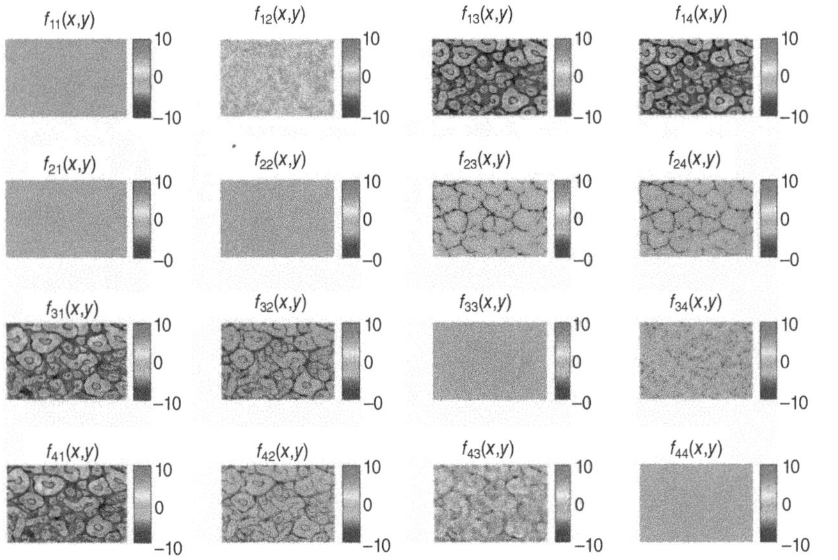

Figure 7.37 Maps of the elements of the first-order differential matrix for a histological brain tissue section ($z = 60\ \mu\text{m}$; $\tau = 0.21$; $\Lambda = 43\%$).

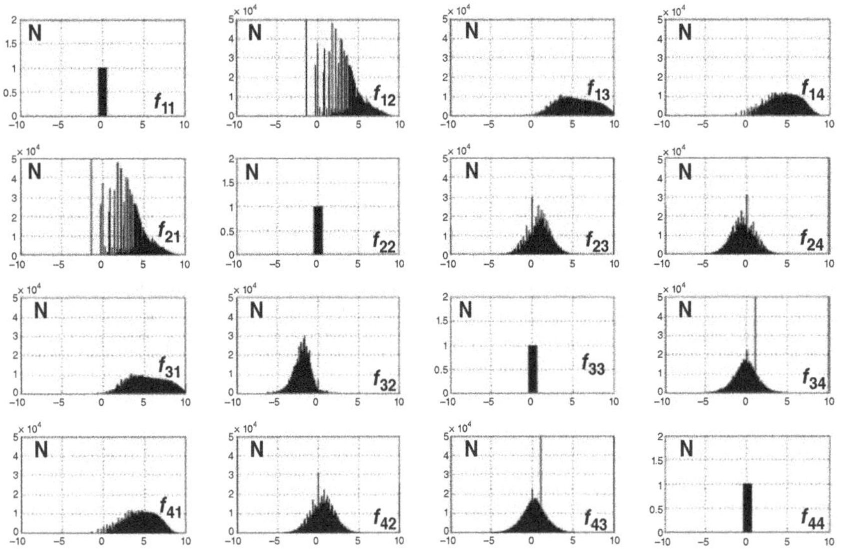

Figure 7.38 Histograms of the distributions of the values of the first-order differential matrix elements for a histological brain tissue section.

Histological sections were produced using standard freezing microtomy.

The results from Mueller-matrix mapping of the diagonal elements of the second-order differential matrix for diffusive layers of vaginal wall tissue are presented: Figure 7.37 for donors and Figure 7.38 for prolapse patients.

A good $\left(\text{Ac}\left(\Delta f^2_{22}\right) = 88 - 93\%\right)$ and excellent $\left(\text{Ac}\left(\Delta f^2_{33;44}\right) = 95 - 96\%\right)$ level of balanced accuracy was reached for the differential diagnostics of biopsy samples of vaginal wall tissue from donors and patients with prolapse.

The experimental validation of this method was demonstrated, and the parameters of the second-order differential Mueller matrix of rectum wall tissue were determined. Differential diagnosis of the connecting tissue component of the vaginal wall was carried out through statistical analysis of fluctuations in the coordinate distributions of linear birefringence and dichroism in diffusive samples from donors and prolapse patients. High levels of balanced accuracy, including good and excellent, were achieved for these samples (Table 7.30).

Table 7.30 Balanced accuracy in differential Mueller-matrix mapping.

Ac, %	Δf_{22}^2	Δf_{33}^2	Δf_{44}^2
$M_{i=1}$	61%	67%	64%
$M_{i=2}$	72%	74%	71%
$M_{i=3}$	88%	92%	93%
$M_{i=4}$	93%	95%	96%

Table 7.31 Statistical moments of the distributions of anisotropy parameters of the depolarizing brain tissue layer.

$M_{i=1;2;3;4}$	f_{12}	f_{13}	f_{14}	f_{23}	f_{24}	f_{34}
$M_{i=1}$	2.11	3.21	4.07	0.88	0.57	0.34
$M_{i=2}$	2.71	1.98	2.28	0.56	0.39	0.22
$M_{i=3}$	1.44	0.89	1.14	0.32	0.17	0.49
$M_{i=4}$	1.81	0.56	0.41	0.21	0.44	0.35

7.9.3 Mueller-matrix Mapping of Biological Polycrystalline Layers Using Reference Wave

The first section offers a concise theoretical overview of the differential Mueller-matrix approach for partially depolarizing layers. It includes experimentally obtained maps of the first-order differential matrix for a polycrystalline structure in a histological brain tissue section. The statistical moments of orders 1 through 4, which describe the distribution of matrix elements, are outlined. The second section presents statistical analyses of birefringence and dichroism in histological sections from mouse liver tissue-both normal and diabetic. Objective criteria for differential diagnosis of diabetes are also established.

The results of studies of the two-dimensional structure of the elements ($(\langle f_{12}\rangle = \langle f_{21}\rangle)$; $(\langle f_{13}\rangle = \langle f_{31}\rangle)$; $(\langle f_{14}\rangle = \langle f_{41}\rangle)$; $(\langle f_{23}\rangle = -\langle f_{32}\rangle)$; $(\langle f_{24}\rangle = -\langle f_{42}\rangle)$; $(\langle f_{34}\rangle = -\langle f_{43}\rangle)$) of the differential matrix of the first order for a histological brain tissue section are illustrated by a series of dependencies (maps and histograms of f_{ik} distributions), shown in Figures 7.37 and 7.38.

The data indicate that all elements of the first-order differential matrix, except the diagonal $\langle f_{11;22;33;44}\rangle$, have non-zero values, reflecting the optical (phase and amplitude) anisotropy of the brain tissue layer. A strong correlation was observed between the theoretical predictions and experimental results regarding the symmetry of this differential matrix in a partially depolarizing biological layer $(\langle f_{12}\rangle = \langle f_{21}\rangle)$; $(\langle f_{13}\rangle = \langle f_{31}\rangle)$; $(\langle f_{14}\rangle = \langle f_{41}\rangle)$; $(\langle f_{23}\rangle = -\langle f_{32}\rangle)$; $(\langle f_{24}\rangle = -\langle f_{42}\rangle)$; $(\langle f_{34}\rangle = -\langle f_{43}\rangle)$. In other words, the results of experimental measurements of the ensemble of six elements of this matrix operator enables to make a direct analysis of the average parameters of optical anisotropy ($LD = \langle f_{12;21}\rangle$; $CD = \langle f_{13;31}\rangle$; $LD' = \langle f_{14;41}\rangle$; $LB = \langle f_{34;43}\rangle$; $\mathrm{CB} = \langle f_{23;32}\rangle$; $\mathrm{LB}' = \langle f_{24;42}\rangle$) of the multiply scattering biological layer.

Table 7.31 shows the results of the statistical analysis, including the first to fourth moments of the coordinate distributions of the differential Mueller matrix elements of the first order.

This feature served as the baseline. The sensitivity of the value $M_{i=1;2;3;4}$ to the specifics of the coordinate distributions of optical anisotropy parameters in the partially depolarizing brain tissue layer was assessed for the statistical analysis of both birefringence and dichroism.

7.9.4 Clinical Application of Mueller-matrix Mapping of Biological Layers in Differential Diagnostics of Diabetes

Histological sections of mouse liver tissue from two groups of samples were studied.

- Healthy animals – control group 1 (36 samples);
- Diabetes-affected animals – investigated group 2 (36 samples).

Figures 7.39 and 7.40 display the results of differential Mueller-matrix mapping, showing the distributions of linear birefringence (Figure 7.39) and dichroism (Figure 7.40) in histological sections of mouse liver tissue from group 1 (fragments (a), (c)) and group 2 (fragments (b), (d)).

It achieved a good $Ac(\tau) = 86 - 91\%$) and excellent ($Ac(\delta) = 95\%$)) level of balanced accuracy for the differential diagnostics of mouse liver tissue samples from healthy and diabetes-affected animals (Table 7.32).

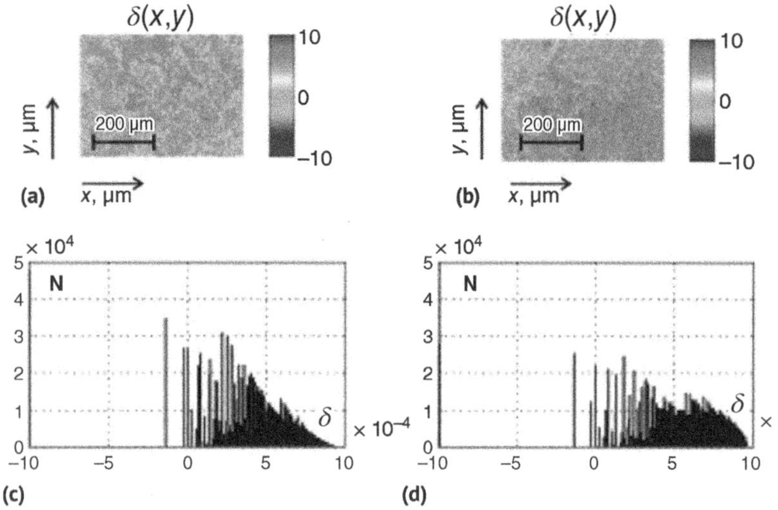

Figure 7.39 Maps (a) and (b) and histograms (c) and (d) of the distribution of the values of linear birefringence for histological sections of mouse liver tissue from healthy ((a), (c)) and diabetes-affected ((b), (d)) animals.

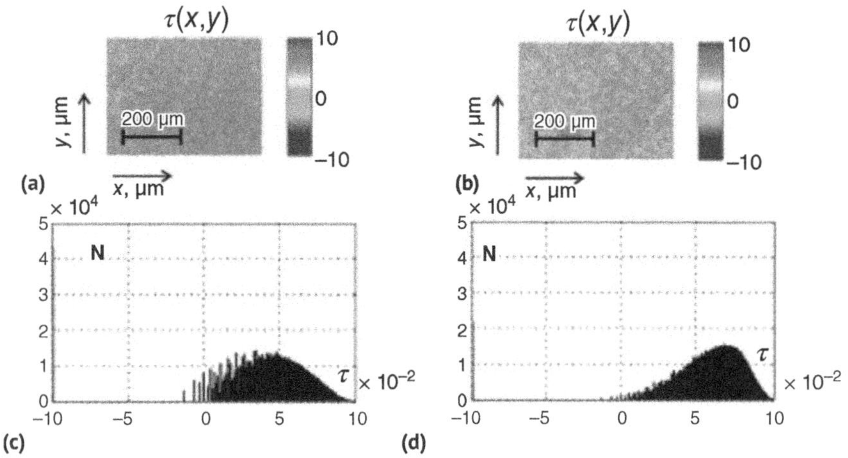

Figure 7.40 Maps (a) and (b) and histograms (c) and (d) of the distribution of the values of linear dichroism for histological sections of mouse liver tissue from healthy ((a), (c)) and diabetes-affected ((b), (d)) animals.

Table 7.32 Balanced accuracy of the method of differential Mueller-matrix mapping.

AC	$\delta(\%)$	$\tau(\%)$
$M_{i=1}$	71	84
$M_{i=2}$	84	81
$M_{i=3}$	90	86
$M_{i=4}$	95	91

7.9.5 Conclusions

A brief overview of the theoretical basis of the differential Mueller-matrix mapping method for analyzing the polycrystalline structure of depolarizing biological layers was provided. The experimental validation results were presented, and the first-order differential Mueller matrix parameters for a brain tissue histological section were identified. Differential diagnostics involved a statistical analysis of the coordinate distributions of linear birefringence and dichroism parameters in mouse liver samples from both healthy and diabetic animals. The study achieved high levels of balanced accuracy, including good and excellent, in differentiating between healthy and diabetic liver tissues.

References

1 Layden D, Ghosh N, Vitkin IA. Quantitative polarimetry for tissue characterization and diagnosis. In Wang RK, Tuchin VV, editors. Advanced Biophotonics: Tissue Optical Sectioning. Boca Raton: CRC Press; 2013. pp. 73–108.
2 Vitkin A, Ghosh N, de Martino A. Tissue polarimetry. In Andrews DL, editor. Photonics: Scientific Foundations, Technology and Applications. John Wiley & Sons, Ltd.; 2015. pp. 239–321.
3 Lee HR, Lotz C, Groeber-Becker FK, Dembski S, Novikova T. Digital histology with Mueller polarimetry and FastDBSCAN. Appl Opt. 2022;61(32):9616–9624.
4 Kim M, Lee HR, Ossikovski R, Malfait-Jobart A, Lamarque D, Novikova T. Optical diagnosis of gastric tissue biopsies with Mueller microscopy and statistical analysis. J Europ Opt Soc Rapid Publ. 2022;18(2):10.
5 Lee HR, Li P, Yoo TS, Lotz C, Groeber-Becker FK, Dembski S, et al. Digital histology with Mueller microscopy: how to mitigate an impact of tissue cut thickness fluctuations. J Biomed Opt. 2019;24(7):076004.
6 Li P, Lee HR, Chandel S, Lotz C, Groeber-Becker FK, Dembski S, et al. Analysis of tissue microstructure with Mueller microscopy: logarithmic decomposition and Monte Carlo modeling. J Biomed Opt. 2020;25(1):015002.
7 Lee HR, Yoo TS, Li P, Lotz C, Groeber-Becker FK, Dembski S, et al. Mueller microscopy of anisotropic scattering media: theory and experiments. In Proc. SPIE 10677 Unconventional Optical Imaging. 2018; 1067718.
8 Ma H, He H, Ramella-Roman JC. Mueller matrix microscopy. In Ramella-Roman JC, Novikova T, editors. Polarized Light in Biomedical Imaging and Sensing. Cham: Springer; 2023. pp. 281–321.
9 He H, Liao R, Zeng N, Li P, Chen Z, Liu X, et al. Mueller matrix polarimetry—an emerging new tool for characterizing the microstructural feature of complex biological specimen. J Lightwave Technol. 2019;37:2534–2548.
10 Oldenbourg R. A new view on polarization microscopy. Nature. 1996;381:811–812.
11 Gurjar RS, Backman V, Perelman LT, Georgakoudi I, Badizadegan K, Itzkan I, et al. Imaging human epithelial properties with polarized light-scattering spectroscopy. Nat Med. 2001;7:1245–1248.
12 Ramella-Roman JC, Saytashev I, Piccini M. A review of polarization-based imaging technologies for clinical and preclinical applications. J Opt. 2020;22:123001.
13 Qi J, Elson DS. Mueller polarimetric imaging for surgical and diagnostic applications: a review. J Biophotonics. 2017;10:950–982.
14 Fujiwara H. Spectroscopic Ellipsometry: Principles and Applications. John Wiley & Sons; 2007.
15 Mendoza-Galván A, Muñoz-Pineda E, Ribeiro SJ, Santos MV, Järrendahl K, Arwin H. Mueller matrix spectroscopic ellipsometry study of chiral nanocrystalline cellulose films. J Opt. 2018;20:024001.
16 Azzam RMA. Stokes-vector and Mueller-matrix polarimetry. J Opt Soc Am A. 2016;33:1396–1408.
17 Azzam RMA. Mueller-matrix ellipsometry: a review. Proc SPIE. 1997;3121:369–405.
18 Song B, Gu H, Zhu S, Jiang H, Chen X, Zhang C, et al. Broadband optical properties of graphene and HOPG investigated by spectroscopic Mueller matrix ellipsometry. Appl Surf Sci 2018;439:1079–1087.
19 Jiang H, Ma Z, Gu H, Chen X, Liu S. Characterization of volume gratings based on distributed dielectric constant model using Mueller matrix ellipsometry. Appl Sci 2019;9:698.
20 Chang JT, He H, Wang Y, Huang Y, Li X, He C, et al. Division of focal plane polarimeter-based 3×4 Mueller matrix microscope: a potential tool for quick diagnosis of human carcinoma tissues. J Biomed Opt 2016;21:056002.
21 Fu YF, Huang Z, He H, Ma H, Wu J. Flexible 3×3 Mueller matrix endoscope prototype for cancer detection. IEEE Trans Instrum Meas. 2018;67:1700–1712.
22 Angelsky PO, Ushenko AG, Dubolazov AV, Sidor MI, Bodnar GB, Koval G, et al. The singular approach for processing polarization-inhomogeneous laser images of blood plasma layers. J Opt. 2013;15(4):044030.
23 Ushenko AG, Dubolazov AV, Ushenko VA, Novakovskaya OY. Statistical analysis of polarization-inhomogeneous Fourier spectra of laser radiation scattered by human skin in the tasks of differentiation of benign and malignant formations.

J Biomed Opt. 2016;21(7):071110.
24 Garazdyuk MS, Bachinskyi VT, Vanchulyak OY, Ushenko AG, Dubolazov OV, Gorsky MP. Polarization-phase images of liquor polycrystalline films in determining time of death. Appl Opt. 2016;55(l):B67–B71.
25 Angelskii OV, Ushenko OG, Burkovets DN, Arkhelyuk OD, Ushenko YA. Polarization-correlation studies of multifractal structures in biotissues and diagnostics of their pathologic changes. Laser Phys. 2000;10(5):1136–1142.
26 Yermolenko S, Ushenko A, Ivashko P, Goudail F, Gruia I, Gavrilă C, et al. Spectropolarimetry of cancer change of biotissues. In Ninth International Conference on Correlation Optics. SPIE. 2009;7388:404–410.
27 Jóźwicki R, Patorski K, Angelsky OV, Ushenko AG, Burkovets DN, Ushenko YA. Automatic polarimetric system for early medical diagnosis by biotissue testing. Opt Appl. 2002;32(4):603–612.
28 Ushenko AG, Dubolazov OV, Ushenko VA, Novakovskaya OY, Olar OV. Fourier polarimetry of human skin in the tasks of differentiation of benign and malignant formations. Appl Opt. 2016;55(12):B56–B60.
29 Ushenko AG, Pishak VP. Coherent-domain optical methods: biomedical diagnostics, environmental and material science. In Laser Polarimetry of Biological Tissue: Principles and Applications. 2004.
30 Ushenko AG, Fediv AI, Marchuk YF. Correlation and fractal structure of Jones matrices of human bile secret. In European Conference on Biomedical Optics. 2009;7368:73681Q.
31 Ushenko AG. Depolarization of a laser emission field as the correlation development of its polarization structure. In International Conference on Correlation Optics. 1997;3317:331–339.
32 Bachinskyi V, Boychuk T, Ushenko A. Laser Polarimetry of Biological Tissues and Fluids: Chapter 4. Methods of Spatial-frequency, Singular and Correlation Analysis of Biological Layers Object Fields. LAP Lambert Academic Publishing; 2018.
33 Ushenko AG, Burkovets DN, Yermolenko SB, Arkhelyuk AD, Pishak VP, Yuzko AM, et al. Stokes polarimetry of biotissues. In Fourth International Conference on Correlation Optics. SPIE. 1999;3904:527–533.
34 Ushenko VA, Dubolazov AV, Pidkamin LY, Sakchnovsky MY, Bodnar AB, Ushenko YA, et al. Mapping of polycrystalline films of biological fluids utilizing the Jones-matrix formalism. Laser Phys. 2018;28(2):025602.
35 Ushenko AG, Angelsky PO, Sidor M, Marchuk YF, Andreychuk DR, Pashkovskaya NV. Spatial-frequency selection of complex degree of coherence of laser images of blood plasma in diagnostics and differentiation of pathological states of human organism of various nosology. Appl Opt. 2014;53(10):B172–B180.
36 Olar EI, Ushenko AG, Ushenko YA. Polarization correlation measurements of the phase tomograms of optically anisotropic biofractals. Laser Phys. 2004;14(8):1115–1121.
37 Guminetskiy SH, Ushenko OG, Polyanskiy IP, Motrych AV, Grynchuk FV. The optical method for investigation of the peritonitis progressing process. In Eighth International Conference on Correlation Optics. 2008;7008:626–632.
38 Ushenko AG, Fediv AI, Marchuk YF. Singular structure of polarization images of bile secret in diagnostics of human physiological state. In: European Conference on Biomedical Optics. Optica Publishing Group; 2009.
39 Ushenko Y, Ushenko A, Dubolazov A, Soltys I, Litvinenko O, Wanchuliak O, et al. Mueller-matrix interferometric multifractal scaling of optically anisotropic architectonics of diffuse blood facies: fundamental and applied aspects. J Biophotonics. 2025;18(3):e202400412. https://doi.org/10.1002/jbio.202400412.
40 Ushenko A, Soltys I, Dubolazov A, Ushenko Y, Bilookyi V, Bilookyi O, et al. 3D Jones-matrix thesiography of biological fluid facies. J Innov Opt Health Sci. 2025;18(2):2443002. https://doi.org/10.1142/S1793545824430028.
41 Ushenko A, Pavlyukovich N, Khukhlina O, Pavlyukovich O, Gorsky M, Soltys I, et al. Layer-by-layer multifractal scanning of optically anisotropic architectonics of blood plasma films: fundamental and applied aspects. Photonics. 2025;12:215. https://doi.org/10.3390/photonics12030215.
42 Ushenko A, Pavlyukovich N, Khukhlina O, Pavlyukovich O, Soltys I, Dubolazov A, et al. Blood Plasma Film Multifractal Scanning in COVID-19 Consequences Diagnostics. J Biophotonics. 2024;17(11):e202400356. https://doi.org/10.1002/jbio.202400356.
43 Ushenko A, Dubolazov A, Zheng J, Litvinenko A, Gorsky M, Ushenko Y, et al. 3D polarization-interference holographic histology for wavelet-based differentiation of biological tissues with different necrotic states. J Biomed Opt. 2024;29(5):052920. https://doi.org/10.1117/1.JBO.29.5.052920.
44 Ushenko O, Bilookyi O, Zheng J, Dubolazov A, Olar O, Ushenko Y, et al. 3D digital holographic polarimetry of laser speckle fields formed by polycrystalline blood films: a tool for differential diagnosis of thyroid pathology. Front Phys. 2024;12:1426469. https://doi.org/10.3389/fphy.2024.1426469.
45 Wójcik W, Hu Z, Ushenko Y, Smolarz A, Soltys I, Dubolazov O, et al. Optical sensor system for 3D Jones matrix reconstruction of optical anisotropy maps of self-assembled polycrystalline soft matter films. Sensors. 2024;24(5):1589. https://doi.org/10.3390/s24051589.

46 Ushenko A, Zheng J, Gorsky M, Dubolazov A, Ushenko Y, Soltys I, et al. 3D digital holographic polarimetry of diffuse optically anisotropic biological tissue object fields. Front Phys. 2023;11:1288935. https://doi.org/10.3389/fphy.2023.1288935.

47 Dubolazov A, Ushenko V, Trifonyuk L, Stashkevich A, Soltys I, Ushenko Y, et al. Polarization-singular approach to imaging Mueller-matrix polarimetry in differential diagnosis of histological sections of biopsy of tumors of the uterus and prostate. Front Phys. 2021;9:711212. https://doi.org/10.3389/fphy.2021.711212.

48 Ushenko VO, Trifonyuk L, Ushenko YA, Dubolazov OV, Gorsky MP, Ushenko AG. Polarization singularity analysis of Mueller-matrix invariants of optical anisotropy of biological tissue samples in cancer diagnostics. J Opt. 2021;23(6):064004. https://doi.org/10.1088/2040-8986/abf97a.

49 Ushenko YA, Koval GD, Ushenko AG, Dubolazov OV, Ushenko VA, Novakovskaia OY. Mueller-matrix of laser-induced autofluorescence of polycrystalline films of dried peritoneal fluid in diagnostics of endometriosis. J Biomed Opt. 2016;21(7):071116. https://doi.org/10.1117/1.JBO.21.7.071116.

8

Wavelet-enhanced Polarimetry for the Analysis of Different Necrotic and Pathological States

Oleksandr Ushenko[1,2], Iryna Soltys[2], Oleksandr Dubolazov[2], Olexandra Litvinenko[3], Olexander Bilookyi[4], Oleh Wanchuliak[3], and Yurii Ushenko[5,6]

[1] *Institute of Zhejiang University – Taizou, Taizhou, China*
[2] *Optics and Publishing Department, Chernivtsi National University, Chernivtsi, Ukraine*
[3] *Department of Forensic Medicine and Medical Law, Bucovinian State Medical University, Chernivtsi, Ukraine*
[4] *Department of Surgery, Bucovinian State Medical University, Chernivtsi, Ukraine*
[5] *Computer Science Department, Chernivtsi National University, Chernivtsi, Ukraine*
[6] *Department of Physics, Shaoxing University, Shaoxing, Zhejiang, China*

8.1 Introduction

Optical methods for diagnosing various polycrystalline objects (including layers of biological tissues) and visualizing their structure occupy a prominent place due to their high informativeness and the possibility of multifunctional monitoring of the investigated environment (photometric, spectral, polarization, and correlation monitoring). However, at present, there is no unified methodological approach for diagnosing such structures [1–14].

A significant advancement in biomedical polarimetric diagnostics has been the development of Mueller-matrix techniques, which has led to the establishment of Mueller-matrix microscopy (MMM) as a separate field. The fundamental results obtained within MMM for biological tissues are also promising for applied clinical and preclinical studies of their polycrystalline structure.

In parallel, methods and systems of laser polarimetry for biological tissues and liquid samples have been intensively developed, employing a wide range of analytical algorithms (statistical, correlation, fractal, wavelet, Fourier, etc.) for quantitative data processing [15–23].

At present, two major problems remain unresolved in biomedical laser polarimetric diagnostics. The first is the dependence of experimental data on the degree of depolarization of the laser radiation, which not only integrates the polarization distributions but also compromises the unambiguous relationships between polarization-inhomogeneous object fields and the parameters of polycrystalline biological layers. The second major informational limitation of polarimetric mapping is the integration of data across all geometric scales of the optically anisotropic architectonics of biological layers.

This section focuses on the development and experimental validation of methods for 3D scale-selective polarimetry of multiply scattered fields in diffuse layers, aimed at eliminating the distorting influence of a high level of depolarized background and increasing the sensitivity and accuracy of histological differential diagnosis of the causes of necrotic changes in the optically anisotropic architectonics of biological tissues [23–31].

8.1.1 Objects of Investigations

To carry out a comprehensive study of the histological sections of myocardium samples, the following groups were formed:

- coronary heart disease (CHD) – Group 1, 12 samples;
- acute coronary insufficiency (ACI) – Group 2, 12 samples.

The following two groups consisted of histological sections of lung tissue obtained from individuals who died from myocardial infarction and had the following concomitant pathologies:

- bronchial asthma (BA) – Group 3, 15 samples;
- pulmonary fibrosis (PF) – Group 4, 15 samples.

Figure 8.1 presents microscopic images of all histological sample groups: myocardium (CHD, panel (a); ACI, panel (b)) and lung tissue (BA, panel (c); PF, panel (d)).

Analysis of myocardial images from both groups reveals the presence of the fibrillar structure of myosin fibers (Figure 8.1a, b). A tissue sample with ACI is characterized by a well-spatially ordered fibrillar network with average geometric dimensions of 15–20 microns (Figure 8.1b). The myocardial fibrillar network in CHD is disordered and consists of thinner (10 microns) myosin fibrils (Figure 8.1a).

Analysis of lung tissue images with BA reveals a weakly structured parenchymal morphological organization (Figure 8.1c). The tissue with PF demonstrates a well-formed fibrillar connective tissue component with an average fiber size of 5–10 microns (Figure 8.1d).

This selection of objects relates to both the fundamental and applied components of our research.

The fundamental component: from a physical point of view, the selected tissues have different optically anisotropic architectures. The myocardium is characterized by spatially structured, linearly birefringent networks of myosin fibrils. The lung tissue is predominantly parenchymal, with slight linear birefringence of the collagen fibrillar networks in the connective tissue component.

Cases of CHD and ACI lead to various necrotic changes in the optical anisotropy of myosin fibers and their spatially structured networks. In CHD, myosin fibers thin, and the spatial order of the fibrillar networks decreases (Figure 8.1a). As a result, the magnitude of structural anisotropy decreases. In ACI, myosin fibers are broken in some areas, but the spatial ordering of the fibrillar network is maintained (Figure 8.1b). Therefore, the birefringence level remains comparable to the anisotropy parameter of healthy tissue.

BA does not significantly alter the architecture of the collagen fibrillar networks in the connective tissue component (Figure 8.1c). PF, however, is accompanied by significant collagen fiber growth and an increase in birefringence (Figure 8.1d).

Therefore, a comparative physical analysis of the results obtained will allow us to determine the capabilities of our method for detecting changes in the optically anisotropic architecture of biological tissue samples with different morphologies.

The applied component. An important and not yet entirely solved forensic task using light microscopy methods is determining whether the cause of myocardial death is natural (CHD) or violent (ACI). The main problem in histological diagnosis is the high level of diffuse background in images of myocardial histological sections.

The optical and geometric parameters of the diffuse histological section samples are presented in Table 8.1.

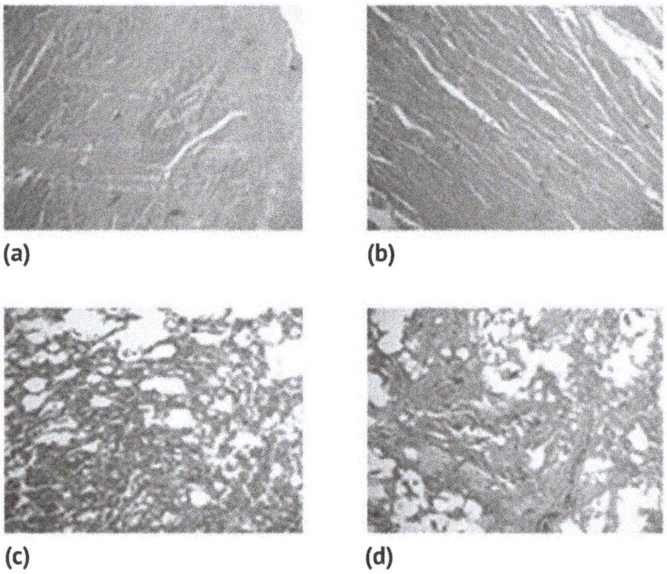

Figure 8.1 Microscopic images (×40) of histological samples of myocardium and lung tissue: a, CHD; b, ACI; c, BA; d, PF.

8.1.2 Wavelet Differentiation of Azimuth Polarization Map of the Myocardium

By the method of phase scanning, we have determined the phase plane of a single scattering. All further discussion refers to the experimental results obtained in this plane, were $\delta_t^\star \leq \pi/8$.

Figure 8.2 shows maps and histograms of the distributions of random values of the azimuth of polarization in the object fields of the myocardium from Groups 1 and 2.

As an initial reference for the azimuths of polarization $(\alpha^{j=1}(\delta_t^\star = \pi/8))$, we used the condition $\otimes \leftrightarrow \alpha^{j=1} = 0$.

The analysis of the obtained data revealed a coordinate-heterogeneous (Figure 8.2a, c) and statistically distributed (Figure 8.2b, d) structure of polarization azimuth maps in the object fields of myocardial samples from both groups.

Topologically, the azimuthal polarization maps are formed by local domains of varying geometric dimensions (Figure 8.2a, c). The distribution of the sizes of such polarization domains differs across the object fields of myocardial samples from different groups. This can be explained by the high degree of correlation, in the phase plane of single scattering, between the scales of the optically anisotropic fibrillar structures of the myocardium and those of the polarization domains.

At the same time, the histograms of the distributions (Figure 8.2b, d) are quite similar in the localization of extremes and in the ranges of variation of the random values of the polarization azimuth. In other words, statistical distributions of polarization parameters do not provide information about the large-scale (topological) structure of necrotic changes in myocardial architectonics.

Table 8.1 Optical and geometric parameters of histological tissue section samples of both types.

Parameters	Myocardium	Lung's tissue
Geometrical thickness, h, μm	50–60	50–60
Coefficient of attenuation, τ, μm	0.18–0.22	0.19–0.23
Depolarization degree, Δ, %	47–53	54–59

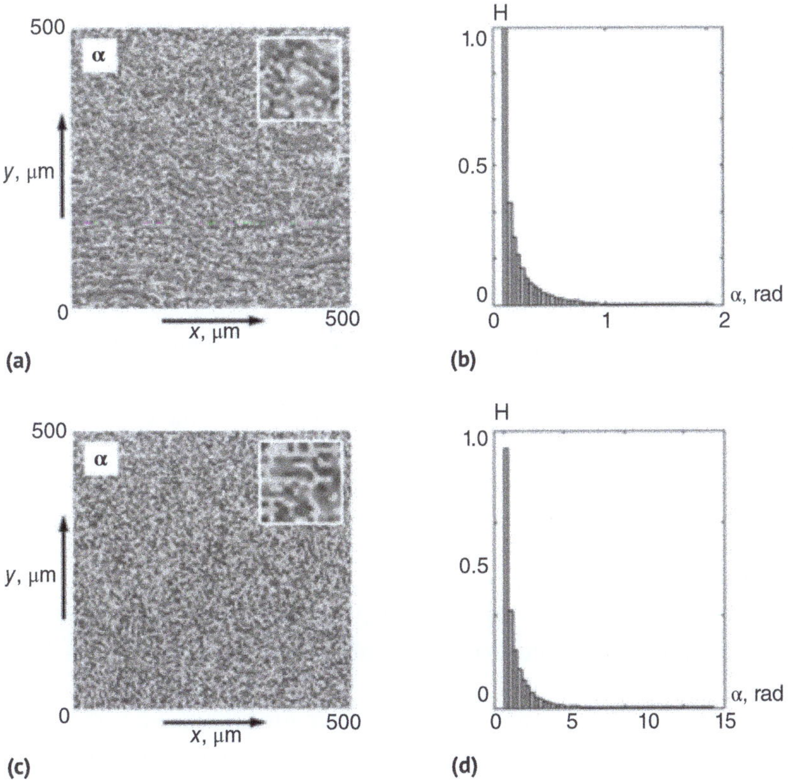

Figure 8.2 Coordinate (a), (b) and probabilistic (c), (d) distributions of random values of the azimuth of polarization in the object fields of myocardial samples from individuals deceased due to ACI (a), (c) and CHD (b), (d).

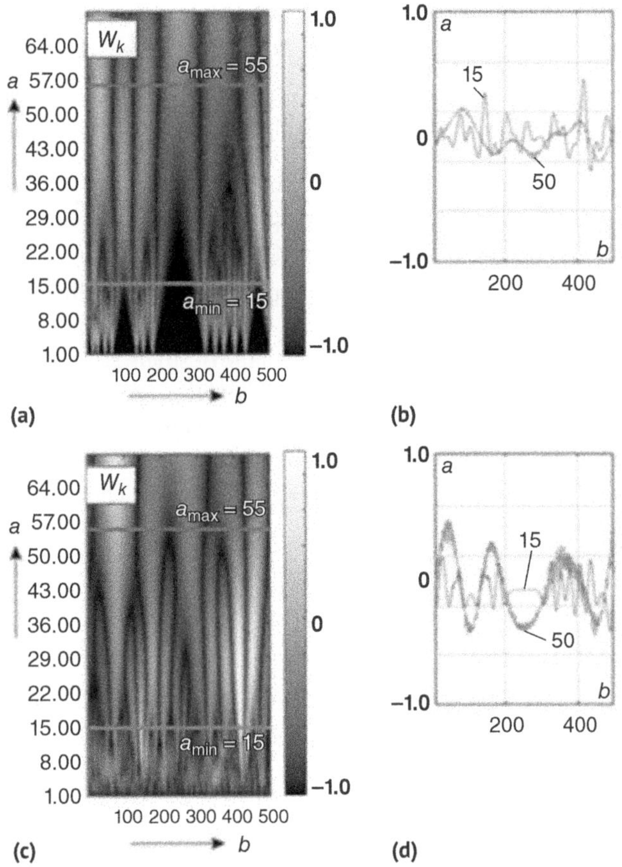

Figure 8.3 Maps (a, c) and multiscale linear cross-sections (b, d) of the wavelet coefficients $W(a, b)$ of the polarization azimuth $\alpha(\delta_t^\star = \pi/8, m, n)$ for histological sections of the myocardium from individuals deceased due to CHD (a, b) and ACI (c, d).

Figure 8.3 presents the results of wavelet analysis of the polarization azimuth maps $\alpha(\delta_t^\star \leq \pi/8, m, n)$ obtained from microscopic images of histological sections of the myocardium. This figure consists of two parts.

The first part (Figure 8.3a, c) shows a two-dimensional array of wavelet coefficient amplitude values. It was obtained as a result of algorithmic line-by-line scanning of MHAT distributions using a variable-scale function $\Psi\left(\frac{x-b}{a}\right)$.

The second part (Figure 8.3b, d) provides examples of the linear distributions of the wavelet coefficient amplitudes for two scales of the MHAT function.

The data show that the two-dimensional amplitude distributions of the wavelet coefficients form a fluctuating surface (Figure 8.3a, c).

The amplitudes and oscillation periods vary for different scales a of the MHAT function $\Psi\left(\frac{x-b}{a}\right)$. As the scale increases ($a \uparrow$), the oscillation period of the linear distributions of the wavelet coefficients W_b also increases [32–37].

The amplitude modulation depth of the wavelet coefficients W_b is specific to each scale (Figure 8.3b, d). The revealed regularities can be explained as follows. The magnitude of the wavelet coefficient amplitude at each scanning point is determined by the degree of mutual correlation between the geometric parameters of the polarization domains. The greater the cross-correlation, the higher the amplitude of the wavelet coefficient. At small scales of the polarization maps, such wavelet correlations can be observed for the fine-structured elements of the myocardial fibrillar architectonics. At larger scales, variations in the amplitude of the wavelet coefficients are associated with the degree of ordering of large-scale fibers within the fibrillar network.

In cases of death due to ACI, the fibrillar myosin network is more ordered and broadly structured compared to the architectonics of the myocardium in individuals who died from CHD (Figure 8.1a, b). Therefore, as a result of the wavelet transform of the polarization maps of the myocardial object field for the samples from Group 2, higher amplitude values and a greater modulation depth of their variations are observed (Figure 8.3a–d).

Quantitatively, the differences between the polarization maps of the myocardial object fields were determined (Table 8.2, Figure 8.4).

From the analysis of the obtained data, it can be seen that the maximum accuracy of the differential diagnosis between CHD and ACI has reached only a satisfactory level $Ac(M_{3,4} = 79.2\%)$.

Based on the obtained statistical analysis, the scales at which the differences were most pronounced were identified. The differences between $M_1(W_{a,b})$ and $M_2(W_{a,b})$ were maximal (Table 8.3). In our case, these scales correspond to (Figure 8.3).

The data also revealed individual modulation of the amplitude values $W_b(a_{min} = 15)$ and $W_b(a_{max} = 55)$.

In the case of CHD, the maximum amplitude and variations in the wavelet $W_{a,b}$ coefficients occur on large scales of the wavelet function scanning in the phase plane $\delta_t^\star = \pi/8$ of the $\alpha(\delta_t^\star = \pi/8, m, n)$ map.

For the case of ACI, the maximum amplitude and variations in the wavelet coefficients $W_{a,b}$ occur at small scales of the wavelet function scanning in the phase plane $\delta_t^\star = \pi/8$ of the $\alpha(\delta_t^\star = \pi/8, m, n)$ map.

Analysis of statistical estimation of fluctuations in the amplitudes of multiscale wavelet coefficients $W_{a,b}$ distributions $\alpha(\delta_t^\star = \pi/8, m, n)$ of the polarization azimuth revealed the following equations $M_{1,2}(\text{ACI}, W_b(a_{min} = 15)) \ll M_{1,2}(\text{CHD}, W_b(a_{min} = 15))$ and $M_{1,2}(\text{ACI}, W_b(a_{max} = 55)) < M_{1,2}(\text{CHD}, W_b(a_{max} = 55)$. As a result, the following levels of differential diagnostic accuracy for CHD and ACI were determined: for small scales $a_{min} = 15$- very good $Ac(M_1) = 91.7\%$ and excellent $Ac(M_2) = 95.8\%$ are equal. For large scales, $a_{max} = 50$ is a good level of $Ac(M_{1,2}) = 87.5\%$.

Table 8.2 Statistical moments of the polarization azimuth map $\alpha(\delta_t^\star = \pi/8, m, n)$.

$M_{1,2,3,4}$	M	ACI	Ac, %
M_1	0.07 ± 0.005	0.09 ± 0.006	70.8
M_2	0.04 ± 0.003	0.05 ± 0.003	75
M_3	1.73 ± 0.095	1.29 ± 0.067	79.2
M_4	2.39 ± 0.13	2.11 ± 0.12	79.2

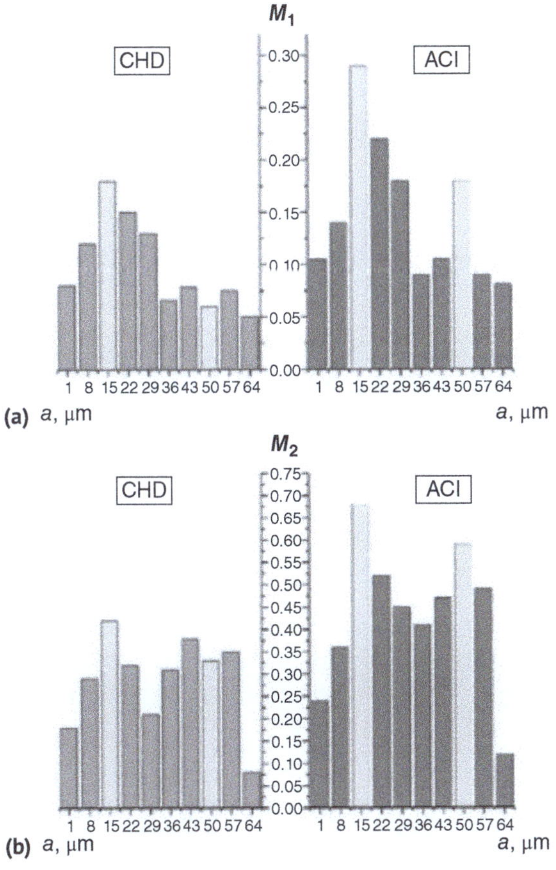

Figure 8.4 Scale dependences of the mean M_1 (a) and the variance M_2 (b) of the wavelet coefficients $W_{a,b}$ of the $\alpha\left(\delta_t^\star = \frac{\pi}{8}, m, n\right)$.

8.1.3 Wavelet Differentiation of the Polarization Ellipticity Maps of the Myocardium

Figure 8.5 shows maps and histograms of the distributions of random values of the polarization ellipticity magnitude in the object fields of histological sections of the myocardium from groups 1 and 2 [29–34].

As an initial reference for the polarization ellipticity $(\beta^{j=1}(\delta_t^\star = \pi/8))$, we used the condition $\otimes \leftrightarrow \beta^{j=1} = 0$. Similar to the polarization azimuth maps (Figure 8.2), the coordinate distributions of ellipticity were found to be spatially inhomogeneous (Figure 8.5a, c) and statistically distributed (Figure 8.5b, d).

The sets of polarization domains in the ellipticity maps are specific to the object fields of myocardial samples from different groups. At the same time, the statistical distributions of random values of the polarization ellipticity magnitude do not provide information about the large-scale (topological) structure of necrotic changes in the architectonics of the birefringent fibrillar networks of the myocardium.

Table 8.3 Mean and variance of the wavelet coefficients $W_{a,b}$ of the $\alpha(\delta_t^\star = \pi/8, m, n)$.

	$a_{min} = 15$		
M_i	CHD	ACI	Ac, %
M_1	0.18 ± 0.01	0.29 ± 0.016	91.7
M_2	0.46 ± 0.025	0.68 ± 0.036	95.8
	$a_{max} = 50$		
M_i	CHD	ACI	Ac, %
M_1	0.09 ± 0.005	0.12 ± 0.007	87.5
M_2	0.43 ± 0.023	0.55 ± 0.029	87.5

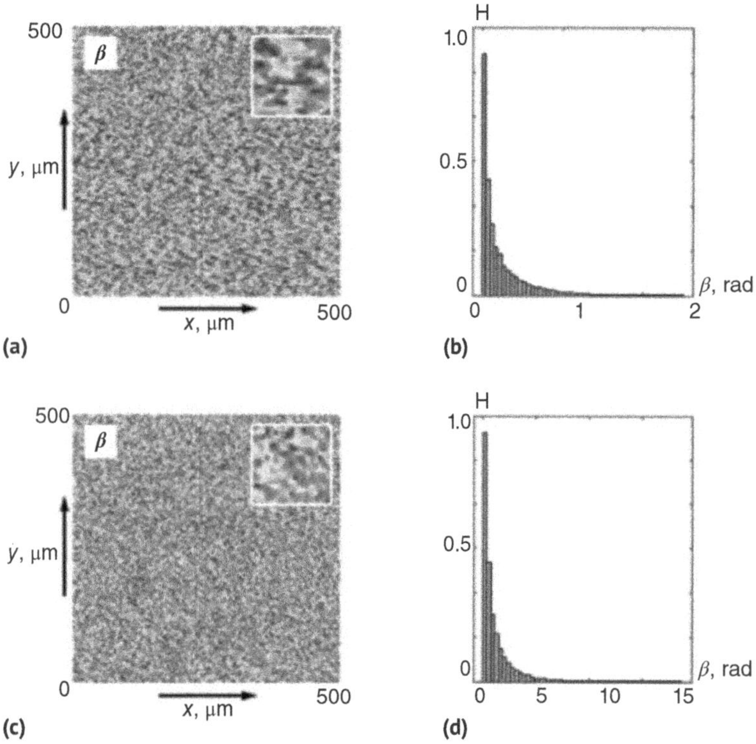

Figure 8.5 Coordinate (a), (c) and probabilistic (b), (d) distributions of random values of the polarization ellipticity magnitude in the object fields of myocardial samples from individuals who died due to CHD (a), (b) and ACI (c), (d).

Table 8.4 presents the results of the statistical analysis of the polarization ellipticity distributions for myocardial samples from the CHD and ACI groups.

It was found that the maximum accuracy of differential diagnostics of CHD and ACI exceed a satisfactory level $Ac(M_4 = 83.3\%)$.

Figure 8.6 shows the results of the application of the wavelet analysis technique to $\beta(\delta_t^\star = \pi/8, m, n)$.

We obtained fluctuating surfaces of the amplitudes of the wavelet coefficients $W_{a,b}$ (Figure 8.6a, c). The amplitudes and the oscillation period of the linear distributions of the wavelet coefficients increase as the scale a of the MHAT function $\Psi\left(\frac{x-b}{a}\right)$ increases. The depth of modulation of the amplitude of the wavelet coefficients of the polarization ellipticity maps (Figure 8.5) is specific to each scale (Figure 8.6b, d). At all scales of the polarization ellipticity maps, the wavelet correlations are determined by the level of structural anisotropy or birefringence. The value of this optical anisotropy parameter is related to the degree of spatial ordering of the fibers within the fibrillar network. Therefore, in cases of death resulting from ACI, higher amplitude values are observed, and the depth of modulation of their variations is greater (Figure 8.3a–d).

Table 8.4 Statistical moments of polarization ellipticity maps of the $\beta(\delta_t^\star = \pi/8, m, n)$.

$M_{1,2,3,4}$	$Z_{1,2,3,4}$	CHD	ACI	Ac, %
M_1	Z_1	0.11 ± 0.007	0.13 ± 0.008	70.8
M_2	Z_2	0.07 ± 0.005	0.09 ± 0.006	75
M_3	Z_3	1.21 ± 0.069	0.93 ± 0.057	79.2
M_4	Z_4	1.87 ± 0.099	1.36 ± 0.079	83.3

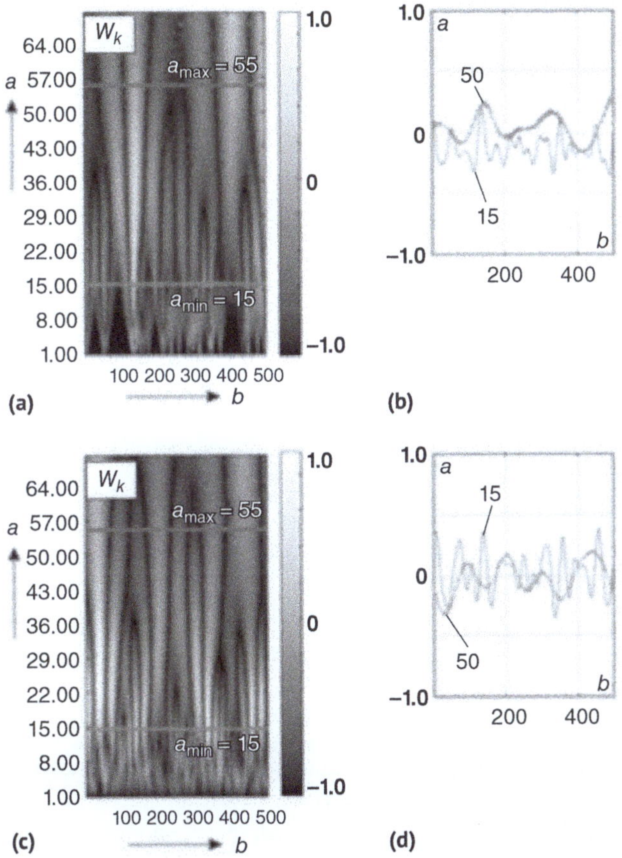

Figure 8.6 Maps (a, c) and multiscale linear cross-sections (b, d) of the polarization ellipticity $\beta(\delta_t^\star = \pi/8, m, n)$ wavelet coefficients $W_{a,b}$ for histological sections of the myocardium from individuals who died of CHD (a, b) and ACI (c, d).

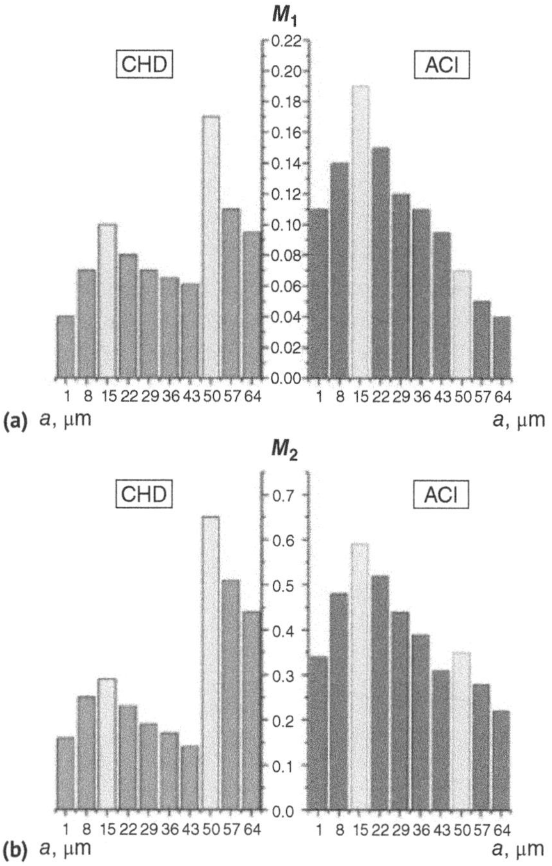

Figure 8.7 Scale dependences of the mean M_1 (a) and the variance M_2 (b) of the wavelet coefficients $W_{a,b}$ of the $\beta(\delta_t^\star = \pi/8, m, n)$.

Table 8.5 Mean M_1 and variance M_2 of wavelet coefficients of ellipticity maps $\beta(\delta_t^\star = \pi/8, m, n)$.

		$a_{min} = 15$	
M_i	CHD	ACI	Ac, %
M_1	0.11 ± 0.006	0.18 ± 0.01	95.8
M_2	0.31 ± 0.016	0.59 ± 0.031	100
		$a_{max} = 50$	
M_i	CHD	ACI	Ac, %
M_1	0.08 ± 0.005	0.14 ± 0.008	91.7
M_2	0.39 ± 0.021	0.62 ± 0.033	91.7

We quantified these amplitude fluctuations of the wavelet coefficients $W_{a,b}$ by calculating statistical parameters $(M_1(W_{a,b})$ and $M_2(W_{a,b})$ for all scales a of the MHAT function $\Psi\left(\frac{x-b}{a}\right)$ (Figure 8.7, Table 8.5).

High levels of differential diagnosis accuracy have been established. For small scales $a_{min} = 15$ an excellent level of $Ac(M_1) = 95.8\%$ and $Ac(M_2) = 100\%$ was achieved. For large scales $a_{max} = 50$ is a very good level of $Ac(M_{1,2}) = 91.7\%$ was obtained.

8.1.3.1 Wavelet-based Differentiation of Polarization Azimuth Maps of Lung Tissue

Figure 8.8 shows maps and histograms of the distributions of random values of the polarization azimuth of object fields in histological sections of lung tissue from Groups 3 and 4 [35–39].

Figure 8.9 presents the results of wavelet analysis of polarization maps $\alpha(\delta_t^\star = \pi/8, m, n)$ for BA and PF cases.

The presence of individual fluctuations in the amplitude of the wavelet coefficients at all scales of the salt-like MHAT function was observed (Figure 8.9a, c). As in the case of wavelet decomposition of polarization maps of myocardial samples, the amplitude at each scanning point is determined by the degree of mutual correlation of the geometric parameters of the polarization domain (Figure 8.8a, c) and the scale of the MHAT function "window."

In the case of BA, the optical birefringence of the pulmonary parenchyma is negligible (Figure 8.1c). For PF, the level of structural anisotropy increases due to the proliferation of connective tissue (Figure 8.1d). Consequently, as a result of the wavelet transform of the polarization maps of the sample from group 3, small amplitude values with shallow modulation depth are observed, in contrast to the higher amplitude values in the sample from Group 4 (Figure 8.9a–d).

We quantified such fluctuations in the amplitude of the wavelet coefficients $W_{a,b}$ by calculating statistical parameters $(M_1(W_{a,b}))$ and variation $M_2(W_{a,b})$ for all scales a of the MHAT function $\Psi\left(\frac{x-b}{a}\right)$ (Figure 8.10). The diagnostically optimal scales were found to be $a_{\min} = 22$ and $a_{\max} = 43$ (Table 8.6).

High levels of differential diagnosis accuracy were established. For small scales, $a_{\min} = 22$ good and very good level of $\mathrm{Ac}(M_1) = 90\%$ and $\mathrm{Ac}(M_2) = 93.3\%$ were achieved. For large scales, $a_{\max} = 43$ very good and excellent levels were observed: $\mathrm{Ac}(M_1) = 93.3\%$ and $\mathrm{Ac}(M_2) = 96.7\%$ level. These results can be physically related to changes across all scales in the geometric dimensions of parenchymal connective tissue birefringence fibrils in the lung parenchyma in cases of fibrosis. Consequently, at all scales, there is maximal modulation of the polarization azimuth $\alpha(\delta_t^\star = \pi/8, m, n)$ and, accordingly, variations in the amplitudes of the wavelet coefficients $W_{a,b}$. In contrast, for BA, the modulation of the wavelet coefficients is minimal.

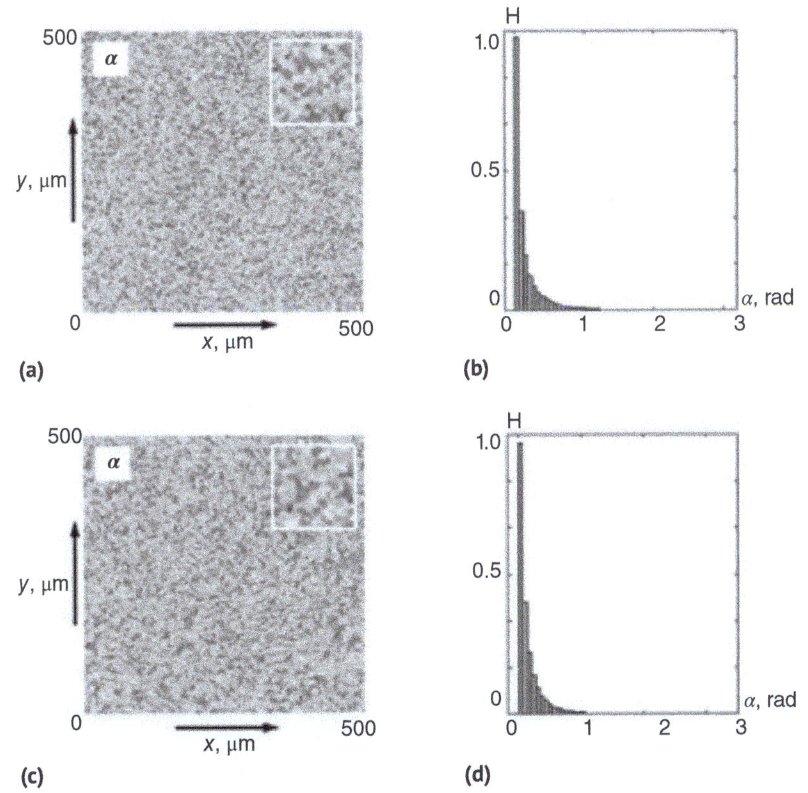

Figure 8.8 Coordinate (a), (c) and probabilistic (b), (d) distributions of random values of the polarization azimuth of object fields in lung tissue samples with BA (a), (b) and PF (c), (d).

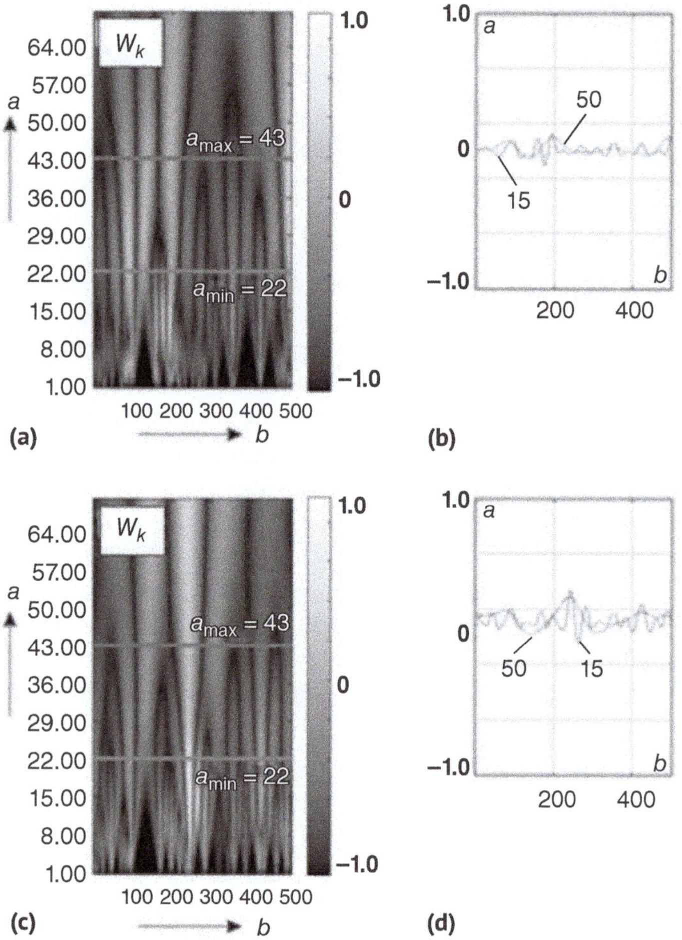

Figure 8.9 Maps (a, c) and multiscale linear cross-sections (b, d) of the polarization azimuth $\alpha(\delta_t^\star = \pi/8, m, n)$ wavelet coefficients $W_{a,b}$ of lung tissue histological sections those who died from BA (a, b) and PF (c, d).

8.1.3.2 Wavelet Differentiation of Ellipticity Polarization Maps of Lung Tissue

Figure 8.11 shows maps and histograms of the distributions of random values of the polarization ellipticity in histological sections of lung tissue from Groups 3 and 4.

As with the polarization azimuth maps (Figure 8.10), the coordinate distributions of the polarization ellipticity in object fields of lung tissue samples were found to be spatially inhomogeneous (Figure 8.11a, c). The histograms of the random-value distributions of the polarization ellipticity in samples from Groups 3 and 4 are quite similar (Figure 8.11b, d).

As a result, the differences between the values of the first to fourth-order statistical moments $Z_{1,2,3,4}$ are 25–35% (Table 8.7). The maximum accuracy of differential diagnosis between BA and PF corresponds to a satisfactory level of $\mathrm{Ac}(M_{1,2,3,4}) = 80 - 83.3\%$. At the same time, the differences in the topological structure of the polarization ellipticity maps are more pronounced in the fields of histological sections from Groups 3 and 4. Figure 8.12 shows wavelet maps $W(a, b)$ of $\beta(\delta_t^\star = \pi/8, m, n)$ for BA and PF cases.

As shown in Table 8.8, Figure 8.13, the mean and variance of the linear dependencies of the wavelet coefficients of ellipticity in maps of lung tissue samples from Group 4 (PF) are 3–4 times greater than the corresponding statistical parameters for BA.

Thus, high levels of accuracy in the differential diagnosis of BA and PF cases were established. An excellent accuracy level of 96.7–100% was achieved across all scales a of the MHAT function $\Psi\left(\frac{x-b}{a}\right)$.

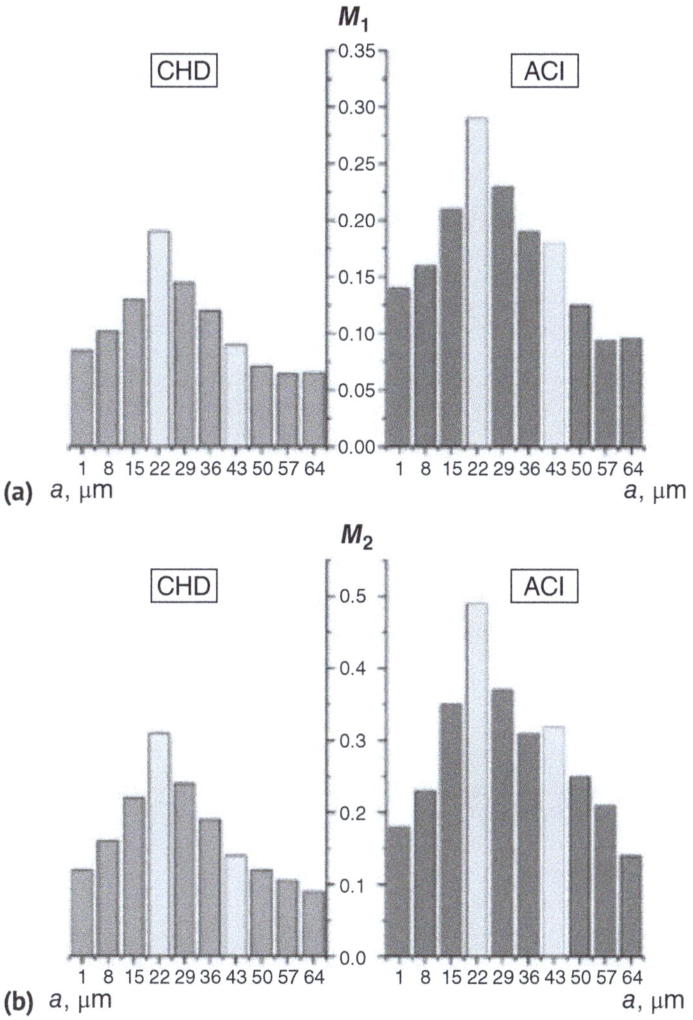

Figure 8.10 Scale dependences of the mean M_1 (a) and the variance M_2 (b) of the wavelet coefficients distributions for polarization azimuth maps $\alpha(\delta_t^\star - \pi/8, m, n)$.

Table 8.6 Mean M_1 and the variance M_2 of the wavelet coefficients distributions for polarization azimuth maps $\alpha(\delta_t^\star = \pi/8, m, n)$.

	$a_{min} = 22$		
M_i	**BA**	**PF**	**Ac, %**
M_1	0.19 ± 0.01	0.29 ± 0.016	90
M_2	0.31 ± 0.017	0.49 ± 0.027	93.3
	$a_{max} = 43$		
M_i	**BA**	**PF**	**Ac, %**
M_1	0.09 ± 0.005	0.16 ± 0.09	93.3
M_2	0.14 ± 0.08	0.32 ± 0.018	96.7

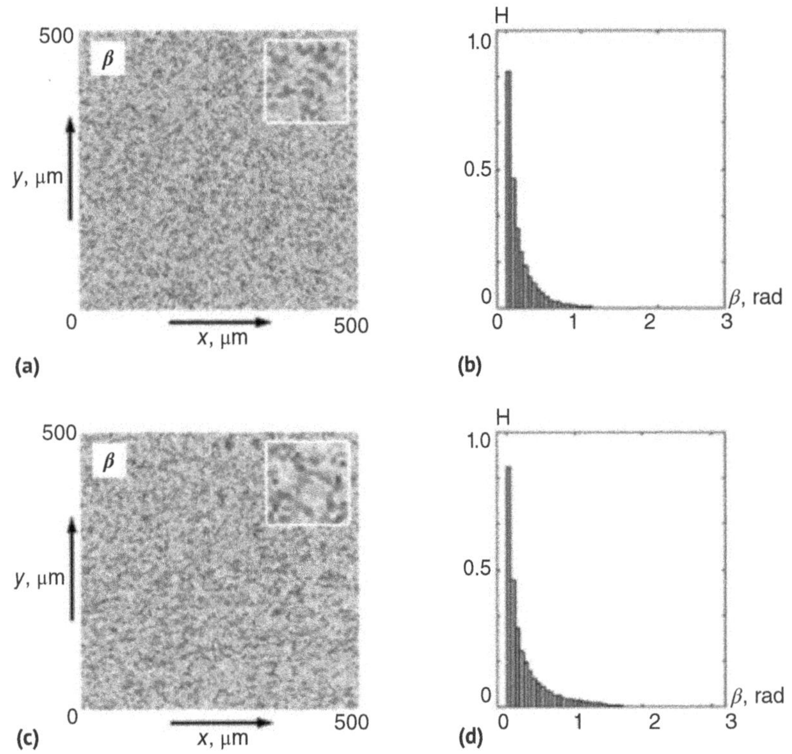

Figure 8.11 Coordinate (a), (c) and probabilistic (b), (d) distributions of random values of the magnitude of the ellipticity of polarization of object fields of lung tissue samples of deceased as a result of BA (a), (b) and PF (c), (d).

Table 8.7 Statistical moments $M_{i=1,2,3,4}$ of polarization ellipticity maps of the $\beta(\delta_t^\star = \pi/8, m, n)$.

$M_{1,2,3,4}$	BA	PF	Ac, %
M_1	0.05 ± 0.003	0.08 ± 0.006	80
M_2	0.03 ± 0.002	0.05 ± 0.003	80
M_3	1.44 ± 0.075	1.03 ± 0.057	83.3
M_4	2.06 ± 0.11	1.61 ± 0.089	83.3

8.1.4 Conclusions

1. The polarization–interference method for mapping and phase selection of diffuse layers of biological tissues with different morphological structures, scattered by varying numbers of polarization-inhomogeneous components in the object field, has been analytically substantiated and experimentally validated.
2. Using digital holographic reconstruction with phase scanning of complex amplitude distributions, polarization maps of a single scattered component in the object field of diffuse, depolarizing histological sections of fibrillar myocardium and parenchymal lung tissue with various types of pathology were obtained algorithmically:
   - myocardium – "coronary heart disease (CHD) – acute coronary insufficiency (ACI)";
   - lung tissue – "asthma (BA) – fibrosis (PF)."
3. Using the scanning soliton-shaped MHAT function, a scale-selective wavelet decomposition of a series of polarization maps of the polarization azimuth and ellipticity was performed. The geometric scales of the structural elements in the polarization maps of a single scattered component within diffuse fields were determined and physically justified for differentiating pathological changes in the optically anisotropic architecture of myocardial and lung tissue samples.
4. For the determined scales, within the framework of statistical analysis of the linear distributions of wavelet coefficient amplitudes, differential diagnostic markers were identified for CHD–ACI and BA–PF.

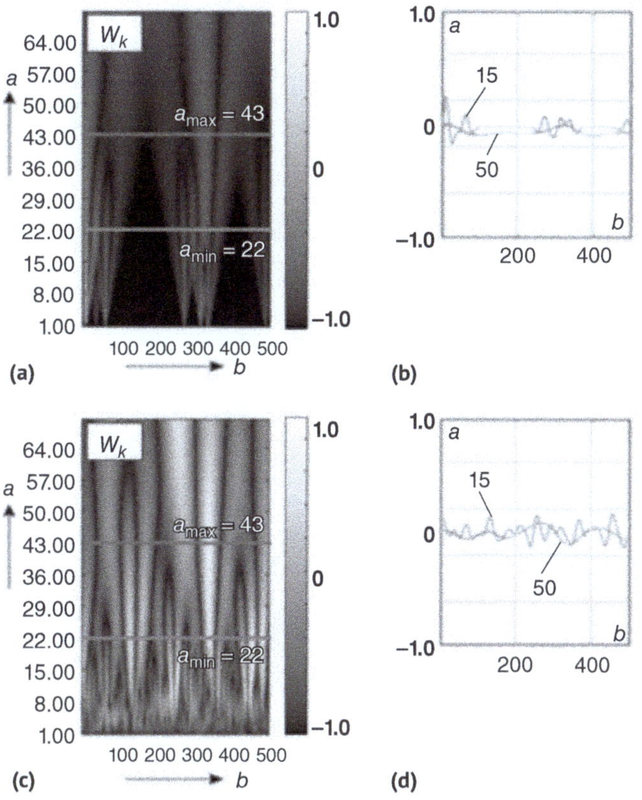

Figure 8.12 Maps $\beta(\delta_t^\star = \pi/8, m, n)$ (a, c) and multiscale linear cross-sections (b, d) of the polarization ellipticity wavelet coefficients $W_{a,b}$ of lung tissue histological sections those who died from BA (a, c) and PF (b, d).

Table 8.8 Mean M_1 and the variance M_2 of the wavelet coefficients distributions for polarization azimuth maps $\beta(\delta_t^\star = \pi/8, m, n)$.

	$a_{min} = 22$		
M_i	BA	PF	Ac, %
M_1	0.13 ± 0.008	0.17 ± 0.009	96.7
M_2	0.36 ± 0.021	0.64 ± 0.038	100
	$a_{max} = 43$		
M_i	BA	PF	Ac, %
M_1	0.078 ± 0.042	0.135 ± 0.007	100
M_2	0.19 ± 0.011	0.56 ± 0.031	100

5. Diagnostic relationships between the first and fourth-order statistical moments, which characterize the distributions of wavelet coefficients of polarization azimuth and ellipticity maps, and the accuracy levels of differentiating various pathological conditions were established. For the "CHD-ACI" cases, a very good accuracy level was obtained for $Ac(Z_{1,2}; a_{min} = 15) = 91.7\%$ with an excellent level of $Ac(M_1) = 95.8\%$ and $Ac(M_2; a_{max} = 50) = 100\%$. For the BA–PF cases, a very good accuracy was observed for $Ac(M_1; a_{min} = 22) = 92.3\%$ and excellent level of $Ac(M_2; a_{min} = 22; a_{max} = 43) = 100\%$.

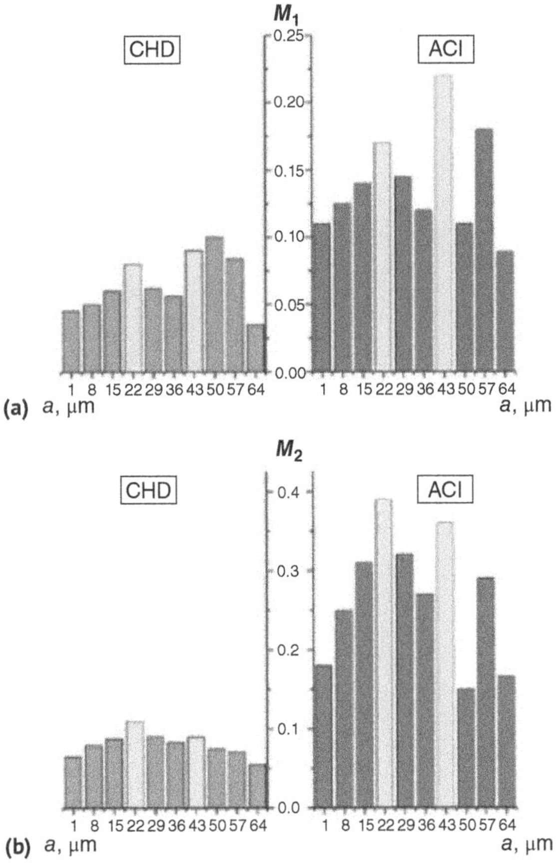

Figure 8.13 Scale dependences of the mean M_1 (a) and the variance M_2 (b) of wavelet coefficients distributions for $\beta(\delta_l^\star = \pi/8, m, n)$.

8.2 Mueller-matrix Interferometric Multifractal Scaling of the Optically Anisotropic Architecture of Diffuse Blood Films: Fundamental and Applied Aspects

8.2.1 Introduction

Among the numerous biomedical diagnostic applications of soft matter structures, Mueller-matrix polarimetry (MMP) of the optical anisotropy of biological tissues occupies a prominent position. In this area of polarimetric biomedicine, two main objectives can be distinguished [34–45]:

- Fundamental: the search for and physical substantiation of relationships between the coordinate distributions of matrix elements and the spatial-angular parameters of the polycrystalline components of biological layers.
- Applied: the identification of objective parameters characterizing the distribution of matrix elements, which can serve as diagnostic markers of the physiological or pathological state of biological tissues.

8.2.2 Optical Model of Blood Films

Here, we provide a brief theoretical overview. A significant achievement in this field has been the successful integration of polarimetric and interferometric mapping methods, which enabled the development of various approaches to 3D polarimetry of biological objects and environments [46–58].

Polarization-sensitive optical coherence tomography (PSOCT) techniques have emerged as particularly promising in this context. This laser polarization–interference technique allows not only the acquisition of layered images from shallow depths (up to 2 mm) of biological tissues but also the extraction of coordinate distributions of elements of the Jones and Mueller matrices, as well as maps of optical anisotropy.

At the same time, the diagnostic capabilities of Mueller-matrix polarimetry (MMP) are limited by significant depolarization of light within the volume of biological objects. The principal solution to this problem has been the generalization of the MMP technique by combining multichannel polarization interferometry with digital holographic reconstruction of embossed depolarization maps. For the first time, this approach enables layer-by-layer mapping of depolarization in biological tissues using the Mueller matrix method.

Methodological development of this approach involved defining reconstruction algorithms and applying scale-selective wavelet analysis to oncologic and necrotic changes (related to myocardial death and duration of mechanical injury) in layer-by-layer coordinate distributions of the "birefringence–dichroism" properties of human tissue samples. However, preparation of biological tissue samples requires invasive biopsy procedures, which are not always feasible.

The relevance of this area of MMP extends beyond the practical accessibility of biological fluids to fundamental considerations. Biological fluids reflect homeostasis and provide a convenient medium for studying the dynamics of physiological and pathological processes. Complete dehydration of a test drop leads to the formation of a films – a self-assembled, generally not only statistical but also scale-self-similar (multifractal) macro portrait of molecular interactions in the fluid. Consequently, both the statistical and topographic structure of the films are associated with the development and progression of pathological processes and carry valuable information for clinical diagnostics [49–63].

Therefore, 3D MMP methods, combined with statistical and multifractal analyses of the layered topographic structure, are promising for further study of diffuse blood films and the architectonics of supramolecular networks self-assembled during dehydration.

Based on this, the fundamental goal of our study is to develop and experimentally validate a new method using a Mueller matrix interferometer with phase extraction to map optical anisotropy in a single-scattering supramolecular layer of diffuse, depolarizing blood films.

The applied objective of our study is a clinically and forensically relevant application of the polarization-interference Mueller-matrix technique for rapid and objective determination of blood loss volume, which can be associated with mortality rates of up to 54% due to hemorrhagic shock. To date, a wide range of direct (gravimetric and colorimetric) and indirect (biochemical and biophysical methods have been developed for assessing blood loss.

The direct gravimetric method is simple and accessible, calculating blood loss by determining the difference in mass of surgical materials before and after use. However, it has a significant error margin of 10–12%. The colorimetric method is more accurate, based on quantitative determination of blood cells and subsequent calculation of the lost volume, but it is labor-intensive and time-consuming. Indirect biochemical methods rely on analysis of hemodynamic parameters and blood hemoconcentration, but they require stationary laboratory facilities and costly reagents. Biophysical methods utilize a wide range of objective indicators, including dyes, radioactive isotopes, bioimpedance measurements via electroplethysmography, and assessment of vascular filling through rheographic recordings of tissue resistance changes. Thermodilution, ultrasound, Doppler imaging, and echocardiography are also commonly employed.

The main limitation of indirect biochemical and biophysical methods is their multiparametric complexity, high cost, and relatively large error rates (10–15%) in estimating blood loss.

In this context, Jones-matrix reconstruction (tomography) algorithms have been developed to map the distribution of optical anisotropy parameters in polycrystalline films of biological fluids. A film of any biological fluid is a complex, spatially inhomogeneous, optically anisotropic structure formed by various biochemical and molecular crystalline complexes.

In this section, we will focus on dehydrated blood polycrystalline films, which consist of two main components:

- blood plasma;
- blood particles (shaped elements).

Blood plasma is a heterogeneous complex-structured liquid, essentially a colloidal-polymer solution. Traditionally, its molecular components – albumin (~60%), globulin (~40%), fibrinogen, and hemoglobin – are effectively studied using biochemical methods. However, for investigating supramolecular optically anisotropic protein structures (dimers, trimers, monomers) ranging from 100 to 2000 nm in size, there are currently insufficiently effective research techniques [64–72].

In a dried blood plasma drop, three zones of different thicknesses h can be distinguished:

- *edge zone:* polycrystalline albumin layer exhibiting linear birefringence and dichroism, with a thickness of approximately $h \sim 50 \div 60\,\mu m$;
- *transitional zone:* outer and inner layers of optically isotropic cubic NaCl crystals, with an intermediate globulin layer showing circular birefringence and dichroism, thickness $h \sim 25\,\mu m \div 30\,\mu m$;
- *central zone:* outer layer of cubic crystals of NaCl crystals, thickness $h \sim 5 \div 15\,\mu m$.

Shaped elements – erythrocytes, platelets, and leukocytes – exhibit circular birefringence and dichroism.

8.2.3 Basic Analytical Relations

A polycrystalline film of any biological fluid can be represented as a sequence of partial layers exhibiting six types of optical anisotropy: linear and circular birefringence $\left(LB_{(0,90);(45,135)}, CB_{\otimes,\oplus}\right)$ and linear and circular dichroism $\left(LD_{(0,90);(45,135)}, CD_{\otimes,\oplus}\right)$[72–81].

In our study, we used the red spectral range (wavelength $\lambda = 0.6328\,\mu\text{m}$) of He–Ne laser probing radiation, where the absorption of blood films proteins is minimal. Therefore, without loss of generality, we limit our consideration to the phase anisotropy mechanisms $LB_{0,90}$, $LB_{45,135}$ and $CB_{\otimes;\oplus}$ corresponding to linearly polarized light (0° ÷ 90° and 45° ÷135°) and circularly right- (⊕) and left- (⊖) polarized components

$$LB_{0;90} = \frac{2\pi}{\lambda}(n_0 - n_{90})h; \tag{8.1}$$

$$LB_{45;135} = \frac{2\pi}{\lambda}(n_{45} - n_{135})h; \tag{8.2}$$

$$CB_{\oplus;\ominus} = \frac{2\pi}{\lambda}(n_{\oplus} - n_{\ominus})h, \tag{8.3}$$

where $\left(n_{0;90;45;135;\oplus;\ominus}\right)$ is the refractive indices for orthogonal amplitude components; h is the geometrical layer thickness.

The mechanisms of linear birefringence (structural anisotropy $LB_{(0,90);(45,135)}$) are inherent in the polycrystalline networks of albumin and fibrin. Molecular complexes of globulin and blood corpuscles have optical activity $\left(CB_{\oplus;\ominus}\right)$. Crystals of Na—Cl salt have a cubic crystal lattice and are optically isotropic.

Our theoretical consideration of the interaction of polarized laser radiation with optically anisotropic polycrystalline architectonics of supramolecular networks of dehydrated blood films (films) is based on the well-known Jones-Mueller matrix formalism.

According to this approach, the Jones complex matrix $||J||$ of supramolecular blood films networks exhibiting linear and circular birefringence can be expressed as follows

$$||J|| = \begin{Vmatrix} j_{11} & j_{12} \\ j_{21} & j_{22} \end{Vmatrix} = \begin{Vmatrix} \left(\cos 0.5H - \frac{iLB_{0-90}}{H}\sin 0.5H\right) & \left(\frac{CB - iLB_{45-135}}{H}\sin 0.5H\right) \\ \left(\frac{-CB + iLB_{45-135}}{H}\sin 0.5H\right) & \left(\cos 0.5H + \frac{iLB_{0-90}}{H}\sin 0.5H\right) \end{Vmatrix}, \tag{8.4}$$

where

$$H = \sqrt{LB_{0-90}^2 + LB_{0-90}^2 + CB^2}. \tag{8.5}$$

To derive the explicit form of the Mueller matrix $||F||$ from the measured set of Stokes vector parameters VS, the Kronecker equation was used

$$||F|| = K(||J|| \oplus ||J^*||)K^{-1}. \tag{8.6}$$

Here, ⊕ denotes the Kronecker operator

$$K = \frac{1}{\sqrt{2}}\begin{Vmatrix} 1 & 0 & 0 & 1 \\ 1 & 0 & 0 & -1 \\ 0 & 1 & 1 & 0 \\ 0 & i & -i & 0 \end{Vmatrix}. \tag{8.7}$$

From equations (8.4)–(8.7), the Mueller matrix of the birefringent blood films can be expressed as follows

$$\|F\| = \left\|\begin{matrix} 1 & 0 & 0 & 0 \\ 0 & f_{22} & f_{32} & f_{42} \\ 0 & f_{32} & f_{33} & f_{34} \\ 0 & f_{42} & f_{43} & f_{44} \end{matrix}\right\| = \left\|\begin{matrix} 1 & 0 & 0 & 0 \\ 0 & (\cos\gamma + \alpha \mathrm{LB}^2_{0-90}) & (\alpha \mathrm{LB}_{0-90}\mathrm{LB}_{45-135} + \beta \mathrm{CB}) & (-\alpha \mathrm{LB}_{0-90}\mathrm{CB} + \beta \mathrm{LB}_{45-135}) \\ 0 & (\alpha \mathrm{LB}_{0-90}\mathrm{LB}_{45-135} - \beta \mathrm{CB}) & (\cos\gamma + \alpha \mathrm{LB}^2_{45-135}) & (-\alpha \mathrm{LB}_{45-135}\mathrm{CB} - \beta \mathrm{LB}_{0-90}) \\ 0 & (-\alpha \mathrm{LB}_{0-90}\mathrm{CB} - \beta \mathrm{LB}_{45-135}) & (-\alpha \mathrm{LB}_{45-135}\mathrm{CB} + \beta \mathrm{LB}_{0-90}) & (\cos\gamma + \alpha \mathrm{CB}^2) \end{matrix}\right\| \tag{8.8}$$

Here, $\alpha = \frac{(1-\cos\gamma)}{\gamma^2}$; $\beta = \frac{\sin\gamma}{\gamma}$ and $\gamma = \sqrt{\mathrm{LB}^2_{0-90} + \mathrm{LB}^2_{45-135} + \mathrm{CB}^2}$.

By jointly solving the set of partial Equations (8.8), the following analytical relationships between the parameters of LB and CB and the corresponding set of Mueller matrix elements f_{pl} are obtained

$$\mathrm{LB}_{0-90} = \frac{1 - \left(\frac{f_{24}+f_{42}}{f_{34}+f_{43}}\right)^2}{2\left(\frac{f_{22}-f_{33}}{f_{23}+f_{32}}\right)}; \tag{8.9}$$

$$\mathrm{LB}_{45-135} = \frac{\left(\frac{f_{34}+f_{43}}{f_{24}+f_{42}}\right) - \left(\frac{f_{24}+f_{42}}{f_{34}+f_{43}}\right)}{2\left(\frac{f_{22}-f_{33}}{f_{23}+f_{32}}\right)}; \tag{8.10}$$

$$\mathrm{LB} = \sqrt{\mathrm{LB}^2_{0-90} + \mathrm{LB}^2_{45-135}}; \tag{8.11}$$

$$\mathrm{CB} = \left(\frac{1 - \left(\frac{f_{24}+f_{42}}{f_{34}+f_{43}}\right)^2}{2\left(\frac{f_{22}-f_{33}}{f_{23}+f_{32}}\right)}\right)\left(\frac{f_{34}+f_{43}}{f_{23}+f_{32}}\right). \tag{8.12}$$

In the following, Equations (8.9)–(8.12) will be referred to as the "information parameters" Y which are defined as

$$\mathrm{Y} = \begin{pmatrix} \mathrm{LB}(f_{pl}) \\ \mathrm{CB}(f_{pl}) \end{pmatrix}. \tag{8.13}$$

It should be noted that only under conditions of single scattering (when the number of interactions between the laser radiation and birefringent domains is $k = 1$) the anisotropy parameters (LB and CB) become uniquely related to the elements f_{pl} of the Mueller matrix $\|F\|$. In this case, a direct functional relationship exists between the parameters of the Stokes vector $S_{j=1,2,3,4}$ and the complex amplitudes parameters U_x and U_y of the object field of the blood films

$$f_{pl} = \Psi_j \begin{pmatrix} S_1 \\ S_2 \\ S_3 \\ S_4 \end{pmatrix} = \Psi_j \begin{pmatrix} |U_x|^2 + |U_y|^2 \\ |U_x|^2 - |U_y|^2 \\ 2|U_x U_y| \cos\xi \\ 2|U_x U_y| \sin\xi \end{pmatrix}. \tag{8.14}$$

Here, ξ is the value of the total phase shift between U_x and U_y

$$\xi = \delta(\text{LB}) + \varepsilon(\text{CB}) = \frac{2\pi}{\lambda}(\Delta n(\text{LB}) + \Delta n(\text{CB}))h. \tag{8.15}$$

For the case of single scattering, the magnitude of the phase shift $\xi^{k=1}$ can be estimated using the following optical-geometric conditions

$$\begin{pmatrix} k = 1 \\ (\Delta n(\text{LB}) + \Delta n(\text{CB})) \sim 10^{-3} \\ \lambda = 0.6328\,\mu\text{m} \\ h \sim 30\,\mu\text{m} \div 50\,\mu\text{m} \end{pmatrix} \Longrightarrow \xi^{k=1} \sim \frac{\pi}{10} \div \frac{\pi}{6}. \tag{8.16}$$

Thus, the condition for the experimental implementation of the theoretical Equations (8.4)–(8.13) corresponds to the magnitude of the phase shift

$$\xi^{k=1} \leq \frac{\pi}{10}. \tag{8.17}$$

Similarly, we determined the scattering multiplicity $(\xi^* = \xi^{k=N})$, at which a complete reversal of the phase shift period is achieved

$$\begin{pmatrix} \xi^* = \xi^{k=N} \\ (\Delta n(\text{LB}) + \Delta n(\text{CB})) \sim 10^{-3} \\ \lambda = 0,6328\,\mu\text{m} \\ h \sim 30\,\mu\text{m} \div 50\,\mu\text{m} \end{pmatrix} \Longrightarrow 2\pi(N \sim 12 \div 20) \Longrightarrow \begin{pmatrix} \cos\gamma = 1 \\ \alpha = 0 \\ \beta = 0 \end{pmatrix}. \tag{8.18}$$

When condition (8.18) is satisfied within the volume of the blood films, all off-diagonal elements of the Mueller matrix $\|F(\xi^* = \xi^{k=N})\|$ are "minimized" and eventually "disappear," – $f_{pl;p\neq l} = 0$.

As a result, the symmetry of the Mueller matrix F (8.8) of the blood films is transformed into a diagonal operator corresponding to a diffuse depolarizer

$$\|F(\xi^* = \xi^{k=N})\| \Longrightarrow \begin{Vmatrix} f_{11} & 0 & 0 & 0 \\ 0 & f_{22} & 0 & 0 \\ 0 & 0 & f_{33} & 0 \\ 0 & 0 & 0 & f_{44} \end{Vmatrix}, \tag{8.19}$$

in this regime, due to multiple scattering (ξ^*), the polarization states are statistically averaged, which resulting from the amplitude–phase superposition of partial laser waves (8.14).

In other words, at the phase shift level $\xi^* = \xi^{k=N}$, the additive addition of intensities $\sum_{l=1}^{N}\left(|U_x|^2 + |U_y|^2\right)_l$ becomes predominant $\sum_{l=1}^{N}\begin{pmatrix} U_x \\ U_y \end{pmatrix}_l$, for which the following analytical conditions are satisfied

$$f_{11} \gg f_{22}; f_{33}; f_{44}; \tag{8.20}$$

$$\xi^* \Longrightarrow \begin{pmatrix} \langle U_x \rangle \to 0 \\ \langle U_y \rangle \to 0 \end{pmatrix} \Longrightarrow \begin{pmatrix} f_{22} \to 0 \\ f_{33} \to 0 \\ f_{44} \to 0 \end{pmatrix}. \tag{8.21}$$

Thus, isolating the single – scattered component $(\xi^{k=1})$ from the diffuse object field of the blood films is an important diagnostic task. To address this problem, we employed the phase-scanning method combined with digital holographic reconstruction of the complex amplitude fields of optically anisotropic histological sections of biological tissue samples.

8.2.4 Biological Samples

For the determination of blood loss volume in deceased individuals, blood samples were collected for the following cases:

- Blood loss volume V = 0, – control group 1 ($q = 28$ samples);
- Blood loss volume V = 500 mL, – research group 2 ($q = 28$ samples);
- Blood loss volume V = 1000 mL, – research group 3 ($q = 28$ samples).

A 30 μL drop of blood was applied onto an optically homogeneous glass substrate heated to body temperature (36.6 °C). The complete drying (dehydration) process lasted 45–50 minutes.

Table 8.9 presents the optical parameters of the polycrystalline blood film samples for each group.

8.2.5 Diagnostic Algorithm

The procedure for diagnostically assessing differences between blood films samples with varying volumes of blood loss was as follows:

- In the absence of phase scanning, for the object field diffusely scattered by the entire volume of the experimental samples, optical anisotropy maps were measured $Y(U^{\Theta}_{0;90}(\xi^{k=N}), (m,n))$;
- by applying the phase scanning, a single scattered component $(\xi^{k=1} \leq \pi/10)$ was extracted from the general diffuse background (8.16, 8.17) and algorithmically (8.9–8.12, 8.22–8.24) birefringence maps were restored $\begin{pmatrix} LB(f_{pl}(U^{\Theta}_{0;90}(\xi^{k=1}))) \\ CB(f_{pl}(U^{\Theta}_{0;90}(\xi^{k=1}))) \end{pmatrix}(m,n)$ optically thin layer of polycrystalline architectonics of blood films;
- as part of the statistical analysis, the central moments $M_{g=1;2;3;4}(Y)$ were calculated, which characterize distributions $Y(\xi^{k=1}, m, n)$;
- within the framework of fractal analysis, the multifractal spectra $F^{LB;CB}(h)$, which characterize the distributions $Y(\xi^{k=1}, m, n)$, were algorithmically determined;
- the central statistical moments of the first to fourth orders were calculated, which characterize the distributions (spectra $F^{LB,CB}$) of fractal dimensions of n optical anisotropy maps $LB(f_{ik}(U^{\Theta}_{0;90}(\xi^{k=1})), m, n)$ and $CB(f_{ik}(U^{\Theta}_{0;90}(\xi^{k=1})), m, n)$;
- within the framework of information analysis to determine the accuracy (Ac) of differential diagnosis of blood loss volume sequentially for markers $M_{g=1;2;3;4}(Y)$ and $Z_{g=1;2;3;4}(F(h))$ the sensitivity $\left(Se_{12;13;23} = \frac{a_{12;13;23}}{a_{12;13;23}+b_{12;13;23}} 100\%\right)$, specificity $\left(Sp_{12;13;23} = \frac{c_{12;13;23}}{c_{12;13;23}+d_{12;13;23}} 100\%\right)$ and balanced accuracy $(Ac_{12;13;23} = 0.5(Se_{12;13;23} + Sp_{12;13;23}))$ were also calculated. Here, $a_{12;13;23}$ and $b_{12;13;23}$ are the number of correct and incorrect diagnoses within "Group 1 – Group 2"; "Group 1 – Group 3" and "Group 2 – Group 3."

8.2.6 Statistical and Multifractal Structure of Anisotropy Maps of Blood Diffuse Films

A series of fragments of Figures 8.14 and 8.15 present the results of the "diffuse" $(\xi^* = \xi^{k=N})$ algorithmic map reconstruction $Y(U^{\Theta}_{0;90}(\xi^{k=N}), (m,n))$ (a)–(c) and histograms $H(Y(U^{\Theta}_{0;90}(\xi^{k=N})))$ (d)–(f) of distributions linear $LB(U^{\Theta}_{0;90}(\xi^{k=N}), (m,n))$ (Figure 8.14) and circular $CB(U^{\Theta}_{0;90}(\xi^{k=N}), (m,n))$ (Figure 8.15) of supramolecular optically anisotropic polycrystalline networks of blood films from deceased individuals with different volumes of blood loss – Group 1 (Figure 8.15a, d); Group 2 (Figure 8.15b, e); and Group 3 (Figure 8.15c, f).

Table 8.9 Optical parameters of the polycrystalline blood film samples from all groups ($q = 18$).

Parameter	Group 1	Group 2	Group 3
Attenuation coefficient τ, cm^{-1}	0.67 ± 0.038	0.65 ± 0.036	0.62 ± 0.034
Depolarization degree Λ, %	59.0 ± 0.87	54.0 ± 0.85	51.0 ± 0.82

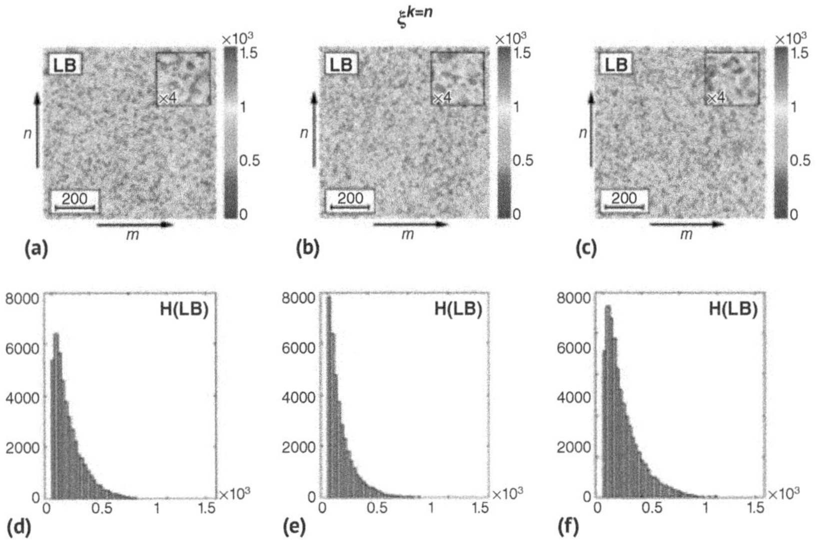

Figure 8.14 Coordinate (a)–(c) and probabilistic (d)–(f) distributions of the magnitude of linear birefringence of "diffuse" maps $\mathrm{LB}(U^{\Theta}_{0;90}(\xi^{k=N}), (m, n))$ of blood films from deceased individuals with different volume of blood loss.

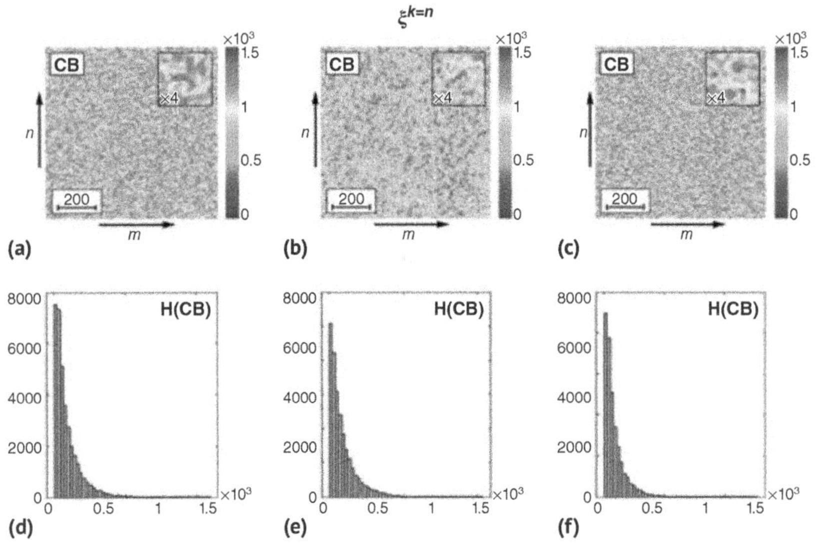

Figure 8.15 Coordinate (a)–(c) and probabilistic (d)–(f) distributions of the magnitude of the circular birefringence of "diffuse" maps $\mathrm{CB}(U^{\Theta}_{0;90}(\xi^{k=N}), (m, n))$ of blood films from deceased individuals with different volume of blood loss.

The results of statistical analysis anisotropy maps $\left(\mathrm{Y}(U^{\Theta}_{0;90}(\xi^{k=N}), (m, n))\right)$ for all groups of diffuse blood films samples are presented in Table 8.10.

Table 8.11 illustrates the accuracy Ac of statistical differential diagnosis of diffuse blood films samples from all groups of deceased individuals.

The analysis of the obtained data showed the following:

- probabilistic distributions of random values of birefringence parameters (Figures 8.14 and 8.15a–c) of supramolecular architectonics of diffuse blood films samples are complex (all central statistical moments $M_{g=1;2;3;4} \neq 0$, Table 8.10) and quite similar (Figures 8.14 and 8.3d–f) by statistical structure ($M_{g=1;2;3;4}$(Group 1) $\approx M_{g=1;2;3;4}$(Group 2) $\approx M_{g=1;2;3;4}$(Group 3), Table 8.10;
- intergroup differences in the values of the central statistical moments of the first to fourth orders, which characterize $\mathrm{Y}(U^{\Theta}_{0;90}(\xi^{k=N}), (m, n))$, are relatively small and do not exceed 25–30% (Table 8.10);

Table 8.10 Statistical parameters of the "integral" anisotropy maps of blood films.

$LB(\xi^{k=N}, m, n)$			
$M_{l=1;2;3;4}$	**Group 1**	**Group 2**	**Group 3**
$M_{l=1} \times 10^{-3}$	0.68 ± 0.037	0.63 ± 0.035	0.59 ± 0.033
$M_{l=2} \times 10^{-3}$	0.24 ± 0.014	0.21 ± 0.013	0.19 ± 0.011
$M_{l=3}$	0.82 ± 0.044	0.88 ± 0.048	0.93 ± 0.049
$M_{l=4}$	1.21 ± 0.072	1.32 ± 0.072	1.39 ± 0.077
$CB(\xi^{k=N}, m, n)$			
$M_{l=1;2;3;4}$	**Group 1**	**Group 2**	**Group 3**
$M_{l=1} \times 10^{-3}$	0.56 ± 0.031	0.53 ± 0.029	0.49 ± 0.028
$M_{l=2} \times 10^{-3}$	0.14 ± 0.009	0.11 ± 0.006	0.09 ± 0.005
$M_{l=3}$	1.18 ± 0.063	1.29 ± 0.069	1.37 ± 0.073
$M_{l=4}$	2.02 ± 0.11	2.13 ± 0.13	2.36 ± 0.14

Table 8.11 Statistical parameters of multifractal spectra of "integral" anisotropy maps of blood films.

$F\{h, LB(\xi^{k=N}, m, n)\}$			
$M_{l=1;2;3;4}$	**Group 1**	**Group 2**	**Group 3**
$M_{l=1} \times 10^{-3}$	1.57 ± 0.091	1.66 ± 0.093	1.74 ± 0.092
$M_{l=2} \times 10^{-3}$	0.21 ± 0.012	0.16 ± 0.009	0.12 ± 0.007
$M_{l=3}$	2.77 ± 0.15	3.08 ± 0.17	3.45 ± 0.19
$M_{l=4}$	3.22 ± 0.18	3.56 ± 0.19	4.06 ± 0.22
$F\{h, CB(\xi^{k=N}, m, n)\}$			
$M_{l=1;2;3;4}$	**Group 1**	**Group 2**	**Group 3**
$M_{l=1} \times 10^{-3}$	1.71 ± 0.098	1.79 ± 0.11	1.84 ± 0.12
$M_{l=2} \times 10^{-3}$	0.18 ± 0.11	0.14 ± 0.008	0.105 ± 0.006
$M_{l=3}$	2.91 ± 0.16	3.31 ± 0.18	3.84 ± 0.21
$M_{l=4}$	3.52 ± 0.19	4.16 ± 0.23	4.53 ± 0.26

- the accuracy (Table 8.11) of differential diagnosis of blood films samples by the using the statistical method was unsatisfactory $\left(Ac(Y(U^{\Theta}_{0;90}(\xi^{k=N})) \sim 64.7\% \div 70.6\%\right)$.

From a physical point of view, the results obtained can be explained by the following considerations [75–84].

Under conditions of multiple volumetric scattering by blood films samples, statistical averaging of polarization manifestations occurs $\left(U^{\Theta}_{0;90}(\xi^{k=N})\right)$ along with the topographic distributions of linear $LB(\xi^{k=N}, (m, n))$ and circular $CB(\xi^{k=N}, (m, n))$ birefringence of supramolecular polycrystalline architectonics of blood films.

At a high level of repeatedly scattered depolarized background, statistical manifestations of linear (structural anisotropy $LB_{(0,90);(45,135)}$ of the polycrystalline networks of albumin and fibrin) and circular (molecular complexes of globulin and blood corpuscles $CB_{\oplus;\ominus}$) birefringence, as well as the spatial–angular variations in the supramolecular networks of biological crystals in blood drops (Groups 1–3) are too small to enable reliable differentiation of blood loss volume. The integral scale-self-similar structure (topographic) of the optical anisotropy map, Table 8.11, illustrates the accuracy Ac of statistical differential diagnosis of diffuse blood films samples from all groups of deceased peoples of diffuse films of blood is formed as a result of the superposition of layered fractal measures $Y^r_i(d)$ of linear and circular birefringence $Y(U^{\Theta}_{0;90}(\xi^{k=N}), (m, n))$.

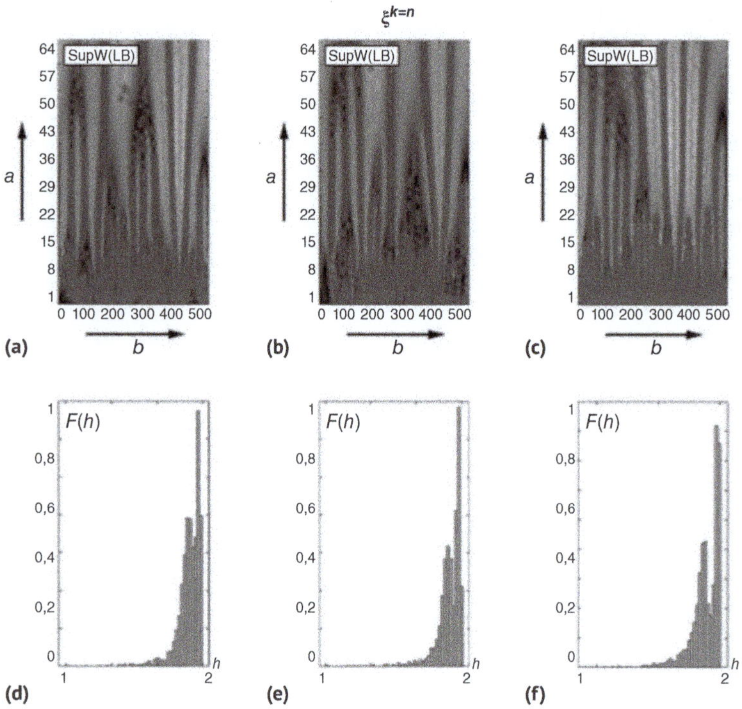

Figure 8.16 Skeletons $\sup|W(a^*, x_i(a^*))|$ (a), (c), (e) and multifractal spectra $F\{h, \mathrm{LB}(\xi^{k=N}, m, n)\}$ (b), (d), (f) distributions of the magnitude of linear birefringence of "diffuse" maps $\mathrm{LB}(U^{\Theta}_{0;90}(\xi^{k=N}), (m, n))$ films of the blood of the deceased with a different volume of blood loss.

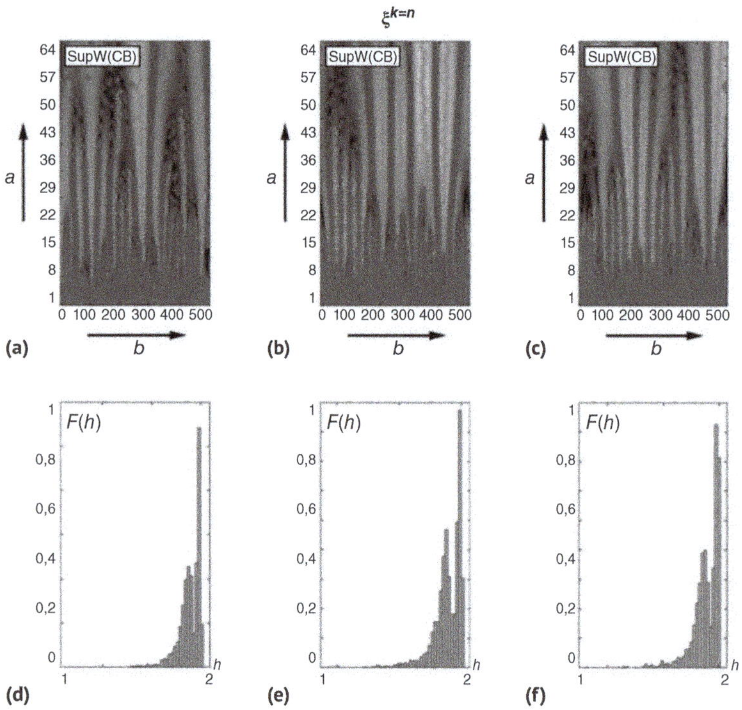

Figure 8.17 Skeletons $\sup|W(a^*, x_i(a^*))|$ (a), (c), (e) and multifractal spectra $F\{h, \mathrm{CB}(\xi^{k=N}, m, n)\}$ (b), (d), (f) distributions of the magnitude of the circular birefringence of "diffuse" maps $\mathrm{CB}(U^{\Theta}_{0;90}(\xi^{k=N}), (m, n))$ films of the blood of the deceased with a different volume of blood loss.

Figures 8.16 and 8.17 present the results of a multifractal analysis of "diffuse" optical anisotropy maps of the entire volume of blood films samples using the WTMM method. Here $\sup|W(a^*, x_i(a^*))|$ denotes the skeletons of the maxima of the wavelet coefficient modules $W^{Y}(a, b)$ (panels (a), (c), (e)) and the multifractal spectra represent the distributions of linear birefringence $LB(U^{\Theta}_{0;90}(\xi^{k=N}), (m, n))$ (Figure 8.14) and circular $CB(U^{\Theta}_{0;90}(\xi^{k=N}), (m, n))$ (Figure 8.15) birefringence of supramolecular optically anisotropic polycrystalline networks of films of the blood of the deceased with different volume of blood loss – Group 1 (Figure 8.15b); group 2 (Figure 8.15d) and group 3 (Figure 8.15f).

Results of multifractal analysis anisotropy maps $(Y(U^{\Theta}_{0;90}(\xi^{k=N}), (m, n)))$ for all groups of diffuse blood films samples are presented in Table 8.12.

Table 8.13 illustrates the accuracy Ac of multifractal differential diagnosis of diffuse blood films samples from all groups of deceased individuals.

Table 8.12 Accuracy of statistical differential diagnostics of "integral" of blood films anisotropy maps.

	$\mathrm{Ac}\{\mathrm{LB}(\xi^{k=N}, m, n)\}$, 100%		
$M_{l=1;2;3;4}$	Group 1–2	Group 1–3	Group 2–3
$M_{l=1} \times 10^{-3}$	64.3	66.1	62.5
$M_{l=2} \times 10^{-3}$	64.3	66.1	62.5
$M_{l=3}$	66.1	69.7	64.3
$M_{l=4}$	69.7	71.5	66.1
	$\mathrm{Ac}\{\mathrm{CB}(\xi^{k=N}, m, n)\}$, 100%		
$M_{l=1;2;3;4}$	Group 1–2	Group 1–3	Group 2–3
$M_{l=1} \times 10^{-3}$	64.3	67.9	64.3
$M_{l=2} \times 10^{-3}$	64.3	69.7	64.3
$M_{l=3}$	66.1	71.5	64.3
$M_{l=4}$	69.7	71.5	66.1

Table 8.13 Accuracy of multifractal differential diagnosis of "integral" blood films anisotropy maps.

	$\mathrm{Ac}\{F[h, \mathrm{LB}(\xi^{k=N}, m, n)]\}$, 100%		
$M_{l=1;2;3;4}$	Group 1–2	Group 1–3	Group 2–3
$M_{l=1} \times 10^{-3}$	71.5	69.7	67.9
$M_{l=2} \times 10^{-3}$	76.8	78.6	75
$M_{l=3}$	83.9	83.9	82.1
$M_{l=4}$	83.9	85.7	82.1
	$\mathrm{Ac}\{F[h, \mathrm{CB}(\xi^{k=N}, m, n)]\}$, 100%		
$M_{l=1;2;3;4}$	Group 1–2	Group 1–3	Group 2–3
$M_{l=1} \times 10^{-3}$	71.5	69.7	67.9
$M_{l=2} \times 10^{-3}$	76.8	78.6	75
$M_{l=3}$	82.1	82.1	80.4
$M_{l=4}$	82.1	83.9	80.4

The analysis of the obtained data showed the following:

- skeletons $\sup|W(a^*, x_i(a^*))|$ of the wavelet maps series $W^{Y}(a, b)$ (Figures 8.16 and 8.17a, c, e) exhibit a pronounced network of lines corresponding to the maxima of the wavelet coefficient modules, representing the transformation of the topographic distributions of the optical anisotropy parameters of the polycrystalline blood films from all groups of deceased individuals:
- the wavelet transformations $W^{Y}(a, b)$ algorithmically reconstructed from skeletons $\sup|W(a^*, x_i(a^*))|$ multifractal spectra $F^{Y}(h_r)$ series of optical anisotropy maps $Y(U^{\Theta}_{0;90}(\xi^{k=N}), (m, n))$ (Figures 8.14 and 8.15, panels a–c – Figures 8.16 and 8.17, panels b, d, f) a wide set of partial fractal dimensions h_r characterizes them.
- intergroup differences between the third and fourth central statistical moments, which characterize multifractal spectra $F^{Y}(h_r)$, increased to 30–40% (Table 8.12);
- multifractal analysis (Table 8.13) of the topographic structure of anisotropy maps $Y(U^{\Theta}_{0;90}(\xi^{k=N}), (m, n))$ of diffuse blood films samples provided a satisfactory level of accuracy $(Ac(Y(U^{\Theta}_{0;90}(\xi^{k=N})) \sim 80.4\% \div 83.3\%)$ for differentiating the volume of blood loss.

From a physical point of view, the results of the multifractal analysis of blood films anisotropy maps can be explained as follows: each measure $Y^r_i(d)$ is correlated with the topographic arrangement of partial supramolecular networks of biological crystals, which are characterized by specific values of the local fractal dimension h_r.

From a biochemical point of view, the process of blood loss is accompanied by a predominant concentration loss of optically active formed elements of the blood (a decrease in circular birefringence CB) and changes in the tertiary and quaternary structure of protein compounds (LB). Consequently, during the dehydration of such blood samples, a modified topographic structure of supramolecular networks is formed. A characteristic manifestation of blood loss is the predominance of large-scale (tens of micrometers) albumin-globulin networks compared to the "small-scale" (several to ten micrometers) optically active polycrystalline structures of formed elements. The reconstructed optical anisotropy maps $\left(\begin{pmatrix} LB(f_{pl}(U^{\Theta}_{0;90}(\xi^{k=N}))) \\ CB(f_{pl}(U^{\Theta}_{0;90}(\xi^{k=N}))) \end{pmatrix}(m, n)\right)$ provide an informational "topographic portrait" of this process.

Wavelet analysis allows for a quantitative assessment of changes in the topographic structure of anisotropy maps. The result of the scale-selective transformation of optical anisotropy parameter distributions in supramolecular networks of blood films during blood loss is manifested as differences in the amplitudes of wavelet coefficients at different scales $(1 \le a \le 70)$ of the MHAT function $\Pi\left(\frac{x-b}{a}\right)$. The amplitudes $W_{a,b}$ of large scales $(a \ge 20)$, corresponding to large-scale linearly birefringent albumin–globulin structures, decrease less compared to the amplitudes at small scales $(a \le 20)$, which represent small-scale optically active structures. These differences lead to a reduction in the range of variation of the fractal dimensions h_r of the topographic structure of optical anisotropy maps due to a decrease in the probability of formation of partial CB and small-scale albumin-globulin complexes components with small values of fractal dimensions h_r. Such a transformation of the singularity spectra $F^{Y}(h_r)$ as a result of blood loss is accompanied by a decrease in dispersion and an increase in the value of the central statistical moments of higher orders $(M_{3;4}(F^{Y}(h_r)))$, which characterize the skewness and kurtosis of fractal dimensions distributions h_r. which represent small-scale optically active structures.

Quantitatively (Table 8.13), this is reflected in an increase in differences – up to (up to 45–55%) between the values of the first to fourth-order central statistical moments characterizing the distributions of fractal dimensions of the optical anisotropy map topography $Y(U^{\Theta}_{0;90}(\xi^{k=N}), (m, n))$, and in a corresponding increase in the accuracy of differential diagnosis of blood loss volume, with $Ac(Y(U^{\Theta}_{0;90}(\xi^{k=N})) \sim 80.4\% \div 85.7\%$.

8.2.7 Statistical and Multifractal Structure of Anisotropy Maps of "Phase" Optically Thin Blood Films

A series of fragments in Figures 8.18 and 8.19 show the results of the "phase" $(\xi^{k=1})$ algorithmic map $Y(U^{\Theta}_{0;90}(\xi^{k=1}), (m, n))$ (panels (a)–(c)) and histograms $H(Y((\xi^{k=1}), (m, n))$ (panels (d)–(f)) reconstruction of the linear (Figure 8.18) and circular (Figure 8.19) birefringence of optically thin supramolecular layer of blood films – Group 1 (panels (a), (d)); Group 2 (panels (b), (e)) and Group 3 (panels (c), (f)).

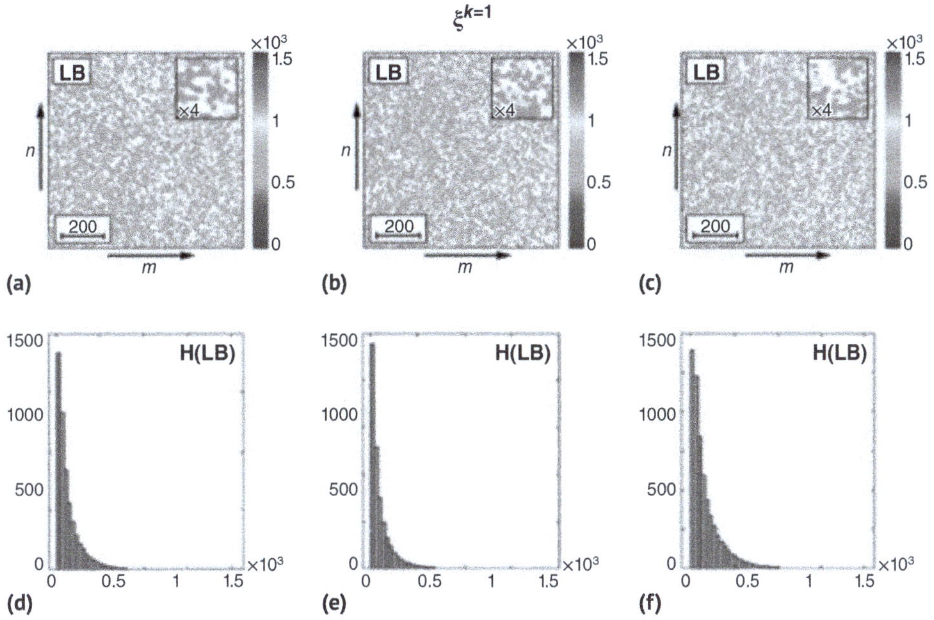

Figure 8.18 Coordinate (a)–(c) and probabilistic (d)–(f) distributions of the magnitude of linear birefringence of "phase" maps $\mathrm{LB}(U^{\Theta}_{0;90}(\xi^{k=1}), (m, n))$ films of the blood of the deceased with a different volume of blood loss.

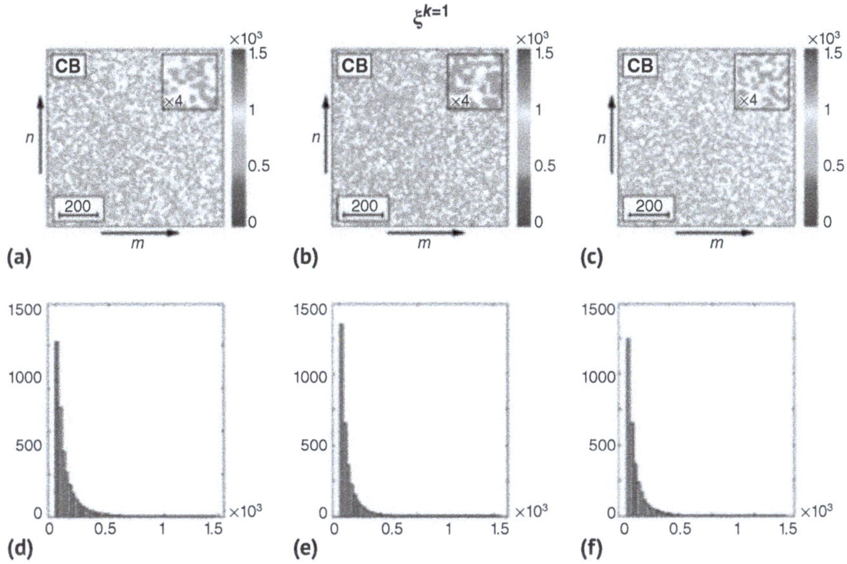

Figure 8.19 Coordinate (a)–(c) and probabilistic (d)–(f) distributions of the magnitude of the circular birefringence of "phase" maps $\mathrm{CB}(U^{\Theta}_{0;90}(\xi^{k=1}), (m, n))$ films of the blood of the deceased with a different volume of blood loss.

Figures 8.20 and 8.21 illustrate the results of the multifractal analysis of the "phase" maps of the optical anisotropy of the non-depolarizing $(\xi^{k=1})$ layer of blood films samples. Here, $\sup|W(a^*, x_i(a^*))|$ are the skeletons of the maxima of the modules of the wavelet coefficients (panels (a),(c),(e)) and multifractal spectra of topographic maps of linear and circular birefringence of optically thin layers of blood films – Group 1 (panel (b)); Groups 2 (panel (d)) and 3 (panel (f)).

The results of the statistical analysis of "optically thin" blood films anisotropy maps are presented in Table 8.14. Table 8.15 provides the statistical parameters of the multifractal spectra of optically thin 'phase' maps of blood films with different volumes of blood loss. Tables 8.17 and 8.18 illustrate the accuracy of statistical and multifractal differential diagnostics of the optically thin phase layers of blood films samples from all groups of deceased individuals.

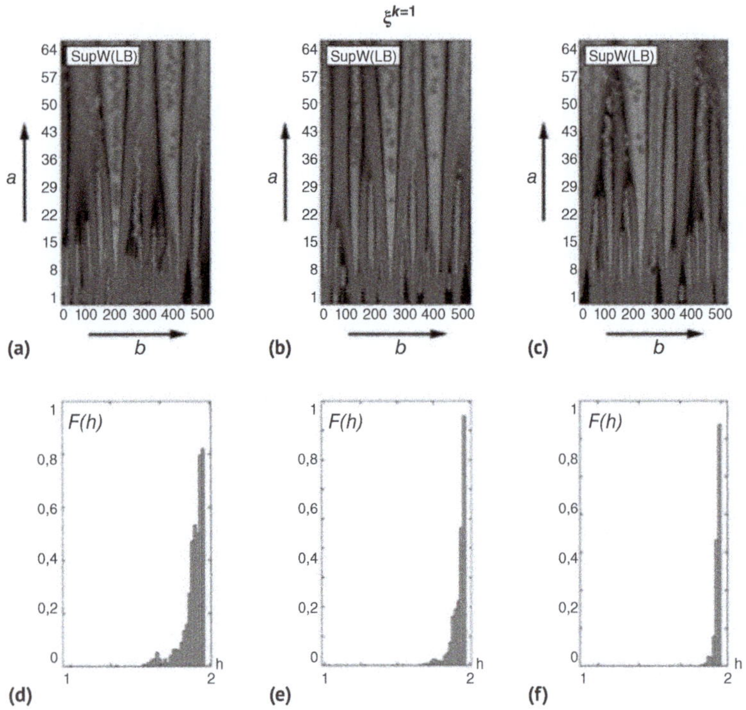

Figure 8.20 Skeletons $\sup|W(a^*, x_i(a^*))|$ (a), (c), (e)and multifractal spectra $F\{h, \mathrm{LB}(\xi^{k=1}, m, n)$ (b), (d), (f) distributions of the magnitude of linear birefringence of "phase" maps $\mathrm{LB}(U^{\Theta}_{0;90}(\xi^{k=1}), (m, n))$ films of the blood of the deceased with a different volume of blood loss.

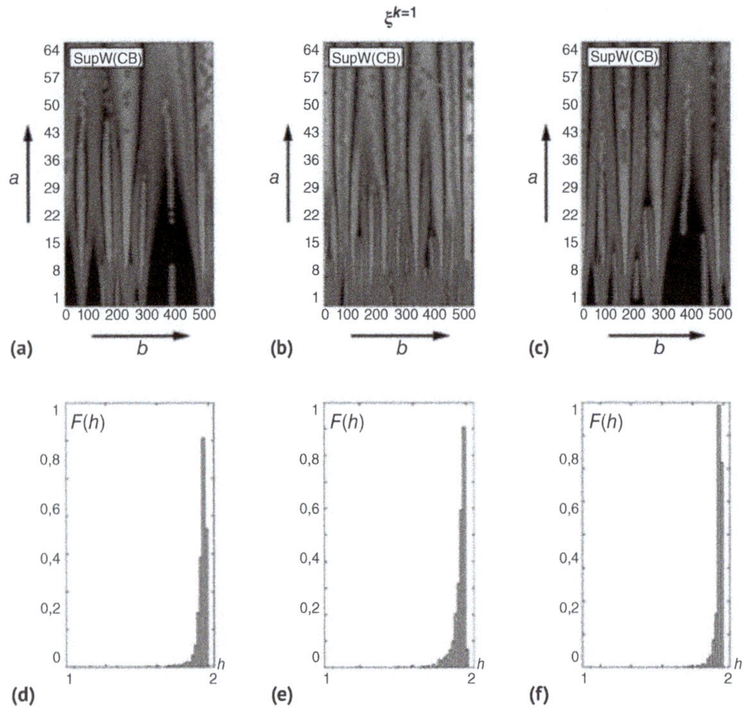

Figure 8.21 Skeletons $\sup|W(a^*, x_i(a^*))|$ (a), (c), (e) and multifractal spectra $F\{h, \mathrm{CB}(\xi^{k=1}, m, n)$ (b), (d), (f) distributions of the magnitude of the circular birefringence of "phase" maps $\mathrm{CB}(U^{\Theta}_{0;90}(\xi^{k=1}), (m, n))$ films of the blood of the deceased with a different volume of blood loss.

Table 8.14 Statistical parameters of "phase" maps of blood films anisotropy.

$M_{l=1;2;3;4}$	Group 1	Group 2	Group 3
$LB(\xi^{k=1}, m, n)$			
$M_{l=1} \times 10^{-3}$	0.27 ± 0.017	0.22 ± 0.012	0.18 ± 0.009
$M_{l=2} \times 10^{-3}$	0.103 ± 0.006	0.088 ± 0.0055	0.079 ± 0.005
$M_{l=3}$	1.49 ± 0.088	1.65 ± 0.098	1.76 ± 0.087
$M_{l=4}$	2.11 ± 0.13	2.42 ± 0.14	2.64 ± 0.14
$CB(\xi^{k=1}, m, n)$			
$M_{l=1;2;3;4}$	Group 1	Group 2	Group 3
$M_{l=1} \times 10^{-3}$	0.16 ± 0.009	0.13 ± 0.007	0.11 ± 0.006
$M_{l=2} \times 10^{-3}$	0.077 ± 0.0004	0.064 ± 0.0004	0.057 ± 0.0031
$M_{l=3}$	1.84 ± 0.11	2.26 ± 0.14	2.57 ± 0.14
$M_{l=4}$	2.43 ± 0.14	2.94 ± 0.18	3.36 ± 0.19

Table 8.15 Statistical parameters of multifractal spectra of "phase" maps of blood films anisotropy.

$M_{l=1;2;3;4}$	Group 1	Group 2	Group 3
$F\{h, LB(\xi^{k=1}, m, n)\}$			
$M_{l=1} \times 10^{-3}$	1.67 ± 0.097	1.73 ± 0.11	1.79 ± 0.11
$M_{l=2} \times 10^{-3}$	0.16 ± 0.009	0.11 ± 0.006	0.068 ± 0.0004
$M_{l=3}$	3.17 ± 0.18	4.19 ± 0.23	5.84 ± 0.32
$M_{l=4}$	4.33 ± 0.23	5.16 ± 0.29	6.59 ± 0.36
$F\{h, CB(\xi^{k=1}, m, n)\}$			
$M_{l=1;2;3;4}$	Group 1	Group 2	Group 3
$M_{l=1} \times 10^{-3}$	1.61 ± 0.092	1.66 ± 0.091	1.73 ± 0.099
$M_{l=2} \times 10^{-3}$	0.12 ± 0.007	0.072 ± 0.0004	0.057 ± 0.0004
$M_{l=3}$	3.61 ± 0.21	4.89 ± 0.27	6.24 ± 0.36
$M_{l=4}$	4.73 ± 0.27	5.81 ± 0.33	7.75 ± 0.41

Analysis of the obtained data on the phase extraction of a single scattered component of the diffuse object field of blood films showed:

- probabilistic distributions of random values of birefringence parameters of an optically thin layer of supramolecular architectonics of blood films samples have an individual statistical structure ($M_{g=1;2;3;4}$ (Group 1) $\neq$ $M_{g=1;2;3;4}$ (Group 2) $\neq M_{g=1;2;3;4}$ (Group 3, – Table 8.14);
- intergroup differences between the values of the central statistical moments of the third and fourth orders, which characterize the skewness $(M_{g=3})$ and the kurtosis $(M_{g=4})$ of the coordinate distributions of "phase" maps $\Upsilon(U^{\Theta}_{0;90}(\xi^{k=1}), (m, n))$, are significant and reach 1.65–1.8 times (Table 8.14);
- algorithmically reconstructed from skeletons $\sup|W(a^*, x_i(a^*))|$ wavelet transformations $W^{\Upsilon}(a, b)$ multifractal spectra $F^{\Upsilon}(h_r)$ series of optical anisotropy "phase" maps have an individual structure and they are characterized by a much smaller set of partial fractal dimensions (h_r) in comparison with the data obtained for diffuse blood films samples;
- with an increase in the volume of blood loss of the half-width of the multifractal spectra $F^{LB,CB}(h_r)$ of optical anisotropy "phase" maps decreases;
- the accuracy (Table 8.16) of differential diagnostics of "phase" birefringence maps of optically thin blood layers increases to a good $(Ac(\Upsilon(U^{\Theta}_{0;90}(\xi^{k=1})) \sim 83.9\%)$ and very good $(Ac(\Upsilon(U^{\Theta}_{0;90}(\xi^{k=1})) \sim 89.3\%)$ levels;

- multifractal analysis (Table 8.17) of the topographic structure of optical anisotropy maps provided the maximum excellent level of accuracy $\left(Ac\left(Y\left(U^{\Theta}_{0;90}\left(\xi^{k=1}\right)\right) \sim 100\%\right)\right)$ differentiation of blood loss volume.

From a physical point of view, the following considerations can explain the results obtained using layered phase scanning of a holographically reconstructed field of complex amplitudes. Under conditions of single scattering $\left(\xi^{k=1}\right)$ by blood films samples, direct relationships are formed between the Mueller-matrix images and the topographic distributions of linear and circular birefringence parameters $\xi^{k=1}$ of birefringence supramolecular polycrystalline architectonics of blood films. As a result, the statistical manifestations of spatial-angular changes in the supramolecular networks of biological crystals (albumin, globulin, fibrinogen, elastin, etc.) in blood drops (Groups 1–3) are maximally pronounced, allowing reliable differentiation of blood loss volume. Consequently, the accuracy of differential diagnosis of blood loss volume increases to good (80.4%) and very good (89.5%) levels.

The topography of the "phase" maps of birefringence of blood films is maximally correlated with the fractal measures of linear and circular birefringence of an optically thin layer of polycrystalline architectonics.

Therefore, the transformation of the scale-self-similar structure of the supramolecular optically fine architectonics of blood films is highly sensitive (Table 8.17). Quantitatively, this manifests as an increase (up to 2.1–2.3 times) in the differences between the values of the central statistical moments of the third and fourth orders, which characterize the

Table 8.16 Accuracy of statistical differential diagnostics of "phase" maps of blood films anisotropy.

	$Ac\{LB(\xi^{k=1}, m, n)\}$, 100%		
$M_{l=1;2;3;4}$	**Group 1–2**	**Group 1–3**	**Group 2–3**
$M_{l=1} \times 10^{-3}$	78.6	80.4	76.8
$M_{l=2} \times 10^{-3}$	80.4	82.1	78.6
$M_{l=3}$	83.9	87.5	83.9
$M_{l=4}$	85.7	89.3	83.9
	$Ac\{CB(\xi^{k=1}, m, n)\}$, 100%		
$M_{l=1;2;3;4}$	**Group 1–2**	**Group 1–3**	**Group 2–3**
$M_{l=1} \times 10^{-3}$	76.8	78.6	73.3
$M_{l=2} \times 10^{-3}$	78.6	80.4	75
$M_{l=3}$	80.4	83.9	78.6
$M_{l=4}$	82.1	85.7	80.4

Table 8.17 Accuracy of multifractal differential diagnostics of "phase" maps of blood films anisotropy.

	$Ac\{F[h, LB(\xi^{k=1}, m, n)]\}$, 100%		
$M_{l=1;2;3;4}$	**Group 1–2**	**Group 1–3**	**Group 2–3**
$M_{l=1} \times 10^{-3}$	78.6	82.1	76.8
$M_{l=2} \times 10^{-3}$	91.1	96.4	89.3
$M_{l=3}$	94.6	100	91.1
$M_{l=4}$	96.4	100	92.9
	$Ac\{F[h, CB(\xi^{k=1}, m, n)]\}$, 100%		
$M_{l=1;2;3;4}$	**Group 1-2**	**Group 1-3**	**Group 2-3**
$M_{l=1} \times 10^{-3}$	78.6	82.1	76.8
$M_{l=2} \times 10^{-3}$	85.7	89.3	89.3
$M_{l=3}$	89.3	94.6	91.1
$M_{l=4}$	91.1	96.4	91.1

skewness and kurtosis of the distribution of fractal dimensions of the topography of the optical anisotropy "phase" maps $Y(U^{\Theta}_{0;90}(\xi^{k=1}), (m, n))$. As a result, the accuracy of differential diagnosis of blood loss volume reaches the maximum level of 100%.

8.2.8 Conclusions

1. The method of Mueller-matrix interferometry with layered phase scanning of object fields, reconstructed via digital holography, has been analytically developed and experimentally validated for studying the optically anisotropic architectonics of supramolecular networks in blood films.
2. Within the single-scattering approximation, analytical relationships (algorithms) were established between the coordinate distributions of Mueller-matrix elements in an optically thin layer of polycrystalline architectonics and the parameters of linear and circular birefringence of supramolecular networks of biological crystals self-assembled during dehydration.
3. A novel approach is proposed for the objective evaluation of optical anisotropy maps of blood films, based on statistical analysis of the maxima of wavelet coefficient modules and multifractal spectra of the linear and circular birefringence distributions of polycrystalline supramolecular networks.
4. Using phase scanning, a single scattered component of the object field from diffuse blood films of deceased individuals with varying blood loss volumes was isolated, and a set of birefringence parameter maps for optically thin layers of the supramolecular architectonics was algorithmically reconstructed.
5. Comparative statistical and multifractal analysis of the "phase" maps of linear and circular birefringence enabled the identification of diagnostic markers – skewness and kurtosis of the fractal dimension distribution – achieving up to 92.9% accuracy in differentiating blood films samples from deceased individuals with varying blood loss volumes (up to 1000 mL).

8.3 Three-dimensional Digital Holographic Scanning and Multifractal Scaling Polarization Maps Biological Fluids Films

8.3.1 Introduction

This section investigates the fundamental relationship between the scale-self-similar structure of optical anisotropy maps (multifractal spectra) and the architectonics of dehydrated films of biological fluids. This is accomplished through multifractal scaling of holographically reconstructed layered polarization maps of optically anisotropic films, exhibiting dendritic (blood plasma – FBP) and dendritic-spherulite (bile – FB) networks. The elucidated relationships are subsequently applied to a practical forensic task: differentiation of the nature of mechanical skin injuries using the developed method of polarization-interference mapping of rat blood films [77–88].

8.3.2 Objects of Investigations

In this section we will consider two types of BL films:

- the blood plasma – ***FBP***;
- cystic bile – ***FB***.

The choice of these objects is motivated by both fundamental and applied considerations. From a *fundamental perspective*, the biochemical composition of blood and bile (BL) differs. Consequently, during the dehydration of BL films, molecular self-assembly leads to the formation of polycrystalline networks with distinct architectonics and optical anisotropy. The data obtained on the scale-self-similar structure of such networks can be used to establish the relationships between the topology of optical anisotropy maps and the spatial-angular characteristics of biological crystals.

From an *applied perspective*, the quantitative parameters (first- and fourth-order statistical moments), that characterize the multifractal spectra $F(h_r)$ of the ellipticity polarization maps $\beta(\varphi^*_k, u, t)$ can serve as markers for the differential diagnosis of human organs under various pathological conditions.

8.3.2.1 Blood Plasma Films

Blood plasma is a heterogeneous liquid of complex structure, a colloidal polymer solution. Traditionally, its molecular components – albumin (~60%), globulin (~40%), fibrinogen, hemoglobin are effectively studied using biochemical methods.

However, there are currently insufficient methods to study the supramolecular optically anisotropic protein structures (dimers, trimers, and monomers) of blood plasma, which range in size from 100 to 2000 nm. Within the structure of the blood plasma surface, three zones of differing thickness h can be distinguished:

- edge (polycrystalline albumin layer with linear birefringence and dichroism, $h \sim 50 - 60\,\mu m$);
- transitional (outer and inner layers of optically isotropic cubic crystals of Na—Cl salt, intermediate globulin layer with circular birefringence and dichroism, $h \sim 25 - 30\,\mu m$);
- the central (outer layer of cubic crystals of salt Na-Cl, $h \sim 5 - 15\,\mu m$).

This type of plasma structure predominantly exhibits a dendritic supramolecular polycrystalline network, characterized by structural anisotropy (LB) and minimal optical activity (CB). Hereafter, this network will be referred to as dendritic.

8.3.2.2 Bile Films

The main components of bile are bile acids, which exist as conjugates with glycine, taurine, and other paired acids. In addition, bile contains proteins (immunoglobulins), cholesterol, bilirubin, and metals such as copper, zinc, lead, indium, magnesium, mercury, and others. The formation of fatty acid crystals and monohydrate cholesterol produces a dendritic polycrystalline network in FB, resulting in structural anisotropy (LB).

Another, equally developed, FB architecture is formed by optically active crystals of calcium bilirubinate and calcium microspherolites, which are characterized by circular birefringence (CB). Thus, the polycrystalline component of FB exhibits comparable manifestations of structural anisotropy and optical activity. In this context, the network of such crystals will be referred to as dendritic-spherulite.

All films were prepared by applying a drop of the liquid onto a glass substrate at room temperature ($t = 22\,°C$) and allowing it to dry for 24 hours.

The optical-geometric parameters of BL films are presented in Table 8.18.

8.3.3 Polarization Mapping of Films Microscopic Images

8.3.3.1 Ellipticity Polarization Maps of Blood Plasma Films

Figure 8.22 shows the results of polarization-interference mapping of microscopic images of blood plasma films (FBP), obtained with layered digital holographic reconstruction of the coordinate distributions of polarization ellipticity. The results confirm the adequacy of the formation of polarization distributions according to the proposed Jones matrix model of the optical anisotropy of supramolecular networks of biological FBP crystals.

The integral object field FBP (Figure 8.22a is characterized by a superposition of all components of scattered radiation $\begin{Bmatrix} Q(\beta_l^{\odot}) \\ Q(\beta_l^{\circledast}) \\ R(\beta^{\boxdot}) \end{Bmatrix}$. As a result, fairly equally probable distributions are formed $M(\beta^*)$ (Figure 8.22b).

As the phase scanning parameter φ_k progressively thinner optically layers with lower scattering multiplicity are distinguished. Consequently, both the average value $\overline{\beta_l}$ and the range of $\Delta\beta_l$ changes in the random values of the ellipticity of polarization β_i decrease. progressively thinner optically layers with lower scattering multiplicity are distinguished. Consequently, both the average value $\beta(u, t)$ and $\beta(\varphi = \pi/4, u, t)$; $\beta(\varphi = \pi/8, u, t)$ of these FBP object field components.

Table 8.18 Optical- geometric parameters of BL films.

Parameters	FBP (23 samples)
Attenuation coefficient τ, cm^{-1}	0.17 ± 0.011
Depolarization degree Λ, %	15 ± 0.12
Parameters	**FB (23 samples)**
Attenuation coefficient τ, cm^{-1}	0.21 ± 0.014
Depolarization degree Λ, %	21 ± 0.14

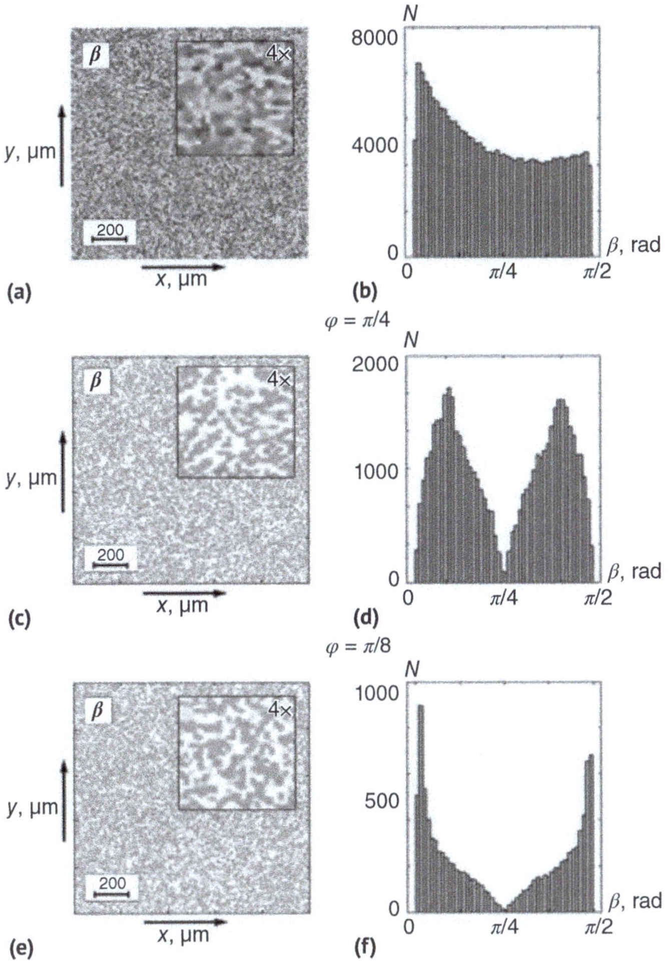

Figure 8.22 Maps (a), (c), (e) and histograms (b), (d), (f) distributions $N(\beta)$ of ellipticity of polarization $\beta(u,t)$ (a) and $\beta(\varphi = {}^{\pi}/_{4}, u, t)$; $\beta(\varphi = {}^{\pi}/_{8}, u, t)$ (c), (e) in the FBP microscopic image.

In other words, the probability distributions of the random values of polarization ellipticity β_l increasingly reflect the true polarization maps of the supramolecular network of biological FBP crystals

$$\beta(\varphi = {}^{\pi}/_{4}, u, t) \Longrightarrow \begin{pmatrix} \beta_{11}^{\odot} & \cdots & \beta_{1t}^{\odot} \\ \vdots & \ddots & \vdots \\ \beta_{u1}^{\odot} & \cdots & \beta_{ut}^{\odot} \end{pmatrix} + \begin{pmatrix} \beta_{11}^{\circledast} & \cdots & \beta_{1t}^{\circledast} \\ \vdots & \ddots & \vdots \\ \beta_{u1}^{\circledast} & \cdots & \beta_{ut}^{\circledast} \end{pmatrix}; \tag{8.22}$$

$$\beta(\varphi = {}^{\pi}/_{8}, u, t) \Longrightarrow \begin{pmatrix} \beta_{11}^{\odot} & \cdots & \beta_{1t}^{\odot} \\ \vdots & \ddots & \vdots \\ \beta_{u1}^{\odot} & \cdots & \beta_{ut}^{\odot} \end{pmatrix}. \tag{8.23}$$

All experimentally obtained polarization maps $\beta(m,n)$ and $\beta(\varphi = {}^{\pi}/_{4}, u, t)$; $\beta(\varphi = {}^{\pi}/_{8}, u, t)$ have a coordinate (Figure 8.22a, c, e) and probabilistic (Figure 8.23b, d, f) the structure of random values of ellipticity.

A comparative analysis of a series of layered polarization maps (φ, u, t) in various phase sections $\varphi = {}^{\pi}/_{4}; {}^{\pi}/_{8}$ of the FBP object field revealed the following transformations of the distributions of random values of the ellipticity of polarization β.

8.3.3.2 Ellipticity Polarization Maps of Bile Films

The fragments of Figure 8.23 illustrate the results of polarization-interference mapping of the microscopic image of the bile films (*FB*) with layered digital holographic reconstruction of the coordinate distributions of the ellipticity of polarization $\beta(u,t)$ and $\beta(\varphi = {}^{\pi}/_{4}, u, t)$; $\beta(\varphi = {}^{\pi}/_{8}, u, t)$.

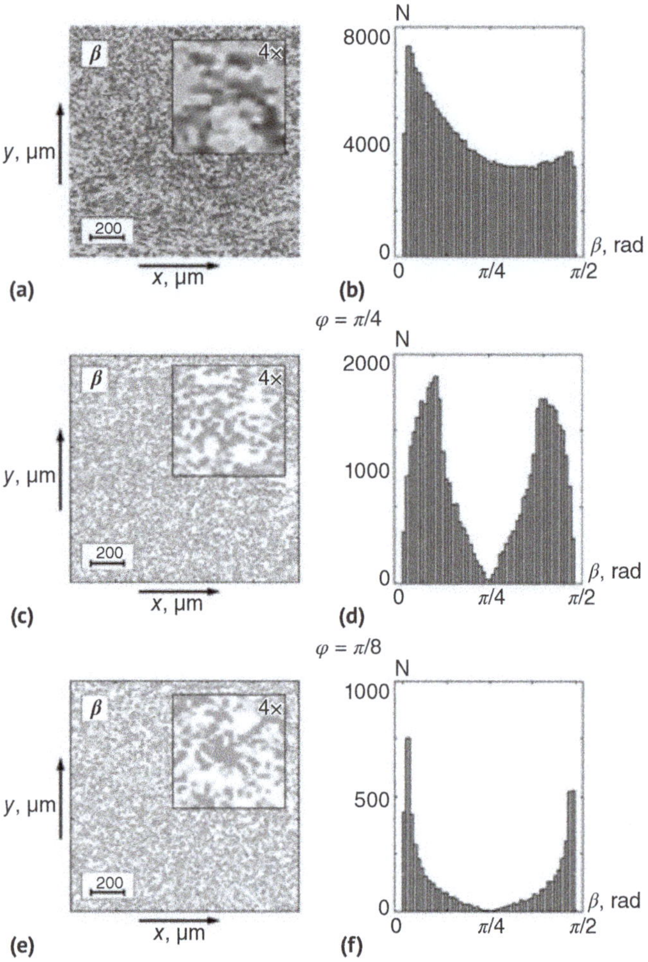

Figure 8.23 Maps (a), (c), (e) and histograms (b), (d), (f) distributions $N(\beta)$ of ellipticity of polarization $\beta(u, t)$ (a) and $\beta(\varphi = \pi/4, u, t)$; $\beta(\varphi = \pi/8, u, t)$ (c), (e) in the ***FB*** microscopic image.

The presence of polarization inhomogeneity of the object field for the films of a biological fluid with different architectonics of a polycrystalline supramolecular network has been experimentally established.

The revealed fact confirms the presence of "developed" probable distributions of $N(\beta)$ random values of the magnitude of the ellipticity of polarization β both in the field itself and in a series of its phase sections with a lower light scattering multiplicity (Figure 8.23b, d, f).

On the other hand, a comparative analysis of the obtained polarization maps $\beta(u, t)$ and $\beta(\varphi = \pi/4, u, t)$; $\beta(\varphi = \pi/8, u, t)$ for FB and FBP, found not only the adequacy of the Jones matrix model of optical anisotropy of supramolecular networks of biological crystals, but also statistical similarity (position and magnitude of extremes, range of variation of random values) the formation of polarization distributions $N(\beta)$ and $N(\varphi_k, \beta)$ all components of the object field films of biological fluids, – Figures 8.22 and 8.23b, d, f).

More detailed and objective distribution data $N(\beta)$ and $N(\varphi_k, \beta)$ illustrate the results of statistical analysis polarizing maps for FB and FBP.

Table 8.19 shows the values of statistical moments of the first and fourth orders, which characterize a series of distributions $\beta(u, t)$ and $\beta(\varphi = \pi/4, u, t)$; $\beta(\varphi = \pi/8, u, t)$.

Despite the sufficient similarity of histograms $N(\beta)$ and $N(\varphi_k, \beta)$ for FB and FBP comparison of statistical analysis of distributions $\beta(u, t)$ and $\beta(\varphi = \pi/4, u, t)$; $\beta(\varphi = \pi/8, u, t)$ it also revealed distinctive features. Thus, the most sensitive to the specifics of the probability distributions of the ellipticity of polarization there were statistical moments of higher orders $M_{p=3;4}(\varphi_k, \beta)$.

Table 8.19 Statistical parameters of distributions $\beta(u, t)$ and $\beta(\varphi = \pi/4, u, t)$; $\beta(\varphi = \pi/8, u, t)$ object fields FB and FBP.

$M_{l=1;2;3;4}$	FBP (23 samples)	FB (23 samples)
$M_{l=1}$	0.63 ± 0.034	0.71 ± 0.039
$M_{l=2}$	0.44 ± 0.026	0.53 ± 0.031
$M_{l=3}$	0.37 ± 0.021	0.22 ± 0.014
$M_{l=4}$	0.51 ± 0.028	0.39 ± 0.023
$M_{l=1;2;3;4}$	$\varphi = \pi/4$	
$M_{l=1}$	0.31 ± 0.018	0.36 ± 0.021
$M_{l=2}$	0.19 ± 0.011	0.23 ± 0.014
$M_{l=3}$	0.53 ± 0.028	0.41 ± 0.023
$M_{l=4}$	0.71 ± 0.039	0.57 ± 0.032
$M_{l=1;2;3;4}$	$\varphi = \pi/8$	
$M_{l=1}$	0.12 ± 0.007	0.15 ± 0.009
$M_{l=2}$	0.07 ± 0.004	0.09 ± 0.006
$M_{l=3}$	0.77 ± 0.041	0.61 ± 0.034
$M_{l=4}$	0.93 ± 0.048	0.72 ± 0.038

The range of changes in these statistical parameters within the range of changes in the phase sections of the object field is for FBP – 1.82 – 2.08 times; for FB – 1.85 – 2.77 times.

Intergroup (FB and FBP) differences in values $Z_{3;4}(\varphi_k, \beta)$ make up 1.26 – 1.33 times.

Statistical moments of the first and second orders that characterize the mean and variance of distributions $\beta(u, t)$ and $\beta(\varphi = \pi/4, u, t)$; $\beta(\varphi = \pi/8, u, t)$, differ slightly for FBP and FB, – within 10–15%.

Thus, the results of statistical analysis of probability distributions $N(\beta)$ and $N(\varphi_k, \beta)$ (Figures 8.22 and 8.23b, d, f) we have demonstrated certain possibilities in their objective differentiation.

On the other hand, the topographic distributions of polarizing maps were out of the field of view of such an analysis $\beta(u, t)$ and $\beta(\varphi = \pi/4, u, t)$; $\beta(\varphi = \pi/8, u, t)$, (Figures 8.22 and 8.23a–c).

Even high-quality evaluation of the specifics of such coordinate distributions, including the ellipticity of polarization of object fields, FBP and FB, reveals significant differences between them.

Therefore, to identify additional markers of the optically anisotropic properties of biological fluid films, we used a different analytical approach.

This is a multifractal analysis of the scale-self-similar structure of polarization maps $\beta(u, t)$ and $\beta(\varphi = \pi/4, u, t)$; $\beta(\varphi = \pi/8, u, t)$ FBP and FB, which are formed by supramolecular polycrystalline networks of biological crystals with different architectonics.

8.3.3.3 Method MMWT of Microscopic Images

In Figures 8.24 and 8.25 the lines of maximal modules (skeletons –denote maxima, "and minima of the wavelet coefficient amplitudes) of the wavelet coefficient maps $W^{FBP,FB}(a, b)$ (panels (a), (b), (c)) are shown, along with multifractal spectra $F(h_i)$ of FBP and FB polarization maps $\beta(u, t)$ and $\beta(\varphi = \pi/4, u, t)$; $\beta(\varphi = \pi/8, u, t)$ (panels (b), (d), (f)).

A comparative analysis using the MWTM method revealed that:

- the polarization maps of optically anisotropic supramolecular films networks of both types of liquids (FBP and FB) are characterized by a complex scale – like structure, which is characterized by individual multifractal spectra $F^{FBP}(\varphi_k, h_r)$ and $F^{FB}(\varphi_k, h_r)$, Figures 8.24 and 8.25b, d, f).
- The multifractal spectra $F^{FBP;FB}(\varphi_k, h_r)$ of $\beta(u, t)$ and $\beta(\varphi = \pi/4, u, t)$; $\beta(\varphi = \pi/8, u, t)$ have an asymmetric structure with a clearly distinct extremum and a range of fractal dimension variations $1.45 \lesssim h_r \lesssim 1.91$.
- For FBP ellipticity polarization maps the multifractal spectra $F^{FBP}(\varphi_k, h_r)$ are "narrower" and more asymmetric (Figure 8.24b, d, f) than similar multifractal spectral dependences $F^{BP}(\varphi_k, h_r)$ for FB polarization inhomogeneous field (Figure 8.25b, d, f).

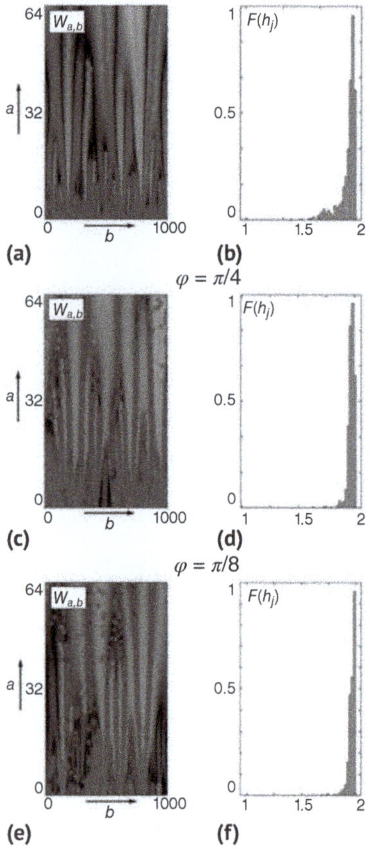

Figure 8.24 Skeletons of wavelet transform (a), (c), (e) and multifractal spectra (b), (d), (f) of polarization maps $\beta(u, t)$ (a) and $\beta(\varphi = {}^{\pi}/_{4}, u, t)$; $\beta(\varphi = {}^{\pi}/_{8}, u, t)$ of the FBP microscopic image.

The main extremes of multifractal spectra $F^{\text{FBP}}(\varphi_k, h_r)$ and $F^{\text{FB}}(\varphi_k, h_r)$ are localized in different parts of the fractal dimension scale: $\begin{cases} \max(\text{F}^{\text{FBP}}) \Leftrightarrow h_i = 1.76; \\ \max(\text{F}^{\text{FB}}) \Leftrightarrow h_i = 1.88. \end{cases}$

The fractal dimensions range Δh_r is larger for the spectrum $F^{\text{FBP}}(\varphi_k, h_r)$ than for $F^{\text{FB}}(\varphi_k, h_r) - \Delta h_r(\text{FB}) > \Delta h_r(\text{FBP})$.

From a physical point of view, the obtained results can be interpreted in terms of the following considerations.

First, each local optically anisotropic degree formed during FBP dehydration corresponds to the corresponding fractal measure in the polarization maps is $\beta(u, t)$ and $\beta(\varphi = {}^{\pi}/_{4}, u, t)$; $\beta(\varphi = {}^{\pi}/_{8}, u, t)$, which has its own dimension h_r.

The magnitude and probability of $F(h_r)$ such fractal dimension is determined by the specifics self-assembled supramolecular crystallite complexes.

Second, from the fractal geometry theory point of view, in the extreme case, the supramolecular needle crystals of FBP dendritic network spatial organization architectonics can lead to a straight line.

For such a boundary situation, the multifractal spectrums $F^{\text{FBP}}(\varphi_k, h_r)$ and $F^{\text{FB}}(\varphi_k, h_r)$ degenerates into a delta function with the base value of the fractal dimension $h_{\text{FBP}} \Rightarrow 1.0$.

Third, a different situation is realized for spherulite supramolecular networks of FB that form fractal measures in polarization maps. In this case, the limiting geometric configuration corresponds to polycrystalline architectonics in the form of a plane, associated with another characteristic fractal dimension $h_{\text{FB}} \Rightarrow 2.0$ of polarization maps $\beta(u, t)$ and $\beta(\varphi = {}^{\pi}/_{4}, u, t)$; $\beta(\varphi = {}^{\pi}/_{8}, u, t)$.

For a real physical situation, such trends are manifested in the different localization of the main extrema of the multifractal spectra $(\max(F^{\text{FBP}}(\varphi_k, h_r))$ and $\max(F^{\text{FB}}(\varphi_k, h_r))$.

In this case, the fractal dimensions variation range Δh_i is larger for the spectrum $F^{\text{FB}}(\varphi_k, h_r)$ than for $F^{\text{FBP}}(\varphi_k, h_r) - \Delta h_r(\text{FB}) > \Delta h_r(\text{FBP})$.

A quantitative statistical estimate of multifractal spectra $F^{\text{FBP}}(\varphi_k, h_r)$ and $F^{\text{FB}}(\varphi_k, h_r)$ is illustrated in Table 8.20.

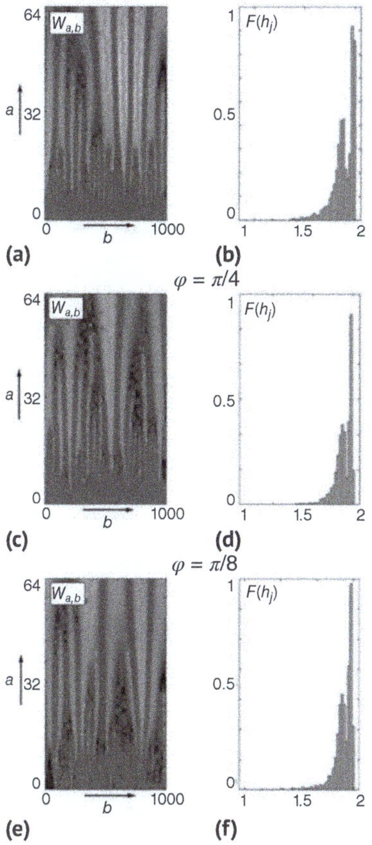

Figure 8.25 Skeletons of wavelet transform (a), (c), (e) and multifractal spectra (b), (d), (f) of polarization maps $\beta(u, t)$ (a) and $\beta(\varphi = \pi/4, u, t)$; $\beta(\varphi = \pi/8, u, t)$ of the FB microscopic image.

Comparison results of physical and statistical analysis of multifractal structure of $\beta(u, t)$ and $\beta(\varphi = \pi/4, u, t)$; $\beta(\varphi = \pi/8, u, t)$ revealed a correlation between them. Namely, $M_{p=1;2}(F^{\mathrm{FBP}}(\varphi_k, h_r)) < M_{p=1;2}(F^{\mathrm{FB}}(\varphi_k, h_r))$, which is due to the different localization $(h_r(\mathrm{FB}) > h_r(\mathrm{FBP}))$ of the extrema $F^{\mathrm{FBP}}(\varphi_k, h_r)$ and $F^{\mathrm{FB}}(\varphi_k, h_r)$, as well as differences between the fractal dimensions ranges – $\Delta h_r(\mathrm{FB}) > \Delta h_r(\mathrm{FBP})$.

The reverse trend $M_{p=3;4}(F^{\mathrm{FBP}}(\varphi_k, h_r)) > M_{p=3;4}(F^{\mathrm{FB}}(\varphi_k, h_r))$ is realized for the third and fourth-order statistical moments, which characterize the multifractal spectra distributions skewness and kurtosis. Mathematically, this regularity is associated with the inversely proportional dependencies $M_3 \sim 1/M_2^3$ and $M_4 \sim 1/M_2^4$.

The parameters most sensitive to changes in the scale self-similarity of various phase components of the object field were the third and fourth-order statistical moments, which characterize the skewness and kurtosis of the distributions $\boldsymbol{\beta(u, t)}$ and $\boldsymbol{\beta(\varphi = \pi/4, u, t)}$; $\boldsymbol{\beta(\varphi = \pi/8, u, t)}$. The range of variation in the value of these multifractal markers is: for FBP, 1.73–1.81 times, and for FB, 1.68–2.16 times. The intergroup differences reach values in the range of 1.29–1.62.

8.4 Biomedical Applications Examples

8.4.1 Multifractal Diagnostics of Sepsis in Mice with Different Durations of Mechanical Skin Injury

The objects of the study were blood films from three groups of rats:

- Healthy rats – control group 1, consisting of 22 species.
- Rats with septic infection 6 hours after mechanical skin injury – experimental group 2, consisting of 22 species.
- Rats with septic infection 18 hours after mechanical skin injury – experimental group 3, consisting of 22 species.

The results of statistical and informational analysis of multifractal spectra $F^{\mathrm{FB}}(\varphi_k, h_r)$ for BLF, are presented in Table 8.21.

Table 8.20 Statistical characteristics of multifractal spectra $F^{FBP}(\varphi_k, h_r)$ and $F^{FB}(\varphi_k, h_r)$.

	"Integral"	
$M_{l=1;2;3;4}$	$F^{FBP}(\varphi_k, h_r)$	$F^{FB}(\varphi_k, h_r)$
$M_{l=1}$	1.39 ± 0.065	1.65 ± 0.076
$M_{l=2}$	0.11 ± 0.008	0.14 ± 0.011
$M_{l=3}$	1.47 ± 0.074	0.91 ± 0.047
$M_{l=4}$	2.03 ± 0.092	1.46 ± 0.076
	$\varphi_k = \pi/4$	
$M_{l=1;2;3;4}$	$F^{FBP}(\varphi_k, h_r)$	$F^{FB}(\varphi_k, h_r)$
$M_{l=1}$	1.35 ± 0.061	1.61 ± 0.072
$M_{l=2}$	0.09 ± 0.006	0.11 ± 0.008
$M_{l=3}$	2.07 ± 0.11	1.39 ± 0.087
$M_{l=4}$	2.99 ± 0.16	1.86 ± 0.106
	$\varphi_k = \pi/8$	
$M_{l=1;2;3;4}$	$F^{FBP}(\varphi_k, h_r)$	$F^{FB}(\varphi_k, h_r)$
$M_{l=1}$	1.31 ± 0.055	1.59 ± 0.067
$M_{l=2}$	0.07 ± 0.005	0.09 ± 0.006
$M_{l=3}$	2.54 ± 0.15	1.98 ± 0.11
$M_{l=4}$	3.63 ± 0.19	2.34 ± 0.14

Table 8.21 Statistical characteristics of $F^{FBP}(\varphi_k, h_j)$ and $F^{FB}(\varphi_k, h_j)$ for BLF.

			"Integral"			
$\beta(u,t)$	Group 1	Group 2	Group 3	$Ac_{1;2}$, %	$Ac_{1;3}$, %	$Ac_{2;3}$, %
M_3	1.34 ± 0.079	1.08 ± 0.083	0.84 ± 0.054	83.3	85.7	76.1
M_4	2.41 ± 0.14	2.03 ± 0.11	1.63 ± 0.092	83.3	88.1	78.6
			$\varphi_k = \pi/4$			
$\beta(u,t)$	Group 1	Group 2	Group 3	$Ac_{1;2}$, %	$Ac_{1;3}$, %	$Ac_{2;3}$, %
M_3	2.19 ± 0.12	1.58 ± 0.067	1.25 ± 0.054	83.3	92.3	85.7
M_4	3.01 ± 0.18	2.24 ± 0.16	1.73 ± 0.12	83.3	92.3	88.1
			$\varphi_k = \pi/8$			
$\beta(u,t)$	Group 1	Group 2	Group 3	$Ac_{1;2}$, %	$Ac_{1;3}$, %	$Ac_{2;3}$, %
M_3	3.02 ± 0.18	2.08 ± 0.12	1.57 ± 0.097	92.3	94.6	90.5
M_4	4.17 ± 0.22	3.04 ± 0.15	2.13 ± 0.13	94.6	96.3	92.3

The information analysis data presented in Table 8.21 demonstrate the high efficiency of MWTM algorithms for blood films. High accuracy was achieved not only in diagnostics ($Ac_{1;3} = 94.6 - 96.3\%$) but also in differentiating the severity of sepsis in rats with different extents of mechanically induced injury ($Ac_{2;3} = 90.5\% - 92.3\%$).

Table 8.22 Balanced accuracy of differential diagnosis of thyroid pathology.

	Skewness, SM_3	
$Ac_{12} = 98\%$ $Ac_{23} = 97.1\%$	$Ac_{13} = 98\%$ $Ac_{24} = 98\%$	$Ac_{14} = 99\%$ $Ac_{34} = 99\%$
	Kurtosis, SM_4	
$Ac_{12} = 100\%$ $Ac_{23} = 99\%$	$Ac_{13} = 100\%$ $Ac_{24} = 99\%$	$Ac_{14} = 100\%$ $Ac_{34} = 99\%$

8.4.2 Multifractal Diagnostics of Thyroid Pathology

Four groups of patients were studied:

1. Control group 1 – healthy donors (51 patients).
2. Research group 2 – patients with nodular goiter (51 patients).
3. Research group 3 – patients with autoimmune thyroiditis (51 patients).
4. Research group 4 – patients with papillary cancer (51 patients).

The following levels of balanced accuracy of differential diagnosis of various pathologies were demonstrated:
Diagnosis of thyroid pathology:

- "normal – nodular goiter" ~98–100% – "excellent" level;
- "normal autoimmune thyroiditis" ~98–100% – "excellent" level;
- "normal – papillary cancer" ~99–100% – "excellent" level.

Differential diagnosis (Table 8.22):

- "nodular goiter – autoimmune thyroiditis" ~97–99% – "excellent" level;
- "nodular goiter – papillary cancer" ~98–99% – "excellent" level;
- "autoimmune thyroiditis – papillary cancer" ~99% – "excellent" level.

8.4.3 Conclusions

1. A Jones-matrix model providing an analytical description of the formation of polarization-inhomogeneous object fields by optically anisotropic supramolecular networks of biological crystals is proposed, and the physical relationships between the birefringence parameters of the polycrystalline architectonics of biological fluid films and the polarization maps of the object field components with different light-scattering multiplicities are established.
2. A method of 3D polarization-interference mapping with digital holographic phase scanning of scattered radiation fields is proposed and experimentally validated. Using this approach, layered maps of polarization ellipticity $\beta(u, t)$ and $\beta(\varphi = \pi/4, u, t)$; $\beta(\varphi = \pi/8, u, t)$ are algorithmically determined for the diffuse object fields of biological fluid films with different structures of optically anisotropic polycrystalline architectonics – dendritic supramolecular networks of blood plasma and dendritic-spherulitic supramolecular networks of bile.
3. Within the framework of statistical analysis of the probability distributions of random values of polarization ellipticity, the parameters most sensitive to the structural features of the supramolecular networks of biological crystals are identified and physically justified – namely, the third and fourth-order statistical moments, which characterize the skewness and kurtosis of the distributions $\beta(u, t)$ and $\beta(\varphi = \pi/4, u, t)$; $\beta(\varphi = \pi/8, u, t)$. The range of variation of these statistical parameters across the phase sections of the object field is: FBP – 1.82 – 2.08 times; for FB – 1.85 – 2.77 times. Intergroup differences (FB and FBP) in the values $M_{3;4}(\varphi_k, \beta)$ amount to 1.26 – 1.33 times.
4. A multifractal approach, based on the analysis of the distributions of maxima of the amplitude modules of the wavelet transform, was employed for an additional assessment of another information parameter: the topographic scale-self-similar structure of the polarization maps $\beta(u, t)$ and $\beta(\varphi = \pi/4, u, t)$; $\beta(\varphi = \pi/8, u, t)$ of the object fields of blood plasma and bile films.

5. Multifractal spectra $F^{FBP}(\varphi_k, h_r)$ and $F^{FB}(\varphi_k, h_r)$ of the scale–self-similar structure of the polarization maps were obtained and statistically analyzed across different phase sections of the object fields of blood plasma and bile films. As a result of physical analysis of the obtained data, new parameters most sensitive to changes in the scale self-similarity of various phase components of the object field were identified – namely, the third and fourth orders statistical moments, which characterize the skewness and kurtosis of the $\beta(u, t)$ and $\beta(\varphi = {}^{\pi}/_{4}, u, t)$; $\beta(\varphi = {}^{\pi}/_{8}, u, t)$. The range of variation in these multifractal markers is: for FBP, 1.73 – 1.81 times and for FB, 1.68 – 2.16 times. The intergroup differences reach a magnitude of 1.29 – 1.62.
6. The fundamental results of this multifractal analysis were successfully applied to a practical forensic task – differentiation of mechanical skin injury using the developed method of polarization-interference mapping of rat blood films, achieving excellent accuracy $\left(\mathrm{Ac}_{1;3} = 94.6 - 96.3\%\right)$.

References

1 Daubechies I. Ten Lectures on Wavelets. Society for Industrial and Applied Mathematics; 1992.
2 Cohen A, Daubechies I, Vial P. Wavelets on the interval and fast wavelet transforms. Appl Comput Harmon Anal. 1993;1(1):54–81.
3 Meyer Y. Wavelets: Algorithms & Applications. Philadelphia: Society for Industrial and Applied Mathematics (SIAM); 1993.
4 Mallat S. A theory for multiresolution signal decomposition: the wavelet representation. IEEE Trans Pattern Anal Mach Intell. 1989;11(7):674–693.
5 Daubechies I. Wavelets on the interval. In Progress in Wavelet Analysis and Applications: Proceedings of the International Conference "Wavelets and Applications", Toulouse, France. 1992.
6 Farge M. Wavelet transforms and their applications to turbulence. Annu Rev Fluid Mech. 1992;24(1):395–458.
7 Ma H, He H, Ramella-Roman JC. Mueller matrix microscopy. In Ramella-Roman JC, Novikova T, editors. Polarized Light in Biomedical Imaging and Sensing. Cham: Springer; 2023. 281–321.
8 He H, Liao R, Zeng N, Li P, Chen Z, Liu X, et al. Mueller matrix polarimetry – an emerging new tool for characterizing the microstructural feature of complex biological specimen. J Lightwave Technol. 2019;37:2534–2548.
9 Wang Y, He H, Chang J, Zeng N, Liu S, Li M, et al. Differentiating characteristic microstructural features of cancerous tissues using Mueller matrix microscope. Micron. 2015;79:8–15.
10 Abbasian V, Moradi A-R. Microsphere-assisted super-resolved Mueller matrix microscopy. Opt Lett. 2020;45:4336–4339.
11 Zhang Y, Lee SY, Zhang Y, Furst D, Fitzgerald J, Ozcan A. Wide-field imaging of birefringent synovial fluid crystals using lens-free polarized microscopy for gout diagnosis. Sci Rep. 2016;6:28793.
12 de Boer JF, Milner TE. Review of polarization sensitive optical coherence tomography and Stokes vector determination. J Biomed Opt. 2002;7(3):359–371. https://doi.org/10.1117/1.1483879.
13 Ku S, Mahato KK, Mazumder N. Polarization-resolved Stokes-Mueller imaging: a review of technology and applications. Lasers Med Sci. 2019;34(7):1283–1293. https://doi.org/10.1007/s10103-019-02752-1.
14 He Y, Li K, Li W, Qiu Y, Li D, Wang C, et al. Polarization coherency matrix tomography. J Biophotonics. 2023;16(9): e202300093. https://doi.org/10.1002/jbio.202300093.
15 Yang D, Yuan Z, Hu M, Liang Y. Zebrafish brain and skull imaging based on polarization-sensitive optical coherence tomography. J Biophotonics. 2022;15(12):e202200112. https://doi.org/10.1002/jbio.202200112.
16 Park JE, Xin Z, Kwon DY, Kim SW, Lee H, Jung MJ, et al. Application of polarization sensitive-optical coherence tomography to the assessment of phase retardation in subpleural cancer in rabbits. Tissue Eng Regen Med. 2021;18(1):61–69. https://doi.org/10.1007/s13770-020-00318-9.
17 Willemse J, Gräfe MGO, Verbraak FD, de Boer JF. In vivo 3D determination of peripapillary scleral and retinal layer architecture using polarization-sensitive optical coherence tomography. Transl Vis Sci Technol. 2020;9(11):21. https://doi.org/10.1167/tvst.9.11.21.
18 Killeen AA, Ossina N, McGlennen RC, Minnerath S, Borgos J, Alexandrov V, et al. Protein self-organization patterns in dried serum reveal changes in B-Cell disorders. Mol Diag Ther. 2006;10:371–380.
19 Brutin D, Sobac B, Loquet B, Sampol J. Pattern formation in drying drops of blood. J Fluid Mech. 2011;667:85–95.
20 Heckbert SR, Vedder NB, Hoffman W, Winn RK, Hudson LD, Jurkovich GJ, et al. Outcome after hemorrhagic shock in trauma patients. J Trauma Acute Care Surg. 1998; 45(3):545–549.

21 Johar RS, Smith RP. Assessing gravimetric estimation of intraoperative blood loss. J Gynecol Surg. 1993;9(3):151–154. https://doi.org/10.1089/gyn.1993.9.151.

22 Convertino VA, Lye KR, Koons NJ, Joyner MJ. Physiological comparison of hemorrhagic shock and V̇ O2max: A conceptual framework for defining the limitation of oxygen delivery. Exp Biol Med (Maywood). 2019;244(8):690–701. https://doi.org/10.1177/1535370219846425.

23 Lemmer JH, Despotis GJ. Antithrombin III concentrate to treat heparin resistance in patients undergoing cardiac surgery. J Thorac Cardiovasc Surg. 2002;123:213–217. https://doi.org/10.1067/mtc.2002.119060.

24 Verma AK, Xu D, Garg A, Blaber AP, Tavakolian K. Effect of aging on muscle-pump baroreflex of individual leg muscles during standing. Front Physiol. 2017;8:767. https://doi.org/10.3389/fphys.2017.00767.

25 Swan HJ, Ganz W, Forrester J, Marcus H, Diamond G, Chonette D. Catheterization of the heart in man with use of a flow-directed balloon-tipped catheter. N Engl J Med. 1970;283(9):447–451. https://doi.org/10.1056/NEJM197008272830902.

26 Bercovitz RS, Shewmake AC, Newman DK, Niebler RA, Scott JP, Stuth E, et al. Validation of a definition of excessive postoperative bleeding in infants undergoing cardiac surgery with cardiopulmonary bypass. J Thorac Cardiovasc Surg. 2018;155(5):2112–2124. https://doi.org/10.1016/j.jtcvs.2017.12.038.

27 Jones CR. New calculus for the treatment of optical systems. VII. Properties of the N-matrices. J Opt Soc Am. 1948;38:671–685.

28 Azzam RMA. Propagation of partially depolarized light through anisotropic media with or without depolarization. A differential 4 × 4 matrix calculus. J Opt Soc Am. 1978;68:1756–1767.

29 Arteaga O, Canillas A. Analytic inversion of the Mueller-Jones polarization matrices for homogeneous media. Opt Lett. 2010;35:559–561.

30 Barakat R. Exponential versions of the Jones and Mueller–Jones polarization matrices. J Opt Soc Am A. 1996;13:158–163.

31 Bacry E, Muzy JF, Arneodo A. Singularity spectrum of fractal signals from wavelet analysis: exact results. J Stat Phys. 1993;70(3-4):635–674.

32 Lu SY, Chipman RA. Interpretation of Mueller matrices based on polar decomposition. J Opt Soc Am A. 1996;13:1106–1113.

33 Arteaga O, Canillas A. Pseudopolar decomposition of the Jones and Mueller-Jones exponential polarization matrices. J Opt Soc Am A. 2009;26:783–793.

34 Ossikovski R, De Martino A, Guyot S. Forward and reverse product decompositions of depolarizing Mueller matrices. Opt Lett. 2007;32:689–691.

35 Ortega-Quijano N, Arce-Diego JL. Mueller matrix differential decomposition. Opt Lett. 2011;36:1942–1944.

36 Wood MF, Ghosh N, Moriyama EH, Wilson BC, Vitkin IA. Proof-of-principle demonstration of a Mueller matrix decomposition method for polarized light tissue characterization in vivo. J Biomed Opt. 2009;14:014029.

37 Wang Y, He H, Chang J, He C, Liu S, Li M, et al. Mueller matrix microscope: a quantitative tool to facilitate detections and fibrosis scorings of liver cirrhosis and cancer tissues. J Biomed Opt. 2016;21:071112.

38 Liu T, Lu M, Chen B, Zhong Q, Li J, He H, et al. Distinguishing structural features between Crohn's disease and gastrointestinal luminal tuberculosis using Mueller matrix derived parameters. J Biophotonics. 2019;12:e201900151.

39 Gil JJ. Invariant quantities of a Mueller matrix under rotation and retarder transformations. J Opt Soc Am A. 2016;33:52–58.

40 Iqbal M, Ahmad I, Khaliq A, Khan S. Comparative study of Mueller matrix transformation and polar decomposition for optical characterization of turbid media. Optik. 2020;224:165508.

41 Sun T, Liu T, He H, Wu J, Ma H. Distinguishing anisotropy orientations originated from scattering and birefringence of turbid media using Mueller matrix derived parameters. Opt Lett. 2018;43:4092–4095.

42 Swami MK. Manhas S, Buddhiwant P, Ghosh N, Uppal A, Gupta PK. Polar decomposition of 3 × 3 Mueller matrix: a tool for quantitative tissue polarimetry. Opt Express. 2006;14:9324–9337.

43 Wang Y, Guo Y, Zeng N, Chen D, He H, Ma H. Study on the validity of 3 × 3 Mueller matrix decomposition. J Biomed Opt. 2015;20:065003.

44 Morio J, Goudail F. Influence of the order of diattenuator, retarder, and polarizer in polar decomposition of Mueller matrices. Opt Lett. 2004;29:2234–2236.

45 Ghosh N, Wood MFG, Vitkin IA. Influence of the order of the constituent basis matrices on the Mueller matrix decomposition-derived polarization parameters in complex turbid media such as biological tissues. Opt Commun. 2010;283:1200–1208.

46 Ghosh N, Wood MFG, Vitkin IA. Mueller matrix decomposition for extraction of individual polarization parameters from complex turbid media exhibiting multiple scattering, optical activity, and linear birefringence. J Biomed Opt. 2008;13:044036.

47 Gil JJ. Characteristic properties of Mueller matrices. J Opt Soc Am A. 2000;17:328–334.

48 Ossikovski R. Analysis of depolarizing Mueller matrices through a symmetric decomposition. J Opt Soc Am A. 2009;26:1109–1118.

49 Vizet J, Ossikovski R. Symmetric decomposition of experimental depolarizing Mueller matrices in the degenerate case. Appl Opt. 2018;57:1159–1167.

50 Oldenbourg R. A new view on polarization microscopy. Nature. 1996;381:811–812.

51 Gurjar RS, Backman V, Perelman LT, Georgakoudi I, Badizadegan K, Itzkan I, et al. Imaging human epithelial properties with polarized lightscattering spectroscopy. Nat Med. 2001;7:1245–1248.

52 Ramella-Roman JC, Saytashev I, Piccini M. A review of polarization-based imaging technologies for clinical and preclinical applications. J Opt. 2020;22:123001.

53 Qi J, Elson DS. Mueller polarimetric imaging for surgical and diagnostic applications: a review. J Biophotonics. 2017; 10:950–982.

54 Fujiwara H. Spectroscopic Ellipsometry: Principles and Applications. John Wiley & Sons; 2007.

55 Mendoza-Galván A, Muñoz-Pineda E, Ribeiro SJ, Santos MV, Järrendahl K, Arwin H. Mueller matrix spectroscopic ellipsometry study of chiral nanocrystalline cellulose films. J Opt. 2018;20:024001.

56 Azzam RMA. Stokes-vector and Mueller-matrix polarimetry. J Opt Soc Am A. 2016;33:1396–1408.

57 Azzam RMA. Mueller-matrix ellipsometry: a review. Proc SPIE. 1997;3121:369–405.

58 Song B, Gu H, Zhu S, Jiang H, Chen X, Zhang C, et al. Broadband optical properties of graphene and HOPG investigated by spectroscopic Mueller matrix ellipsometry. Appl Surf Sci. 2018;439:1079–1087.

59 Jiang H, Ma Z, Gu H, Chen X, Liu S. Characterization of volume gratings based on distributed dielectric constant model using Mueller matrix ellipsometry. Appl Sci. 2019;9:698.

60 Chang J, He H, Wang Y, Huang Y, Li X, He C, et al. Division of focal plane polarimeter-based 3 × 4 Mueller matrix microscope: a potential tool for quick diagnosis of human carcinoma tissues. J Biomed Opt. 2016;21:056002.

61 Angelsky PO, Ushenko AG, Dubolazov AV, Sidor MI, Bodnar GB, Koval G, et al. The singular approach for processing polarization-inhomogeneous laser images of blood plasma layers. J Opt. 2013;15(4):044030.

62 Ushenko AG, Dubolazov AV, Ushenko VA, Novakovskaya OY. Statistical analysis of polarization-inhomogeneous Fourier spectra of laser radiation scattered by human skin in the tasks of differentiation of benign and malignant formations. J Biomed Opt. 2016;21(7):071110.

63 Garazdyuk MS, Bachinskyi VT, Vanchulyak OY, Ushenko AG, Dubolazov OV, Gorsky MP. Polarization-phase images of liquor polycrystalline films in determining time of death. Appl Opt. 2016;55(12):B67–B71.

64 Angelskii OV, Ushenko OG, Burkovets DN, Arkhelyuk OD, Ushenko YA. Polarization-correlation studies of multifractal structures in biotissues and diagnostics of their pathologic changes. Laser Phys. 2000;10(5):1136–1142.

65 Yermolenko S, Ushenko A, Ivashko P, Goudail F, Gruia I, Gavrilă C, et al. Spectropolarimetry of cancer change of biotissues. In Ninth International Conference on Correlation Optics. SPIE. 2009;7388:404–410.

66 Jóźwicki R, Patorski K, Angelsky OV, Ushenko AG, Burkovets DN, Ushenko YA. Automatic polarimetric system for early medical diagnosis by biotissue testing. Opt Appl. 2002;32(4):603–612.

67 Ushenko AG, Dubolazov OV, Ushenko VA, Novakovskaya OY, Olar OV. Fourier polarimetry of human skin in the tasks of differentiation of benign and malignant formations. Appl Opt. 2016;55(12):B56–B60.

68 Ushenko AG, Pishak VP. Coherent-domain optical methods: biomedical diagnostics, environmental and material science. In Laser polarimetry of biological tissue: Principles and applications. Kluwer Academic Publishers. 2004.

69 Ushenko AG, Fediv AI, Marchuk YF. Correlation and fractal structure of Jones matrices of human bile secret. In European Conference on Biomedical Optics. 2009;7368:73681Q.

70 Ushenko AG. Depolarization of a laser emission field as the correlation development of its polarization structure. In International Conference on Correlation Optics. 1997;3317:331–339.

71 Bachinskyi V, Boychuk T, Ushenko A. Laser Polarimetry of Biological Tissues and Fluids: Chapter 4. Methods of Spatial-frequency, Singular and Correlation Analysis of Biological Layers Object Fields. LAP Lambert Academic Publishing; 2018.

72 Ushenko AG, Burkovets DN, Yermolenko SB, Arkhelyuk AD, Pishak VP, Yuzko AM, et al. Stokes polarimetry of biotissues. In Fourth International Conference on Correlation Optics. SPIE. 1999;3904:527–533.

73 Ushenko VA, Dubolazov AV, Pidkamin LY, Sakchnovsky MY, Bodnar AB, Ushenko YA, et al. Mapping of polycrystalline films of biological fluids utilizing the Jones-matrix formalism. Laser Phys. 2018;28(2):025602.

74 Ushenko AG, Angelsky PO, Sidor M, Marchuk YF, Andreychuk DR, Pashkovskaya NV. Spatial-frequency selection of complex degree of coherence of laser images of blood plasma in diagnostics and differentiation of pathological states of human organism of various nosology. Appl Opt. 2014;53(10):B172–B180.

75 Olar EI, Ushenko AG, Ushenko YA. Polarization correlation measurements of the phase tomograms of optically anisotropic biofractals. Laser Phys. 2004;14(8):1115–1121.

76 Guminetskiy SH, Ushenko OG, Polyanskiy IP, Motrych AV, Grynchuk FV. The optical method for investigation of the peritonitis progressing process. In Eighth International Conference on Correlation Optics. 2008;7008:626–632.

77 Ushenko AG, Fediv AI, Marchuk YF. Singular structure of polarization images of bile secret in diagnostics of human physiological state. In: European Conference on Biomedical Optics. Optica Publishing Group; 2009.

78 Ushenko Y, Ushenko A, Dubolazov A, Soltys I, Litvinenko O, Wanchuliak O, et al. Mueller-matrix interferometric multifractal scaling of optically anisotropic architectonics of diffuse blood facies: fundamental and applied aspects. J Biophotonics. 2025;18(3):e202400412. https://doi.org/10.1002/jbio.202400412.

79 Ushenko A, Soltys I, Dubolazov A, Ushenko Y, Bilookyi V, Bilookyi O, et al. 3D Jones-matrix thesiography of biological fluid facies. J Innov Opt Health Sci. 2025;18(2):2443002. https://doi.org/10.1142/S1793545824430028.

80 Ushenko A, Pavlyukovich N, Khukhlina O, Pavlyukovich O, Gorsky M, Soltys I, et al. Layer-by-layer multifractal scanning of optically anisotropic architectonics of blood plasma films: fundamental and applied aspects. Photonics. 2025;12:215. https://doi.org/10.3390/photonics12030215.

81 Ushenko A, Pavlyukovich N, Khukhlina O, Pavlyukovich O, Soltys I, Dubolazov A, et al. Blood plasma film multifractal scanning in COVID-19 consequences diagnostics. J Biophotonics. 2024;17(11):e202400356. https://doi.org/10.1002/jbio.202400356.

82 Ushenko A, Dubolazov A, Zheng J, Litvinenko A, Gorsky M, Ushenko Y, et al. 3D polarization-interference holographic histology for wavelet-based differentiation of biological tissues with different necrotic states. J Biomed Opt. 2024;29(5):052920. https://doi.org/10.1117/1.JBO.29.5.052920.

83 Ushenko O, Bilookyi O, Zheng J, Dubolazov A, Olar O, Ushenko Y, et al. 3D digital holographic polarimetry of laser speckle fields formed by polycrystalline blood films: a tool for differential diagnosis of thyroid pathology. Front Phys. 2024;12:1426469. https://doi.org/10.3389/fphy.2024.1426469.

84 Wójcik W, Hu Z, Ushenko Y, Smolarz A, Soltys I, Dubolazov O, et al. Optical sensor system for 3D Jones matrix reconstruction of optical anisotropy maps of self-assembled polycrystalline soft matter films. Sensors. 2024;24(5):1589. https://doi.org/10.3390/s24051589.

85 Ushenko A, Zheng J, Gorsky M, Dubolazov A, Ushenko Y, Soltys I, et al. 3D digital holographic polarimetry of diffuse optically anisotropic biological tissue object fields. Front Phys. 2023;11:1288935. https://doi.org/10.3389/fphy.2023.1288935.

86 Dubolazov A, Ushenko V, Trifonyuk L, Stashkevich A, Soltys I, Ushenko Y, et al. Polarization-singular approach to imaging Mueller-matrix polarimetry in differential diagnosis of histological sections of biopsy of tumors of the uterus and prostate. Front Phys. 2021;9:711212. https://doi.org/10.3389/fphy.2021.711212.

87 Ushenko VO, Trifonyuk L, Ushenko YA, Dubolazov OV, Gorsky MP, Ushenko AG. Polarization singularity analysis of Mueller-matrix invariants of optical anisotropy of biological tissue samples in cancer diagnostics. J Opt. 2021;23(6):064004. https://doi.org/10.1088/2040-8986/abf97a.

88 Ushenko YA, Koval GD, Ushenko AG, Dubolazov OV, Ushenko VA, Novakovskaia OY. Mueller-matrix of laser-induced autofluorescence of polycrystalline films of dried peritoneal fluid in diagnostics of endometriosis. J Biomed Opt. 2016;21(7):071116. https://doi.org/10.1117/1.JBO.21.7.071116.

9

Holographic Scanning of Myocardial Layers

Yurii Ushenko[1,2], Weidong Dou[1], Oleksandr Ushenko[2,3], Oleksandr Dubolazov[2], Iryna Soltys[2], Viktor Bachinskiy[4], Oleh Vanchuliak[4], Marta Garazdyuk[4], Nataliia Pavlyukovich[4], Oleksandr Pavlyukovich[4], Vitaliy Rozhko[4], and Oleksandr Olar[2]

[1] Shaoxing University, Shaoxing, China
[2] Yurii Fedkovych Chernivtsi National University, Chernivtsi, Ukraine
[3] Taizhou Institute of Zhejiang University, Taizhou, China
[4] Bukovinian State Medical University, Chernivtsi, Ukraine

9.1 Three-dimensional Digital Polarization-holographic Wavelet Histology in Determining the Duration of Mechanical Damage to the Myocardium

9.1.1 Objects of Investigations

The following groups of histological myocardium samples were created [1–7]:

- Duration of mechanical injury (6 hours) – Group 1, 21 samples.
- Duration of mechanical injury (18 hours) – Group 2, 21 samples.

From a morphological perspective, the myocardium features spatially organized, linearly birefringent networks of myosin fibrils. The duration of mechanical injury leads to different necrotic effects on these fibers and their structured networks. In cases of prolonged injury (Group 2), myosin fibers become thinner, and their organized network decreases. Conversely, short-term injury (Group 1) causes rupture in specific areas of the myosin fibers while maintaining the overall spatial order of the network.

The optical–geometric parameters of the histological section samples are shown in Table 9.1.

9.1.2 Wavelet Differentiation of the Myocardium Histological Sections Azimuth Polarization Map

Figure 9.1 shows the results of the wavelet analysis of $W(\omega_k, b)$ maps of the polarization azimuth $\alpha(\theta, m, n)$ in the phase $(\theta = \pi/8)$ myocardium histological sections [8–12].

The individual modulation depth of the value of amplitudes $\omega_k(a = 15)$ and $\omega_k(a = 57)$ of the wavelet coefficients ω_k of the myocardial microscopic image's polarization map $\alpha(\theta, m, n)$ (phase cross-section $(\theta = \pi/8)$) from both groups was revealed (Table 9.1).

For the case of long-term injury (18 hours), the minimum amplitude values of the wavelet coefficients ω_k and correspond to all scales of MHAT, which scans the polarization map $\alpha(\theta, m, n)$ in the phase plane $\theta = \pi/8$.

For the case of a "short-term" injury (6 hours), the inverse tendency is established – the presence of the amplitude maximum values and spread of the wavelet coefficients ω_k values in all scales in the phase plane $\theta = \pi/8$.

Figure 9.2 and Table 9.2 show the dependences of the mean values $M_1(\omega_k)$ and the variance $M_1(\omega_k)$, which characterize the multiscale dependences of the wavelet coefficients amplitudes $\omega_k(a = 15)$ and $\omega_k(a = 57)$ for myocardium histological sections maps $\alpha(\theta = {}^{\pi}/_{8}, m, n)$ the from different groups.

Table 9.1 Optical and geometric parameters of myocardium histological section samples.

Parameters	Myocardium
Geometrical thickness, h, μm	40–45
Coefficient of attenuation, τ μm	0.13–0.17
Depolarization degree, Δ, %	27–33

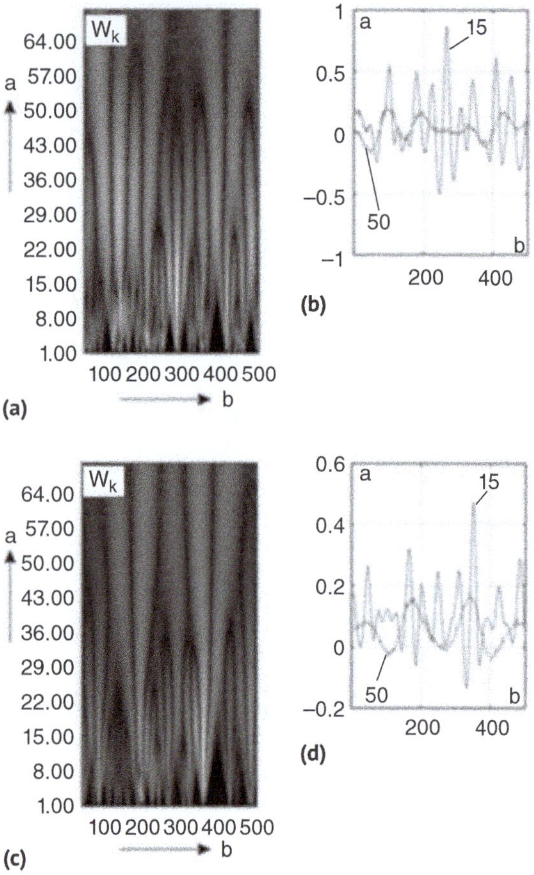

Figure 9.1 Maps (a,c) and multiscale linear cross-section distributions (b,d) of polarization azimuth wavelet coefficients in the phase section $\theta = \pi/8$ of microscopic images of myocardium histological section samples with different mechanical injury durations (6 and 18 hours).

The following differential diagnostic accuracy values have been established [13–19]:

- for small scales, $a_{min} = 15$ is a very good level of $Ac(M_1) = 92.3\%$ and excellent $Ac(M_2) = 100\%$ level;
- for large scales, $a_{max} = 75$ is the maximum accuracy level of $Ac(M_{1,2}) = 100\%$.

The results are likely due to morphological alterations in the myocardium. Mechanical injury causes spatial disorganization of myosin fibers, which is further amplified by necrotic changes over time at various scales of the birefringent fibrils.

Optically, the transformations are indicated according to the model representations developed by us they are manifested in an increase in the orientation range of the optical axes $\rho(m, n)$, which is the "predominant" factor in the formation of single-fold $\Lambda(\alpha_j; \rho_j)$ and "small-fold" $\Lambda^n\left(\sum_{j=1}^{n} q(\rho_j)\right)$ scattered components of the object field of samples of histological sections of the traumatically damaged myocardium [20–28].

Consequently, in myocardial samples subjected to prolonged mechanical injury, the largest changes in polarization azimuth occur, along with corresponding variations in wavelet amplitudes of the ω_k coefficients occur at all fibrillar grid scales.

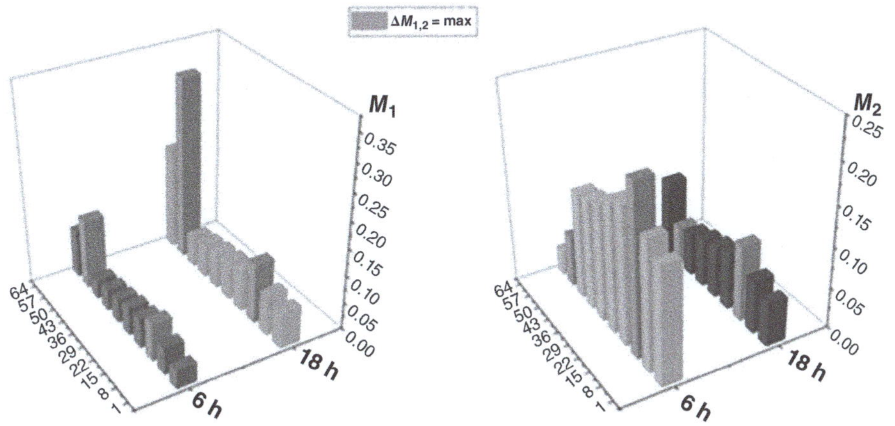

Figure 9.2 Scale dependences of the mean value M_1 and the variance M_2 of microscopic images polarization azimuth wavelet coefficients distributions of myocardium histological sections in phase section $\theta = \pi/8$.

Table 9.2 Mean M_1 and variance M_2 of the polarization azimuth multiscale wavelet coefficients in the phase section $\theta = \pi/8$.

	$a_{min} = 15$		
M_i	Group 1	Group 2	Ac, %
M_1	0.07 ± 0.004	0.11 ± 0.006	92.3
M_2	0.094 ± 0.005	0.21 ± 0.011	100
	$a_{max} = 57$		
M_i	Group 1	Group 2	Ac, %
M_1	0.064 ± 0.004	0.13 ± 0.007	100
M_2	0.095 ± 0.005	0.34 ± 0.018	100

Statistically, this scenario leads to significant differences between the mean and variance values and can be characterized by the following equations $\begin{cases} M_1(T = 6 \text{ hours}; \omega_k(\alpha_i; \rho_i)) < M_1(T = 18 \text{ hours}; \omega_k(\alpha_i; \rho_i)); \\ M_2(T = 6 \text{ hours}; \omega_k(\alpha_i; \rho_i)) < M_2(T = 18 \text{ hours}; \omega_k(\alpha_i; \rho_i)). \end{cases}$

9.1.3 Wavelet Differentiation of the Myocardium Histological Sections Polarization Ellipticity Map

Figure 9.3 shows distributions $\beta(\theta, m, n)$ wavelet maps $W(\omega_k, b)$ in the phase ($\theta = \pi/8$) plane of the myocardium histological sections microscopic images from both groups.

In the phase plane $\theta = \pi/8$ of the polarization ellipticity map $\beta(\theta, m, n)$ for myocardial samples from Group 2, the minimum level of amplitude and depth of modulation of the wavelet coefficients ω_k at all scales of the scanning MHAT function is established. Within the framework of forensic information analysis, excellent accuracy levels (96.7–100%) of differential diagnosis of the duration of mechanical myocardial injury were established for all scanning MHAT function scales a_i [29–37].

The results of the polarization maps $\beta(\theta = \pi/8, m, n)$ wavelet differentiation are related to how mechanical damage to the myocardium causes not only spatial disorganization of myosin fibers, but also stretching and thinning of birefringence fibrils.

Statistical analysis of the wavelet coefficients amplitudes $\omega_k(a = 15)$ and $\omega_k(a = 57)$ in the phase maps $\beta(\theta = \pi/8, m, n)$ are shown in Figure 9.4 and in Table 9.3.

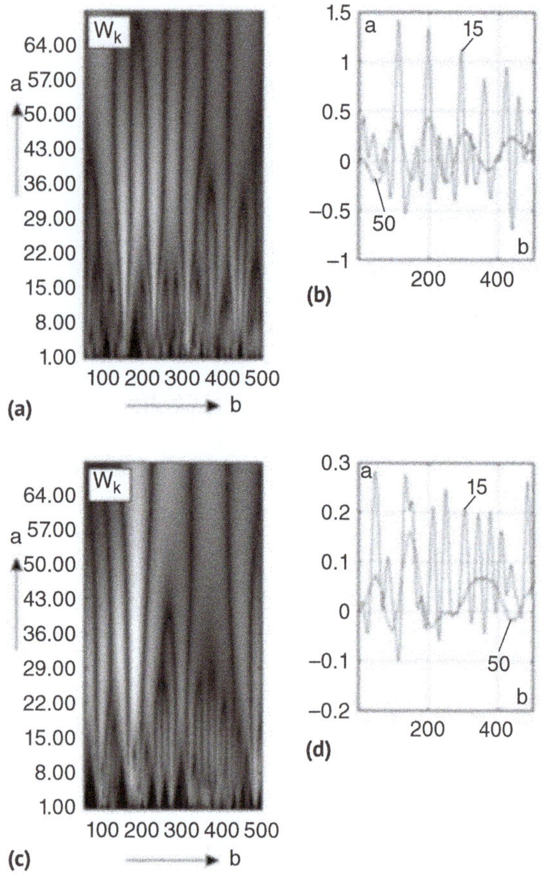

Figure 9.3 Maps (a,c) and multiscale linear cross-section distributions (b,d) of polarization ellipticity wavelet coefficients in the phase section $\theta = \pi/8$ of microscopic images of myocardium histological section samples with different mechanical injury durations (6 and 18 hours).

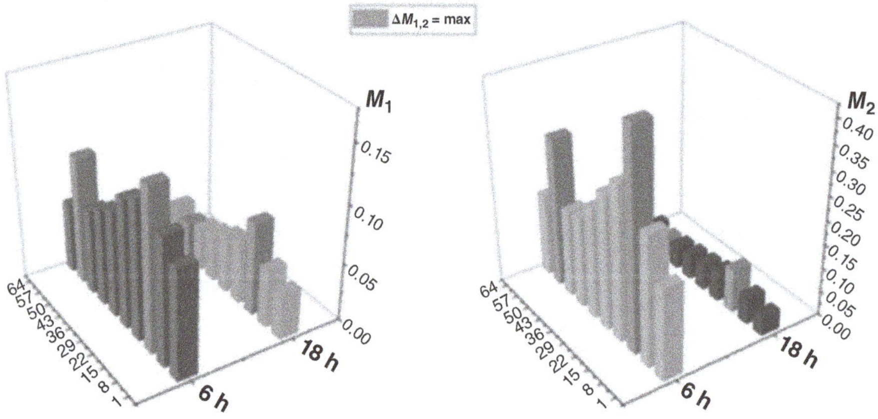

Figure 9.4 Scale dependences of the mean value M_1 and the variance M_2 of microscopic images polarization ellipticity wavelet coefficients distributions of myocardium histological sections in phase section $\theta = \pi/8$.

Optically, changes in the morphological structure of myocardial fibrillar networks are evident, and based on our model representations, these changes are reflected in a reduction of the phase modulation depth $\delta(m, n)$ between the orthogonal components of the E_x and E_y amplitudes for myocardial samples with a longer duration of mechanical injury [38–44].

As a result, for myocardial samples from Group 2, at all fibrillar grid scales, there is a minimal modulation of the polarization ellipticity $\beta(\theta = \pi/8, m, n)$ and, accordingly, variations in the wavelet coefficients ω_k amplitudes.

Table 9.3 Mean M_1 and variance M_2 in the phase section $\theta = \pi/8$ of microscopic images ellipticity polarization wavelet coefficients distributions of myocardium histological sections.

	$a_{min} = 15$		
M_i	**Group 1**	**Group 2**	**Ac, %**
M_1	0.14 ± 0.008	0.08 ± 0.0005	92.3
M_2	0.45 ± 0.026	0.09 ± 0.005	100
	$a_{max} = 57$		
M_i	**Group 1**	**Group 2**	**Ac, %**
M_1	0.11 ± 0.006	0.04 ± 0.0002	96.7
M_2	0.31 ± 0.016	0.05 ± 0.0003	100

Statistically, this scenario leads to significant differences between the mean and variance values and can be characterized by the following equations $\begin{cases} M_1(T = 6\text{ hours}; \omega_k(\beta_i; \delta_i)) > M_1(T = 18\text{ hours}; \omega_k(\beta_i; \delta_i)); \\ M_2(T = 6\text{ hours}; \omega_k(\beta_i; \delta_i)) > M_2(T = 18\text{ hours}; \omega_k(\beta_i; \delta_i)). \end{cases}$

9.2 Three-dimensional Polarization–Interference Holographic Histology for Wavelet-based Differentiation of the Polycrystalline Component of Biological Tissues with Different Necrotic States

9.2.1 Objects of Investigations

To carry out the detailed study of myocardium histological section samples [45–49]:

- coronary heart disease (CHD) – Group 1, 12 samples;
- acute coronary insufficiency (ACI) – Group 2, 12 samples.

The second two groups included lung tissue histological sections from individuals who died of myocardial infarction and had the following concurrent pathologies:

- bronchial asthma (BA) – Group 3, 15 samples;
- pulmonary fibrosis (PF) – Group 4, 15 samples.

Figure 9.5 displays microscopic images of all groups of histological samples, including myocardium (CHD, Figure 9.5a; ACI, Figure 9.5b) and lung tissue (BA, Figure 9.5c; PF, Figure 9.5d).

Analysis of myocardial images from both groups shows the fibrillar structure of myosin fibers (Figure 9.5a, b). A tissue sample with ACI displays a well-organized fibrillar network with average geometric dimensions of 15 to 20 microns (Figure 9.5b).

The myocardial fibrillar network with CHD is disorganized and composed of thinner (10 microns) myosin fibrils (Figure 9.5a).

Analysis of lung tissue images with BA shows a weakly organized parenchymal structure (Figure 9.5c). The tissue image with PF displays a formed fibrillar connective tissue component with an average fiber size of 5–10 microns (Figure 9.5d).

This selection of objects relates to both the basic and applied aspects of this research.

The fundamental component varies physically, as the chosen tissues exhibit different levels of optical anisotropy. The myocardium features spatially organized, linearly birefringent networks of myosin fibrils. Meanwhile, lung tissue is mainly parenchymal, with weak linear birefringence present in the collagen fibrillar networks of its connective tissue [50–54].

Cases of CHD and ACI cause various necrotic changes in the optical anisotropy of myosin fibers and their structured networks. In CHD, myosin fibers become thinner, and the organization of the fibrillar networks declines (Figure 9.5a), leading to a decrease in structural anisotropy. In ACI, myosin fibers rupture in specific areas, but the overall organization

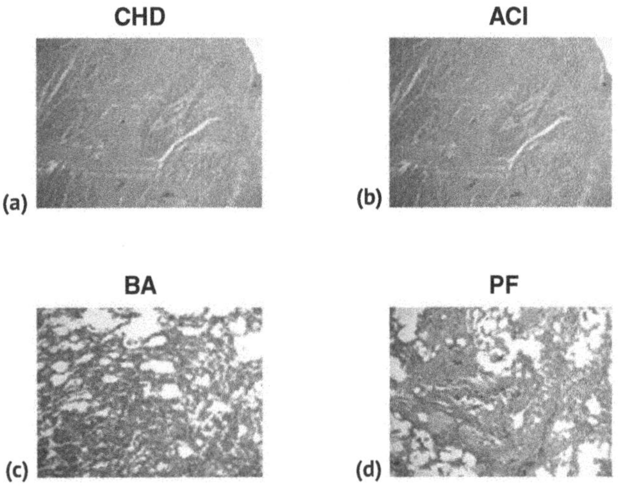

Figure 9.5 Microscopic images (×40) of histological samples of myocardium and lung tissue. (a) CHD; (b) ACI; (c) BA; (d) PF.

Table 9.4 Optical and geometric parameters of histological section tissues samples of both types.

Parameters	Myocardium	Lung tissue
Geometrical thickness, h, μm	50–60	50–60
Coefficient of attenuation, τ μm	0.18–0.22	0.19–0.23
Depolarization degree, Δ, %	47–53	54–59

of the myosin network remains intact (Figure 9.5b). Consequently, the level of birefringence stays similar to the anisotropy parameter of healthy tissue. BA does not significantly change the architecture of the collagen fibrillar networks in the connective tissue component (Figure 9.5c). PF is characterized by substantial growth of collagen fibers and an increase in birefringence (Figure 9.5d).

Therefore, a comparative physical analysis of the results will help assess this method's ability to identify changes in the optical anisotropy of biological tissue samples with varying morphologies.

The component in question is the determination of whether myocardial death was due to a natural cause, such as CHD, or a violent cause like ACI, which remains an important yet unresolved forensic challenge when using light microscopy. The primary difficulty in histological diagnosis is the high level of diffuse background present in images of myocardial tissue sections.

Table 9.4 shows the optical and geometric parameters of the diffuse histological section samples.

9.2.2 Wavelet Differentiation of the Azimuth Polarization Map of the Myocardium

Figure 9.6 shows maps and histograms of the distributions of random azimuth values of the polarization of object fields in the myocardium from Groups 1 and 2.

The data analysis revealed a coordinate-heterogeneous (Figure 9.6a, c) and statistically distributed (Figure 9.6b, d) structure of polarization azimuth maps for the object fields of myocardial samples from both groups. Topologically, these maps consist of local domains with varying geometric sizes (Figure 9.6a, c). The scales of these polarization domains differ across the object fields from different groups, likely due to the high correlation between the scales of myocardial fibrillar structures and polarization domains in the phase plane. Although the histograms of distribution (Figure 9.6b, d) are quite similar in their extremal points and the ranges of polarization azimuth variation, the statistical distributions do not reveal information about the large-scale (topological) structure of necrotic changes in myocardial architecture [55–61].

Figure 9.7 presents the results of wavelet analysis of the polarization azimuth $\alpha(\delta_t^\star \leq {}^{\pi}/_{8}, m, n)$ maps of the myocardium histological sections microscopic images. This drawing consists of two parts.

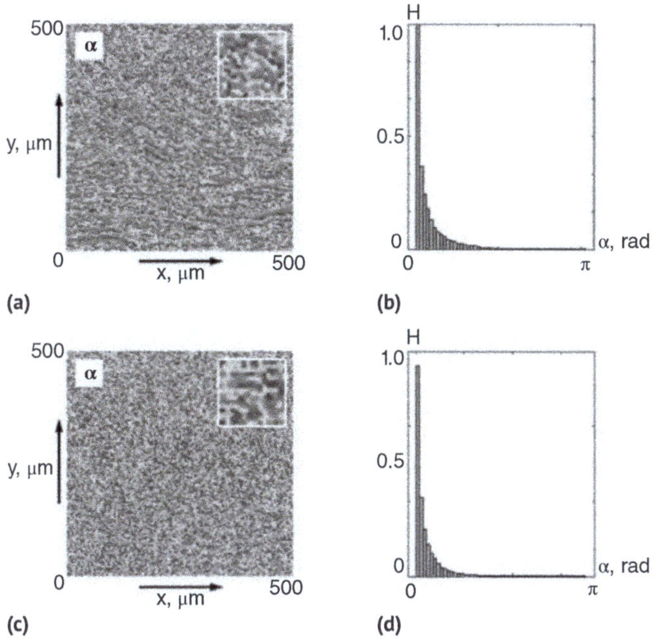

Figure 9.6 Coordinate (a), (b) and probabilistic (c), (d) distributions of random values of the azimuth of polarization of object fields of myocardial samples of deceased as a result of ACI (a), (c) and CHD (b), (d).

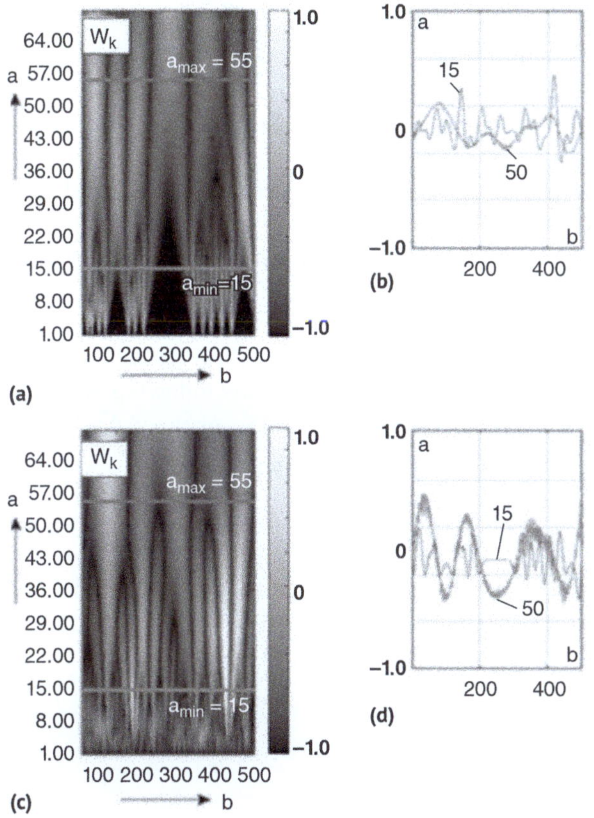

Figure 9.7 Maps (a,c) and multiscale linear cross-sections (b,d) of the polarization azimuth $\alpha(\delta_i^\star = \pi/8, m, n)$ wavelet coefficients $W_{a,b}$ of myocardium histological sections of those who died from CHD (a,b and c,d) and ACI (c,d).

The first part (panels (a), (c)) represents a two-dimensional (2D) array of values of the amplitudes of the wavelet coefficients. It is formed as a result of the algorithmic (ratios (24)–(26)) line-by-line scanning of MHAT distributions by a variable-scale function $\Psi\left(\frac{x-b}{a}\right)$.

The second part (panels (b), (d)) illustrates the linear distributions of wavelet coefficient amplitudes at two scales of the MHAT function.

The data show that the 2D amplitude distributions of the wavelet coefficients represent a fluctuating surface (Figure 9.7a, c). The amplitudes and the oscillation period are distributed differently for different scales a of the MHAT function $\Psi\left(\frac{x-b}{a}\right)$. As the scale increases ($a \uparrow$), the oscillation period of the linear distributions of the wavelet coefficients W_b increases. The amplitude modulation depth of the wavelet coefficients W_b is individual for different scales (Figure 9.7b, d).

The observed patterns can be explained by considering that the wavelet coefficient amplitude at each scan point depends on the mutual correlation of the geometric parameters of the polarization domain (Figure 9.6a, c) and the size of the MHAT function's "window" $\Psi\left(\frac{x-b}{a}\right)$. The greater the cross-correlation, the higher the amplitude of the wavelet coefficient.

At small scales on the polarization maps, wavelet correlations are evident for the detailed elements of myocardial fibrillar architecture. At larger scales, changes in the wavelet coefficient amplitudes reflect the degree of organization of the large-scale fibers within the fibrillar network.

In cases of death caused by ACI, the fibrillar myosin network appears more ordered and extensively structured compared to the myocardium architecture seen in individuals who died of CHD (Figure 9.6a, b). As a result, the wavelet transform of the polarization maps of the myocardial field for samples from group 2 shows higher amplitude values and increased modulation depth of their variations (Figure 9.7a–d).

The differences between the polarization maps of the myocardial object fields were quantified in two ways. The first involved directly calculating statistical moments that describe the distributions (see Table 9.5).

From the analysis of the data obtained, it can be seen that the maximum accuracy of the differential diagnosis of CHD and ACI practically achieved only a satisfactory level $\text{Ac}(M_{3,4} = 79.2\%)$.

The second is a method for estimating fluctuations in the amplitude of the wavelet coefficients by calculating statistical parameters (mean $M_1(W_{a,b})$ and variation $M_2(W_{a,b})$) for all scales of the MHAT function (Figure 9.8).

The differences between $M_1(W_{a,b})$ and $M_2(W_{a,b})$ were maximal (Table 9.6).

From the obtained data, it was revealed that individual modulation of the amplitude values $W_b(a_{\min} = 15)$ and $W_b(a_{\max} = 55)$.

For the case of CHD, the maximum amplitude and variations in the wavelet value of the $W_{a,b}$ coefficients occur on large scales of the wavelet function scanning in the phase plane $\delta_t^\star = {}^{\pi}/_{8}$ of the $\alpha\left(\delta_t^\star = {}^{\pi}/_{8}, m, n\right)$ map.

For the case of ACI, the maximum amplitude and variations in the wavelet value of the $W_{a,b}$ coefficients occur on small scales of the wavelet function scanning in the phase plane $\delta_t^\star = {}^{\pi}/_{8}$ of the $\alpha\left(\delta_t^\star = {}^{\pi}/_{8}, m, n\right)$ map.

Analysis of statistical estimation of fluctuations in the amplitudes of multiscale wavelet coefficients $W_{a,b}$ distributions of the polarization azimuth $\alpha\left(\delta_t^\star = {}^{\pi}/_{8}, m, n\right)$ revealed next equations $M_{1,2}(\text{ACI}, W_b(a_{\min} = 15)) \ll M_{1,2}(\text{CHD}, W_b(a_{\min} = 15))$ and $M_{1,2}(\text{ACI}, W_b(a_{\max} = 55)) < M_{1,2}(\text{CHD}, W_b(a_{\max} = 55)$. As a result, the following levels of accuracy of differential diagnosis of cases were determined CHD and ACI. For small scales, $a_{\min} = 15$- very good $\text{Ac}(M_1) = 91.7\%$ and excellent $\text{Ac}(M_2) = 95.8\%$ are equal. For large scales, $a_{\max} = 50$ is a good level of $\text{Ac}\left(M_{1,2}\right) = 87.5\%$.

9.2.3 Wavelet Differentiation of the Ellipticity Polarization Maps of Myocardium

Figure 9.9 displays maps and histograms of the distributions of random values of the ellipticity magnitude of the polarization of object fields in histological sections of the myocardium from Groups 1 and 2 [62–66].

As an initial reference for the ellipticity of polarization $\left(\beta^{j=1}\left(\delta_t^\star = {}^{\pi}/_{8}\right)\right)$, we used the equation $\otimes \leftrightarrow \beta^{j=1} = 0$.

Table 9.5 Statistical moments of the polarization azimuth maps $\alpha\left(\delta_t^\star = {}^{\pi}/_{8}, m, n\right)$.

$M_{1,2,3,4}$	CHD	ACI	Ac, %
M_1	0.07 ± 0.005	0.09 ± 0.006	70.8
M_2	0.04 ± 0.003	0.05 ± 0.003	75
M_3	1.73 ± 0.095	1.29 ± 0.067	79.2
M_4	2.39 ± 0.13	2.11 ± 0.12	79.2

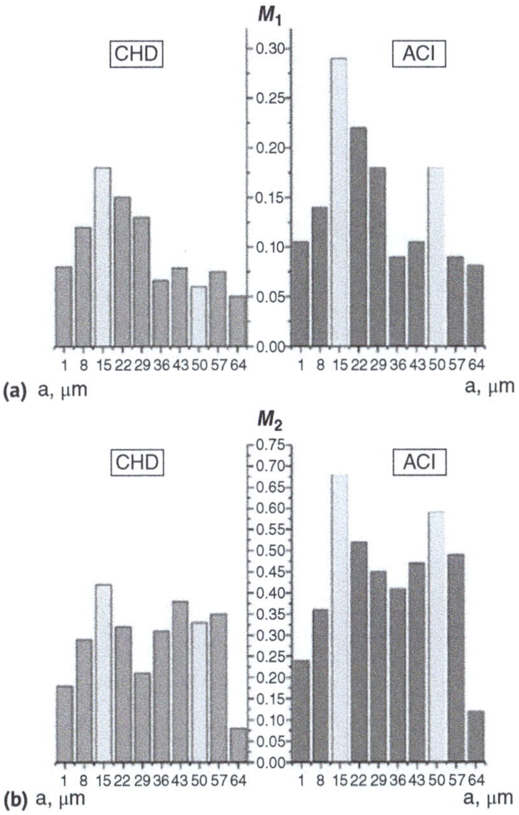

Figure 9.8 Scale dependences of the mean M_1 (a) and the variance M_2 (b) of the wavelet coefficients $W_{a,b}$ of the $\alpha(\delta_t^\star = \pi/8, m, n)$.

Table 9.6 Mean M_1 and variance M_2 of the wavelet coefficients $W_{a,b}$ of the $\alpha(\delta_t^\star = \pi/8, m, n)$.

	$a_{\min} = 15$		
M_i	**CHD**	**ACI**	**Ac, %**
M_1	0.18 ± 0.01	0.29 ± 0.016	91.7
M_2	0.46 ± 0.025	0.68 ± 0.036	95.8
	$a_{\max} = 50$		
M_i	**CHD**	**ACI**	**Ac, %**
M_1	0.09 ± 0.005	0.12 ± 0.007	87.5
M_2	0.43 ± 0.023	0.55 ± 0.029	87.5

Similar to the polarization azimuth maps (Figure 9.6), the coordinate distributions of polarization ellipticity were found to be uneven across the coordinates (Figure 9.9a, b) and exhibit a statistical pattern (Figure 9.9b, d). The polarization domain sets in the ellipticity maps are unique to the object fields of myocardial samples from different groups. However, the statistical distributions of the random polarization ellipticity values do not reveal information about the large-scale (topological) structure of necrotic changes in the organization of the myocardium's birefringent fibrillar networks.

Figure 9.10 shows the results of the application wavelet analysis technique $W(a, b)$ for maps of the ellipticity $\beta(\delta_t^\star = \pi/8, m, n)$.

Were obtained fluctuating surfaces of the amplitudes of the wavelet coefficients $W_{a,b}$ (Figure 9.10a, c). The amplitudes and the oscillation period of the linear distributions of the wavelet coefficients increase as the scale a of the MHAT function $\Psi\left(\frac{x-b}{a}\right)$ increases.

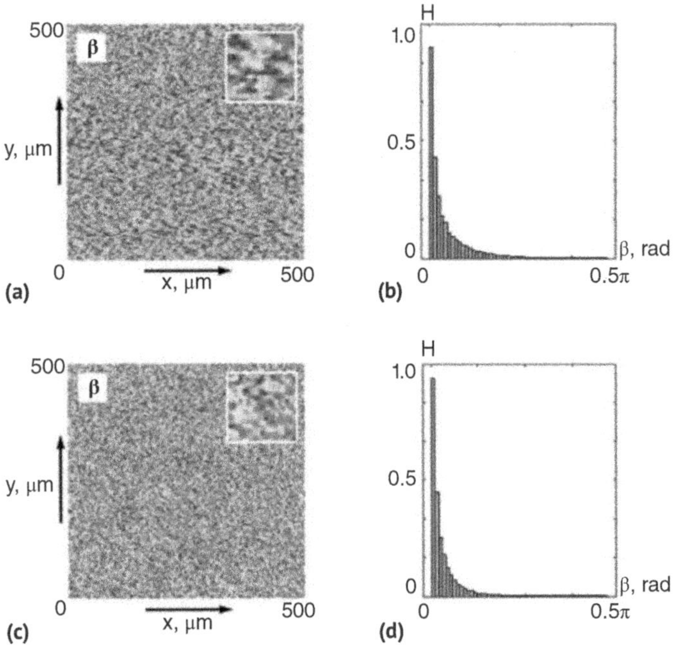

Figure 9.9 Coordinate (a), (c) and probabilistic (b), (d) distributions of random values of the magnitude of the ellipticity of polarization of object fields of myocardial samples that died as a result of CHD (a), (b) and ACI (c), (d).

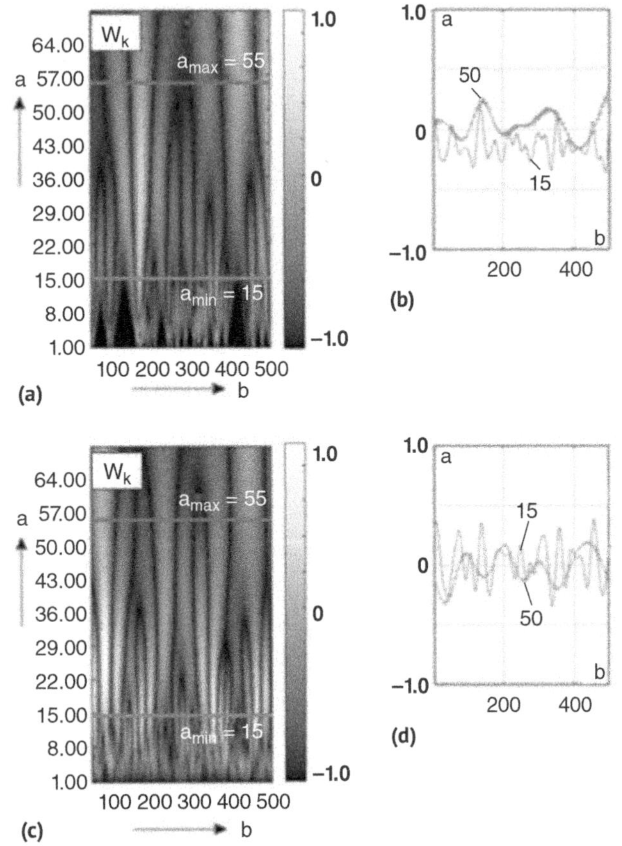

Figure 9.10 Maps (a,c) and multiscale linear cross-sections (b,d) of the polarization ellipticity $\beta(\delta_t^\star = \pi/8, m, n)$ wavelet coefficients $W_{a,b}$ of myocardium histological sections of those who died from CHD (a,b) and ACI (c,d).

The extent of modulation in the amplitudes of wavelet coefficients within the polarization ellipticity maps (Figure 9.10) varies across different scales (see Figure 9.10b, d).

Table 9.7 shows the results of a statistical analysis of the polarization ellipticity distributions in myocardial samples from the CHD and ACI groups.

It was found that the maximum accuracy of differential diagnostics of CHD and ACI exceed a satisfactory level $Ac(M_4 = 83.3\%)$.

At all levels of the polarization ellipticity maps, the wavelet correlations depend on the degree of structural anisotropy or birefringence. The magnitude of this optical anisotropy parameter relates to how well the fibers are spatially organized within the fibrillar network. Therefore, in cases of death caused by ACI, more amplitude values are produced, and the extent of their variation modulation is greater (Figure 9.7a–d).

Was quantified such amplitude fluctuations of the wavelet coefficients $W_{a,b}$ by calculating statistical parameters $(M_1(W_{a,b})$ and $M_2(W_{a,b})$ for all scales a of the MHAT function $\Psi\left(\frac{x-b}{a}\right)$ (Figure 9.11, Table 9.8).

High levels of differential diagnosis accuracy have been established for small scales $a_{min} = 15$ is an excellent level of $Ac(M_1) = 95.8\%$ and $Ac(M_2) = 100\%$.

For large scales $a_{max} = 50$ is a very good level of $Ac(M_{1,2}) = 91.7\%$.

Table 9.7 Statistical moments of polarization ellipticity maps of the $\beta(\delta_t^\star = \pi/8, m, n)$.

$M_{1,2,3,4}$	CHD	ACI	Ac, %
M_1	0.11 ± 0.007	0.13 ± 0.008	70.8
M_2	0.07 ± 0.005	0.09 ± 0.006	75
M_3	1.21 ± 0.069	0.93 ± 0.057	79.2
M_4	1.87 ± 0.099	1.36 ± 0.079	83.3

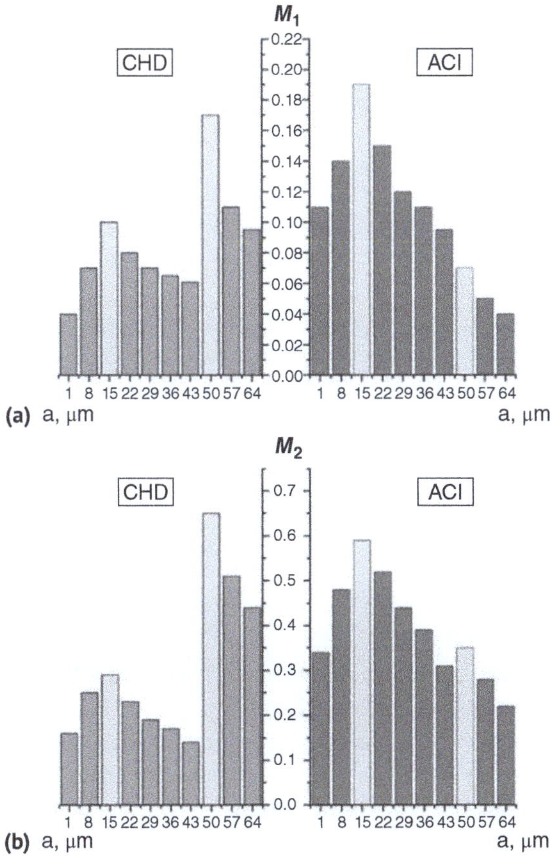

Figure 9.11 Scale dependences of the mean M_1 (a) and the variance M_2(b) of the wavelet coefficients $W_{a,b}$ of the $\beta(\delta_t^\star = \pi/8, m, n)$.

Table 9.8 Mean M_1 and variance M_2 of wavelet coefficients of polarization ellipticity maps $\beta(\delta_i^{\star} = \pi/8, m, n)$.

$a_{min} = 15$			
M_i	CHD	ACI	Ac, %
M_1	0.11 ± 0.006	0.18 ± 0.01	95.8
M_2	0.31 ± 0.016	0.59 ± 0.031	100
$a_{max} = 50$			
M_i	CHD	ACI	Ac, %
M_1	0.08 ± 0.005	0.14 ± 0.008	91.7
M_2	0.39 ± 0.021	0.62 ± 0.033	91.7

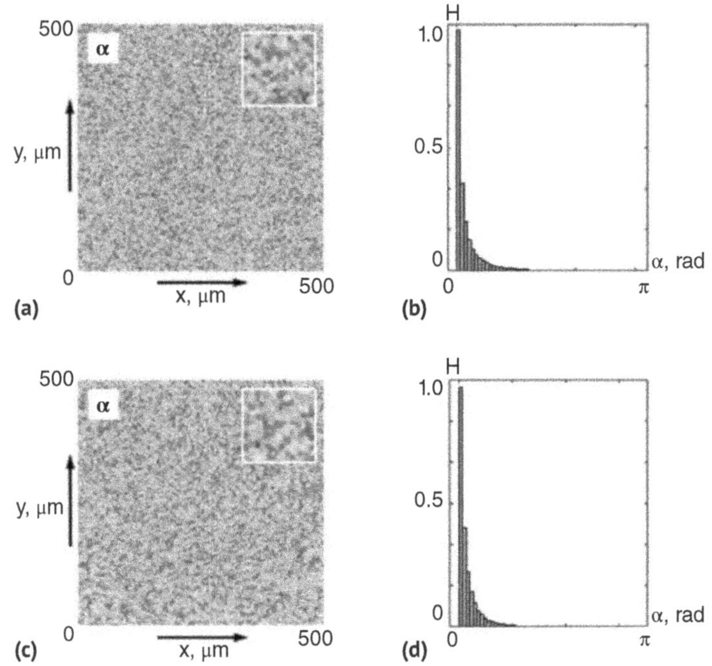

Figure 9.12 Coordinate (a), (c) and probabilistic (b), (d) distributions of the random values of polarization azimuth in the object fields of lung tissue samples with BA (a), (b) and PF (c), (d).

9.2.4 Wavelet Differentiation of the Polarization Azimuth Maps of Lung Tissue

Figure 9.12 presents maps and histograms illustrating the distributions of random polarization azimuth values in the object fields of histological sections of parenchymal lung tissue from both Groups 3 and 4.

The data show that the polarization azimuth maps for the lung tissue samples from both groups have a coordinate-inhomogeneous structure (see Figure 9.12a, c) and are statistically distributed (see Figure 9.12b, d).

Similar to myocardial object fields (Figures 9.6 and 9.9), azimuth polarization maps consist of local domains with varying geometric sizes (Figure 9.12a, c). The histograms of these distributions (Figure 9.12b, d) show comparable patterns in the placement of extreme values and the ranges of variation for the random polarization azimuth values.

Table 9.9 Statistical moments of the polarization azimuth maps of the $\alpha(\delta_t^\star = \pi/8, m, n)$.

$M_{1;2;3;4}$	BA	PF	Ac, %
M_1	0.09 ± 0.01	0.11 ± 0.016	66.7
M_2	0.07 ± 0.025	0.08 ± 0.036	63.3
M_3	1.12 ± 0.005	0.94 ± 0.007	73.3
M_4	1.71 ± 0.023	1.49 ± 0.029	76.7

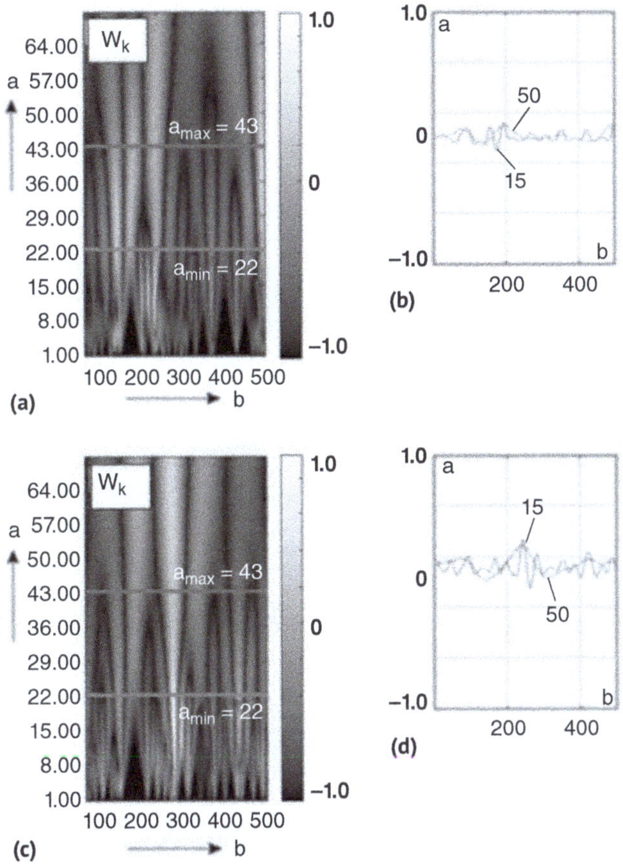

Figure 9.13 Maps (a,c) and multi-scale linear cross-sections (b,d) of the polarization azimuth $\alpha(\delta_t^\star = \pi/8, m, n)$ wavelet coefficients $W_{a,b}$ of lung tissue histological sections those who died from BA (a,b) and PF (c,d).

As a result, the differences between the values of the statistical moments of the first and fourth orders $M_{1;2;3;4}$ are insignificant and do not exceed 15–25% (Table 9.9).

Figure 9.13 presents the results of wavelet analysis of polarization maps $\alpha(\delta_t^\star = \pi/8, m, n)$ for BA and PF cases.

Analysis of the obtained data shows that the maximum accuracy of differential diagnosis between CHD and ACI does not reach a satisfactory level, with $\mathrm{Ac}(M_{1,2,3,4} < 80\%)$. Individual fluctuations in the amplitude of the wavelet coefficients at all scales of the salt-like MHAT function were observed (Figure 9.13a, c).

Similar to wavelet decomposition of myocardial polarization maps, the amplitude at each scan point is based on the mutual correlation of the polarization domain's geometric parameters (Figure 9.12a, c) and the size of the MHAT function's "window."

In BA, the pulmonary parenchyma shows minimal optical birefringence (Figure 9.12c). In contrast, PF exhibits increased structural anisotropy because of connective tissue proliferation (Figure 9.12d).

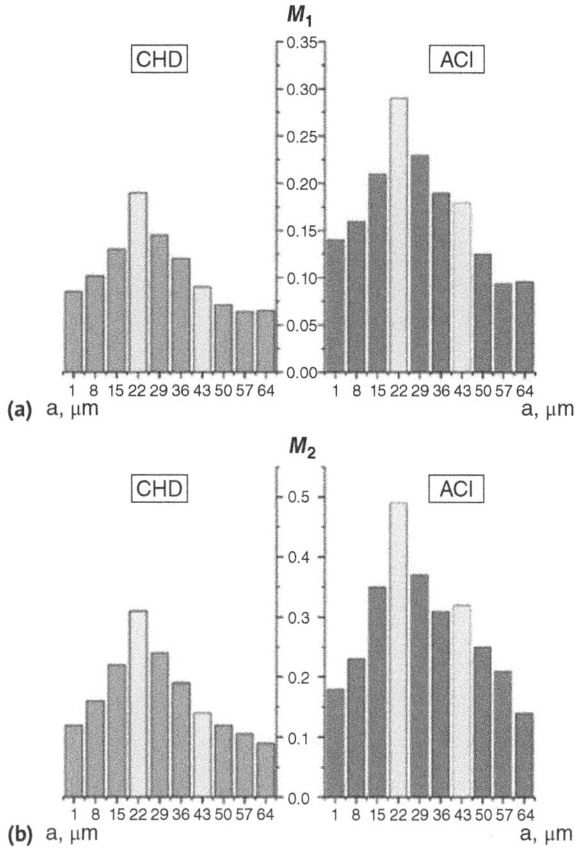

Figure 9.14 Coordinate (a), (c) and probabilistic (b), (d) distributions of random values of the magnitude of the ellipticity of polarization of object fields of lung tissue samples of deceased as a result of BA (a), (b) and PF (c), (d).

Table 9.10 Mean M_1 and the variance M_2 of the wavelet coefficients distributions for polarization azimuth maps $\alpha(\delta_t^\star = \pi/8, m, n)$.

	$a_{min} = 22$		
M_i	BA	PF	Ac, %
M_1	0.19 ± 0.01	0.29 ± 0.016	90
M_2	0.31 ± 0.017	0.49 ± 0.027	93.3
	$a_{max} = 43$		
M_i	BA	PF	Ac, %
M_1	0.09 ± 0.005	0.16 ± 0.09	93.3
M_2	0.14 ± 0.08	0.32 ± 0.018	96.7

As a result of the wavelet transform applied to the polarization maps of the sample from Group 3, smaller amplitude values with lower modulation depth are observed compared to those in the data for the sample from Group 4 (Figure 9.13a–d). These fluctuations in the amplitude of the wavelet coefficients $W_{a,b}$ were quantified by calculating the statistical parameters $(M_1(W_{a,b}))$ and variance $M_2(W_{a,b})$ for all scales a of the MHAT function $\Psi\left(\frac{x-b}{a}\right)$ (Figure 9.14).

The scales proved to be diagnostically optimal $a_{min} = 22$ and $a_{max} = 43$ (Table 9.10).

High levels of differential diagnosis accuracy have been established. For small scales, $a_{min} = 22$ is a good and very good level of $Ac(M_1) = 90\%$ and $Ac(M_2) = 93.3\%$. For large scales, $a_{max} = 43$ is a very good $Ac(M_1) = 93.3\%$ and excellent $Ac(M_2) = 96.7\%$ level.

The results can be physically explained by changes across all scales of the birefringent fibrils' geometric dimensions within the lung parenchyma's connective tissue in fibrosis cases. Consequently, at all scales, there is maximum modulation of the polarization azimuth $\alpha(\delta_t^\star = \pi/8, m, n)$ and, accordingly, variations in the amplitudes of the wavelet coefficients $W_{a,b}$. In BA, the modulation of the wavelet coefficients is minimal.

9.2.5 Wavelet Differentiation of Ellipticity Polarization Maps of Lung Tissue

Figure 9.15 presents maps and histograms illustrating the distributions of random polarization ellipticity magnitude values in the object fields of lung tissue histological sections from Groups 3 and 4 [66–68].

Similar to the polarization azimuth maps (Figure 9.12), the spatial distributions of ellipticity in lung tissue samples showed coordinate inhomogeneity (Figure 9.15a, c). The histograms of the ellipticity random-value distributions for samples from Groups 3 and 4 are quite similar (Figure 9.15b, d). Consequently, the differences between the values of the first and fourth statistical moments $M_{i=1,2,3,4}$ amount to approximately 25–35% (Table 9.11).

It can be seen that the maximum accuracy of the differential diagnosis between BA and PF achieves a satisfactory level, with $\mathrm{Ac}(M_{1,2,3,4}) = 80 - 83.3\%$. At the same time, differences in the topological structure of the polarization ellipticity maps are more pronounced in the object fields of lung tissue histological sections from Groups 3 and 4.

In Figure 9.16 shows wavelet maps $W(a, b)$ of $\beta(\delta_t^\star = \pi/8, m, n)$ for BA and PF cases.

As shown in Table 9.12, the mean and variance of the linear dependencies of the wavelet coefficients of ellipticity in the maps of lung tissue samples from Group 4 (PF) are three to four times higher than the corresponding statistical parameters for BA. Thus, high accuracy in the differential diagnosis of BA and PF cases has been demonstrated. An excellent accuracy level of $96.7 - 100\%$ was obtained on all a scales of MHAT function $\Psi\left(\frac{x-b}{a}\right)$.

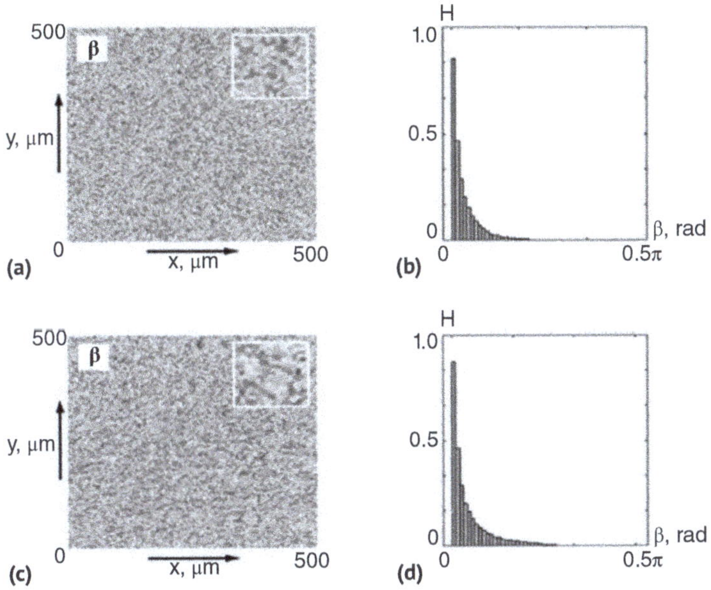

Figure 9.15 Maps $\beta(\delta_t^\star = \pi/8, m, n)$ (a,c) and multiscale linear cross-sections (b,d) of the polarization ellipticity wavelet coefficients $W_{a,b}$ of lung tissue histological sections those who died from BA (a,b) and PF (c,d).

Table 9.11 Statistical moments $M_{i=1,2,3,4}$ of polarization ellipticity maps of the $\beta(\delta_t^\star = \pi/8, m, n)$.

$M_{1,2,3,4}$	BA	PF	Ac, %
M_1	0.05 ± 0.003	0.08 ± 0.006	80
M_2	0.03 ± 0.002	0.05 ± 0.003	80
M_3	1.44 ± 0.075	1.03 ± 0.057	83.3
M_4	2.06 ± 0.11	1.61 ± 0.089	83.3

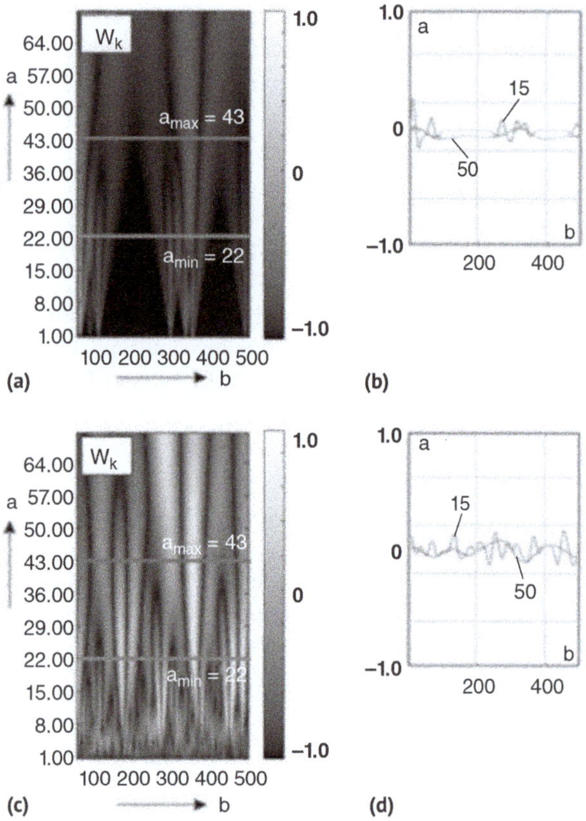

Figure 9.16 Maps (a,c) and multiscale linear cross-sections (b,d) of the polarization azimuth $\alpha(\delta_t^\star = \pi/8, m, n)$ wavelet coefficients $W_{a,b}$ of lung tissue histological sections those who died from BA (a,b) and PF (c,d).

Table 9.12 Mean M_1 and the variance M_2 of the wavelet coefficients distributions for polarization azimuth maps $\beta(\delta_t^\star = \pi/8, m, n)$.

	$a_{min} = 22$		
M_i	BA	PF	Ac,%
M_1	0.13 ± 0.008	0.17 ± 0.009	96.7
M_2	0.36 ± 0.021	0.64 ± 0.038	100
	$a_{max} = 43$		
M_i	BA	PF	Ac,%
M_1	0.078 ± 0.042	0.135 ± 0.007	100
M_2	0.19 ± 0.011	0.56 ± 0.031	100

9.3 Conclusions

1. Scenarios for creating multiscale wavelet coefficients that describe the coordinate polarization distributions of the multiscattered object field in myocardial samples within the phase section $\theta = \pi/8$ were determined and physically justified, under conditions where the influence of the diffuse component is minimized.
2. In the context of statistical analysis, diagnostic equations linking the mean and variance were identified, describing the multiscale wavelet coefficient amplitude distributions of myocardium microscopic image azimuth and ellipticity maps. Additionally, diagnostic accuracy levels for distinguishing the duration of mechanical injury were determined.

3. Polarization maps of azimuth $\alpha(\theta = \pi/8, m, n)$:
   - for small scales $a_{min} = 15$ – excellent level $Ac(M_1) = 95.8\%$ and $Ac(M_2) = 100\%$.
   - for large scales $a_{max} = 50$ – very good level $Ac(M_{1,2}) = 91.7\%$ and excellent level $Ac(M_1) = 95.8\%$.
4. Polarization maps of ellipticity $\beta(\theta = \pi/8, m, n)$ for all scales excellent level $Ac(M_2) = 96.7 - 100\%$.
5. The polarization–interference method for mapping and selecting phases of diffuse layers in biological tissues with diverse morphological structures – scattered with various polarization-inhomogeneous components – has been both theoretically justified and experimentally confirmed.
6. Using digital holographic reconstruction through phase scanning of complex amplitude distributions, polarization maps of the single-scattered component in the object field of diffuse depolarizing histological sections of fibrillar myocardium and parenchymal lung tissue, with the following types of pathology were generated algorithmically.
   myocardium – "coronary heart disease (CHD)–acute coronary insufficiency (ACI)";
   lung tissue – "asthma (BA) – fibrosis (PF)."
7. A scale-selective wavelet decomposition was carried out on a series of polarization maps of azimuth and ellipticity, using a scanning soliton-shaped MHAT function. This process identified and physically justified the geometric scales of structural elements within the polarization maps of the single-scattered component in diffuse fields. As a result, it allowed for distinguishing pathological changes in the optically anisotropic structure of myocardial and lung tissue samples.
8. Using statistical analysis of the linear distribution of wavelet coefficient amplitudes, markers for differential diagnosis-namely CHD-ACI and BA-PF-are identified for the obtained scales.
9. Diagnostic equations linking the first to fourth order statistical moments-describing the wavelet coefficient distributions in azimuth and ellipticity maps-and the diagnostic accuracy levels for distinguishing different pathological conditions were developed. For the "CHD-ACI" cases, a very high accuracy of $Ac(M_{1,2}; a_{min} = 15) = 91.7\%$ was achieved, with an excellent level for $Ac(M_1) = 95.8\%$ and $Ac(M_2; a_{max} = 50) = 100\%$. For "BA - PF" cases, a very good accuracy of $Ac(M_1; a_{min} = 22) = 92.3\%$ was observed, and an excellent level of $Ac(M_2; a_{min} = 22; a_{max} = 43) = 100\%$ was obtained.

References

1 Huang D, Swanson EA, Lin CP, Schuman JS, Stinson WG, Chang W, et al. Optical coherence tomography. Science. 1991;254(5035):1178–1181.

2 Hee MR, Huang D, Swanson EA, Fujimoto JG. Polarization-sensitive low-coherence reflectometer for birefringence characterization and ranging. J Opt Soc Am B. 1992;9(6):903–908.

3 Park BH, de Boer JF. Polarization sensitive optical coherence tomography. In Drexler W, Fujimoto JG, editors. Optical Coherence Tomography - Technology and Applications. Cham: Springer; 2015. pp. 1055–1101.

4 Hitzenberger CK, Pircher M. MUW approach of PS OCT. In Drexler W, Fujimoto JG, editors. Optical Coherence Tomography - Technology and Applications. Cham: Springer; 2015. pp. 1103–1136.

5 Yasuno Y, Ju MJ, Hong YJ, Makita S, Lim Y, Yamanari M. Jones matrix based polarization sensitive optical coherence tomography. In Drexler W, Fujimoto JG, editors. Optical Coherence Tomography - Technology and Applications. Cham: Springer; 2015. pp. 1137–1162.

6 de Boer JF, Milner TE. Review of polarization sensitive optical coherence tomography and Stokes vector determination. J Biomed Opt. 2002;7(3):359–371.

7 Pircher M, Hitzenberger CK, Schmidt-Erfurth U. Polarization Sensitive optical coherence tomography in the human eye. Prog Retin Eye Res. 2011;30(6):431–451.

8 Stifter D. Beyond biomedicine: a review of alternative applications and developments for optical coherence tomography. Appl Phys B. 2007;88(3):337–357.

9 Jones RC. A new calculus for the treatment of optical systems I. Description and discussion of the calculus. J Opt Soc Am A. 1941;31(7):488–493.

10 Gil JJ, Bernabeu E. Obtainment of the polarizing and retardation parameters of a non-depolarizing optical system from the polar decomposition of its Mueller matrix. Optik. 1987;76:67–71.

11 Bohren CF, Huffman DR. Absorption and Scattering of Light by Small Particles. New York, NY: Wiley Interscience; 1983.

12 Shurcliff WA, Ballard SS. Polarized Light. New York: Van Nostrand; 1964.

13 Park BH, Saxer C, Srinivas SM, Nelson JS, de Boer JF. In vivo burn depth determination by highspeed fiber-based polarization sensitive optical coherence tomography. J Biomed Opt. 2001;6(4):474–479.

14 Park BH, Pierce MC, Cense B, de Boer JF. Real-time multi-functional optical coherence tomography. Opt Express. 2003;11(7):782–793.

15 Saxer CE, de Boer JF, Park BH, Zhao Y, Chen Z, Nelson JS. High-speed fiber based polarization-sensitive optical coherence tomography of in vivo human skin. Opt Lett. 2000;25(18):1355–1357.

16 Park BH, Pierce MC, Cense B, de Boer JF. Jones matrix analysis for a polarization-sensitive optical coherence tomography system using fiber-optic components. Opt Lett. 2004;29(21):2512–2514.

17 Park BH, Pierce MC, Cense B, de Boer JF. Optic axis determination accuracy for fiber-based polarization-sensitive optical coherence tomography. Opt Lett. 2005;30(19):2587–2589.

18 Kemp NJ, Park J, Zaatari HN, Rylander HG, Milner TE. High-sensitivity determination of birefringence in turbid media with enhanced polarization-sensitive optical coherence tomography. J Opt Soc Am A. 2005;22(3):552–560.

19 Kemp NJ, Zaatari HN, Park J, Rylander III HG, Milner TE. Form-biattenuance in fibrous tissues measured with polarization-sensitive optical coherence tomography (PS-OCT). Opt Express. 2005;13(12):4611–4628.

20 Götzinger E, Pircher M, Geitzenauer W, Ahlers C, Baumann B, Michels S, et al. Retinal pigment epithelium segmentation by polarization sensitive optical coherence tomography. Opt Express. 2008;16(21):16410–16422.

21 Haskell RC, Carlson FD, Blank PS. Form birefringence of muscle. Biophys J. 1989;56(2):401–413.

22 van Blokland GJ. Ellipsometry of the human retina in vivo: preservation of polarization. J Opt Soc Am A. 1985;2(1):72–75.

23 Klein Brink HB, van Blokland GJ. Birefringence of the human foveal area assessed in vivo with Mueller-matrix ellipsometry. J Opt Soc Am A. 1988;5(1):49–57.

24 Todorović M, Jiao S, Wang LV, Stoica G. Determination of local polarization properties of biological samples in the presence of diattenuation by use of Mueller optical coherence tomography. Opt Lett. 2004;29(20):2402–2404.

25 Schmitt JM, Xiang SH. Cross-polarized backscatter in optical coherence tomography of biological tissue. Opt Lett. 1998;23(13):1060–1062.

26 Baumann B, Baumann SO, Konegger T, Pircher M, Götzinger E, Schlanitz F, et al. Polarization sensitive optical coherence tomography of melanin provides intrinsic contrast based on depolarization. Biomed Opt Express. 2012;3(7):1670–1683.

27 Makita S, Yamanari M, Yasuno Y. Generalized Jones matrix optical coherence tomography: performance and local birefringence imaging. Opt Express. 2010;18(2):854–876.

28 Chipman RA. Polarization analysis of optical systems. Opt Eng, 1989;28:280290.

29 Lu SY, Chipman RA. Interpretation of Mueller matrices based on polar decomposition. J Opt Soc Am A. 1996;13(5): 1106–1113.

30 Jiao S, Yao G, Wang LV. Depth-resolved two-dimensional stokes vectors of backscattered light and Mueller matrices of biological tissue measured with optical coherence tomography. Appl Opt. 2000;39(34):6318–6324.

31 Adie SG, Hillman TR, Sampson DD. Detection of multiple scattering in optical coherence tomography using the spatial distribution of Stokes vectors. Opt Express. 2007;15(26):18033–18049.

32 Lee SW, Kang JH, Yoo JY, Kang MS, Oh JT, Kim BM. Quantification of scattering changes using polarization-sensitive optical coherence tomography. J Biomed Opt. 2008;13(5):054032.

33 Götzinger E, Baumann B, Pircher M, Hitzenberger CK. Polarization maintaining fiber based ultra-high resolution spectral domain polarization sensitive optical coherence tomography. Opt Express. 2009;17(25):22704–22717.

34 Götzinger E, Pircher M, Baumann B, Schmoll T, Sattmann H, Leitgeb RA, et al. Speckle noise reduction in high speed polarization sensitive spectral domain optical coherence tomography. Opt Express. 2011;19(15):14568–14585.

35 Daubechies I. Ten Lectures on Wavelets. Society for Industrial and Applied Mathematics; 1992.

36 Cohen A, Daubechies I, Vial P. Wavelets on the interval and fast wavelet transforms. Appl Comput Harmon Anal. 1993;1(1):54–81.

37 Meyer Y. Wavelets: Algorithms & Applications. Philadelphia: Society for Industrial and Applied Mathematics (SIAM); 1993.

38 Mallat S. A theory for multiresolution signal decomposition: the wavelet representation. IEEE Trans Pattern Anal Mach Intell. 1989;11(7):674–693.

39 Daubechies I. Wavelets on the interval. In Progress in Wavelet Analysis and Applications: Proceedings of the International Conference. Toulouse, France: Wavelets and Applications; 1992.

40 Farge M. Wavelet transforms and their applications to turbulence. Annu Rev Fluid Mech. 1992;24(1):395–458.

41 Angelsky PO, Ushenko AG, Dubolazov AV, Sidor MI, Bodnar GB, Koval G, et al. The singular approach for processing polarization-inhomogeneous laser images of blood plasma layers. J Opt. 2013;15(4):044030.

42 Ushenko AG, Dubolazov AV, Ushenko VA, Novakovskaya OY. Statistical analysis of polarization-inhomogeneous Fourier spectra of laser radiation scattered by human skin in the tasks of differentiation of benign and malignant formations. J Biomed Opt. 2016;21(7):071110.

43 Garazdyuk MS, Bachinskyi VT, Vanchulyak OY, Ushenko AG, Dubolazov OV, Gorsky MP. Polarization-phase images of liquor polycrystalline films in determining time of death. Appl Opt. 2016;55(12):B67–B71.

44 Angelskii OV, Ushenko OG, Burkovets DN, Arkhelyuk OD, Ushenko YA. Polarization-correlation studies of multifractal structures in biotissues and diagnostics of their pathologic changes. Laser Phys. 2000;10(5):1136–1142.

45 Yermolenko S, Ushenko A, Ivashko P, Goudail F, Gruia I, Gavrilă C, et al. Spectropolarimetry of cancer change of biotissues. In Ninth International Conference on Correlation Optics. SPIE. 2009;7388:404–410.

46 Jóźwicki R, Patorski K, Angelsky OV, Ushenko AG, Burkovets DN, Ushenko YA. Automatic polarimetric system for early medical diagnosis by biotissue testing. Opt Appl. 2002;32(4):603–612.

47 Ushenko AG, Dubolazov OV, Ushenko VA, Novakovskaya OY, Olar OV. Fourier polarimetry of human skin in the tasks of differentiation of benign and malignant formations. Appl Opt. 2016;55(12):B56–B60.

48 Ushenko AG, Pishak VP. Coherent-domain optical methods: Biomedical diagnostics, environmental and material science. In Laser Polarimetry of Biological Tissue: Principles and Applications. 2004.

49 Ushenko AG, Fediv AI, Marchuk YF. Correlation and fractal structure of Jones matrices of human bile secret. In European Conference on Biomedical Optics. 2009;7368:73681Q.

50 Ushenko AG. Depolarization of a laser emission field as the correlation development of its polarization structure. In International Conference on Correlation Optics. 1997;3317:331–339.

51 Bachinskyi V, Boychuk T, Ushenko A. Laser Polarimetry of Biological Tissues and Fluids: Chapter 4. Methods of Spatial-frequency, Singular and Correlation Analysis of Biological Layers Object Fields. LAP Lambert Academic Publishing; 2018.

52 Ushenko AG, Burkovets DN, Yermolenko SB, Arkhelyuk AD, Pishak VP, Yuzko AM, et al. Stokes polarimetry of biotissues. In Fourth International Conference on Correlation Optics. SPIE. 1999;3904:527–533.

53 Ushenko VA, Dubolazov AV, Pidkamin LY, Sakchnovsky MY, Bodnar AB, Ushenko YA, et al. Mapping of polycrystalline films of biological fluids utilizing the Jones-matrix formalism. Laser Phys. 2018;8(2):025602.

54 Ushenko AG, Angelsky PO, Sidor M, Marchuk YF, Andreychuk DR, Pashkovskaya NV. Spatial-frequency selection of complex degree of coherence of laser images of blood plasma in diagnostics and differentiation of pathological states of human organism of various nosology. Appl Opt. 2014;53(10):B172–B180.

55 Olar EI, Ushenko AG, Ushenko YA. Polarization correlation measurements of the phase tomograms of optically anisotropic biofractals. Laser Phys. 2004;14(8):1115–1121.

56 Guminetskiy SH, Ushenko OG, Polyanskiy IP, Motrych AV, Grynchuk FV. The optical method for investigation of the peritonitis progressing process. In Eighth International Conference on Correlation Optics. 2008;7008:626–632.

57 Ushenko AG, Fediv AI, Marchuk YF. Singular structure of polarization images of bile secret in diagnostics of human physiological state. In: European Conference on Biomedical Optics. Optica Publishing Group; 2009.

58 Ushenko Y, Ushenko A, Dubolazov A, Soltys I, Litvinenko O, Wanchuliak O, et al. Mueller-matrix interferometric multifractal scaling of optically anisotropic architectonics of diffuse blood facies: fundamental and applied aspects. J Biophotonics. 2025;18(3):e202400412. https://doi.org/10.1002/jbio.202400412.

59 Ushenko A, Soltys I, Dubolazov A, Ushenko Y, Bilookyi V, Bilookyi O, et al. 3D Jones-matrix thesiography of biological fluid facies. J Innov Opt Health Sci. 2025;18(2):2443002. https://doi.org/10.1142/S1793545824430028.

60 Ushenko A, Pavlyukovich N, Khukhlina O, Pavlyukovich O, Gorsky M, Soltys I, et al. Layer-by-layer multifractal scanning of optically anisotropic architectonics of blood plasma films: fundamental and applied aspects. Photonics. 2025;12:215. https://doi.org/10.3390/photonics12030215.

61 Ushenko A, Pavlyukovich N, Khukhlina O, Pavlyukovich O, Soltys I, Dubolazov A, et al. Blood plasma film multifractal scanning in COVID-19 consequences diagnostics. J Biophotonics. 2024;17(11):e202400356. https://doi.org/10.1002/jbio.202400356.

62 Ushenko A, Dubolazov A, Zheng J, Litvinenko A, Gorsky M, Ushenko Y, et al. 3D polarization-interference holographic histology for wavelet-based differentiation of biological tissues with different necrotic states. J Biomed Opt. 2024;29(5):052920. https://doi.org/10.1117/1.JBO.29.5.052920.

63 Ushenko O, Bilookyi O, Zheng J, Dubolazov A, Olar O, Ushenko Y, et al. 3D digital holographic polarimetry of laser speckle fields formed by polycrystalline blood films: a tool for differential diagnosis of thyroid pathology. Front Phys. 2024;12:1426469. https://doi.org/10.3389/fphy.2024.1426469.

64 Wójcik W, Hu Z, Ushenko Y, Smolarz A, Soltys I, Dubolazov O, et al. Optical sensor system for 3D Jones matrix reconstruction of optical anisotropy maps of self-assembled polycrystalline soft matter films. Sensors. 2024;24(5):1589. https://doi.org/10.3390/s24051589.

65 Ushenko A, Zheng J, Gorsky M, Dubolazov A, Ushenko Y, Soltys I, et al. 3D digital holographic polarimetry of diffuse optically anisotropic biological tissue object fields. Front Phys. 2023;11:1288935. https://doi.org/10.3389/fphy.2023.1288935.

66 Dubolazov A, Ushenko V, Trifonyuk L, Stashkevich A, Soltys I, Ushenko Y, et al. Polarization-singular approach to imaging Mueller-matrix polarimetry in differential diagnosis of histological sections of biopsy of tumors of the uterus and prostate. Front Phys. 2021;9:711212. https://doi.org/10.3389/fphy.2021.711212.

67 Ushenko VO, Trifonyuk L, Ushenko YA, Dubolazov OV, Gorsky MP, Ushenko AG. Polarization singularity analysis of Mueller-matrix invariants of optical anisotropy of biological tissue samples in cancer diagnostics. J Opt. 2021;23(6):064004. https://doi.org/10.1088/2040-8986/abf97a.

68 Ushenko YA, Koval GD, Ushenko AG, Dubolazov OV, Ushenko VA, Novakovskaia OY. Mueller-matrix of laser-induced autofluorescence of polycrystalline films of dried peritoneal fluid in diagnostics of endometriosis. J Biomed Opt. 2016;21(7):071116. https://doi.org/10.1117/1.JBO.21.7.071116.

10

Three-dimensional Layer-by-layer Digital Multifractal Scanning of Polarization Maps of Biological Tissue Architectonics for Mechanical Trauma Diagnostics

Yurii Ushenko[1,2], Weidong Dou[1], Oleksandr Ushenko[2,3], Oleksandr Dubolazov[2], Iryna Soltys[2], Viktor Bachinskiy[4], Oleh Vanchuliak[4], Marta Garazdyuk[4], Nataliia Pavlyukovich[4], Oleksandr Pavlyukovich[4], Vitaliy Rozhko[4], and Oleksandr Olar[2]

[1] *Shaoxing University, Shaoxing, People's Republic of China*
[2] *Yurii Fedkovych Chernivtsi National University, Chernivtsi, Ukraine*
[3] *Taizhou Institute of Zhejiang University, Taizhou, People's Republic of China*
[4] *Bukovinian State Medical University, Chernivtsi, Ukraine*

10.1 Three-dimensional Polarimetry Phase Scanning Method

An example of the implementation of a polarization–interference technique with digital holographic reconstruction of the three-dimensional (3D distribution of azimuths of polarization $Az(r)$ of the object field of an optically thick histological section of the traumatically intact liver of a deceased from acute coronary insufficiency (Figure 10.1) [1–6].

Figure 10.1a–c illustrates partial distributions ("polarization measures") of polarization azimuths, which are formed by various physical mechanisms:

- a is a polarization measure, which is formed mainly by acts of single scattering and interference $\left(Az(r) \longleftrightarrow \begin{pmatrix} Az^{\odot}(\vartheta_k, m, n) \\ Az^{\circledast}_{l=2}(\vartheta_k, m, n) \\ Az^{\star}_{l=1,2}(\vartheta_k, m, n) \end{pmatrix}\right)$. The range of variation in the magnitude of the azimut a is a polarization measure, which is formed mainly by acts of single scattering and interference $\left(Az(r) \longleftrightarrow \begin{pmatrix} Az^{\odot}(\vartheta_k, m, n) \\ Az^{\circledast}_{l=2}(\vartheta_k, m, n) \\ Az^{\star}_{l=1,2}(\vartheta_k, m, n) \end{pmatrix}\right)$. The range of variation in the magnitude of the azimuth of polarization is insignificant – $\Delta\alpha(r) \cong \pi/10$.
- h of polarization is insignificant – $\Delta\alpha(r) \cong \pi/10$.
- b is a polarization measure, which is formed by acts of scattering and interference of average multiplicity $\left(Az(r) \longleftrightarrow \begin{pmatrix} Az^{\circledast}_{l=3,\dots p}(\vartheta_k, m, n) \\ Az^{\star}_{l=3,\dots,p}(\vartheta_k, m, n) \end{pmatrix}\right)$. The range of variation in the magnitude of the azimuth of polarization increases – $\Delta\alpha(r) \cong \pi/4$.

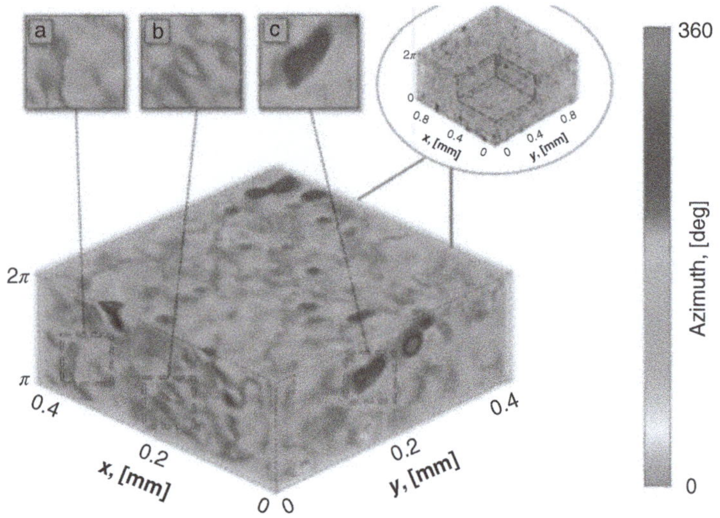

Figure 10.1 Three-dimensional distribution of polarization azimuths of the diffuse object field of the histological section of the traumatically intact liver.

- c is a polarization measure, which is formed mainly by acts of multiple scattering and interference $\left(Az(r) \longleftrightarrow \begin{pmatrix} Az^{\circledast}_{l=p,\ldots,w}(\vartheta_k, m, n) \\ Az^{\star}_{l=p,\ldots,w}(\vartheta_k, m, n) \end{pmatrix}\right)$. The range of variation in the magnitude of the azimuth of polarization increases as much as possible $\Delta\alpha(r) \cong \pi/2$.

Apply the wavelet schedule $W(a, b)$ for each polarization parameters coordinate distributions $\begin{pmatrix} Az^{\odot}(\vartheta_k, m, n) \\ Az^{\circledast}_{l=2,\ldots,w}(\vartheta_k, m, n) \\ Az^{\star}_{l=1,2,\ldots,w}(\vartheta_k, m, n) \end{pmatrix}$.

Determine the wavelet amplitudes maxima modules skeletons sup $\left|W(a^*, x_j(a^*))\right|$ for biological tissue polarization maps $\left(\begin{pmatrix} Az^{\odot}(m, n) \\ Az^{\circledast}_{l=2,\ldots,w}(m, n) \\ Az^{\star}_{l=1,2,\ldots,w}(m, n) \end{pmatrix} \text{and} \begin{pmatrix} Az^{\odot}(\vartheta_k, m, n) \\ Az^{\circledast}_{l=2,\ldots,w}(\vartheta_k, m, n) \\ Az^{\star}_{l=1,2,\ldots,w}(\vartheta_k, m, n) \end{pmatrix}\right)$, which serve as the basis for calculating $(F^{\alpha}(h_z))$ multifractal spectra by the WTMM method, – $\text{FM}(z, d, \sup|W(a^*, x_j(a^*))|) \Longleftrightarrow \ln\text{FM}(z, d)/\ln d$.

The multifractal spectrum $F^{\alpha}(\vartheta_k, h_z)$ resulting was analyzed using the following algorithms to calculate mean (M_1), variance (M_2), skewness (M_3), and kurtosis (M_4) values [7–16].

10.2 Objects and Their Characteristics

To conduct a comprehensive study of histological sections of the liver, the following cases were examined [17–24]:

- mechanical injury occurring 6 hours prior to sampling – Group 1, 19 samples;
- mechanical injury occurring 12 hours prior to sampling – Group 2, 19 samples.

The microscopic structure of the liver resembles a honeycomb, where each unit corresponds to a hepatic lobule with a size ranging from 1 to 2.5 mm. Rows of hepatocytes are arranged radially, resembling the spokes of a wheel. Numerous such rows form layered hepatic laminae. The structural support for hepatocytes is provided by delicate reticulin fibers composed of type III collagen.

Table 10.1 The liver histological section samples optical-geometric parameters.

Parameters	Liver
Geometrical thickness, h, μm	50–60
Coefficient of attenuation, τ μm	0.21–0.26
Depolarization degree, Δ, %	49–55

The optical anisotropy of liver tissue is primarily determined by circular birefringence arising from its main structural elements – the hepatic lobules, which account for 60–80% of the total liver mass. Each lobule is composed of radial rows of hepatocytes forming polygonal or hexagonal parenchymal cells. These structures are optically active and induce rotation of the polarization plane of probing laser radiation.

In addition to circular birefringence, liver tissue exhibits minor linear birefringence, which generates elliptically polarized waves and originates from fibrillar collagen networks of loose connective tissue.

Based on these structural and optical properties, it can be assumed that polarization azimuth maps are the most sensitive indicators of morphological alterations in liver tissue, including those resulting from mechanical injury [25–28].

The liver histological section samples optical-geometric parameters are presented in Table 10.1.

10.3 Experimental Results and Their Discussion

10.3.1 Stokes Polarimetric Differentiation of the Prescription of Mechanical Injury to the Liver

The series of Figures 10.2–10.4 shows the polarization maps $\alpha(m,n)$; $\alpha(\vartheta_k, m, n)$ (a), (b) and histograms of distributions of random values of the azimuth of polarization $N(\alpha)$ (c), (d) of microscopic images of histological sections of the liver of deceased with different prescription of mechanical trauma liver – 6 hours (a), (c) and 12 hours (b), (d).

The analysis of the obtained data revealed the presence of coordinate-inhomogeneous polarization maps $\alpha(m,n)$ and $\alpha(\vartheta_k, m, n)$ (Figures 10.2–10.4, (a), (b)) in all phase sections ϑ_k of microscopic images of histological sections of the liver with different duration of mechanical injury. This fact confirms the adequacy of proposed model analysis of the processes of formation of various polarization measures $\begin{pmatrix} Az^{\odot}(m,n) \\ Az^{\circledast}_{l=2}(m,n) \\ Az^{\star}_{l=1,2}(m,n) \end{pmatrix} \begin{pmatrix} Az^{\odot}(\vartheta_k, m,n) \\ Az^{\circledast}_{l=2}(\vartheta_k, m,n) \\ Az^{\star}_{l=1,2}(\vartheta_k, m,n) \end{pmatrix}$ in the structure of object fields of optically anisotropic liver parenchyma.

For repeatedly scattered diffuse fields $\begin{pmatrix} Az^{\odot}(m,n) \\ Az^{\circledast}_{l=2}(m,n) \\ Az^{\star}_{l=1,2}(m,n) \end{pmatrix}$, almost equally probable distributions $N(\alpha^*)$ are formed random values of the polarization azimuth α^* (Figure 10.2c, d).

As the influence of the light scattering multiplicity decreases by phase scanning $\vartheta_k = \delta_x^{\otimes} - \delta_y^{\otimes}$ of the holographically reconstructed field of complex amplitudes $E_{x;y}$; E_{0°; E_{90°, the structure of histograms $N(\alpha)$ and $N(\vartheta_k, \alpha)$ changes – local extremes are formed, ranges of variation of random values of the azimuth of polarization decrease (Figures 10.3 and 10.4, (c), (d)).

A comparison of the coordinate distributions (m,n) of the polarization characteristics $\left(Az(r) \longleftrightarrow \begin{pmatrix} Az^{\circledast}_{l=p,\dots,w}(\vartheta_k, m,n) \\ Az^{\star}_{l=p,\dots,w}(\vartheta_k, m,n) \end{pmatrix} \right)$ of "integral" repeatedly scattered microscopic images of diffuse liver samples with different injury duration did not reveal significant differences between them (Figure 10.2 a–d).

Algorithmic phase scanning ϑ_k provides the possibility of obtaining large coordinate $m \times n$ (Figures 10.3 and 10.4, (a), (b)) and statistical differences $N(\alpha)$ (Figures 10.3 and 10.4, (c), (d)) between phase polarization maps $\alpha(m,n)$ of samples with different duration of mechanical injury.

From a physical point of view, the results obtained can be explained by the following considerations.

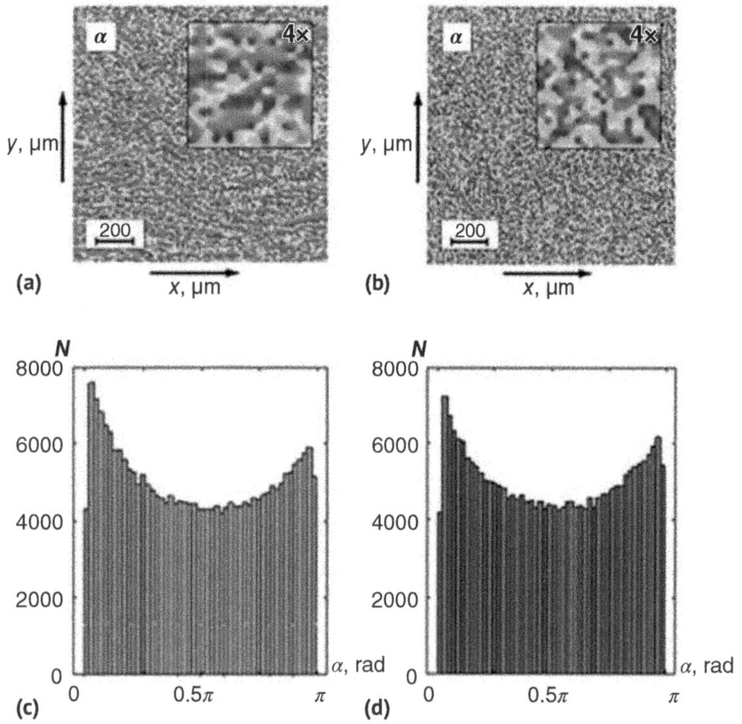

Figure 10.2 Polarization maps $\alpha(m, n)$ (a), (b) and histograms of $N(\alpha)$ (c), (d) microscopic images of histological sections of the liver with a probability of mechanical injury of 6 hours (a), (c) and 12 hours (b), (d).

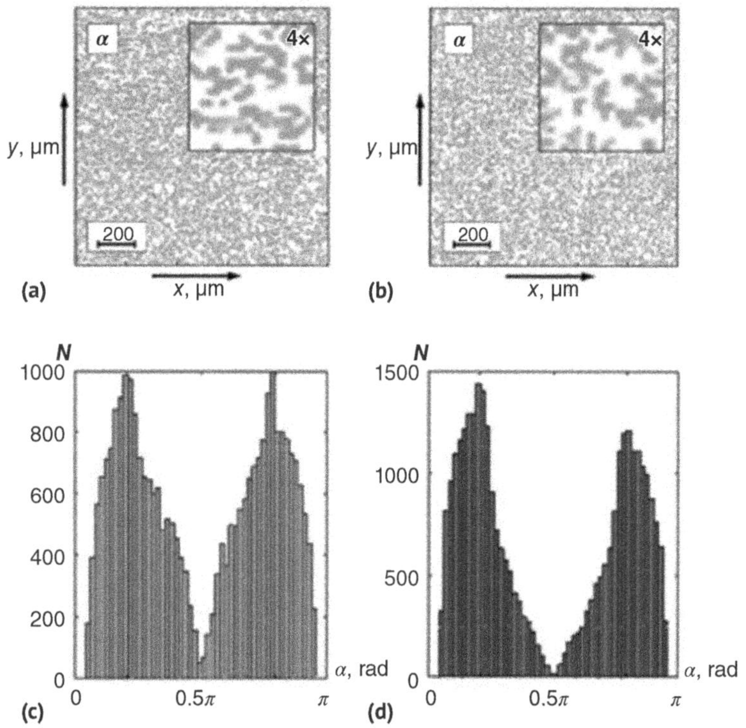

Figure 10.3 Phase section $\vartheta_k = \pi/4$ – polarization maps $\alpha(m, n)$ (a), (b) and histograms of $N(\alpha)$ (c), (d) microscopic images of histological sections of the liver with a probability of mechanical injury of 6 hours (a), (c) and 12 hours (b), (d) [29–36].

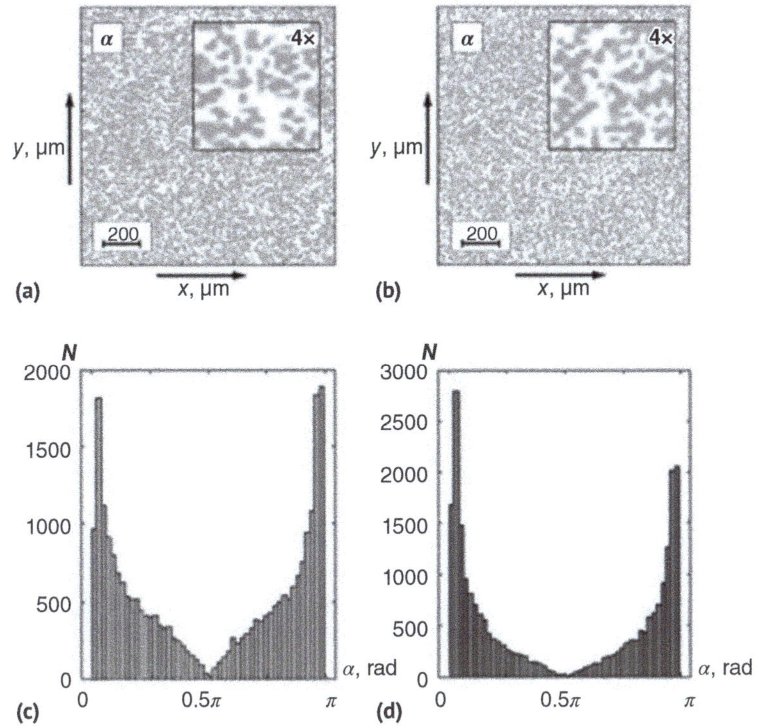

Figure 10.4 Phase section $\vartheta_k = \pi/8$ – polarization maps $\alpha(m, n)$ (a), (b) and histograms of $N(\alpha)$ (c), (d) microscopic images of histological sections of the liver with a probability of mechanical injury of 6 hours (a), (c) and 12 hours (b), (d).

First, traumatic injury induces a mechanical transformation of the optically anisotropic architectonics of liver tissue. These morphological alterations generally manifest at the macroscale level relative to the size of individual liver lobules or their larger structural complexes. Therefore, the overall degree of mechanical disruption of optical birefringence remains insignificant.

This is especially valid under conditions of multiple scattering, as observed in the polarization measure c presented in Figure 10.1 – $\left(Az(r) \longleftrightarrow \begin{pmatrix} Az^{\circledast}_{l=p,\dots,w}(\vartheta_k, m, n) \\ Az^{\star}_{l=p,\dots,w}(\vartheta_k, m, n) \end{pmatrix} \right)$ within the volume of the diffuse parenchymal layer.

Holographic reconstruction of the components of the object field with a lower light scattering multiplicity (like "polarization measures a and b" in Figure 10.1 – $\begin{pmatrix} Az^{\odot}(\vartheta_k, m, n) \\ Az^{\circledast}_{l=2}(\vartheta_k, m, n) \\ Az^{\star}_{l=1,2}(\vartheta_k, m, n) \end{pmatrix}$ and $\begin{pmatrix} Az^{\circledast}_{l=3,\dots p}(\vartheta_k, m, n) \\ Az^{\star}_{l=3,\dots,p}(\vartheta_k, m, n) \end{pmatrix}$ makes it possible to increase the sensitivity (Ac) of the polarimetry method to a traumatic change in the phase anisotropy of more optically thin layers of liver parenchyma.

As a result, the degree of difference between coordinate $(m \times n)$ and statistical $N(\alpha)$ increases as the scattered components are isolated a few times – multiple scattering $\begin{pmatrix} Az^{\circledast}_{l=2,\dots,w} \\ Az^{\star}_{l=2,\dots,w} \end{pmatrix}$ and acts of scattering and interference of average $\begin{pmatrix} Az^{\circledast}_{l=3,\dots p}(\vartheta_k = \pi/4, m, n) \\ Az^{\star}_{l=3,\dots,p}(\vartheta_k = \pi/4, m, n) \end{pmatrix}$ or once $\begin{pmatrix} Az^{\odot}(\vartheta_k = \pi/8, m, n)) \\ Az^{\star}_{l=1}(\vartheta_k = \pi/8, m, n)) \end{pmatrix}$ multiplicity in a diffuse object field.

Second, in parallel, the process of "traumatic" reduction of birefringence is influenced by the temporary necrotic destruction of optically active $\left(\left\|\begin{matrix} q_{11}(\gamma;\delta;\xi) & \cdots & q_{14}(\gamma;\delta;\xi) \\ \vdots & \ddots & \vdots \\ q_{41}(\gamma;\delta;\xi) & \cdots & q_{44}(\gamma;\delta;\xi) \end{matrix}\right\|\right)$ lobules of the liver.

As the prescription of mechanical injury increases, the influence of this factor increases.

Therefore, the liver parenchyma with a more long-standing mechanical injury has less birefringence. As a result, the average value and ranges of variation of random values of the azimuth of polarization, which are described by the corresponding histograms $N(\alpha)$, should decrease [37–42].

The results of statistical and informational analysis of the effectiveness of the Stokes polarimetric differential diagnosis of the prescription of mechanical injury to the liver is given in Table 10.2.

The results of the statistical analysis of the polarization maps $\alpha(m, n)$ and $\alpha(\vartheta_k, m, n)$ correlate well with the proposed model representations about the optical anisotropy of the liver parenchyma, as well as with the physical substantiation of the Stokes polarimetry data.

As can be seen, the differences between the first and second order statistical moments, which characterize the mean M_1 and variance M_2 of distributions, do not exceed 10–15%.

As a result, the accuracy of the differential diagnosis of the prescription of mechanical injury is unsatisfactory, – Ac < 70%.

The level of accuracy increases somewhat (by 10–12% in the "optically thin" phase section $\vartheta_k = \pi/8$) for statistical moments of the third (M_3) and fourth (M_4) orders, which mathematically are more sensitive to variations in the magnitude of the mean and variance, $-M_3 \sim 1/M_2^3$ and $M_4 \sim 1/M_2^4$. As a result, the level of accuracy of differential diagnostics increases to satisfactory $\text{Ac}(M_3; M_4) = 80.1 - 82.3\%$.

It should be noted that in the framework of statistical analysis ($M_{i=1;2;3;4}$) of polarization maps ($\alpha(m, n)$ and $\alpha(\vartheta_k, m, n)$) of microscopic images of liver samples, only quantitative or concentration mechanical changes in the optically anisotropic architectonics of the liver parenchyma are detected.

Table 10.2 Statistical characteristics of the polarization maps $\alpha(m, n)$ and $\alpha(\vartheta_k, m, n)$ the accuracy Ac of the differential diagnosis of the prescription of mechanical injury.

Parameters	Group 1	Group 2	Accuracy
$M_{i=1;2;3;4}$			
M_1	0.12 ± 0.007	0.105 ± 0.007	<70
M_2	0.38 ± 0.022	0.31 ± 0.018	<70
M_3	0.21 ± 0.013	0.27 ± 0.016	71.4
M_4	0.31 ± 0.017	0.39 ± 0.022	73.8
$\vartheta_k = \frac{\pi}{4}$			
Parameters	**Group 1**	**Group 2**	**Accuracy**
$M_{i=1;2;3;4}$			
M_1	0.09 ± 0.006	0.075 ± 0.005	<70
M_2	0.18 ± 0.011	0.16 ± 0.008	<70
M_3	0.53 ± 0.029	0.67 ± 0.044	76.2
M_4	0.71 ± 0.041	0.88 ± 0.053	75.6.6
$\vartheta_k = \frac{\pi}{8}$			
Parameters	**Group 1**	**Group 2**	**Accuracy**
$M_{i=1;2;3;4}$			
M_1	0.06 ± 0.004	0.047 ± 0.003	<70
M_2	0.11 ± 0.006	0.098 ± 0.006	<70
M_3	0.71 ± 0.043	0.86 ± 0.056	80.1
M_4	1.07 ± 0.062	1.21 ± 0.079	82.3

Such changes are evaluated by the corresponding set of central statistical moments of the first to fourth orders $M_{t=1;2;3;4}(\vartheta_k, \alpha)$.

At the same time, an unused information resource is the architectonic construction of the most optically anisotropic parenchyma, which can be estimated by the multifractal parameters h_z or their spectra $F(h_z)$.

This approach may be more sensitive to destructive mechanical changes in the optical anisotropy of the liver parenchyma with a short duration of injury, which are experimentally evaluated by the coordinate distributions of the azimuth of polarization, – $\alpha(m, n)$ and $\alpha(\vartheta_k, m, n)$.

10.3.2 Multifractal Differentiation of the Prescription of Mechanical Liver Injury

Figures 10.5–10.7 presents the results of a multifractal analysis of polarization maps of $\alpha(m, n)$ and $\alpha(\vartheta_k, m, n)$ liver samples using the WTMM method.

Analysis of a series of obtained fractal dimension spectra $F(h_j)$ revealed a scale-self-similar structure of polarization maps $\alpha(m, n)$ and $\alpha(\vartheta_k, m, n)$ of the object field of histological sections of the liver with mechanical trauma of various ages, Figure 10.5b, d [43–47].

Each multifractal spectrum ($F(h_z)$ and $F(\vartheta_k, h_z)$) is characterized by a localized main extremum $F(h_z)$ and a relatively small range of changes in fractal dimensions $h_{\min} \leq h \leq h_{\max}$. Note that the "boundary" values of topographic distributions h_z of polarization maps are theoretically limited to $h_{\min} = 1$ (straight line) and $h_{\max} = 2$ (plane).

Algorithmic phase scanning (ϑ_k) of the object field of complex amplitudes $H_{x,y}$ provides the possibility of estimating the multifractal structure of polarization measures with a small scattering multiplicity (like measures $b - \begin{pmatrix} Az^{\circledast}_{l=3,\ldots p}(\vartheta_k, m, n) \\ Az^{\star}_{l=3,\ldots,p}(\vartheta_k, m, n) \end{pmatrix}$ or a on Figure 10.1 – $\begin{pmatrix} Az^{\odot}(\vartheta_k, m, n) \\ Az^{\circledast}_{l=2}(\vartheta_k, m, n) \\ Az^{\star}_{l=1,2}(\vartheta_k, m, n) \end{pmatrix}$ in a diffuse field $\begin{pmatrix} Az^{\odot}(\vartheta_k, m, n) \\ Az^{\circledast}_{l=2,\ldots,w}(\vartheta_k, m, n) \\ Az^{\star}_{l=1,2,\ldots,w}(\vartheta_k, m, n) \end{pmatrix}$.

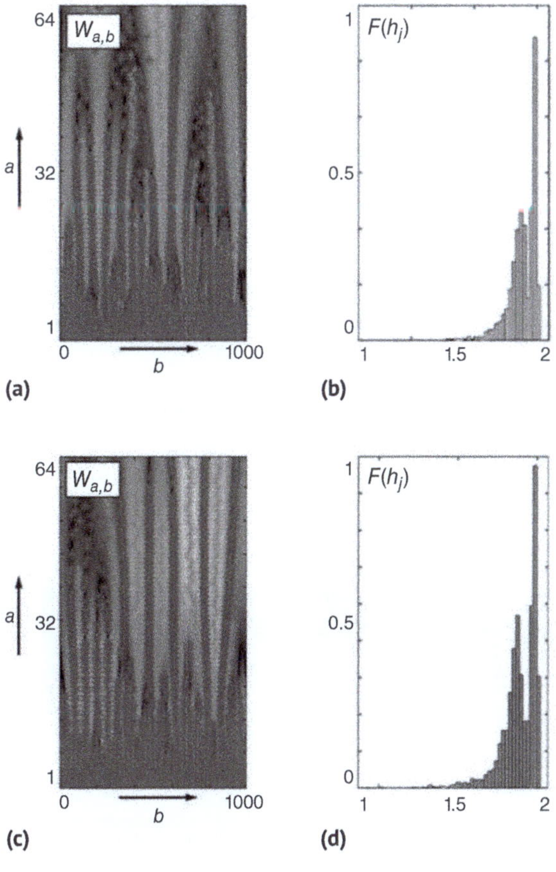

Figure 10.5 Skeletons of wavelet maps $W(a, b)$ (a), (c) and multifractal spectra (b), (d) of polarization azimuth maps $\alpha(m, n)$ of microscopic images of histological sections of the liver with a duration of mechanical injury of 6 hours (a), (c) and 12 hours (b), (d).

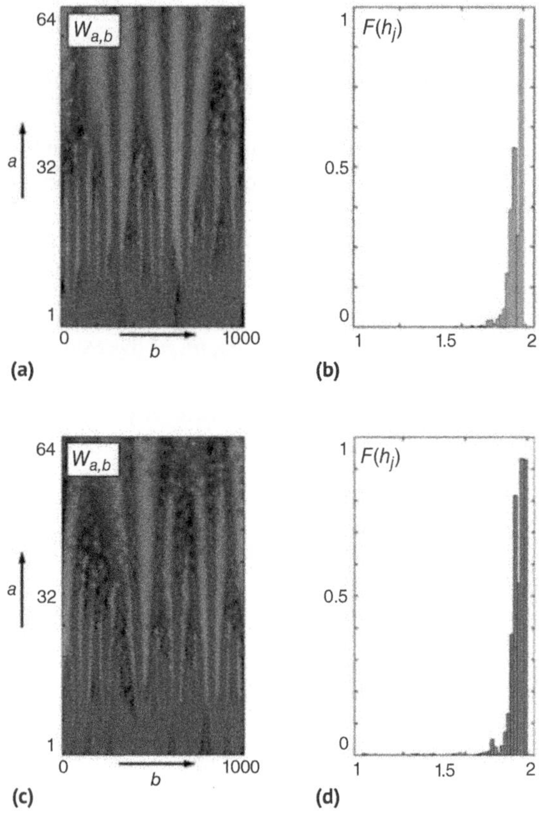

Figure 10.6 Phase section $\vartheta_k = \pi/4$ – skeletons of wavelet maps $W(a, b)$ (a), (c) and multifractal spectra (b), (d) of polarization azimuth maps $\alpha(m, n)$ of microscopic images of histological sections of the liver with a duration of mechanical injury of 6 hours (a), (c) and 12 hours (b), (d).

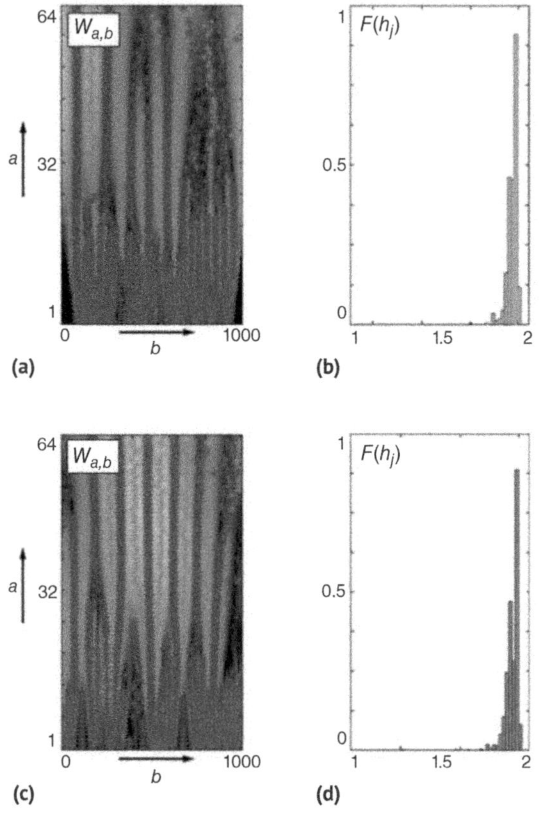

Figure 10.7 Phase section $\vartheta_k = \pi/8$ – skeletons of wavelet maps $W(a, b)$ (a), (c) and multifractal spectra (b), (d) of polarization azimuth maps $\alpha(m, n)$ of microscopic images of histological sections of the liver with a duration of mechanical injury of 6 hours (a), (c) and 12 hours (b), (d).

The coordinate and topographic structure of the polarization maps of such components turns out to be quite unambiguously related to the optically anisotropic architectonics $\left(\left\|\begin{matrix} q_{11}(\gamma;\delta;\xi) & \cdots & q_{14}(\gamma;\delta;\xi) \\ \vdots & \ddots & \vdots \\ q_{41}(\gamma;\delta;\xi) & \cdots & q_{44}(\gamma;\delta;\xi) \end{matrix}\right\|\right)$ of the liver parenchyma.

Therefore, the transformation of the multifractal spectra $F(h_z)$ and $F(\vartheta_k, h_z)$ of polarization maps $\alpha(m, n)$ and $\alpha(\vartheta_k, m, n)$ is most sensitive to mechanical traumatic changes with small limitations.

Experimentally, such scenarios manifest themselves in the formation of more "narrow" multifractal spectra $F(h_z)$ and $F(\vartheta_k, h_z)$ for short-term mechanical trauma (Figures 10.6 and 10.7, (b)) in comparison with similar fractal dimension spectra for longer-term injuries (Figures 10.6 and 10.7, (d)).

Quantitatively, the scenarios of traumatic transformation of the scale-self-similar topographic structure of polarization maps $\alpha(m, n)$ and $\alpha(\vartheta_k, m, n)$ of the object field of liver samples and its phase sections illustrate the results of statistical analysis $M_{i=1;2;3;4}$ of a series of multifractal spectra $F(h_z)$ and $F(\vartheta_k, h_z)$, Table 10.3.

Statistical analysis of multifractal spectra $F(h_z)$ obtained by MWTM method revealed the most sensitive markers for differential diagnosis of the prescription of mechanical liver injury.

Such markers were the statistical moments of the third and fourth orders, which characterize the skewness M_3 and the kurtosis M_4 of the distributions of fractal dimensions h_z of layered (ϑ_k) maps of the azimuth of polarization $\alpha(\vartheta_k, m, n)$ of the object field of liver samples [44–51].

For laser radiation repeatedly scattered by diffuse liver samples, a good accuracy of differential diagnosis of the prescription of mechanical injury has been demonstrated, $\text{Ac}(M_3; M_4) = 85.7 - 85.6.1\%$.

In various phase sections ϑ_k of the object field, an even greater increase in the accuracy of differential diagnostics has been achieved to very good $(Ac(\vartheta_k = \pi/4; M_3; M_4) = 90.4\%)$ and excellent $(\text{Ac}(\vartheta_k = \pi/8; M_3; M_4) = 95.2\%$ levels.

10.3.3 Conclusions

Layer-by-layer maps of the polarization azimuth $\alpha(\vartheta_k, m, n)$ of microscopic images of liver samples with different durations of mechanical injury were experimentally determined and interpreted within the framework of the proposed optical anisotropy model.

Statistical and informational analysis of a series of layered polarization distributions $\alpha(\vartheta_k, m, n)$, repeatedly scattered by histological sections of the liver, demonstrated satisfactory accuracy of differential diagnosis of the timing of mechanical injury, reaching Ac = 76.2–82.3%.

Table 10.3 Statistical characteristics of multifractal spectra $F(h_z)$ and $F(\vartheta_k, h_z)$ and accuracy Ac of differential diagnosis of mechanical injury prescription.

Parameters	Group 1	Group 2	Accuracy
M_1	1.72 ± 0.03	1.81 ± 0.04	<70
M_2	0.14 ± 0.008	0.16 ± 0.009	<70
M_3	1.88 ± 0.099	1.12 ± 0.069	85.7
M_4	3.21 ± 0.19	2.33 ± 0.13	85.6.1
$\vartheta_k = \pi/4$			
M_1	1.69 ± 0.088	1.78 ± 0.091	<70
M_2	0.09 ± 0.005	0.13 ± 0.007	85.7
M_3	2.47 ± 0.15	1.85 ± 0.11	90.4
M_4	4.03 ± 0.24	3.11 ± 0.17	90.4
$\vartheta_k = \pi/8$			
M_1	1.66 ± 0.089	1.76 ± 0.093	<70
M_2	0.05 ± 0.003	0.09 ± 0.006	85.6.1
M_3	3.16 ± 0.18	2.11 ± 0.13	95.2
M_4	4.83 ± 0.27	3.41 ± 0.19	95.2

The shading indicates different levels of diagnostic accuracy; darker shades correspond to higher diagnostic performance. It is used only for visual differentiation and has no additional statistical meaning.

Using multifractal analysis based on the maxima of the modulus of the wavelet transform coefficients, a series of spectra of fractal dimensions h_z of spatial distributions of the polarization azimuth $\alpha(\vartheta_k, m, n)$ was obtained and interpreted for different phase sections ϑ_k of the object field for liver samples with different durations of mechanical injury.

The parameters most sensitive to changes in the structure of the multifractal spectra $F(h_z)$ were identified: the statistical moments of the third and fourth orders, characterizing the skewness M_3 and kurtosis M_4 of the distributions of fractal dimensions of the topographic structure (m, n) of layered polarization maps $\alpha(\vartheta_k, m, n)$.

By applying phase scanning ϑ_k of the components of the scattered object field, the dynamics of the multifractal increase in the accuracy of differential diagnosis of the timing of mechanical injury to the liver were investigated, ranging from good performance in the diffuse field (Ac = 85.7%) to excellent performance in the optically thin phase layer ($\vartheta_k = \pi/8$, Ac = 95.2%).

10.4 Traditional Histological Methods of Diagnosing the Genesis and Duration of Hemorrhages in the Human Brain

Verification of the cause of death (CD) resulting from ischemic cerebral infarction, traumatic, and nontraumatic hemorrhages makes it possible to exclude a violent origin of death. Furthermore, determining the precise timing of injury enables forensic investigators to significantly narrow the range of potential suspects in criminal cases. In many autopsy procedures, macroscopic evaluation alone is insufficient for an expert to accurately determine the genesis of a hemorrhage. Therefore, additional sampling for forensic histological examination is required [50–58].

In general, the histological method represents the gold standard for diagnosing the cause of death and establishing the vitality and temporal progression of pathological processes in biological tissues in forensic medical practice. Ischemic cerebral infarction and intracerebral hemorrhages remain among the most frequent pathological entities studied in forensic histology.

In this section, three validated histochemical methods were used for the objective diagnosis of the nature, origin, and temporal characteristics of damage to brain tissue. Each method plays an important role in the differential diagnosis of ischemic cerebral infarction (ICI), traumatic-origin hemorrhages (HTG), and nontraumatic hemorrhages (HNG), as well as in the temporal stratification of these processes.

To address the objectives defined by the study design, within the framework of the traditional histological approach to diagnosing the genesis and temporal development of hemorrhages in human brain tissue, the following tasks were formulated:

- to determine the morphological characteristics of ischemic cerebral infarction at different stages of its formation; to characterize the dynamics of changes in the HBS in cases of traumatic and nontraumatic hemorrhages; to compare histological changes between study groups based on Nissl, Spielmeyer, and Perls staining;
- and to evaluate the diagnostic informativeness of each histochemical technique, as well as the histological method as a whole, for determining the duration and origin of ischemic cerebral infarction and traumatic and nontraumatic hemorrhages.

The main research method used was light microscopy of histological brain preparations at a magnification of 400×, with corresponding digital photographs of microscopic objects obtained [59–65].

Histological evaluation of HBS samples of the main groups depending on the time from the moment of injury demonstrated changes in tissue architecture characteristic of ischemic, hemorrhagic, and traumatic lesions during the first two weeks after infarction or traumatic brain injury. The studies covered key time stages of the pathological process – from early changes occurring in the first hours after injury to later reactive and reparative transformations of HBS.

This provided the opportunity for intergroup comparison, verification of diagnostic signs of the duration of the lesion, relevant in forensic practice and attempts to establish histological criteria for the genesis of hemorrhages.

10.4.1 Nissl Technique

Chromatophilic substance or Nissl bodies (also called "tigroid" in the literature) are large basophilic granules contained in neurons. By their nature, these granules are cisterns of the granular endoplasmic reticulum, containing a large number of ribosomes, and are the site of protein synthesis and, together with neurofibrils, belong to the special components of nerve cells. As a result of Nissl staining, basophilic RNA granules are stained blue.

This staining method is used to determine the localization of the perikaryon, since the chromatophilic substance is contained only in the perikaryon and dendrites and is absent in the axon and axon hillock. During certain pathological processes, this substance may disappear – the so-called chromatolysis occurs, indicating the exhaustion of the nerve cell.

Histological assessment, the results of microscopy of histological preparations of HBS samples (stained by the Nissl method) of the main study groups, depending on the time from the moment of injury, demonstrate changes in tissue architecture characteristic of ischemic, hemorrhagic, and traumatic lesions (Figure 10.8–10.10).

In ICI observations (1–14 days), there was fragmentation of Nissl substance in neurons near areas of destruction of nervous tissue; in foci of necrosis, rounding of neurons and disappearance of Nissl substance were detected, as shown in Figure 10.8.

In the study group of hemorrhages of nontraumatic genesis (1–12 days), fragmentation of the Nissl substance and/or its absence were noted in individual surviving neurons in the necrosis zone (Figure 10.9).

In brain tissue samples from the area of hemorrhages of traumatic genesis (1–10 days), the brain substance was semi-destroyed, rounded neurons were identified, with homogenization of the Nissl substance, as shown in Figure 10.10.

10.4.2 Spielmeyer's Method

The next step was to perform microscopy of HBS preparations stained according to Spielmeyer to visualize myelinated nerve fibers. Myelinated nerve fibers provide fast and efficient transmission of electrical signals in the nervous system. To act as an electrical insulator, the myelin sheath is a multilamellar membrane structure formed by the spiral wrapping and subsequent compaction of the oligodendroglia plasma membrane around the axons of the central nervous system. Current evidence indicates that the myelin sheath is more than an inert insulating membrane structure. Oligodendrocytes are metabolically active and functionally coupled to the adjacent axon through cytoplasm-rich myelin channels for the movement of macromolecules into and out of the internodal periaxonal space beneath the myelin sheath.

Microscopy of histological preparations stained using the Spielmeyer method revealed the following morphological changes in the examined HBS samples with hemorrhages.

In cases of ICI (2–14 days), the absence of myelin fibers is noted in the area of necrosis, along the periphery of necrosis there are signs of destruction of myelin structures, in the preserved areas there are myelin fibers (Figure 10.11).

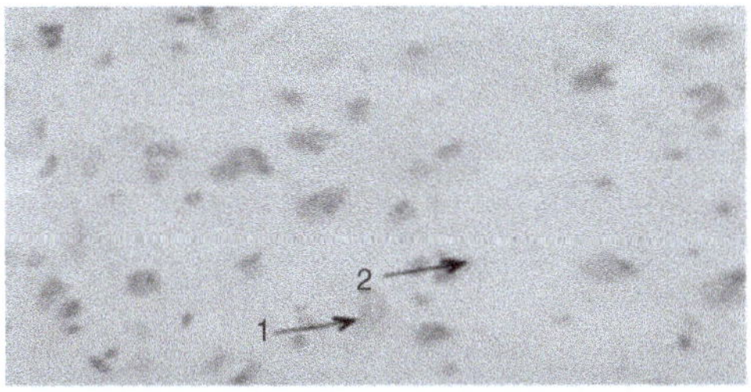

Figure 10.8 Histological section of the ischemic cerebral infarction area (5 days). Staining: histochemical Nissl method. Microphotograph. Magnification: objective 40×, eyepiece 10×: 1, rounded neuron; 2, disappearance of Nissl substance in neuron processes.

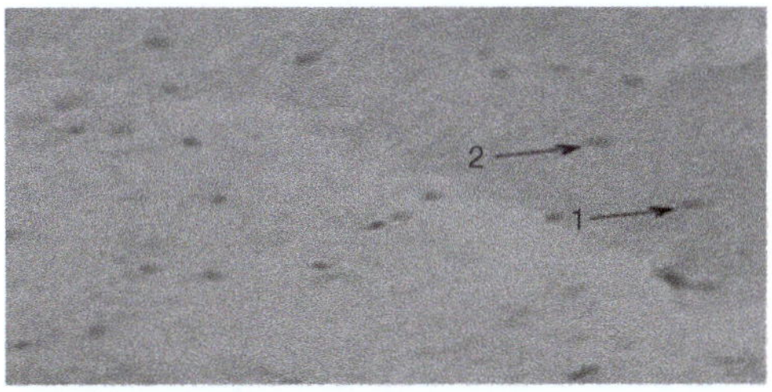

Figure 10.9 Histological section of human brain substance from the area of nontraumatic hemorrhage (5 days). Staining: histochemical Nissl method. Microphotograph. Magnification: objective 40×, eyepiece 10×: 1, rounded neuron; 2, disappearance of Nissl substance in processes.

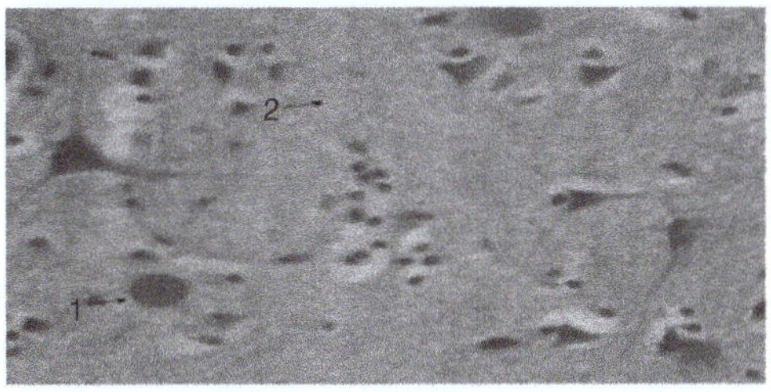

Figure 10.10 Histological section of the human brain from the area of hemorrhage of traumatic genesis (4 days). Staining: histochemical method according to Nissl. Microphotograph. Magnification: objective 40×, eyepiece 10×: 1, rounded neuron; 2, homogenization of Nissl substance.

Figure 10.11 Histological section of the ischemic cerebral infarction area (5 days). Staining: histochemical method according to Spielmeyer. Microphotograph. Magnification: objective 40×, eyepiece 10×: 1, absence of myelin fibers in the necrotic area; 2, signs of destruction of myelin fibers along the periphery of the necrotic area; 3, myelin fibers present in the preserved areas.

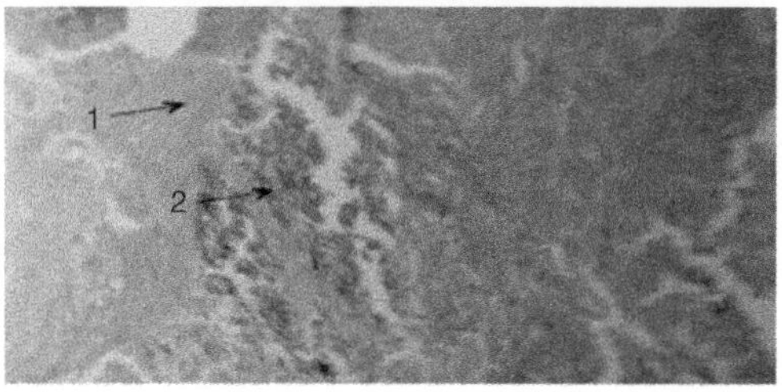

Figure 10.12 Histological section of human brain tissue from a nontraumatic hemorrhage zone (13 days). Staining: histochemical method according to Spielmeyer. Microphotograph. Magnification: objective 40×, eyepiece 10×: 1, absence of myelin fibers in the hemorrhage zone; 2, remains of myelin fiber fragments.

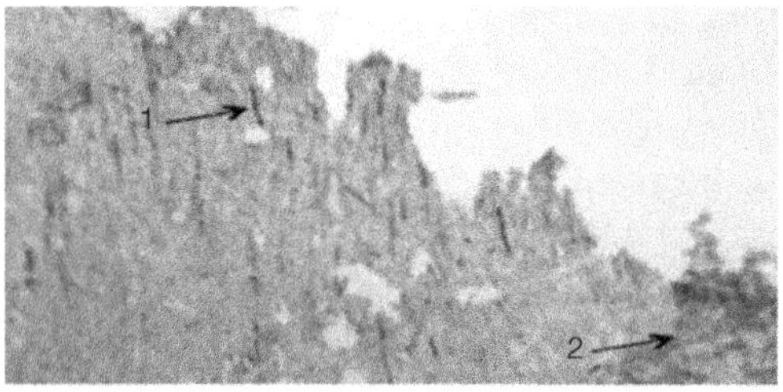

Figure 10.13 Histological section of the human brain from the area of hemorrhage of traumatic genesis (2–15 days). Staining: histochemical method according to Spielmeyer. Microphotograph. Magnification: objective 40×, eyepiece 10×: 1, single fragments of the remains of myelin fibers in the area of destroyed brain tissue with hemorrhages; 2, hemorrhage.

Evaluation of observations of human brain matter from HNG zones (2–16 days) demonstrated complete destruction of myelin fibers in areas of necrosis and hemorrhage, with in some places minor remnants of short fragments of myelin fibers (Figure 10.12).

In the HTG study groups (2–15 days), single fragments of myelin fiber remnants were found in the area of destroyed brain tissue with hemorrhages, as shown in Figure 10.13.

Analysis of the obtained data from the histological study of morphological changes in the tissue elements of the human brain with different genesis of hemorrhage formation did not reveal stable relationships between changes in the structure of the nervous tissue and the cause of the hemorrhage.

Therefore, the classical method of histological differential diagnostics of the genesis and time of hemorrhage formation in HBS is not effective and cannot be used to find constant and stable relationships between the cause of hemorrhage formation and changes in the structure of cells and tissue of the deceased.

Thus, given the nonspecificity of degenerative changes in the structural elements of the brain substance, it can be concluded that standard methods of histological studies of brain preparations do not provide accurate and objective information about the genesis of cerebral hemorrhages.

At the same time, the possibility of diagnosing degenerative changes in HBS and their fairly detailed descriptive characteristics has been proven. Therefore, it is important to continue searching for other, objective methods for diagnosing changes in the structure of HBS depending on the cause of death.

10.4.3 Perls Staining Technique

During microscopic examination of histological samples stained using the Perls histochemical method, iron salts were visualized as blue deposits due to their interaction with blood salts. The intense bright blue coloration of Prussian blue is likely associated with the presence of iron in multiple oxidation states (ferrous and ferric forms). The presence of atoms of a single element in different oxidation states within a compound often results in the appearance or enhancement of color intensity.

The formation of hemorrhage in the hematoencephalic barrier structure (HBS), followed by extensive red blood cell disintegration, leads to the accumulation of iron-containing products at hemorrhagic sites. Consequently, iron appears in tissues where it is normally absent.

Analysis of histological samples of human brain tissue with hemorrhages of different origins and varying times of formation, examined using light microscopy and stained with the Perls method, revealed a number of characteristic alterations in tissue architecture associated with ischemic, hemorrhagic and traumatic lesions.

In observations of nontraumatic hemorrhages (HNG) in samples collected 3–15 days after injury, the reaction to hemosiderin was strongly positive in areas of hemorrhage and necrosis. Large accumulations of hemosiderin and individual deep pigment inclusions were detected in the perifocal regions, as demonstrated in Figure 10.14.

Hemosiderin was present in 24 of 30 samples with varying degrees of brightness of Prussian blue.

In micropreparations of 6- and 15-day hemorrhages, the pigment was visualized as strongly stained, in 10- and 13-day samples from the moment of hemorrhage formation, it had a weakly stained color (Figure 10.15).

In HBS samples with HNG aged 3 days, hemosiderin was not found at all, as was the case in the 7-day sample.

In HTG observations (samples 2–12 days), the reaction to hemosiderin in HBS was assessed as weakly positive – in the foci of hemorrhage, small lumps of intensely stained hemosiderin were observed in places, in particular in the 10-day preparation.

In 2-, 3-, 6- and 11-day samples, pigment was absent. This fact may have been due to the relatively small volume of hemorrhage. It is entirely logical that in the overwhelming majority of observations of ischemic infarctions, the Perls reaction was negative.

However, it should be noted that in individual samples from the ischemic infarction study group, usually after 7 days, a weakly positive Perls reaction was observed, indicating late iron deposition (Figure 10.16).

These findings are obviously related to secondary hemorrhagic infiltration of infarct foci.

The described changes are shown using Figure 10.17.

It is quite logical that in the observations of the control group with IHD (samples 2–14 days), there were no traces of hemosiderin in HBS.

A statistical analysis of the presence of hemosiderin in HBS samples was performed depending on the time of hemorrhage onset. When comparing, no statistically probable difference was found ($\chi^2 = 6.04$; df $= 5$; $P = 0.112$) between the HNG and HTG groups, which confirms the data obtained in the preliminary analysis of the samples.

Considering the fact that it is not always possible to detect pigment in a sample and the lack of a direct relationship between the appearance of hemosiderin and the age of hemorrhage, it can be concluded that the histological method of studying brain matter does not provide accurate and objective information regarding the age of hemorrhage.

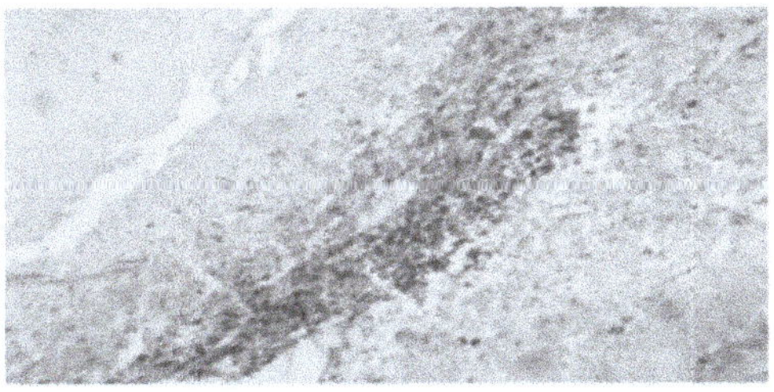

Figure 10.14 Hemorrhage into the human brain tissue of nontraumatic genesis (13 days). Staining: Perls histochemical method. Microphotograph. Magnification: objective 40×, eyepiece 10×: reaction to hemosiderin is intensely positive in areas of hemorrhage and necrosis (arrow).

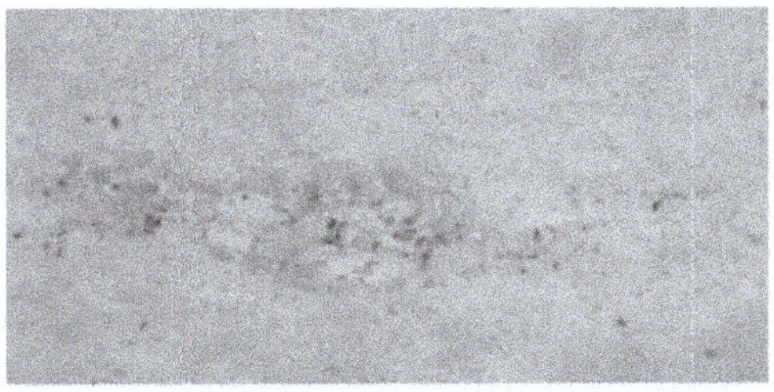

Figure 10.15 Histological section of human brain matter from a site of nontraumatic hemorrhage (10 days). Staining: Perls histochemical method. Microphotograph. Magnification: objective 40×, eyepiece 10×: weakly positive reaction to hemosiderin in the hemorrhage sites (the arrow indicates individual hemosiderin accumulations).

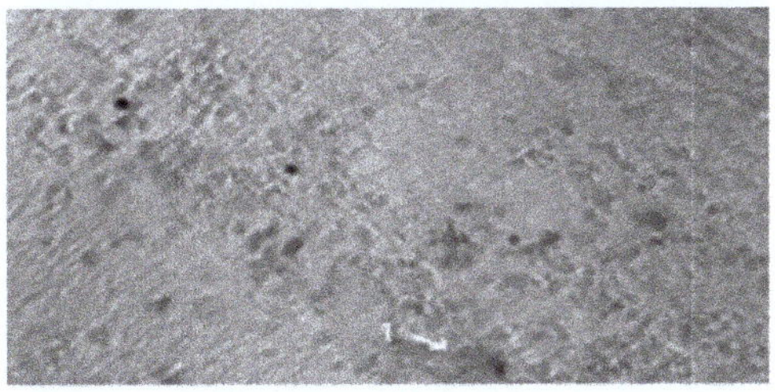

Figure 10.16 Hemorrhage of traumatic genesis (10 days) in a histological section of human brain matter. Staining: Perls histochemical method. Microphotograph. Magnification: objective 40×, eyepiece 10×: 1, small lumps of hemosiderin in hemorrhage foci.

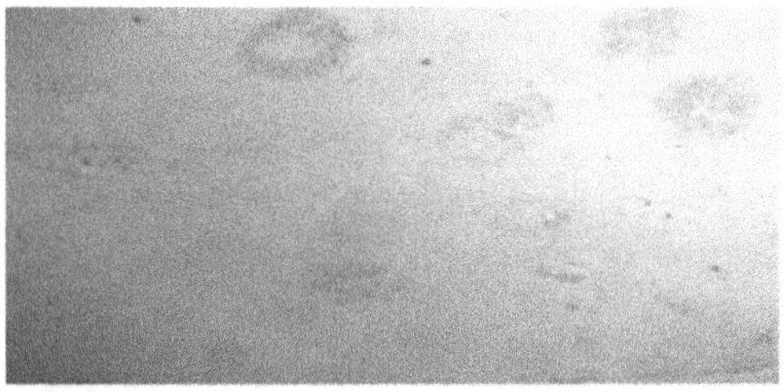

Figure 10.17 Ischemic infarction (7 days) in a histological section of human brain matter. Staining: Perls histochemical method. Microphotograph. Magnification: objective 40×, eyepiece 10×: arrow points to small foci of hemosiderin.

Thus, the performed histological studies of the evolution of cerebral ischemic infarctions and hemorrhages of traumatic and nontraumatic genesis revealed temporal patterns in all groups that have diagnostic value for establishing the chronology of HBS damage.

10.4.4 Operational Characteristics of the Effectiveness of Differential Diagnostics of Injuries of Various Genesis

It was established that for each of the cases – control group 1 (myocardial infarction – 60 samples), experimental group 2 (hemorrhages of traumatic genesis – 60 samples), experimental group 3 (cerebral infarction of ischemic genesis – 60 samples) and experimental group 4 (cerebral infarction of hemorrhagic genesis – 60 samples) as a result of histological analysis the following indicators (sensitivity Se, specificity Sp, balanced accuracy Ac) of the diagnostic power of the method are achieved:

- Groups "1 (60 samples) – 2 + 3 + 4 (180 samples)" – sensitivity Se = 88.9% (number of false positive conclusions $b = 20$), specificity Sp = 85% (number of false negative conclusions $d = 4$), balanced accuracy Ac = 87.9%;
- Groups "2 (60 samples) – 3 (60 samples)" – sensitivity Se = 88.3% (number of false positive conclusions $b = 7$), specificity Sp = 85% (number of false negative conclusions $d = 9$), balanced accuracy Ac = 86.7%;
- Groups "2 (60 samples) – 4 (60 samples)" – sensitivity Se = 86.7% (number of false positive conclusions $b = 8$), specificity Sp = 83.3% (number of false negative conclusions $d = 10$), balanced accuracy Ac = 85%;
- Groups "3 (60 samples) – 4 (60 samples)" – sensitivity Se = 83.3% (number of false positive conclusions $b = 10$), specificity Sp = 81.7% (number of false negative conclusions $d = 11$), balanced accuracy Ac = 82.5%.

The established parameters of the diagnostic power of the traditional method of histological research were used as reference, comparative values with the data on the diagnostic efficiency of a combination of polarization, Mueller-matrix, polarization-correlation, and tomographic methods (Table 10.4).

10.4.5 Results of Application of Traditional Histological Methods of Diagnostics of Genesis and Duration of Formation of Hemorrhages in the Substance of Human Brain

The Nissl technique demonstrated the degree of preservation of neuronal integrity and glial reactivity in response to damage of various genesis. In cases of ischemic infarction of human brain matter, early manifestations (1–3 days) included pronounced cytoplasmic eosinophilia and nuclear pyknosis, which evolved into widespread chromatolysis and vacuolization by 4–7 days, and led to almost complete loss of neurons with pronounced astroglial proliferation by 8–14 days. In observations of nontraumatic hemorrhages in HBS, a delayed neuronal reaction was established: the structure of neurons, preserved in the early stages, was gradually replaced by neuronal loss and perifocal gliosis. Traumatic hemorrhages had the characteristics of a sudden and heterogeneous lesion, with pronounced neuronal destruction, microvascular ruptures and an intense inflammatory response. Axonal delamination is particularly characteristic, particularly in the deep white matter and at the border of the cortex and subcortex. Staining according to Spielmeyer turned out to be especially informative in these cases: already on the 1–3 day after the injury, fragmented, and swollen myelin fibers were detected. On the 4–7 day, foci of demyelination and vacuolization of the white matter were observed, which by the 8–14 day progressed with fusion into foci of complete loss of myelin. These changes were in sharp contrast to the more diffuse and delayed demyelination observed in ischemic infarctions and HNG, highlighting the diagnostic value of this method for recognizing traumatic injuries.

The Perls histochemical method allowed us to visualize iron metabolism in hemorrhage foci quite accurately. In both types of hemorrhage – traumatic and nontraumatic genesis – hemosiderin deposits were noticeable between the fourth and seventh days, with subsequent increase in intensity and spread up to 8–14 days. Iron-positive macrophages and perivascular hemosiderin accumulation were particularly pronounced in HBS samples with HTG. It should be noted that in some observations of ischemic infarctions, usually after 7 days, a weakly positive result of the Perls reaction was noted, which

Table 10.4 Specificity, sensitivity, accuracy of histological evaluation of microscopic images of brain preparations.

Groups "1 (60 samples) – 2 + 3 + 4 (180 samples)"			
Parameters	**Sensitivity, Se,**	**Specificity, Sp, %**	**Accuracy, Ac, %**
	$a = 160; b = 20$ 88.9	$c = 51; d = 9$ 85	$n = 240$ 87.9
Groups "2 (60 samples) – 3 (60 samples)"			
Parameters	**Sensitivity, Se, %**	**Specificity, Sp, %**	**Accuracy, Ac, %**
	$a = 53; b = 7$ 88.3	$c = 51; d = 9$ 85	$n = 120$ 86.7
Groups "2 (60 samples) – 4 (60 samples)"			
Parameters	**Sensitivity, Se, %**	**Specificity, Sp, %**	**Accuracy, Ac, %**
	$a = 52; b = 8$ 86.7	$c = 50; d = 10$ 83.3	$n = 120$ 85
Groups "3 (60 samples) – 4 (60 samples)"			
Parameters	**Sensitivity, Se, %**	**Specificity, Sp, %**	**Accuracy, Ac, %**
	$a = 50; b = 10$ 83.3	$c = 49; d = 11$ 81.7	$n = 360$ 82.5

indicated late iron deposition. These findings are probably associated with secondary hemorrhagic infiltration of infarction foci. The possibility of delayed changes in infarcts and perivascular localization of iron should be taken into account to differentiate the timing and mechanisms of HBS damage during complex histological analysis. Based on the analysis of the results of histochemical studies, it was established that the corresponding time dependencies of the duration of hemorrhage are not linear, but, on the contrary, have fluctuations in their own values – uneven appearance of pigment in HBS with a gradual increase in time from the moment of formation of hemorrhages of various genesis.

However, the conducted studies indicate the presence of changes in the structure of HBS. Therefore, it is important to continue the search for other, objective methods for diagnosing the genesis, and duration of hemorrhage in HBS.

1. The Nissl method made it possible to identify the dynamics of neuronal and glial changes in ischemic infarctions and hemorrhages:
   - ICI – characterized by progressive chromatolysis, vacuolization, and almost complete loss of neurons up to 8–14 days, which is accompanied by astroglial proliferation;
   - HNG – demonstrate a delayed neuronal response with substitution by gliosis;
   - HTG – have a sudden heterogeneous lesion with pronounced destruction of neurons, micro-ruptures of vessels, inflammation, and axonal delamination in the white matter.
2. The Spielmeyer staining method proved to be highly informative for the differentiation of traumatic injuries:
   - fragmented and edematous myelin fibers are observed already on the first to third day;
   - by the fourth to seventh days, foci of demyelination are formed, which by the 8th–14th day turn into zones of complete myelin loss;
   - these changes differ from the more gradual and diffuse demyelination in ischemic infarctions.
3. The Perls reaction revealed the peculiarities of iron metabolism:
   - HNG and HTG: intensive hemosiderin deposition from 4 to 7 days, with a peak up to 8 to 14 days.
   - ICI – weak positive reaction after 7 days, which may be associated with secondary hemorrhagic infiltration.
     - Iron-positive macrophages and perivascular hemosiderin are especially pronounced in traumatic injuries; however, the dynamics of pigment accumulation and structural changes is oscillatory, which makes it difficult to more accurately determine the age of the injury.

4. Histochemical methods (Nissl, Spielmeyer, Perls) are effective for detecting morphological changes in various types of brain damage, but these changes are nonspecific and have limitations in terms of the accuracy of determining the age. To improve diagnostics, these methods need to be integrated with laser tomographic approaches.

10.5 Forensic Medical Assessment of Cerebral Infarction, Hemorrhages of Traumatic and Nontraumatic Genesis Using 3D Mueller-matrix Microscopy

This section contains the following materials and results:

1. Developed structural-logical scheme and research design of the latest method of 3D Mueller-matrix microscopy of the layer-by-layer structure of the polycrystalline component of depolarizing hilogical sections of the brain;
2. Differential diagnostics of hemorrhage formation of traumatic genesis, cerebral infarction of ischemic and hemorrhagic genesis using 3D Mueller-matrix microscopy:
   - layered azimuthally invariant Mueller matrix invariant of linear birefringence (MMI LB) of histological sections of the brain and operational characteristics of the method of their statistical analysis;
   - layered azimuthally invariant Mueller matrix invariant of optical activity (circular birefringence) (MMI OA) of histological sections of the brain and operational characteristics of the method of their statistical analysis.

10.5.1 Structural and Logical Scheme of 3D Mueller-matrix Microscopy of Histological Sections of the Brain

The design of the study of the volumetric polycrystalline structure of histological sections of the brain of deceased people of all groups includes the following experimental and algorithmic stages (Table 10.5):

1. The use of an additional reference laser wave superimposed on a polarization-inhomogeneous microscopic image of a depolarizing histological section of the brain;
2. Digital holographic reproduction of a set of layer-by-layer distributions of the magnitude of azimuthally invariant Mueller-matrix parameters characterizing volumetric distributions of linear and circular birefringence; and
3. Identification of additional forensic criteria for improving the effectiveness of the Mueller-matrix microscopy method for the objective differentiation of brain tissue samples of patients who died from ischemic heart disease (control group 1), traumatic hemorrhage (experimental group 2), cerebral infarction of ischemic (study group 3) and hemorrhagic (study group 4) genesis.

10.5.2 Differential Diagnostics of Hemorrhage Formation of Traumatic Genesis, Cerebral Infarction of Ischemic and Hemorrhagic Genesis Using 3D Mueller-matrix Microscopy

This part of the work presents the results of statistical analysis of layer-by-layer maps of the distribution of the magnitude of the Mueller matrix invariant (MMI), obtained using the method of phase scanning and digital holographic reconstruction of the distribution of complex amplitudes of the laser field transformed within the optically anisotropic volume of histological sections of nervous tissue of the brain.

Table 10.5 Structural and logical scheme of 3D Mueller-matrix microscopy of histological sections of the brain.

Histological sections of the brain of deceased people			
Control group of deceased (Group 1)	Hemorrhage of traumatic genesis (Group 2)	Cerebral infarction of ischemic genesis (Group 3)	Cerebral infarction of hemorrhagic genesis (Group 4)
3D Mueller-matrix mapping			
Layered maps of optical activity (circular birefringence) (MMI OA)		Layered maps of linear birefringence (MMI LB)	
Statistical analysis of layered maps of MMI OA, MMI LB			
Average values and standard deviations of the magnitude of statistical moments of the third to fourth orders characterizing coordinate distributions MMI OA, MMI LB			
Criteria for differential diagnosis of histological brain sections of deceased persons from Groups 1–4			

It should be noted that phase scanning enables algorithmic selection and reconstruction of cross-sections representing polarization manifestations of the object's properties under different scattering multiplicities, ranging from minimal to maximal. The minimum phase corresponds to nearly single scattering, which is analogous to an optically thin layer of biological tissue. The maximum phase reflects multiple scattering within the volume of a real sample under study.

10.5.2.1 Layered Azimuth-invariant MMI OA of Histological Brain Sections

The results of the study of layer-by-layer coordinate distributions of the Mueller matrix invariant of circular birefringence (MMI CB) of protein complexes of nervous tissue of samples of histological sections of the brain of deceased people from different groups using the 3D Mueller-matrix microscopy method are presented in a series of fragments (Figures 10.18–10.20).

We used the following phase sections of the 3D distribution of the field of complex amplitudes for digital holographic reproduction of the MMI CB:

- $\Delta = 0.4$ rad – almost single scattering;
- $\Delta = 0.8$ rad – medium multiplicity scattering;
- $\Delta = 1.2$ rad – multiple scattering in the volume of a real sample.

From the analysis of the obtained data of 3D Mueller-matrix mapping of layer-by-layer polarization manifestations of optical activity of a set of histological sections of the brain for cases of ischemic heart disease (Group 1), traumatic hemorrhage (Group 2), hemorrhagic cerebral infarction (Group 3), and ischemic (Group 4) genesis, it was established:

1. Phase section $\Delta = 0.4$ rad:
   - the individual topographic structure of all MMI CB maps of this phase section of the optically anisotropic component of histological sections of the nervous tissue of the brain of deceased persons from all groups (Figure 10.18a, c, e, g);
   - for 3D distributions of the optical activity values of brain samples from all (control 1 and experimental 2–4) groups, significant values of the average and a spread of values are characteristic (see Figure 10.18b, d, f, h).

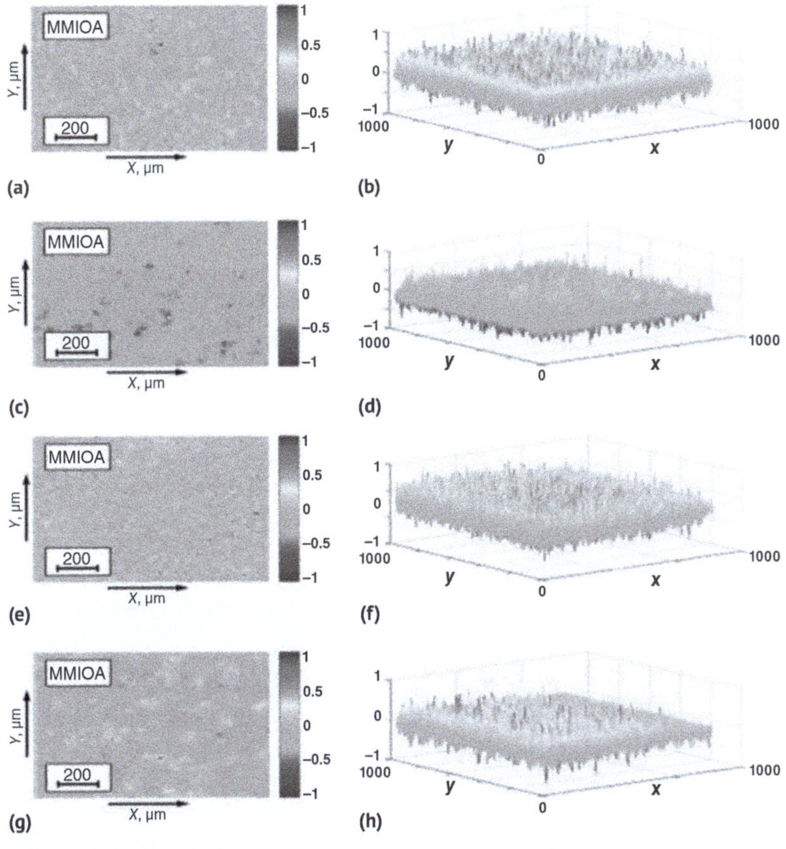

Figure 10.18 Two-dimensional (a), (c), (e), (g) and 3D (b), (d), (f), (h) distribution maps of the MMI CB value of histological sections of the brains of deceased individuals from Group 1 (a), (b), Group 2 (c), (d), Group 3 (e), (f), and Group 4 (g), (h) for the phase section $\Delta = 0.4$ rad.

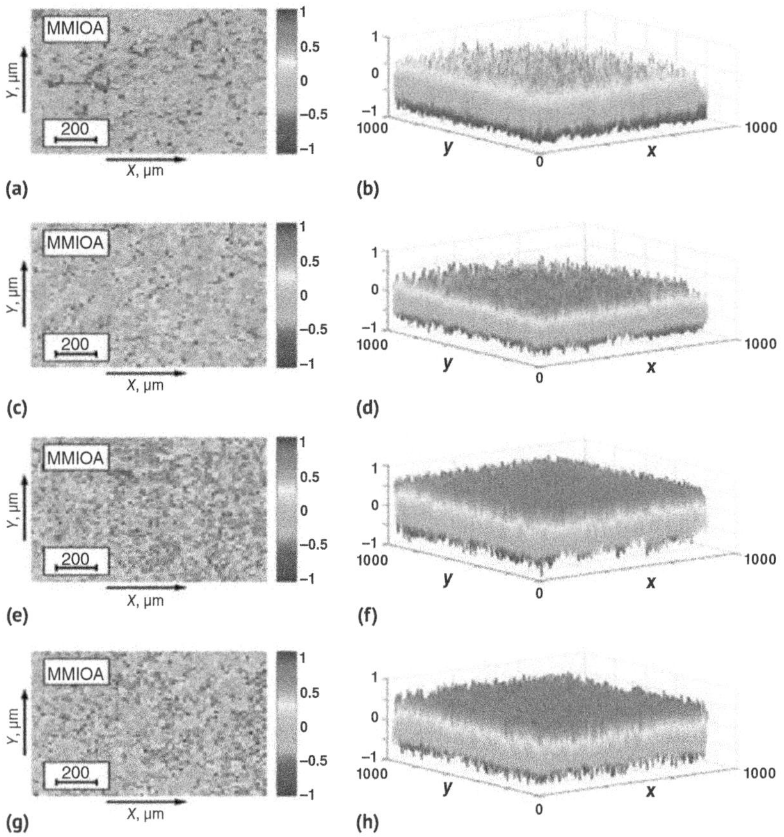

Figure 10.19 Two-dimensional (a), (c), (e), (g) and 3D (b), (d), (f), (h) distribution maps of the MMI CB value of histological sections of the brains of deceased individuals from Group 1 (a), (b), Group 2 (c), (d), Group 3 (e), (f), and Group 4 (g), (h) for the phase section $\Delta = 0.8$ rad.

2. Phase section $\Delta = 0.8$ rad: for the 3D layered map of MMI CB in the phase section with an intermediate scattering multiplicity, samples from all (control 1 and experimental 2–4) groups are characterized by an increase in the spread of values (see Figure 10.19).
3. Phase section $\Delta = 1.2$ rad: close to the results of direct Mueller-matrix polarimetry of all integral maps of the MMI CB histological sections of the nervous tissue of the brain of deceased people from all groups topological structure of the distributions of the values of circular birefringence in the phase plane formed by multiple scattering of laser radiation throughout the entire volume of the sample.

The series of Tables 10.6–10.8 presents the average values and errors of representativeness of the determination of a set of statistical moments of the first and fourth order, characterizing the layer-by-layer Mueller-matrix images of polarization manifestations of circular birefringence of optically active protein complexes of nervous tissue.

A comparative analysis of the obtained data showed:

1. Phase cross-section $\Delta = 0.4$ rad (see Table 10.5):
   Maximum intergroup differences in the value of the set of statistical moments of the first to fourth orders and statistical significance of differentiation ($p < 0.05$) of cases of ischemic heart disease (control group 1), traumatic hemorrhage (experimental group 2), cerebral infarction of ischemic (study group 3), and hemorrhagic (study group 4) genesis.
2. Phase cross-section $\Delta = 0.8$ rad (see Table 10.7):
   Significant intergroup differences in average values within representative samples of histological brain sections of the value of the set of statistical moments of the first to fourth orders and the statistical significance of differentiation of cases from the control group 1) and research groups 2–4.
3. Phase cross-section $\Delta = 1.2$ rad (see Table 10.8):
   the possibility of statistically reliable differentiation, as for the results of direct Mueller-matrix polarimetry of all integral maps of the MMI CB of histological sections of the nervous tissue of the brain of deceased persons from all groups, is realized only between Group 1 and Groups 2–4.

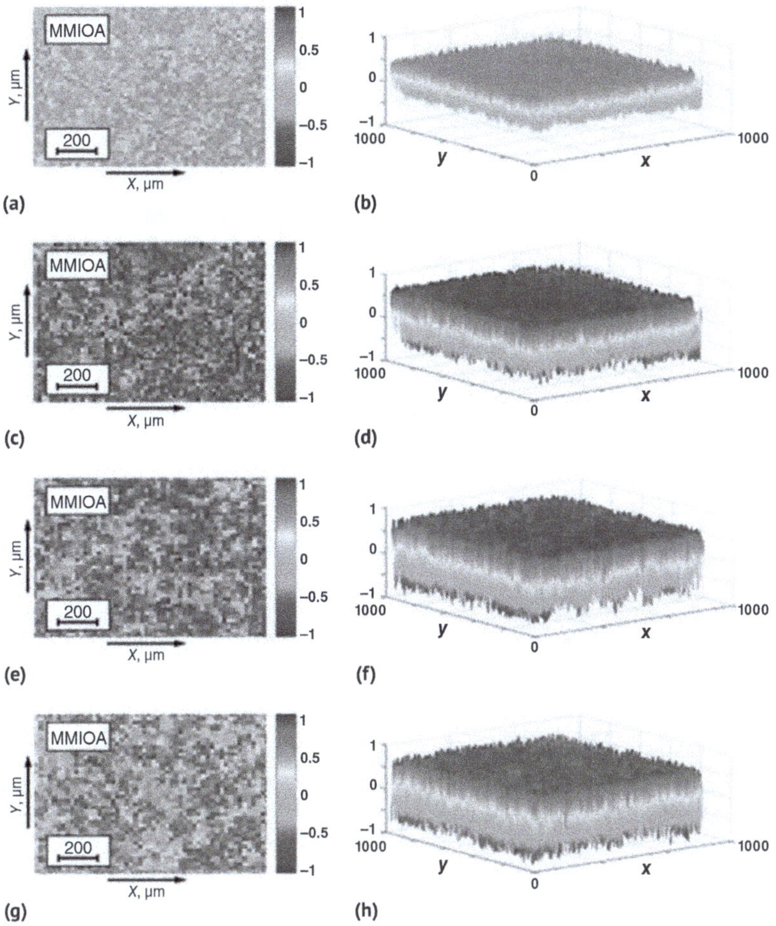

Figure 10.20 Two-dimensional (a), (c), (e), (g) and 3D (b), (d), (f), (h) distribution maps of the MMI CB value of histological sections of the brains of deceased individuals from Group 1 (a), (b), Group 2 (c), (d), Group 3 (e), (f), and Group 4 (g), (h) for the phase section $\Delta = 1.2$ rad.

Table 10.6 Statistical moments of first and fourth orders characterizing the distribution of the magnitude of the MMI OA of histological sections of the brain of Groups 1–4 for the phase section $\Delta = 0.4$ rad.

Parameters	Group 1	Group 2	Group 3	Group 4
$\overline{M_3} \pm m_3$	0.87 ± 0.035	1.12 ± 0.059	1.57 ± 0.063	1.98 ± 0.089
p_{ik}	$p_{1(2;3;4)}(t = 8.74) < 0.001$; $p_{23}(t = 5.214) < 0.001$; $p_{24}(t = 8.054) < 0.001$; $p_{34}(t = 3.76) < 0.001$			
$\overline{M_4} \pm m_4$	1.88 ± 0.087	2.56 ± 0.11	3.09 ± 0.13	3.89 ± 0.15
p_{ik}	$p_{1(2;3;4)}(t = 8.311) < 0.001$; $p_{23}(t = 3.112) < 0.001$; $p_{24}(t = 7.15) < 0.05$; $p_{34}(t = 4.3) < 0.001$			

Table 10.7 Statistical moments of first to fourth orders characterizing the distribution of the magnitude of the MMI OA of histological sections of the brain of Groups 1–4 for the phase section $\Delta = 0.8$ rad.

Parameters	Group 1	Group 2	Group 3	Group 4
$\overline{M_3} \pm m_3$	0.68 ± 0.026	0.82 ± 0.035	1.05 ± 0.043	1.29 ± 0.052
p_{ik}	$p_{1(2;3;4)}(t = 7.388) < 0.001$; $p_{23}(t = 4.148) < 0.001$; $p_{24}(t = 7.498) < 0.001$; $p_{34}(t = 3.557) < 0.001$			
$\overline{M_4} \pm m_4$	1.18 ± 0.047	1.45 ± 0.066	1.89 ± 0.083	2.28 ± 0.094
p_{ik}	$p_{1(2;3;4)}(t = 7.404) < 0.001$; $p_{23}(t = 4.149) < 0.001$; $p_{24}(t = 7.226) < 0.001$; $p_{34}(t = 3.11) < 0.001$			

Table 10.8 Statistical moments of first to fourth orders characterizing the distribution of the magnitude of the MMI OA of histological sections of the brain of groups 1–4 for the phase section $\Delta = 1.2$ rad.

Parameters	Group 1	Group 2	Group 3	Group 4
$\overline{M_3} \pm m_3$	0.56 ± 0.019	0.62 ± 0.023	0.75 ± 0.028	0.82 ± 0.033
p_{ik}	$p < 0.05$	$p > 0.05$		
$\overline{M_4} \pm m_4$	0.71 ± 0.028	0.84 ± 0.036	1.09 ± 0.044	1.12 ± 0.058
p_{ik}	$p < 0.05$	$p > 0.05$		

Table 10.9 Balanced accuracy of the method of statistical analysis of MMI OA maps of histological brain sections for the phase section $\Delta = 0.4$ rad.

Parameters	Balanced accuracy Ac, %		
Groups	"2"–"4"	"2"–"3"	"3"–"4"
M_3	95 $a = 98; b = 2;$ $c = 92; d = 8$	92 $a = 94; b = 6;$ $c = 90; d = 10$	85 $a = 88; b = 12;$ $c = 82; d = 18$
M_4	95 $a = 98; b = 2;$ $c = 92; d = 8$	92 $a = 94; b = 6;$ $c = 90; d = 10$	85 $a = 88; b = 12;$ $c = 82; d = 18$

Table 10.10 Balanced accuracy of the method of statistical analysis of MMI OA maps of histological brain sections for the phase section $\Delta = 0.8$ rad.

Parameters	Balanced accuracy Ac, %		
Groups	"2"–"4"	"2"–"3"	"3"–"4"
M_3	88 $a = 90; b = 10;$ $c = 86; d = 14$	85 $a = 88; b = 12;$ $c = 82; d = 18$	78 $a = 80; b = 20;$ $c = 76; d = 24$
M_4	88 $a = 90; b = 10;$ $c = 86; d = 14$	86 $a = 88; b = 12;$ $c = 84; d = 16$	76 $a = 78; b = 22;$ $c = 74; d = 26$

Table 10.11 Balanced accuracy of the method of statistical analysis of MMI OA maps of histological brain sections for the phase section $\Delta = 1.2$ rad.

Parameters	Balanced accuracy Ac, %		
Groups	"2"–"4"	"2"–"3"	"3"–"4"
M_3	76 $a = 78; b = 22;$ $c = 74; d = 26$	74 $a = 76; b = 24;$ $c = 72; d = 28$	68 $a = 70; b = 30;$ $c = 66; d = 34$
M_4	80 $a = 82; b = 18;$ $c = 78; d = 22$	76 $a = 78; b = 22;$ $c = 74; d = 26$	72 $a = 74; b = 26;$ $c = 70; d = 30$

10.5.2.2 Operational Characteristics of the Method of Statistical Analysis of Layered Maps of MMI CB of Histological Sections of the Brain

The series Tables 10.9–10.11 presents the results of the information analysis of the determination of the strength (balanced accuracy) of the 3D Mueller-matrix mapping method in different phase planes of the distributions of the MMI CB value.

The possibility and effectiveness of differentiation were considered:

- control group 1 – all experimental groups 2+3+4;
- experimental group 2 – experimental group 4;
- experimental group 2 – experimental group 3;
- experimental group 3 – experimental group 4.

A comparative analysis of the obtained data on the balanced accuracy of differentiation of cases of traumatic hemorrhage, cerebral infarction of hemorrhagic and ischemic genesis using the 3D Mueller-matrix mapping MMI CB method revealed:

1. Phase cross-section $\Delta = 0.4$ rad (see Table 10.9):
   The maximum level of balanced accuracy of intergroup differentiation by calculating the statistical moments of the 3rd and 4th order, characterizing the asymmetry and excess of the distribution of the MMI CB value of histological brain sections:
   - "2" – "4": excellent accuracy 95–96% (the number of false positive and false negative conclusions is 2 and 8, respectively);
   - "2" – "3": good accuracy 92–93% (the number of false positive and false negative conclusions is 6 and 10, respectively);
   - "3" – "4": satisfactory accuracy 85–86% (the number of false positive and false negative conclusions is 12 and 18, respectively).
2. Phase cross-section $\Delta = 0.8$ rad (see Table 10.10):
   A significant reduction (by 10–15%) to a satisfactory level of balanced accuracy of intergroup differentiation by calculating the asymmetry and excess of the distributions of the MMI CB value of histological brain sections:
   - "2" – "4": satisfactory accuracy 89–90% (the number of false positive b and false negative d conclusions is 10 and 14, respectively);
   - "2" – "3": satisfactory accuracy 89–90% (the number of false positive b and false negative d conclusions is 12 and 18, respectively);
   - "3"–"4": unsatisfactory accuracy 76–78% (the number of false positive b and false negative d conclusions is 20 and 24, respectively).
3. Phase cross-section $\Delta = 1.2$ rad (see Table 10.11):
   Critical reduction to an unsatisfactory level of balanced accuracy of intergroup differentiation by calculating the asymmetry and excess of the distributions of the value of the MMI CB of histological brain sections:
   - "2"–"4": unsatisfactory accuracy 78–82% (the number of false positive and false negative conclusions is 18 and 22, respectively);
   - "2"–"3": unsatisfactory accuracy 75%–80% (the number of false positive and false negative conclusions is 22 and 26, respectively);
   - "3"–"4": unsatisfactory accuracy of 75–76% (the number of false positive and false negative conclusions is 26 and 30, respectively).

10.5.3 Layered Azimuth-invariant MMI LB of Histological Brain Sections

The series of fragments in Figures 10.21–10.23 present the results of a study of holographically reproduced layer-by-layer coordinate distributions of the value of the MMI LB of spatially structured fibrillar networks of tissue from samples of histological sections of the brains of deceased people from different groups using the 3D Mueller-matrix microscopy method.

Analysis of 3D Mueller-matrix mapping data obtained by step-by-step phase scanning of the volume distribution of the field of complex amplitudes, layered maps of Mueller matrix invariants characterizing the polarization manifestations of structural anisotropy (linear birefringence) of the networks of protein fibrils and fibers of the nervous tissue of a set of histological sections of the brain for cases of ischemic heart disease (Group 1), traumatic hemorrhage (Group 2), cerebral infarction of hemorrhagic (Group 3) and ischemic (Group 4) genesis revealed:

1. Phase cross-section $\Delta = 0.4$ rad:
   1. coordinate-inhomogeneous, individual for each case (Groups 1–4), topographic structure of all two-dimensional MMI LB maps of a given phase section, corresponding to the conditions of single scattering of the optically anisotropic component of histological sections of the nervous tissue of the brain of deceased people from all groups (Figure 10.21);

   2. for a series of 3D distributions of the magnitude of linear birefringence, in comparison with the data of the study of the optical activity of brain samples (Figure 10.21) of the control 1 and experimental 2–4 groups, large values of the average and a spread of values are characteristic.
2. Phase cross-section $\Delta = 0.8$ rad:
   1. due to the increase in the multiplicity of light scattering, a more uniform, in comparison with the results of polarization-holographic 3D Mueller-matrix polarimetry of the distributions of the magnitude of linear birefringence in the optically thin phase plane ($\Delta = 0.4$ rad, topological structure of MMI LB samples of nervous tissue of the brain of deceased people from all groups (Figure 10.22);
   2. the statistical structure of the MMI LB maps is transformed – the distribution of the value of linear birefringence in the phase section with an intermediate scattering multiplicity of samples from the control 1 and experimental 2–4 groups is characterized by an increase in the spread of values (Figure 10.22).
3. Phase cross-section $\Delta = 1.2$ rad:
   1. the similarity of the results of the methods of direct Mueller-matrix polarimetric microscopy of all integral maps of the MMI LB of histological sections of the nervous tissue of the brain of deceased people of all groups the topological structure of the distributions of the magnitude of linear birefringence in the phase plane, which is formed by multiple scattering of laser radiation throughout the entire volume of the sample;
   2. 3D maps of MMI LB samples of histological sections of the brain from all have the maximum, but close in size, average and dispersion (Figure 10.23).

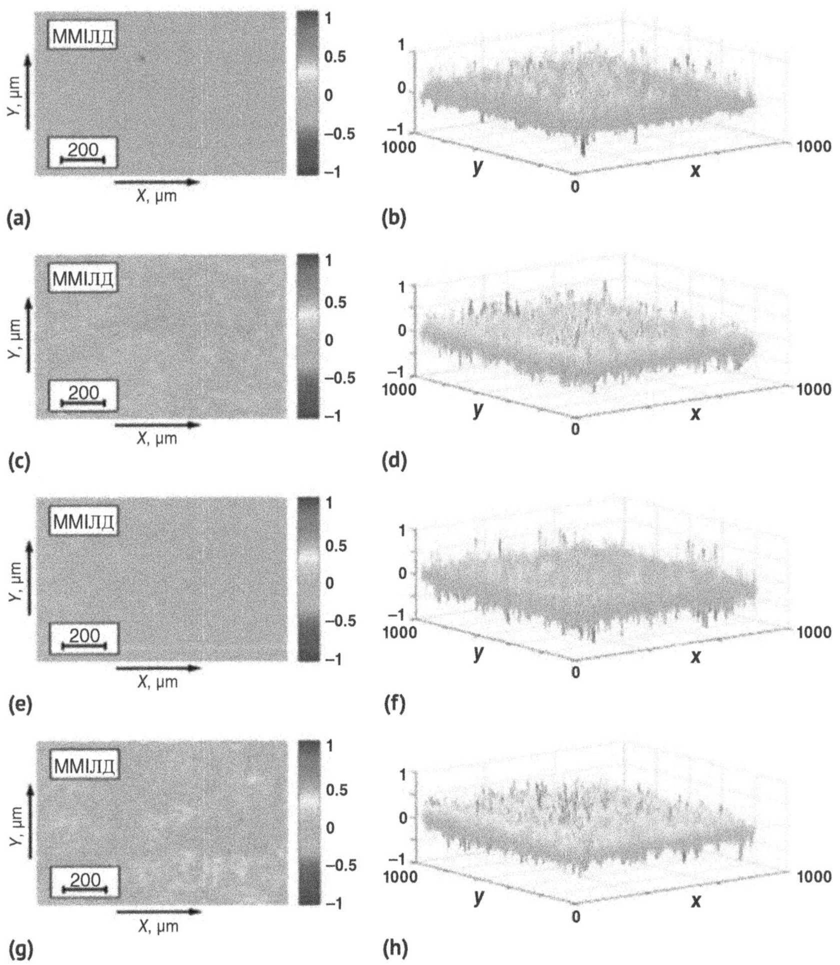

Figure 10.21 Two- (a), (c), (e), (g) and three-dimensional (b), (d), (f), (h) distribution maps of the MMI LB value of histological sections of the brains of deceased individuals from Group 1 (a), (b), Group 2 (c), (d), Group 3 (e), (f), and Group 4 (g), (h) for the phase section $\Delta = 0.4$ rad.

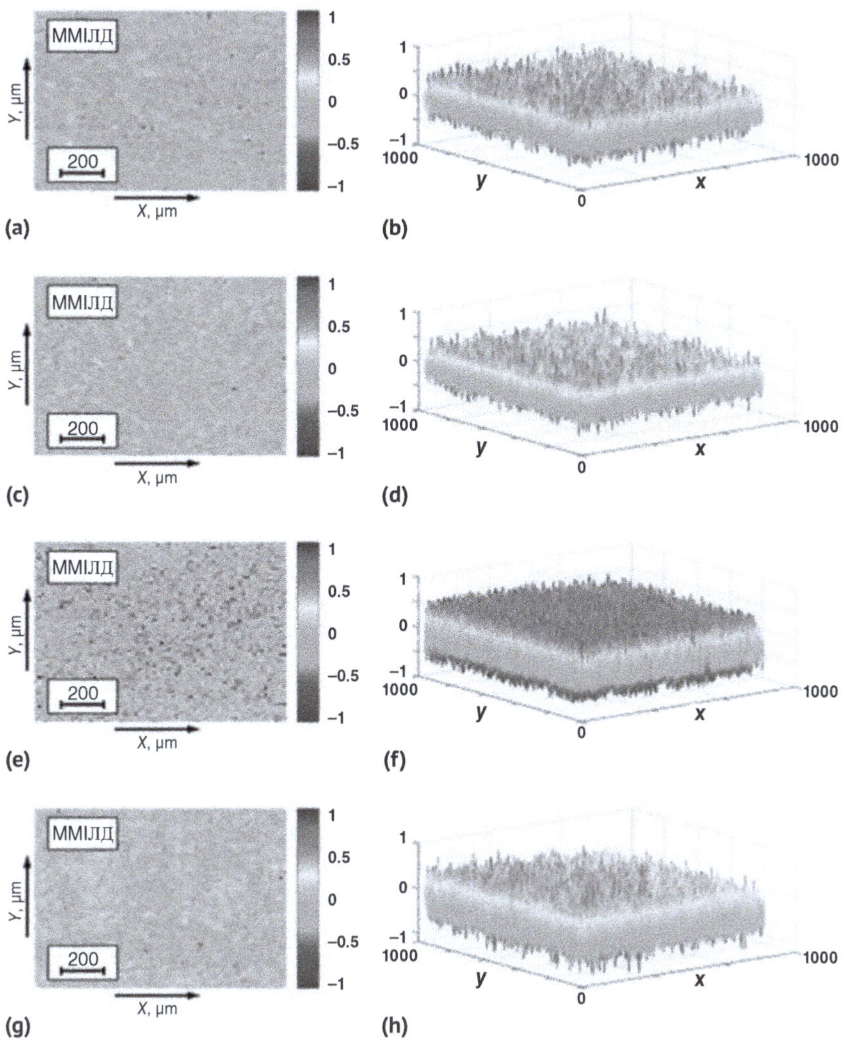

Figure 10.22 Two (a), (c), (e), (g) and three-dimensional (b), (d), (f), (h) distribution maps of the MMI LB value of histological sections of the brains of deceased individuals from Group 1 (a), (b), Group 2 (c), (d), Group 3 (e), (f) and Group 4 (g), (h) for the phase section $\Delta = 0.8$ rad.

The series of Tables 10.12–10.14 presents the average values and representativeness errors for determining a set of statistical moments of the first to fourth orders.

It is established:

1. Phase section $\Delta = 0.4$ rad (see Table 10.12):

   For all the studied groups, there are maximum intergroup differences in the value of the set of statistical moments of the first to fourth orders and statistical reliability of differentiation of cases of ischemic heart disease (control group 1), traumatic hemorrhage (experimental group 2), cerebral infarction of ischemic (experimental group 3) and hemorrhagic (experimental group 4) genesi.
2. Phase section $\Delta = 0.8$ rad (see Table 10.13):

   Significant intergroup differences in average values within representative samples of histological brain sections of the value of the set of statistical moments of the first to fourth orders and statistical reliability of differentiation of cases from the control group 1 and experimental groups 2–4.
3. Phase section $\Delta = 1.2$ rad (see Table 10.14):

   In contrast to the data from the statistical analysis of the MMI LB data, for statistical moments of the third and fourth orders, the possibility is realized.

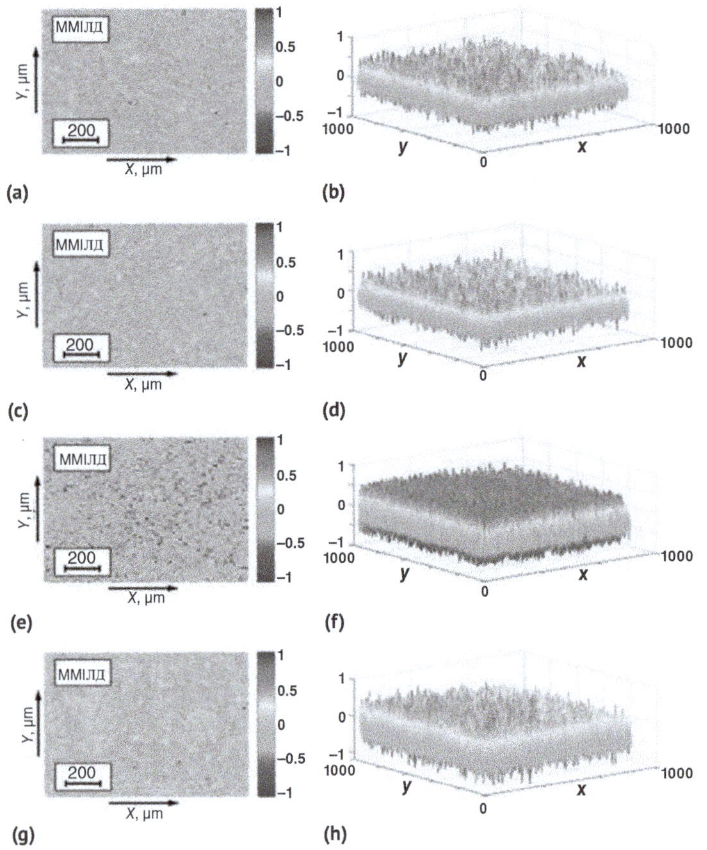

Figure 10.23 Two- (a), (c), (e), (g) and three-dimensional (b), (d), (f), (h) distribution maps of the MMI LB value of histological sections of the brains of deceased individuals from Group 1 (a), (b), Group 2 (c), (d), Group 3 (e), (g), and Group 4 (f), (h) for the phase section $\Delta = 1.2$ rad.

Table 10.12 Statistical moments of first to fourth orders characterizing the distribution of the MMI LB value of histological sections of the brain of Groups 1–4 for the phase section $\Delta = 0.4$ rad.

Parameters	Group 1	Group 2	Group 3	Group 4
$\overline{M_3} \pm m_3$	1.18 ± 0.042	1.82 ± 0.078	2.53 ± 0.11	3.19 ± 0.13
p_{ik}	$p_{1(2;3;4)}(t = 11.694) < 0.001$; $p_{23}(t = 5.265) < 0.001$; $p_{24}(t = 9.037) < 0.001$; $p_{34}(t = 3.876) < 0.001$			
$\overline{M_4} \pm m_4$	2.33 ± 0.105	3.29 ± 0.13	4.11 ± 0.17	5.48 ± 0.21
p_{ik}	$p_{1(2;3;4)}(t = 9.826) < 0.001$; $p_{23}(t = 3.831) < 0.001$; $p_{24}(t = 8.867) < 0.001$; $p_{34}(t = 5.071) < 0.001$			

Table 10.13 Statistical moments of first to fourth orders characterizing the distribution of the MMI LB value of histological sections of the brain of Groups 1–4 for the phase section $\Delta = 0.8$ rad.

Parameters	Group 1	Group 2	Group 3	Group 4
$\overline{M_3} \pm m_3$	0.81 ± 0.036	1.12 ± 0.047	1.55 ± 0.061	2.01 ± 0.092
p_{ik}	$p_{1(2;3;4)}(t = 9.899) < 0.001$; $p_{23}(t = 5.584) < 0.001$; $p_{24}(t = 8.615) < 0.001$; $p_{34}(t = 4.167) < 0.001$			
$\overline{M_4} \pm m_4$	1.78 ± 0.077	2.19 ± 0.089	2.71 ± 0.11	3.14 ± 0.14
p_{ik}	$p_{1(2;3;4)}(t = 6.582) < 0.001$; $p_{23}(t = 3.675) < 0.001$; $p_{24}(t = 5.727) < 0.001$; $p_{34}(t = 2.415) < 0.001$			

Table 10.14 Statistical moments of first to fourth orders characterizing the distribution of the MMI LB value of histological sections of the brain of Groups 1–4 for the phase section $\Delta = 1.2$ rad.

Parameters	Group 1	Group 2	Group 3	Group 4
$\overline{M_3} \pm m_3$	0.65 ± 0.025	0.74 ± 0.028	0.85 ± 0.036	1.01 ± 0.045
p_{ik}	$p_{1(2;3;4)}(t = 4.913) < 0.001$; $p_{23}(t = 2.412) < 0.001$; $p_{24}(t = 5.094) < 0.001$; $p_{34}(t = 2.776) < 0.001$			
$\overline{M_4} \pm m_4$	0.77 ± 0.031	0.89 ± 0.037	0.97 ± 0.041	1.14 ± 0.049
p_{ik}	$p_{1(2;3;4)}(t = 4.383) < 0.001$; $p_{23}(t = 1.449) < 0.001$; $p_{24}(t = 4.072) < 0.001$; $p_{34}(t = 2.661) < 0.001$			

Table 10.15 Balanced accuracy of the method of statistical analysis of MMI LB maps of histological brain sections for the phase section $\Delta = 0.4$ rad.

Parameters	Balanced accuracy Ac, %		
Groups	"2"–"4"	"2"–"3"	"3"–"4"
M_3	98 $a = 100$; $b = 0$; $c = 96$; $d = 4$	95 $a = 97$; $b = 3$; $c = 93$; $d = 7$	92 $a = 94$; $b = 6$; $c = 90$; $d = 10$
M_4	97 $a = 99$; $b = 1$; $c = 95$; $d = 5$	96 $a = 98$; $b = 2$; $c = 94$; $d = 6$	91.5 $a = 93$; $b = 7$; $c = 90$; $d = 10$

Table 10.16 Balanced accuracy of the method of statistical analysis of MMI LB maps of histological brain sections for the phase section $\Delta = 0.8$ rad.

Parameters	Balanced accuracy Ac, %		
Groups	"2"–"4"	"2"–"3"	"3"–"4"
M_3	92.5 $a = 94$; $b = 6$; $c = 91$; $d = 9$	89 $a = 92$; $b = 8$; $c = 86$; $d = 4$	85 $a = 89$; $b = 1$; $c = 81$; $d = 9$
M_4	92 $a = 94$; $b = 6$; $c = 90$; $d = 10$	87 $a = 89$; $b = 11$; $c = 85$; $d = 15$	86 $a = 88$; $b = 12$; $c = 84$; $d = 16$

10.5.4 Operational Characteristics of the Method of Statistical Analysis of Layered Maps of MMI LB of Histological Sections of the Brain

The series of Tables 10.15–10.17 presents the results of the information analysis of the determination of the power (balanced accuracy) of the 3D Mueller-matrix mapping method of topographic distributions of the magnitude of linear birefringence of spatially structured fibrillar networks of the nervous tissue of the brain in the differentiation of cases of traumatic hemorrhages, cerebral infarction of ischemic and hemorrhagic genesis.

From the analysis of the obtained data on the balanced accuracy of differentiation of cases of traumatic hemorrhage (experimental group 2), cerebral infarction of hemorrhagic (experimental group 3) and ischemic (experimental group 4) genesis using the method of 3D Mueller-matrix mapping with polarization-holographic reproduction of layered MMI LB maps, it follows:

1. Phase section $\Delta = 0.4$ rad (see Table 10.15):
   The maximum level of balanced accuracy of intergroup differentiation by calculating statistical moments of the third and fourth orders, characterizing the asymmetry and excess of the distribution of the MMI LB value of histological brain sections:
   - "2"–"4": excellent accuracy 96–98% (number of false positives $b = 0$, number of false negatives $d = 4$);
   - "2"–"3": good accuracy 92–95% (number of false positives $b = 3$, number of false negatives $d = 7$);
   - "3"–"4": satisfactory accuracy 85–86% (number of false positives $b = 6$, number of false negatives $d = 10$).

Table 10.17 Balanced accuracy of the method of statistical analysis of MMI LB maps of histological brain sections for the phase section $\Delta = 1.2$ rad.

Parameters		Balanced accuracy Ac, %	
Groups	"2"–"4"	"2"–"3"	"3"–"4"
M_3	83 $a = 86; b = 14;$ $c = 80; d = 20$	79 $a = 81; b = 19;$ $c = 77; d = 23$	75 $a = 77; b = 23;$ $c = 73; d = 27$
M_4	82.5 $a = 85; b = 15;$ $c = 80; d = 20$	76 $a = 78; b = 22;$ $c = 74; d = 26$	74 $a = 76; b = 24;$ $c = 72; d = 28$

2. Phase section $\Delta = 0.8$ rad (see Table 10.16):
   Some reduction (up to 15%) of the level of balanced accuracy between group differentiation by calculating the asymmetry and excess of the distributions of the value of the set of MMI LB maps of histological brain sections:
   - "2"–"4": good accuracy 92–93% (the number of false positive and false negative conclusions is 6 and 9, respectively);
   - "2"–"3": satisfactory accuracy 87–89% (the number of false positive and false negative conclusions is 8 and 14, respectively);
   - "3"–"4": unsatisfactory accuracy 85–87% (the number of false positive and false negative conclusions is 12 and 16, respectively).
3. Phase section $\Delta = 1.2$ rad (see Table 10.17):
   Unsatisfactory accuracy level – balanced accuracy (79–83%).

10.5.5 Conclusions

1. A structural and logical scheme was developed, and the design of the 3D Mueller-matrix mapping method was experimentally validated to enhance the diagnostic potential of forensic differentiation of cases of cerebral infarction, hemorrhagic lesions of traumatic origin, and determination of the time of their formation using digital holography and azimuth-invariant Mueller-matrix microscopy.
2. It was demonstrated that, for each phase section of the volumetric distributions of the complex amplitude field, the values of sensitivity (Se, %), specificity (Sp, %), and balanced accuracy (Ac, %) of the statistical analysis of the spatial distributions of the MMI-CB and MMI-LB reach their maximum at small phase shifts corresponding to the level of single scattering.
3. The highest level of balanced accuracy of intergroup differentiation was obtained by calculating the statistical moments of the third and fourth orders, characterizing the skewness and kurtosis of the distributions of the MMI-CB values of histological brain sections:
   - "2"–"4": excellent accuracy (95–96%);
   - "2"–"3": good accuracy (92–93%);
   - "3"–"4": satisfactory accuracy (85–86%).
4. An excellent level of balanced accuracy in intergroup differentiation was achieved by calculating the statistical moments of the third and fourth orders, characterizing the skewness and kurtosis of the distributions of the magnitude of linear birefringence of fibrillar networks in nervous tissue of histological brain sections:
   - "2"–"4": excellent accuracy (100%);
   - "2"–"3": good accuracy (95–96%);
   - "3"–"4": satisfactory accuracy (91–92%).

References

1 Lu SY, Chipman RA. Interpretation of Mueller matrices based on polar decomposition. J Opt Soc Am A. 1996;13:1106–1113.
2 Arteaga O, Canillas A. Pseudopolar decomposition of the Jones and Mueller-Jones exponential polarization matrices. J Opt Soc Am A. 2009;26:783–793.

3 Ossikovski R, De Martino A, Guyot S. Forward and reverse product decompositions of depolarizing Mueller matrices. Opt Lett. 2007;32:689–691.

4 Ortega-Quijano N, Arce-Diego JL. Mueller matrix differential decomposition. Opt Lett. 2011;36:1942–1944.

5 Wu PJ, Walsh Jr. JT. Stokes polarimetry imaging of rat-tail tissue in a turbid medium using incident circularly polarized light. Lasers Surg Med. 2005;37:396–406.

6 Macdonald C, Meglinski I. Backscattering of circular polarized light from a disperse random medium influenced by optical clearing. Laser Phys Lett. 2011;8:324–328.

7 Qi J, He H, Lin J, Dong Y, Chen D, Ma H, et al. Assessment of tissue polarimetric properties using Stokes polarimetric imaging with circularly polarized illumination. J Biophotonics. 2018;11:e201700139.

8 Kunnen B, Macdonald C, Doronin A, Jacques S, Eccles M, Meglinski I. Application of circularly polarized light for non-invasive diagnosis of cancerous tissues and turbid tissue-like scattering media. J Biophotonics. 2015;8:317–323. Xu M, Alfano RR. Circular polarization memory of light. Phys Rev E. 2005;72:065601. Macdonald CM, Jacques SL, Meglinski IV. Circular polarization memory in polydisperse scattering media. Phys Rev E. 2015;91:033204.

9 Wood MF, Ghosh N, Moriyama EH, Wilson BC, Vitkin IA. Proof-of-principle demonstration of a Mueller matrix decomposition method for polarized light tissue characterization in vivo. J Biomed Opt. 2009;14:014029.

10 Ramella-Roman JC, Saytashev I, Piccini M. A review of polarization-based imaging technologies for clinical and preclinical applications. J Opt. 2020;22:123001.

11 Chang J, He H, Wang Y, Huang Y, Li X, He C, et al. Division of focal plane polarimeter-based 3×4 Mueller matrix microscope: a potential tool for quick diagnosis of human carcinoma tissues. J Biomed Opt. 2016;21:056002.

12 Fu Y, Huang Z, He H, Ma H, Wu J. Flexible 3 × 3 Mueller matrix endoscope prototype for cancer detection. IEEE Trans Instrum Meas. 2018;67:1700–1712.

13 De Boer JF, Hitzenberger CK, Yasuno Y. Polarization sensitive optical coherence tomography—a review. Biomed Opt Express. 2017;8:1838–1873.

14 Pezzaniti JL, Chipman RA. Mueller matrix imaging polarimetry. Opt Eng. 1995;34:1558–1568.

15 Azzam RMA. Photopolarimetric measurement of the Mueller matrix by Fourier analysis of a single detected signal. Opt Lett. 1978;2:148–150.

16 Goldstein DH. Mueller matrix dual-rotating retarder polarimeter. Appl Opt. 1992;31:6676–6683.

17 Smith MH. Optimization of a dual-rotating-retarder Mueller matrix polarimeter. Appl Opt. 2002;41:2488–2493.

18 Dubreuil M, Rivet S, Le Jeune B, Cariou J. Snapshot Mueller matrix polarimeter by wavelength polarization coding. Opt Express. 2007;15:13660–13668.

19 Sun M, He H, Zeng N, Du E, Guo Y, Liu S, et al. Characterizing the microstructures of biological tissues using Mueller matrix and transformed polarization parameters. Biomed Opt Express. 2014;5:4223–4234. Li P, Lv D, He H, Ma H. Separating azimuthal orientation dependence in polarization measurements of anisotropic media. Opt Express. 2018;26:3791–3800.

20 Li P, Lee HR, Chandel S, Lotz C, Groeber-Becker FK, Dembski S, et al. Analysis of tissue microstructure with Mueller microscopy: logarithmic decomposition and Monte Carlo modeling. J Biomed Opt. 2020;25:015002.

21 Li P, Tariq A, He H, Ma H. Characteristic Mueller matrices for direct assessment of the breaking of symmetries. Opt Lett. 2020;45:706–709.

22 Vizet J, Rehbinder J, Deby S, Roussel S, Nazac A, Soufan R, et al. In vivo imaging of uterine cervix with a Mueller polarimetric colposcope. Sci Rep. 2017;7:2471.

23 Zhanghao K, Liu W, Li M, Wu Z, Wang X, Chen X, et al. High-dimensional super-resolution imaging reveals heterogeneity and dynamics of subcellular lipid membranes. Nat Commun. 2020;11:5890.

24 Ku S, Mahato KK, Mazumder N. Polarization-resolved Stokes–Mueller imaging: a review of technology and applications. Lasers Med Sci. 2019;34:1283–1293.

25 Yang B, Lesicko J, Sharma M, Hill M, Sacks MS, Tunnell JW. Polarized light spatial frequency domain imaging for nondestructive quantification of soft tissue fibrous structures. Biomed Opt Express. 2015;6:1520–1533.

26 Clancy NT, Arya S, Qi J, Stoyanov D, Hanna GB, Elson DS. Polarised stereo endoscope and narrowband detection for minimal access surgery. Biomed Opt Express. 2014;5:4108–4117.

27 Manhas S, Vizet J, Deby S, Vanel JC, Boito P, Verdier M, et al. Demonstration of full 4 × 4 Mueller polarimetry through an optical fiber for endoscopic applications. Opt Express. 2015;23:3047–3054.

28 Wood TC, Elson DS. Polarization response measurement and simulation of rigid endoscopes. Biomed Opt Express. 2010;1:463–470.

29 Vizet J, Manhas S, Tran J, Validire P, Benali A, Garcia-Caurel E, et al. Optical fiber-based full Mueller polarimeter for endoscopic imaging using a two-wavelength simultaneous measurement method. J Biomed Opt. 2016;21:071106.
30 Rivet S, Bradu A, Podoleanu A. 70 kHz full 4 × 4 Mueller polarimeter and simultaneous fiber calibration for endoscopic applications. Opt Express. 2015;23:23768–23786.
31 Forward S, Gribble A, Alali S, Lindenmaier AA, Vitkin IA. Flexible polarimetric probe for 3 × 3 Mueller matrix measurements of biological tissue. Sci Rep. 2017;7:11958.
32 Muzy JF, Bacry E, Arneodo A. The multifractal formalism revisited with wavelets. Int J Bifurcat Chaos. 1994;4:245.
33 Grassberger P. Generalized dimensions of strange attractors. Phys Lett A. 1983;97:227. Grassberger P, Procaccia I. Measuring the strangeness of strange attractors. Physica D. 1983;9:189.
34 Hentschel HG, Procaccia I. The infinite number of generalized dimensions of fractals and strange attractors. Physica D. 1983;8:435.
35 Daubechies I. Wavelets on the interval (In Meyer Y, Roques S, editors, Progress in Wavelet Analysis and Applications). In Proceedings of the International Conference "Wavelets and Applications," Toulouse, France June 1992;95–107. Atlantica Seguier Frontieres; 1993.
36 Farge M. Wavelet transforms and their applications to turbulence. Annu Rev Fluid Mech. 1992;24(1):395–458.
37 Mescher AL. Junqueira's Basic Histology: Text and Atlas. 13th ed. New York: Mcgraw-Hill Education; 2013. ISBN 978-0-07-180720-3, 0-07-180720-9.
38 Angelsky PO, Ushenko AG, Dubolazov AV, Sidor MI, Bodnar GB, Koval G, et al. The singular approach for processing polarization-inhomogeneous laser images of blood plasma layers. J Opt. 2013;15(4):044030.
39 Ushenko AG, Dubolazov AV, Ushenko VA, Novakovskaya OY. Statistical analysis of polarization-inhomogeneous Fourier spectra of laser radiation scattered by human skin in the tasks of differentiation of benign and malignant formations. J Biomed Opt. 2016;21(7):071110.
40 Garazdyuk MS, Bachinskyi VT, Vanchulyak OY, Ushenko AG, Dubolazov OV, Gorsky MP. Polarization-phase images of liquor polycrystalline films in determining time of death. Appl Opt. 2016;55(12):B67–B71.
41 Angelskii OV, Ushenko OG, Burkovets DN, Arkhelyuk OD, Ushenko YA. Polarization-correlation studies of multifractal structures in biotissues and diagnostics of their pathologic changes. Laser Phys. 2000;10(5):1136–1142.
42 Yermolenko S, Ushenko A, Ivashko P, Goudail F, Gruia I, Gavrilă C, et al. Spectropolarimetry of cancer change of biotissues. In Ninth International Conference on Correlation Optics. SPIE. 2009;7388:404–410.
43 Jóźwicki R, Patorski K, Angelsky OV, Ushenko AG, Burkovets DN, Ushenko YA. Automatic polarimetric system for early medical diagnosis by biotissue testing. Opt Appl. 2002;32(4):603–612.
44 Ushenko AG, Dubolazov OV, Ushenko VA, Novakovskaya OY, Olar OV. Fourier polarimetry of human skin in the tasks of differentiation of benign and malignant formations. Appl Opt. 2016;55(12):B56–B60.
45 Ushenko AG, Pishak VP. Coherent-domain optical methods: biomedical diagnostics, environmental and material science. In Laser Polarimetry of Biological Tissue: Principles and Applications. Kluwer Academic Publisher; 2004.
46 Ushenko AG, Fediv AI, Marchuk YF. Correlation and fractal structure of Jones matrices of human bile secret. In European Conference on Biomedical Optics. 2009;7368:73681Q.
47 Ushenko AG. Depolarization of a laser emission field as the correlation development of its polarization structure. In International Conference on Correlation Optics. 1997;3317:331–339.
48 Bachinskyi V, Boychuk T, Ushenko A. Laser Polarimetry of Biological Tissues and Fluids: Chapter 4. Methods of Spatial-frequency, Singular and Correlation Analysis of Biological Layers Object Fields. LAP Lambert Academic Publishing; 2018.
49 Ushenko AG, Burkovets DN, Yermolenko SB, Arkhelyuk AD, Pishak VP, Yuzko AM, et al. Stokes polarimetry of biotissues. In Fourth International Conference on Correlation Optics. SPIE. 1999;3904:527–533.
50 Ushenko VA, Dubolazov AV, Pidkamin LY, Sakchnovsky MY, Bodnar AB, Ushenko YA, et al. Mapping of polycrystalline films of biological fluids utilizing the Jones-matrix formalism. Laser Phys. 2018;28(2):025602.
51 Ushenko AG, Angelsky PO, Sidor M, Marchuk YF, Andreychuk DR, Pashkovskaya NV. Spatial-frequency selection of complex degree of coherence of laser images of blood plasma in diagnostics and differentiation of pathological states of human organism of various nosology. Appl Opt. 2014;53(10):B172–B180.
52 Olar EI, Ushenko AG, Ushenko YA. Polarization correlation measurements of the phase tomograms of optically anisotropic biofractals. Laser Phys. 2004;14(8):1115–1121.
53 Guminetskiy SH, Ushenko OG, Polyanskiy IP, Motrych AV, Grynchuk FV. The optical method for investigation of the peritonitis progressing process. In Eighth International Conference on Correlation Optics. 2008;7008:626–632.

54 Ushenko AG, Fediv AI, Marchuk YF. Singular structure of polarization images of bile secret in diagnostics of human physiological state. In: European Conference on Biomedical Optics. Optica Publishing Group; 2009.

55 Ushenko Y, Ushenko A, Dubolazov A, Soltys I, Litvinenko O, Wanchuliak O, et al. Mueller-matrix interferometric multifractal scaling of optically anisotropic architectonics of diffuse blood facies: fundamental and applied aspects. J Biophotonics. 2025;18(3):e202400412. https://doi.org/10.1002/jbio.202400412.

56 Ushenko A, Soltys I, Dubolazov A, Ushenko Y, Bilookyi V, Bilookyi O, et al. 3D Jones-matrix thesiography of biological fluid facies. J Innov Opt Health Sci. 2025;18(2):2443002. https://doi.org/10.1142/S1793545824430028.

57 Ushenko A, Pavlyukovich N, Khukhlina O, Pavlyukovich O, Gorsky M, Soltys I, et al. Layer-by-layer multifractal scanning of optically anisotropic architectonics of blood plasma films: fundamental and applied aspects. Photonics. 2025;12:215. https://doi.org/10.3390/photonics12030215.

58 Ushenko A, Pavlyukovich N, Khukhlina O, Pavlyukovich O, Soltys I, Dubolazov A, et al. Blood Plasma Film Multifractal Scanning in COVID-19 Consequences Diagnostics. J Biophotonics. 2024;17(11):e202400356. https://doi.org/10.1002/jbio.202400356.

59 Ushenko A, Dubolazov A, Zheng J, Litvinenko A, Gorsky M, Ushenko Y, et al. 3D polarization-interference holographic histology for wavelet-based differentiation of biological tissues with different necrotic states. J Biomed Opt. 2024;29(5):052920. https://doi.org/10.1117/1.JBO.29.5.052920.

60 Ushenko O, Bilookyi O, Zheng J, Dubolazov A, Olar O, Ushenko Y, et al. 3D digital holographic polarimetry of laser speckle fields formed by polycrystalline blood films: a tool for differential diagnosis of thyroid pathology. Front Phys. 2024;12:1426469. https://doi.org/10.3389/fphy.2024.1426469.

61 Wójcik W, Hu Z, Ushenko Y, Smolarz A, Soltys I, Dubolazov O, et al. Optical sensor system for 3D Jones matrix reconstruction of optical anisotropy maps of self-assembled polycrystalline soft matter films. Sensors. 2024;24(5):1589. https://doi.org/10.3390/s24051589.

62 Ushenko A, Zheng J, Gorsky M, Dubolazov A, Ushenko Y, Soltys I, et al. 3D digital holographic polarimetry of diffuse optically anisotropic biological tissue object fields. Front Phys. 2023;11:1288935. https://doi.org/10.3389/fphy.2023.1288935.

63 Dubolazov A, Ushenko V, Trifonyuk L, Stashkevich A, Soltys I, Ushenko Y, et al. Polarization-singular approach to imaging Mueller-matrix polarimetry in differential diagnosis of histological sections of biopsy of tumors of the uterus and prostate. Front Phys. 2021;9:711212. https://doi.org/10.3389/fphy.2021.711212.

64 Ushenko VO, Trifonyuk L, Ushenko YA, Dubolazov OV, Gorsky MP, Ushenko AG. Polarization singularity analysis of Mueller-matrix invariants of optical anisotropy of biological tissue samples in cancer diagnostics. J Opt. 2021;23(6):064004. https://doi.org/10.1088/2040-8986/abf97a.

65 Ushenko YA, Koval GD, Ushenko AG, Dubolazov OV, Ushenko VA, Novakovskaia OY. Mueller-matrix of laser-induced autofluorescence of polycrystalline films of dried peritoneal fluid in diagnostics of endometriosis. J Biomed Opt. 2016;21(7):071116. https://doi.org/10.1117/1.JBO.21.7.071116.

11

Imaging of Blood Film Microstructure

Yurii Ushenko[1,2], Weidong Dou[1], Oleksandr Ushenko[2,3], Oleksandr Dubolazov[2], Iryna Soltys[2], Viktor Bachinskiy[4], Oleh Vanchuliak[4], Marta Garazdyuk[4], Nataliia Pavlyukovich[4], Oleksandr Pavlyukovich[4], Vitaliy Rozhko[4], and Oleksandr Olar[2]

[1] Shaoxing University, Shaoxing, China
[2] Yurii Fedkovych Chernivtsi National University, Chernivtsi, Ukraine
[3] Taizhou Institute of Zhejiang University, Taizhou, China
[4] Bukovinian State Medical University, Chernivtsi, Ukraine

11.1 Three-dimensional Digital Holographic Polarimetry of Laser Speckle Fields Formed by Polycrystalline Blood Films: A Tool for Differential Diagnosis of Thyroid Pathology

11.1.1 Introduction

Timely detection of thyroid disease is a global health issue. According to statistics from the World Health Organization, 10–30% of adults worldwide have thyroid conditions, marked by both gland dysfunction and structural changes. Additionally, this percentage increases by 5% each year.

Traditionally, the gold standard for diagnosing thyroid disease is biopsy or surgical removal, with tissue sections stained with hematoxylin and eosin and examined under a microscope. However, early diagnosis remains challenging due to variability in interpreting tissue morphology. Additionally, cancerous changes can be small and barely noticeable [1–12].

These methods have become the focus of digital pathology, utilizing new quantitative image analysis capabilities.

Thin sections of the thyroid nodule capsule, along with microscopic images, histograms, and texture analysis, were utilized. Based on these, the study demonstrated the potential for differentiating thyroid nodules as benign (follicular adenoma) or malignant (papillary carcinoma).

Pathologists can enhance thyroid pathology diagnosis by utilizing new objective quantitative techniques that offer high spatial resolution. Among these, polarization-sensitive second harmonic generation (PSHG) microscopy methods stand out.

PSHG microscopy methods are employed to distinguish cancerous or diseased tissues by examining circular dichroism and adjusting laser polarization states. This allows for the measurement of ultrastructural alterations in collagen frameworks within organ tissues at a submicron scale. The technique has been applied to investigate collagen organization in both benign and malignant fibrillar capsules of the thyroid [13–29].

The development and integration of PSHG microscopy represent advanced techniques in non-invasive polarization-sensitive optical coherence tomography (PSOCT). By employing principles of direct interference recording of the structure of speckle fields in the object, it becomes possible to measure the coordinate distributions of Mueller-matrix elements or images (MMI) layer-by-layer at different depths within biological samples. This approach enables quantitative assessment of optical anisotropy for each biological tissue layer. Consequently, this facilitates precise diagnosis of fibrosis and the differentiation of tumor regions with minimal fibrosis. Nevertheless, the sensitivity of these systems is constrained by a high level of depolarized laser speckle background.

Biological tissue object fields used in phase-scanning polarization–interference methods with complex-amplitude distributions and layer-by-layer digital reconstruction can reduce background interference and improve the diagnostic sensitivity of PSHG and PSOCT techniques [30–37].

These methods enable the phase separation of a non-depolarized, singly scattered component in a laser speckle field. The complex amplitude of the biological layer's non-depolarized component is uniquely linked to the parameters of the optically anisotropic component. Consequently, within a statistical analysis framework, this allows for the most precise detection of pathological changes in human tissue biopsy samples, including polycrystalline structures. Based on this, the techniques have shown potential for diagnosing endometriosis, distinguishing between benign and malignant prostate tumors, and identifying necrotic or traumatic myocardium changes.

Expanding the capabilities of polarization–interference phase-scanning methods could involve studying different types of objects, such as biological fluids. These objects are readily accessible and do not require invasive biopsy surgery. They are among the most vital systems necessary for maintaining the body's essential functions, consisting of biological fluids involved in metabolic processes. In biological fluids (BF), rapid changes occur in molecular composition and interactions among various components of physiological systems, especially under extreme and pathological conditions.

Any physiological or pathological process in a living organism involves specific alterations in proteins and other organic molecules. These changes are highly informative for studying homeostasis at the molecular level and can form the foundation for early disease diagnosis.

Modern laboratory and clinical diagnostics mainly emphasize the many quantitative indicators in BF determination, which reflect the body's essential functions.

Morphological research in biology and medicine primarily focuses on cellular components, whereas BF morphological analysis – the fundamental system supporting life – has only recently become accessible for clinical use. A BF drop resting on a horizontal surface is a self-organizing system and serves as a practical model for examining physical–chemical processes. Its characteristics depend on the dissolved substances in the liquid, external dehydration conditions, and the substrate material. When a BF drop dehydrates, it forms a dry film called a "facies." The structure of this facies contains information about the composition and interactions of the dissolved substances within BF. Analyzing these objects at the macroscopic level of self-organization provides insights into the molecular-level system.

One of the most informative biological fluids about the state of the human body is blood, which contains about hundreds of proteins. Conventionally (according to differences in solubility), they are divided into five main fractions: albumins (55–65%), $\alpha 1$ – (2–4%), $\alpha 2$ – (6–12%), β – (8–12%), and γ – (2–4%) globulins.

Blood albumin levels are high, and its molecules are small, resulting in an extensive surface area of albumin molecules. This large surface makes them well equipped to carry out transport functions, including bilirubin, urobilin, fatty acids, bile salts, and lipid hormones. The albumin fraction also contains transthyretin (prealbumin), which, along with thyroxine-binding globulin (TBG) and albumin, helps transport the thyroid hormone thyroxine and its metabolite iodothyronine.

Globulins are a blood protein, in the molecules of which polypeptide chains are tightly coiled into compact spherical structures – globules (tertiary protein structures). One globulin type is thyroglobulin (Tg). This is a protein produced by the thyroid gland follicular cells, accumulating in its structural and functional unit – the follicle – as a colloid. Patients with Hashimoto's thyroiditis or Graves' disease often have elevated antithyroglobulin antibodies. This indicator is used for diagnostic purposes. Tg is a differentiated (papillary or follicular) thyroid cancer reliable prognostic postoperative marker. In addition, Tg levels may be elevated in Graves' disease.

The optical anisotropic structure and polarization properties of these objects remain underexplored. Only a limited number of studies have examined polarization interferometry of biological fluids for diagnosing pathological changes in human organs [38–49].

This section concentrates on discovering new objective markers for the minimally invasive differential diagnosis of thyroid conditions. It employs comprehensive polarization–interference mapping and digital holography to reconstruct layer-by-layer maps of the ellipticity of speckle fields in blood films. To maximize the method's effectiveness, statistical analysis of polarization-ellipticity maps of the singly scattered component of the speckle field has been performed. This analysis includes blood samples from donors and patients with nodular goiter, autoimmune thyroiditis, and papillary cancer, marking the first application of this approach.

11.1.2 Biological Samples

Four representative groups of dehydration blood films were formed:

1. ***Control group***: Healthy donors aged 30–60 years included 51 individuals, with patients consisting of 48 women and 3 men. Inclusion criteria involved undergoing a puncture biopsy of the thyroid gland without any confirmed pathological processes, along with normal levels of thyroid hormones and autoantibodies. Exclusion criteria were the confirmation of pathological processes through a puncture biopsy of the thyroid gland.
2. ***Experimental group***: A total of 51 patients aged 30–60 (46 women, 5 men) for nodular goiter.

   Inclusion criteria include a puncture biopsy of the thyroid gland confirming a pathological process. Patients must have one or more thyroid colloid nodules, with normal thyroid hormone and autoantibody levels. Exclusion criteria consist of abnormal thyroid hormone levels (either increased or decreased) and elevated autoantibodies. Additionally, individuals with other conditions such as diabetes, hypertension, atherosclerosis, chronic adrenal insufficiency, rheumatoid arthritis, chronic active hepatitis, or systemic lupus erythematosus are excluded.

   A violation of the principle of negative feedback endocrine function regulation characterizes this group.
3. ***Experimental group***: A total of 51 patients aged 30–60 (49 women, 2 men) with autoimmune thyroiditis.

   Inclusion criteria: Patients must have a puncture biopsy confirming stages 1–3 thyroid pathology – such as hypertrophic thyroiditis (Hashimoto's goiter), atrophic thyroiditis, or fibrous invasive Riedel's goiter. They should also show elevated autoantibodies exceeding 1000 IU/mL and have hypothyroidism. Exclusion criteria: Patients with normal thyroid hormone and thyroperoxidase autoantibody levels are ineligible. The presence of other conditions like diabetes, hypertension, atherosclerosis, chronic adrenal insufficiency, rheumatoid arthritis, chronic active hepatitis, or systemic lupus erythematosus disqualifies them. Damage to the hematothyroid barrier is typical for this group.
4. ***Experimental group***: A total of 51 patients aged 30–60 (44 women, 7 men) with papillary cancer thyroid gland.

   Inclusion criteria include a puncture biopsy of the thyroid gland confirming the presence of papillary epithelial cancer cells, with no evidence of infiltration beyond the capsule or lymph node invasion. Normal levels of autoantibodies are also required. Exclusion criteria include abnormal thyroid hormone levels and elevated autoantibodies to thyroperoxidase. The condition may be associated with other diseases such as diabetes, hypertension, atherosclerosis, chronic adrenal insufficiency, rheumatoid arthritis, chronic active hepatitis, or systemic lupus erythematosus. This group is characterized by uncontrolled, atypical thyroid cell growth, forming papillary structures, nuclear inclusions, tumor progression, and a history of frequent exposure to Chernobyl accident effects.

Table 11.1 presents the polycrystalline blood films optical parameters.

11.1.3 Experimental Results

Currently, the tendencies of pathological changes in the polarization maps of optically anisotropic components of thyroid biopsies have been experimentally studied:

- Nodular goiter involves increased optical activity (1, 2) of thyroglobulin protein complexes, caused by follicle enlargement and the development of structural birefringence (3, 4) resulting from nodular connective tissue proliferation.
- Autoimmune thyroiditis results from squamous cell metaplasia of follicular cells; the increase in cell size leads to manifestations of optical activity of thyroglobulin complexes and birefringence of the fibrous cord network.
- Papillary cancer – mainly caused by the structural birefringence resulting from hyperplasia in the form of elongated papillary formations.

Analysis and diagnostics of the results rely on a two-component amorphous anisotropic model of the thyroid gland's optical characteristics. In this model, the amorphous component-comprising parenchyma made of epithelial tissue – is

Table 11.1 The blood films optical parameters.

Parameter	Group 1	Group 2	Group 3	Group 4
Attenuation coefficient τ, cm^{-1}	0.64 ± 0.04	0.61 ± 0.04	0.63 ± 0.04	0.62 ± 0.04
Depolarization degree Λ, %	41 ± 0.9	38 ± 0.8	42 ± 0.8	40 ± 0.7

polarization-inactive and does not modify the laser radiation's polarization state. The anisotropic component includes both optically active elements, such as globulin protein complexes (follicles containing thyroglobulin protein colloid), and structural elements, like the supramolecular fibrillar protein networks of connective tissue within the parenchyma (e). The optically active component predominantly forms maps of polarization azimuth $\alpha(m, n)$. The structural or polycrystalline component predominantly forms maps of polarization ellipticity.

During the blood drop dehydration process, protein molecules naturally self-assemble into a supramolecular network. The structural features of these proteins influence blood film formation. The atoms' arrangement within the molecules determines their structure, which, in three dimensions, results in polymers or polypeptides. Each polypeptide monomer is a sequence of α-amino acids, with all except glycine possessing an asymmetric carbon atom (threonine and isoleucine have two). These amino acids exhibit optical activity or circular birefringence.

The protein structure hierarchy, following the levels, is distinguished.

The primary structure of monomers, such as nucleotides or amino acids, determines their sequence. A key characteristic of amino acids' primary structures is the near-random distribution of their residues. In biophysics, circular birefringence arises at this level.

The main polypeptide protein chain conformational arrangement forms the secondary structure, regardless of the side chains' conformation. At this level, the linear birefringence optical axis direction is formed.

The tertiary structure of a protein molecule is its three-dimensional shape, made up of a single chain. Based on how the polypeptide chain is arranged in space, proteins are classified as globular or fibrillar. Globular proteins usually have an ellipsoid shape and mainly show circular birefringence. Fibrillar (thread-like) proteins have an elongated form and display linear birefringence.

The quaternary structure consists of the spatial arrangement of individual polypeptide chains, which mostly share the same primary, secondary, and tertiary structures.

A unified blood film supramolecular network is formed, combining both structural and functional aspects, and exhibits phase anisotropy, including circular and linear birefringence. The unique polycrystalline supramolecular structure of blood film proteins, as well as the relationship between optical anisotropy values, is demonstrated by polypeptide fragments that possess specific conformational independence and are part of these proteins.

Based on this, we can anticipate changes in the circular and linear birefringence parameters of blood film supramolecular networks, depending on thyroid pathology presence. In our case, a specific prognostic scenario is likely: an increase in thyroglobulin and the albumin that transports it may lead to heightened blood phase anisotropy of supramolecular protein networks, as seen in ill patients.

Our main attention will be focused on the object field single-scattered component polarization maps $\beta(x, y)$ of blood films from healthy donors and patients with various thyroid pathologies.

11.1.4 Statistical Analysis of Polarization Ellipticity Maps

The series of Figures 11.1–11.4 presents the integral maps (panels (a)) and ellipticity values distributions histograms (panels (b)) of donors (control group 1) and patients with thyroid gland various pathologies (experimental groups 2–4).

The analysis of the obtained results revealed a correlation between the experimental data (Figures 11.1–11.4) and the proposed model of blood films (patients in the control and study groups) speckle fields $H_{\text{blood}}(x, y)$ polarization structure formation. It was found that for all the considered pathologies (nodular goiter, autoimmune thyroiditis, papillary cancer), the blood films samples object speckle fields have similar polarization ellipticity maps $\beta_{\text{blood}}(\vartheta_{xy})$ (panels (a), Figures 11.1–11.4). This fact is illustrated by the histograms $N(\beta)$ of the polarization ellipticity value distributions (panels (b), Figures 11.1–11.4). Each blood films speckle field polarization map $\beta_{\text{blood}}(\vartheta_{xy})$ is characterized by a sufficiently wide range of change in the polarization ellipticity value ($0 \leq \beta \leq 0.5\pi$). Moreover, the probabilities of different ellipticity values formation of the donors and sick patients' blood films samples speckle field are quite close in magnitude. At first glance, this experimental does not correspond to the proposed model scenario of the polarization structure $\beta_{\text{object}}(\delta_{xy})$ formation and pathological transformation of the "object field" $E_{\text{object}}(x, y) \equiv \begin{pmatrix} E_x \\ E_y \\ \delta_{xy} \end{pmatrix}$ of blood films optically anisotropic supramolecular networks. We have already noted that the architectonics $(\rho(x, y); \delta_{xy}(x, y))$ of such networks have a multilevel structure – nucleotide monomers, polypeptide protein chains, tertiary and quaternary structure of globular and

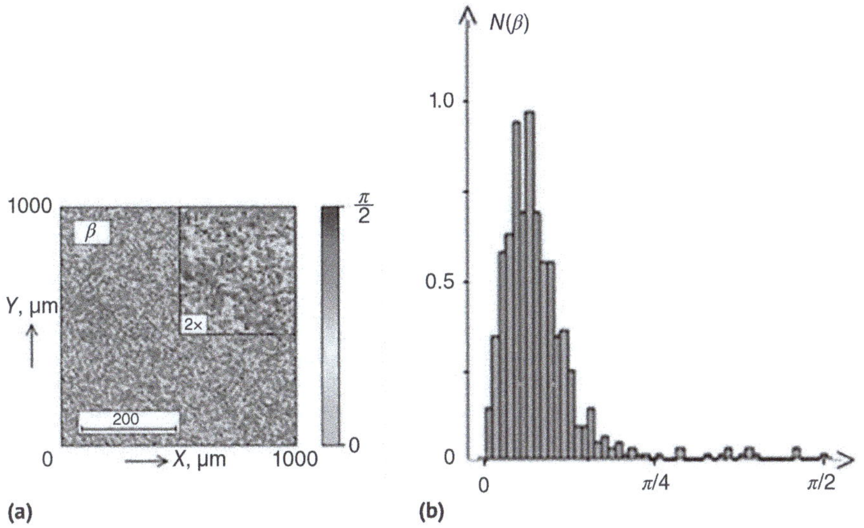

Figure 11.1 The blood film polarization map (a) and histogram (b) for a patient from control group 1 (healthy donors).

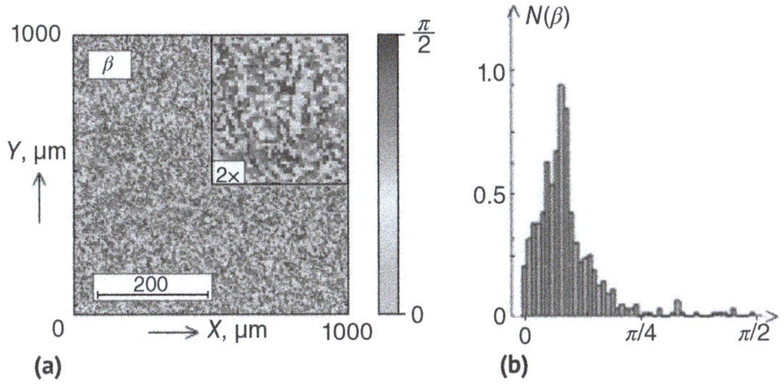

Figure 11.2 The blood film polarization map (a) and histogram (b) for a patient from experimental group 2 (patients with nodular goiter).

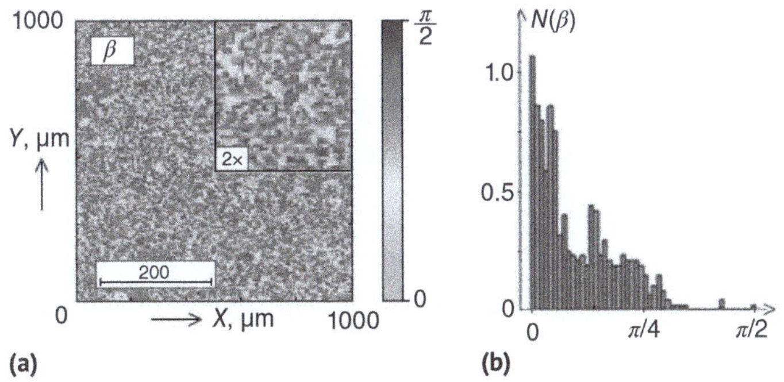

Figure 11.3 The blood film polarization map (a) and histogram (b) for a patient from experimental group 3 (patients with autoimmune thyroiditis).

fibrillar protein chains. When pathology occurs, the blood films optical anisotropy increases due to biochemical (Tg concentration increases) and conformational (structuring of globular and fibrillar protein networks). In terms of polarization, this should manifest itself in an increase in the magnitude and an increase in the range of change in the ellipticity values $\beta_{\text{object}}(\delta_{xy})$. However, in the multiple scattering presence in the blood films volume, regardless of the presence and

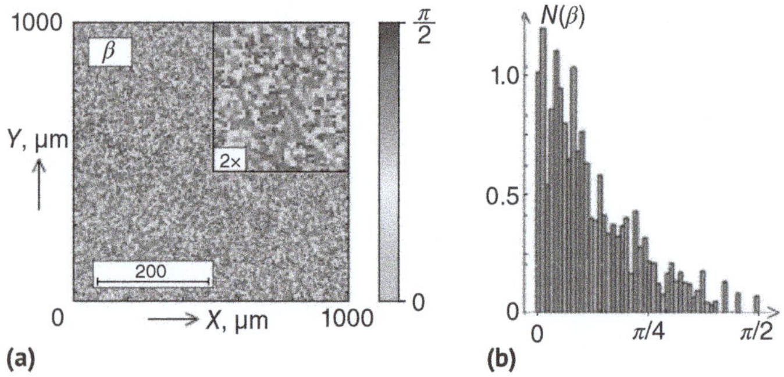

Figure 11.4 The blood film polarization map (a) and histogram (b) for a patient from experimental group 4 (patients with papillary cancer).

Table 11.2 Donors and patients with various thyroid gland pathologies, blood films polarization ellipticity maps, and statistical parameters.

Groups	**Group 1 (51 samples)**	**Group 2 (51 samples)**	**Group 3 (51 samples)**	**Group 4 (51 samples)**
Average, M_1	0.187 ± 0.01	0.205 ± 0.01	0.212 ± 0.01	0.22 ± 0.01
p_{ik}		$p_{12;13;14} > 0.05$ $p_{23;24;34} > 0.05$		
Variance, M_2	0.082 ± 0.005	0.093 ± 0.006	0.114 ± 0.007	0.123 ± 0.008
p_{ik}		$p_{12;13;14} > 0.05$ $p_{23;24;34} > 0.05$		
Skewness, M_3	0.89 ± 0.05	0.78 ± 0.04	0.71 ± 0.04	0.66 ± 0.04
p_{ik}		$p_{12;13;14} > 0.05$ $p_{23;24;34} > 0.05$		
Kurtosis, M_4	1.47 ± 0.09	1.18 ± 0.06	0.94 ± 0.06	0.83 ± 0.05
p_{ik}		$p_{12;13;14} < 0.05$ $p_{23;24;34} > 0.05$		

pathology type, along with the "object" $E_{\text{object}}(x, y) \equiv \begin{pmatrix} E_x \\ E_y \\ \delta_{xy} \end{pmatrix}$ component, a "speckle" $U_{\text{speckle}}(x, y) \equiv \begin{pmatrix} U_x \\ U_y \\ \varphi_{xy} \end{pmatrix}$ component is also formed. The interference mechanisms of such a component formation transform and average the polarization ellipticity value $\beta_{\text{object}}(\delta_{xy})$ "object" distributions, individual for each pathology type. Due to such averaging, diagnostically significant differences between the polarization ellipticity $\beta_{\text{object}}(x, y, \delta_{xy})$ "object" maps are leveled. The fact is that each act of scattered waves $\widetilde{E}_{xj}(r)$ and $\widetilde{E}_{yj}(r)$ interference addition changes the polarization ellipticity $\widetilde{\beta}(x, y, \varphi_{x,y})$ value. Moreover, the "interference" phase shifts $\varphi_{x,y}$ values can significantly exceed similar "object" parameters $\delta_{x,y}$. Due to this, the secondary interference influence on the donors and sick patients' blood films polarization ellipticity $\beta_{\text{speckle}}(x, y, \varphi_{xy})$ maps formation becomes decisive. The random ellipticity values distributions statistical analysis results illustrate the quantitative similarity of the blood films speckle fields polarization structure for all groups (Table 11.2).

A Mann–Whitney U test was employed to compare the polarization ellipticity random values across independent samples. The comparisons included: group 1 versus group 2, group 1 versus group 3, group 1 versus group 4, group 2 versus group 3, group 2 versus group 4, and group 3 versus group 4.

The statistical analysis results revealed:

- Statistically insignificant ($p_{12;13;14} > 0.05$; $p_{23;24;34} > 0.05$) increase in the average M_1 of the distribution $\beta(m, n)$ statistical value within individual representative samples the from 0.187 to 0.22.
- The $\beta(m, n)$ distributions variance M_2 value also turned out to be statistically insignificant ($p_{12;13;14} > 0.05$; $p_{23;24;34} > 0.05$) in the range from 0.082 to 0.123.

- The third-order M_3 statistical moment, which characterizes the $\beta(m, n)$ distributions skewness and varies from 0.66 to 0.89, was also found to be statistically insignificant for differential diagnosis.
- The blood films polarization maps sensitivity to intergroup changes has kurtosis M_4 of the ellipticity random values distributions.
- Kurtosis M_4 varies statistically significantly $\left(p_{12;13;14} \leq 0.05\right)$ from 0.83 to 1.47 and can be used as a marker for detecting the presence of thyroid gland pathology.

Therefore, for different pathologies ("normal – nodular goiter – autoimmune thyroiditis – papillary cancer"), the polarization ellipticity $\beta(m, n)$ distributions mean and variance change slightly. Also, the –third and fourth-order statistical moments, which change by 30–40% across intergroup differences, turned out to be insufficiently sensitive.

The thyroid gland pathology differential diagnosis, balanced accuracy results, are summarized in Table 11.3.

Here a is the correct results total number among all samples according to control group 1 and sequentially to experimental groups 2 (nodular goiter), 3 (autoimmune thyroiditis), and 4 (papillary cancer); b is the false positives total number among all samples according to all groups.

The information analysis results demonstrated a slight diagnostic capacity for distinguishing thyroid pathology (control group 1 and the set of all experimental groups "2–4") – none of the statistical markers achieved a balanced accuracy above 80%.

Table 11.3 Thyroid pathology differential diagnosis balanced accuracy.

	Average, M_1	
$a = 70$	$a = 71$	$a = 72$
$b = 32$	$b = 31$	$b = 30$
$Ac_{12} = 68.6\%$	$Ac_{13} = 69.6\%$	$Ac_{14} = 70.6\%$
$a = 70$	$a = 70$	$a = 71$
$b = 32$	$b = 32$	$b = 31$
$Ac_{23} = 68.6\%$	$Ac_{24} = 68.6\%$	$Ac_{34} = 69.6\%$
	Variance, M_2	
$a = 72$	$a = 73$	$a = 74$
$b = 30$	$b = 29$	$b = 28$
$Ac_{12} = 70.6\%$	$Ac_{13} = 71.5\%$	$Ac_{14} = 72.5\%$
$a = 71$	$a = 72$	$a = 73$
$b = 31$	$b = 30$	$b = 29$
$Ac_{23} = 69.6\%$	$Ac_{24} = 70.6\%$	$Ac_{34} = 71.6\%$
	Skewness, M_3	
$a = 75$	$a = 76$	$a = 77$
$b = 27$	$b = 26$	$b = 25$
$Ac_{12} = 73.5\%$	$Ac_{13} = 74.5\%$	$Ac_{14} = 75.5\%$
$a = 73$	$a = 74$	$a = 75$
$b = 29$	$b = 28$	$b = 27$
$Ac_{23} = 71.5\%$	$Ac_{24} = 72.5\%$	$Ac_{34} = 73.5\%$
	Kurtosis, M_4	
$a = 76$	$a = 77$	$a = 78$
$b = 26$	$b = 25$	$b = 24$
$Ac_{12} = 74.5\%$	$Ac_{13} = 75.5\%$	$Ac_{14} = 76.5\%$
$a = 74$	$a = 75$	$a = 76$
$b = 28$	$b = 27$	$b = 26$
$Ac_{23} = 72.5\%$	$Ac_{24} = 73.5\%$	$Ac_{34} = 74.5\%$

A different situation is realized when using blood films speckle fields $\begin{pmatrix} H_x; \\ H_y; \\ \vartheta_{xy} \end{pmatrix}$ polarization–interference mapping with phase scanning and polarization ellipticity $\beta_{blood}(\Delta\theta_{xy} \to \vartheta_{xy})$ layered maps digital algorithmic reconstruction.

The series Figures 11.5–11.8 illustrates the algorithmically obtained in the phase plane $(\theta_k = \pi/8)$ "object" component $\begin{pmatrix} E_x; \\ E_y; \\ \delta_{xy} \end{pmatrix}$ polarization ellipticity $\beta_{object}(\theta_k, x, y)$ maps in donors and sick patients' blood films speckle fields $\begin{pmatrix} H_x; \\ H_y; \\ \vartheta_{xy} \end{pmatrix}$.

The value $(\theta_k = \pi/8)$ is established by step-by-step $(\Delta\theta_{xy})$ phase scanning and algorithmically reconstructed polarization ellipticity value distributions $\beta_{speckle}(\Delta\theta_{xy}, x, y)$ statistical analysis. The phase plane $(\theta_k = \pi/8)$ corresponds to the statistical criterion $M_{g=1;2;3;4}(\beta(\Delta\theta_{xy} \leq \theta_k)) \approx$ const. for determining the single-scattered "object" component $\begin{pmatrix} E_x; \\ E_y; \\ \delta_{xy} \end{pmatrix}$. "Object" polarization maps $\beta_{object}(\theta_k, x, y)$ (panel (a), Figures 11.5–11.8) comparative analysis revealed a significant decrease in the interval ellipticity value β_{object} change in comparison with "speckle" distributions $\beta_{speckle}(\vartheta_{xy}, x, y)$, panel (a), Figures 11.1–11.4. This confirms the algorithmic elimination of the scattered component $\beta_{speckle}(\varphi_{xy}, x, y)$ distorting influence on the "object"

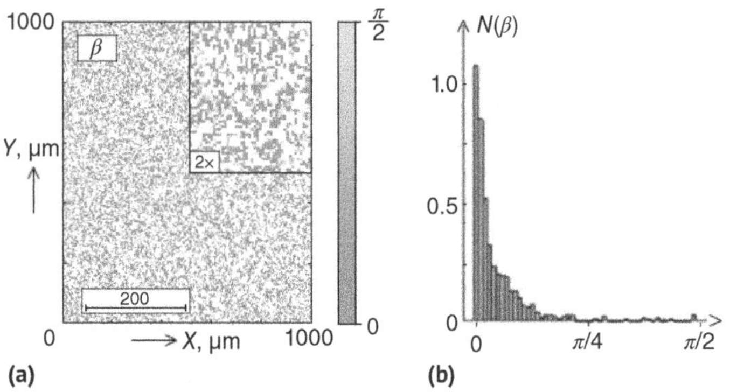

Figure 11.5 The blood film phase $(\theta_k = \pi/8)$ polarization map (a) and histogram (b) for a patient from control group 1 (healthy donors).

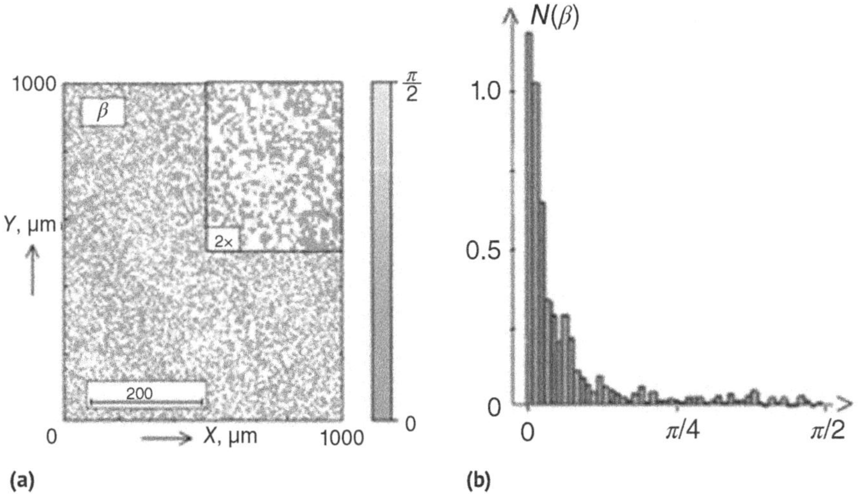

Figure 11.6 The blood film phase $(\theta_k = \pi/8)$ polarization map (a) and histogram (b) for a patient from experimental group 2 (patients with nodular goiter).

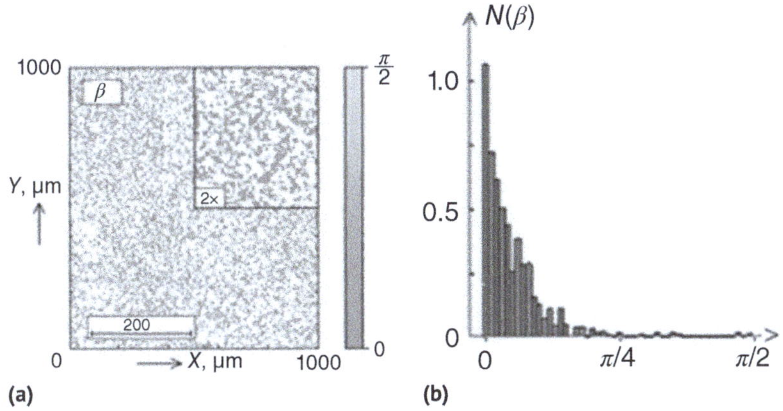

Figure 11.7 The blood film phase ($\theta_k = \pi/8$) polarization map (a) and histogram (b) for a patient from experimental group 3 (patients with autoimmune thyroiditis).

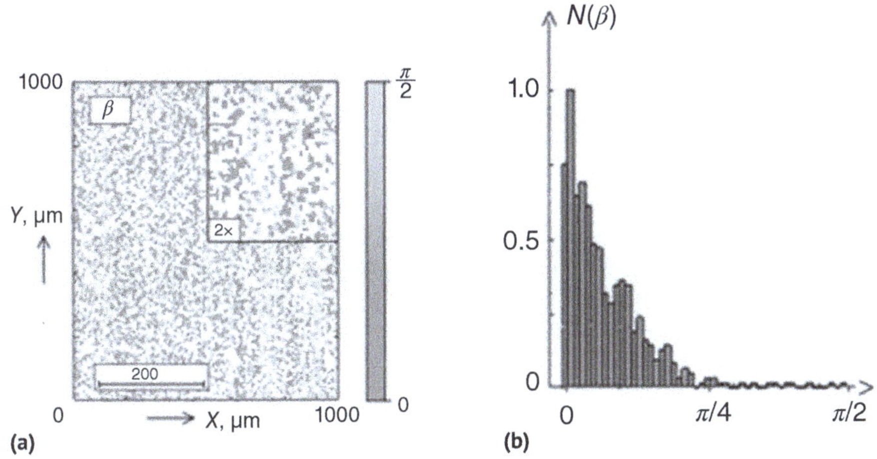

Figure 11.8 The blood film phase ($\theta_k = \pi/8$) polarization map (a) and histogram (b) for a patient from experimental group 4 (patients with papillary cancer).

ellipticity coordinate distributions. In accordance with the presented blood films optical anisotropy model concepts, it is possible to estimate the most probable changes in the singly scattered component $\begin{pmatrix} E_x; \\ E_y; \\ \delta_{xy} \end{pmatrix}$ polarization ellipticity value.

The protein structures' birefringence value corresponds to the level $(\Delta n \sim 1.5 \times 10^{-3})$; the blood films geometric thickness l varies within the range from 30 to 100 μm; the laser radiation wavelength $\lambda = 0.6328$ μm. Assuming that the blood films' entire volume is birefringent, the "object" phase shift value δ_{xy} fluctuates in variation range within the limits from $\pi/6$ to $\pi/2$. Based on this and the single-scattered laser waves $\begin{pmatrix} E_x; \\ E_y; \\ \delta_{xy} \end{pmatrix}$ formation conditions, the maximum $\pi/4$ ellipticity β_{object} level formation is probable. However, in experiments with real blood film samples, polarization states with much lower ellipticity values (shown in panel (b), Figures 11.5–11.8) are more likely. This can be explained by the fact that optically anisotropic supramolecular networks are spatially organized and do not fully occupy the blood film volume.

As a result, the birefringent globular and fibrillar protein chains effective geometric dimensions $l_{effective}$ decrease. As a result, the "object" phase shifts δ_{xy} and ellipticity β_{object} values are concentrated in a "narrower" range – from 0 to $\pi/4$.

The ellipticity distributions $\beta_{\text{object}}(\theta_k, x, y)$ histograms $N(\beta)$ comparative analysis in the phase plane ($\theta_k = \pi/8$) revealed a tendency for its value to consistently increase for the blood films of donors (Figure 11.5b) and patients with nodular goiter (Figure 11.6b), autoimmune thyroiditis (Figure 11.7b), and papillary cancer (Figure 11.8b).

This fact is indicated by a decrease in the histogram kurtosis (SM_4) due to a reduction in the polarization ellipticity β_{object} minimum values. At the same time, the probability distributions $N(\beta)$ of the polarization parameter become more symmetrical – skewness value SM_3 decreases. The revealed scenario of "pathological" transformation of blood films speckle field $\begin{pmatrix} H_x; \\ H_y; \\ \vartheta_{xy} \end{pmatrix}$ "object" component $\begin{pmatrix} E_x; \\ E_y; \\ \delta_{xy} \end{pmatrix}$ ellipticity maps $\beta_{\text{object}}(\theta_k, x, y)$ correlates with the proposed optical anisotropy model of supramolecular networks.

According to the specified model, pathological conditions – ranging from nodular goiter and autoimmune thyroiditis to papillary cancer – are characterized by a consistent rise in ellipticity value and a broader range of variation, driven by biochemical (increasing Tg concentration) and conformational (structuring of globular and fibrillar protein networks).

Within the framework of the statistical approach, such pathological transformations should be accompanied by an increase in the mean (M_1) and variance (M_2) values, which characterize the polarization maps $\beta_{\text{object}}(\theta_k, x, y)$. The higher-order statistical moments $M_{3;4}$ values, $M_3 \sim 1/(M_2)^3$ and $M_4 \sim 1/(M_2)^4$, should decrease.

This scenario of the difference's formation between the blood films polarization "object" maps $\beta_{\text{object}}(\theta_k, x, y)$ from all groups is confirmed by the first- and fourth-order central statistical moments $SM_{g=1;2;3;4}$ calculation (Table 11.4).

Polarization ellipticity maps $\beta_{\text{object}}(\theta_k, x, y)$ statistical analysis revealed.

Statistical significance for diagnostics $(p_{12;13;14})$ and differentiation $(p_{23;24;34})$ of entire set thyroid pathology markers.

The mean M_1 and variance M_2 are statistically significant ($p_{ik} < 0.05$) and consistently (from control group 1 to experimental groups 2–4) increase within – M_1 ~ from 0.15 to 0.194 and M_2 ~ from 0.069 to 0.11.

Skewness M_3 and kurtosis M_4 have statistically reliable diagnostic $(p_{12;13;14} < 0.001)$ and differentiation $(p_{23;24;34} < 0.05)$ sensitivity to changes in the layered maps $\beta_{\text{object}}(\theta_k, x, y)$ structure.

The polarization ellipticity β_{object} values distributions skewness M_3 magnitude vary from 1.21 to 0.81.

The kurtosis SM_4, which characterizes the peak sharpness of the blood films speckle field $\begin{pmatrix} H_x; \\ H_y; \\ \vartheta_{xy} \end{pmatrix}$ single scattered component $\begin{pmatrix} E_x; \\ E_y; \\ \delta_{xy} \end{pmatrix}$ polarization distributions $\beta_{\text{object}}(\theta_k, x, y)$, consistently decreases from 2.07 to 1.22.

Table 11.4 Blood films polarization ellipticity phase maps statistical parameters.

Groups	Group 1 (51 samples)	Group 2 (51 samples)	Group 3 (51 samples)	Group 4 (51 samples)
Average, M_1	0.15 ± 0.008	0.17 ± 0.009	0.185 ± 0.01	0.194 ± 0.01
p_{ik}		$p_{12;13;14} \leq 0.05$; $p_{23;24;34} \leq 0.05$		
Variance, M_2	0.069 ± 0.004	0.081 ± 0.005	0.096 ± 0.006	0.11 ± 0.007
p_{ik}		$p_{12;13;14} \leq 0.05$; $p_{23;24;34} \leq 0.05$		
Skewness, M_3	1.21 ± 0.07	0.99 ± 0.05	0.88 ± 0.05	0.81 ± 0.04
p_{ik}		$p_{12;13;14} \leq 0.001$; $p_{23;24;34} \leq 0.05$		
Kurtosis, M_4	2.07 ± 0.1	1.79 ± 0.1	1.54 ± 0.08	1.22 ± 0.07
p_{ik}		$p_{12;13;14} \leq 0.001$; $p_{23;24;34} \leq 0.05$		

Table 11.5 Thyroid pathology differential diagnosis balanced accuracy.

	Average, M_1	
$a = 85$	$a = 88$	$a = 89$
$b = 17$	$b = 14$	$b = 13$
$Ac_{12} = 84.3\%$	$Ac_{13} = 86.3\%$	$Ac_{14} = 87.3\%$
$a = 80$	$a = 82$	$a = 83$
$b = 22$	$b = 20$	$b = 19$
$Ac_{23} = 78.4\%$	$Ac_{24} = 80.4\%$	$Ac_{34} = 81.4\%$
	Variance, M_2	
$a = 85$	$a = 88$	$a = 89$
$b = 17$	$b = 14$	$b = 13$
$Ac_{12} = 84.3\%$	$Ac_{13} = 86.3\%$	$Ac_{14} = 87.3\%$
$a = 80$	$a = 82$	$a = 83$
$b = 22$	$b = 20$	$b = 19$
$Ac_{23} = 78.4\%$	$Ac_{24} = 80.4\%$	$Ac_{34} = 81.4\%$
	Skewness, M_3	
$a = 93$	$a = 94$	$a = 95$
$b = 9$	$b = 8$	$b = 7$
$Ac_{12} = 91.2\%$	$Ac_{13} = 92{,}1\%$	$Ac_{14} = 93.1\%$
$a = 89$	$a = 90$	$a = 91$
$b = 13$	$b = 12$	$b = 11$
$Ac_{23} = 87.3\%$	$Ac_{24} = 88.2\%$	$Ac_{34} = 89.2\%$
	Kurtosis, M_4	
$a = 93$	$a = 94$	$a = 95$
$b = 9$	$b = 8$	$b = 7$
$Ac_{12} = 91.2\%$	$Ac_{13} = 92.1\%$	$Ac_{14} = 93.1\%$
$a = 89$	$a = 90$	$a = 91$
$b = 13$	$b = 12$	$b = 11$
$Ac_{23} = 87.3\%$	$Ac_{24} = 88.2\%$	$Ac_{34} = 89.2\%$

The analysis results (Table 11.5) showed a marked improvement in sensitivity, rising from "unsatisfactory" to "very good" and "excellent" when compared to integral ellipticity polarization mapping.

The following levels of balanced accuracy were shown for distinguishing different thyroid gland conditions:

1. Thyroid gland pathology diagnosis (control group 1 and all experimental groups "2–4"):
   - "normal – nodular goiter" $Ac_{12}(M_3; M_4) \sim 91.2\%$ – "excellent" level;
   - "normal – autoimmune thyroiditis" $Ac_{13}(M_3; M_4) \sim 92.1\%$ – "excellent" level;
   - "normal – papillary cancer" $Ac_{14}(M_3; M_4) \sim 93.1\%$ – "excellent" level.
2. Differential diagnosis:
   - "nodular goiter – autoimmune thyroiditis" $Ac_{23}(M_3; M_4) \sim 87\%$ – "very good" level;
   - "nodular goiter – papillary cancer" $Ac_{24}(M_3; M_4) \sim 88\%$ – "very good" level;
   - "autoimmune thyroiditis – papillary cancer" $Ac_{34}(M_3; M_4) \sim 89\%$ – "very good" level.

11.1.5 Conclusions

1. A minimally invasive laser method was developed and tested for the first time to map dehydration blood films layer-by-layer using polarization–interference, aiding in the differential diagnosis of thyroid diseases.
2. A model describing the polycrystalline structure of supramolecular networks in blood films was developed for the first time, using linear and circular birefringence approximations, to provide a physical basis and prognostic analysis of speckle field polarization structures.

3. No statistically significant markers were found using direct polarization–interference mapping and statistical analysis of the measured "integral" maps of the ellipticity of the diffuse component of speckle fields in blood films. Consequently, the accuracy of differential diagnosis for thyroid pathology did not surpass 80%.
4. Digital phase scanning of algorithmically reconstructed complex amplitude fields allowed for measuring polarization ellipticity maps of the singly scattered component in blood film speckle fields. This led to the identification and physical validation of new, statistically significant markers for minimally invasive differential diagnosis of thyroid diseases.
5. The following outcomes of thyroid pathology differential diagnosis have been obtained:
   5.1. Thyroid gland pathology diagnosis (control group 1 and all experimental groups "2–4"):
      - "normal – nodular goiter" $Ac_{12}(M_3; M_4) \sim 91.2\%$ - "excellent" level;
      - "normal – autoimmune thyroiditis" $Ac_{13}(M_3; M_4) \sim 92.1\%$ - "excellent" level;
      - "normal – papillary cancer" $Ac_{14}(M_3; M_4) \sim 93.1\%$ - "excellent" level.

   5.2. Differential diagnosis:
      - "nodular goiter – autoimmune thyroiditis" $Ac_{23}(M_3; M_4) \sim 87\%$ - "very good" level;
      - "nodular goiter – papillary cancer" $Ac_{24}(M_3; M_4) \sim 88\%$ – "very good" level;
      - "autoimmune thyroiditis – papillary cancer" $Ac_{34}(M_3; M_4) \sim 89\%$ – "very good" level.

11.2 Polarization-interference Jones Matrix Sensors of Layer-by-layer Scanning of Polycrystalline Dehydrated Blood Films

11.2.1 Introduction

This section aims to analyze algorithms and conduct experimental testing of a novel multifunctional method for polarization–interference visualization of the Jones matrix. It also involves digitally reconstructing optical anisotropy maps of supramolecular networks in dehydrated blood films layer by layer. The focus is on identifying statistically significant markers that characterize these reconstructed maps of birefringence and dichroism, which are useful for early diagnosis (stage 1) and distinguishing between stages (stages 1 and 2) of papillary thyroid cancer [50–62].

11.2.2 Objects of Investigations

Blood films (BF) samples were obtained by applying a liquid drop to a glass substrate at room temperature ($t = 22\,°\text{C}$) and drying for 24 hours. We reduced the volume of a liquid drop to 0.035–0.004 μL. As a result of blood dehydration, blood films were formed with an average height of up to 60 μm and a diameter of up to 3–4 mm.

Optical-geometric parameters of samples BF are presented in Table 11.6.

11.2.3 Jones Matrix Mapping of Phase Anisotropy of Polycrystalline Blood Films

Figure 11.9 shows the experimental results of the polarization–interference Jones matrix mapping method of linear $LB(a, b)$ and circular $CB(a, b)$ birefringence of BF:

- coordinate distributions ($a \times b = 1000 \times 1000$ and $a \times b = 100 \times 100$) of integral maps $T^{LB}(a, b)$ and $T^{CB}(a, b)$, Figure 11.9a, b;
- histograms of probability distributions of $N(LB(1000 \times 1000))$ and $N(CB(1000 \times 1000))$ integral values of linear LB and circular LB birefringence of polycrystalline architectonics, Figure 11.9c, d;

Table 11.6 BF samples optical parameters.

Parameters	BF
Attenuation coefficient τ, cm^{-1}	0.38 ± 0.021
Depolarization degree Λ, %	29 ± 0.18

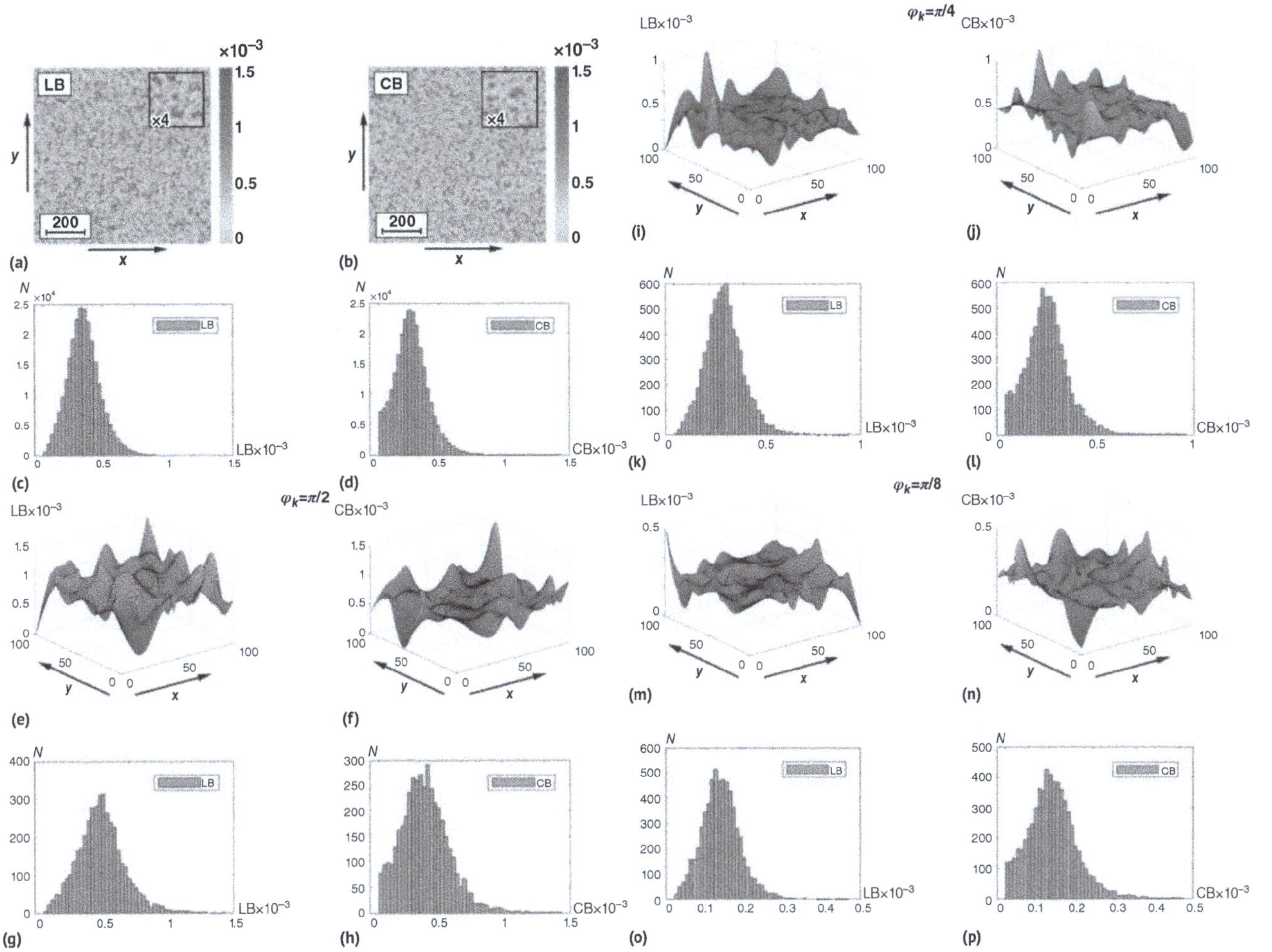

Figure 11.9 Two-dimensional distributions and histograms of integral (a)–(d) and layer-by-layer ($\varphi_k = \pi/2$; $\varphi_k = \pi/4$; $\varphi_k = \pi/8$) (e)–(p)) birefringence (BF) maps: linear birefringence LB ((a), (c), (e), (g), (i), (k), (m) and (o); circular birefringence CB (b), (d), (f), (h), (j), (l), (n), and (p).

- coordinate distributions ($a \times b = 100 \times 100$) of reconstructed layer-by-layer maps $T^{\mathrm{LB}}(\varphi_k, (a, b))$ (panels (e), (i), (m)) and $T^{\mathrm{CB}}(\varphi_k, (a, b))$ (panels (f), (j), (n) in different phase sections ($\varphi_k = \pi/2$, panels (e),(f)), $\varphi_k = \pi/4$, (panels (i), (j)), $\varphi_k = \pi/8$, panels (m), (n));
- histograms $N(\mathrm{LB}(\varphi_k, 100 \times 100))$ and $N(\mathrm{CB}(\varphi_k, 100 \times 100))$ of layer-by-layer distributions of linear and circular birefringence values of polycrystalline architectonics in different phase planes in different phase sections ($\varphi_k = \pi/2$, panels (g), (h)), $\varphi_k = \pi/4$, panels (k), (l)), $\varphi_k = \pi/8$, panels (15), (16)).

Integral maps $T^{\mathrm{LB;CB}}(a, b)$ were obtained by polarization–interference mapping (Figure 11.9) of the coordinate distributions of the elements of the Jones matrix of the blood films with subsequent analytical reconstruction of the LB (1000×1000) and CB (1000×1000) maps.

Reconstruction of the layered maps $T^{\mathrm{LB;CB}}(\varphi_k, 100 \times 100)$ in different phase planes ($\varphi_k = \pi/2$; $\varphi_k = \pi/4$; $\varphi_k = \pi/8$) of the object field of complex amplitudes was carried out by step-by-step phase scanning. The analysis of the maps of linear $T^{\mathrm{LB}}(a, b)$ and circular $T^{\mathrm{CB}}(a, b)$ birefringence confirmed the adequacy of the proposed model representations about the optical anisotropy of the BF polycrystalline architectonics.

Comparison of coordinate and probabilistic (Figure 11.9) distributions of phase anisotropy parameters (LB (1000×1000) and CB (1000×1000)) of BF revealed:

- individual structure of algorithmically reconstructed integral and layer-by-layer maps $T^{\mathrm{LB;CB}}(1000 \times 1000)$ of linear LB and circular CB birefringence of optically anisotropic architectonics, panels (a), (c) and (b), (d);

- dependence of the average level and the magnitude of the range of change of fluctuations of random values of phase anisotropy parameters on the magnitude of the phase parameter φ_k in the reconstructed object field of complex amplitudes, panels (e), (g), (i), (k), (m), (o) and (f), (h), (j), (l), (n), (p);
- the tendency to decrease the magnitude of fluctuations of random values of linear LB and circular CB birefringence in the phase planes $\varphi_k = \pi/2$; $\varphi_k = \pi/4$; $\varphi_k = \pi/8$) of the field of complex amplitudes with smaller values of the scanning parameter φ_k, panels (e), (i), (m) and (f), (j), (n).

The discovered features of the Jones transformation of matrix reconstructed maps $T^{\text{LB;CB}}(\varphi_k, 100 \times 100)$ can be associated with the influence on the calculated value of the birefringence parameter of different multiplicities (q) of light scattering in the considered phase planes ($\varphi_k = \pi/2$; $\varphi_k = \pi/4$; $\varphi_k = \pi/8$)) of the BF object field.

For integral maps $T^{\text{LB}}(a, b)$ and $T^{\text{CB}}(a, b)$ the scattering factor of laser radiation in the volume of polycrystalline blood films is maximal. Each local ($q = 1$) act of interaction of laser radiation with an optically anisotropic domain of polycrystalline architectonics corresponds to "true" values of the set of elements of the Jones matrix w_{ik}. It is possible to reconstruct the Maps $T^{\text{LB;CB}}(a, b)$ of the real 3D polycrystalline structure of the blood films. This "ideal" situation is distorted (multiplied) by multiple (q) acts of interaction – scattering. As a result, the Jones algorithms for reconstructing phase anisotropy maps recreate coordinate distributions with significantly greater values of linear LB and circular CB birefringence. As the value of the phase scanning parameter φ_k decreases (↓), components with a lower multiplicity ($p < q$) of light scattering are extracted from the object field of complex amplitudes. This enables Jones matrix reconstruction of the BF phase anisotropy maps $T^{\text{LB;CB}}(a, b)$ are closer to real ones.

Finally, starting from a certain level of the phase scanning parameter φ_k, the most informative component with a minimum (mainly single $q \to 1$) multiplicity of light scattering in the volume of the blood films is "extracted" from the object field. In this situation, the maximum sensitivity of the reconstructed maps $T^{\text{LB;CB}}(a, b)$ to pathological changes in the BF optically anisotropic architectonics is realized. The considered scenario of layer-by-layer polarization–interference Jones matrix reconstruction of phase anisotropy parameters quantitatively characterizes the results of statistical analysis of the distributions of LB(a, b) and CB(a, b), Table 11.7.

Established:

1. individual and different from Gaussian $\left(M_{3;4}(\text{LB, CB}) = 0\right)$ statistics of holographically reproduced BF maps LB(a, b) and CB(a, b) – $M_{i=1;2;3;4}(\text{LB, CB}) \neq 0$;
2. the third and fourth-order statistical moments $Z_{i=3;4}(\text{LB, CB}(a, b))$ were the most sensitive to changes in the topographic structure of BF phase anisotropy maps;
3. twofold range of phase φ_k change (for different conditions of light scattering in the BF volume) of $M_{i=3;4}(\text{LB}(a, b))$ and $M_{i=3;4}(CB(a, b))$ values.

The results may be related to the following physical considerations, which are based on the model analysis of the BF optical anisotropic properties.

We have previously observed that the biochemical composition of blood can be approximately depicted as a "two-component mixture"-comprising plasma and cellular elements.

Table 11.7 First and fourth-order statistical moments, that characterize the LB(a, b) and CB(a, b) maps.

M_i	LB	LB($\varphi = \pi/2$)	LB($\varphi = \pi/4$)	LB($\varphi = \pi/8$)
$M_1 \times 10^{-3}$	0.42 ± 0.082	0.27 ± 0.071	0.16 ± 0.053	0.093 ± 0.0048
$M_2 \times 10^{-3}$	0.31 ± 0.016	0.28 ± 0.015	0.24 ± 0.014	0.21 ± 0.012
M_3	0.35 ± 0.018	0.41 ± 0.023	0.53 ± 0.027	0.65 ± 0.036
M_4	0.51 ± 0.028	0.67 ± 0.037	0.78 ± 0.042	0.91 ± 0.046
M_i	CB	CB($\varphi = \pi/2$)	CB($\varphi = \pi/4$)	CB($\varphi = \pi/8$)
$M_1 \times 10^{-3}$	0.32 ± 0.067	0.13 ± 0.055	0.094 ± 0.0046	0.089 ± 0.0045
$M_2 \times 10^{-3}$	0.29 ± 0.017	0.23 ± 0.013	0.19 ± 0.011	0.17 ± 0.009
M_3	0.56 ± 0.029	0.71 ± 0.037	0.82 ± 0.043	0.99 ± 0.047
M_4	0.71 ± 0.038	0.94 ± 0.046	0.17 ± 0.064	0.38 ± 0.073

The magnitude of linear birefringence LB is mainly determined by the degree of needle crystals optical axes spatial orientation (γ) of the marginal albumin layer, which can be quite large – $0 \leq \gamma \leq 2\pi$. As a result, the first and second-order statistical moments, which characterize the mean and variance of the distributions $\mathrm{LB}(a, b)$, have a small value. The values of higher orders statistical moments $M_{i=3;4}(\mathrm{LB}(a, b))$, which determine the skewness and kurtosis of polycrystalline dendritic network BF, are quite large (Table 11.7).

The magnitude of BF circular birefringence, observed in globulin crystals and various shaped elements such as erythrocytes, platelets, and leukocytes, is quite substantial. Additionally, in a blood sample, the high concentration of spherulites causes the overall optical activity parameter to be nearly equal to the linear birefringence parameter LB across the entire volume of the needle dendritic network of albumin crystals. Accordingly, the statistical moments $M_{i=1;2}(\mathrm{CB}) \approx M_{i=1;2}(\mathrm{LB})$ and $M_{i=3;4}(\mathrm{CB}) \approx M_{i=3;4}(\mathrm{LB})$ are given in Table 11.7.

In digital holographic phase scanning of volumetric depth BF, different types of polycrystalline networks are consistently identified, characterized by optically anisotropic parameter distributions such as the orientations of optical axes and phase differences between orthogonal amplitude components. As a result, individual orientation distributions ($\gamma_{\min} \leq \gamma \leq \gamma_{\max}$) with a certain random variance of their values are reproduced for each phase plane. Additionally, when small or single scattering conditions are met, the range of variations in the random phase shifts narrows. Approaching the conditions of an optically thin layer, the statistics of the BF phase anisotropy ($\mathrm{LB}(a, b)$ and $\mathrm{CB}(a, b)$) changes in the following scenarios,

$$- \varphi_k \downarrow \Longrightarrow \begin{pmatrix} M_{1;2}(\mathrm{LB}, \mathrm{CB}) \downarrow; \\ M_{3;4}(\mathrm{LB}, \mathrm{CB}) \uparrow \end{pmatrix}.$$

11.2.4 Jones Matrix Mapping of the Amplitude Anisotropy of Polycrystalline Blood Films

Figure 11.10 shows, similar to the polarization–interference Jones matrix phase mapping data in Figure 11.9, the coordinate (panels (a), (b), (e), (f), (i), (j), (m), (n)) and probability (panels (c), (d), (g), (h), (k), (l), (o), (p)) distributions of linear LB (panels (a), (c), (e), (g), (i), (k), (m), (o)) and circular CB (panels (b), (d), (f), (h), (j), (l), (n), (p)) dichroism of BF polycrystalline architectonics. Integral maps $T^{\mathrm{LD}}(a, b)$ and $T^{\mathrm{CD}}(a, b)$ of amplitude anisotropy were reconstructed.

From the analysis of the obtained maps $T^{\mathrm{LD;CD}}(a, b)$ and $T^{\mathrm{LD;CD}}(\varphi_k, a, b)$ of amplitude anisotropy (Figure 11.10) it is evident that for each coordinate distribution of random values of the magnitude of linear and circular dichroism of the BF polycrystalline architectonics, an individual coordinate (panels (a), (e), (i), (m) and panels (b), (f), (j), (n)) and probabilistic (panels (c), (g), (k), (o) and panels (d), (h), (l), (p)) structure is characteristic. A comparison of a series of layered, aps $T^{\mathrm{LD}}(\varphi_k, 100 \times 100)$ of amplitude anisotropy (Figure 11.10e, g, i, k, m, o) and $T^{;\mathrm{CD}}(\varphi_k, 100 \times 100)$ (Figure 11.10e, g, i, k, m, p) demonstrated a tendency, similar to the phase maps $T^{\mathrm{LB}}(\varphi_k, 100 \times 100)$ and $T^{\mathrm{CB}}(\varphi_k, 100 \times 100)$ (Figure 11.9), of decreasing the magnitude of fluctuations of the parameters of optically anisotropic absorption in the phase planes ($\varphi_k = \pi/2$; $\varphi_k = \pi/4$; $\varphi_k = \pi/8$), respectively) of the components of the object field of complex amplitudes with a low multiplicity of light scattering.

Quantitively phase transformation of the $\mathrm{LD}(a, b)$ and $\mathrm{CD}(a, b)$ maps illustrate the first to fourth-order central statistical moments (Table 11.8).

Detected:

- difference from zero (for all phase films depths) values of all central statistical moment's $M_{i=1;2;3;4}(\mathrm{LD}, \mathrm{CD}) \neq 0$;
- dependence of the magnitude of statistical parameters $Z_{i=1;2;3;4}(\mathrm{LD}, \mathrm{CD}) \leftrightarrow q(\varphi_k)$ from the phase scan φ_k depth;
- the most sensitive markers for changes in amplitude anisotropy maps are third- and fourth-order statistical moments $M_{i=3;4}(\mathrm{LD}, \mathrm{CD}(a, b))$;
- as the phase scanning parameter $\varphi_k \uparrow$ value increases (approaching single scattering conditions), the statistical parameters $M_{i=3;4}(\mathrm{LD}(a, b))$ values increase to 85%, and $M_{i=3;4}(\mathrm{CD}, (a, b))$ increase by 60–80%.

From a physical perspective, the results can be attributed to the fact that the values of linear dichroism and linear birefringence LD at all phase depths are determined by the degree of spatial orientation of the optical axes in needle-shaped protein crystals (albumin, fibrin, and elastin). Under conditions of phase approximation for an optically thin layer, the changes in the orientations of the needle crystals' optical axes, as well as the magnitude of phase shifts, follow a scenario where the changes in the 1st and 4th order statistical moments are observed – $\varphi_k \downarrow \Longrightarrow \begin{pmatrix} M_{1;2}(\mathrm{LD}, \mathrm{CD}) \downarrow; \\ M_{3;4}(\mathrm{LD}, \mathrm{CD}) \uparrow \end{pmatrix}$, – Table 11.8.

Along with this, the value of circular dichroism CD combined "shaped elements – albumin – globulin grids" polycrystalline

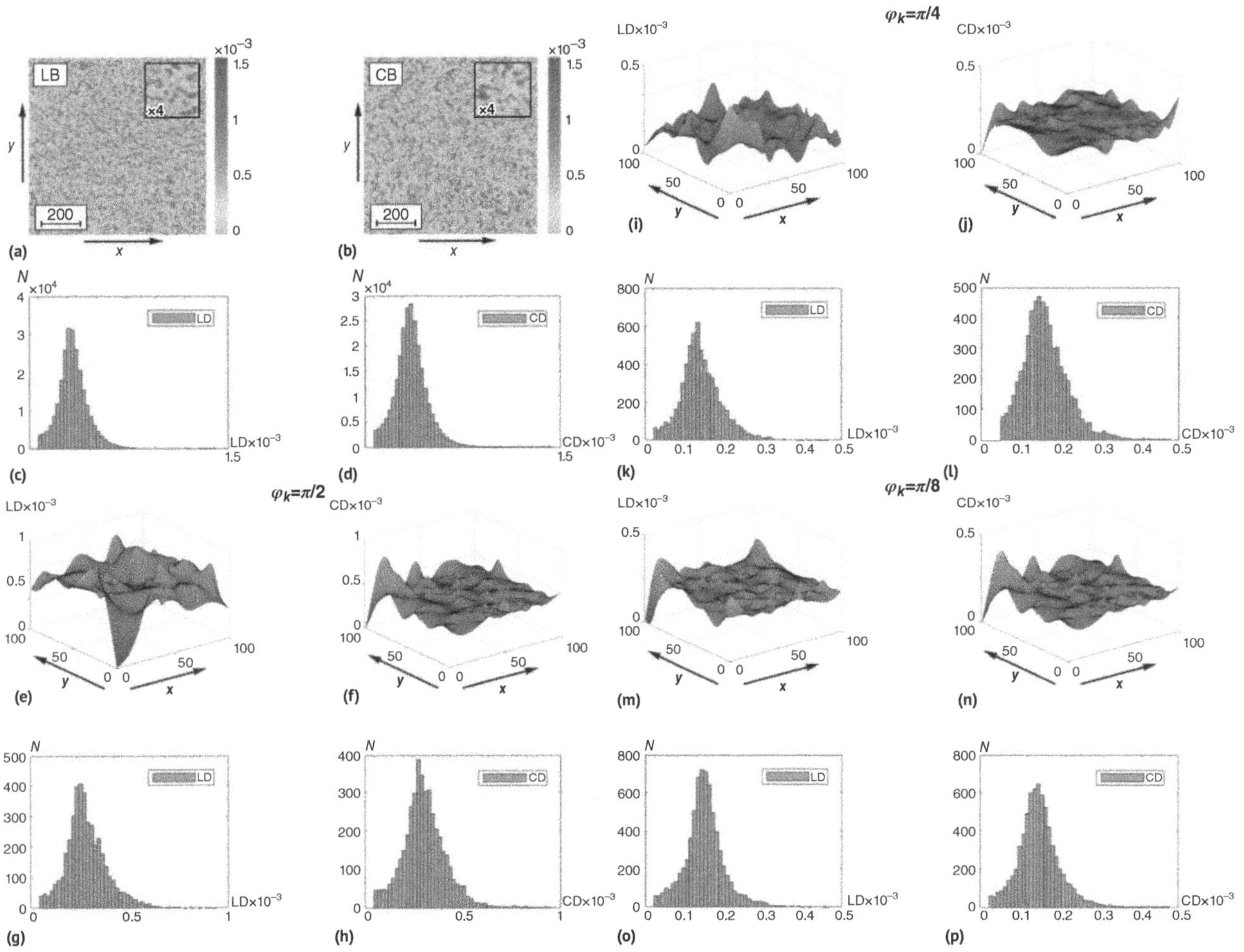

Figure 11.10 Two-dimensional distributions and histograms of integral (a)–(d) and layer-by-layer ($\varphi_k = \pi/2$; $\varphi_k = \pi/4$; $\varphi_k = \pi/8$) (e)–(p) linear LD (a), (c), (e), (g), (i), (k), (m), (o) and circular CD (a), (c), (e), (g), (i), (k), (m), (p) birefringence BF maps.

Table 11.8 First- to fourth-order statistical moments of the BF linear and circular dichroism.

M_i	LD	LD($\varphi = \pi/2$)	LD($\varphi = \pi/4$)	LD($\varphi = \pi/8$)
$M_1 \times 10^{-3}$	0.24 ± 0.069	0.19 ± 0.064	0.07 ± 0.058	0.06 ± 0.054
$M_2 \times 10^{-3}$	0.28 ± 0.015	0.24 ± 0.013	0.22 ± 0.013	0.18 ± 0.009
M_3	0.41 ± 0.022	0.56 ± 0.029	0.68 ± 0.036	0.77 ± 0.039
M_4	0.62 ± 0.033	0.79 ± 0.041	0.92 ± 0.046	0.09 ± 0.051
M_i	**CD**	**CD($\varphi = \pi/2$)**	**CD($\varphi = \pi/4$)**	**CD($\varphi = \pi/8$)**
$M_1 \times 10^{-3}$	0.095 ± 0.0055	0.093 ± 0.0049	0.088 ± 0.0046	0.082 ± 0.0045
$M_2 \times 10^{-3}$	0.21 ± 0.011	0.19 ± 0.011	0.17 ± 0.009	0.15 ± 0.008
M_3	0.65 ± 0.034	0.81 ± 0.043	0.97 ± 0.051	0.16 ± 0.066
M_4	0.83 ± 0.043	0.05 ± 0.054	0.31 ± 0.072	0.54 ± 0.087

network. Thus, the cycle of fundamental polarization–interference studies of BF optical anisotropy in polycrystalline structures, mapped layer-by-layer, provides the basis for implementing applied Jones matrix mapping.

The most sensitive markers ($M_{3;4}$(LB, CB, LD, CD)) to changes in the optical anisotropy of BF supramolecular networks, identified through statistical analysis of layered Jones matrix maps of blood film supramolecular networks, were used in practical studies. These studies assessed the diagnostic effectiveness of the polarization–interference method for early detection of papillary thyroid cancer and for distinguishing its stages.

11.2.5 Diagnostic Capabilities of Jones Matrix Mapping of Polycrystalline Blood Films

This section contains three data arrays from experimental blood film studies [63–73].

The first part focuses on a comparative analysis of the statistical structure of Jones matrix maps measuring linear and circular birefringence and dichroism in blood films. These samples are from donors (control group 1, 26 samples) and patients with stage 1 papillary cancer (experimental group 2, 26 samples).

The second part focuses on evaluating the diagnostic effectiveness of layer-by-layer Jones matrix mapping of BF for detecting early-stage papillary thyroid carcinoma.

The third section presents a comparative analysis of diagnostic effectiveness (group 1) and the differentiation between stage 1 (group 2) and stage 2 (group 3, 26 samples) thyroid cancer using Jones matrix mapping and Mueller-matrix tomography of blood films.

The diagnostic methodology of the developed polarization–interference method includes the following stages:

- polarization–interference mapping of object fields of the set of samples of blood films of Group 1–3;
- calculation of the integral elements of the Jones matrix and the matrix operator of "single" (phase plane $\varphi_k = {}^{\pi}/_{8}$) scattering;
- reconstruction of integral $T^{\text{LB,CB,LD,CD}}(a, b)$ and layer-by-layer $T^{\text{LB,CB,LD,CD}}(\varphi_k = {}^{\pi}/_{8}, a, b)$ maps of BF phase and amplitude anisotropy;
- statistical analysis of maps and determination of the most sensitive parameters (markers) $M_i(\text{LB, CB, LD, CD})$ and $M_i(\varphi_k = {}^{\pi}/_{8}, \text{LB, CB, LD, CD})$ to pathological changes in the BF optically anisotropic architecture;

Figures 11.11 and 11.12 show integral (a, b) (panels (a), (b)) and layer-by-layer $(\varphi_k, (a, b))$ (panels (e), (f)) linear $\text{LB}(\varphi_k, (a, b))$ (Figure 11.11) and circular $\text{CB}(\varphi_k, (a, b))$ (Figure 11.12) birefringence maps, as well as histograms of $G(\text{LB, CB})$ distributions (panels (c), (d), (g), (h)) from control group 1 (panels (a), (c), (e), (g)) and experimental group 2 (panels (b), (d), (f), (h)).

Quantitative phase transformation and intergroup differences between the $\text{LB}(\varphi_k, (a, b))$ and $\text{CB}(\varphi_k, (a, b))$ BF maps are illustrated by the statistical analysis results (Table 11.9).

Statistical analysis of the phase anisotropy maps $\text{LB}(a, b)$ and $\text{CB}(a, b)$ revealed:

- for all phase depths, the central first to fourth- order statistical moments are non-zero-$M_{i=1;2;3;4}(\text{LB, CB}) \neq 0$
- the third- and fourth-order statistical moments $M_{i=3;4}(\text{LB}(a, b), \text{CB}(a, b))$ are the most sensitive markers (especially in the phase plane $\varphi_k = \pi/8$) to changes in BF birefringence;
- for maps $\text{LB}(\varphi_k = \pi/8; (a, b))$, the differences $\Delta M_{i=3;4}(\text{LB}(a, b))$ between the values of BF control and experimental groups diagnostic markers are 30–40%;
- for maps $\text{CB}(\varphi_k = \pi/8; (a, b))$, the differences $\Delta M_{i=3;4}(\text{CB}(a, b))$ between the values of diagnostic markers increase and reach 45–55%.

From a physical perspective, the results obtained can be linked to the following characteristics of BF architectonics in polycrystalline networks' pathological changes.

In earlier classical vector-parametric studies of optical anisotropy, blood films, and plasma were shown to primarily have their linear birefringence value LB influenced by the spatial orientation of the optical axes of albumin, fibrin, and elastin needle crystals.

The orientation changes in the directions of the optical axes in the dendritic grid are very large and almost equidistant – $0 \leq \gamma \leq 2\pi$. As a result, the values of the first- and second-order statistical moments, which characterize $\text{LB}(a, b)$ distributions mean and variance, are quite large. The values of higher-order statistical moments $Z_{i=3;4}(\text{LB}(a, b))$, on the contrary, are insignificant.

During the cancer process, blood crystallization increases. Due to this, for patients' blood droplets dehydrated films from experimental group 2, the range of changes decreases. As a result, the parameters $M_{i=1;2}(\text{LB}(a, b)) \downarrow$ decrease and, conversely, the third and fourth orders $M_{i=3;4}(\text{LB}(a, b)) \uparrow$ statistical moments increase.

Second, the value of the globulin crystals and shaped elements (erythrocytes, platelets, leukocytes, and monocytes) spherulite optically anisotropic network circular birefringence CB], is commensurate with the linear birefringence LB values. In cancer, inflammation increases the concentration of globulin and white blood cell molecules. This increases the

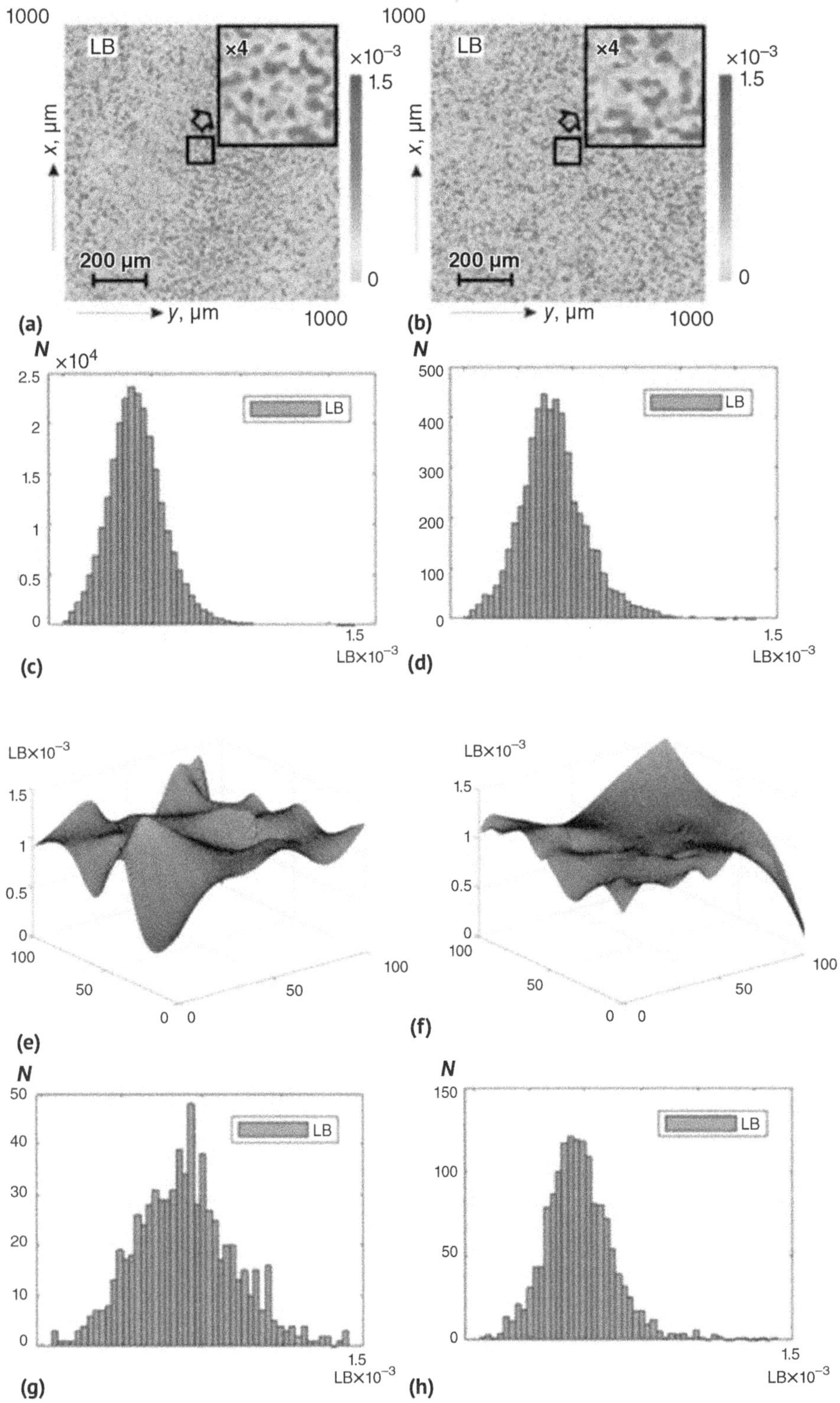

Figure 11.11 BF maps $\{\mathrm{LB}(a, b)\}$ and histograms $G\{\mathrm{LB}(a, b)\}$ of healthy donors ((a), (c), (e), and (g)) and patients with papillary thyroid carcinoma ((b), (d), (f), and (h)).

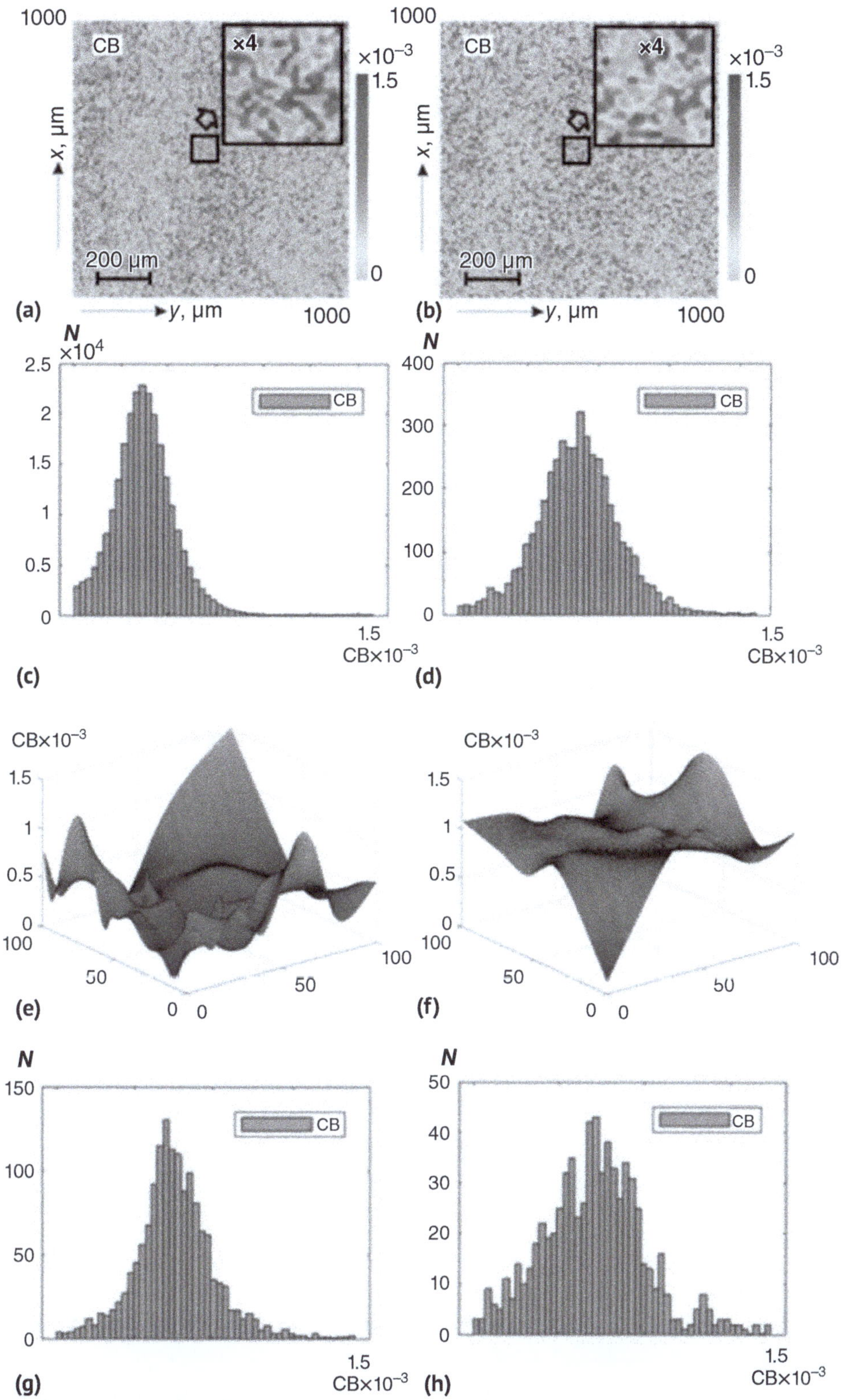

Figure 11.12 BF maps $\{\mathrm{CB}(a, b)\}$ and histograms $G\{\mathrm{CB}(a, b)\}$ of healthy donors ((a), (c), (e), and (g)) and patients with papillary thyroid carcinoma ((b), (d), (f), and (h)).

Table 11.9 Statistical parameters of LB(φ_k, (a, b)) and CB(φ_k, (a, b)).

M_i	LB		CB	
$M_1 \times 10^{-3}$	Group 1	Group 2	Group 1	Group 2
$M_2 \times 10^{-3}$	0.41 ± 0.081	0.34 ± 0.081	0.02 ± 0.062	0.11 ± 0.064
M_3	0.33 ± 0.015	0.28 ± 0.015	0.27 ± 0.015	0.32 ± 0.016
M_4	0.31 ± 0.017	0.42 ± 0.021	0.59 ± 0.025	0.49 ± 0.027
M_i	0.45 ± 0.026	0.53 ± 0.027	0.75 ± 0.042	0.64 ± 0.036
M_i	$\{LB(\pi/8)\}$		$\{CB(\pi/8)\}$	
$M_1 \times 10^{-3}$	Group 1	Group 2	Group 1	Group 2
$M_2 \times 10^{-3}$	0.89 ± 0.046	0.94 ± 0.049	0.88 ± 0.049	0.95 ± 0.046
M_3	0.21 ± 0.011	0.24 ± 0.013	0.16 ± 0.008	0.21 ± 0.011
M_4	0.68 ± 0.035	0.57 ± 0.037	0.69 ± 0.039	0.87 ± 0.048
M_i	0.89 ± 0.044	0.81 ± 0.048	0.82 ± 0.044	1.07 ± 0.059

optical activity of the chiral molecular domains in the dehydrated films of blood droplets from patients with papillary thyroid carcinoma.

By this, first- to fourth-order statistical moments, which characterize the circular birefringence random values distributions, undergo the following changes – $M_{i=1;2}$(CB) ↑ and, conversely, $M_{i=3;4}$(CB) ↓.

Third, as the scan φ_k ↓ phase shift decreases, a more optically thin rat blood-dehydrated film volume is released.

As a result, the albumin, fibrin, and elastin needle crystals' optical axes directions distribution are transformed into an asymmetric one ($\gamma_{\min} \leq \gamma \leq \gamma_{\max}$) with a value certain random variance. In addition, the range of changes in random values of phase shifts between laser radiation amplitudes orthogonal components is narrowed. These changes in the phase anisotropy maps statistics of dehydrated blood drop films LB(a, b) and CB(a, b), – $\begin{pmatrix} M_{1;2}(\text{LB, CB}) \downarrow; \\ M_{3;4}(\text{LB, CB}) \uparrow \end{pmatrix}$. Against this background, cancer differences in the optically anisotropic network's polycrystalline structure of blood samples from experimental group 2 are more clearly revealed.

Figures 11.13 and 11.14 show integral (a, b) (panels (a), (b)) and layer-by-layer (φ_k, (a, b)) (panels (e),(f)) LD(φ_k, (a, b)) (Figure 11.13) and CD(φ_k, (a, b)) (Figure 11.14) maps, as well as histograms of G(LD, CD) (panels (c), (d), (g), (h)).

Control group 1 (panels (a), (c), (e), (g)) and experimental group 2 (panels (b), (d), (f), (h)).

Quantitative phase transformation and intergroup differences between the LD(φ_k, (a, b)) and CD(φ_k, (a, b)) BF maps are illustrated by the statistical analysis results (Table 11.10).

Analysis of BF amplitude anisotropy maps for all phase depths revealed:

- difference all first- to fourth- order statistical moments value from zero $M_{i=1;2;3;4}$(LD, CD) $\neq 0$;
- dependence of statistical distributions $M_{i=1;2;3;4}$(LD, CD) $\leftrightarrow q(\varphi_k)$ on the value of the phase scanning parameter φ_k of BF polycrystalline architectonics;
- the most sensitive diagnostic markers are third- and fourth- order statistical moments $M_{i=3;4}$(LD(a, b), CD(a, b));
- for linear dichroism LD(a, b) maps, the differences $\Delta M_{i=3;4}$(LD(a, b)) between the diagnostic marker's values of the control and experimental BF samples groups are 25–30%;
- for CD(a, b)circular dichroism maps, the differences $\Delta M_{i=3;4}$(CD(a, b)) between the diagnostic marker's values increase and reach 35–45%.

Physically, the results are attributed to phase scanning's (φ_k ↓) ability to isolate a more optically thin segment of BF. This causes the topographic structure of BF maps to closely match the distribution of needle crystal (linear dichroism) optical axes and phase shift values (circular dichroism). Consequently, this allows for the highest accuracy in identifying polycrystalline differences between BF rat samples from both groups.

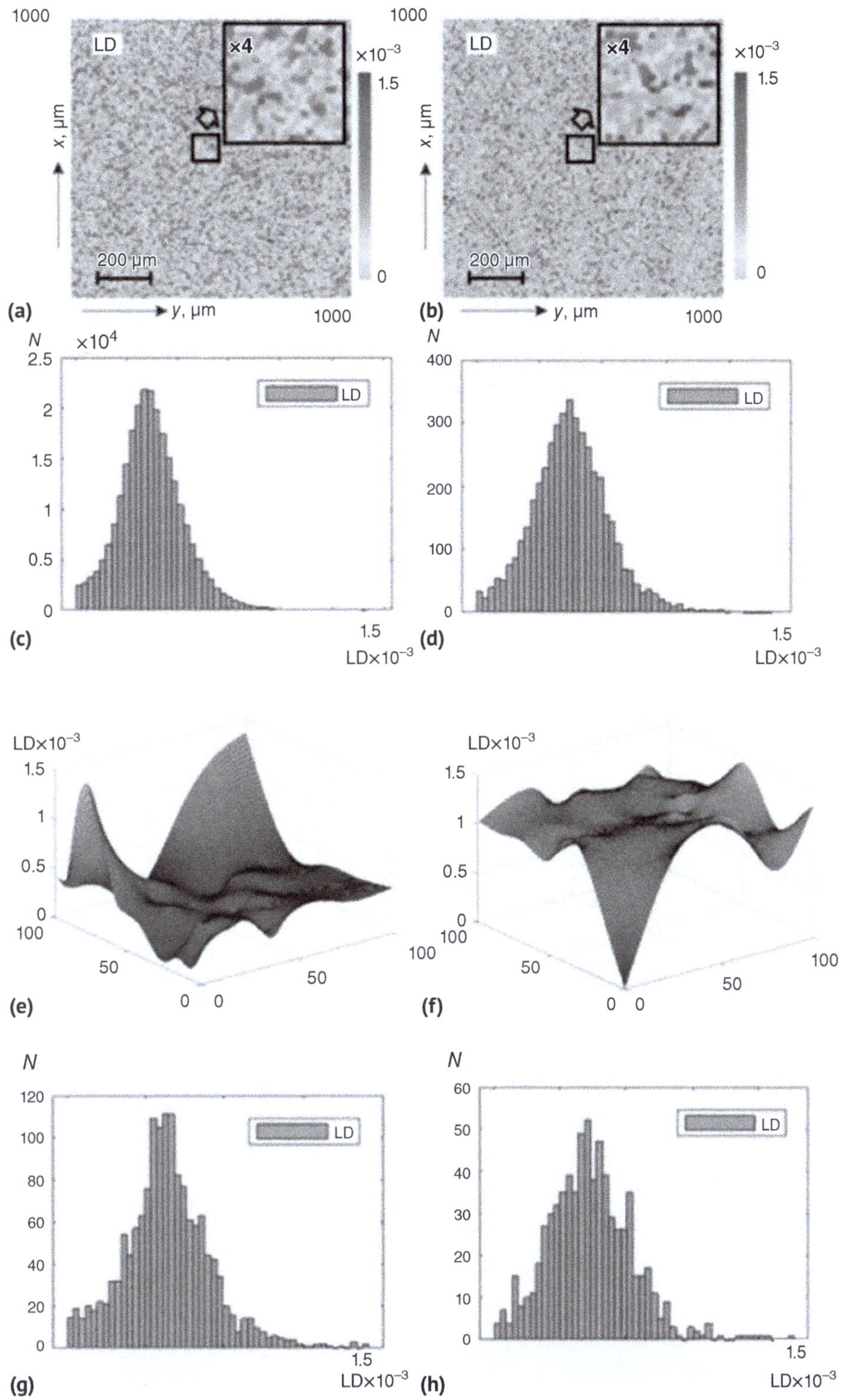

Figure 11.13 BF LD maps $\{\mathrm{LD}(a, b)\}$ and histograms $G\{\mathrm{LD}(a, b)\}$ of healthy donors ((a), (c), (e), and (g)) and patients with papillary thyroid carcinoma ((b), (d), (f), and (h)).

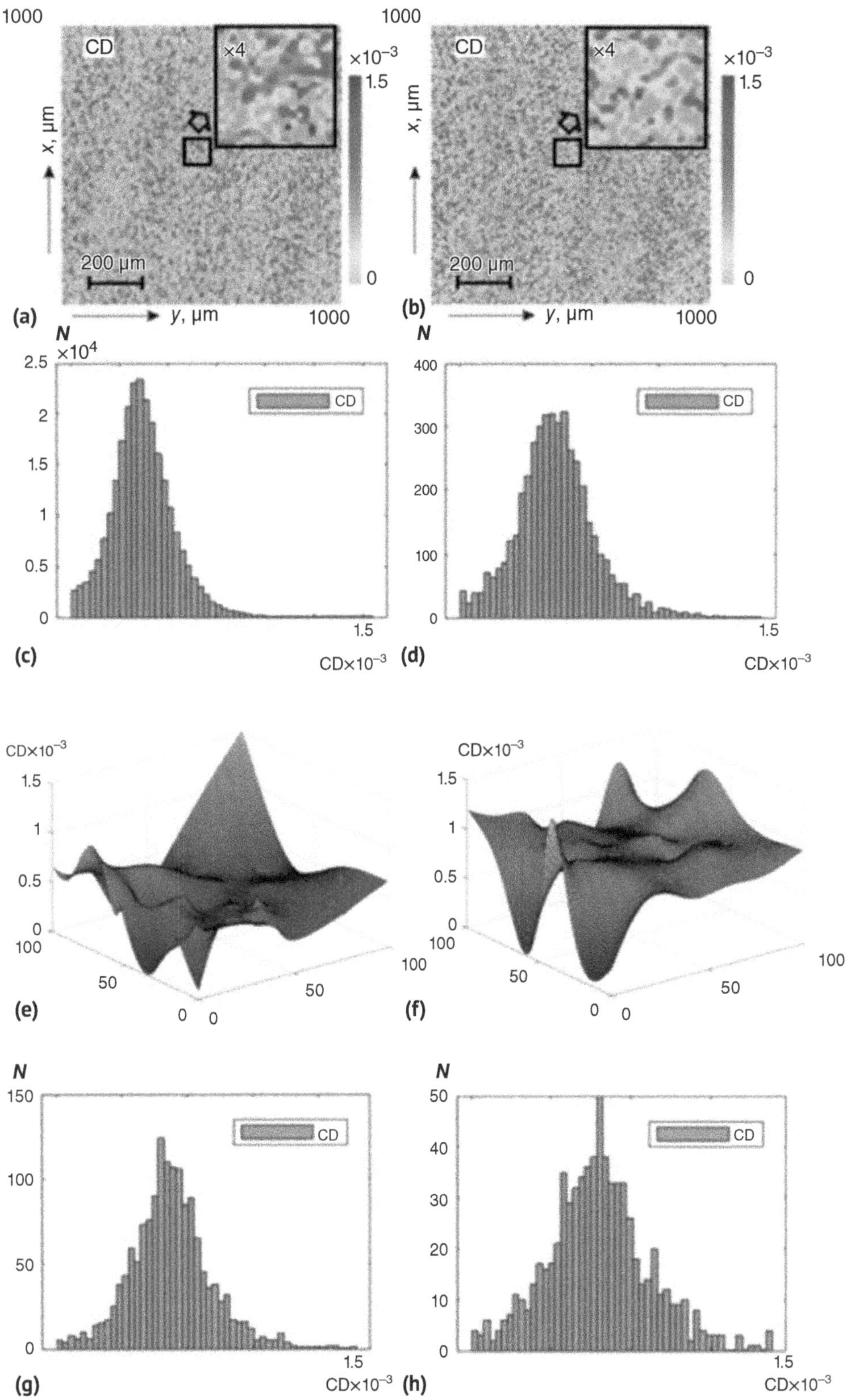

Figure 11.14 BF maps $\{\mathrm{CD}(a, b)\}$ and histograms $G\{\mathrm{CD}(a, b)\}$ of healthy donors ((a), (c), (e), and (g)) and patients with papillary thyroid carcinoma ((b), (d), (f), and (h)).

Table 11.10 Statistical parameters of LD(φ_k, (a, b)) and CD(φ_k, (a, b)).

M_i	LD		CD	
$M_1 \times 10^{-3}$	Group 1	Group 2	Group 1	Group 2
$M_2 \times 10^{-3}$	0.21 ± 0.063	0.29 ± 0.068	0.89 ± 0.051	0.94 ± 0.054
M_3	0.27 ± 0.014	0.33 ± 0.016	0.18 ± 0.011	0.21 ± 0.011
M_4	0.51 ± 0.023	0.43 ± 0.023	0.76 ± 0.036	0.63 ± 0.036
M_i	0.76 ± 0.032	0.64 ± 0.034	0.93 ± 0.047	0.81 ± 0.044
M_i	{LD($\pi/8$)}		{CD($\pi/8$)}	
$M_1 \times 10^{-3}$	Group 1	Group 2	Group 1	Group 2
$M_2 \times 10^{-3}$	0.94 ± 0.054	0.89 ± 0.051	0.81 ± 0.046	0.88 ± 0.043
M_3	0.18 ± 0.009	0.15 ± 0.008	0.14 ± 0.007	0.18 ± 0.009
M_4	0.79 ± 0.038	0.91 ± 0.045	0.19 ± 0.068	0.91 ± 0.067
M_i	0.98 ± 0.052	0.21 ± 0.062	0.44 ± 0.089	0.15 ± 0.089

A slightly lower accuracy (based on the analysis of phase anisotropy maps) may be linked to a reduced level of anisotropic absorption compared to laser scattering by BF samples. Consequently, the sensitivity of the BF polycrystalline component matrix mapping decreases.

It is worth noting that the sensitivity of the polarization–interference Jones matrix mapping method for blood films, used in diagnosing early papillary thyroid cancer, can be enhanced by further refining the algorithm. This improvement would help reduce the effect of the depolarized background during the reconstruction of anisotropy parameters in polycrystalline structures.

11.2.6 Comparative Information Analysis of Jones Matrix Mapping and Mueller-matrix Tomography Data

This section presents the results of a comparative analysis involving a donor group (Group 1) to evaluate the diagnostic effectiveness of Jones matrix mapping and Mueller-matrix tomography of blood films in early detection of stage 1 (where the formation develops inside the gland, the organ capsule remains intact, and symptoms are not yet visible, Group 2). It also assesses the detection of stage 2 (characterized by proliferation of nodes and capsule deformation, Group 3, with 26 samples) of thyroid cancer and differentiates between cancer stages (Groups 2 and 3).

The studies were conducted in the most sensitive phase plane $\varphi = \pi/8$, Table 11.11.

Here, JT – Jones matrix mapping; MMT – Mueller-matrix tomography.

The maximum levels of balanced accuracy for diagnosing papillary thyroid cancer have been set.

1. Jones matrix mapping (stage 1):
   - linear birefringence LB(a, b) – good level 88.5% (Se = 88.5%; Sp = 88.5%);
   - circular birefringence CB(a, b) – very good level 90.4% (Se = 92.3%; Sp = 88.5%);
   - linear dichroism LD(a, b) – good level 82.7% (Se = 80.8%; Sp = 84.6%);
   - circular dichroism CD(a, b) – good level 88.5% (Se = 84.6%; Sp = 92.3%).
2. Mueller-matrix tomography (stage 1) – unsatisfactory accuracy Ac < 80% for tomograms of linear and circular birefringence and dichroism.
3. Jones matrix mapping (stage 2):
   - linear birefringence, LB(a, b) – very good level 94.2% (Se = 96.2%; Sp = 88.5%);
   - circular birefringence, CB(a, b) – excellent level 96.2% (Se = 96.2%; Sp = 96.2%);
   - linear dichroism, LD(a, b) – very good level 92.3% (Se = 96.2%; Sp = 88.5%);
   - circular dichroism, CD(a, b) – very good level 94.2% (Se = 92.3%; Sp = 96.2%).
4. Mueller-matrix tomography (stage 2):
   - linear birefringence LB(a, b) – good level 86.5% (Se = 88.5%; Sp = 84.6%);
   - circular birefringence CB(a, b) – good level 88.2% (Se = 96.2%; Sp = 96.2%);
   - linear LD(a, b) and circular dichroism CD(a, b) – unsatisfactory level (Se; Sp; Ac < 80%).

Table 11.11 Operational characteristics of the diagnostic capabilities of Jones matrix mapping and Mueller-matrix tomography methods for blood films.

$M_{3;4}(\pi/8)$	LB(*a*, *b*)		
Groups	"1" "2"	"1" "3"	"2" "3"
Se(JT), %	$N = 23; n = 3$; **88.5**	$N = 25; n = 1$; **96.2**	$N = 24; n = 2$; **92.3**
Se(MMT), %	$N = 20; n = 6$; **76.9**	$N = 23; n = 3$; **88.5**	$N = 21; n = 5$; **80.8**
Sp(JT), %	$H = 23; h = 3$; **88.5**	$H = 24; h = 2$; **88.5**	$H = 24; h = 2$; **92.3**
Sp(MMT), %	$H = 20; h = 6$; **76.9**	$H = 22; h = 4$; **84.6**	$H = 22; h = 4$; **84.6**
Ac(JT), %	**88.5**	**94.2**	**92.3**
Ac(MMT), %	**76.9**	**86.5**	**82.7**

$M_{3;4}(\pi/8)$	CB(*a*, *b*)		
Groups	"1" "2"	"1" "3"	"2" "3"
Se(JT), %	$N = 24; n = 2$; **92.3**	$N = 25; n = 1$; **96.2**	$N = 25; n = 1$; **96.2**
Se(MMT), %	$N = 21; n = 5$; **80.8**	$N = 24; n = 2$; **92.3**	$N = 23; n = 3$; **88.5**
Sp(JT), %	$H = 23; h = 3$; **88.5**	$H = 25; h = 1$; **96.2**	$H = 24; h = 2$; **92.3**
Sp(MMT), %	$H = 20; h = 6$; **76.9**	$H = 22; h = 4$; **84.6**	$H = 21; h = 5$; **80.8**
Ac(JT), %	**90.4**	**96.2**	**94.2**
Ac(MMT), %	**78.8**	**88.5**	**84.6**

$M_{3;4}(\pi/8)$	LD(*a*, *b*)		
Groups	"1" "2"	"1" "3"	"2" "3"
Se(JT), %	$N = 22; n = 4$; **84.6**	$N = 25; n = 1$; **96.2**	$N = 23; n = 3$; **88.5**
Se(MMT), %	$N = 18; n = 8$; **69.2**	$N = 20; n = 6$; **76.9**	$N = 17; n = 9$; **65.4**
Sp(JT), %	$H = 21; h = 5$; **80.8**	$H = 23; h = 3$; **88.5**	$H = 24; h = 2$; **92.3**
Sp(MMT), %	$H = 19; h = 7$; **73.1**	$H = 20; h = 6$; **76.9**	$H = 16; h = 10$; **61.5**
AC(JT), %	**82.7**	**92.3**	**90.4**
AC(MMT), %	**71.2**	**76.9**	**63.5**

$M_{3;4}(\pi/8)$	CD(*a*, *b*)		
Groups	"1" "2"	"1" "3"	"2" "3"
Se(JT), %	$N = 23; n = 3$; **88.5**	$N = 24; n = 2$; **92.3**	$N = 24; n = 2$; **92.3**
Se(MMT), %	$N = 19; n = 7$; **73.1**	$N = 20; n = 6$; **76.9**	$N = 19; n = 7$; **73.1**
Sp(JT), %	$H = 23; h = 3$; **88.5**	$H = 25; h = 1$; **96.2**	$H = 24; h = 2$; **92.3**
Sp(MMT), %	$H = 18; h = 8$; **69.2**	$H = 19; h = 7$; **73.1**	$H = 18; h = 8$; **69.2**
AC(JT), %	**88.5**	**94.2**	**92.3**
Ac(MMT), %	**71.2**	**75**	**71.2**

The following maximum levels of balanced accuracy of differentiation of papillary thyroid cancer stages were established.

1. Jones matrix mapping (stages 2 and 3):
   - linear birefringence LB(a, b) – very good level 92.3% (Se = 92.3%; Sp = 92.3%);
   - circular birefringence CB(a, b) – very good level 94.2% (Se = 96.2%; Sp = 92.3%);
   - linear dichroism LD(a, b) – very good level 90.4% (Se = 88.5%; Sp = 92.3%);
   - circular dichroism CD(a, b) – very good level 92.3% (Se = 92.3%; Sp = 92.3%).
2. Mueller-matrix diffuse tomography (stages 2 and 3):
   - linear birefringence LB(a, b) – good level 82.7% (Se = 80.8%; Sp = 84.6%);
   - circular birefringence CB(a, b) – good level 84.6% (Se = 88.5%; Sp = 80.8%);
   - linear LD(a, b) and circular dichroism CD(a, b) – unsatisfactory level (Se; Sp; Ac < 80%).

11.2.7 Conclusions

The primary key finding of this research is a novel polarization–interference multifunctional technique for laser Jones matrix mapping of the polycrystalline structure in supramolecular networks of dehydrated blood films. This approach enhances sensitivity and broadens the functional scope of Mueller-matrix tomography, facilitating minimally invasive early detection of asymptomatic stage 1 papillary cancer and distinguishing between its first and second stages.

An equally significant result is the potential for a more expressive four-dimensional approach, compared to the eight-dimensional Mueller matrix, for directly extracting Jones matrix images from phase-stationary speckle fields through polarization–interference.

As a result, a set of blood films polycrystalline component phase (birefringence – LB, CB) and amplitude (dichroism – LD, CD) anisotropy integral and layered maps (a, b) was obtained and physically analyzed.

Regularities and statistical scenarios of phase transformations in the structure of holographically reconstructed maps of optical anisotropy of blood films have been established – $(\varphi_k \uparrow) \Longrightarrow \begin{pmatrix} M_{i=1;2}(\text{LB, CB, LD, CD}(a, b)) \downarrow; \\ M_{i=3;4}(\text{LB, CB, LD, CD}(a, b)) \uparrow \end{pmatrix}$.

The most sensitive markers to changes in the layer-by-layer topographic structure of blood films phase and amplitude anisotropy are established – the third and fourth-order $M_{i=3;4}$(LB, CB, LD, CD(a, b)) statistical moments.

The main advantages of the Jones matrix method in comparison with the Mueller-matrix diffuse tomography technique are shown:

- high accuracy of early diagnostics (stage 1: JM — 90.4% and MMT — 78.8%) of papillary cancer at its asymptomatic stage;
- excellent detection rate of stage 2 papillary cancer: JM — 96.2% and MMT — 88.5%) of cancer diagnostics
- very good accuracy (JM — 94.2% and MMT — 82.7%) of differentiation of papillary thyroid cancer stages.

11.3 Three-dimensional Jones Matrix Tomography Phase and Amplitude Anisotropy Polycrystalline Films of Blood and Its Plasma in the Differential Diagnosis of Prostate Tumors

Three representative groups comprising blood films and their plasma were created:

- Group 1 consisted of $k = 36$ polycrystalline film samples of blood and its plasma taken from patients with adenoma;
- Group 2 consisted of $k = 36$ polycrystalline film samples of blood and its plasma taken from patients with well-differentiated adenocarcinoma (ISUP grade 1);
- Group 3 consisted of $k = 36$ polycrystalline film samples of blood and its plasma taken from patients with moderately differentiated adenocarcinoma (ISUP grades 2–3).

Table 11.12 presents the optical parameters of the samples of polycrystalline blood films and their plasma from each of the groups.

Comparative analysis of the optical parameters of polycrystalline films of both types of liquids revealed a high level of absorption (τ) and depolarization of light (Λ) by blood samples. Their value is 4–5 times higher than that of similar indicators for polycrystalline films of blood plasma.

In other words, blood film samples show more pronounced dichroism, while blood plasma films exhibit birefringence.

Table 11.12 Optical parameters of polycrystalline blood films and its plasma.

Parameter	Group 1	Group 2	Group 3
Blood plasma			
Attenuation coefficient τ, cm^{-1}	0.21 ± 0.011	0.23 ± 0.013	0.22 ± 0.012
Depolarization degree Λ, %	9 ± 0.08	11 ± 0.11	10 ± 0.105
Blood			
Attenuation coefficient τ, cm^{-1}	0.91 ± 0.021	0.93 ± 0.023	0.92 ± 0.022
Depolarization degree Λ, %	48 ± 0.08	52 ± 0.11	50 ± 0.105

11.3.1 Reconstruction of the Polycrystalline Structure of Biological Fluid Films

The results of Jones matrix reproduction of topographic maps $JT(\varphi_k, m \times n)$ of optical anisotropy of polycrystalline films of blood plasma taken from patients with prostate tumors are presented in a series of Figure 11.15 $\left(LB\left(\varphi_j^*, m \times n\right), \langle CB\rangle\left(\varphi_j^*, m \times n\right)\right)$ – Figure 11.16 $\left(LD\left(\varphi_j^*, m \times n\right), \langle CD\rangle\left(\varphi_j^*, m \times n\right)\right)$:

- blood plasma samples taken from patients with benign prostate adenoma, panels (a), (d);
- blood plasma samples taken from patients with well-differentiated adenocarcinoma (ISUP grade 1), panels (b), (e);

blood plasma samples taken from patients with moderately differentiated adenocarcinoma (ISUP grade 2–3), panels (c), (f).

Comparative analysis of the obtained data (Figures 11.15 and 11.16) revealed two opposite tendencies in the transformation of Jones matrix tomograms of linear $\left(LB\left(\varphi_j^*, m \times n\right), LD\left(\varphi_j^*, m \times n\right)\right)$ and circular $\left(\langle CB\rangle\left(\varphi_j^*, m \times n\right), \langle CD\rangle\left(\varphi_j^*, m \times n\right)\right)$ birefringence and dichroism:

- sequential increase in the average level and fluctuations of values $\langle CB\rangle$ and $\langle CD\rangle$ of samples of polycrystalline blood plasma films taken from patients with of adenoma, adenocarcinoma (ISUP grade 1) and adenocarcinoma (ISUP grades 2–3), Figures 11.15 and 11.16 (panels d–f);
- sequential decrease in the average level and fluctuations of values LB and LD of samples of polycrystalline films of blood plasma taken from patients with adenoma, adenocarcinoma (ISUP grade 1) and adenocarcinoma (ISUP grades 2–3), Figure 11.16a–c.

From a physical perspective, the results are linked to shifts in the albumin-to-globulin ratio in the blood plasma of patients with prostate tumors. Malignant conditions typically lead to an increased concentration of optically active globulin molecules. Consequently, there is primarily an increase in both the magnitude and variability of circular birefringence values $\langle CB\rangle$ of samples of polycrystalline blood plasma films taken from patients with adenocarcinoma (ISUP grade 1) and adenocarcinoma (ISUP grades 2–3) in comparison with similar polarization maps of benign adenoma. At the same time, dynamic changes in the anisotropy of circular absorption $\langle CD\rangle$, the extremum of which falls on the ultraviolet and blue regions of the spectrum, are less expressive. Oncological decrease in the concentration of needle albumin molecules affects the decrease in the level of linear birefringence LB and dichroism LD of blood plasma films taken from patients with malignant prostate tumors.

Such processes are quantitatively characterized by a set of statistical moments $\boldsymbol{Z_i}$ characterizing tomograms of linear $\left(LB\left(\varphi_j^*, m \times n\right), LD\left(\varphi_j^*, m \times n\right)\right)$ and circular $\left(\langle CB\rangle\left(\varphi_j^*, m \times n\right), \langle CD\rangle\left(\varphi_j^*, m \times n\right)\right)$ birefringence and dichroism of anisotropy of polycrystalline blood plasma films (Table 11.13).

The data showed a high sensitivity (indicated by gray shading) to variations in the polycrystalline structure of blood plasma films across the following statistical moments:

- **Group 1–2:** $\begin{cases} \Delta M_{3;4}(\langle CB\rangle) = 1.45 \div 1.56; \\ \Delta M_{3;4}(LB) = 1.28 \div 1.29; \end{cases}$

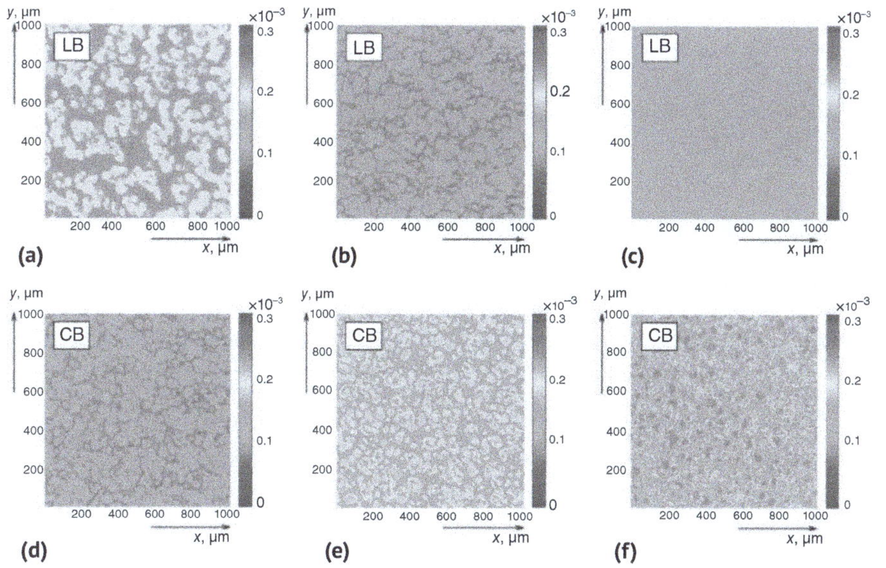

Figure 11.15 Topographic maps $\mathrm{LB}\left(\varphi_j^*, m \times n\right)$ ((a)–(c)) and $\langle\mathrm{CB}\rangle\left(\varphi_j^*, m \times n\right)$ ((d)–(f)) of polycrystalline blood plasma films taken from patients with benign and malignant prostate tumors.

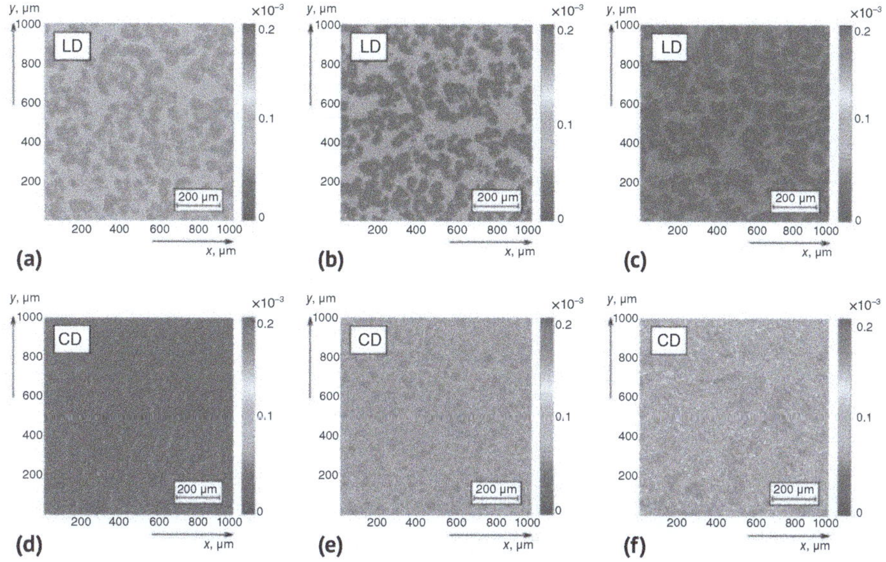

Figure 11.16 Topographic maps $\mathrm{LD}\left(\varphi_j^*, m \times n\right)$ ((a)–(c)) and $\langle\mathrm{CD}\rangle\left(\varphi_j^*, m \times n\right)$ ((d)–(f)) of polycrystalline blood plasma films taken from patients with benign and malignant prostate tumors.

- **Group 1–3:** $\begin{cases} \Delta M_{3;4}(\langle\mathrm{CB}\rangle) = 1.91 \div 1.92; \\ \Delta M_{3;4}(\mathrm{LB}) = 1.51 \div 1.55; \end{cases}$

- **Group 2–3:** $\begin{cases} \Delta M_{3;4}(\langle\mathrm{CB}\rangle) = 1.23 \div 1.32; \\ \Delta M_{3;4}(\mathrm{LB}) = 1.17 \div 1.21. \end{cases}$

Table 11.14 shows the values of sensitivity Se, specificity Sp and balanced accuracy Ac of the method of layered Jones matrix mapping of polycrystalline films of blood plasma taken from patients with benign and malignant prostate tumors.

From the analysis of the data obtained (Table 11.14), it follows that the maximum efficiency of differential diagnosis of adenoma (group 1) and adenocarcinoma (group 2 + group 3) by the method of Jones matrix tomography of circular birefringence $\langle\mathrm{CB}\rangle\left(\varphi_j^*, m \times n\right)$ of polycrystalline blood plasma films reaches an excellent level $\mathrm{Ac}_{1;2;3}(\langle\mathrm{CB}\rangle) = 94.4\% \div 95.8\%$.

At the same time, the accuracy of differentiation of the severity of the oncological process (groups 2–3) is also high – $\mathrm{Ac}_{2;3}(\langle\mathrm{CB}\rangle) > 90\%$.

Table 11.13 Statistical parameters of Jones matrix tomograms JT of polycrystalline blood plasma films.

JT	$LB(\varphi_j^*, m \times n)$			$LD(\varphi_j^*, m \times n)$		
M_i	Group 1	Group 2	Group 3	Group 1	Group 2	Group 3
$M_1 \times 10^{-3}$	0.31 ± 0.016	0.27 ± 0.013	0.24 ± 0.011	0.18 ± 0.008	0.15 ± 0.007	0.13 ± 0.006
$M_2 \times 10^{-3}$	0.23 ± 0.011	0.21 ± 0.011	0.19 ± 0.009	0.12 ± 0.006	0.105 ± 0.005	0.09 ± 0.004
M_3	1.02 ± 0.058	1.31 ± 0.064	1.58 ± 0.079	1.34 ± 0.068	1.49 ± 0.071	1.62 ± 0.083
M_4	1.25 ± 0.065	1.61 ± 0.085	1.89 ± 0.092	1.99 ± 0.11	2.15 ± 0.12	2.31 ± 0.13
JT	$\langle CB \rangle(\varphi_j^*, m \times n)$			$\langle CD \rangle(\varphi_j^*, m \times n)$		
M_i	Group 1	Group 2	Group 3	Group 1	Group 2	Group 3
$M_1 \times 10^{-3}$	0.23 ± 0.016	0.29 ± 0.019	0.35 ± 0.021	0.13 ± 0.012	0.16 ± 0.014	0.19 ± 0.016
$M_2 \times 10^{-3}$	0.19 ± 0.011	0.24 ± 0.012	0.31 ± 0.013	0.11 ± 0.008	0.12 ± 0.009	0.15 ± 0.011
M_3	1.42 ± 0.058	0.91 ± 0.044	0.74 ± 0.039	1.64 ± 0.068	1.41 ± 0.051	1.29 ± 0.043
M_4	1.76 ± 0.065	1.21 ± 0.052	0.92 ± 0.042	2.09 ± 0.089	1.76 ± 0.068	1.52 ± 0.053

The shading indicates different levels of diagnostic accuracy; darker shades correspond to higher diagnostic performance. It is used only for visual differentiation and has no additional statistical meaning.

Table 11.14 Operational characteristics of the diagnostic power of the Jones matrix tomography method of polycrystalline blood plasma films.

JT	$LB(\varphi_j^*, m \times n)$			$LD(\varphi_j^*, m \times n)$		
Groups	Group 1–2	Group 1–3	Group 2–3	Group 1–2	Group 1–3	Group 2–3
Se(%)	86.1	88.9	80.5	77.8	83.3	72.2
Sp(%)	83.3	88.9	77.8	77.8	80.5	69.4
Ac(%)	84.7	88.9	79.15	77.8	81.9	70.8
JT	$\langle CB \rangle(\varphi_j^*, m \times n)$			$\langle CD \rangle(\varphi_j^*, m \times n)$		
Groups	Group 1–2	Group 1–3	Group 2–3	Group 1–2	Group 1–3	Group 2–3
Se(%)	94.4	97.2	91.7	86.1	88.9	83.3
Sp(%)	94.4	94.4	88.9	83.3	88.9	80.5
Ac(%)	94.4	95.8	90.3	84.7	88.9	81.9

The shading indicates different levels of diagnostic accuracy; darker shades correspond to higher diagnostic performance. It is used only for visual differentiation and has no additional statistical meaning.

11.3.2 Polycrystalline Blood Films

In Figures 11.17 and 11.18 shows the results of Jones matrix reproduction of topographic maps $JT(\varphi_k, m \times n)$ of optical anisotropy of polycrystalline blood films taken from patients with prostate tumors.

Comparative analysis of Jones matrix tomography of polycrystalline blood films (Figures 11.17 and 11.18) showed the greatest dynamics of oncological changes in the parameters (mean and value of fluctuations) linear $LD(\varphi_j^*, m \times n)$ (decrease) and circular $\langle CD \rangle(\varphi_j^*, m \times n)$ (growth). Jones matrix tomograms of linear $LB(\varphi_j^*, m \times n)$ and circular $\langle CB \rangle(\varphi_j^*, m \times n)$ birefringence undergo similar, but less expressive changes.

From a physical perspective, the results can be explained by the much higher absorption of laser radiation by whole blood compared to plasma (see Table 11.11). Additionally, within the blood film volume, a considerable portion of the formed elements exhibits circular dichroism. This leads to oncological alterations in the blood's biochemical makeup – such as increased globulin levels and decreased albumin and fibrin levels – being most distinctly indicated by changes in the topographic structure of Jones matrix tomograms of the dichroism in its polycrystalline films.

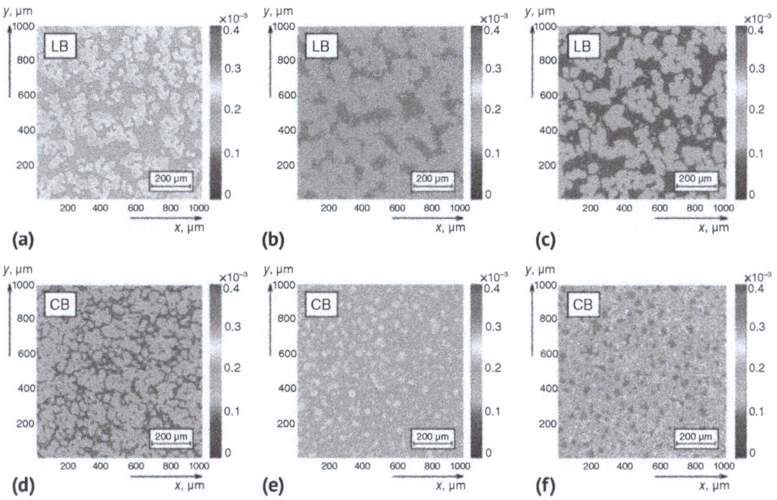

Figure 11.17 Topographic maps $\mathrm{LB}\left(\varphi_j^*, m \times n\right)$ (a)–(c) and $\langle\mathrm{CB}\rangle\left(\varphi_j^*, m \times n\right)$ (d)–(f) of polycrystalline blood films taken from patients with benign and malignant prostate tumors.

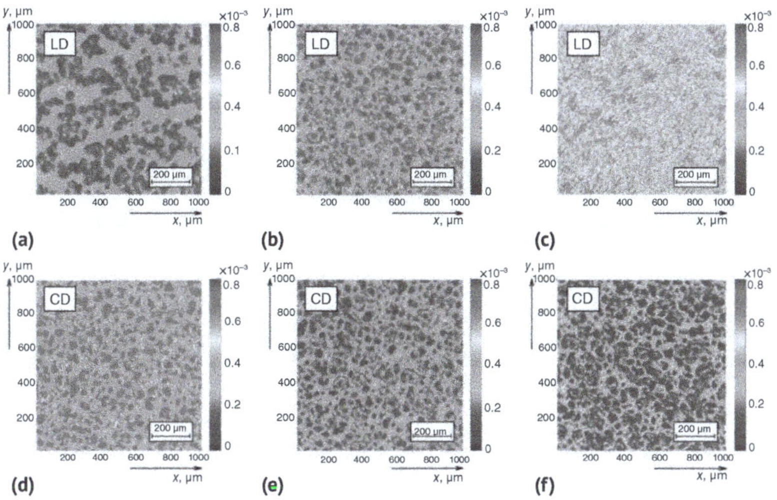

Figure 11.18 Topographic maps $\mathrm{LD}\left(\varphi_j^*, m \times n\right)$ (a)–(c) and $\langle\mathrm{CD}\rangle\left(\varphi_j^*, m \times n\right)$ (d)–(f) of polycrystalline blood films taken from patients with benign and malignant prostate tumors.

The results of statistical analysis of tomograms of linear $\left(\mathrm{LB}\left(\varphi_j^*, m \times n\right), \mathrm{LD}\left(\varphi_j^*, m \times n\right)\right)$ and circular $\left(\langle\mathrm{CB}\rangle\left(\varphi_j^*, m \times n\right), \langle\mathrm{CD}\rangle\left(\varphi_j^*, m \times n\right)\right)$ birefringence and dichroism of polycrystalline blood films are presented in Table 11.15.

Unlike the statistical analysis of Jones matrix tomograms of linear and circular birefringence in blood plasma films (Table 11.13), the highest sensitivity (marked in gray shading) to changes in the polycrystalline structure of blood plasma was observed in the statistical moments that describe the parameters of optically anisotropic absorption.

- **Group 1–2:** $\begin{cases} \Delta M_{3;4}(\langle\mathrm{CD}\rangle) = 1.43 \div 1.56; \\ \Delta M_{3;4}(\mathrm{LD}) = 1.2 \div 1.44; \end{cases}$

- **Group 1–3:** $\begin{cases} \Delta M_{3;4}(\langle\mathrm{CD}\rangle) = 2 \div 2.09; \\ \Delta M_{3;4}(\mathrm{LD}) = 1.41 \div 1.82; \end{cases}$

- **Group 2–3:** $\begin{cases} \Delta M_{3;4}(\langle\mathrm{CD}\rangle) = 1.28 \div 1.46; \\ \Delta M_{3;4}(\mathrm{LD}) = 1.17 \div 1.27. \end{cases}$

Table 11.15 Statistical parameters of Jones matrix tomograms JT of polycrystalline blood films.

JT	$LB(\varphi_j^*, m \times n)$			$LD(\varphi_j^*, m \times n)$		
M_i	Group 1	Group 2	Group 3	Group 1	Group 2	Group 3
$M_1 \times 10^{-3}$	0.37 ± 0.018	0.32 ± 0.015	0.28 ± 0.013	0.68 ± 0.038	0.55 ± 0.027	0.43 ± 0.016
$M_2 \times 10^{-3}$	0.28 ± 0.014	0.24 ± 0.012	0.21 ± 0.011	0.52 ± 0.026	0.41 ± 0.021	0.39 ± 0.014
M_3	0.72 ± 0.038	0.93 ± 0.044	1.15 ± 0.059	0.34 ± 0.018	0.49 ± 0.021	0.62 ± 0.033
M_4	0.85 ± 0.045	1.06 ± 0.055	1.28 ± 0.062	0.59 ± 0.31	0.71 ± 0.34	0.83 ± 0.41
JT	$\langle CB \rangle(\varphi_j^*, m \times n)$			$\langle CD \rangle(\varphi_j^*, m \times n)$		
M_i	Group 1	Group 2	Group 3	Group 1	Group 2	Group 3
$M_1 \times 10^{-3}$	0.29 ± 0.014	0.34 ± 0.017	0.38 ± 0.018	0.53 ± 0.032	0.66 ± 0.034	0.79 ± 0.036
$M_2 \times 10^{-3}$	0.19 ± 0.011	0.26 ± 0.012	0.32 ± 0.014	0.31 ± 0.014	0.42 ± 0.022	0.55 ± 0.025
M_3	0.84 ± 0.048	0.69 ± 0.034	0.57 ± 0.029	0.64 ± 0.038	0.41 ± 0.022	0.32 ± 0.015
M_4	1.07 ± 0.055	0.85 ± 0.042	0.72 ± 0.034	1.09 ± 0.059	0.76 ± 0.036	0.52 ± 0.023

The shading indicates different levels of diagnostic accuracy; darker shades correspond to higher diagnostic performance. It is used only for visual differentiation and has no additional statistical meaning.

Table 11.16 Operational characteristics of the diagnostic power of the Jones matrix tomography method of polycrystalline blood films.

JT	$LB(\varphi_j^*, m \times n)$			$LD(\varphi_j^*, m \times n)$		
Groups	Group 1–2	Group 1–3	Group 2–3	Group 1–2	Group 1–3	Group 2–3
Se(%)	80.5	86.1	75	86.1	88.9	80.5
Sp(%)	77.8	83.3	72.2	83.3	88.9	80.5
Ac(%)	79.15	84.7	73.6	84.7	88.9	80.5
JT	$\langle CB \rangle(\varphi_j^*, m \times n)$			$\langle CD \rangle(\varphi_j^*, m \times n)$		
Groups	Group 1–2	Group 1–3	Group 2–3	Group 1–2	Group 1–3	Group 2–3
Se(%)	86.1	88.9	80.5	94.4	94.4	86.1
Sp(%)	83.3	86.1	77.8	91.7	94.4	83.3
Ac(%)	84.7	87.5	79.15	93.05	94.4	84.7

The shading indicates different levels of diagnostic accuracy; darker shades correspond to higher diagnostic performance. It is used only for visual differentiation and has no additional statistical meaning.

Table 11.16 shows the values of sensitivity Se, specificity Sp and balanced accuracy Ac of the method of layered Jones matrix mapping of polycrystalline films of blood taken from patients with benign and malignant prostate tumors.

The excellent quality $(Ac_{1;2;3}(\langle CD \rangle) = 93.05\% \div 94.4\%)$ of differential diagnosis of adenoma (group 1) and adenocarcinoma (group 2 + group 3) of the prostate by the method of Jones matrix tomography of circular dichroism $\langle CD \rangle(\varphi_j^*, m \times n)$ of polycrystalline blood films was established. At the same time, the accuracy of differentiation of the severity of the oncological process (groups 2–3) is less high than in the case of studies of polycrystalline blood plasma films – $Ac_{2;3}(\langle CB \rangle) \sim 85\%$.

11.3.3 Conclusions

The analytic foundation is established for the 3D Jones matrix mapping method, which uses digital holographic reconstruction to create layered maps of linear and circular birefringence and dichroism in polycrystalline blood films and their plasma.

This method, based on polarization interferometry, was experimentally tested for rapid (under 15 minutes) differential diagnosis of varying severity in adenoma and adenocarcinoma.

A representative sample of polycrystalline blood films and their plasma was studied. Using the phase scanning method of the polycrystalline film's object field, the optimal cross sections were determined. In these cross sections, the maximum differences $\left(\Delta\overline{M}_{i=3;4} \rightarrow \max\right)$ between the values of the statistical moments $M_{i=1;2;3;4}$, which characterize the maps of optical anisotropy of polycrystalline blood films and its plasma samples of Groups 1–3, were realized.

The most sensitive statistical parameters identified are the skewness and kurtosis of the coordinate distributions of linear and circular birefringence and dichroism. The operational characteristics (sensitivity Se, specificity Sp and balanced accuracy Ac) have been determined, which demonstrate the diagnostic power of the 3D Jones matrix mapping of biological films with digital holographic reconstruction of layered maps of optical anisotropy.

The use of algorithms for layer-by-layer Jones matrix tomography of circular birefringence in polycrystalline blood plasma films achieved excellent accuracy $\left(Ac_{1;2;3}(\langle CB\rangle) = 94.4\% \div 95.8\%\right)$ of differentiation of benign and malignant tumors of prostate tissue.

In addition, a high level of balanced accuracy $\left(Ac_{2;3}(\langle CB\rangle) > 90\%\right)$ was achieved in the differential diagnosis of the severity of the oncological process.

A similar result was achieved by Jones matrix reproduction of circular dichroism maps of polycrystalline whole blood films. $\left(Ac_{1;2;3}(\langle CD\rangle) = 93.05\% \div 94.4\%\right)$. However, the accuracy of differential diagnosis of the severity of the oncological process was less high – $Ac_{2;3}(\langle CB\rangle) \sim 85\%$.

References

1 Eftimie LG, Glogojeanu RR, Tejaswee A, Gheorghita P, Stanciu SG, Chirila A, et al. Differential diagnosis of thyroid nodule capsules using random forest guided selection of image features. Sci Rep. 2022;12:21636.

2 Yang WT, Ma BY, Chen Y. A narrative review of deep learning in thyroid imaging: current progress and future prospects. Quant Imaging Med Surg. 2024;14(2):2069–2088. https://doi.org/10.21037/qims-23-908.

3 Tokarz D, Cisek R, Joseph A, Asa SL, Wilson BC, Barzda V. Characterization of pathological thyroid tissue using polarization-sensitive second harmonic generation microscopy. Lab Invest. 2020;100(10):1280–1287. https://doi.org/10.1038/s41374-020-0475-7.

4 Cisek R, Joseph A, Harvey M, Tokarz D. Polarization-sensitive second harmonic generation microscopy for investigations of diseased collagenous tissues. Front Phys. 2021;9:726996. https://doi.org/10.3389/fphy.2021.726996.

5 Hristu R, Eftimie LG, Stanciu SG, Tranca DE, Paun B, Sajin M, et al. Quantitative second harmonic generation microscopy for the structural characterization of capsular collagen in thyroid neoplasms. Biomed Opt Express. 2018;9(8):3923–3936. https://doi.org/10.1364/BOE.9.003923.

6 de Boer JF, Milner TE. Review of polarization sensitive optical coherence tomography and Stokes vector determination. J Biomed Opt. 2002;7(3):359–371. https://doi.org/10.1117/1.1483879.

7 Ku S, Mahato KK, Mazumder N. Polarization-resolved Stokes-Mueller imaging: a review of technology and applications. Lasers Med Sci. 2019;34(7):1283–1293. https://doi.org/10.1007/s10103-019-02752.

8 He Y, Li K, Li W, Qiu Y, Li D, Wang C, Tang Q, Li Z. Polarization coherency matrix tomography. J Biophotonics. 2023;16(9):e202300093. https://doi.org/10.1002/jbio.202300093.

9 Yang D, Yuan Z, Hu M, Liang Y. Zebrafish brain and skull imaging based on polarization-sensitive optical coherence tomography. J Biophotonics. 2022;15(12):e202200112. https://doi.org/10.1002/jbio.202200112.

10 Park JE, Xin Z, Kwon DY, Kim SW, Lee H, et al. Application of polarization sensitive-optical coherence tomography to the assessment of phase retardation in subpleural cancer in rabbits. Tissue Eng Regen Med. 2021;18(1):61–69. https://doi.org/10.1007/s13770-020-00318-9.

11 Willemse J, Gräfe MGO, Verbraak FD, de Boer JF. In vivo 3D determination of peripapillary scleral and retinal layer architecture using polarization-sensitive optical coherence tomography. Transl Vis Sci Technol. 2020;9(11):21. https://doi.org/10.1167/tvst.9.11.21.

12 Baumann B, Augustin M, Lichtenegger A, Harper D, Muck M, Eugui P, et al. Polarization-sensitive optical coherence tomography imaging of the anterior mouse eye. J Biomed Opt. 2018;23(8):1–12. https://doi.org/10.1117/1.JBO.23.8.086005.

13 Hariri LP, Adams DC, Applegate MB, Miller AJ, Roop BW, Villiger M, et al. Distinguishing tumor from associated fibrosis to increase diagnostic biopsy yield with polarization-sensitive optical coherence tomography. Clin Cancer Res. 2019;25(17): 5242–5249. https://doi.org/10.1158/1078-0432.CCR-19-0566.

14 Tao K, Sun K, Ding Z, Ma Y, Kuang H, Zhao H, et al. Catheter-based polarization sensitive optical coherence tomography using similar Mueller matrix method. IEEE Trans Biomed Eng. 2020;67(1):60–68. https://doi.org/10.1109/TBME.2019.2908031.

15 de Boer JF, Hitzenberger CK, Yasuno Y. Polarization sensitive optical coherence tomography. Biomed Opt Express. 2017;8(3):1838–1873.

16 Ushenko A, Zheng J, Gorsky M, Dubolazov A, Ushenko Y, Soltys I, et al. 3D digital holographic polarimetry of diffuse optically anisotropic biological tissue object fields. Front Phys. 2023;11:1288935. https://doi.org/10.3389/fphy.2023.1288935.

17 Ushenko A, Zheng J, Litvinenko A, Gorsky M, Wanchuliak O, Dubolazov A, et al. 3D digital polarization-holographic wavelet histology in determining the duration of mechanical damage to the myocardium. J Biophotonics. 2024;17(3):e202300372.

18 Ushenko A, Dubolazov A, Zheng J, Litvinenko A, Gorsky M, Ushenko Y, et al. 3D polarization-interference holographic histology for wavelet-based differentiation of the polycrystalline component of biological tissues with different necrotic states. J Biomed Opt. 2024;29(5):052920. https://doi.org/10.1117/1.JBO.29.5.052920.

19 Hunter R. Foundations of Colloid Science. New York: Oxford University Press; 2004. p. 806.

20 Chen X, Chen PG, Ouazzani J, Liu Q. Numerical simulations of sessile droplet evaporating on heated substrate. Eur Phys J Spec Top. 2017;226(6):1325–1335

21 Dash S, Garimella SV. Droplet evaporation on heated hydrophobic and superhydrophobic surfaces. Phys Rev E. 2014;89:042402.

22 Gleason K, Putnam SA. Microdroplet evaporation with a forced pinned contact line. Langmuir. 2014;30(34):10548–10555.

23 Hu H, Larson RG. Evaporation of a sessile droplet on a substrate. J Phys Chem B. 2020;106:1334–1344.

24 Vladmir Z, Wang JB, Yan XH. Human blood plasma crystal and molecular biocolloid textures – dismetabolism and genetic breaches. Natur Sci J Xiangtan Univ. 2001;23:118–127.

25 Abe S, Tanaka J, Kojima M, Kanamaru S, Hirata K, Yamashita K, et al. Cell-free protein crystallization for nanocrystal structure determination. Sci Rep. 2022;12:16031.

26 Van den Beld AW, Kaufman JM, Zillikens MC, Lamberts SWJ, Egan JM, van der Lely AJ. The physiology of endocrine systems with ageing. Lancet Diabetes Endocrinol. 2018;6(8):647–658. https://doi.org/10.1016/S2213-8587(18)30026-3.

27 Asa SL, Erickson LA, Rindi G. The spectrum of endocrine pathology. Endocr Pathol. 2023;34(4):368–381. https://doi.org/10.1007/s12022-023-09758-0.

28 Brix K, Szumska J, Weber J, Qatato M, Venugopalan V, Al-Hashimi A, et al. Auto-regulation of the thyroid gland beyond classical pathways. Exp Clin Endocrinol Diabetes. 2020;128(6-7):437–445. https://doi.org/10.1055/a-1080-2969.

29 Grieco G, Wang T, Delcorte O, Spourquet C, Janssens V, Strickaert A, et al. Class III PI3K Vps34 controls thyroid hormone production by regulating thyroglobulin iodination, lysosomal proteolysis, and tissue homeostasis. Thyroid. 2020;30(1):133–146. https://doi.org/10.1089/thy.2019.0182.

30 Boron WF, Boulpaep EL. Medical Physiology: A Cellular and Molecular Approach. Elsevier/Saunders; 2003. p. 1300. ISBN 1-4160-2328-3.

31 Delom F, Mallet B, Carayon P, Lejeune PJ. Role of extracellular molecular chaperones in the folding of oxidized proteins. Refolding of colloidal thyroglobulin by protein disulfide isomerase and immunoglobulin heavy chain-binding protein. J Biol Chem. 2001;276(24):21337–21342. https://doi.org/10.1074/jbc.M101086200.

32 Wagner I, Musso H. New naturally occurring amino acids. Angew Chem Int Ed Engl. 1983;22(11):816–828. https://doi.org/10.1002/anie.198308161.

33 Stoker HS. Organic and Biological Chemistry. Cengage Learning; 2001. 371. ISBN 978-1-305-68645-8.

34 Yildirim Simsir I, Cetinkalp S, Kabalak T. Review of factors contributing to nodular goiter and thyroid carcinoma. Med Princ Pract. 2020;29(1):1–5. https://doi.org/10.1159/000503575.

35 Knobel M. Etiopathology, clinical features, and treatment of diffuse and multinodular nontoxic goiters. J Endocrinol Invest. 2016;39(4):357–373. https://doi.org/10.1007/s40618-015-0391-7.

36 Mallya M, Ogilvy-Stuart AL. Thyrotropic hormones. Best Pract Res Clin Endocrinol Metab. 2018;32(1):17–25. https://doi.org/10.1016/j.beem.2017.10.006.

37 Delitala AP, Capobianco G, Cherchi PL, Dessole S, Delitala G. Thyroid function and thyroid disorders during pregnancy: a review and care pathway. Arch Gynecol Obstet. 2019;299(2):327–338. https://doi.org/10.1007/s00404-018-5018-8.

38 Mohler, J., Bahnson, R. R., Boston, B., Busby, J. E., D'Amico, A., Eastham, J. A. et al. Prostate Cancer. J Natl Compr Cancer Netw. 2010;8:162–200.

39 Grönberg H. Prostate cancer epidemiology. Lancet. 2003;361(9360):859–864.

40 Ilic D, Neuberger MM, Djulbegovic M, Dahm P. Screening for prostate cancer. Cochrane Database Syst Rev. 2013;2013:1.

41 Albertsen PC, Hanley JA, Fine J. 20-Year outcomes following conservative management of clinically localized prostate cancer. J Am Med Assoc. 2005;293:2095–2101.

42 Killeen AA, Ossina N, McGlennen RC, Minnerath S, Borgos J, Alexandrov V, et al. Protein self-organization patterns in dried serum reveal changes in B-Cell disorders. Mol Diag Ther. 2006;10:371–380.

43 Brutin D, Sobac B, Loquet B, Sampol J. Pattern formation in drying drops of blood. J Fluid Mech. 2011;667:85–95.

44 Jones CR. New calculus for the treatment of optical systems.VII. Properties of the N-matrices. J Opt Soc Am. 1948;38:671–685.

45 Azzam RMA. Propagation of partially depolarized light through anisotropic media with or without depolarization. A differential 4x4 matrix calculus. J Opt Soc Am. 1978;68:1756–1767.

46 Arteaga O, Canillas A. Analytic inversion of the Mueller-Jones polarization matrices for homogeneous media. Opt Lett. 2010;35:559–561.

47 Barakat R. Exponential versions of the Jones and Mueller–Jones polarization matrices. J Opt Soc Am A. 1996;13:158–163.

48 Angelsky PO, Ushenko AG, Dubolazov AV, Sidor MI, Bodnar GB, Koval G, et al. The singular approach for processing polarization-inhomogeneous laser images of blood plasma layers. J Opt. 2013;15(4):044030.

49 Ushenko AG, Dubolazov AV, Ushenko VA, Novakovskaya OY. Statistical analysis of polarization-inhomogeneous Fourier spectra of laser radiation scattered by human skin in the tasks of differentiation of benign and malignant formations. J Biomed Opt. 2016;21(7):071110.

50 Garazdyuk MS, Bachinskyi VT, Vanchulyak OY, Ushenko AG, Dubolazov OV, Gorsky MP. Polarization-phase images of liquor polycrystalline films in determining time of death. Appl Opt. 2016;55(12):B67–B71.

51 Angelskii OV, Ushenko OG, Burkovets DN, Arkhelyuk OD, Ushenko YA. Polarization-correlation studies of multifractal structures in biotissues and diagnostics of their pathologic changes. Laser Phys. 2000;10(5):1136–1142.

52 Yermolenko S, Ushenko A, Ivashko P, Goudail F, Gruia I, Gavrilă C, et al. Spectropolarimetry of cancer change of biotissues. In Ninth International Conference on Correlation Optics. SPIE. 2009;7388:404–410.

53 Jóźwicki R, Patorski K, Angelsky OV, Ushenko AG, Burkovets DN, Ushenko YA. Automatic polarimetric system for early medical diagnosis by biotissue testing. Opt Appl. 2002;32(4):603–612.

54 Ushenko AG, Dubolazov OV, Ushenko VA, Novakovskaya OY, Olar OV. Fourier polarimetry of human skin in the tasks of differentiation of benign and malignant formations. Appl Opt. 2016;55(12):B56–B60.

55 Ushenko AG, Pishak VP. Coherent-domain optical methods: biomedical diagnostics, environmental and material science. In Laser Polarimetry of Biological Tissue: Principles and Applications. Kluwer Academic Publisher; 2004.

56 Ushenko AG, Fediv AI, Marchuk YF. Correlation and fractal structure of Jones matrices of human bile secret. In European Conference on Biomedical Optics. 2009;7368:73681Q.

57 Ushenko AG. Depolarization of a laser emission field as the correlation development of its polarization structure. In International Conference on Correlation Optics. 1997;3317:331–339.

58 Bachinskyi V, Boychuk T, Ushenko A. Laser Polarimetry of Biological Tissues and Fluids: Chapter 4. Methods of Spatial-frequency, Singular and Correlation Analysis of Biological Layers Object Fields. LAP Lambert Academic Publishing; 2018.

59 Ushenko AG, Burkovets DN, Yermolenko SB, Arkhelyuk AD, Pishak VP, Yuzko AM, et al. Stokes polarimetry of biotissues. In Fourth International Conference on Correlation Optics. SPIE. 1999;3904:527–533. SPIE.

60 Ushenko VA, Dubolazov AV, Pidkamin LY, Sakchnovsky MY, Bodnar AB, Ushenko YA, et al. Mapping of polycrystalline films of biological fluids utilizing the Jones-matrix formalism. Laser Phys. 2018;28(2):025602.

61 Ushenko AG, Angelsky PO, Sidor M, Marchuk YF, Andreychuk DR, Pashkovskaya NV. Spatial-frequency selection of complex degree of coherence of laser images of blood plasma in diagnostics and differentiation of pathological states of human organism of various nosology. Appl Opt. 2014;53(10):B172–B180.

62 Olar EI, Ushenko AG, Ushenko YA. Polarization correlation measurements of the phase tomograms of optically anisotropic biofractals. Laser Phys. 2004;14(8):1115–1121.

63 Guminetskiy SH, Ushenko OG, Polyanskiy IP, Motrych AV, Grynchuk FV. The optical method for investigation of the peritonitis progressing process. In Eighth International Conference on Correlation Optics. 2008;7008:626–632.

64 Ushenko AG, Fediv AI, Marchuk YuF. Singular structure of polarization images of bile secret in diagnostics of human physiological state. In: European Conference on Biomedical Optics. Optica Publishing Group; 2009.

65 Ushenko Y, Ushenko A, Dubolazov A, Soltys I, Litvinenko O, Wanchuliak O, et al. Mueller-matrix interferometric multifractal scaling of optically anisotropic architectonics of diffuse blood facies: fundamental and applied aspects. J Biophotonics. 2025;18(3):e202400412. https://doi.org/10.1002/jbio.202400412.

66 Ushenko A, Soltys I, Dubolazov A, Ushenko Y, Bilookyi V, Bilookyi O, et al. 3D Jones-matrix thesiography of biological fluid facies. J Innov Opt Health Sci. 2025;18(2):2443002. https://doi.org/10.1142/S1793545824430028.

67 Ushenko A, Pavlyukovich N, Khukhlina O, Pavlyukovich O, Gorsky M, Soltys I, et al. Layer-by-layer multifractal scanning of optically anisotropic architectonics of blood plasma films: fundamental and applied aspects. Photonics. 2025;12:215. https://doi.org/10.3390/photonics12030215.

68 Ushenko A, Pavlyukovich N, Khukhlina O, Pavlyukovich O, Soltys I, Dubolazov A, et al. Blood plasma film multifractal scanning in COVID-19 consequences diagnostics. J Biophotonics. 2024;17(11):e202400356. https://doi.org/10.1002/jbio.202400356.

69 Ushenko O, Bilookyi O, Zheng J, Dubolazov A, Olar O, Ushenko Y, et al. 3D digital holographic polarimetry of laser speckle fields formed by polycrystalline blood films: a tool for differential diagnosis of thyroid pathology. Front Phys. 2024;12:1426469. https://doi.org/10.3389/fphy.2024.1426469.

70 Wójcik W, Hu Z, Ushenko Y, Smolarz A, Soltys I, Dubolazov O, et al. Optical sensor system for 3D Jones matrix reconstruction of optical anisotropy maps of self-assembled polycrystalline soft matter films. Sensors. 2024;24(5):1589. https://doi.org/10.3390/s24051589.

71 Dubolazov A, Ushenko V, Trifonyuk L, Stashkevich A, Soltys I, Ushenko Y, et al. Polarization-singular approach to imaging Mueller-matrix polarimetry in differential diagnosis of histological sections of biopsy of tumors of the uterus and prostate. Front Phys. 2021;9:711212. https://doi.org/10.3389/fphy.2021.711212.

72 Ushenko VO, Trifonyuk L, Ushenko YA, Dubolazov OV, Gorsky MP, Ushenko AG. Polarization singularity analysis of Mueller-matrix invariants of optical anisotropy of biological tissue samples in cancer diagnostics. J Opt. 2021;23(6):064004. https://doi.org/10.1088/2040-8986/abf97a.

73 Ushenko YA, Koval GD, Ushenko AG, Dubolazov OV, Ushenko VA, Novakovskaia OY. Mueller-matrix of laser-induced autofluorescence of polycrystalline films of dried peritoneal fluid in diagnostics of endometriosis. J Biomed Opt. 2016;21(7):071116. https://doi.org/10.1117/1.JBO.21.7.071116.

12

Mapping of Biological Fluid Films

Yurii Ushenko[1,2], *Weidong Dou*[1], *Oleksandr Ushenko*[2,3], *Oleksandr Dubolazov*[2], *Iryna Soltys*[2], *Vitaliy Rozhko*[4], *Oleksandr Olar*[2], *and Vladyslava Sholota*[5]

[1] *Shaoxing University, Shaoxing, China*
[2] *Yurii Fedkovych Chernivtsi National University, Chernivtsi, Ukraine*
[3] *Taizhou Institute of Zhejiang University, Taizhou, China*
[4] *Bukovinian State Medical University, Chernivtsi, Ukraine*
[5] *Vinnitsa National Technical University, Vinnitsa, Ukraine*

12.1 Laser Medical Jones Matrix Imaging of Polycrystalline Bile Films

12.1.1 Real Components of Jones Matrix Images of Dehydrated Bile Films

Figures 12.1 and 12.2 present the results of polarization–interference Jones matrix mapping of the $Rej_{ik}(m,n)$ maps of bile films (BF) [1–12].

Analysis of the obtained data revealed:

- The presence of an individual coordinate-inhomogeneous topographic structure in the Jones matrix images $Rej_{ik}(m,n)$ of dendritic polycrystalline networks for each matrix element (Figure 12.1a, b, e, and f).
- An asymmetrical structure of histograms $G(Rej_{ik}(m,n))$ by the localization of extrema and by the ranges of variation of the random values $Rej_{ik}(m,n)$ (Figure 12.1c, d, g, and h).
- A transformation of the magnitude and position of the extrema and the ranges of change of the histograms $G(Rej_{ik}(m,n))$ depending on the magnitude of the phase shift φ_k during the scanning of the algorithmically reproduced field of complex amplitudes (panels (c), (d), and (g), (h)).
- A sequential growth of the correlation between the model representations and the experimental Jones matrix images $Rej_{ik}(m,n)$, $Rej_{ik}(\varphi_k=\pi/4,m,n)$, and $Rej_{ik}(\varphi_k=\pi/8,m,n)$ of supramolecular polycrystalline networks of the BF sample – $Rej_{11}(m,n)\approx Rej_{22}(m,n)$ and $Rej_{12}(m,n)\approx Rej_{21}(m,n)$.

Table 12.1 shows the quantitative statistical scenarios $(\widetilde{M}_{1;2;3;4}\pm\sigma_{1;2;3;4})$ of the phase transformation of Jones matrix images $j_{ik}(m,n)$ of a representative sample ($q=21$) of dehydrated BF samples. Here, $\widetilde{M}_{1;2;3;4}$ is the average value within the set of dehydrated BF samples, and $\pm\sigma_{1;2;3;4}$ is the standard deviation.

It was revealed that:

- The predicted superiority of the Jones matrix manifestations of the circular birefringence (CB) mechanisms $Rej_{12;21}(\mathrm{CB}^2{}_{\otimes;\oplus})$ over the manifestations of structural anisotropy $Rej_{11;22}(\mathrm{LB}^2+\mathrm{CB}^2{}_{\otimes;\oplus})$ of supramolecular bile networks – $\widetilde{M}_{i=1;2}(Rej_{11;22}(\mathrm{CB}^2{}_{\otimes;\oplus}))>\widetilde{M}_{i=1;2}(Rej_{12;21}(\mathrm{LB}^2+\mathrm{CB}^2{}_{\otimes;\oplus}))$.
- Opposite trends in the change (a decrease ↓) of the mean $(\widetilde{M}_{i=1})$ and variance $(\widetilde{M}_{i=2})$ and (an increase ↑) of the skewness $(\widetilde{M}_{i=3})$ and kurtosis $(\widetilde{M}_{i=4})$, which characterize the transformation of the Jones matrix images of the set of real component elements $Rej_{ik}(m,n)$ in different phase planes $(\varphi_k(m,n))$ of the polycrystalline BF $\begin{pmatrix}\widetilde{M}_{i=1;2}\downarrow\\ M_{i=3,4}\uparrow\end{pmatrix}$.

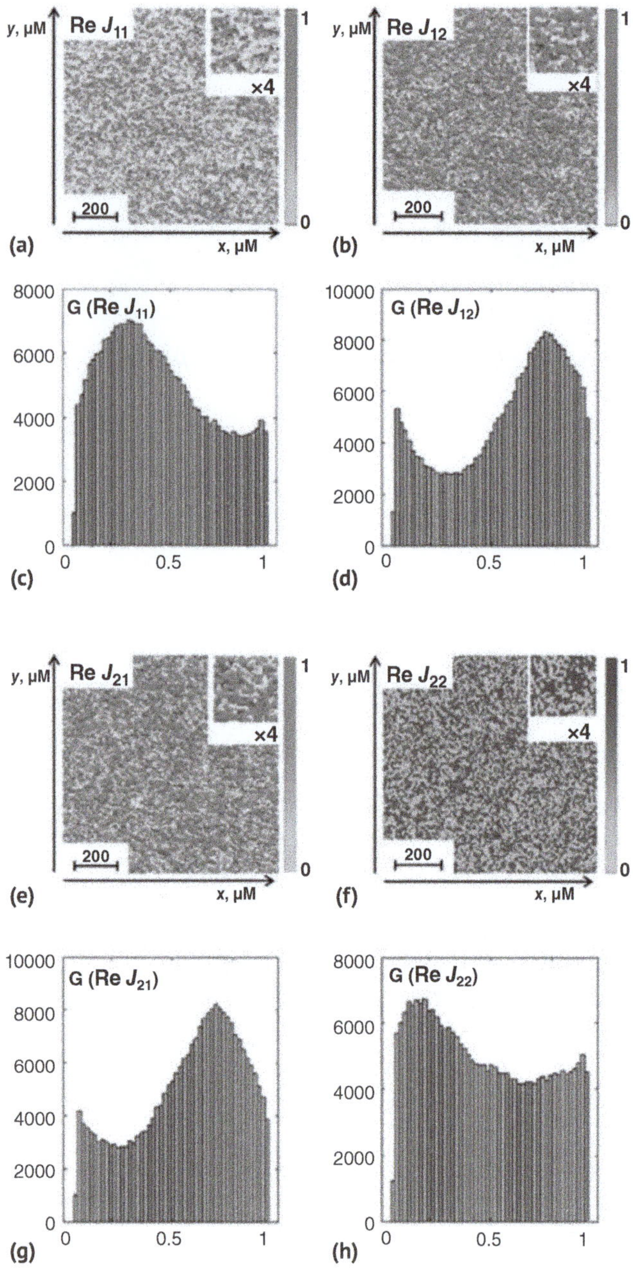

Figure 12.1 Integral coordinate (a), (b), (e), (f) and statistical (c), (d), (g), (h) structure of the real parts of the Jones matrix images $Rej_{ik}(m,n)$ of dehydrated bile films.

- The values of the mean and variance $\widetilde{M}_{i=1;2}(Rej_{ik}(m,n))$ decrease (by 30–50%) with an increase in the phase-scan parameter φ_k.
- The most sensitive markers were the statistical moments $\widetilde{M}_{i=3;4}\left(Rej_{12;21}(m,n)\right)$, which characterize the skewness and kurtosis of the distributions of the real component values of the matrix elements $Rej_{12;21}\left(\mathrm{CB}^2{}_{\otimes;\oplus}\right)$, – with an increase in the phase-scan parameter $\varphi_k \uparrow$ the values of $\widetilde{M}_{i=3;4}$ increased significantly, up to threefold.
- The range of phase variation of the magnitude of the higher-order statistical moments $\widetilde{M}_{i=3;4}\left(Rej_{11;22}\left(\mathrm{LB}^2+\mathrm{CB}^2{}_{\otimes;\oplus}\right)\right)$ is somewhat smaller, reaching up to twofold.

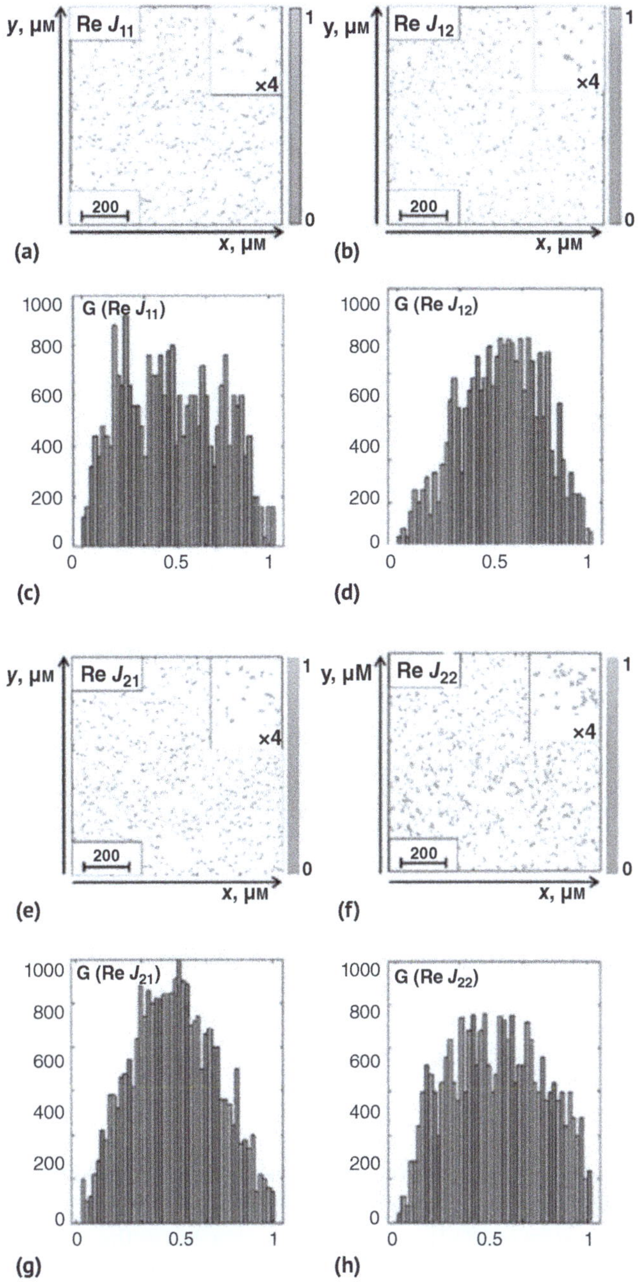

Figure 12.2 Layered ($\varphi_k = \pi/8$) coordinate (a), (b), (e), (f) and statistical (c), (d), (g), (h) structure of the real parts of the Jones matrix images $Rej_{ik}(m, n)$ of dehydrated bile films.

Table 12.1 Central statistical moments of the first to fourth orders characterizing the coordinate distributions of the real parts of the Jones matrix images $Rej_{ik}(m, n)$ of the bile films.

	$Rej_{ik}(m, n)$			
$\widetilde{M}_{1;2;3;4} \pm \sigma_{1;2;3;4}$	$Rej_{11}(m, n)$	$Rej_{12}(m, n)$	$Rej_{21}(m, n)$	$Rej_{22}(m, n)$
$\widetilde{M}_1 \pm \sigma_1$	0.52 ± 0.027	0.47 ± 0.025	0.43 ± 0.022	0.48 ± 0.024
$\widetilde{M}_2 \pm \sigma_2$	0.31 ± 0.018	0.33 ± 0.017	0.29 ± 0.014	0.28 ± 0.016
$\widetilde{M}_3 \pm \sigma_3$	0.21 ± 0.011	0.22 ± 0.012	0.18 ± 0.011	0.23 ± 0.13
$\widetilde{M}_4 \pm \sigma_4$	0.33 ± 0.016	0.31 ± 0.018	0.25 ± 0.013	0.35 ± 0.018

(Continued)

Table 12.1 (Continued)

		$\varphi_k = \pi/4$		
$\widetilde{M}_{1;2;3;4} \pm \sigma_{1;2;3;4}$	$Rej_{11}(m,n)$	$Rej_{12}(m,n)$	$Rej_{21}(m,n)$	$Rej_{22}(m,n)$
$\widetilde{M}_1 \pm \sigma_1$	0.41 ± 0.022	0.39 ± 0.021	0.35 ± 0.017	0.38 ± 0.019
$\widetilde{M}_2 \pm \sigma_2$	0.22 ± 0.011	0.22 ± 0.012	0.19 ± 0.09	0.21 ± 0.013
$\widetilde{M}_3 \pm \sigma_3$	0.44 ± 0.029	0.41 ± 0.023	0.39 ± 0.019	0.46 ± 0.024
$\widetilde{M}_4 \pm \sigma_4$	0.57 ± 0.032	0.51 ± 0.029	0.49 ± 0.026	0.61 ± 0.035
		$\varphi_k = \pi/8$		
$\widetilde{M}_{1;2;3;4} \pm \sigma_{1;2;3;4}$	$Rej_{11}(m,n)$	$Rej_{12}(m,n)$	$Rej_{21}(m,n)$	$Rej_{22}(m,n)$
$\widetilde{M}_1 \pm \sigma_1$	0.33 ± 0.018	0.29 ± 0.015	0.31 ± 0.017	0.35 ± 0.018
$\widetilde{M}_2 \pm \sigma_2$	0.19 ± 0.011	0.16 ± 0.009	0.18 ± 0.01	0.21 ± 0.011
$\widetilde{M}_3 \pm \sigma_3$	0.53 ± 0.023	0.59 ± 0.028	0.54 ± 0.028	0.49 ± 0.027
$\widetilde{M}_4 \pm \sigma_4$	0.69 ± 0.034	0.89 ± 0.045	0.82 ± 0.039	0.72 ± 0.041

12.1.2 Imaginary Components of Jones Matrix Images of Dehydrated BF

Figures 12.3 and 12.4 show the results of layer-by-layer Jones matrix mapping of the maps $\text{lm}j_{ik}(m,n)$ of the optically anisotropic architectonics of supramolecular networks of dehydrated BF [13–20].

The analysis of the obtained data revealed the individual structure of the coordinate distributions $\text{Im}j_{ik}(m,n)$ (panels (a), (b) and (e), (f)) and the probability distributions $(G(\text{Im}j_{ik}(m,n)))$ (panels (c), (d) and (g), (h)), which characterize the imaginary components of the Jones matrix images $\text{Im}j_{11;22}(\text{LB}_{0;90})$ and $\text{Im}j_{12;21}(\text{LB}_{45;135})$.

- Phase layer-by-layer scanning of the algorithmically reproduced object field of complex amplitudes of dehydrated BF demonstrated a significant transformation in the magnitude and position of the extrema and in the ranges of variation of the histograms $G(\text{Im}j_{ik}(m,n))$ (Figures 12.4 and 12.5, (c), (d) and (g), (h)). At the same time, there is a consistent increase in the correlation between the model representations and the experimental data on the structure of the set of imaginary Jones matrix images $\text{Im}j_{ik}(m,n)$, $\text{lm}j_{ik}(\varphi_k = \pi/4, m, n)$ and $\text{lm}j_{ik}(\varphi_k = \pi/8, m, n)$ of the supramolecular polycrystalline networks of the BF sample – $\text{lm}j_{11}(m,n) \approx \text{Im}j_{22}(m,n)$ and $\text{lm}j_{12}(m,n) \approx \text{Im}j_{21}(m,n)$ [21–26].

Table 12.2 presents the statistical characteristics of the phase transformation of the imaginary Jones matrix images $\text{lm}j_{ik}(m,n)$ of the set of dehydrated BF samples.

Comparative analysis of the Jones matrix images $\text{lm}j_{ik}(m,n)$ of dehydrated BF demonstrated a statistical difference in the coordinate distributions of the magnitude of the real component of the matrix elements $\widetilde{M}_{i=1\div4}(\text{lm}j_{11;22}(\text{LB}_{0;90})) \neq \widetilde{M}_{i=1\div4}(\text{lm}j_{12;21}(\text{LB}_{45;135}))$, which characterize the polarization manifestations of structural anisotropy. The most sensitive markers were the statistical moments $\widetilde{M}_{i=3;4}(\text{lm}j_{12;21}(\text{LB}_{45;135}))$, which characterize the skewness and kurtosis of the distributions of the magnitude of the imaginary component of the matrix elements $\text{lmj}_{12;21}(\text{LB}_{45;135})$. During digital phase scanning of the object field of complex amplitudes, the magnitudes of the higher-order statistical moments $\widetilde{M}_{i=3;4}(\varphi_k, \text{lm}j_{12;21}(\text{LB}_{45;135}))$ increased up to fivefold. The range of variation of other markers $M_{i=3;4}(\varphi_k, \text{lm}j_{11;22}(\text{LB}_{0;90}))$ was much smaller and did not exceed 1.6 times.

From a physical point of view, the obtained results can be associated with the following features of the spatial–angular symmetry of the optically anisotropic architectonics of supramolecular networks self-assembled during the dehydration of a bile drop [27–33].

With a decrease in the phase scan shift $\varphi_k \downarrow$ of the algorithmically reproduced field of complex amplitudes, a more optically thin volume of the dehydrated BF is distinguished. As a result, two scenarios of changes in the birefringence parameters of the polycrystalline architectonics are realized.

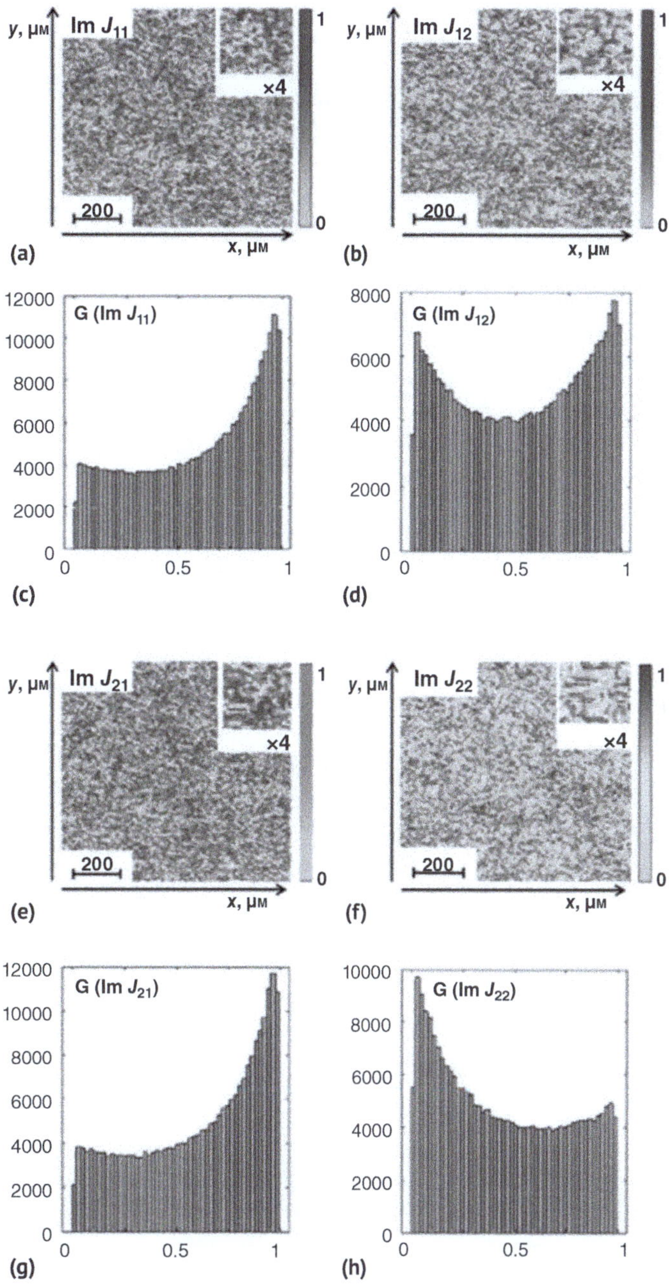

Figure 12.3 Coordinate (a), (b), (e), (f) and statistical (c), (d), (g), (h) structure of the imaginary parts of the Jones matrix images $\mathrm{Im}j_{ik}(m, n)$ of dehydrated bile films.

The first is that the spatial–angular structure of the distribution of the directions of the optical axes of the needle-shaped crystals of cholesterol (CC) and fatty acids (FA) narrows from $(0 \leq \rho \leq 2\pi)$ to $(\rho_{\min} \leq \rho \leq \rho_{\max})$, and the structural anisotropy of the supramolecular network formed in the phase-separated volume of the BF decreases, LB(ρ) ↓. At the same time, the degree of "degradation" of LB for different orientation directions ($\ll 0° - 90° \gg$ and $\ll 45° - 135° \gg$) is different. Due to this, different values of the parameters $\mathrm{LB}_{0;90}(\rho)$ and $\mathrm{LB}_{45;135}(\rho)$ are formed.

Second, the concentration of optically active microliths (M) consisting of Ca salts, mucus, cholesterol, and calcium bilirubinate (CaB) decreases. Accordingly, the value of optical activity $(\mathrm{CB}_{\otimes;\oplus} \downarrow)$ of the spherulite component of the supramolecular network of the BF also changes. As a result of the above-mentioned optical–physical transformations, the values of the real constituent elements of the Jones matrix also decrease $(r_{ik} \downarrow)$. Moreover, as predicted (Section 2.1), the matrix elements $Re\,j_{11;22}(\mathrm{LB}^2 + \mathrm{CB}^2{}_{\otimes;\oplus})$ decrease more rapidly compared to $Re\,j_{12;21}(\mathrm{CB}^2{}_{\otimes;\oplus})$.

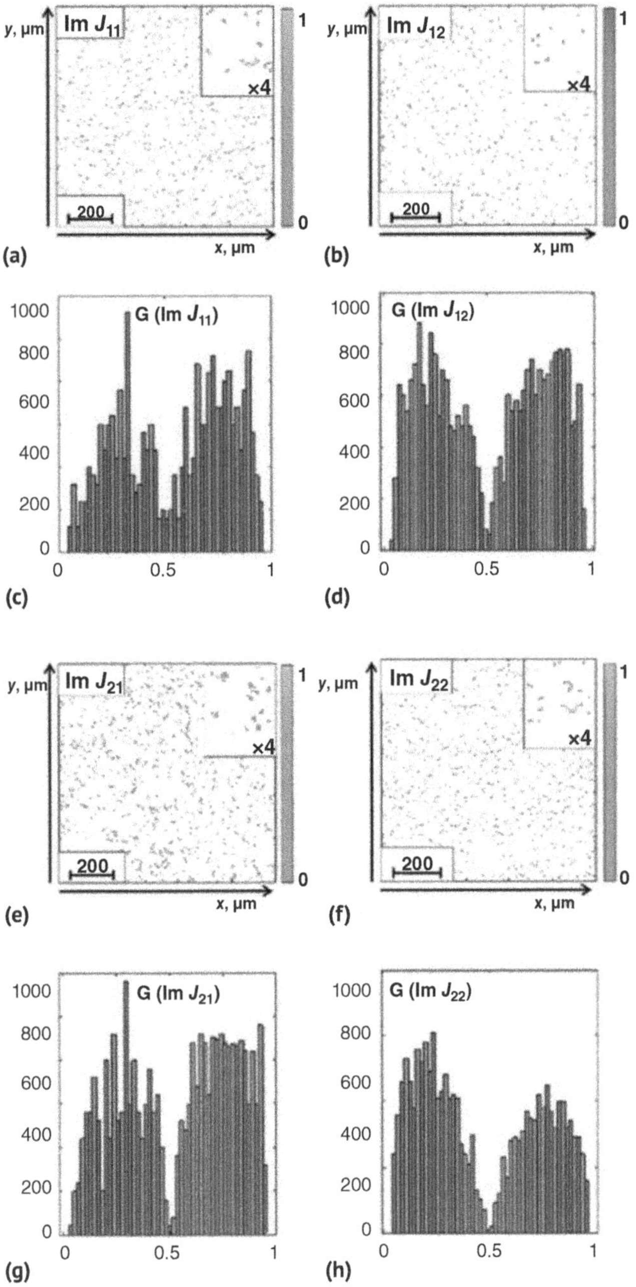

Figure 12.4 Layered ($\varphi_k = \pi/8$) coordinate (a), (b), (e), (f) and statistical (c), (d), (g), (h) structure of the imaginary parts of the Jones matrix images $\text{Im}j_{ik}(m,n)$ of dehydrated bile films.

Table 12.2 Central statistical moments of the first to fourth orders characterizing the coordinate distributions of the imaginary parts of the Jones matrix images $\text{Im}j_{ik}(m,n)$ of the bile films.

	$\text{Im}j_{ik}(m,n)$			
$\widetilde{M}_{1;2;3;4} \pm \sigma_{1;2;3;4}$	$\text{Im}j_{11}(m,n)$	$\text{Im}j_{12}(m,n)$	$\text{Im}j_{21}(m,n)$	$\text{Im}j_{22}(m,n)$
$\widetilde{M}_1 \pm \sigma_1$	0.45 ± 0.024	0.51 ± 0.027	0.53 ± 0.031	0.41 ± 0.022
$\widetilde{M}_2 \pm \sigma_2$	0.29 ± 0.016	0.34 ± 0.019	0.36 ± 0.021	0.27 ± 0.016
$\widetilde{M}_3 \pm \sigma_3$	0.17 ± 0.009	0.12 ± 0.007	0.104 ± 0.006	0.15 ± 0.008
$\widetilde{M}_4 \pm \sigma_4$	0.23 ± 0.012	0.17 ± 0.009	0.15 ± 0.009	0.21 ± 0.012

		$\varphi_k = \pi/4$		
$\widetilde{M}_{1;2;3;4} \pm \sigma_{1;2;3;4}$	$\mathrm{Im}j_{11}(m,n)$	$\mathrm{Im}j_{12}(m,n)$	$\mathrm{Im}j_{21}(m,n)$	$\mathrm{Im}j_{22}(m,n)$
$\widetilde{M}_1 \pm \sigma_1$	0.39 ± 0.021	0.44 ± 0.025	0.48 ± 0.025	0.39 ± 0.021
$\widetilde{M}_2 \pm \sigma_2$	0.22 ± 0.013	0.25 ± 0.014	0.29 ± 0.016	0.21 ± 0.012
$\widetilde{M}_3 \pm \sigma_3$	0.29 ± 0.015	0.49 ± 0.022	0.56 ± 0.029	0.27 ± 0.014
$\widetilde{M}_4 \pm \sigma_4$	0.33 ± 0.017	0.68 ± 0.029	0.79 ± 0.036	0.35 ± 0.018
		$\varphi_k = \pi/8$		
$\widetilde{M}_{1;2;3;4} \pm \sigma_{1;2;3;4}$	$\mathrm{Im}j_{11}(m,n)$	$\mathrm{Im}j_{12}(m,n)$	$\mathrm{Im}j_{21}(m,n)$	$\mathrm{Im}j_{22}(m,n)$
$\widetilde{M}_1 \pm \sigma_1$	0.31 ± 0.019	0.29 ± 0.016	0.33 ± 0.017	0.29 ± 0.016
$\widetilde{M}_2 \pm \sigma_2$	0.15 ± 0.008	0.15 ± 0.008	0.17 ± 0.009	0.13 ± 0.008
$\widetilde{M}_3 \pm \sigma_3$	0.28 ± 0.015	0.61 ± 0.037	0.54 ± 0.029	0.31 ± 0.017
$\widetilde{M}_4 \pm \sigma_4$	0.37 ± 0.016	0.81 ± 0.041	0.78 ± 0.043	0.39 ± 0.028

Due to the predominant influence of the optical activity of microliths (M) consisting of Ca salts, mucus, cholesterol, and calcium bilirubinate (CaB), the statistical moments $\widetilde{M}_{i=3;4}(Rej_{12;21}(\mathrm{CB}^2{}_{\otimes;\oplus}))$ were found to be quantitatively the most sensitive, in terms of the phase range of eigenvalue variations, to the transformation of the polycrystalline architectonics of the BF.

12.1.3 Biomedical Application

Two representative groups ($q = 21$) of dehydrated BF samples were studied: healthy donors (control group 1) and patients with a verified diagnosis of calculous cholecystitis (study group 2). Within both groups, the mean values $(M_{3,4})$, standard deviations $(\pm\sigma_{3,4})$ and statistical significance (p) of the most sensitive detected markers were calculated (Tables 12.1 and 12.2). These markers characterize the skewness and kurtosis of the coordinate distributions of the value $Rej_{ik}(m,n)$ and $\mathrm{Im}j_{ik}$.

Figures 12.5 and 12.6 present integral (panels (a), (b)) and layer-by-layer $\varphi_k = \pi/8$ (panels (e), (f)) maps $Rej_{11}(\mathrm{LB}^2 + \mathrm{CB}^2{}_{\otimes;\oplus})$ and $Rej_{12}(\mathrm{CB}^2{}_{\otimes;\oplus})$ as well as histograms $G(Rej_{11;21})$ of the distributions (panels (c), (d), (g), (h)) of the values of the real component of the matrix elements r_{11} and r_{12} for the set of dehydrated BF from Group 1 (panels (c), (g)) and Group 2 (panels (d), (h)).

Figures 12.7 and 12.8 show similar dependences for the imaginary components of the Jones matrix elements $\mathrm{Im}j_{11}(\mathrm{LB}_{0;90})$ and $\mathrm{Im}j_{12}(\mathrm{LB}_{45;135})$, which characterize the polarization manifestations of the structural anisotropy of supramolecular networks of dehydrated BF.

The results of the statistical and informational analysis of the Jones matrix images $Rej_{11;12}(m,n)$ and $\mathrm{Im}j_{11;12}(m,n)$ of dehydrated BF from healthy donors and patients with calculous cholecystitis are presented in Tables 12.3 and 12.4.

Comparative analysis of the data on the statistical structure of the real part of the integral $(Rej_{11;12}(m,n))$ and layer-by-layer $(Rej_{11;12}(\varphi_k, m, n))$ Jones matrix images of dehydrated BF samples demonstrated the predicted trends in the set of azimuthally invariant diagnostic markers $M_{i=3;4}(Rej_{11;12}(m,n))$ – a decrease in the values of $M_{i=3;4}(Rej_{11}(m,n))$ and $M_{i=3;4}(Rej_{12}(m,n))$, which were calculated for the case of calculous cholecystitis. At the same time, there is a predominant increase in the parameters $M_{1;2}(Rej_{12;21}(\mathrm{CB}^2{}_{\otimes;\oplus}))$ compared to $M_{1;2}(Rej_{11;22}(\mathrm{LB}^2 + \mathrm{CB}^2{}_{\otimes;\oplus}))$. As a result, the differences between the diagnostic markers $M_{i=3;4}(Rej_{11})$ and $M_{i=3;4}(Rej_{12})$ reach 1.6 and 1.7 times, respectively (Table 12.3). Consequently, rather high levels of balanced accuracy of differential diagnosis of calculous cholecystitis are achieved – good $\mathrm{Ac}(M_{i=3;4}(Rej_{11})) = 85.7\%$ and very good $\mathrm{Ac}(M_{i=3;4}(Rej_{12})) = 90.5\%$ (highlighted in gray). At the same time, the depolarizing effect of multiple scattering within the volume of dehydrated BF significantly averages the polarization differences between the polycrystalline structures of the supramolecular networks of BF from healthy donors and patients with calculous cholecystitis. Eliminating this effect by digital phase scanning of the algorithmically reconstructed field of

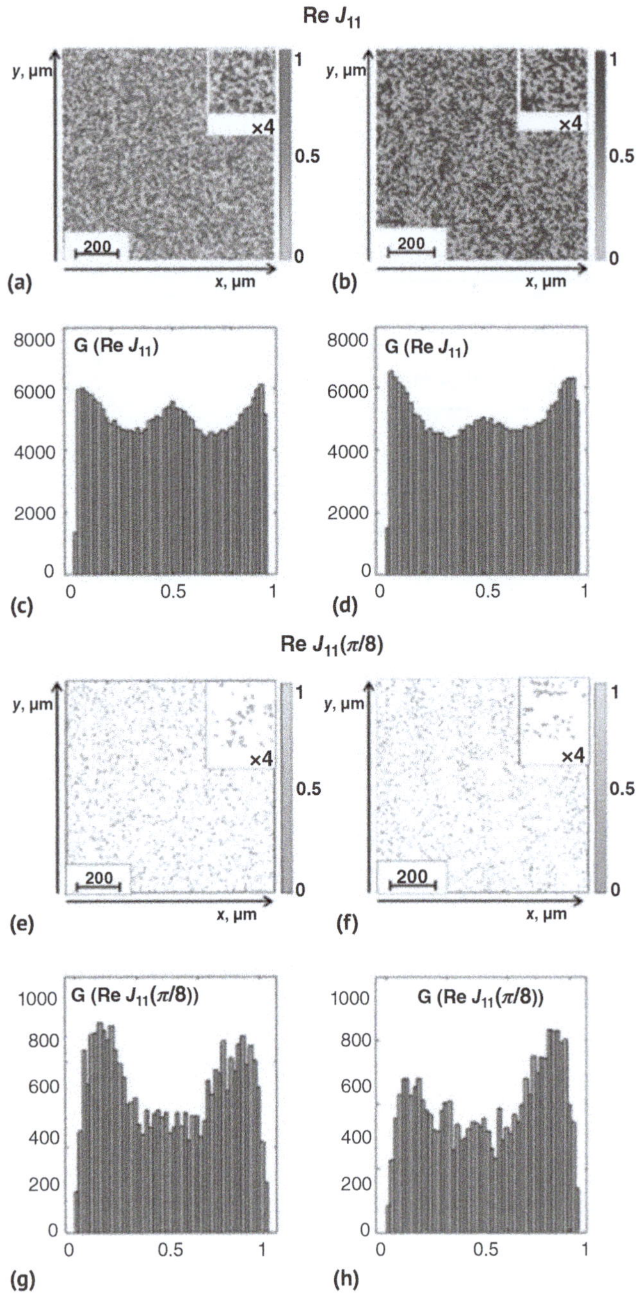

Figure 12.5 Jones matrix images $Rej_{11}(m,n)$ (a), (b), (e), (f) and histograms $G(Rej_{11}(m,n))$ (c), (d), (g), (h) of the distributions of the value Rej_{11} for dehydrated bile films of healthy donors (a), (c), (e), (g) and patients with calculous cholecystitis (b), (d), (f), (h).

complex amplitudes significantly increases the sensitivity of the layer-by-layer Jones matrix mapping method. In particular, in the phase plane ($\varphi_k = \pi/8$) of the object field component with the minimum multiplicity of light scattering, the statistical structure of the Jones matrix maps $Rej_{11;12}(\varphi_k = \pi/8, a, b)$ exhibits maximum differences between the diagnostic markers $M_{i=3;4}(Rej_{11;12})$, 1.9 and 2.3 times, respectively. As a result, the accuracy of differential diagnosis reaches a very good level $\text{Ac}(M_{i=3;4}(\varphi_k = \pi/8, Rej_{11})) = 90.5\%$ (highlighted in gray) and an excellent level $\text{Ac}(M_{i=3;4}(\varphi_k = \pi/8, Rej_{12})) = 95.2\%$ (highlighted in gray) – Table 12.3.

A different pattern is observed for the results of mapping the imaginary components of the Jones matrix elements $\text{Im}j_{11}(\text{LB}_{0;90})$ and $\text{Im}j_{12}(\text{LB}_{45;135})$, which characterize the azimuthally dependent ($\text{LB}_{0;90}$ and $\text{LB}_{45;135}$) structural anisotropy of the supramolecular networks of dehydrated BF. Comparison of the statistical analysis of the coordinate distributions of

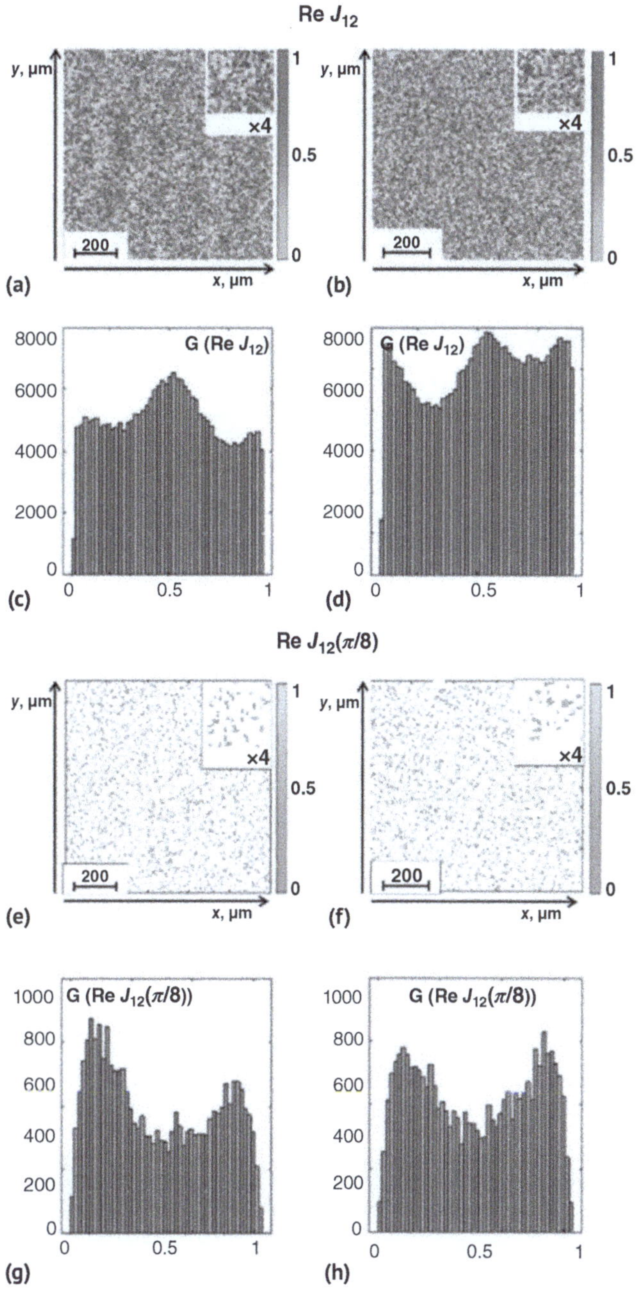

Figure 12.6 Jones matrix images $Rej_{12}(m,n)$ (a), (b), (e), (f) and histograms $G(Rej_{12}(m,n))$ (c), (d), (g), (h) of the distributions of the value Rej_{12} for dehydrated bile films of healthy donors (a), (c), (e), (g) and patients with calculous cholecystitis (b), (d), (f), (h).

the values $\text{Im}j_{11}(\text{LB}_{0;90})$ and $\text{Im}j_{12}(\text{LB}_{45;135})$ revealed a correlation between them and the prognostic scenarios of pathological changes in LB. At the same time, the differences between the diagnostic marker values $M_{i=3;4}(\text{Im}j_{11})$ and $M_{i=3;4}(\text{Im}j_{12})$ are considerably smaller (1.3 and 1.2 times, respectively), Table 12.4.

As a result, the balanced accuracy of differential diagnosis decreases to a satisfactory level — $\text{Ac}(M_{i=3;4}(\text{Im}j_{11})) = 83.3\% - 85.7\%$ and to an unsatisfactory level – $\text{Ac}(M_{i=3;4}(\text{Im}j_{12})) = 78.6\%$.

Phase scanning of the complex amplitude field and algorithmic reconstruction of the Jones matrix images $\text{Im}j_{11;12}(\varphi_k = \pi/8, m, n)$ of an optically thin layer of dehydrated BF resulted in a slight increase in the differences between the diagnostic markers $M_{i=3;4}(\text{Im}j_{11;12})$ — up to 1.4 and 1.3 times, respectively. As a result, a moderate improvement in

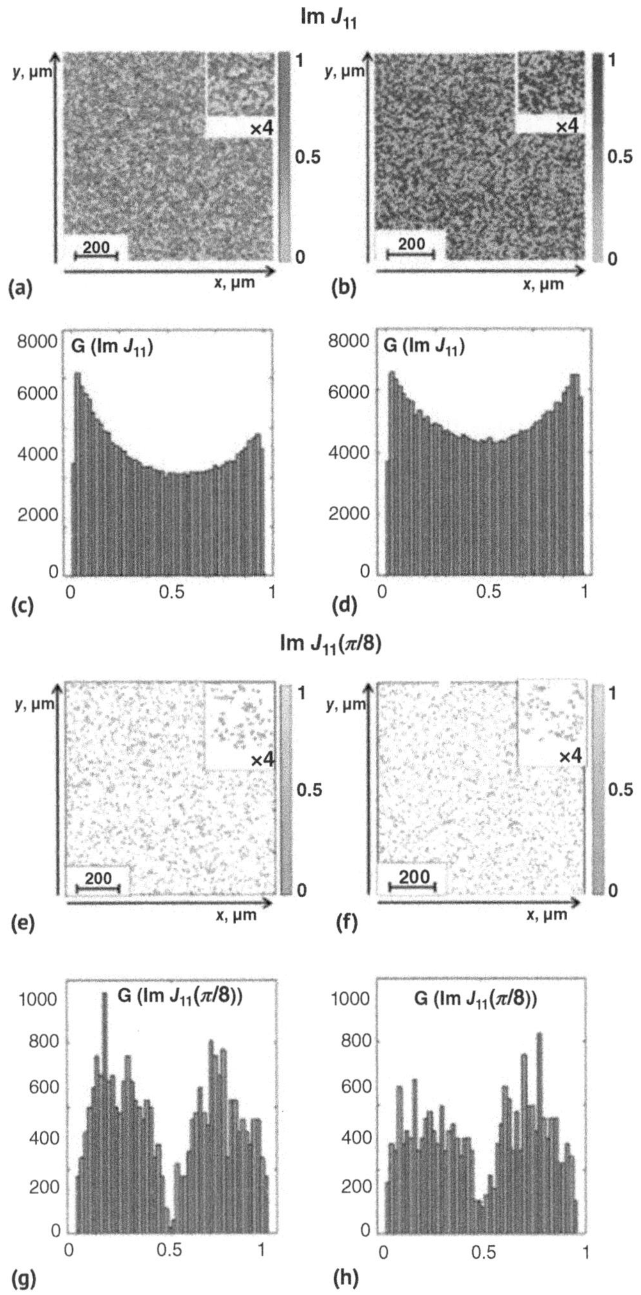

Figure 12.7 Jones matrix images $\text{Im}j_{11}(m,n)$ (a), (b), (e), (f) and histograms $G(\text{Im}j_{11}(m,n))$ (c), (d), (g), (h)) of the distributions of the value $\text{Im}j_{11}$ for dehydrated bile films of healthy donors (a), (c), (e), (g)) and patients with calculous cholecystitis (b), (d), (f), (h).

the balanced accuracy of differential diagnosis to a satisfactory level was achieved – $\text{Ac}(M_{i=3;4}(\varphi_k = \pi/8, \text{Im}j_{11})) = 81\%$ and $\text{Ac}(M_{i=3;4}(\varphi_k = \pi/8, \text{Im}j_{12})) = 83.3\% - 85.7\%$, Table 12.4.

From a physical point of view, the results obtained can be explained by the fact that the values of $\text{Im}j_{11}(\text{LB}_{0;90})$ and $\text{Im}j_{12}(\text{LB}_{45;135})$ in all phase planes φ_k are determined by the orientational features of the distributions of needle crystal orientations ($\rho_{\min} \le \rho(\varphi_k) \le \rho_{\max}$) within the supramolecular network. As a result of serial intra-group measurements for individual samples of dehydrated BF, different angular conditions $\rho(\varphi_k)$ and, consequently, different conditions for the formation of the values of the imaginary components of the Jones matrix elements $\text{Im}j_{11}(\text{LB}_{0;90})$ and $\text{Im}j_{12}(\text{LB}_{45;135})$ are

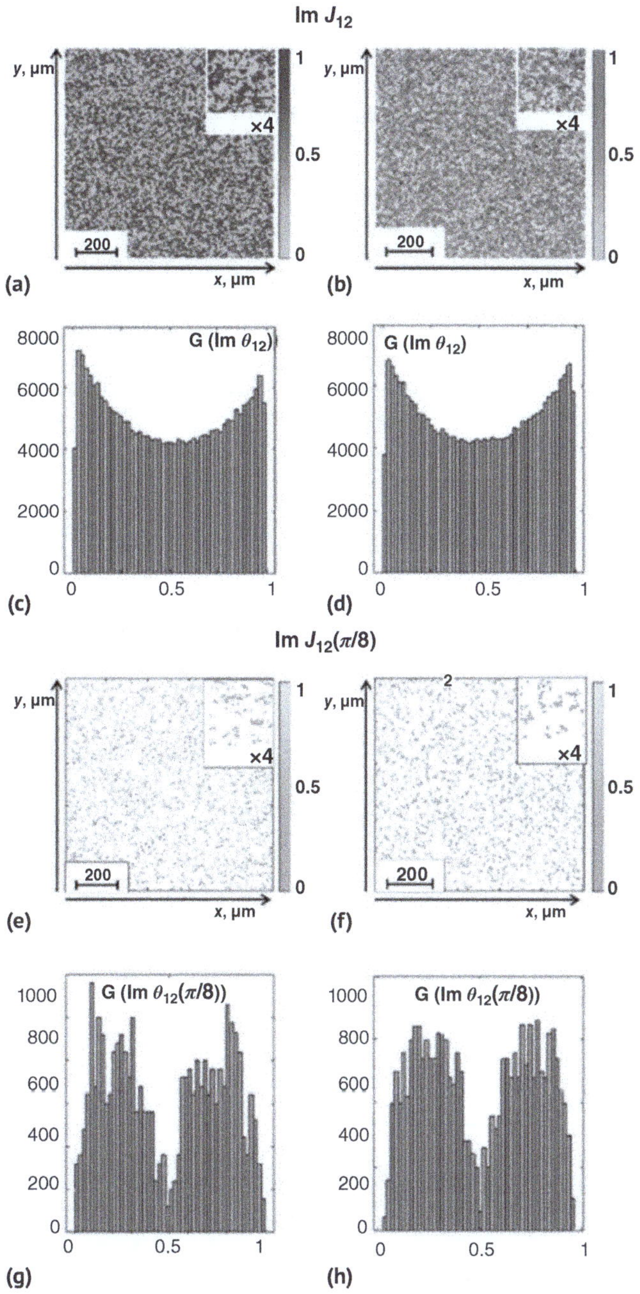

Figure 12.8 Jones matrix images $\text{Im}j_{12}(m, n)$ (a), (b), (e), (f) and histograms $G(\text{Im}j_{12}(m, n))$ (c), (d), (g), (h) of the distributions of the value $\text{Im}j_{12}$ for dehydrated bile films of healthy donors (a), (c), (e), (g) and patients with calculous cholecystitis (b), (d), (f), (h).

realized. Based on this, the overall group-averaged ($q = 21$) values of the elements $\text{Im}j_{11;12}(q)$ are obtained. Consequently, the differences between the statistical moments $M_{i=3;4}(\text{Im}j_{11}(\text{LB}_{0;90}))$ and $M_{i=3;4}(\text{Im}j_{12}(\text{LB}_{45;135}))$, which characterize the orientational distributions of structural anisotropy $\text{LB}_{0;90}$ and $\text{LB}_{45;135}$ are reduced. As a result, the balanced accuracy of differential diagnosis of calculous cholecystitis using the markers $M_{i=3;4}(\text{Im}j_{11;12})$ is significantly decreased:

- $\text{Ac}(M_{i=3;4}(\text{Im}j_{11})) = 81\% = 83.3\%$ – satisfactory level;
- $\text{Ac}(M_{i=3;4}(\varphi_k = \pi/8, \text{Im}j_{11})) = 83.3\% - 85.7\%$ – satisfactory level;
- $\text{Ac}(M_{i=3;4}(\text{Im}j_{12})) = 78.6\%$ – unsatisfactory level;
- $\text{Ac}(M_{i=3;4}(\varphi_k = \pi/8, \text{Im}j_{12})) = 81\%$ – satisfactory level.

Table 12.3 Statistical markers $\widetilde{M}_{3;4}$ and accuracy of differential diagnosis of calculous cholecystitis based on Jones matrix images $Rej_{11;12}(m, n)$ of dehydrated bile films.

	$Rej_{11}(m, n)$			$Rej_{11}(\varphi_k = \pi/8, m, n)$		
$\widetilde{M}_{3;4} \pm \sigma_{3;4}$	**Group 1**	**Group 2**	**Ac, %**	**Group 1**	**Group 2**	**Ac, %**
$M_3 \pm \sigma_3$	0.34 ± 0.018	0.21 ± 0.011	85.7	0.84 ± 0.043	0.43 ± 0.021	**90.5**
p_{12}	$p_{12} < 0.05$		$c = 2$; $d = 4$	$p_{12} < 0.001$		$c = 2$; $d = 2$
$M_4 \pm \sigma_4$	0.59 ± 0.029	0.33 ± 0.017	85.7	0.98 ± 0.053	0.59 ± 0.031	**90.5**
p_{12}	$p_{12} < 0.05$		$c = 3$; $d = 3$	$p_{12} < 0.001$		$c = 1$; $d = 3$
	$Rej_{12}(m, n)$			$Rej_{12}(\varphi_k = \pi/8, m, n)$		
$\widetilde{M}_{3;4} \pm \sigma_{3;4}$	**Group 1**	**Group 2**	**Ac, %**	**Group 1**	**Group 2**	**Ac, %**
$M_3 \pm \sigma_3$	0.36 ± 0.01	0.22 ± 0.012	**90.5**	0.93 ± 0.047	0.41 ± 0.028	**95.2**
p_{12}	$p_{12} < 0.001$		$c = 1$; $d = 3$	$p_{12} < 0.001$		$c = 1$; $d = 1$
$M_4 \pm \sigma_4$	0.52 ± 0.031	0.34 ± 0.018	**90.5**	1.34 ± 0.074	0.77 ± 0.041	**95.2**
p_{12}	$p_{12} < 0.001$		$c = 2$; $d = 2$	$p_{12} < 0.001$		$c = 1$; $d = 1$

Table 12.4 Statistical markers $\widetilde{M}_{3;4}$ and accuracy of differential diagnosis of calculous cholecystitis based on Jones matrix images $\mathrm{Im}j_{11;12}(m, n)$ of dehydrated bile films.

	$\mathrm{Im}j_{11}(m, n)$			$\mathrm{Im}j_{11}(\varphi_k = \pi/8, m, n)$		
$\widetilde{M}_{3;4} \pm \sigma_{3;4}$	**Group 1**	**Group 2**	**Ac, %**	**Group 1**	**Group 2**	**Ac, %**
$M_3 \pm \sigma_3$	0.22 ± 0.015	0.17 ± 0.009	81	0.38 ± 0.019	0.27 ± 0.015	83.3
p_{12}	$p_{12} < 0.05$		$c = 3$; $d = 5$	$p_{12} < 0.05$		$c = 3$; $d = 4$
$M_4 \pm \sigma_4$	0.28 ± 0.016	0.21 ± 0.012	83.3	0.33 ± 0.017	0.37 ± 0.019	85.7
p_{12}	$p_{12} < 0.05$		$c = 3$; $d = 4$	$p_{12} < 0.05$		$c = 2$; $d = 4$
	$\mathrm{Im}j_{12}(m, n)$			$\mathrm{Im}j_{12}(\varphi_k = \pi/8, m, n)$		
$\widetilde{M}_{3;4} \pm \sigma_{3;4}$	**Group 1**	**Group 2**	**Ac, %**	**Group 1**	**Group 2**	**Ac, %**
$M_3 \pm \sigma_3$	0.14 ± 0.009	0.12 ± 0.007	78.6	0.81 ± 0.042	0.61 ± 0.037	81
p_{12}	$p_{12} < 0.05$		$c = 4$; $d = 5$	$p_{12} < 0.05$		$c = 3$; $d = 5$
$M_4 \pm \sigma_4$	0.23 ± 0.012		78.6	1.09 ± 0.059	0.81 ± 0.043	81
p_{12}	$p_{12} < 0.05$		$c = 4$; $d = 5$	$p_{12} < 0.05$		$c = 4$; $d = 4$

12.1.4 Conclusions

1. An optical model of the polycrystalline structure of supramolecular networks in dehydrated BF has been proposed, and a set of analytical relationships has been established between the phase anisotropy parameters (linear and circular birefringence) of the polycrystalline architectonics and the real and imaginary components of the Jones matrix elements.

2. The results of experimental testing of the polarization-interference mapping method with digital Fourier reconstruction of the complex amplitude distributions of object fields, followed by algorithmic calculation of the coordinate distributions of the magnitudes of the real and imaginary components of the Jones matrix elements of dehydrated BF in a series of phase planes, are presented and physically analyzed.
3. The scenarios of phase transformation in the statistical structure of the imaginary and real components of the Jones matrix images of supramolecular networks of dehydrated BF – $\widetilde{M}_{i=1;2}\left\{\varphi_k \downarrow, \begin{pmatrix} Re j_{ik}; \\ \text{Im} j_{ik} \end{pmatrix}\right\} \downarrow$ and $\widetilde{M}_{i=3;4}\left(\left\{\varphi_k \downarrow, \begin{pmatrix} Re j_{ik}; \\ \text{Im} j_{ik} \end{pmatrix}\right\}\right) \uparrow$ have been established and physically substantiated.
4. A set of diagnostic markers most sensitive to the transformation of supramolecular polycrystalline networks of dehydrated BF has been identified – the third- and fourth-order statistical moments that characterize the skewness and kurtosis of the coordinate distributions of the real and imaginary components of the Jones matrix elements, $\widetilde{M}_{i=3;4}\begin{pmatrix} Re j_{12}(\text{CB}^2{}_{\otimes;\oplus}); \\ Re j_{11}(\text{LB}^2 + \text{CB}^2{}_{\otimes;\oplus}) \end{pmatrix}$.
5. An excellent accuracy of $\text{Ac}(M_{i=3;4}(\varphi_k = \pi/8, Re j_{12})) = 95.2\%$ was demonstrated for the layer-by-layer Jones matrix mapping method of dehydrated BF in the differential diagnosis of calculous cholecystitis – the early (latent) stage of gallstone disease.

12.2 Methods and Means of Laser Layer-by-layer Jones Matrix Mapping of Polycrystalline Films of Synovial Fluid

12.2.1 Synovial Fluid Films

Synovial fluid consists of two main components: liquid and protein–polysaccharide. Its main biochemical element, which provides the viscoelastic properties of synovial fluid, is hyaluronan – a polysaccharide from the group of glycosaminoglycans.

From the point of view of the model concepts developed by us, these microstructures exhibit optical activity – circular birefringence (CB) [34–39].

Hyaluronan stabilizes the structure of proteoglycans, which together form complex molecules of high molecular weight. These molecules are deposited within the collagen matrix of cartilage, providing it with elasticity. Spatially elongated macromolecular structures described herein give rise to structural anisotropy or LB.

During the dehydration of a drop of synovial fluid, the hydrated shells of biochemical molecules and their structures are lost. As result, polycrystalline supramolecular networks or optically anisotropic architectonics are formed. The substances that most frequently crystallize in synovial fluid include:

1. Sodium urate – crystals most often have a needle-like shape with two sharp ends and, in polarized light, exhibit strong negative (left-handed) birefringence.
2. Calcium pyrophosphate dihydrate – forms rhomboid-shaped crystals or thick needles with a blunt end, which exhibit weak positive (right-handed) birefringence.
3. Hydroxyapatite – crystals have a spindle-like shape.
4. Cholesterol – crystals have a plate-like shape and exhibit strong positive (right-handed) LB.

Therefore, the polycrystalline network of the synovial fluid films formed during dehydration represents dendritic, optically active (CB) structures that, at the supramolecular level, exhibit predominant LB.

12.2.2 Real Jones Matrix Images of Polycrystalline Synovial Fluid Films

Figures 12.9–12.11 present the results of polarization–interference Jones matrix mapping of the $Re(j_{ik}(m, n))$ distributions of the synovial fluid films.

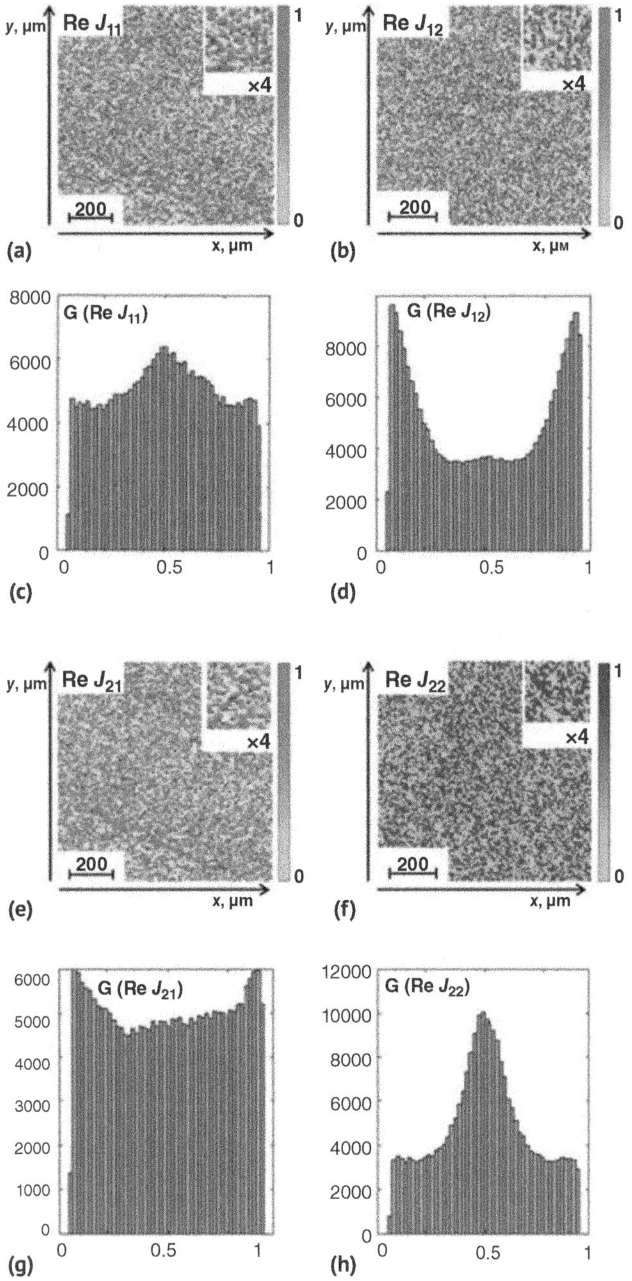

Figure 12.9 Integral coordinate (a), (b), (e), (f) and statistical (c), (d), (g), (h) structure of the real parts of the Jones matrix images $Re j_{ik}(m, n)$ of the dendritic polycrystalline network of the dehydrated synovial fluid film.

Analysis of the obtained data revealed:

- The presence of an individual coordinate-inhomogeneous topographic structure of the Jones matrix images $Re(j_{ik}(m, n))$ of dendritic polycrystalline networks for each matrix element, Figure 12.9a, b and e, f.
- An asymmetrical structure of the histograms $G(Re(j_{ik}))$ in terms of the localization of extrema and the ranges of variation of the random values $Re(j_{11;12;21;22})$, Figure 12.9c, d and g, h.
- Transformation of the magnitude and position of the extrema, as well as the ranges of variation of the histograms $G(Re(j_{ik}))$, depending on the value of the phase shift φ_k during the scanning of the algorithmically reconstructed complex amplitude field (Figures 12.10 and 12.11, c, d and g, h).

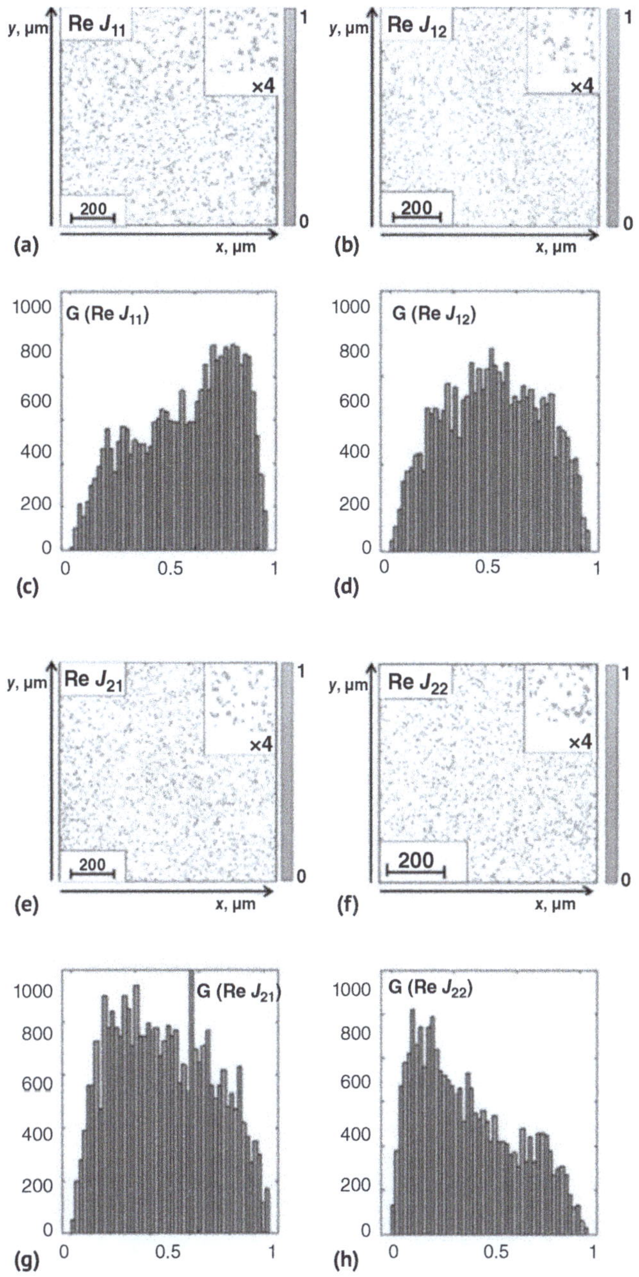

Figure 12.10 Layered ($\varphi_k = \pi/4$) coordinate (a), (b), (e), (f) and statistical (c), (d), (g), (h) structure of the real parts of the Jones matrix images $Rej_{ik}(m, n)$ of the dendritic polycrystalline network of the dehydrated synovial fluid film.

- A consistent increase in the correlation between the model representations and the experimental data on the structure of the set of Jones matrix images $Re(fi = \pi/2, j_{ik}(m, n)$, $Re(\varphi_k = \pi/4, j_{ik}(m, n)$ and $Re(\varphi_k = \pi/8, j_{ik}(m, n))$ of the supramolecular polycrystalline networks of a synovial fluid films sample – $Re(\varphi_k, j_{11}(m, n)) \approx Re(\varphi_k, j_{22}(m, n))$ and $Re(\varphi_k, j_{12}(m, n)) \approx Re(\varphi_k, j_{21}(m, n))$, – Figures 12.10 and 12.11a, b and e, f.

Table 12.5 presents the quantitative statistical parameters $(\widetilde{M}_{1;2;3;4} \pm \sigma_{1;2;3;4})$ of the phase transformation of the Jones matrix images $Re(j_{ik}(m, n))$ for 21 samples of polycrystalline synovial fluid films.

Here, $\widetilde{M}_{1;2;3;4}$ is the mean value within the group of synovial fluid polycrystalline film samples, and $\pm\sigma_{1;2;3;4}$ is the standard deviation.

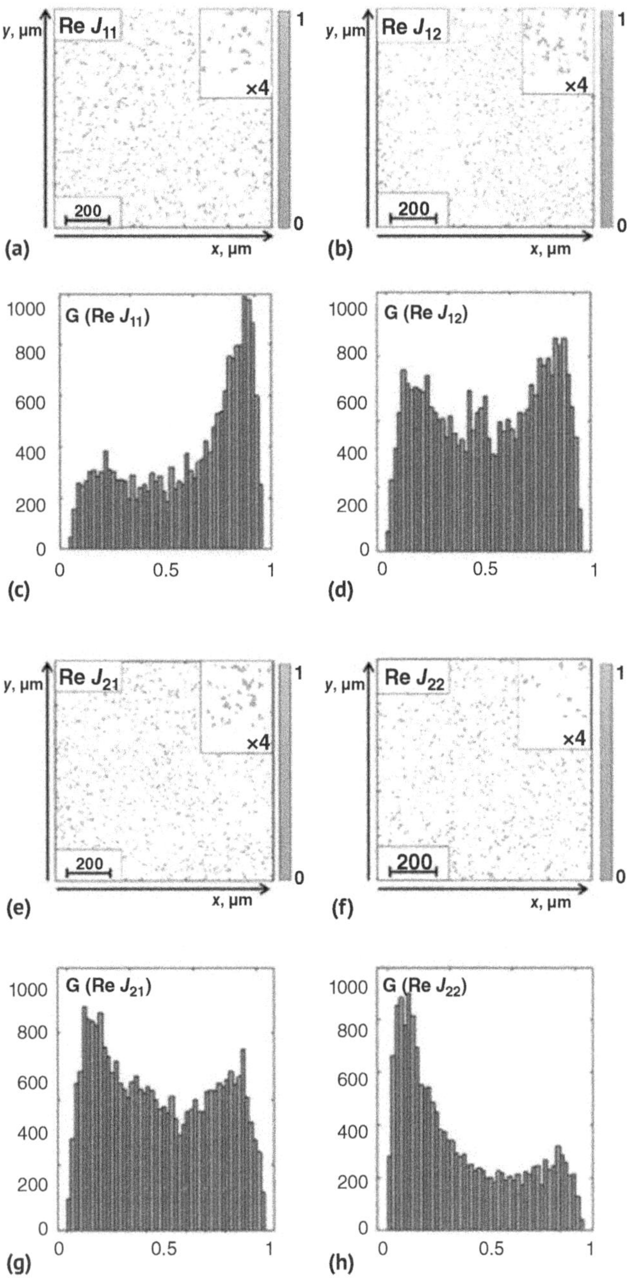

Figure 12.11 Layered ($\varphi_k = \pi/8$) coordinate (a), (b), (e), (f) and statistical (c), (d), (g), (h) structure of the real parts of the Jones matrix images $Re\,j_{ik}(m,n)$ of the dendritic polycrystalline network of the dehydrated synovial fluid film.

The following was revealed:

- Statistical similarity of the coordinate distributions (m, n) of the magnitude of the real components of the matrix elements – $M_{i=1\div4}(Re(J_{11})) \approx M_{i=1\div4}(Re(J_{11}))$ and $M_{i=1\div4}(Re(J_{12})) \approx M_{i=1\div4}(Re(J_{21}))$; the maximum differences between the values of the statistical moments $M_{i=1;2;3;4}(Re(j_{ik}(m,n)))$ do not exceed 10%.
- The predicted predominance of the Jones matrix manifestations of the LB mechanisms $(Re(j_{11;22}(m,n))$ – over the optical activity $(Re(j_{12;21}(m,n)$ – of the supramolecular networks of synovial fluid: $\widetilde{M}_{i=1;2}(Re(J_{11;22})) > \widetilde{M}_{i=1;2}(Re(J_{12;21}))$.
- Opposite trends were observed in the changes – a decrease (↓) in the mean $(\widetilde{M}_{i=1})$ and variance $(\widetilde{M}_{i=2})$ and an increase (↑) in the skewness $(\widetilde{M}_{i=3})$ and kurtosis $(\widetilde{M}_{i=4})$ – which characterize the transformation of the Jones matrix images $Re(j_{ik}(a,b))$ in different phase planes $(\varphi_k(m,n))$ of the polycrystalline synovial fluid film – $\begin{pmatrix} \widetilde{M}_{i=1;2} \downarrow \\ M_{i=3,4} \uparrow \end{pmatrix}$.

Table 12.5 Central statistical moments of the first to fourth orders characterizing the distributions of the real parts of the Jones matrix images $j_{ik}(m, n)$.

		$Re\{j_{ik}\}$		
$\widetilde{M}_{1;2;3;4} \pm \sigma_{1;2;3;4}$	$Re\{j_{11}\}$	$Re\{j_{12}\}$	$Re\{j_{21}\}$	$Re\{j_{22}\}$
$\widetilde{M}_1 \pm \sigma_1$	0.62 ± 0.034	0.38 ± 0.021	0.43 ± 0.023	0.58 ± 0.027
$M_2 \pm \sigma_2$	0.43 ± 0.023	0.21 ± 0.012	0.33 ± 0.017	0.38 ± 0.021
$M_3 \pm \sigma_3$	0.18 ± 0.011	0.32 ± 0.017	0.28 ± 0.015	0.36 ± 0.019
$M_4 \pm \sigma_4$	0.29 ± 0.016	0.41 ± 0.025	0.37 ± 0.019	0.58 ± 0.031
		$\varphi_k = \pi/4$		
$\widetilde{M}_{1;2;3;4} \pm \sigma_{1;2;3;4}$	$Re\{j_{11}\}$	$Re\{j_{12}\}$	$Re\{j_{21}\}$	$Re\{j_{22}\}$
$\widetilde{M}_1 \pm \sigma_1$	0.42 ± 0.023	0.29 ± 0.016	0.31 ± 0.019	0.38 ± 0.012
$M_2 \pm \sigma_2$	0.21 ± 0.011	0.18 ± 0.011	0.21 ± 0.012	0.19 ± 0.008
$M_3 \pm \sigma_3$	0.59 ± 0.032	0.44 ± 0.024	0.39 ± 0.022	0.67 ± 0.039
$M_4 \pm \sigma_4$	0.78 ± 0.041	0.63 ± 0.034	0.54 ± 0.029	0.89 ± 0.046
		$\varphi_k = \pi/8$		
$\widetilde{M}_{1;2;3;4} \pm \sigma_{1;2;3;4}$	$Re\{j_{11}\}$	$Re\{j_{12}\}$	$Re\{j_{21}\}$	$Re\{j_{22}\}$
$\widetilde{M}_1 \pm \sigma_1$	0.33 ± 0.014	0.19 ± 0.011	0.21 ± 0.012	0.28 ± 0.011
$M_2 \pm \sigma_2$	0.19 ± 0.012	0.14 ± 0.008	0.16 ± 0.009	0.17 ± 0.009
$M_3 \pm \sigma_3$	0.68 ± 0.035	0.53 ± 0.026	0.55 ± 0.031	0.72 ± 0.039
$M_4 \pm \sigma_4$	0.96 ± 0.047	0.77 ± 0.041	0.73 ± 0.037	1.01 ± 0.051

- The values of the mean and variance $\widetilde{M}_{i=1;2}(Re(j_{ik}(m, n)))$ decrease slightly (within the range of 30–50%) with an increase in the phase scanning parameter $\varphi_k \uparrow$.
- The most sensitive markers are the statistical moments $\widetilde{M}_{i=3;4}(Re(j_{11;22}(m, n))$, which characterize the skewness and kurtosis of the distributions of the real components of the matrix elements $Re(j_{11;22}(\mathrm{LB}^2 + \mathrm{CB}^2{}_{\otimes;\oplus}))$, – with an increase in the phase scanning parameter $\varphi_k \uparrow$, the values of $\widetilde{Z}_{i=3;4}$ increase significantly – up to threefold.
- The range of phase variation in the magnitude of the higher-order statistical moments $\widetilde{M}_{i=3;4}(Re(j_{12;21}(\mathrm{CB}^2{}_{\otimes;\oplus}))$, which characterize the manifestations of circular birefringence in the supramolecular network of the synovial fluid films, is somewhat smaller and does not exceed a twofold increase.

12.2.3 Imaginary Components of Jones Matrix Images of Polycrystalline Synovial Fluid Films

Figures 12.12–12.14 present the results of Jones matrix mapping of the $\mathrm{Im}(j_{ik}(m, n))$ distributions of polycrystalline synovial fluid films.

Analysis of the obtained data revealed both similarities and certain differences compared to the results of layer-by-layer polarization–interference mapping of the Jones matrix images of the real parts of the matrix elements $Re(j_{ik}(m, n))$ of supramolecular dendritic networks.

A similar feature is the presence, for each imaginary part of the matrix elements, of an individual coordinate-inhomogeneous topographic structure of the Jones matrix images $\mathrm{Im}(j_{ik}(a, b))$ of the dendritic polycrystalline networks

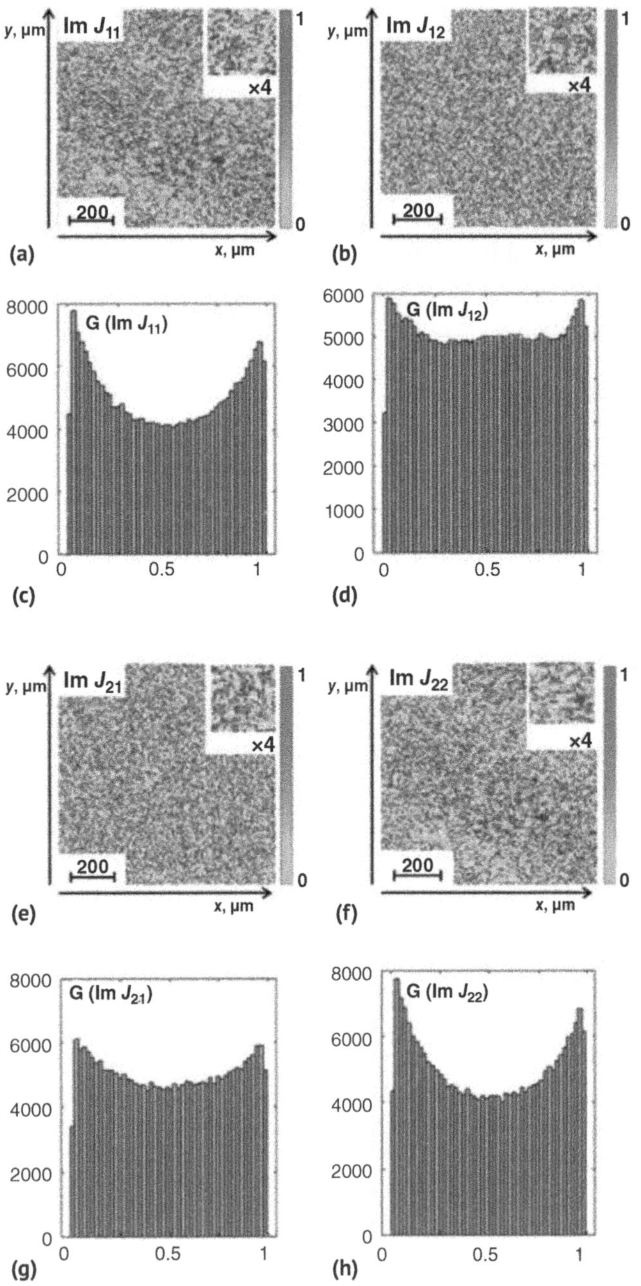

Figure 12.12 Coordinate (a), (b), (e), (f) and statistical (c), (d), (g), (h) structure of the imaginary parts of the Jones matrix images $\mathrm{lm}j_{ik}(m, n)$ of the dendritic polycrystalline network of the dehydrated synovial fluid film.

of the synovial fluid films – Figures 12.12–12.14, panels (a), (b) and (e), (f). The structure of the set of histograms $G(\mathrm{lm}(J_{11;12;21;22}))$ of the probability distributions of the random values $\mathrm{lm}(J_{11;12;21;22})$ of the dehydrated polycrystalline synovial fluid film is asymmetric in terms of the position and magnitude of the extrema – panels (c), (d) and (g), (h).

During phase layer-by-layer scanning of the object field formed within the volume of an optically anisotropic polycrystalline synovial fluid film, a significant dependence is observed – namely, the transformation of the magnitude and position of the extrema and the ranges of variation of the histograms $G(\mathrm{lm}(J_{ik}))$ – on changes in the phase shift value φ_k (panels (c), (d) and (g), (h)). At the same time, there is a consistent increase in the correlation between the model representations and the experimental data on the structure of the set of Jones matrix images $\mathrm{lm}(j_{ik}(m, n))$,

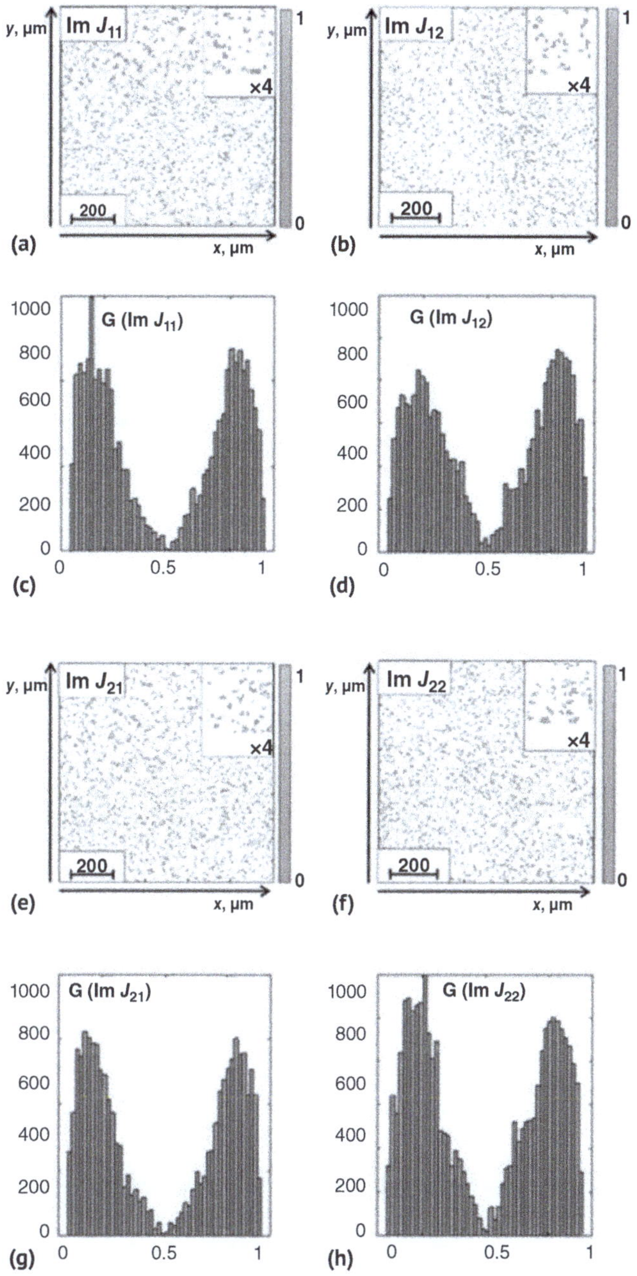

Figure 12.13 Layered ($\varphi_k = \pi/4$) coordinate (a), (b), (e), (f) and statistical (c), (d), (g), (h) structure of the imaginary parts of the Jones matrix images $\mathrm{Im}j_{ik}(m, n)$ of the dendritic polycrystalline network of the dehydrated synovial fluid film.

$\mathrm{Im}(\varphi_k = \pi/4, j_{ik}(m, n))$ and $\mathrm{Im}(\varphi_k = \pi/8, j_{ik}(m, n))$ of the supramolecular polycrystalline networks of a synovial fluid films sample – $\mathrm{Im}(\varphi_k, j_{11}(m, n)) \approx \mathrm{Im}(\varphi_k, j_{22}(m, n))$ and $\mathrm{Im}(\varphi_k, j_{12}(m, n)) \approx \mathrm{Im}(\varphi_k, j_{21}(m, n))$ (panels (a), (b) and (e), (f)).

Table 12.6 presents the quantitative statistical characteristics of the phase transformation of the Jones matrix images $\mathrm{Im}(j_{ik}(m, n))$ for a set of polycrystalline synovial fluid film samples.

Statistical analysis of the Jones matrix images $\mathrm{Im}(j_{ik}(m, n))$ of polycrystalline synovial fluid films revealed the following:

- Statistical similarity of the coordinate distributions of the magnitude of the imaginary components of the matrix elements – $\widetilde{M}_{i=1\div4}(\mathrm{Im}(J_{11})) \approx \widetilde{M}_{i=1\div4}(\mathrm{Im}(J_{11}))$ and $\widetilde{M}_{i=1\div4}(\mathrm{Im}(J_{12})) \approx \widetilde{M}_{i=1\div4}(\mathrm{Im}(J_{21}))$; the maximum differences between the values of the statistical moments $\widetilde{M}_{i=1;2;3;4}(\mathrm{Im}(j_{ik}(m, n)))$ do not exceed 10–15%.

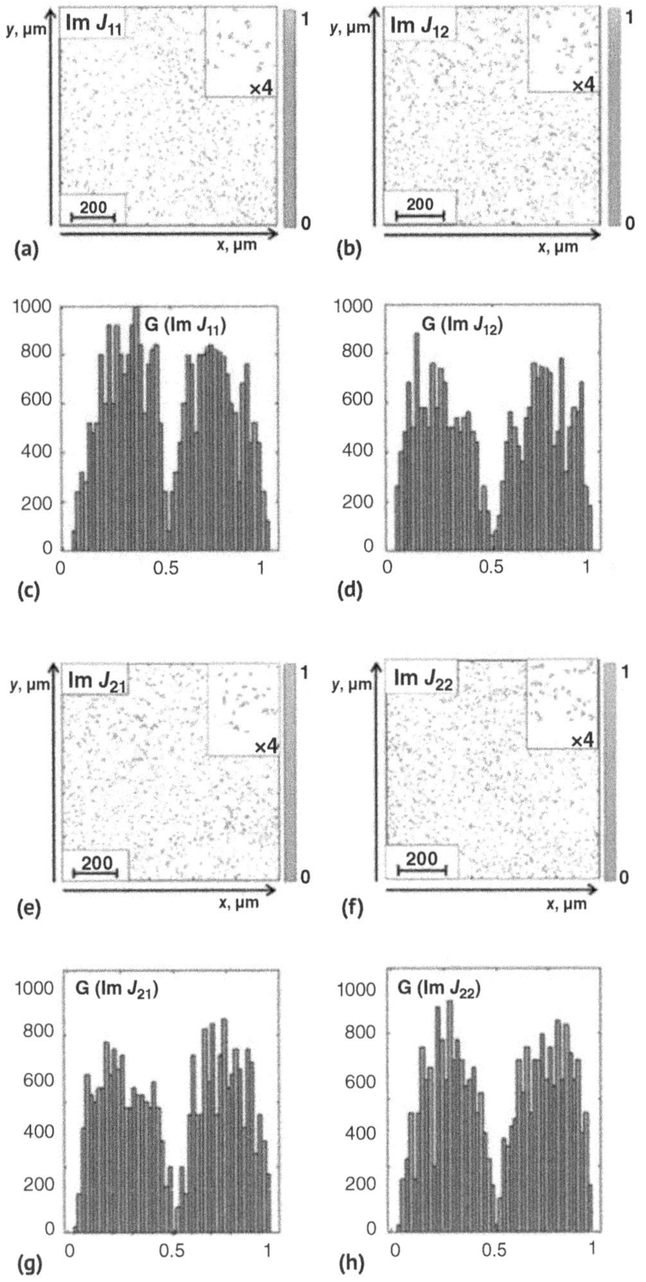

Figure 12.14 Layered ($\varphi_k = \pi/8$) coordinate (a), (b), (e), (f) and statistical (c), (d), (g), (h) structure of the imaginary parts of the Jones matrix images $\mathrm{Im}j_{ik}(m, n)$ of the dendritic polycrystalline network of the dehydrated synovial fluid film.

- Opposite trends were observed in the changes – a decrease (↓) in the mean $(\widetilde{M}_{i=1})$ and variance $(\widetilde{M}_{i=2})$ and an increase (↑) in the skewness $(\widetilde{M}_{i=3})$ and kurtosis $(\widetilde{M}_{i=4})$ – which characterize the transformation of the Jones matrix images $\mathrm{Im}(j_{ik}(a, b))$ in different phase planes $(\varphi_k(a, b))$ of the polycrystalline BF $\begin{pmatrix} \widetilde{M}_{i=1;2} \downarrow \\ M_{i=3,4} \uparrow \end{pmatrix}$.
- The most sensitive markers are the statistical moments $\widetilde{M}_{i=3;4}(\mathrm{Im}(j_{ik}(a, b))$, which characterize the skewness and kurtosis of the distributions of the magnitude of the imaginary components of the matrix elements $\mathrm{Im}(j_{ik}(LB_{0;90;45;135}))$, – with an increase in the phase scanning parameter φ_k ↑, the values of $\widetilde{M}_{i=3;4}$ increase by up to fivefold.

From a physical point of view, the obtained results can be associated with the following features of the architectonics of supramolecular dendritic polycrystalline networks self-assembled during the dehydration of a drop of synovial fluid.

Table 12.6 Central statistical moments of the first to fourth orders characterizing the coordinate distributions of the imaginary parts of the Jones matrix images $j_{ik}(m,n)$.

		$\mathrm{Im}\{j_{ik}\}$		
$\widetilde{M}_{1;2;3;4} \pm \sigma_{1;2;3;4}$	$\mathrm{Im}\{j_{11}\}$	$\mathrm{Im}\{j_{12}\}$	$\mathrm{Im}\{j_{21}\}$	$\mathrm{Im}\{j_{22}\}$
$\widetilde{M}_1 \pm \sigma_1$	0.71 ± 0.039	0.56 ± 0.031	0.52 ± 0.028	0.69 ± 0.027
$M_2 \pm \sigma_2$	0.39 ± 0.021	0.44 ± 0.024	0.41 ± 0.022	0.41 ± 0.023
$M_3 \pm \sigma_3$	0.27 ± 0.017	0.18 ± 0.011	0.14 ± 0.008	0.22 ± 0.012
$M_4 \pm \sigma_4$	0.33 ± 0.018	0.26 ± 0.014	0.21 ± 0.011	0.29 ± 0.016
		$\varphi_k = \pi/4$		
$\widetilde{M}_{1;2;3;4} \pm \sigma_{1;2;3;4}$	$\mathrm{Im}\{j_{11}\}$	$\mathrm{Im}\{j_{12}\}$	$\mathrm{Im}\{j_{21}\}$	$\mathrm{Im}\{j_{22}\}$
$\widetilde{M}_1 \pm \sigma_1$	0.41 ± 0.022	0.47 ± 0.026	0.52 ± 0.028	0.49 ± 0.027
$M_2 \pm \sigma_2$	0.26 ± 0.014	0.29 ± 0.017	0.31 ± 0.016	0.25 ± 0.014
$M_3 \pm \sigma_3$	0.39 ± 0.021	0.45 ± 0.024	0.47 ± 0.028	0.41 ± 0.022
$M_4 \pm \sigma_4$	0.58 ± 0.031	0.64 ± 0.034	0.69 ± 0.036	0.63 ± 0.033
		$\varphi_k = \pi/8$		
$\widetilde{M}_{1;2;3;4} \pm \sigma_{1;2;3;4}$	$\mathrm{Im}\{j_{11}\}$	$\mathrm{Im}\{j_{12}\}$	$\mathrm{Im}\{j_{21}\}$	$\mathrm{Im}\{j_{22}\}$
$\widetilde{M}_1 \pm \sigma_1$	0.31 ± 0.017	0.36 ± 0.019	0.41 ± 0.023	0.39 ± 0.021
$M_2 \pm \sigma_2$	0.22 ± 0.013	0.24 ± 0.014	0.26 ± 0.015	0.25 ± 0.013
$M_3 \pm \sigma_3$	0.88 ± 0.045	0.49 ± 0.026	0.62 ± 0.033	0.74 ± 0.039
$M_4 \pm \sigma_4$	1.36 ± 0.077	0.85 ± 0.041	0.89 ± 0.047	1.18 ± 0.061

With a decrease in the phase scanning shift $\varphi_k \downarrow$ of the algorithmically reconstructed complex amplitude field, a more optically thin volume of the dehydrated synovial fluid film is distinguished. As a result, two scenarios of changes in the polycrystalline architectonics are realized.

First, the spatial–angular structure of the distribution of the optical axis orientations of the needle-shaped crystals narrows significantly – from $(0 \leq \gamma \leq 2\pi)$ to $(\gamma_{\min} \leq \gamma \leq \gamma_{\max})$ – and the structural anisotropy formed within the volume of the synovial fluid films decreases $(\mathrm{LB}_{0;90};\ \mathrm{LB}_{45;135} \downarrow)$.

Second, the concentration of optically active polysaccharides (hyaluronans) decreases $(\mathrm{CB}_{\otimes;\oplus} \downarrow)$.

According to these optical–physical transformations, the statistical structure of the real and imaginary components of the Jones matrix images of the optically anisotropic architectonics of supramolecular networks in the synovial fluid films changes as follows:

- $Re\left(j_{11;22}\left(\mathrm{LB}^2 + \mathrm{CB}^2_{\otimes;\oplus}\right)\right)$ and $Re\left(j_{12;21}\left(\mathrm{CB}^2_{\otimes;\oplus}\right)\right) - \begin{pmatrix} \widetilde{M}_{i=1;2} \downarrow; \\ \widetilde{M}_{i=3;4} \uparrow \end{pmatrix}$;
- $\mathrm{Im}\left(j_{11;22}\left(\mathrm{LB}_{0;90}\right)\right)$ and $\mathrm{Im}\left(j_{12;21}\left(\mathrm{LB}_{45;135}\right)\right) - \begin{pmatrix} \widetilde{M}_{i=1;2} \downarrow; \\ \widetilde{M}_{i=3;4} \uparrow \end{pmatrix}$.

Due to the predominant influence of the structural anisotropy of the dendritic networks formed by sodium urate, calcium pyrophosphate dihydrate, hydroxyapatite, and cholesterol crystals, the statistical moments $\widetilde{M}_{i=3;4}\left(\mathrm{Im}\left(j_{ik}\left(\mathrm{LB}_{0;90;45;135}\right)\right)\right)$ were found to be the most quantitatively sensitive to changes in the polycrystalline architectonics of the synovial fluid films.

12.2.4 Biomedical Application

This section presents an example of a study evaluating the effectiveness of the layer-by-layer Jones matrix mapping method for the differential diagnosis of polycrystalline film samples of synovial fluid from the knee joints of injured patients with inflammatory arthritis (Group 1, 19 samples) and septic arthrosis (Group 2, 19 samples) [40–45].

The obtained results are presented in Table 12.7.

Table 12.7 Statistical parameters of the distributions of the real and imaginary parts of the matrix elements j_{11} and j_{12} of polycrystalline synovial fluid films.

	$Re\{j_{11}(\pi/8)\}$			$Re\{j_{12}(\pi/8)\}$		
M_i	Group 1	Group 2	Ac, %	Group 1	Group 2	Ac, %
$\widetilde{M}_3 \pm \sigma_3$	0.68 ± 0.037	0.83 ± 0.043	86.8	0.53 ± 0.031	0.64 ± 0.037	73.7
p	$p < 0.05$			$p < 0.05$		
$M_4 \pm \sigma_4$	0.96 ± 0.054	1.27 ± 0.069	89.5	0.77 ± 0.044	0.94 ± 0.049	73.7
p	$p < 0.05$			$p < 0.05$		
	$\mathrm{Im}\{j_{11}(\pi/8)\}$			$\mathrm{Im}\{j_{12}(\pi/8)\}$		
M_i	Group 1	Group 2	Ac, %	Group 1	Group 2	Ac, %
$\widetilde{M}_3 \pm \sigma_3$	0.88 ± 0.045	1.27 ± 0.065	**92.1**	0.49 ± 0.026	0.81 ± 0.042	**97.4**
p	$p < 0.001$			$p < 0.001$		
$M_4 \pm \sigma_4$	1.36 ± 0.073	1.98 ± 0.11	**92.1**	0.85 ± 0.047	1.38 ± 0.079	**97.4**
p	$p < 0.001$			$p < 0.001$		

The following maximum levels of diagnostic accuracy for the differential diagnosis of the severity of traumatic knee joint injury were established through statistical analysis of the real and imaginary Jones matrix images of polycrystalline synovial fluid films in the phase plane $\varphi_k = \pi/8$:

- distributions $Re\left(j_{11}\left(\mathrm{LB}^2 + \mathrm{CB}^2{}_{\otimes;\oplus}\right)\right)$ – diagnostic accuracy Ac = 86.8%.
- distributions $Re\left(j_{12}\left(\mathrm{CB}^2{}_{\otimes;\oplus}\right)\right)$ – maximum diagnostic accuracy Ac = 73.7%.
- distributions $\mathrm{Im}\left(j_{11}\left(\mathrm{LB}_{0;90}\right)\right)$ – Ac = 92.1%.
- distributions $\mathrm{Im}\left(j_{12}\left(\mathrm{LB}_{45;135}\right)\right)$ – Ac = 97.4%.

From a physical point of view, the obtained results can be explained by the fact that in the case of a septic inflammatory condition, the crystallization process of the synovial fluid is accompanied by the destruction of dendritic birefringent ($\mathrm{LB}_{0;90;45;135} \downarrow$) supramolecular networks, with only a slight change in the concentration of polysaccharides ($\mathrm{CB}_{\otimes;\oplus} \approx$ const.). As a result, the mean and variance values ($\widetilde{M}_{1;2} \downarrow$) characterizing the distributions of $Re\left(j_{11}\left(\mathrm{LB}^2 + \mathrm{CB}^2{}_{\otimes;\oplus}\right)\right)$ and $\mathrm{Im}\left(j_{11}\left(\mathrm{LB}_{0;90}\right)\right)$; $\mathrm{Im}\left(j_{12}\left(\mathrm{LB}_{45;135}\right)\right)$ of polycrystalline synovial fluid films in patients with septic arthritis decrease. Conversely, the values of the most sensitive statistical markers of optical anisotropy parameters $\widetilde{M}_{3;4} \uparrow$ (Tables 12.5 and 12.6) significantly increase. As a result, within both groups of polycrystalline synovial fluid film samples, the maximum differences are observed in the skewness and kurtosis values of the distributions $Re(j_{11})$ and $\mathrm{Im}(j_{11;12})$. Consequently, the highest level (97.4%) of balanced accuracy in the differential diagnosis of the severity of traumatic knee joint pathology is achieved.

Changes in the spherulitic component $\mathrm{CB}_{\otimes;\oplus} \downarrow$ during the crystallization of the polycrystalline supramolecular network of synovial fluid films in patients with various knee joint pathologies were found to be insignificant. This is evidenced by the considerably smaller difference between the statistical markers $\widetilde{M}_{3;4}\left(Re\left\{J_{12}\left(\mathrm{CB}_{\otimes;\oplus}, \pi/8\right)\right\}\right)$ and the relatively low diagnostic accuracy (73.7%) of the differential analysis.

12.2.5 Conclusions

1. A set of analytical relationships has been established between the phase anisotropy parameters (linear and circular birefringence) of the supramolecular networks of the synovial fluid films and the real and imaginary components of the Jones matrix elements.
2. The results of the experimental testing of the polarization–interference mapping method with digital Fourier reconstruction of the complex amplitude distributions of object fields, along with the algorithmic calculation of the coordinate distributions of the real and imaginary components of the Jones matrix elements of dehydrated synovial fluid films across a series of phase planes, are presented and physically analyzed.

3. The scenarios of transformation in the statistical structure of the Jones matrix images within the phase planes of the algorithmically reconstructed complex amplitude field of polycrystalline synovial fluid films – $\widetilde{M}_{i=1;2}\left\{\varphi_k \downarrow, \begin{pmatrix} Re(j_{ik}) \\ \mathrm{Im}(j_{ik}) \end{pmatrix}\right\} \downarrow$ and $\widetilde{M}_{i=3;4}\left\{\varphi_k \downarrow, \begin{pmatrix} Re(j_{ik}) \\ \mathrm{Im}(j_{ik}) \end{pmatrix}\right\} \uparrow$ – have been established and physically substantiated.
4. The set of diagnostic markers most sensitive to changes in the orientational–phase structure of polycrystalline networks in synovial fluid films has been determined – the third- and fourth -order statistical moments characterizing the skewness and kurtosis of the coordinate distributions of the real and imaginary components of the Jones matrix $\widetilde{M}_{i=3;4}\left\{\varphi_k, \begin{pmatrix} \mathrm{Im}(j_{11;22}(\mathrm{LB}_{0;90})) \\ \mathrm{Im}(j_{12;21}(\mathrm{LB}_{45;135})) \end{pmatrix}\right\}$.
5. A high diagnostic accuracy $\left(\mathrm{Ac}\left(Re\left(J_{12}\left(\mathrm{CB}^2{}_{\otimes;\oplus}\right)\right) = 97.4\%\right)\right)$ of the layer-by-layer Jones matrix mapping method for polycrystalline synovial fluid films was demonstrated in the differential diagnosis of the severity of traumatic knee joint pathology.

12.3 3D Jones Matrix Mapping of Biological Fluid Films

12.3.1 Synovial Fluid Films Maps of Phase Anisotropy

We consider two types of films BL:

- knee joint synovial fluid;
- cystic bile.

The choice of such objects is connected with both fundamental and applied aspects.

Fundamental – the BL data biochemical composition is different. Therefore, in the dehydration and molecular self-assembly process, BL films optical anisotropy polycrystalline networks various architectonics are formed. The obtained data on the polarization manifestations of the optical properties of such networks can be used to determine the totality of physical and quantitative relationships between the optical anisotropy maps topology and the spatial-angular structure specifics of biological crystals ensembles.

Applied – quantitative parameters (in our work we used a set of first to fourth-order statistical moments), which characterize the Jones matrix, maps can be used as differential diagnosis markers of human organs various pathological conditions.

12.3.1.1 Synovial Fluid Films

Synovial fluid consists of two main components – liquid and protein-polysaccharide.

The liquid is essentially blood plasma, which is a heterogeneous complex-structured liquid, a heterogeneous colloidal-polymer solution. The molecular components (monomers of albumin, globulin, fibrinogen) in the process of dehydration form a dendritic supramolecular optically anisotropic (dimers, trimers, monomers) protein structures (100 – 2000 nm) polycrystalline network of needle-like crystals, which has linear birefringence (LB) and linear dichroism (LD).

Another component providing viscoelastic properties of synovial fluid is hyaluronan, a polysaccharide, which is a long linear or branched chain (structural anisotropy "LB – LD") of monosaccharide optically active molecular complexes (circular birefringence [CB] – circular dichroism [CD]").

Thus, the synovial fluid films possess a predominantly dendritic polycrystalline network with structural anisotropy ("LB – LD") and insignificant optical activity ("CB – CD"). In the future, such a network will be called dendritic.

12.3.1.2 Bile Films

The main component of bile is bile acids, which are in the form of conjugates (compounds) with glycine and taurine and other so-called paired acids.

In addition, bile contains proteins (immunoglobulins), cholesterol, bilirubin, metals (copper, zinc, lead, indium, magnesium, mercury and others).

In reality, the presence of partial crystals of fatty acids, cholesterol monohydrate, forms a dendritic network of structural anisotropy – "LB – LD".

Table 12.8 Optical parameters of polycrystalline biological films.

Parameters	Synovial Liquid
Attenuation coefficient τ, cm^{-1}	0.28 ± 0.016
Depolarization degree Λ, %	16 ± 0.13
Parameters	**Bile**
Attenuation coefficient τ, cm^{-1}	0.37 ± 0.023
Depolarization degree Λ, %	24 ± 0.15

Another, but less developed, architectonics of BF is formed by optically active crystals – calcium bilirubinate and calcium carbonate microspherulites ("CB – CD").

So, the birefringent architectonics polycrystalline component has comparable manifestations of structural anisotropy and optical activity. In this sense, we will call the network of such crystals – dendritic-spherulite.

BL films were obtained by applying a drop of liquid to a glass substrate at room temperature ($t = 22\,°C$) and drying for 24 hours.

The optical-geometric parameters of films BL are presented in Table 12.8.

Figure 12.15 shows a set linear LB(φ_k, $m \times n$) and circular CB(φ_k, $m \times n$) birefringence integral ($m \times n$) and layer-by-layer (φ_k, $m \times n$) maps and histograms G(LB, CB).

Analysis of the obtained data revealed:

- Individual structure of LB($\varphi_k, m \times n$) and CB($\varphi_k, m \times n$) maps for FSL dendritic polycrystalline network (Figure 12.15a, b, e, f, i, j).
- Asymmetric structure of distributions G(LB, CB) for FSL histograms (Figure 12.15c, d, g, h, k, l).
- Transformation of the magnitude and position of histogram G(LB, CB) extremes during phase scanning φ_k (Figure 12.15g, h, k, l).

Statistical analysis of maps LB(m, n) and CB(m, n) revealed:

- Individual and different from normal $\left(M_{3;4}(\text{LB, CB}) = 0\right)$ statistics of LB(m, n) and CB(m, n) maps, – all central statistical moments $M_{i=1;2;3;4}$(LB, CB) $\neq 0$.
- Dependence of distributions $M_{i=1;2;3;4}$(LB, CB) $\leftrightarrow f(\varphi_k)$ from the phase scanning depth.
- Statistical parameters (markers) are most sensitive to changes in phase anisotropy maps – statistical moments of the third and fourth orders $M_{i=3;4}$(LB, CB(m, n)), which characterize the FSL dendritic network LB and CB distributions skewness and excess.
- As the value of the phase scanning parameter $\varphi_k \uparrow$ increases, the values of $M_{i=3;4}$(LB) increase up to two times, and $M_{i=3;4}$(CB) increase by 50%.

Quantitatively, the dendritic polycrystalline FSL network LB($m \times n$) and CB($m \times n$) maps phase transformation scenarios are shown in Table 12.9.

From a physical point of view, the results obtained can be linked to the following features of the architectonics of dendritic polycrystalline FSL networks.

First, in previous classical vector-parametric studies of biological layers optical anisotropy, it was demonstrated that the LB magnitude is mainly determined by the needle crystals optical axes directions spatial orientation γ degree – in our case, the FSL dendritic network. The variations of this parameter in the films plane are very large and almost equidistant – $0 \leq \gamma \leq 2\pi$. As a result, the first and second orders statistical moments values, which characterize the mean and variance of the LB(m, n) distributions, are quite large. The higher-order statistical moments $M_{i=3;4}$(LB(m, n)) values, on the contrary, are insignificant.

Secondly, the FSL needle crystal network circular birefringence CB value, which characterizes polarization states mutual transformations due to molecular domains optical activity of, is 2–3 times less than the LB value. This circumstance indicates a significant advantage of the needle crystal networks structural anisotropy over the chiral molecular FSL domains optical activity. Accordingly, statistical moments $M_{i=1;2}$(CB) > $M_{i=1;2}$ (LB) and vice versa $M_{i=3;4}$(CB) < $M_{i=3;4}$ (LB).

Third, with a decrease in the phase shift of the scan $\varphi_k \downarrow$, a more optically thin volume of FSL is released as a result of the needle crystals optical axes directions distribution is transformed into an asymmetric one ($\gamma_{\min} \leq \gamma \leq \gamma_{\max}$) with a certain their random value dispersion. In addition, the range of changes in phase shifts random values between laser radiation orthogonal amplitudes components is narrowed. This changes the statistics of Jones matrix reproduced maps LB(m, n) and CB(m, n), $-\begin{pmatrix} M_{1;2}(\text{LB, CB}) \downarrow; \\ M_{3;4}(\text{LB, CB}) \uparrow \end{pmatrix}$.

12.3.2 FSL Maps of Amplitude Anisotropy

Figure 12.16 illustrates integral and layer-by-layer$\varphi_k, m, n)$ maps and histograms $G(\text{LD, CD})$ of linear LD(φ_k, m, n) and circular CD(φ_k, m, n) dichroism.

Statistical scenarios for the transformation of LD(m, n) and CD(m, n) maps illustrate the data, which are shown in Table 12.10.

Obtained data analysis revealed the presence (as in the case of FSL phase introscopy, Figure 12.15) individual structure for Jones matrix reproduced maps LD(φ_k, m, n) and CD(φ_k, m, n).

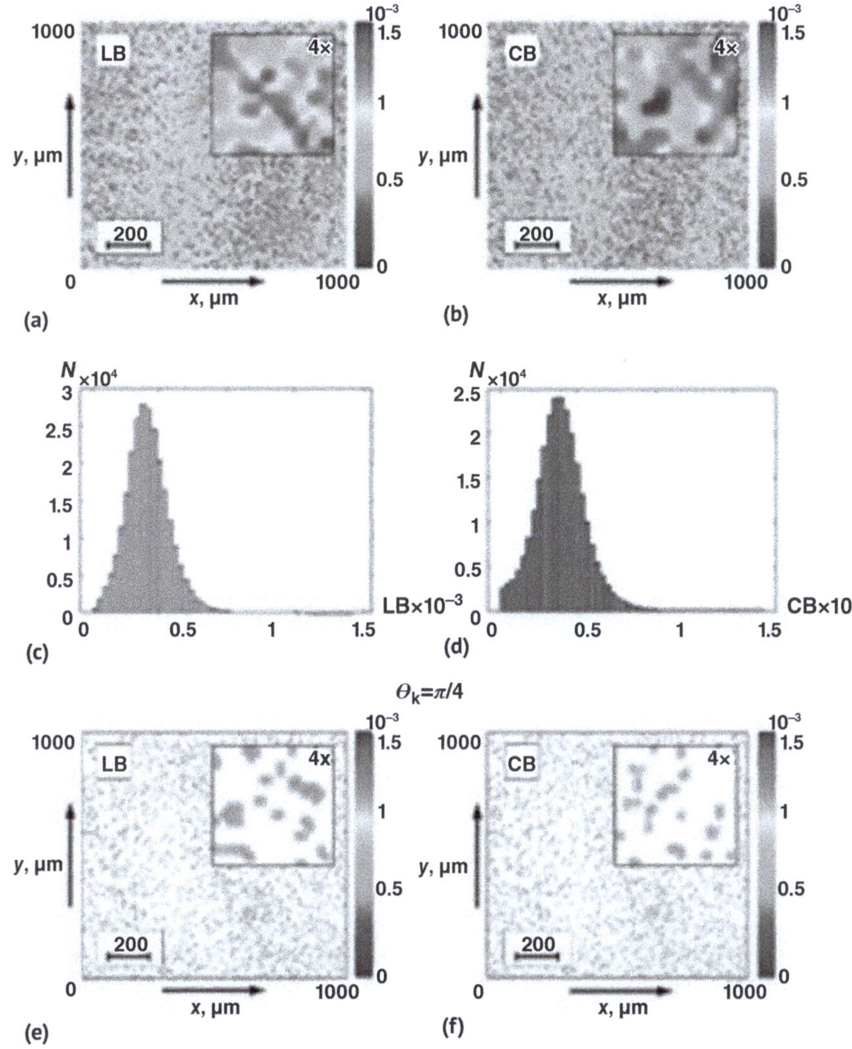

Figure 12.15 Integral (a), (b) and layer-by-layer ($\varphi = \pi/4$, e, f); $\varphi = \pi/8$, (i), (j) maps and histograms of FSL dendritic network. Linear LB (c), (g), (k) and circular CB (d), (h), (l) birefringence.

(Continued)

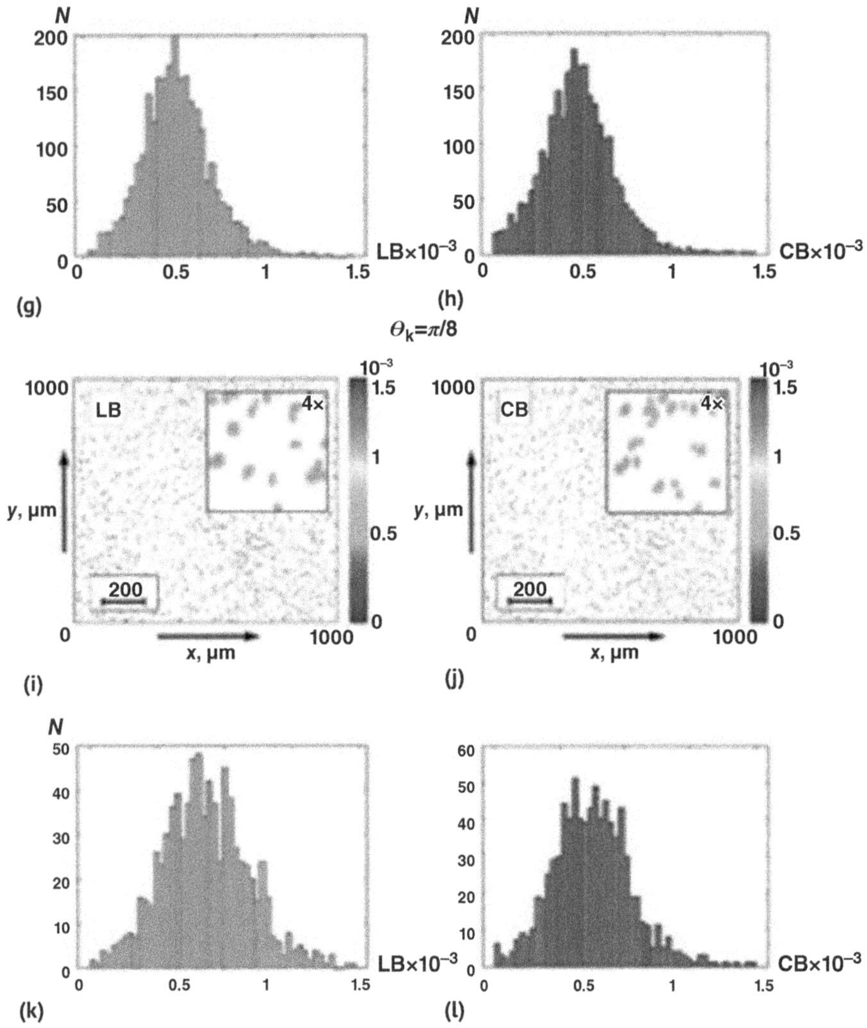

Figure 12.15 **(Cont'd)**

Table 12.9 The first and fourth-order statistical moments the that characterize FSL maps.

M_i	LB	LB($\varphi = \pi/4$)	LB($\varphi = \pi/8$)
$M_1 \times 10^{-3}$	0.93 ± 0.047	0.72 ± 0.038	0.67 ± 0.036
$M_2 \times 10^{-3}$	0.25 ± 0.013	0.16 ± 0.009	0.12 ± 0.007
M_3	0.29 ± 0.016	0.42 ± 0.023	0.53 ± 0.028
M_4	0.41 ± 0.022	0.69 ± 0.037	0.78 ± 0.41
M_i	**CB**	**CB($\varphi = \pi/4$)**	**CB($\varphi = \pi/8$)**
$M_1 \times 10^{-3}$	0.52 ± 0.028	0.39 ± 0.021	0.31 ± 0.016
$M_2 \times 10^{-3}$	0.11 ± 0.006	0.07 ± 0.003	0.05 ± 0.003
M_3	0.69 ± 0.036	0.86 ± 0.044	0.93 ± 0.048
M_4	0.88 ± 0.045	1.11 ± 0.057	1.32 ± 0.073

Depending on the scan phase shift φ_k magnitude take place the magnitude and histograms G(LD, CD) extremes position (Figure 12.16g, h, k, l).

Detected:

- Difference the of all central statistical moments value from zero $\left(M_{i=1;2;3;4}(\text{LD, CD}) \neq 0\right)$.
- Phase (φ_k) dependence of LD(m, n) and CD(m, n) maps statistical distributions $M_{i=1;2;3;4}(\text{LD, CD}) \leftrightarrow q(\varphi_k)$ from the depth phase scanning.
- The third- to fourth-orders statistical moments $M_{i=3;4}(\text{LD, CD}(m, n))$, which characterize the skewness and excess of the LD and CD distributions, were the most sensitive to changes of amplitude anisotropy.
- As the phase scanning parameter $\varphi_k \uparrow$ value increases, the statistical parameters $M_{i=3;4}(\text{LD}(m, n))$ values increase up to two times, and $M_{i=3;4}(\text{CD}(m, n))$ increase by 80%.

From a physical point of view, the obtained results can be related to the fact that the value of LD, as well as LB FSL, is determined by the needle crystals optical axes directions γ spatial orientation degree. Due to the wide range of such axis's orientations, the first- and second-order statistical moments values, which characterize the coordinate distributions LD(m, n) average and variance are quite large. At the same time, the dendritic network of FSL needle-like crystals shows a

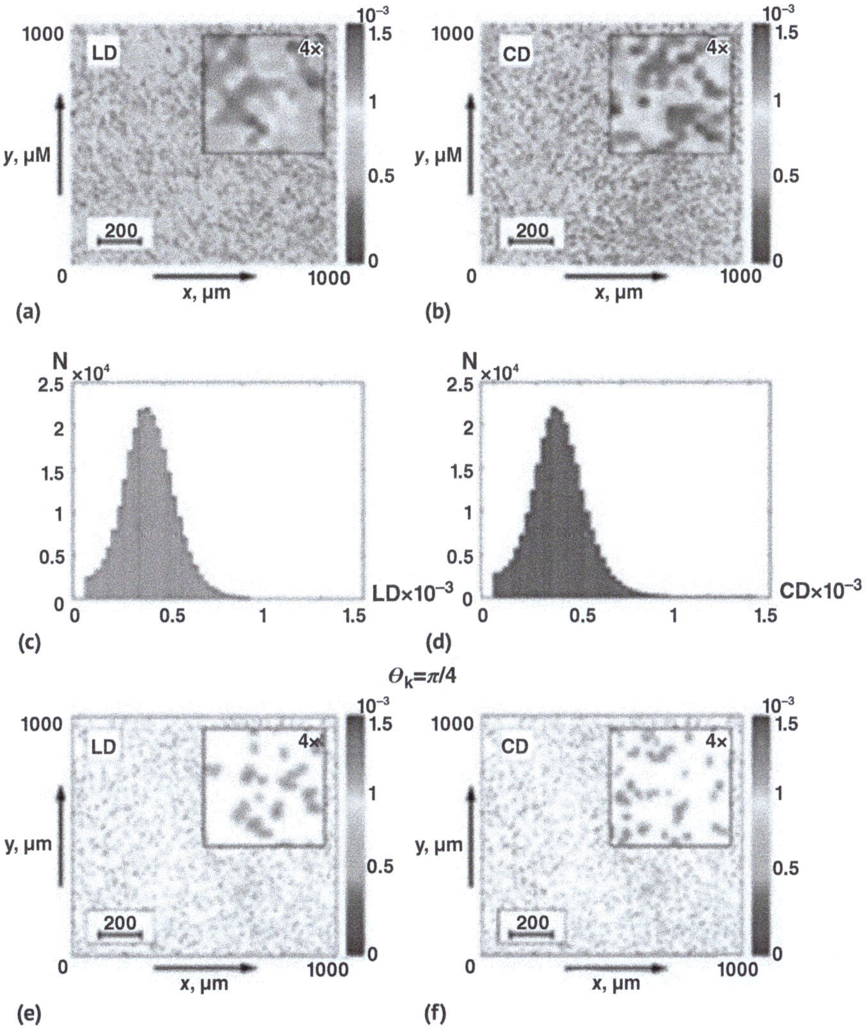

Figure 12.16 Integral (a), (b) and layer-by-layer ($\varphi = \pi/4$, (e), (f); $\varphi = \pi/8$, (i), (j)) maps and histograms of linear LD (c), (g), (k) and circular CD (d), (h), (l) dichroism.

(Continued)

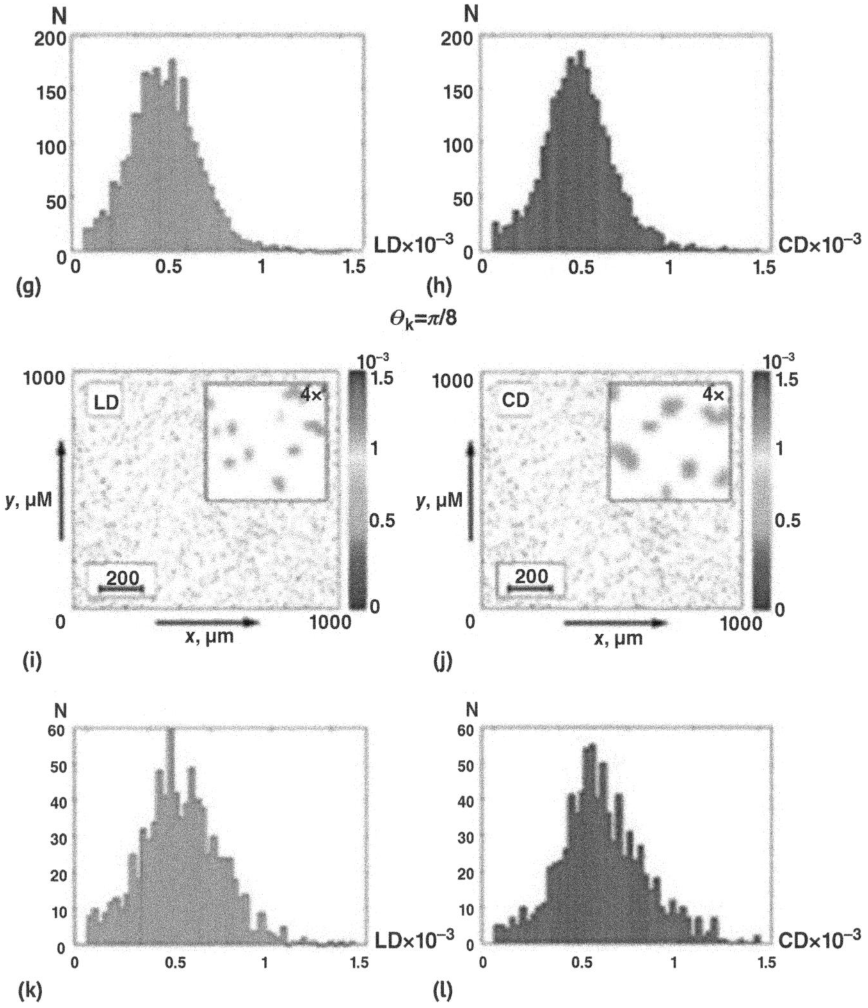

Figure 12.16 (Cont'd)

Table 12.10 The first to fourth-order statistical moments the that characterize the FSL maps of dichroism.

M_i	LD	LD($\varphi = \pi/4$)	LD($\varphi = \pi/8$)
$M_1 \times 10^{-3}$	0.83 ± 0.043	0.67 ± 0.035	0.56 ± 0.029
$M_2 \times 10^{-3}$	0.18 ± 0.009	0.14 ± 0.008	0.11 ± 0.006
M_3	0.34 ± 0.018	0.53 ± 0.028	0.67 ± 0.037
M_4	0.52 ± 0.027	0.74 ± 0.039	0.98 ± 0.046
M_i	CD	CD($\varphi = \pi/4$)	CD($\varphi = \pi/8$)
$M_1 \times 10^{-3}$	0.35 ± 0.019	0.23 ± 0.012	0.18 ± 0.009
$M_2 \times 10^{-3}$	0.105 ± 0.006	0.06 ± 0.003	0.04 ± 0.002
M_3	0.73 ± 0.037	0.96 ± 0.054	1.21 ± 0.068
M_4	0.96 ± 0.051	1.45 ± 0.078	1.69 ± 0.089

CD value that is 2–3 times less than the LD value. Accordingly, statistical moments $M_{i=1;2}(\text{CD}) > M_{i=1;2}(\text{LD})$ and vice versa $M_{i=3;4}(\text{CD}) < M_{i=3;4}(\text{LD})$.

Phase scanning $\varphi_k \downarrow$ provides the ability to isolate a more optically thin volume of FSL. This changes the statistics needle crystals optical axes directions distribution, and also narrows the range of changes in laser radiation phase shifts between amplitude orthogonal components random values. As a result, the FSL amplitude anisotropy maps statistics are transformed, $\begin{pmatrix} M_{1;2}(\text{LD, CD}) \downarrow; \\ M_{3;4}(\text{LD, CD}) \uparrow \end{pmatrix}$.

12.3.3 BF Maps of Phase Anisotropy

Figure 12.17 shows LB and CB vector parametric and layer-by-layer maps and histograms of BF optically anisotropic networks.

Table 12.11 illustrates the first to fourth-order statistical moments that characterize the BF maps of a birefringent dendritic network.

Established:

- As in the case of the FSL needle crystals dendritic network maps studies, individual and different from the Gaussian $(M_{i=1;2;3;4}(\text{LB, CB}) \neq 0)$.

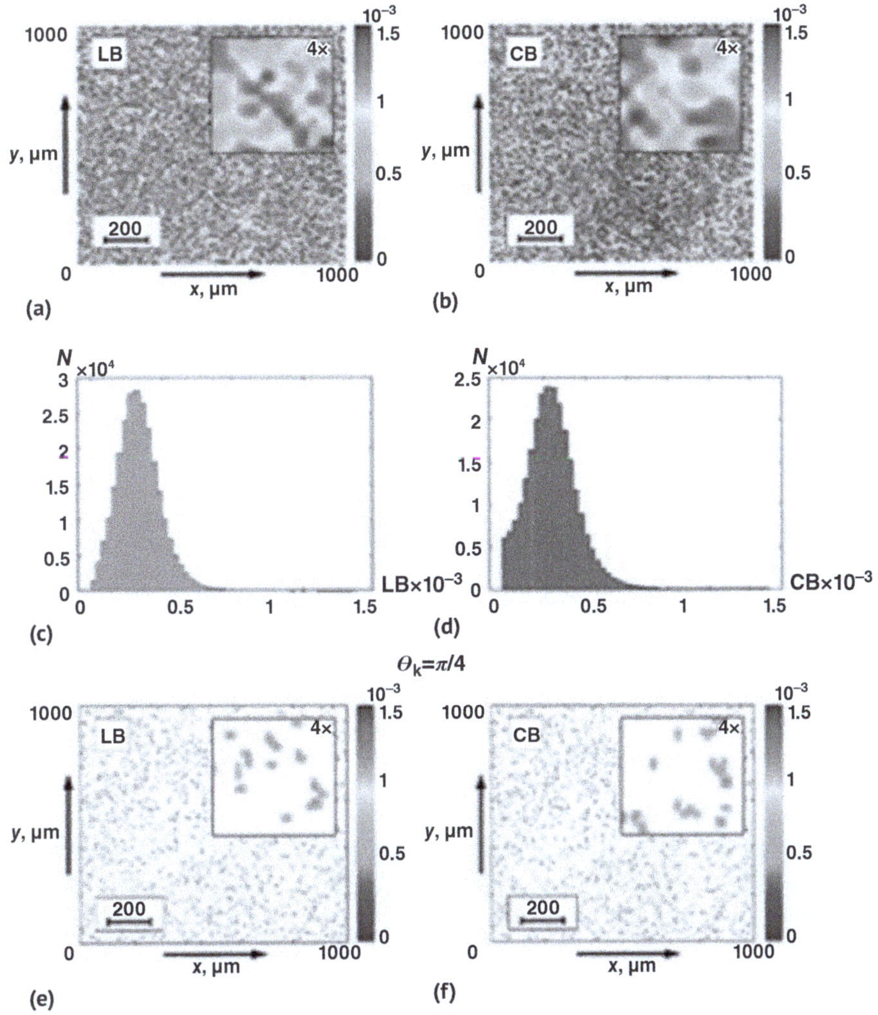

Figure 12.17 Integral (a), (b) and layer-by-layer ($\varphi = \pi/4$, (e), (f); $\varphi = \pi/8$, (i), (j) maps and histograms of BF dendritic network. Linear LB (c), (g), (k) and circular CB (d), (h), (l) birefringence.

(Continued)

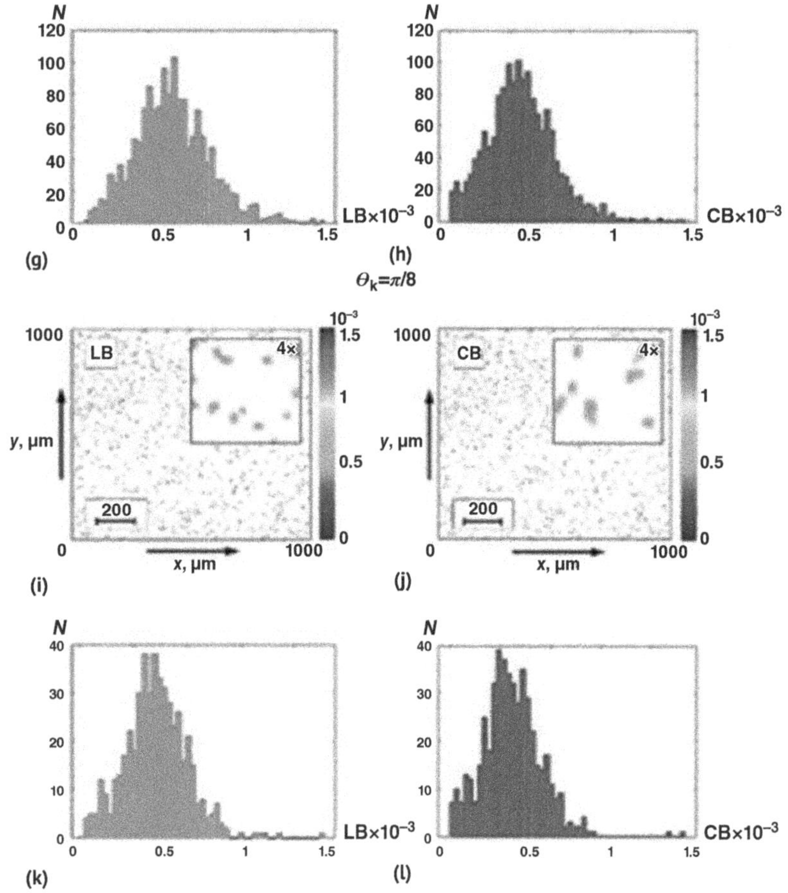

Figure 12.17 (Cont'd)

Table 12.11 The first- to fourth-order statistical moments that characterize BF.

M_i	LB	LB($\varphi = \pi/4$)	LB($\varphi = \pi/8$)
$M_1 \times 10^{-3}$	0.59 ± 0.031	0.45 ± 0.025	0.39 ± 0.021
$M_2 \times 10^{-3}$	0.12 ± 0.007	0.08 ± 0.005	0.065 ± 0.004
M_3	0.43 ± 0.023	0.61 ± 0.034	0.74 ± 0.039
M_4	0.62 ± 0.032	0.86 ± 0.048	0.94 ± 0.051
M_i	CB	CB($\varphi = \pi/4$)	CB($\varphi = \pi/8$)
$M_1 \times 10^{-3}$	0.21 ± 0.011	0.15 ± 0.008	0.12 ± 0.007
$M_2 \times 10^{-3}$	0.34 ± 0.018	0.22 ± 0.012	0.17 ± 0.009
M_3	0.43 ± 0.022	0.69 ± 0.037	0.81 ± 0.046
M_4	0.57 ± 0.032	0.86 ± 0.045	1.03 ± 0.054

- The third- and fourth-order statistical moments $M_{i=3;4}$(LB, CB(m, n)) were the most dynamic and, accordingly, sensitive to changes in the BF samples.
- As the value of the phase scanning parameter $\varphi_k \uparrow$ increases, changes of $M_{i=3;4}$(LB) reach 50–60%, and $M_{i=3;4}$(CB) increase up to two times.

The results obtained can be linked to the following physical considerations.

First, the presence of fatty acids, cholesterol monohydrate needle crystals, which form the BF dendritic optically anisotropic network component, causes the structural LB occurrence. Its value is mainly determined by the needle crystals optical axes directions γ spatial orientation degree. In the limit approximation, the orientation distributions are almost equidistant- $0 \leq \gamma \leq 2\pi$. As a result, the first and second-order statistical moments values, which characterize the mean and variance, as well as the higher orders $M_{i=3;4}(\text{LB}(m, n))$ statistical moments values which determine the skewness and excess of the polycrystalline BF network dendritic component $\text{LB}(m, n)$ distributions are quite large (Table 12.11).

Second, the circular birefringent network with different architectonics (calcium bilirubin crystals and calcium carbonate microspherolites) has quite large CB value. However, due to the spherulites lower concentration, the optical activity parameter CB values are almost two times less than the structural anisotropy parameter – the BF needle dendritic network LB. Accordingly, statistical moments $M_{i=1;2}(\text{CB}) < M_{i=1;2}(\text{LB})$ and vice versa $M_{i=3;4}(\text{CB}) > M_{i=3;4}(\text{LB})$, Table 12.11.

Third, BF volume-depth digital holographic selection consistently highlights various random implementations of optical-axis direction distributions and phase shifts. As a result, the fatty acids and cholesterol monohydrate needle crystals optical axes directions coordinate distribution is transformed into an asymmetric one ($\gamma_{\min} \leq \gamma \leq \gamma_{\max}$) with a certain random their value dispersion. In addition, the range of changes in random values of phase shifts is narrowed not only between orthogonal linearly polarized but also between right and left circularly polarized laser radiation amplitudes, polarized components – calcium bilirubin crystals and calcium carbonate microspherolites. This changes the statistics of BF Jones matrix reproduced phase anisotropy maps $\text{LB}(m, n)$ and $\text{CB}(m, n)$, are illustrated $\begin{pmatrix} M_{1;2}(\text{LB, CB}) \downarrow; \\ M_{3;4}(\text{LB, CB}) \uparrow \end{pmatrix}$.

12.3.4 BF Maps of Amplitude Anisotropy

Figure 12.18 illustrates integral (m, n) and layer-by-layer (φ_k, m, n) maps and histograms of linear $\text{LD}(\varphi_k, m, n)$ and circular $\text{CB}(\varphi_k, m, n)$ distributions.

The results of Jones matrix mapping are demonstrated:

- The polycrystalline BF linear $\text{LD}(\varphi_k, m, n)$ and circular $\text{CD}(\varphi_k, m, n)$ dichroism network are characterized by various statistical distributions (Figure 12.18a, b, e, f, i, j).
- A more asymmetric (than for $G(\text{LB, CB})$) structure of histograms $G(\text{LD, CD})$, Figure 12.18c, d, g, h, k, l.
- A clear dependence of the histograms $G(\text{LD, CD})$ (magnitude and extremes position) from the change in the scanning phase shift φ_k magnitude, is shown in Figure 12.18g, h, k, l.

Statistical scenarios for the phase transformation of maps $\text{LD}(m, n)$ and $\text{CD}(m, n)$ are illustrate the first to fourth-order statistical moments (Table 12.12).

From the analysis of statistical data revealed:

- All first to fourth-order central statistical moments values are difference from zero $M_{i=1;2;3;4}(\text{LD, CD}) \neq 0$.
- Phase dependence of the value $M_{i=1;2;3;4}(\text{LD, CD}) \leftrightarrow q(\varphi_k)$ for the BF samples.
- Statistical parameters (markers) are most sensitive to changes in the maps topographic structure of amplitude anisotropy – third and fourth-order statistical moments $M_{i=3;4}(\text{LD, CD}(m, n))$.
- As the BF polycrystalline component phase scanning parameter $\varphi_k \uparrow$ increases, the statistical parameters $M_{i=3;4}(\text{LD}(m, n))$ values increase up to 2 times, and $M_{i=3;4}(\text{CD}(m, n))$ increase by 60–80%.

From a physical point of view, the results obtained can be related to the fact that the value of LD, as well as LB, is determined by the BF fatty acids, cholesterol monohydrate needle crystals, spatial orientation, and optical axes directions γ degree.

Due to the wide range partial crystals optical axes orientations, as well as the magnitude of the phase shift between the objective laser radiation amplitude orthogonally polarized components, the first to fourth-order statistical moments values, which characterize the $\text{LD}(m, n)$ coordinate distribution, are quite large. Along with this, the compact dendritic-spherulite network (calcium bilirubin crystals and calcium carbonate microspherolites) CD value is two times less than the LD value. According to the end, statistics $M_{i=1;2}(\text{LD}) > M_{i=1;2}(\text{CD})$ and data $M_{i=3;4}(\text{CD}) < M_{i=3;4}(\text{CD})$.

12.3.5 Biomedical Application

In this section, we present brief data from the applied Jones matrix layer-by-layer method testing for biological fluids films.

For the initial applied (clinical) study of the diagnostic capabilities and efficiency of the method of layered polarization–interference Jones matrix mapping of biological fluid films, we examined several types of previously verified pathology of various human organs:

- Inflammation of the gallbladder – bile (diagnosis of cholecystitis was carried out based on general and biochemical blood tests, general urine analysis, microscopic and bacteriological examination of bile, as well as ultrasound introscopy).
- Inflammation of the knee joint – synovial fluid (the verified diagnosis was determined by tomographic, ultrasound, and biochemical laboratory methods).
- Thyroid tumors – blood plasma (the verified diagnosis was determined by analyzing blood for hormones (thyroid-stimulating hormone, triiodothyronine, thyroxine, etc.) and tumor markers, ultrasound, and radionuclide studies).

A statistical test based on the Mann–Whitney test was used for pairwise comparisons (control group 1 vs. research group 2) of optically anisotropic parameters, random values from statistically independent samples.

12.3.5.1 Differential Diagnosis of Cholelithiasis

BF were prepared from two groups of patients:

- Group 1 ($n = 57$ samples) – healthy donors (aged 30–60; 41 women, 16 men);
- Group 2 ($n = 57$ samples) patients with gallstone disease (aged 32–61; 47 women, 10 men)).

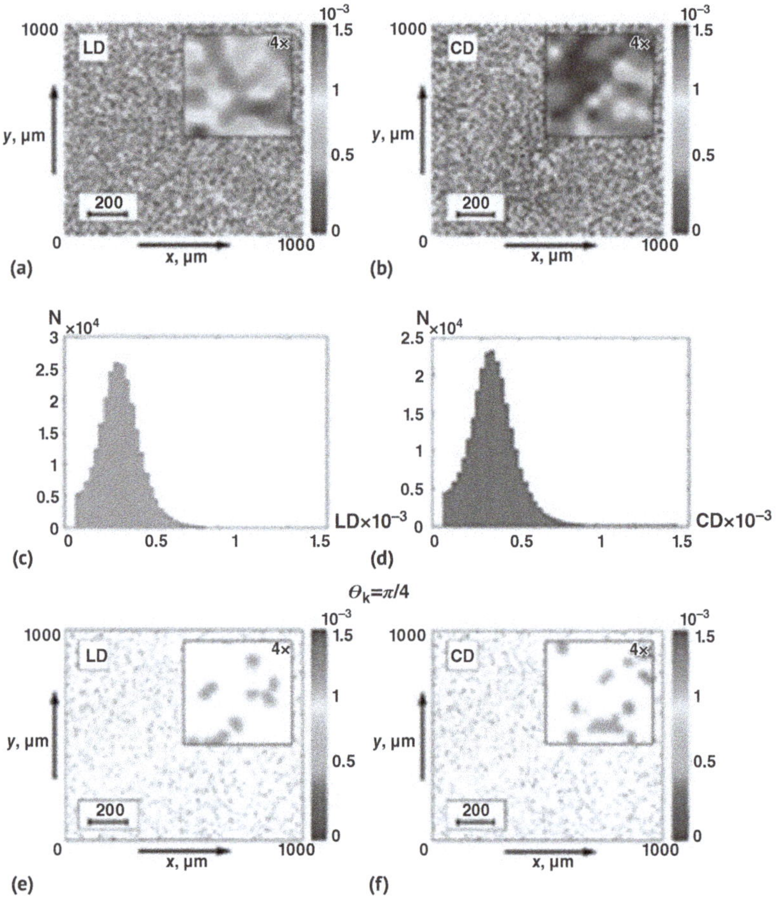

Figure 12.18 Integral (a), (b) and layer-by-layer ($\varphi = \pi/4$, (e), (f)); $\varphi = \pi/8$, (i), (j) BF maps and histograms. Linear LD (c), (g), (k) and circular CD (d), (h), (l) dichroism distributions.

(Continued)

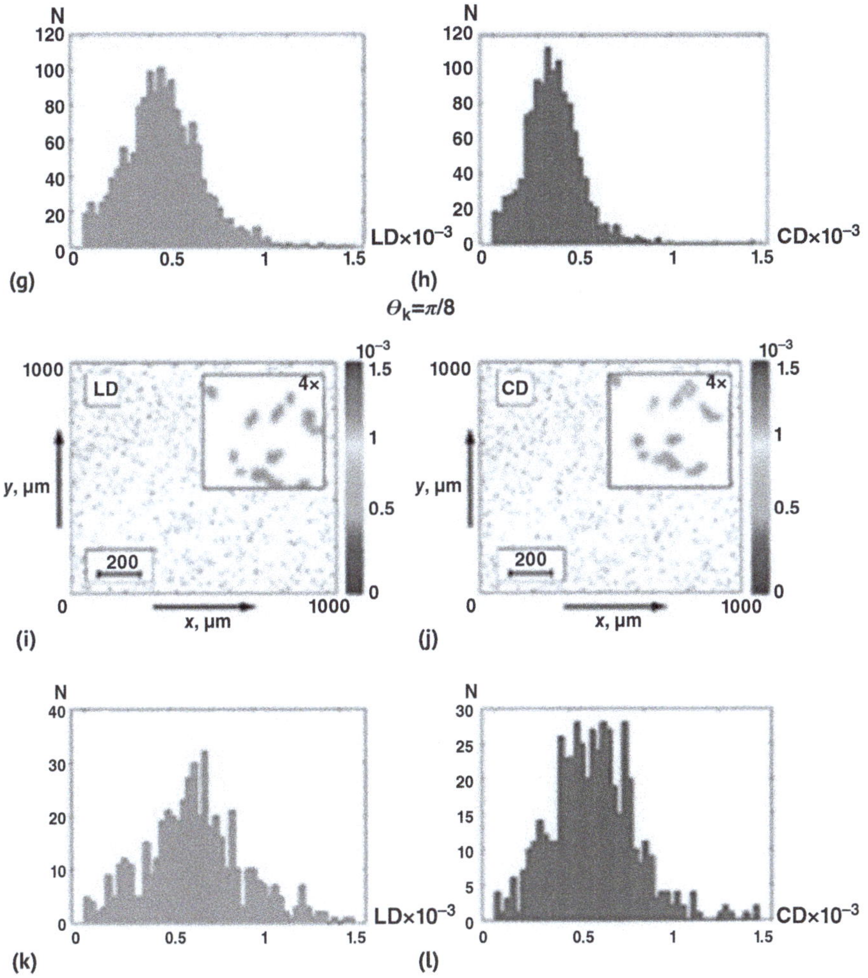

Figure 12.18 (Cont'd)

Table 12.12 The first to fourth-order statistical moments that characterize the BF maps.

M_i	LD	LD($\varphi = \pi/4$)	LD($\varphi = \pi/8$)
$M_1 \times 10^{-3}$	0.99 ± 0.054	0.84 ± 0.044	0.79 ± 0.042
$M_2 \times 10^{-3}$	0.28 ± 0.015	0.19 ± 0.011	0.16 ± 0.009
M_3	0.47 ± 0.025	0.68 ± 0.036	0.77 ± 0.041
M_4	0.65 ± 0.036	0.88 ± 0.049	0.94 ± 0.052
M_i	**CD**	**CD($\varphi = \pi/4$)**	**CD($\varphi = \pi/8$)**
$M_1 \times 10^{-3}$	0.49 ± 0.026	0.36 ± 0.019	0.33 ± 0.018
$M_2 \times 10^{-3}$	0.11 ± 0.006	0.07 ± 0.004	0.06 ± 0.003
M_3	0.19 ± 0.011	0.32 ± 0.017	0.43 ± 0.023
M_4	0.27 ± 0.015	0.48 ± 0.025	0.59 ± 0.033

The differential diagnostics accuracy results are presented in Table 12.13.
Excellent accuracy of differential diagnostics has been achieved $Ac(\varphi_k = {}^{\pi}/_{8}; (LB)) = 95.6 - 97.3\%$.

12.3.5.2 Knee Joint Pathologies Differential Diagnosis

Synovial fluid films from operated patients two groups were prepared:

- Group 1 ($n = 51$ samples) patients with reactive synovitis (aged 41–64; 20 women, 31 men);
- Group 2 ($n = 51$ samples) patients with reactive synovitis (aged 39–67; 27 women, 24 men).

The differential diagnostic accuracy results are presented in Table 12.14.
Very good accuracy of differential diagnostics has been achieved $Ac(\varphi = \pi/8; Z_3; Z_4(LB)) = 94.1\%$.

12.3.5.3 Differential Diagnosis of Benign and Malignant Tumors of the Thyroid Gland

Two groups whole blood films (WBF) samples were formed:

- Group 1 ($n = 53$ samples) patients with follicular adenoma (aged 27–63; 41 women, 22 men);
- Group 2 ($n = 53$ samples) patients with papillary thyroid carcinoma (aged 32–66; 37 women, 26 men).

The differential diagnostics accuracy results are presented in Table 12.15.
Excellent accuracy of differential diagnostics has been achieved $Ac(\varphi = \pi/8; Z_4(LB)) = 95.3\%$.
Thus, we have demonstrated the strong prospects for applying our method in the differential diagnosis of inflammatory and pathological conditions of various human organs.

Table 12.13 Balanced accuracy of differential diagnosis of cholelithiasis.

JT	Ac(LB), %	Ac(CB), %	Ac(LD), %	Ac(CD), %
M_3	86	79.8	82.5	71.1
M_4	88.6	79.8	82.5	71.1
		$\varphi = \pi/4$		
M_3	92.1	83.3	88.6	78.1
M_4	93	83.3	88.6	78.1
		$\varphi = \pi/8$		
M_3	95.6	88.1	92.1	81.6
M_4	97.3	88.1	92.1	82.5

Table 12.14 Differential diagnosis balanced accuracy of knee joint pathology.

JT	Ac(LB), %	Ac(CB), %	Ac(LD), %	Ac(CD), %
M_3	70.5	69.6	65.6	62.7
M_4	71.5	70.5	65.6	63.7
		$\varphi = \pi/4$		
M_3	79.4	78.4	74.5	70.5
M_4	80.4	78.4	75.5	70.5
		$\varphi = \pi/8$		
M_3	94.1	91.1	84.3	81.3
M_4	94.1	91.1	85.3	81.3

Table 12.15 Differential diagnosis balanced accuracy of knee joint pathology.

JT	Ac(LB), %	Ac(CB), %	Ac(LD), %	Ac(CD), %
M_3	82.1	76.4	80.2	76.4
M_4	82.1	77.4	80.2	76.4
		$\varphi = \pi/4$		
M_3	87.7	83.4	83.4	82.1
M_4	89.6	83.4	84.9	82.1
		$\varphi = \pi/8$		
M_3	93.4	91.5	89.6	84.9
M_4	95.3	91.5	89.6	84.9

Note that these are only the first steps. The method of polarization–interference Jones matrix mapping requires broader clinical testing and the collection of a significantly larger volume of statistical data. At the same time, one of its essential advantages may be its high sensitivity (at early preclinical stages of the disease) and the possibility of conducting screening studies.

It should also be noted that mastering this method at this stage requires special training of service personnel. However, this problem can be overcome by the technical improvement of diagnostic polarization-interference equipment.

The proposed technique of polarization–interference Jones matrix mapping of polycrystalline architectonics of films of biological fluids of human organs demonstrated the first encouraging results in the differential diagnostics of several inflammatory (cholelithiasis, reactive synovitis and septic arthritis) and pathological (benign and malignant tumors of the thyroid gland) conditions of various human organs. The prospects for further development of such studies are associated with the fact that, during the dehydration of a drop of biological fluid, a scale-self-similar or multifractal architectonics of optically anisotropic supramolecular networks forms. The features of this process and its relationship with the occurrence and severity of pathology can be quantitatively characterized by the spectra of singularities of the optical anisotropy maps. As a result, one can expect to obtain new, more sensitive markers for early (preclinical) diagnostics of human organ pathologies. However, at the moment we are not aware of any publications in this area of biomedical diagnostics. Therefore, one of our nearest studies will be the development of principles for differentiation of prostate cancer stages by statistical analysis of the spectra of singularities of maps of LB and CB and dichroism of blood films.

Based on this, the next stage of our studies will be the search for and diagnostic use of new relationships between the scale-self-similar (multifractal) structure of maps of self-assembled supramolecular networks of biological fluid films and various pathological conditions of human organs.

12.3.6 Conclusions

1. The layer-by-layer matrix introscopy method – LB, CB and LD, CD coordinate distributions reproduction by polarization–interference mapping of the object fields complex amplitudes, biological fluids films is analytically substantiated and experimentally tested.
2. The dehydrated films polycrystalline component phase (birefringence – LB, CB) and amplitude (dichroism – LD, CD) anisotropy integral and layered maps (m, n) is obtained and physically analyzed:
   - synovial fluid with dendritic polycrystalline network structure;
   - bile with dendritic-spherulite polycrystalline network structure.
3. The relationships between the first- to fourth-orders central statistical moments, which characterize integral and layer-by-layer optical anisotropy maps of biological fluid films, and architectonics polycrystalline structure are determined and physically analyzed.
4. Statistical regularities of phase scanning in biological fluid films are established – $(\varphi_k \downarrow) \Longrightarrow \begin{pmatrix} M_{i=1;2}(\text{LB, CB, LD, CD}(m,n)) \downarrow; \\ M_{i=3;4}(\text{LB, CB, LD, CD}(m,n)) \uparrow \end{pmatrix}$.

5. The most sensitive markers to changes in the layer-by-layer topographic structure of phase and amplitude anisotropy maps are third- to fourth-orders statistical moments $M_{i=3;4}$(LB, CB, LD, CD($m \times n$)), which have maximum ranges of change in value with increasing phase scanning step φ_k:
   - synovial fluid films – in the range from 1.5 to 2 times;
   - BF – from 1.3 to 2 times.
6. The following differential diagnostics balanced accuracy levels have been achieved:
   - "cholelithiasis" – excellent accuracy Ac($\varphi_k = {}^{\pi}/_{8}$(LB)) = 95.2 − 97.6%;
   - "reactive synovitis – septic arthritis" is a very good accuracy Ac(M_4(LB)) = 94.4%;
   - "follicular adenoma – papillary thyroid carcinoma" is a excellent accuracy Ac($\varphi = \pi/8$; Z_4(LB)) = 97.1%.

12.4 Multifractal Analysis of Jones Matrix Images of the Dichroism of the Polycrystalline Structure of BF in the Early Differential Diagnosis of Gallbladder Disease

12.4.1 Polycrystalline Structure of BF

The main component of bile is bile acids (67%). Half of them are primary bile acids: cholic and chenodeoxycholic; the rest are secondary: deoxycholic, lithocholic, allocholic, and ursodeoxycholic acids. Bile acids in bile are present in the form of conjugates (compounds) with glycine and taurine—glycocholic, glycochenodeoxycholic, urocholic, and other so-called paired acids. Bile acids and their conjugates are sometimes referred to as "bile salts."

During the dehydration of bile, the following crystalline formations are formed:

- *Cholesterol crystals* (CC) appear as thin, colorless tetrahedral plates. Such structures form slight structural anisotropy – LB and LD. Under normal conditions, these crystalline formations are found in small quantities only in certain portions of bile. A large amount of cholesterol in bile portions indicates a change in the colloidal stability of bile and occurs in cases of cholelithiasis.
- *Microliths* (M) – dark, compact, rounded, or polyhedral formations composed of Ca salts, mucus, and cholesterol. These structures are optically active and exhibit anisotropic circular birefringence and dichroism. Microliths can be detected in cases of gallstone disease, along with cholesterol, fatty acids, and calcium bilirubinate crystals.
- *Fatty acids* (FA) appear as needle-shaped crystals, sometimes quite deep. Crystallite formations of this type are characterized by combined optical anisotropy – structural anisotropy of LB – dichroism in needle-shaped polycrystalline networks, as well as optical activity and CD of fatty acid spherulites. In pathology (cholecystitis), fatty acids occur together with cholesterol, microliths, and calcium bilirubinate.
- *Calcium bilirubinate* (CaB) appears as small grains – optically active spherulites of golden-yellow and brown color, which precipitate into a sediment visible macroscopically.

Therefore, as a result of the dehydration process of a bile drop, supramolecular optically anisotropic networks with two types of components are formed – linear (CC + FA) and circular (M + CaB + FA) birefringence and dichroism.

The magnitude of LB–LD, or the structural anisotropy of the first component, is azimuthally dependent and determined by the specific features of the angular distribution of orientations ($0 \le \rho \le \pi$) of the optical axes of the needle-shaped crystals of the supramolecular network formed during the dehydration of a bile drop.

Circular birefringence and dichroism of the second component are primarily associated with the concentration of M + CaB + FA crystalline domains.

Given the biochemical composition of bile, a predominant influence of circular birefringence and dichroism can be expected in the formed polycrystalline architectonics of the dehydrated film.

Pathological processes corresponding to the occurrence of calculous cholecystitis, followed by the formation of calculi and the development of gallstone disease, are accompanied by an increase in the overall level of crystallization and in the optical anisotropic absorption of dehydrated bile.

Two representative groups ($q = 21$) of bile samples were studied: healthy donors (control group 1) and patients with a verified diagnosis of calculous cholecystitis (study group 2).

12.4.2 Statistical Analysis of Linear and Circular Dichroism Maps of BF

Figures 12.19 and 12.20 show the integral ((a), (b)) and layer-by-layer ((e), (f)) maps, as well as the histograms $G(\text{LD}, \text{CD})$ of the distributions ((c), (d), (g), (h)) of the linear $\text{LD}(\varphi_k, m, n)$ and circular $\text{CD}(\varphi_k, m, n)$ dichroism values of BF from healthy donors ((a), (c), (e), (g)) and patients with calculous cholecystitis ((b), (d), (f), (h)).

The statistical scenarios of phase transformation for the set of $\text{LD}(m, n)$ and $\text{CD}(m, n)$ maps of polycrystalline supramolecular networks of BF illustrate the values of the statistical moments of the first to fourth orders, which are presented in Tables 12.16 and 12.17.

Analysis of the amplitude anisotropy maps of BF from healthy donors and patients with calculous cholecystitis revealed the following:

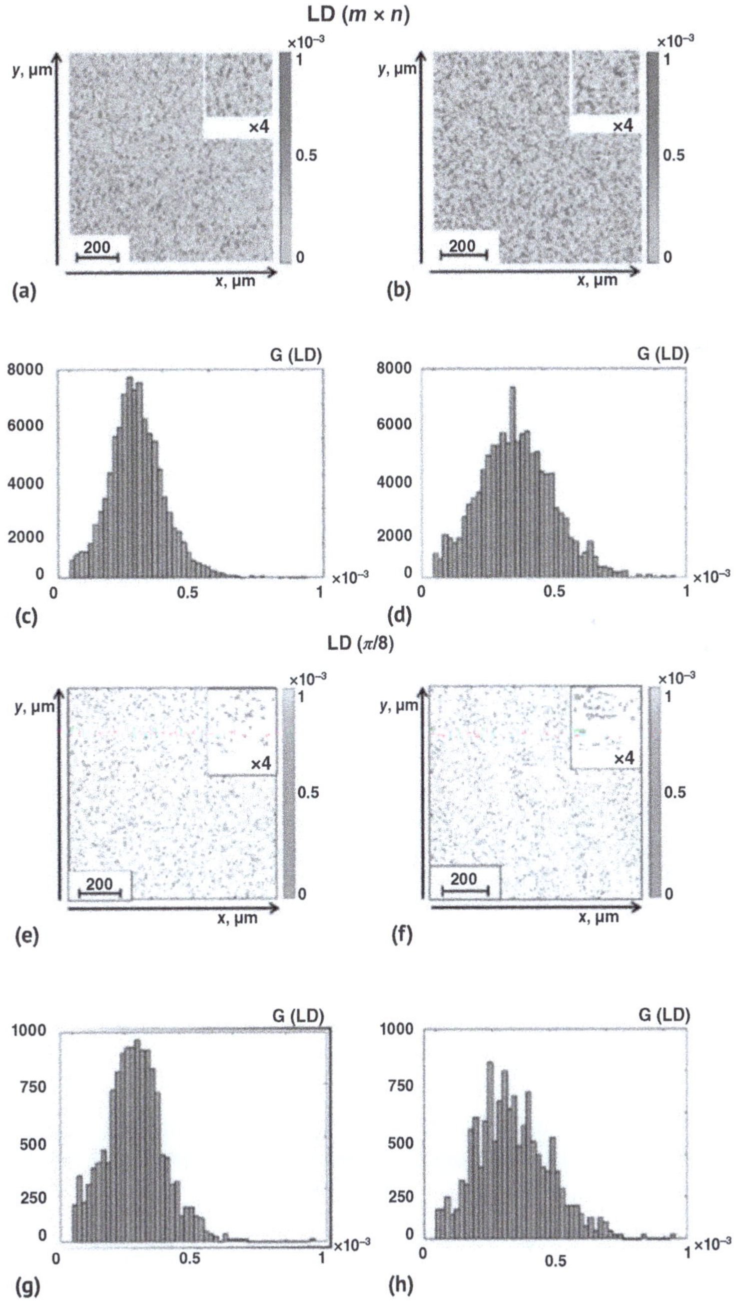

Figure 12.19 Maps and histograms of the distribution of the LD value in bile film samples from healthy individuals and patients with calculous cholecystitis.

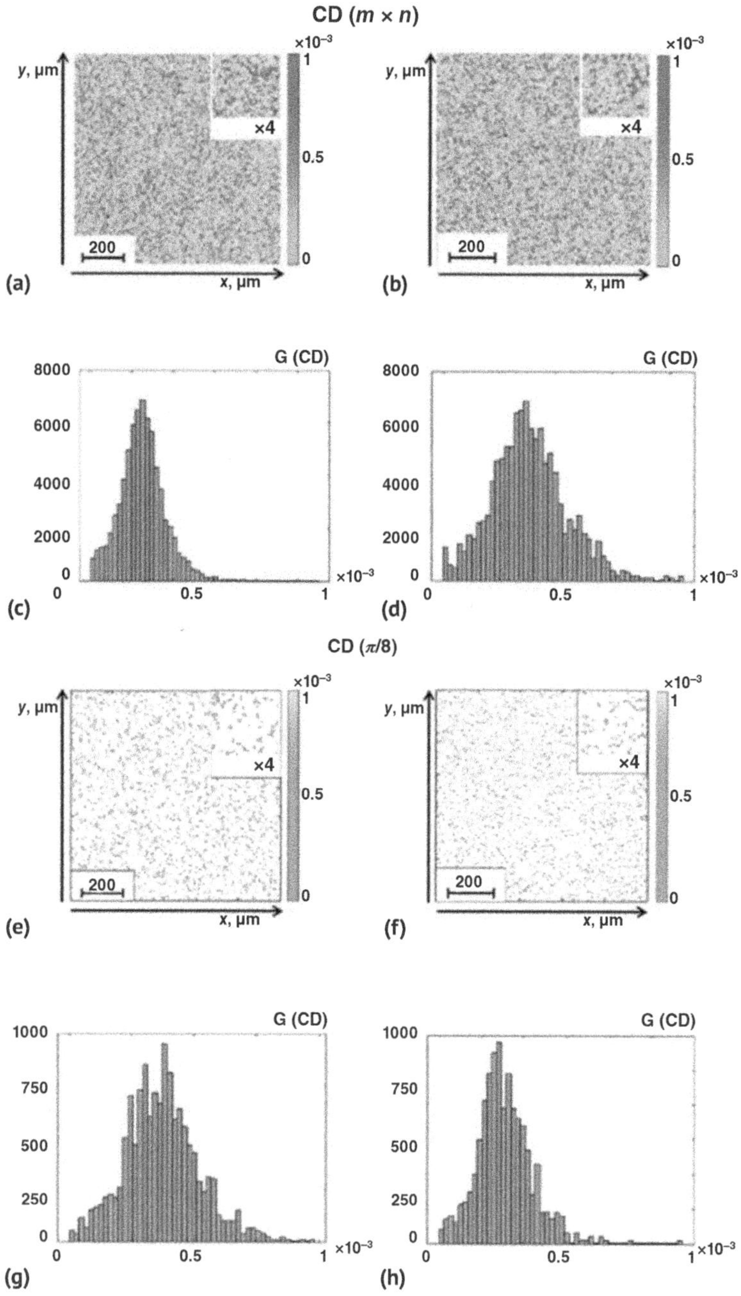

Figure 12.20 Maps and histograms of the distributions of CD in dehydrated polycrystalline bile films from healthy individuals and patients with calculous cholecystitis.

- A deviation from zero was observed for all central statistical moments of the first to fourth orders, $\widetilde{M}_{i=1;2;3;4}(\text{LD, CD}) \neq 0$. which characterize the coordinate distributions of the amplitude anisotropy parameters $\text{LD}(m, n)$ and $\text{CD}(m, n)$.
- Statistical significance was established for all calculated parameters: $\widetilde{M}_i - \widetilde{M}_{i=1;2}(\text{LD, CD}) - p < 0.05$ and $\widetilde{M}_{i=;3;4}(\text{LD, CD}) - p < 0.001$
- A phase (φ_k) dependence of the statistical distributions $\widetilde{M}_{i=1;2;3;4}(\text{LD, CD}) \leftrightarrow q(\varphi_k)$ of the dichroism maps $\text{LD}(m, n)$ and $\text{CD}(m, n)$ was observed, corresponding to the depth of phase scanning of the polycrystalline architectonics of dehydrated BF samples.
- The most sensitive diagnostic markers of changes in the dichroism of supramolecular networks are the statistical moments of the third to fourth orders, $\widetilde{M}_{i=3;4}(\text{LD, CD})$.

Table 12.16 Statistical parameters of the distributions of the LD magnitude and the balanced accuracy of differential diagnosis.

	LD			$\{LD(\varphi_k = \pi/8)\}$		
$\widetilde{M}_{i=1;2;3;4} \pm \sigma_{i=1;2;3;4}$	Group 1	Group 2	Ac, %	Group 1	Group 2	Ac, %
$\widetilde{M}_1 \times 10^{-3}$	0.41 ± 0.022	0.46 ± 0.026	76.1 $c = 6$; $d = 5$	0.32 ± 0.017	0.36 ± 0.018	78.3 $c = 4$; $d = 6$
p	$p < 0.05$			$p < 0.05$		
$\widetilde{M}_2 \times 10^{-3}$	0.11 ± 0.006	0.13 ± 0.006	76.1 $c = 5$; $d = 6$	0.06 ± 0.004	0.08 ± 0.003	78.3 $c = 5$; $d = 5$
p	$p < 0.05$			$p < 0.05$		
$\widetilde{M}_3$	0.26 ± 0.013	0.21 ± 0.011	80.4 $c = 4$; $d = 5$	0.46 ± 0.024	0.35 ± 0.023	84.8 $c = 3$; $d = 4$
p	$p < 0.001$			$p < 0.001$		
$\widetilde{M}_4$	0.37 ± 0.019	0.29 ± 0.015	80.4 $c = 4$; $d = 5$	0.65 ± 0.034	0.52 ± 0.033	84.8 $c = 4$; $d = 3$
p	$p < 0.001$			$p < 0.001$		

Table 12.17 Statistical parameters of the distributions of CD and the balanced accuracy of differential diagnosis.

	CD			$\{CD(\varphi_k = \pi/8)\}$		
$\widetilde{M}_{i=1;2;3;4} \pm \sigma_{i=1;2;3;4}$	Group 1	Group 2	Ac, %	Group 1	Group 2	Ac, %
$\widetilde{M}_1 \times 10^{-3}$	0.44 ± 0.024	0.49 ± 0.025	80.4 $c = 3$; $d = 6$	0.37 ± 0.021	0.42 ± 0.022	84.8 $c = 3$; $d = 4$
p	$p < 0.05$			$p < 0.05$		
$\widetilde{M}_2 \times 10^{-3}$	0.27 ± 0.015	0.31 ± 0.015	80.4 $c = 4$; $d = 4$	0.16 ± 0.009	0.19 ± 0.009	84.8 $c = 3$; $d = 4$
p	$p < 0.05$			$p < 0.05$		
$\widetilde{M}_3$	0.39 ± 0.021	0.49 ± 0.025	82.6 $c = 3$; $d = 4$	0.61 ± 0.034	0.89 ± 0.041	89.1 $c = 2$; $d = 3$
p	$p < 0.001$			$p < 0.001$		
$\widetilde{M}_4$	0.61 ± 0.033	0.75 ± 0.036	82.6 $c = 3$; $d = 4$	0.82 ± 0.045	1.21 ± 0.062	89.1 $c = 1$; $d = 3$
p	$p < 0.001$			$p < 0.001$		

- For the LD maps $LD(m, n)$, the maximum differences $\Delta\widetilde{M}_{i=3;4}(\varphi_k = {}^{\pi}/_{8}, LD)$ between the diagnostic marker values of the control and study groups of BF samples are 25–35%
- For the CD maps $CD(m, n)$, the maximum differences $\Delta\widetilde{M}_{i=3;4}(\varphi_k = {}^{\pi}/_{8}, CD)$ increase and reach 45–55%.

As a result of information analysis, the maximum accuracy of the differential diagnosis of calculous cholecystitis was determined, reaching the following levels:

- $LD(m, n)$ maps – $Ac((\varphi_k = \pi/8)) = 84.8\%$, corresponding to a practically good level;
- $CD(m, n)$ maps – $Ac((\varphi_k = \pi/8)) = 89.1\%$, corresponding to a good level.

From a physical standpoint, the results of the Jones matrix differential diagnosis for early-stage gallstone disease can be explained by the following considerations.

The influence of CD of microlith crystals, fatty acids, and calcium bilirubinate (M + CaB + FA) is predominant compared to the LD of the needle-shaped crystals of cholesterol and fatty acids (CC + FA) in the dehydrated BF. Within the framework of statistical analysis, this corresponds to – $\begin{pmatrix} \widetilde{M}_{i=1;2}(\text{CD}) > \widetilde{M}_{i=1;2}(\text{LD}); \\ \widetilde{M}_{i=3;4}(\text{CD}) < \widetilde{M}_{i=3;4}(\text{LD}) \end{pmatrix}$.

Pathological processes corresponding to the development of calculous cholecystitis, followed by the formation of calculi and the progression of cholelithiasis, are accompanied by an increase in the overall level of crystallization and in the optical anisotropic absorption of dehydrated BF. As a result, for the dichroism maps of BF from patients, an increase in the values of $\widetilde{M}_{i=1;2}\begin{pmatrix}\text{LD}\\ \text{CD}\end{pmatrix} \uparrow$ is observed against the background of a decrease in $\widetilde{M}_{i=3;4}\begin{pmatrix}\text{LD}\\ \text{CD}\end{pmatrix} \downarrow$, as well as an increase in the difference (Δ) between the most sensitive statistical markers determined (Tables 12.16 and 12.17) – $\Delta\widetilde{M}_{i=3;4}\begin{pmatrix}\text{LD}\\ \text{CD}\end{pmatrix} \uparrow$. However, the presence of a multiply scattered depolarized background significantly limits the sensitivity of the laser dichroism introscopy method for studying the optically anisotropic architectonics of supramolecular networks of BF. Consequently, the value of the balanced accuracy does not exceed a satisfactory level: $\text{Ac}(\widetilde{M}_{i=3;4}(\text{LD})) = 80.4\%$ and $\text{Ac}(\widetilde{M}_{i=3;4}(\text{CD})) = 82.6\%$.

Phase scanning $\varphi_k \downarrow$ makes it possible to isolate a more optically thin volume of the dehydrated bile droplet film. As a result, the multiplicity of light scattering decreases, and clearer relationships are established between the algorithmically reconstructed dichroism maps $\text{DT}(\varphi_k, m, n)$ and the statistics of the distribution of the directions of the optical axes of needle-shaped crystals (linear dichroism $\text{LD}(m, n)$), as well as the magnitudes of the phase shifts between the orthogonal components of the laser radiation amplitude ($\text{CD}(m, n)$).

As a result, the differences between the diagnostic markers $\Delta\widetilde{M}_{i=3;4}\left(\varphi_k, \begin{pmatrix}\text{LD}\\ \text{CD}\end{pmatrix}\right) \uparrow\uparrow$, which more clearly characterize the skewness and excess of the distributions of LD and CD in the supramolecular networks of BF of healthy donors and patients, increase. This is confirmed by the improvement in the accuracy of differential diagnosis to a good level: $(\text{Ac}(\widetilde{M}_{i=3;4}(\varphi_k = \pi/8, \text{LD})) = 84.8\%)$ and $(\text{Ac}(\widetilde{M}_{i=3;4}(\varphi_k = \pi/8, \text{CD})) = 89.1\%)$.

At the same time, the statistical analysis of the set of layer-by-layer maps $\text{DT}(\varphi_k, m, n)$ (Figures 12.19 and 12.20) does not make it possible to obtain information about the features of the scale-selective and scale-self-similar structure of the coordinate distributions (maps) of LD and CD of supramolecular networks of biochemical crystals (CC + FA) and (M + CaB + FA) self-assembled during the dehydration process.

Therefore, the next step was to study the scale-self-similar and multifractal structure of the LD and CD maps in order to identify new diagnostic markers that are more sensitive to pathological changes in the polycrystalline structure of BF.

12.4.3 Wavelet Analysis of Jones Matrix Dichroism Maps of Supramolecular Networks of BF

This section presents the results of applying the scale-self-similar wavelet decomposition $W^{\text{LD;CD}}(a, b)$ to the set of dichroism maps $\text{DT}(\varphi_k, m, n)$ of BF samples from healthy donors and patients with calculous cholecystitis.

The results of the wavelet analysis of the set of $\text{DT}(\varphi_k = {}^{\pi}/_{8}, m, n)$ dichroism maps of supramolecular networks of BF are presented in a series of fragments in Figure 12.21 (healthy donors – control group 1) and Figure 12.22 (patients with calculous cholecystitis – study group 2).

The linear coordinate distributions of the amplitudes of the wavelet coefficients) at different scales ($a_{\min} = 15$ – "light gray lines" and $a_{\max} = 55$ – "dark gray lines") of the scanning algorithmically reconstructed layer-by-layer Jones matrix maps $\text{DT}(\varphi_k = {}^{\pi}/_{8}, m, n)$ of the soliton-like MHAT function obtained in the phase plane $\varphi_k = {}^{\pi}/_{8}$ were evaluated within the framework of the statistical analysis algorithms by calculating the mean and variance – $\begin{pmatrix} M_{1;2}\left(W^{\text{LD;CD}}(\varphi_k = {}^{\pi}/_{8}, a_{\min} = 15, b)\right); \\ M_{1;2}\left(W^{\text{LD;CD}}(\varphi_k = {}^{\pi}/_{8}, a_{\max} = 55, b)\right) \end{pmatrix}$ – see Tables 12.18 and 12.19.

Comparative statistical and informational analysis of the amplitude distributions of multiscale wavelet coefficients $W(a, b)$ of the DT(LD, CD) maps of amplitude anisotropy of dehydrated BF from healthy donors and patients with calculous cholecystitis is presented in Table 12.18.

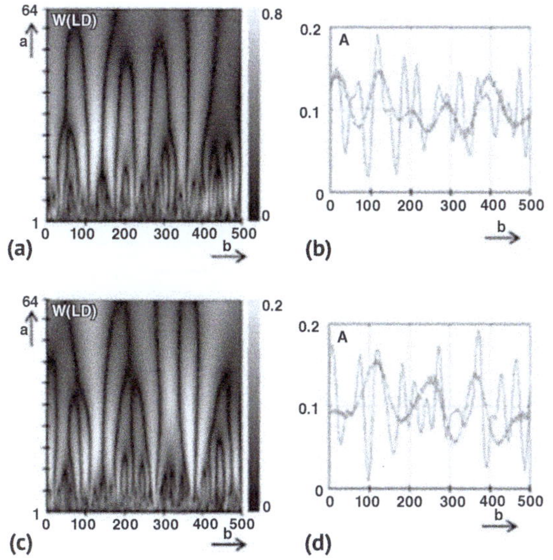

Figure 12.21 Maps (a), (c) and linear dependences (b), (d) of the amplitudes of the wavelet coefficients $W(a, b)$ of the LD maps of bile films from healthy donors (a), (b) and patients with calculous cholecystitis (c), (d).

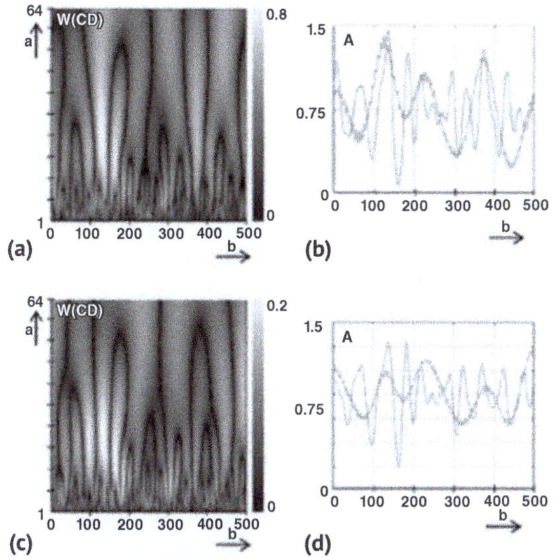

Figure 12.22 Maps (a), (c) and linear dependences (b), (d) of the amplitudes of the wavelet coefficients $W(a, b)$ of the CD maps of bile films from healthy donors (a), (b) and patients with calculous cholecystitis (c), (d).

It was established:

- For BF samples from healthy donors and patients with calculous cholecystitis, the maximum differences (1.35–1.45 times) between the statistical moments of the first and second orders are observed at small scales $a_{\min} = 15$ of the MHAT wavelet function $W^{\text{LD}}(a, b)$ corresponding to the structural dichroism maps of amplitude anisotropy.
- A similar pattern is observed for the relationship between the statistical parameters $M_{i=1;2}$, which characterize the distributions $W^{\text{CD}}(a_{\min} = 15, b)$ of the wavelet coefficients of the CD maps of BF from Groups 1 and 2. The corresponding differences between the statistical moments of the first and second orders reach 55–65%.

As a result, an increase in the maximum accuracy of differential wavelet diagnostics of the early stage of gallstone disease is observed:

- For the wavelet maps $W^{\text{LD}}(a_i, b)$ of the distributions of the LD magnitude – $\text{Ac}(\widetilde{M}_{i=2}(\varphi_k = \pi/8)) = 91.2\%$, which corresponds to a very good diagnostic level;

Table 12.18 Statistical parameters of the amplitude distributions of the wavelet coefficients $W(a, b)$ of the LD and CD maps.

LD						
a_i	$a_{min} = 15$			$a_{max} = 55$		
$\widetilde{M}_i$	Group 1	Group 2	Ac, %	Group 1	Group 2	Ac, %
$\widetilde{M}_1$	0.044 ± 0.003	0.059 ± 0.004	86.9 $c = 4;$ $d = 2$	0.047 ± 0.003	0.061 ± 0.004	84.8 $c = 3;$ $d = 4$
p	$p < 0.001$			$p < 0.001$		
M_2	0.074 ± 0.004	0.094 ± 0.005	86.9 $c = 3;$ $d = 3$	0.063 ± 0.004	0.082 ± 0.005	84.8 $c = 3;$ $d = 4$
p	$p < 0.001$			$p < 0.001$		
CD						
a_i	$a_{min} = 15$			$a_{max} = 55$		
M_i	Group 1	Group 2	Ac, %	Group 1	Group 2	Ac, %
$\widetilde{M}_1$	0.22 ± 0.013	0.34 ± 0.016	93.5 $c = 1;$ $d = 2$	0.18 ± 0.011	0.27 ± 0.013	91.3 $c = 1;$ $d = 3$
p	$p < 0.001$			$p < 0.001$		
$\widetilde{M}_2$	0.34 ± 0.019	0.56 ± 0.029	93.5 $c = 1;$ $d = 2$	0.29 ± 0.018	0.41 ± 0.023	91.3 $c = 2;$ $d = 2$
p	$p < 0.001$			$p < 0.001$		

- For the wavelet maps $W^{CD}(a_i, b)$ of the distributions of the CD magnitude – $\text{Ac}(\widetilde{M}_{i=2}(\varphi = \pi/8)) = 94.2\%$, which also corresponds to a very good diagnostic level.

From a physical point of view, the improvement in the sensitivity of differential diagnosis of calculous cholecystitis can be attributed to the different rates of bile crystallization observed at various geometric scales of biochemical crystal formation. The architectonics of supramolecular networks primarily begins at the fine-scale microcrystallite level. The application of wavelet decomposition to the maps $W^{LD}(a_{min} = 15, b)$ and $W^{CD}(a_{min} = 15, b)$ enables direct statistical analysis of pathological small-scale alterations in the optically anisotropic architectonics of BF in patients with calculous cholecystitis. As a result, due to an increase in the mean $\widetilde{M}_{i=1}(W^{LD;CD}(a_{min} = 15, b)) \uparrow$ and variance $\widetilde{M}_{i=2}(W^{LD;CD}(a_{min} = 15, b)) \uparrow$ of the amplitude values of the wavelet coefficients, the overall sensitivity of the method increases, and the accuracy of differential diagnostics improves by approximately 5–7% compared to the direct statistical evaluation of the set of dichroism maps $\text{DT}(\varphi_k = {}^{\pi}/_{8}, m, n)$ of supramolecular BF networks.

12.4.4 Multifractal Analysis of Layer-by-layer Jones Matrix Dichroism Maps of BF

This section presents the results of applying the wavelet transform Mueller matrix (WTMM) method for the algorithmic determination of the multifractal spectra $F^{LD;CD}(h)$, which characterize the distributions of fractal dimensions h reconstructed in the phase plane $\varphi_k = {}^{\pi}/_{8}$ for the maps $\text{DT}^{LD;CD}(m, n)$ of the polycrystalline optically anisotropic architectonics of BF samples from Groups 1 and 2.

Figures 12.23 and 12.24 show the skeletons $\sup|W(a^*, x_i(a^*))|$ of the wavelet coefficient maps $W^{LD;CD}(a, b)$ (panels (a), (c)) and the multifractal spectra $F^{LD}(h)$ and $F^{CD}(h)$ (panels (b), (d)) of the phase maps of LD, $\text{DT}(\varphi_k = {}^{\pi}/_{8}, \text{LD}(m, n))$ and

Table 12.19 Statistical characteristics of the distributions of the fractal dimensions of LD and CD maps of bile films.

F(h)			
T		**LD**	**Ac, %**
$\widetilde{M}_1$	1.74 ± 0.086	1.59 ± 0.087	84.8
p		$p < 0.05$	$c = 3$; $d = 4$
$\widetilde{M}_2$	0.14 ± 0.008	0.19 ± 0.011	91.3
p		$p < 0.05$	$c = 2$; $d = 2$
$\widetilde{M}_3$	1.49 ± 0.079	0.91 ± 0.048	91.3
p		$p < 0.001$	$c = 1$; $d = 3$
$\widetilde{M}_4$	3.15 ± 0.17	1.88 ± 0.11	93.5
p		$p < 0.001$	$c = 1$; $d = 2$
T		**CD**	**Ac, %**
$\widetilde{M}_1$	1.68 ± 0.086	1.57 ± 0.096	86.9
p		$p < 0.05$	$c = 2$; $d = 4$
$\widetilde{M}_2$	0.16 ± 0.009	0.19 ± 0.011	86.9
p		$p < 0.05$	$c = 3$; $d = 3$
$\widetilde{M}_3$	1.16 ± 0.022	0.64 ± 0.035	95.6
p		$p < 0.001$	$c = 1$; $d = 1$
$\widetilde{M}_4$	2.71 ± 0.14	1.32 ± 0.079	97.8
p		$p < 0.001$	$c = 0$; $d = 1$

CD, DT($\varphi_k = \pi/8$, CD(m, n)) of the supramolecular networks of BF samples self-assembled during the dehydration process for representative samples of healthy donors (panels (a), (b)) and sick patients (panels (c), (d)).

Analysis of the obtained data $F^{\text{LD}}(h)$ and $F^{\text{CD}}(h)$ demonstrated the presence of a multifractal structure in the coordinate distributions of the optically anisotropic absorption parameters DT($\varphi_k = \pi/8$, $LD(m, n)$) and DT($\varphi_k = \pi/8$, CD(m, n)) of the components of supramolecular networks of biochemical crystals self-assembled during the dehydration of BF – linear (needle-shaped crystals of cholesterol and fatty acids – CC + FA) and circular (spherulitic and pit-shaped crystals of microliths, calcium bilirubinate, and fatty acids M + CaB + FA) dichroism. This fact is indicated by the presence of distributions of fractal dimensions $1.6 \leq h_r(\text{LD}) \leq 1.92$ (Figure 12.23b) and $1.55 \leq h_r(\text{CD}) \leq 1.91$ (Figure 12.24b) with narrowly localized (probability $F^{\text{LD;CD}} \sim 90\% - 95\%$) main extrema of the principal fractal dimensions $h_{\max}(\text{LD}) = 1.86$ and $h_{\max}(\text{CD}) = 1.78$.

Pathological changes are manifested in the expansion of the variation ranges of fractal dimensions $1.35 \leq h_r(\text{LD}) \leq 1.91$ (Figure 12.23d) and $1.45 \leq h_r(\text{CD}) \leq 1.93$ (Figure 12.24d), as well as in the increase in the probabilities of additional (relative to the main ($h_{\max}$) fractal dimensions corresponding to the extrema of the $F^{\text{LD;CD}}$ distributions.

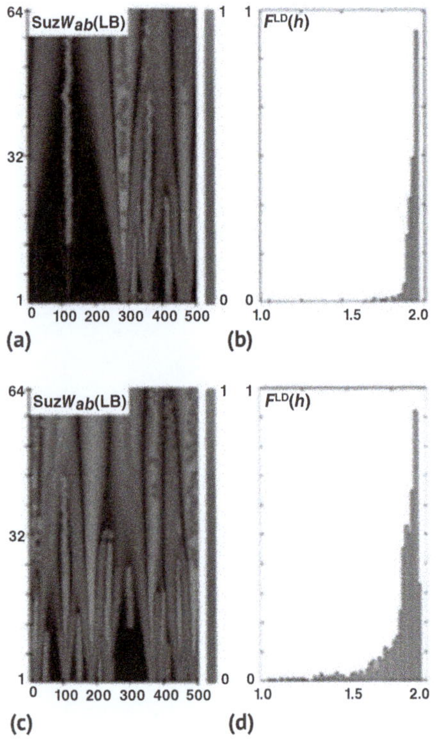

Figure 12.23 Skeletons sup $W_{ab}(\mathrm{LD})$ ((a), (c)) and spectra of fractal dimensions $F^{\mathrm{LD}}(h)$ ((b), (d)) of maps of LD of bile films of healthy ((a), (b)) and sick patients ((c), (d)).

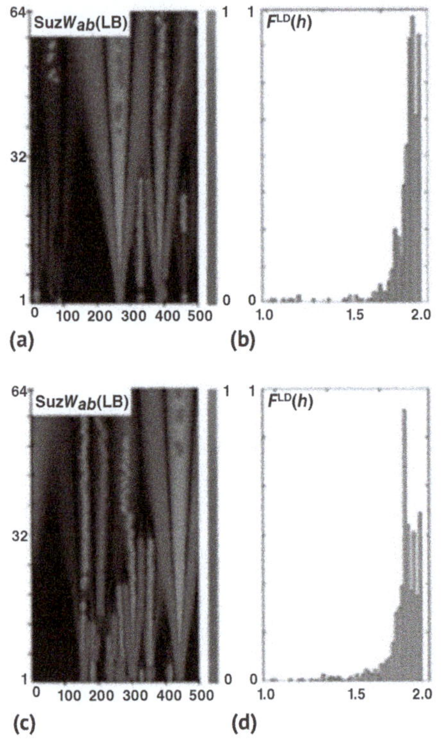

Figure 12.24 Skeletons sup $W_{ab}(\mathrm{CD})$ (a), (c) and spectra of fractal dimensions $F^{\mathrm{CD}}(h)$ (b), (d) of maps of circular CD dichroism of bile films of healthy (a), (b) and sick patients (c), (d).

From a physical point of view, the revealed features of the structure of the multifractal spectra $F^{\mathrm{LD;CD}}(h)$ can be associated with the fractal geometry of the supramolecular bile networks self-assembled during dehydration. As a result, fractal measures $Z(r, d)$ of the optically anisotropic properties of the formed polycrystalline architectonics $\mathrm{DT}(\varphi_k = {}^{\pi}/_{8}, \mathrm{LD}(m, n))$ and $\mathrm{DT}(\varphi_k = {}^{\pi}/_{8}, \mathrm{CD}(m, n))$ are created. Each such measure $\mathrm{DT}_r(m, n)$ corresponds to the fractal distribution $F^{\rho}(h) \leftrightarrow F^{\mathrm{LD}}(h)$

of the orientation of the optical axes of needle crystals (M + CaB + FA) or to the scale self-similarity of the geometric dimensions $F^z(h) \leftrightarrow F^{CD}(h)$ of optically active molecular crystals (CC + FA).

When pathological changes occur during the dehydration of the biochemical structure of bile, additional ("add") fractal measures $DT_{add}(m, n)$ are formed in the geometry of the optically anisotropic architectonics of supramolecular networks. As a result, the range of fractal dimensions ($1.35 \leq h_r(LD) \leq 1.91$ and $1.45 \leq h_r(CD) \leq 1.93$), characterizing the dichroism maps $DT(\varphi_k = \pi/8, LD(m, n))$ and $DT(\varphi_k = \pi/8, CD(m, n))$ expands. Predictively, such changes in the multifractal spectra $F^{LD;CD}(h)$ are manifested by an increase in the half-width $(\widetilde{M}_{i=2}(LD, CD) \uparrow)$ and a decrease in the peak sharpness $(\widetilde{M}_{i=3;4}(LD, CD) \downarrow)$ of the distributions of fractal dimensions $F^{LD;CD}(h)$.

The quantitative results of the statistical analysis of the multifractal spectra $F^{LD;CD}(h)$ of the amplitude anisotropy parameters of polycrystalline bile droplet films are presented in Table 12.19.

It is established:

- The difference from the normal distribution $(\widetilde{M}_{i=3;4} = 0)$ in the individual multifractal spectra $F^{LD;CD}(h)$ of the maps of LD, $DT(\varphi_k = \pi/8, LD(m, n))$ and CD, $DT(\varphi_k = \pi/8, CD(m, n))$ is observed – the values of all central statistical moments of the first to fourth orders are non-zero, $\widetilde{M}_{i=1;2;3;4}(F^{LD;CD}(h)) \neq 0$.
- The statistical significance of the mean $\widetilde{M}_{i=1}$ and variance $\widetilde{M}_{i=2}(p < 0.05)$, as well as skewness $\widetilde{M}_{i=3}$ and kurtosis $\widetilde{M}_{i=4}(p < 0.001)$, which characterize the distributions of fractal dimensions $F^{LD;CD}(h)$ of maps $DT(\varphi_k = \pi/8, LD(m, n))$ and $DT(\varphi_k = \pi/8, CD(m, n))$.
- The maximum differences $(\Delta\widetilde{M}_{i=3;4})$ between the statistical moments of higher orders, which characterize the skewness and kurtosis of the multifractal spectra $F^{LD;CD}(h)$ of the phase maps of dichroism $DT(\varphi_k = \pi/8, LD(m, n))$ and $DT(\varphi_k = \pi/8, CD(m, n))$ of BF from control group 1 and experimental group 2 – $\begin{pmatrix} \Delta\widetilde{M}_{i=3;4}(LD) = 1.6 \div 1.7 \\ \Delta\widetilde{M}_{i=3;4}(CD) = 1.8 \div 2.0 \end{pmatrix}$.

As a result, the maximum balanced accuracy of differential multifractal diagnostics reaches a very high level:

- multifractal spectrum $F(LD, h)$ of LD – $Ac(M_{i=2;4}(\varphi = \pi/8)) = 94.1\%$ – very good level.
- multifractal spectrum $F(CD, h)$ of CD – $Ac(M_{i=2;4}(\varphi = \pi/8)) = 97.2\%$ – excellent level.

12.4.5 Conclusions

1. A method for Jones matrix layer-by-layer introscopy was developed and experimentally tested to analyze the optically anisotropic architecture of supramolecular networks in dehydrated bile layers.
2. Maps of LD and CD in various phase sections of the complex amplitude field of dehydrated bile layers from healthy donors and patients with calculous cholecystitis were experimentally measured and physically analyzed.
3. Within the framework of statistical analysis, the most sensitive diagnostic markers of changes in the dichroism of supramolecular networks of BF were determined – the statistical moments of the third and fourth orders $\widetilde{M}_{i=3;4}(LD, CD)$. On this basis, in the phase plane $\varphi_k = \pi/8$, the maximum accuracy Ac of the differential diagnosis of calculous cholecystitis was determined, reaching the following maximum levels:
   - maps $LD(m, n)$ – $Ac((\varphi_k = \pi/8)) = 84.8\%$ – practically good level.
   - maps $CD(m, n)$ – $Ac((\varphi_k = \pi/8)) = 89.1\%$ – good level.
4. By applying scale-selective wavelet analysis to the set of $DT(\varphi_k = \pi/8, m, n)$ dichroism maps of supramolecular bile networks, the sensitivity and accuracy of differential wavelet diagnostics of the early stage of cholelithiasis were improved:
   - wavelet map $W^{LD}(a_i, b)$ of LD magnitude distributions LD – $Ac(\widetilde{M}_{i=2}(\varphi_k = \pi/8)) = 91.2\%$ – very good level.
   - wavelet map $W^{CD}(a, b)$ of CD magnitude distributions CD – $Ac(\widetilde{M}_{i=2}(\varphi = \pi/8)) = 94.2\%$ – very good level.
5. For the first time, the wavelet transform method of modulus maxima of LD and CD distributions was applied to evaluate the multifractal structure of dehydrated bile layers – specifically, the optically anisotropic supramolecular networks – using layer-by-layer maps. The spectra of fractal dimension distributions of the layered LD and CD maps were algorithmically obtained and analyzed using statistical methods.
6. As a result, the maximum balanced accuracy of differential multifractal diagnostics reached a very high level:
   - multifractal spectrum $F^{LD}(h)$ of LD – $Ac(M_{i=2;4}(\varphi = \pi/8)) = 94.1\%$ – very good level.
   - multifractal spectrum $F^{CD}(h)$ of CD – $Ac(M_{i=2;4}(\varphi = \pi/8)) = 97.2\%$ - excellent level.

References

1 Hunter R. Foundations of Colloid Science. New York: Oxford University Press; 2004. p. 806.
2 Chen X, Chen PG, Ouazzani J, Liu Q. Numerical simulations of sessile droplet evaporating on heated substrate. Eur Phys J Spec Top. 2017;226(6):1325–1335.
3 Dash S, Garimella SV. Droplet evaporation on heated hydrophobic and superhydrophobic surfaces. Phys Rev E. 2014;89:042402.
4 Gleason K, Putnam SA. Microdroplet evaporation with a forced pinned contact line. Langmuir. 2014;30(34):10548–10555.
5 Hu H, Larson RG. Evaporation of a sessile droplet on a substrate. J Phys Chem B. 2002;106:1334–1344.
6 Serra J. Image Analysis and Mathematical Morphology. New York: Academic Press; 1988. p. 411.
7 Shirvaikar M, Trivedi M. Developing texture-based image clutter measures for object detection. Opt Eng. 1992;31(12):2628–2639.
8 Thiele U. Patterned deposition at moving contact lines. Adv Colloid Interface Sci. 2014;206:399–413.
9 Ushenko YA, Ushenko VA, Dubolazov AV, Balanetskaya VO, Zabolotna NI. Mueller-matrix diagnostics of optical properties of polycrystalline networks of human blood plasma. Opt Spectrosc. 2012;112(6):884–892.
10 Ushenko YA, Dubolazov AV, Balanetskaya VO, Karachevtsev AO, Ushenko VA. Wavelet-analysis of polarization maps of human blood plasma. Opt Spectrosc. 2012;113(3):332–343.
11 Ushenko VO. Spatial-frequency polarization phasometry of biological polycrystalline networks. Opt Mem Neural Netw. 2013;22(1):56–64.
12 Ushenko VA, Pavlyukovich ND, Trifonyuk L. Spatial-frequency azimuthally stable cartography of biological polycrystalline networks. Int J Opt. 2013;2013:7.
13 Dubolazov AV, Pashkovskaya NV, Ushenko YA, Marchuk YF, Ushenko VA, Novakovskaya OY. Birefringence images of polycrystalline films of human urine in early diagnostics of kidney pathology. Appl Opt. 2016;55:B85–B90.
14 Ushenko AG, Pashkovskaya NV, Dubolazov OV, Ushenko Y, Marchuk Y, Ushenko V. Mueller matrix images of polycrystalline films of human biological fluids. Rom Rep Phys. 2015;67(4):1467–1479.
15 Sreelal MM, Vinu RV, Singh RK. Jones matrix microscopy from single-shot intensity measurement. Opt Lett. 2017;42:5194–5197.
16 Park K, Yang TD, Seo D, Hyeon MG, Kong T, Kim BM, et al. Jones matrix microscopy for living eukaryotic cells. ACS Photonics. 2021;8:3042.
17 Liu H, Rv V, Chen K, Liao D, Chen Z, Pu J. Real-time Jones matrix synthesis by compact polarization in line holographic microscopy. Laser Photonics Rev. 2024;18(7):2301261.
18 Angelsky PO, Ushenko AG, Dubolazov AV, Sidor MI, Bodnar GB, Koval G, et al. The singular approach for processing polarization-inhomogeneous laser images of blood plasma layers. J Opt. 2013;15(4):044030.
19 Ushenko AG, Dubolazov AV, Ushenko VA, Novakovskaya OY. Statistical analysis of polarization-inhomogeneous Fourier spectra of laser radiation scattered by human skin in the tasks of differentiation of benign and malignant formations. J Biomed Opt. 2016;21(7):071110.
20 Garazdyuk MS, Bachinskyi VT, Vanchulyak OY, Ushenko AG, Dubolazov OV, Gorsky MP. Polarization-phase images of liquor polycrystalline films in determining time of death. Appl Opt. 2016;55(12):B67–B71.
21 Angelskii OV, Ushenko OG, Burkovets DN, Arkhelyuk OD, Ushenko YA. Polarization-correlation studies of multifractal structures in biotissues and diagnostics of their pathologic changes. Laser Phys. 2000;10(5):1136–1142.
22 Yermolenko S, Ushenko A, Ivashko P, Goudail F, Gruia I, Gavrilă C, et al. Spectropolarimetry of cancer change of biotissues. In Ninth International Conference on Correlation Optics. SPIE. 2009;7388:404–410.
23 Jóźwicki R, Patorski K, Angelsky OV, Ushenko AG, Burkovets DN, Ushenko YA. Automatic polarimetric system for early medical diagnosis by biotissue testing. Opt Appl. 2002;32(4):603–612.
24 Ushenko AG, Dubolazov OV, Ushenko VA, Novakovskaya OY, Olar OV. Fourier polarimetry of human skin in the tasks of differentiation of benign and malignant formations. Appl Opt. 2016;55(12):B56–B60.
25 Ushenko AG, Pishak VP. Coherent-domain optical methods: biomedical diagnostics, environmental and material science. In Laser Polarimetry of Biological Tissue: Principles and Applications. Springer, New York, 2004.
26 Ushenko AG, Fediv AI, Marchuk YF. Correlation and fractal structure of Jones matrices of human bile secret. In European Conference on Biomedical Optics. 2009;7368:73681Q.
27 Ushenko AG. Depolarization of a laser emission field as the correlation development of its polarization structure. In International Conference on Correlation Optics. 1997;3317:331–339.

28 Bachinskyi V, Boychuk T, Ushenko A. Laser Polarimetry of Biological Tissues and Fluids: Chapter 4. Methods of Spatial-frequency, Singular and Correlation Analysis of Biological Layers Object Fields. LAP Lambert Academic Publishing; 2018.

29 Ushenko AG, Burkovets DN, Yermolenko SB, Arkhelyuk AD, Pishak VP, Yuzko AM, et al. Stokes polarimetry of biotissues. In Fourth International Conference on Correlation Optics. SPIE. 1999;3904:527–533.

30 Ushenko VA, Dubolazov AV, Pidkamin LY, Sakchnovsky MY, Bodnar AB, Ushenko YA, et al. Mapping of polycrystalline films of biological fluids utilizing the Jones-matrix formalism. Laser Phys. 2018;28(2):025602.

31 Ushenko AG, Angelsky PO, Sidor M, Marchuk YF, Andreychuk DR, Pashkovskaya NV. Spatial-frequency selection of complex degree of coherence of laser images of blood plasma in diagnostics and differentiation of pathological states of human organism of various nosology. Appl Opt. 2014;53(10):B172–B180.

32 Olar EI, Ushenko AG, Ushenko YA. Polarization correlation measurements of the phase tomograms of optically anisotropic biofractals. Laser Phys. 2004;14(8):1115–1121.

33 Guminetskiy SH, Ushenko OG, Polyanskiy IP, Motrych AV, Grynchuk FV. The optical method for investigation of the peritonitis progressing process. In Eighth International Conference on Correlation Optics. 2008;7008:626–632.

34 Ushenko AG, Fediv AI, Marchuk YuF. Singular structure of polarization images of bile secret in diagnostics of human physiological state. In: European Conference on Biomedical Optics. Optica Publishing Group; 2009.

35 Ushenko Y, Ushenko A, Dubolazov A, Soltys I, Litvinenko O, Wanchuliak O, et al. Mueller-matrix interferometric multifractal scaling of optically anisotropic architectonics of diffuse blood facies: fundamental and applied aspects. J Biophotonics. 2025;18(3):e202400412. https://doi.org/10.1002/jbio.202400412.

36 Ushenko A, Soltys I, Dubolazov A, Ushenko Y, Bilookyi V, Bilookyi O, et al. 3D Jones-matrix thesiography of biological fluid facies. J Innov Opt Health Sci. 2025;18(2):2443002. https://doi.org/10.1142/S1793545824430028.

37 Ushenko A, Pavlyukovich N, Khukhlina O, Pavlyukovich O, Gorsky M, Soltys I, et al. Layer-by-layer multifractal scanning of optically anisotropic architectonics of blood plasma films: fundamental and applied aspects. Photonics. 2025;12:215. https://doi.org/10.3390/photonics12030215.

38 Ushenko A, Pavlyukovich N, Khukhlina O, Pavlyukovich O, Soltys I, Dubolazov A, et al. Blood plasma film multifractal scanning in COVID-19 consequences diagnostics. J Biophotonics. 2024;17(11):e202400356. https://doi.org/10.1002/jbio.202400356.

39 Ushenko A, Dubolazov A, Zheng J, Litvinenko A, Gorsky M, Ushenko Y, et al. 3D polarization-interference holographic histology for wavelet-based differentiation of biological tissues with different necrotic states. J Biomed Opt. 2024;29(5):052920. https://doi.org/10.1117/1.JBO.29.5.052920.

40 Ushenko O, Bilookyi O, Zheng J, Dubolazov A, Olar O, Ushenko Y, et al. 3D digital holographic polarimetry of laser speckle fields formed by polycrystalline blood films: a tool for differential diagnosis of thyroid pathology. Front Phys. 2024;12: 1426469. https://doi.org/10.3389/fphy.2024.1426469.

41 Wójcik W, Hu Z, Ushenko Y, Smolarz A, Soltys I, Dubolazov O, et al. Optical sensor system for 3D Jones matrix reconstruction of optical anisotropy maps of self-assembled polycrystalline soft matter films. Sensors. 2024;24(5):1589. https://doi.org/10.3390/s24051589.

42 Ushenko A, Zheng J, Gorsky M, Dubolazov A, Ushenko Y, Soltys I, et al. 3D digital holographic polarimetry of diffuse optically anisotropic biological tissue object fields. Front Phys. 2023;11:1288935. https://doi.org/10.3389/fphy.2023.1288935.

43 Dubolazov A, Ushenko V, Trifonyuk L, Stashkevich A, Soltys I, Ushenko Y, et al. Polarization-singular approach to imaging Mueller-matrix polarimetry in differential diagnosis of histological sections of biopsy of tumors of the uterus and prostate. Front Phys. 2021;9:711212. https://doi.org/10.3389/fphy.2021.711212.

44 Ushenko VO, Trifonyuk L, Ushenko YA, Dubolazov OV, Gorsky MP, Ushenko AG. Polarization singularity analysis of Mueller-matrix invariants of optical anisotropy of biological tissue samples in cancer diagnostics. J Opt. 2021;23(6):064004. https://doi.org/10.1088/2040-8986/abf97a.

45 Ushenko YA, Koval GD, Ushenko AG, Dubolazov OV, Ushenko VA, Novakovskaia OY. Mueller-matrix of laser-induced autofluorescence of polycrystalline films of dried peritoneal fluid in diagnostics of endometriosis. J Biomed Opt. 2016;21(7):071116. https://doi.org/10.1117/1.JBO.21.7.071116.

13

Multifractal Scanning for COVID-19 Diagnostics

Yurii Ushenko[1,2], Weidong Dou[1], Oleksandr Ushenko[2,3], Oleksandr Dubolazov[2], Iryna Soltys[2], Vitaliy Rozhko[4], Oleksandr Olar[2], and Vladyslava Sholota[5]

[1] *Shaoxing University, Shaoxing, China*
[2] *Yurii Fedkovych Chernivtsi National University, Chernivtsi, Ukraine*
[3] *Taizhou Institute of Zhejiang University, Taizhou, China*
[4] *Bukovinian State Medical University, Chernivtsi, Ukraine*
[5] *Vinnitsa National Technical University, Vinnitsa, Ukraine*

13.1 Blood Plasma Films Multifractal Scanning in COVID-19 Consequences Diagnostics

13.1.1 Ellipticity Polarization Maps of Blood Plasma Films – Group 1

We will examine blood plasma films from two groups (Table 13.1):

- Control Group 1 – blood plasma without COVID-19 infection.
- Research Group 2 – blood plasma from COVID-19 patients with a duration of 2 years.

Figure 13.1 presents microscopic images of blood plasma films, polarization–interference mapping results with ellipticity coordinate distributions, and layered digital holographic reconstructions.

The integral object field BPF1 (Figure 13.1a) is characterized by all scattered radiation components of superposition. As a result, fairly equally probable distributions are formed $M(\beta^*)$ (Figure 13.1b).

The phase-scanning parameter value φ_k decreases, more and more optically thin layers with low scattering multiplicity are distinguished. As a result, the average value of $\overline{\beta_l}$ and the range of $\Delta\beta_l$ changes in the ellipticity β_i random values decrease.

13.1.2 Ellipticity Polarization Maps of Blood Plasma Films – Group 2

Figure 13.2 illustrates the blood plasma films BPF2 – group 2 microscopic image polarization–interference mapping digital holographic reconstruction $\beta(u, t)$ and $\beta(\varphi = \pi/4, u, t)$; $\beta(\varphi = \pi/8, u, t)$ results.

The blood plasma film samples from this group also showed polarization-inhomogeneous ellipticity maps in all phase sections, as shown in Figure 13.2a, c and e.

Comparison with the polarization maps $\beta(u, t)$ of the blood plasma samples from group 1 revealed polarization ellipticity random values probability statistical similarity of distributions, Figures 13.1 and 13.2 (panels (b), (d), (f)). However, the polarization maps histograms of blood plasma film samples from group 2 show higher average levels and greater variation in ellipticity values. This is most evident during the “optically thin” phase layer $\beta(\varphi = \pi/4, u, t)$; $\beta(\varphi = \pi/8, u, t)$.

More detailed and objective distribution data $N(\beta)$ and $N(\varphi_k, \beta)$ illustrate the results of statistical analysis polarizing maps [1–24].

Table 13.1 Blood plasma films optical-geometric parameters.

Parameters	BPF1 (23 samples)
Attenuation coefficient τ, cm^{-1}	0.17 ± 0.011
Depolarization degree Λ, %	15 ± 0.12
Parameters	**BPF2 (23 samples)**
Attenuation coefficient τ, cm^{-1}	0.21 ± 0.014
Depolarization degree Λ, %	21 ± 0.14

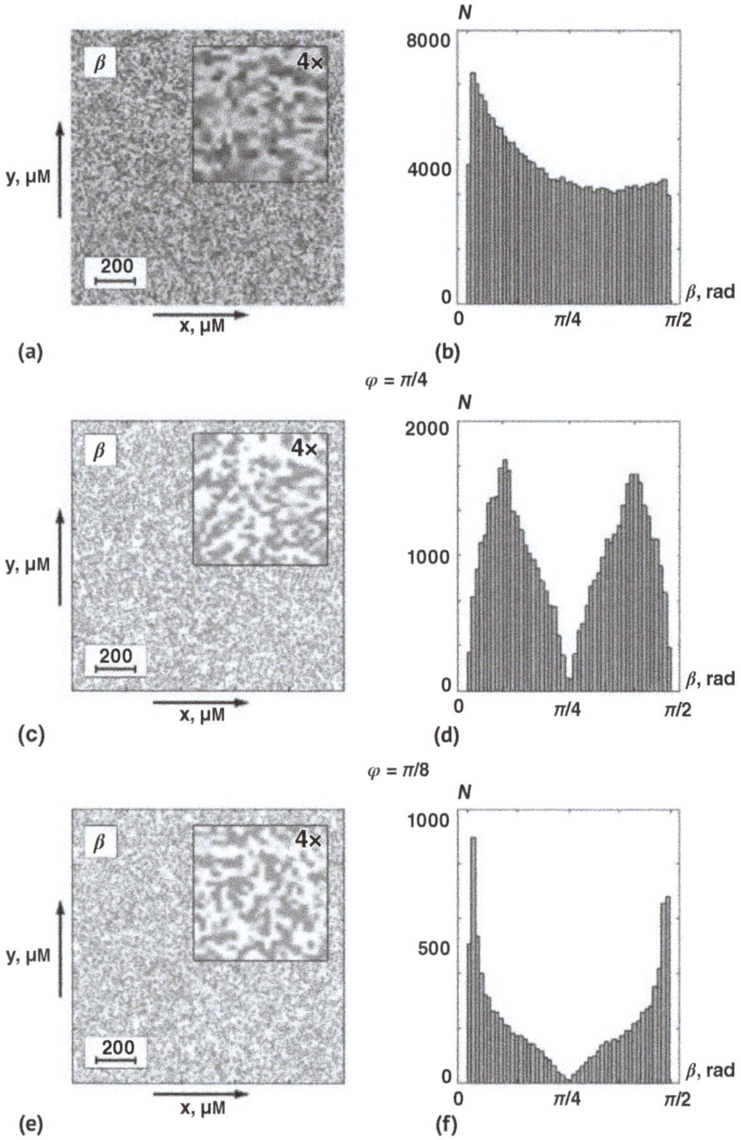

Figure 13.1 BPF1 microscopic image ellipticity polarization maps $\beta(u, t)$; $\beta(\varphi = \pi/4, u, t)$; $\beta(\varphi = \pi/8, u, t)$ (a), (c), (e) and histograms $N(\beta)$ (b), (d), (f).

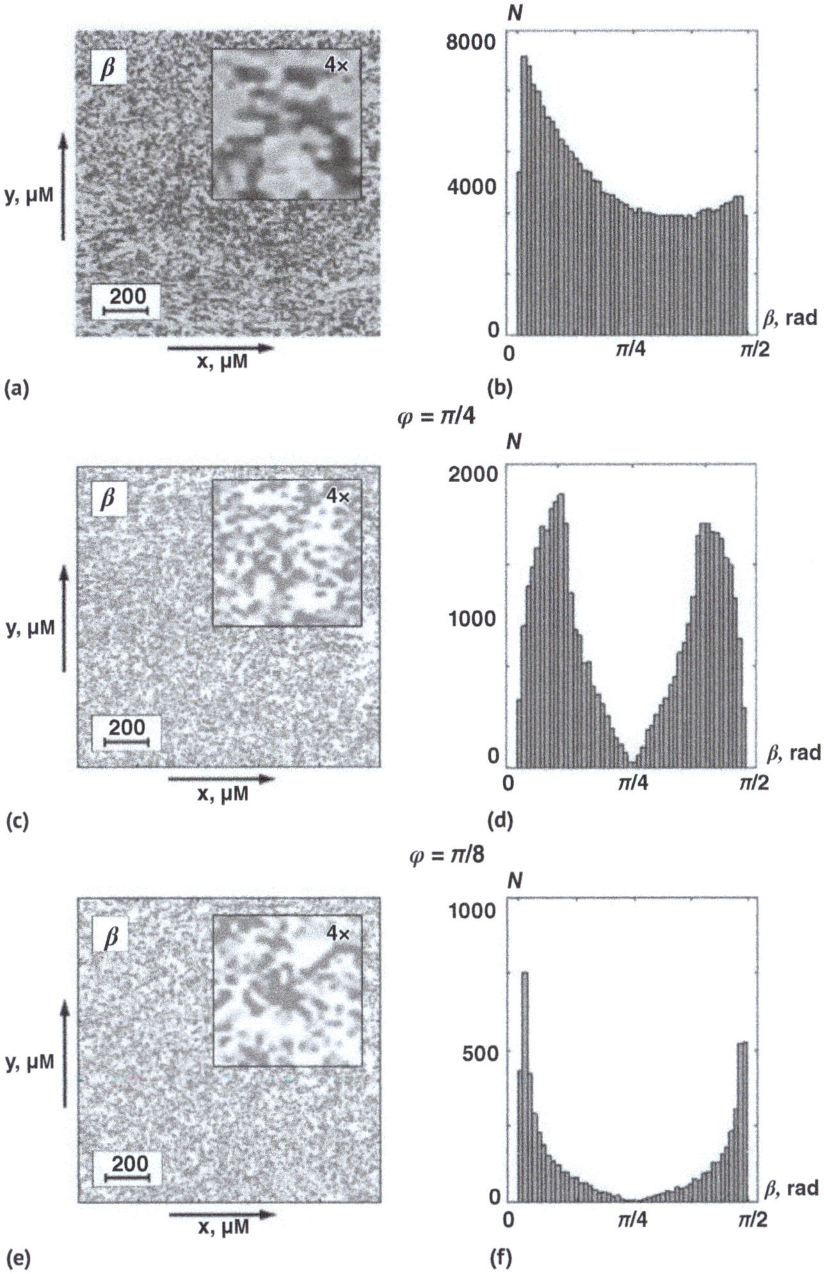

Figure 13.2 BPF2 microscopic image ellipticity polarization maps $\beta(u,t)$; $\beta(\varphi = {}^{\pi}/_{4}, u, t)$; $\beta(\varphi = {}^{\pi}/_{8}, u, t)$ (a), (c), (e) and histograms $N(\beta)$ (b), (d), (f).

13.1.3 Ellipticity Polarization Maps Statistical Analysis

Table 13.2 shows the first- to fourth-order statistical moments values, which characterize $\beta(u,t)$ and $\beta(\varphi = {}^{\pi}/_{4}, u, t)$; $\beta(\varphi = {}^{\pi}/_{8}, u, t)$ distributions.

Despite the histograms $N(\beta)$ and $N(\varphi_k, \beta)$ sufficient similarity groups 1 and 2 samples, distributions $\beta(u,t)$ and $\beta(\varphi = {}^{\pi}/_{4}, u, t)$; $\beta(\varphi = {}^{\pi}/_{8}, u, t)$ comparison statistical analysis also revealed distinctive features. Thus, the most sensitive to the ellipticity probability distributions specifics there were higher orders $M_{p=3;4}(\varphi_k, \beta)$ statistical moments.

The changes range in these statistical parameters within the changes range in the object field phase sections for group 1 is – 1.82–2.08 times; for group 2, 1.85–2.77 times. Intergroup differences in values $M_{3;4}(\varphi_k, \beta)$ make up 1.26–1.33 times.

Table 13.2 BPF1; 2 $\beta(u,t)$ and $\beta(\varphi = {}^{\pi}/_{4}, u, t)$; $\beta(\varphi = {}^{\pi}/_{8}, u, t)$ distributions statistical parameters.

$M_{p=1;2;3;4}$	BPF1 (23 samples)	BPF2 (23 samples)
M_1	0.63 ± 0.034	0.71 ± 0.039
M_2	0.44 ± 0.026	0.53 ± 0.031
M_3	0.37 ± 0.021	0.22 ± 0.014
M_4	0.51 ± 0.028	0.39 ± 0.023
$M_{p=1;2;3;4}$	$\varphi = {}^{\pi}/_{4}$	
M_1	0.31 ± 0.018	0.36 ± 0.021
M_2	0.19 ± 0.011	0.23 ± 0.014
M_3	0.53 ± 0.028	0.41 ± 0.023
M_4	0.71 ± 0.039	0.57 ± 0.032
$M_{p=1;2;3;4}$	$\varphi = {}^{\pi}/_{8}$	
M_1	0.12 ± 0.007	0.15 ± 0.009
M_2	0.07 ± 0.004	0.09 ± 0.006
M_3	0.77 ± 0.041	0.61 ± 0.034
M_4	0.93 ± 0.048	0.72 ± 0.038

The first- and second-orders statistical moments that characterize the distributions $\beta(u,t)$ and $\beta(\varphi = {}^{\pi}/_{4}, u, t)$; $\beta(\varphi = {}^{\pi}/_{8}, u, t)$ mean and variance, differ slightly – within 10–15%.

Thus, the probability distributions $N(\beta)$ and $N(\varphi_k, \beta)$ statistical analysis results (Figures 13.1 and 13.2, panels b, d, and f) we have demonstrated certain possibilities in their objective differentiation.

Note that within the statistical analysis framework, the polarization map's topographic structure is not considered. On the other hand, even a semi-quantitative analysis reveals significant differences in topographic construction between blood plasma samples, microscopic images, and ellipticity coordinate distributions from the control and study groups [25–37].

Therefore, to search for differentiation in blood plasma films' optically anisotropic properties and new additional markers, we used a different analytical approach. This is a polarization maps $\beta(u,t)$ and $\beta(\varphi = {}^{\pi}/_{4}, u, t)$; $\beta(\varphi = {}^{\pi}/_{8}, u, t)$ scale-self-similar structure multifractal analysis.

13.1.4 Multifractal Spectra of BPF1 and BPF2 Polarization Maps

Figures 13.3 and 13.4 show the lines of maximal modules (wavelet coefficient amplitudes skeletons – "dark gray dots" – maxima; "light gray dots" – minima) maps $W^{BPF1,BPF2}(a,b)$ wavelets coefficients (panels a–c) and polarization maps $\beta(u,t)$ and $\beta(\varphi = {}^{\pi}/_{4}, u, t)$; $\beta(\varphi = {}^{\pi}/_{8}, u, t)$ (panels b, d, and f) multifractal spectra $F(h_i)$.

A comparative analysis of MWTM method results revealed that:

- Both types (groups 1 and 2) of optically anisotropic supramolecular film networks have polarization maps characterized by a complex scale-like structure, which is described by individual multifractal spectra $F^{BPF1}(\varphi_k, h_r)$ and $F^{BPF2}(\varphi_k, h_r)$, (panels b, d, and f).
- The multifractal spectra $F^{BPF1;BPF2}(\varphi_k, h_r)$ of $\beta(u,t)$ and $\beta(\varphi = {}^{\pi}/_{4}, u, t)$; $\beta(\varphi = {}^{\pi}/_{8}, u, t)$ have an asymmetric structure with a clearly distinct extremum and a fractal dimension changes range – $1.45 \lesssim h_r \lesssim 1.91$.
- Ellipticity polarization maps and multifractal spectra $F^{BPF1}(\varphi_k, h_r)$ for blood plasma samples (group 1) are "narrower" and more asymmetric (panels b, d, and f) than similar multifractal spectral dependences $F^{BPF2}(\varphi_k, h_r)$ polarization inhomogeneous field (panels b, d, and f).

The main extremes of multifractal spectra $F^{BPF1}(\varphi_k, h_r)$ and $F^{BPF2}(\varphi_k, h_r)$ are localized in different fractal dimension scale parts – $\begin{cases} \max(F^{BPF1}) \Leftrightarrow h_i = 1.76; \\ \max(F^{BPF2}) \Leftrightarrow h_i = 1.88. \end{cases}$

The fractal dimensions range Δh_r is larger for the spectrum $F^{BPF1}(\varphi_k, h_r)$ than for $F^{BPF2}(\varphi_k, h_r)$ – $\Delta h_r(\text{BPF2}) > \Delta h_r(\text{BPF1})$.

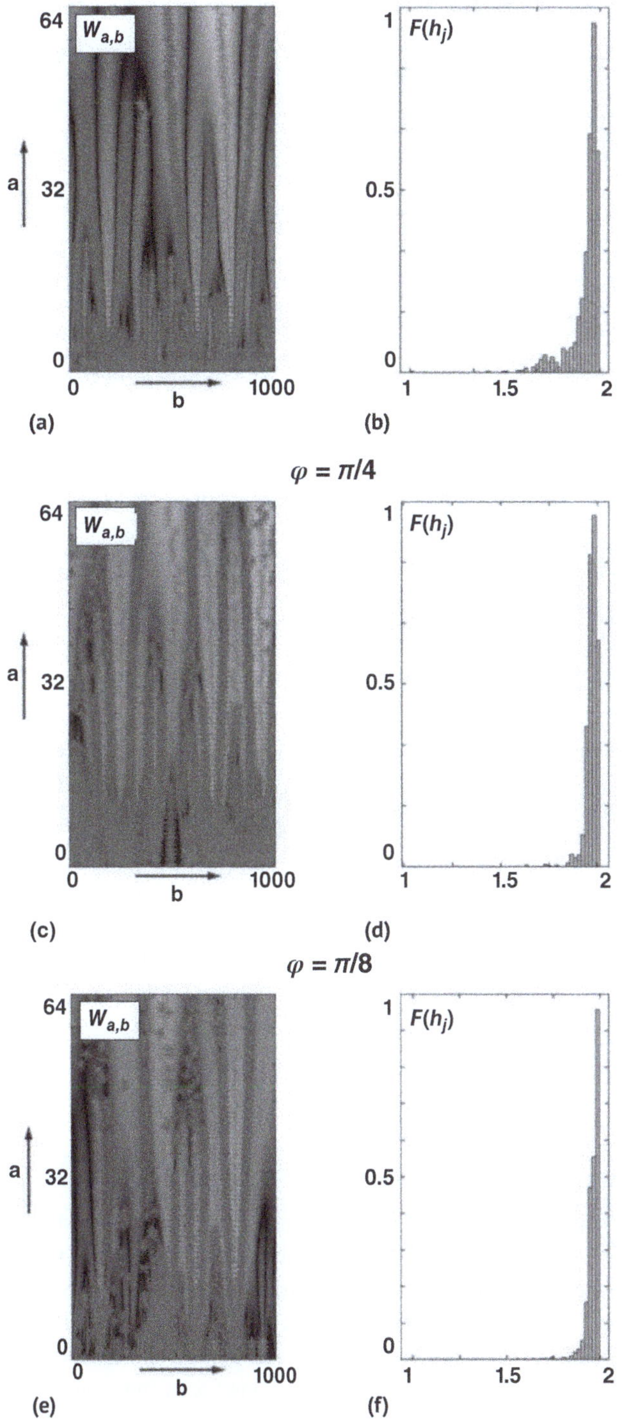

Figure 13.3 Wavelet transform skeletons (a), (c), and (e) and multifractal spectra BPF1 polarization maps $\beta(u, t)$ (a) and $\beta(\varphi = \pi/4, u, t)$; $\beta(\varphi = \pi/8, u, t)$ (b), (d), (f).

From a physical perspective, the results obtained can be connected to the following considerations.

First, each local optically anisotropic degree formed during blood plasma dehydration corresponds to the related fractal measure in the polarization maps is $\beta(u, t)$ and $\beta(\varphi = \pi/4, u, t)$; $\beta(\varphi = \pi/8, u, t)$, which has its own dimension h_r.

The magnitude and probability of $F(h_r)$ such fractal dimension is determined by the specifics self-assembled supramolecular crystallite complexes.

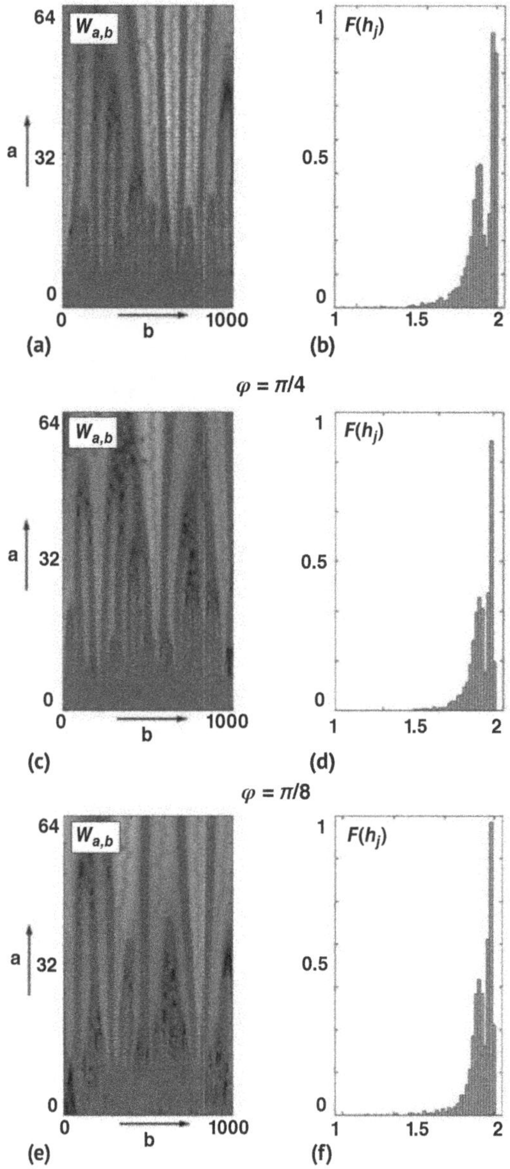

Figure 13.4 Wavelet transform skeletons (a), (c), and (e) and multifractal spectra BPF2 polarization maps $\beta(u, t)$ (a) and $\beta(\varphi = {}^{\pi}/_{4}, u, t)$; $\beta(\varphi = {}^{\pi}/_{8}, u, t)$ (b), (d), and (f).

Second, from the fractal geometry perspective, in the extreme case, the dendritic network's spatial organization, architecture, and supramolecular needle crystals can lead to a straight line.

For such a boundary situation, the multifractal spectrums $F^{BPF1}(\varphi_k, h_r)$ and $F^{BPF2}(\varphi_k, h_r)$ degenerates into a delta function with the fractal dimension base value $h_{BPF1} \Rightarrow 1.0$.

Third, blood plasma films (group 2 – COVID-19, with a duration of 2 years) exhibit a distinct supramolecular network pattern, forming fractal structures visible in the polarization maps.

The long-term effects of COVID-19 are biochemically shown by an increase in the concentration of anti-inflammatory proteins, specifically globulins with tertiary and quaternary structures. During blood plasma dehydration, such changes can lead to a broader formation of fractal measures in supramolecular networks. Consequently, the corresponding ellipticity maps will exhibit a wide range of variations in fractal dimensions. In this context, the boundary geometric condition is polycrystalline architectonics on a plane, which relates to a different value of fractal dimension $h_{BPF2} \Rightarrow 2.0$ of polarization maps $\beta(u, t)$ and $\beta(\varphi = {}^{\pi}/_{4}, u, t)$; $\beta(\varphi = {}^{\pi}/_{8}, u, t)$.

In real physical situations, similar trends are observed in various localization multifractal spectra's main extremes $\max(F^{BPF1}(\varphi_k, h_r))$ and $\max(F^{BPF2}(\varphi_k, h_r))$.

In this case, the range of variation in fractal dimensions Δh_i is larger for the spectrum $F^{BPF2}(\varphi_k, h_r)$ than for $F^{BPF1}(\varphi_k, h_r) - \Delta h_r(BPF2) > \Delta h_r(BPF1)$.

A numerical statistical estimate of multifractal spectra. $F^{BPF1}(\varphi_k, h_r)$ and $F^{BPF2}(\varphi_k, h_r)$ is illustrated in Table 13.3.

Maps $\beta(u, t)$ and $\beta(\varphi = \pi/4, u, t)$; $\beta(\varphi = \pi/8, u, t)$ multifractal structure physical and statistical analysis comparison results revealed a correlation between them. Namely, $M_{p=1;2}(F^{BPF1}(\varphi_k, h_r)) < M_{p=1;2}(F^{BPF2}(\varphi_k, h_r))$, which is due to the different localization ($h_r(BPF2) > h_r(BPF1)$) of the extrema $F^{BPF1}(\varphi_k, h_r)$ and $F^{BPF2}(\varphi_k, h_r)$, as well as differences between the fractal dimensions ranges – $\Delta h_r(BPF2) > \Delta h_r(BPF1)$.

The reverse trend $M_{i=3;4}(F^{BPF1}(\varphi_k, h_r)) > M_{i=3;4}(F^{BPF2}(\varphi_k, h_r))$ appears in the third- and fourth-order statistical moments, which describe the skewness and kurtosis of multifractal spectra distributions. Mathematically, this pattern correlates with inversely proportional relationships $M_3 \sim 1/M_2^3$ and $M_4 \sim 1/M_2^4$.

The most sensitive to changes in the object field were the third- and fourth-order statistical moments of the various phase components' scale self-similarity parameters. The range of variation for the multifractal markers is for BPF1, 1.73–1.81 times, and for BPF2, 1.68–2.16 times. The differences between groups reach a magnitude of 1.29–1.62.

13.1.5 Clinical Application

To evaluate the clinical effectiveness of the layer-by-layer polarization maps multifractal analysis method, we explored its potential for differentiating blood plasma films:

- Control group 1 (25 samples).
- Study group 2 (50 samples) – COVID-19 age one year.
- Study group 3 (50 samples) – COVID-19 is two years old.

Table 13.3 Statistical characteristics of multifractal spectra $F^{BPF1}(\varphi_k, h_r)$ and $F^{BPF2}(\varphi_k, h_r)$.

	"Integral"	
$M_{p=1;2;3;4}$	$F^{BPF1}(h_r)$	$F^{BPF2}(h_r)$
M_1	1.39 ± 0.065	1.65 ± 0.076
M_2	0.11 ± 0.008	0.14 ± 0.011
M_3	1.47 ± 0.074	0.91 ± 0.047
M_4	2.03 ± 0.092	1.46 ± 0.076
	$\varphi_k = \pi/4$	
$M_{p=1;2;3;4}$	$F^{BPF1}(\varphi_k, h_r)$	$F^{BPF2}(\varphi_k, h_r)$
M_1	1.35 ± 0.061	1.61 ± 0.072
M_2	0.09 ± 0.006	0.11 ± 0.008
M_3	2.07 ± 0.11	1.39 ± 0.087
M_4	2.99 ± 0.16	1.86 ± 0.106
	$\varphi_k = \pi/8$	
$M_{p=1;2;3;4}$	$F^{BPF1}(\varphi_k, h_r)$	$F^{BPF2}(\varphi_k, h_r)$
M_1	1.31 ± 0.055	1.59 ± 0.067
M_2	0.07 ± 0.005	0.09 ± 0.006
M_3	2.54 ± 0.15	1.98 ± 0.11
M_4	3.63 ± 0.19	2.34 ± 0.14

Table 13.4 Multifractal method differentiation accuracy.

		"Integral"	
$\beta(u,t)$	$Ac_{1;2}$, %	$Ac_{1;3}$, %	$Ac_{2;3}$, %
M_3	83.3	85.7	76.1
M_4	83.3	88.1	78.6
		$\varphi_k = \pi/4$	
$\beta(u,t)$	$Ac_{1;2}$, %	$Ac_{1;3}$, %	$Ac_{2;3}$, %
M_3	83.3	92.3	85.7
M_4	83.3	92.3	88.1
		$\varphi_k = \pi/8$	
$\beta(u,t)$	$Ac_{1;2}$, %	$Ac_{1;3}$, %	$Ac_{2;3}$, %
M_3	92.3	94.6	90.5
M_4	94.6	96.3	92.3

Within the evidence-based medicine framework, the sensitivity Se, specificity Sp and accuracy Ac of differential diagnosis were determined (Table 13.4) [38–54].

The following differential diagnosis achieved maximum accuracy levels.

COVID-19 pathology diagnosis:

- "duration of COVID-19 one year" $Ac_{1;2} = 94.6\%$;
- "duration of COVID-19 two years" $Ac_{1;3} = 96.4\%$.

Differential diagnosis:

- "duration of COVID-19 one – two years" $Ac_{2;3} = 92.3\%$;

13.1.6 Conclusions

1. A blood plasma films optically anisotropic supramolecular networks Jones matrix model for polarization-inhomogeneous object fields analytical description is proposed.
2. A 3D polarization–interference mapping method with blood plasma films microscopic images digital holographic phase scanning is proposed and experimentally tested.
3. Within the probable ellipticity random values distributions statistical analysis framework the most sensitive parameters to the supramolecular network's specifics are determined and physically justified, third- and fourth-order statistical moments, which characterize $\beta(u,t)$ and $\beta(\varphi = \pi/4, u, t)$; $\beta(\varphi = \pi/8, u, t)$ distributions skewness and kurtosis.
4. The wavelet transforms maxima amplitude modules method was applied and new multifractal markers of differential diagnostics of COVID-19 duration were found – skewness and kurtosis of polarization maps $\beta(u,t)$ and $\beta(\varphi = \pi/4, u, t)$; $\beta(\varphi = \pi/8, u, t)$ multifractal spectra.
5. Polarization maps multifractal spectra $F^{FBP1}(\varphi_k, h_r)$ and $F^{FBP2}(\varphi_k, h_r)$ are obtained and statistically analyzed in different blood plasma films object fields phase sections.
6. New parameters, most sensitive to variations in the scale self-similarity of polarization maps, are identified as the third- and fourth-order statistical moments, which describe the distributions $\beta(u,t)$ and $\beta(\varphi = \pi/4, u, t)$; $\beta(\varphi = \pi/8, u, t)$ skewness and kurtosis.
7. The obtained multifractal analysis fundamental results were successfully used for differentiation long-term COVID-19 consequences with excellent accuracy ($Ac = 92.3\% - 96.4\%$).

13.2 Layer-by-layer Multifractal Analysis of the Optically Anisotropic Structure of Blood Plasma Films: Fundamental and Applied Perspectives

Table 13.5 presents the BPF optical-geometric parameters.

13.2.1 Results and Discussion

This section presents the results of the BPF1 and BPF2 maps, along with comparative statistical and multifractal analyses.

At the first stage, we conducted the traditional statistical analysis of the algorithmically reconstructed maps $JT = \begin{pmatrix} LB(Re\, j_{pk}) \\ CB(Re\, j_{pk}) \end{pmatrix}(\varphi_k, m, n)$ of BPF1 (Figure 13.5, Table 13.6) and BPF2 (Figure 13.6, Table 13.7).

Based on the JT analysis presented in (Figure 13.5, Table 13.6, Figure 13.6, Table 13.7), we observed that both linear (LB) and circular birefringence (CB) distributions deviate from Gaussian, as indicated by all nonzero central statistical moments.

In all phase planes (Tables 13.6 and 13.7), only negligible differences were observed between the values of the central statistical moments that characterize the maps $JT = \begin{pmatrix} LB(Re\, j_{pk}) \\ CB(Re\, j_{pk}) \end{pmatrix}(\varphi_k, m, n)$.

The maximum differences are between 35% and 40%.

Therefore, it can be concluded that the sensitivity of the statistical analysis of the BPF1 and BPF2 phase anisotropy distributions was not sufficiently high.

In the second stage, to analyze the topographic distributions of phase anisotropy parameters (LB, CB), we used the fractal dimension method (h_r), which has been thoroughly tested in earlier research. This technique involves calculating the slope angles (ϑ) of straight-line $H(\vartheta)$ segments in the power spectra (PD) logarithmic dependencies $\ln(PD(JT(m, n))) - \ln L^{-1}$ for distribution $JT(m, n)$.

Figure 13.7 illustrates the fractal analysis results of the BPF1 map – JT^{LB} ((a)) and JT^{CB} ((d)).

Analysis of the data collected (Figure 13.4) showed a consistent decline in the $\ln(PD(JT^{LB,CB}) - \ln L^{-1}$ dependencies over the entire size L_i range changes in BPF1 structural elements $JT^{LB,CB}$.

Linear plots showcasing various inclinations ($\vartheta_i \approx$ const.) angles to $\ln(PD(JT^{LB,CB})) - \ln L^{-1}$ were found, which indicates a complex, generally speaking, the multifractal structure of $JT^{LB,CB}$ maps.

From a physical perspective view, this $\ln(PD(JT^{LB,CB}) - \ln L^{-1}$ structure can be related to the fact that each fractal measure in the distributions $JT^{LB,CB}(m, n)$ corresponds to a certain (partial) scale-self-like (with dimension h_r within $L_{min} \leq L_i \ll L_{max}$) set of the BPF1 biological crystals. Such partial polycrystalline networks form gradually during the dehydration of blood plasma drops.

Table 13.5 BPF optical-geometric parameters.

BPF1	
Parameters	
Attenuation coefficient τ, cm^{-1}	0.28 ± 0.016
Depolarization degree Λ, %	16 ± 0.13
Films diameter D, mm	5.5 ± 0.27
BPF2	
Parameters	
Attenuation coefficient τ, cm^{-1}	0.39 ± 0.022
Depolarization degree Λ, %	24 ± 0.15
Films diameter D, mm	5.5 ± 0.27

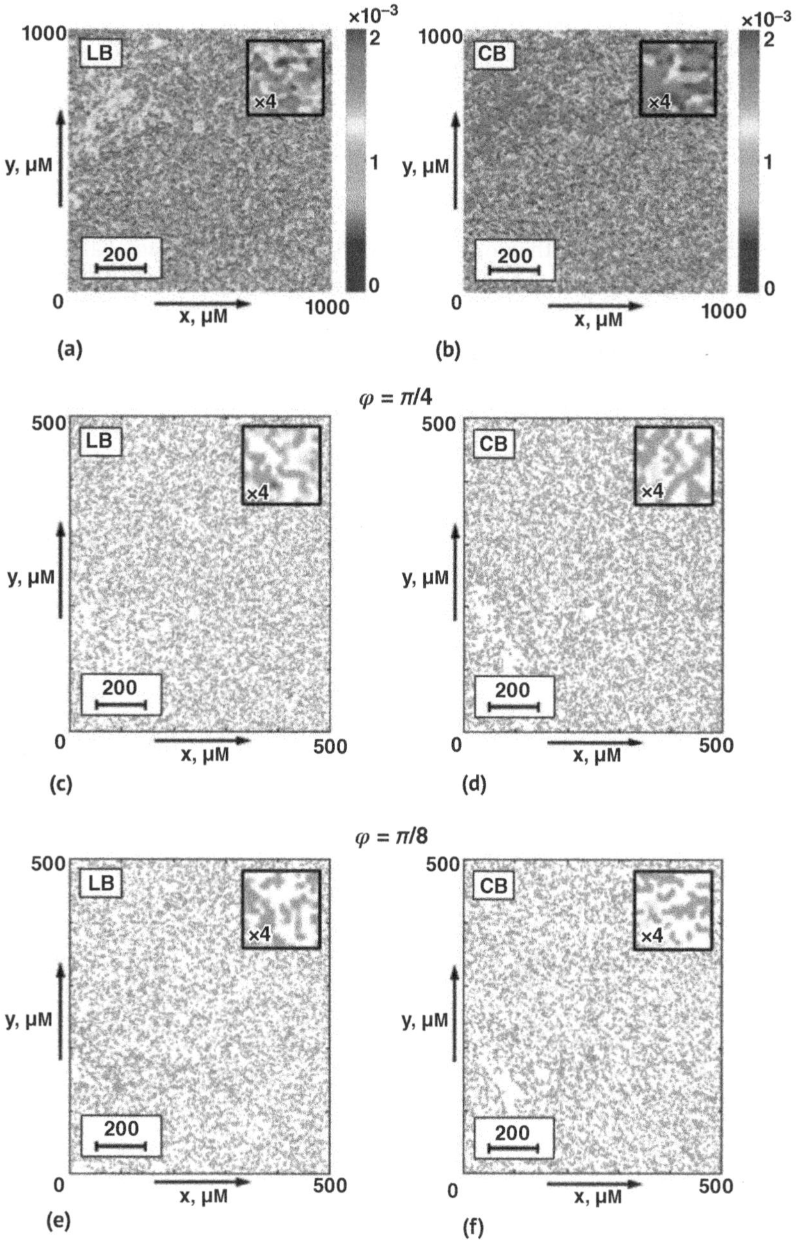

Figure 13.5 Jones-matrix reconstructed integral (a), (b), layer-by-layer (c), (d) and (e), (f) maps of phase anisotropy $JT = \begin{pmatrix} LB(Re\, j_{pk}) \\ CB(Re\, j_{pk}) \end{pmatrix}$ – coordinate distributions of LB (a), (c), (e) and CB (b), (d), (f) of the supramolecular networks of the BPF1 sample.

Typically, this process involves several key (structural) stages:

The initial stage is basic or “prefractal.” The small-scale $(2\,\mu m \leq L_{i(\min)} \leq 10\,\mu m)$ molecular domains crystallization process occurs. The topographic structure of such a measure is $JT^{LB}(L_{i(\min)})$ (albumin molecules crystallites) and circular $JT^{CB}(L_{i(\min)})$ (globulin molecules crystallites). This fact is shown by the lack of a consistent inclination angle of the approximating curve. $H(\vartheta)$ to the distributions $\ln\left(PD(JT^{LB,CB}\right) - \ln L^{-1}$ – Figure 13.7.

The second involves intermediate or “quasi-fractal” BPF1 crystallite molecular birefringence domains that are sequentially combined into larger structures $(20\,\mu m \leq L_{i(\text{middle})} \leq 200\,\mu m \div 400\,\mu m)$ spatially structured needle crystals networks. Thus, a birefringence maps $JT^{LB;CB}(h_r, L_{i(\text{middle})})$ scale-self-like (h_j) measures set is formed.

Table 13.6 The first to fourth central statistical moments that describe the anisotropy maps JT of the BPF1 phase.

JT	LB	CB
$M_1 \times 10^{-3}$	1.21 ± 0.077	1.03 ± 0.059
$M_2 \times 10^{-3}$	0.22 ± 0.014	0.18 ± 0.011
M_3	0.34 ± 0.019	0.42 ± 0.023
M_4	0.48 ± 0.028	0.51 ± 0.028
	$\varphi_k = \pi/4$	
$M_1 \times 10^{-3}$	1.09 ± 0.058	0.94 ± 0.052
$M_2 \times 10^{-3}$	0.17 ± 0.009	0.15 ± 0.008
M_3	0.42 ± 0.025	0.51 ± 0.027
M_4	0.54 ± 0.029	0.63 ± 0.024
	$\varphi_k = \pi/8$	
$M_1 \times 10^{-3}$	0.95 ± 0.052	0.89 ± 0.049
$M_2 \times 10^{-3}$	0.14 ± 0.008	0.12 ± 0.007
M_3	0.53 ± 0.029	0.62 ± 0.037
M_4	0.64 ± 0.035	0.73 ± 0.039

These processes in BPF1 polycrystalline self-assembly demonstrate linear slope plots that approximate $H(\vartheta)$ dependences $\ln\left(PD(JT^{LB,CB}\right) - \ln L^{-1}$ (Figure 13.7).

The third is the final or "multifractal" set. These local fractal measures create maps of biological crystals $JT^{LB;CB}\left(L_{i(middle)}\right)$ is combined into a single macroscale $\left(200\,\mu m \div 400\,\mu m \leq L_{i(max)} \leq 1000\,\mu m\right)$ supramolecular BPF1 network. The structural elements sizes corresponding range in coordinate distributions $JT^{LB;CB}\left(L_{i(max)}\right)$ has the fractal structure – dependence $\ln\left(PD(JT^{LB,CB}\right) - \ln L^{-1}$ is characterized by a single slope angle $(\vartheta_i)\begin{cases} h_1\left(LB, L_{i(max)}\right) \Longleftrightarrow \vartheta_1\left(LB, L_{i(max)}\right) = \text{const.}; \\ h_2\left(CB, L_{i(middle)}\right) \Longleftrightarrow \vartheta_2\left(CB, L_{i(max)}\right) = \text{const.} \end{cases}$

Therefore, it can be concluded that the BPF1 polycrystalline network self-assembly process results in the formation of a mainly multifractal supramolecular optically anisotropic network structure. This BPF1 structure exhibits optical anisotropy $JT^{LB;CB}(L_i, h_r = \text{const.})$ partial scale-self-like (fractal) distributions, which are characterized by the fractal dimension spectra $F(h_r)$ of LB and CB. Moreover, the max $F(h_r)$ extreme values will fall on the supramolecular polycrystalline formation's large scales $\left(L_{i(max)}\right)$, while minimum values will relate to small-scale formations of molecular optical anisotropy.

We employed the WTMM method to examine BPF1's optically anisotropic multifractal structure in a more thorough and objective manner.

Figure 13.7 illustrates the maximum magnitudes (wavelet coefficient amplitudes) of the skeletons – with "dark gray dots" indicating the maxima and "light gray dots" showing the minima – of the wavelet coefficient lines in the maps $W^{LB,CB}(a, b)$ and BPF1 maps JT^{LB} (fragments (b), (c)) and JT^{CB} ((e), (f)) multifractal spectra $F(h_i)$.

A comparative analysis of results from the WTMM method showed that:

Supramolecular BPF1 network exhibits optically anisotropic properties $JT^{LB;CB}$ is characterized by a complex scale-like structure, which is characterized by individual multifractal spectra $F^{LB}(h_r)$ and $F^{CB}(h_r)$ for LB and CB (Figure 13.7c, f).

The BPF1 maps $JT^{LB;CB}$ multifractal spectra $F^{LB;CB}(h_r)$ have an asymmetric structure with a distinct extremum and fractal dimension changes of $1.35 \lesssim h_r \lesssim 1.75$ range.

For maps of BPF1 LB (JT^{LB}) the multifractal spectra $F^{LB}(h_r)$ are "narrower" and more asymmetric (Figure 13.7c) than similar multifractal spectral dependences $F^{CB}(h_r)$, which are calculated for CB (Figure 13.7f) maps JT^{CB}.

The phase anisotropy maps of multifractal spectra $F^{LB}(h_r)$ and $F^{CB}(h_r)$ main extremes are localized in the different parts of the fractal dimension scale – $\begin{cases} \max\left(F^{LB}\right) \Leftrightarrow h_r = 1.56; \\ \max\left(F^{CB}\right) \Leftrightarrow h_r = 1.68 \end{cases}$.

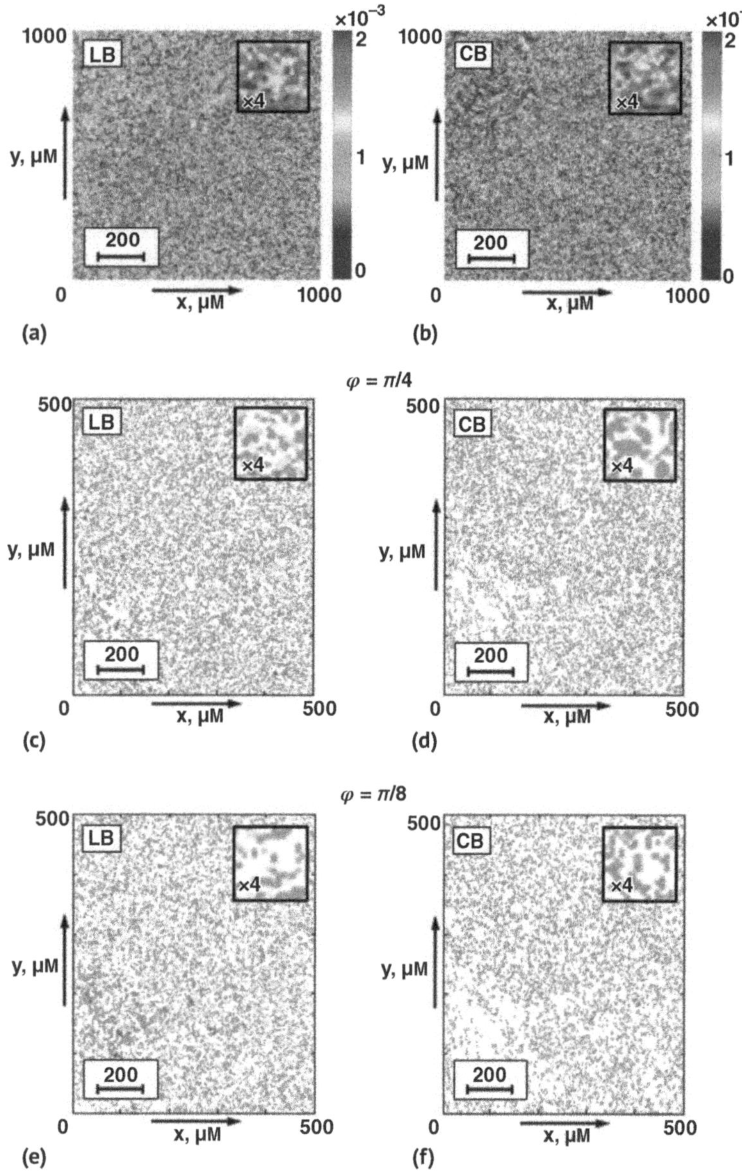

Figure 13.6 Jones-matrix reconstructed integral (a), (b), layer-by-layer (c), (d), and (e), (f) maps of phase anisotropy $T = \begin{pmatrix} \mathrm{LB}(Re\, j_{pk}) \\ \mathrm{CB}(Re\, j_{pk}) \end{pmatrix}$ - coordinate distributions of LB (a), (c), (e) and CB (b), (d), (f) of the supramolecular networks of the BPF1 sample.

The range of fractal dimensions Δh_r is more significant for the spectrum $F^{\mathrm{CB}}(h_r)$ than for $F^{\mathrm{LB}}(h_r) - \Delta h_r(\mathrm{CB}) > \Delta h_r(\mathrm{LB})$.

Physically speaking, the results can be connected to the following points:

First, each local optically anisotropic degree that develops during dehydration in the BPF1 maps is associated with its fractal dimension h_i value, the probability, and magnitude of which $(F(h_r))$ is determined by the specifics of self-assembled supramolecular crystallite complexes.

Second, the highest probability of the fractal dimension is inherent in the BPF1 birefringence distributions LB; CB $(200\,\mu\mathrm{m} \div 400\,\mu\mathrm{m} \leq L_{i(\max)} \leq 1000\,\mu\mathrm{m})$.

Third, from the perspective of fractal geometry theory, in extreme cases, the spatial organization of supramolecular needle crystal dendritic network architectures can form a straight line. In such boundary conditions, the multifractal spectrum reduces to a delta function with the base value of the fractal dimension $h_{\mathrm{LB}} \Rightarrow 1.0$. Spherulite supramolecular networks that create fractal patterns in CB maps exhibit a different scenario. In this case, the boundary geometry is characterized by polycrystalline architectonics in a planar form, which aligns with a different fractal dimension value $h_{\mathrm{LB}} \Rightarrow 2.0$.

Table 13.7 The first to fourth central statistical moments that describe the anisotropy maps JT of the BPF2 phase.

JT	LB	CB
$M_1 \times 10^{-3}$	1.12 ± 0.068	1.23 ± 0.071
$M_2 \times 10^{-3}$	0.19 ± 0.011	0.24 ± 0.014
M_3	0.39 ± 0.022	$0.34 \pm 0.,019$
M_4	0.54 ± 0.029	0.45 ± 0.025
	$\varphi_k = \pi/4$	
$M_1 \times 10^{-3}$	0.99 ± 0.054	1.14 ± 0.069
$M_2 \times 10^{-3}$	0.15 ± 0.008	0.21 ± 0.012
M_3	0.48 ± 0.027	0.44 ± 0.027
M_4	0.63 ± 0.035	0.53 ± 0.029
	$\varphi_k = \pi/8$	
$M_1 \times 10^{-3}$	0.91 ± 0.048	0.98 ± 0.058
$M_2 \times 10^{-3}$	0.11 ± 0.006	0.12 ± 0.007
M_3	0.65 ± 0.036	0.56 ± 0.031
M_4	0.76 ± 0.041	0.67 ± 0.036

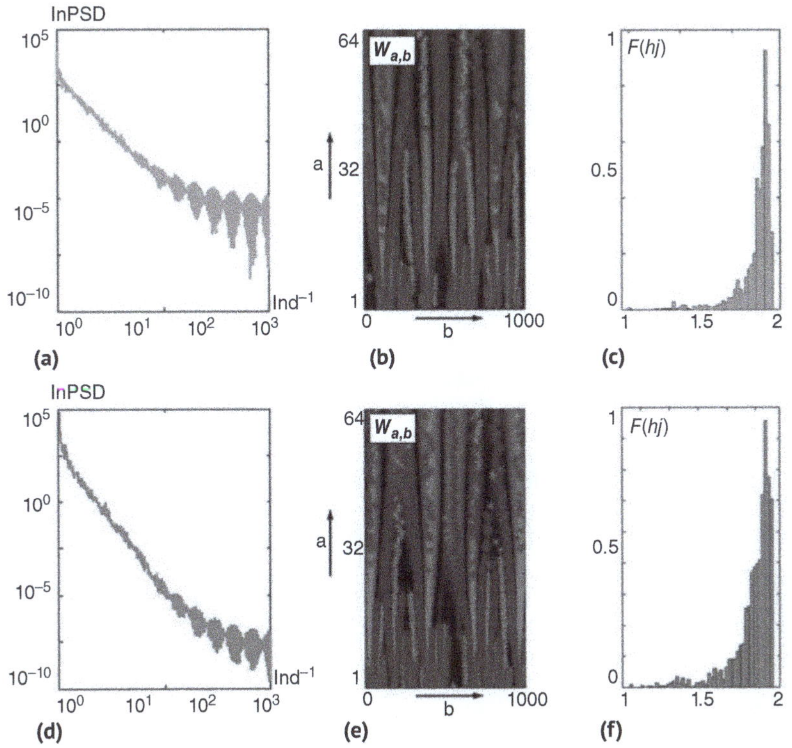

Figure 13.7 Power spectra with logarithmic dependencies (a), (d); maxima skeletons of wavelet transform modules (b), (e); and multifractal spectra of BPF1 phase anisotropy maps (c), (f).

In a real physical situation, such trends appear in various localization multifractal spectra at the main extremes $(\max(F^{CB}(h_r))$ and $\max(F^{LB}(h_r)))$. In this case, the variation range of the fractal dimensions Δh_i is larger for the spectrum $F^{CB}(h_r)$ than for $F^{LB}(h_r) - \Delta h_r(CB) > \Delta h_r(LB)$.

A quantitative estimate of the multifractal spectrum $F^{LB}(h_r)$ and $F^{CB}(h_r)$ is illustrated in Table 13.8.

Comparison of physical and statistical analysis maps $JT^{LB;CB}$ of the multifractal structure revealed a correlation between them. Namely, $M_{i=1;2}(F^{LB}(h_r)) < M_{i=1;2}(F^{CB}(h_r))$, which is due to the extrema $F^{LB}(h_r)$ and $F^{CB}(h_r)$ different localization (h_r(CB) > h_r(LB)), as well as differences between the fractal dimensions ranges – Δh_r(CB) > Δh_r(LB). The reverse trend $M_{i=3;4}(F^{LB}(h_r)) > M_{i=3;4}(F^{CB}(h_r))$ appears in the third and fourth-order statistical moments, which describe the skewness and kurtosis of the LB and CB multifractal spectra distributions.

The fractal analysis of the BPF2 polycrystalline structure revealed that the primary difference from the previously obtained BPF1 dependencies (Figure 13.7a, d) is the presence of a greater number of quasilinear segments (h_r) with a smaller geometric length (d_i) presence. Alongside this, oscillations in the distribution of optical anisotropy parameter power spectra show logarithmic dependencies (see Figure 13.8a, d). These observations suggest differences in the BPF2 maps JT^{LB} and JT^{CB} exist, as well as multifractal (with more fractal measures) construction [55–73].

Figure 13.8 presents the fractal analysis results of the BPF2 maps $JT^{LB;CB}$, specifically in Figure 13.8b, c, e, and f.

The BPF2 maps $F^{LB;CB}(FB, h_r)$ WTMM analysis results confirmed the presence of multifractal spectra, the parameters of which differ from those previously obtained for the supramolecular films network dependencies $F^{LB}(h_r)$ and $F^{CB}(h_r)$ (Figure 13.7).

Table 13.8 The multifractal spectra $F^{LB}(h_r)$ and $F^{CB}(h_r)$ statistical characteristics – BPF1.

$M_{i=1;2;3;4}$	$F^{LB}(h_r)$	$F^{CB}(h_r)$
M_1	1.31 ± 0.065	1.65 ± 0.076
M_2	0.11 ± 0.008	0.21 ± 0.012
M_3	1.47 ± 0.074	0.91 ± 0.047
M_4	2.03 ± 0.092	1.46 ± 0.076

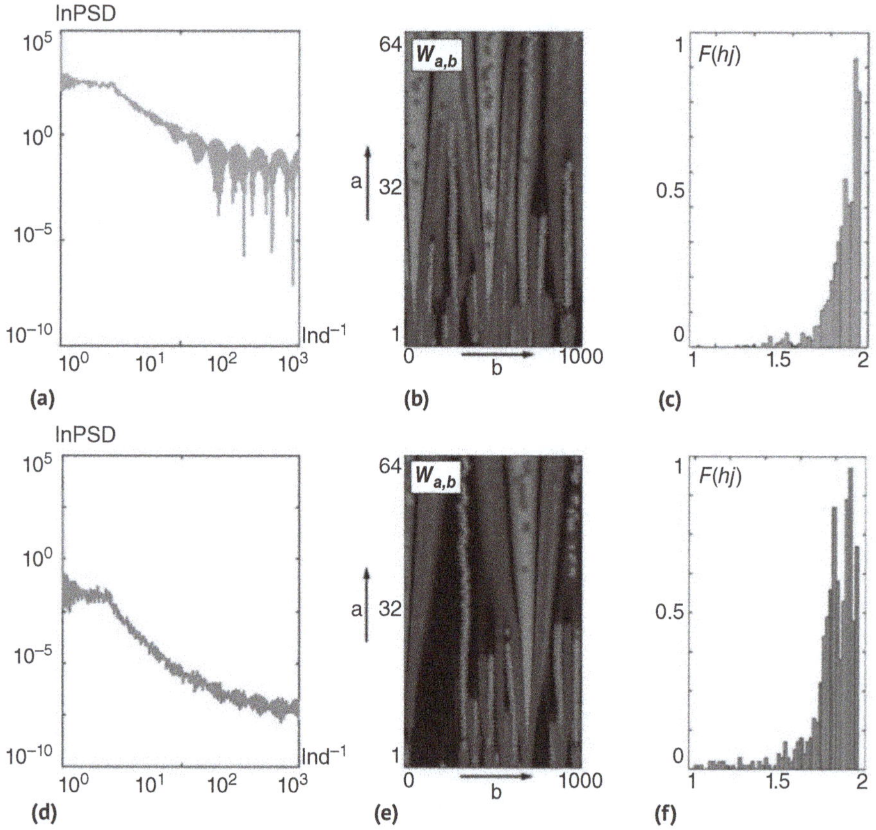

Figure 13.8 Power spectra logarithmic dependences (a), (d); wavelet transform modules' maxima skeletons (b), (e); and multifractal spectra of BPF2 phase anisotropy maps (c), (f).

In particular, the BPF2 maps $JT^{LB;CB}$ multifractal spectra $F^{LB;CB}(h_r)$ are less asymmetric and have greater width of fractal dimensions $1.3 \lesssim h_r \lesssim 1.89$. At the same time, these multifractal spectra $F^{LB}(h_r)$ and $F^{CB}(h_r)$ characterized the differences between main extremes localization – $\begin{cases} \max(F^{LB}) \Leftrightarrow h_r = 1.45; \\ \max(F^{CB}) \Leftrightarrow h_r = 1.83. \end{cases}$

The BPF2 self-assembled supramolecular network's multifractal structure $JT^{LB;CB}$ shows specificity that can be linked to numerous birefringent crystallites, including linear ones like albumin needle crystals and circular ones such as globulin microspherulites.

This causes the number of fractal dimensions in the BPF2 maps, which characterize local scale-like measures, to increase. Consequently, the BPF2 supramolecular crystallite complexes form self-assembled multifractal spectra $F^{LB;CB}(h_r)$ half-width increases.

BPF2 phase anisotropy maps show multifractal spectra. $F^{LB}(h_r)$ and $F^{CB}(h_r)$ quantitative statistical assessment is illustrated in Table 13.9.

The data analysis presented in Table 13.9 confirmed the physical interpretations of BPF2 maps JT^{LB} and JT^{CB} multifractal structure. In particular, small differences were observed between the values of the first- and second-order statistical moments – $M_{i=1;2}(F^{LB}(h_r)) \gtrsim M_{i=1;2}(F^{CB}(h_r))$.

This fact reflects the "statistical weight" of the fractal measures in the BPF2, related to the LB and CB of the crystallite supramolecular network. Consequently, this network can be described as "dendritic-spherulitic," unlike the "dendritic" nature of the BPF2 polycrystalline structure.

Simultaneously, the distinctiveness of dendritic and spherulite fractal measures is reflected in notable differences in the statistical moments of the third and fourth orders, which describe the multifractal spectra $F^{CB}(h_r)$ and $F^{LB}(h_r)$ skewness and kurtosis – $M_{i=3;4}(F^{LB}(h_r)) > M_{i=3;4}(F^{CB}(h_r))$.

Therefore, we have shown that during blood plasma dehydration, the self-assembly of a supramolecular network results in films with a complex, optically anisotropic structure. The polarization characteristics of these films are described by their individual multifractality spectra $F^{CB}(h_r)$ and $F^{LB}(h_r)$ for various blood films types (BPF1 and BPF2).

Figures 13.9 and 13.10 series panels (a)–(f), as well as Tables 13.10 and 13.11, present the BPF2 linear $(JT^{LB}(\varphi_k))$ and circular $(JT^{LB}(\varphi_k))$ birefringence multifractal and statistical analysis results in the phase planes $\varphi_k = \pi/4$ and $\varphi_k = \pi/8$, respectively.

Within the fractal analysis framework (see Figures 13.9 and 13.10, panels (a), (d)) in different phase sections ($\varphi_k = {}^{\pi}/_{4}$ and $\varphi_k = {}^{\pi}/_{8}$), it was found that with a decrease in the phase scanning parameter $\varphi_k \downarrow$ value, the number (h_j) decreases and the linear sections geometric extent (L_i) in dependencies $\ln\left(\mathrm{PD}(JT^{LB,CB})\right) - \ln L^{-1}$ increases. Meanwhile, the decrease in oscillation amplitude suggests that the measures of crystallites are increasing, reflecting a growth in the large-scale, self-similar organization within the supramolecular BPF1 network.

Multifractal structure $F^{LB;CB}(h_i, \varphi_k)$ statistical analysis revealed the following $JT^{LB;CB}(h_r, \varphi_k = {}^{\pi}/_{4}; {}^{\pi}/_{8})$ regularities.

Individual multifractal spectra for linearly and circularly birefringent fractal measures, which are objectively characterized by the first- to fourth-order central statistical moments – $M_{i=1;2;3;4}(F^{LB}(h_r, \varphi_k)) \neq M_{i=1;2;3;4}(F^{CB}(h_r, \varphi_k))$, Tables 13.10 and 13.11.

With phase parameter $\varphi_k \downarrow$ values a decrease ($\downarrow$) has been progressively declining in the spectra $F^{LB;CB}(h_r, \varphi_k)$ half-width, Figures 13.9 and 13.10, panels b, c, e, and f;

The following magnitude $M_{i=1;2;3;4}(F^{LB;CB}(h_r, \varphi_k \downarrow)) \Rightarrow \begin{pmatrix} M_{i=1;2}(F^{LB;CB}(h_r, \varphi_k)) \downarrow \\ M_{i=3;4}(F^{LB;CB}(h_r, \varphi_k)) \uparrow \end{pmatrix}$ phase transformations $\varphi_k \downarrow$, Tables 13.6 and 13.7.

Table 13.9 Statistical characteristics of BPF2 multifractal spectra $F^{LB}(h_r)$ and $F^{CB}(h_r)$.

$M_{i=1;2;3;4}$	$F^{LB}(h_i)$	$F^{CB}(h_i)$
M_1	1.49 ± 0.075	1.76 ± 0.086
M_2	0.15 ± 0.008	0.18 ± 0.009
M_3	1.07 ± 0.005	0.69 ± 0.041
M_4	1.32 ± 0.07	0.99 ± 0.048

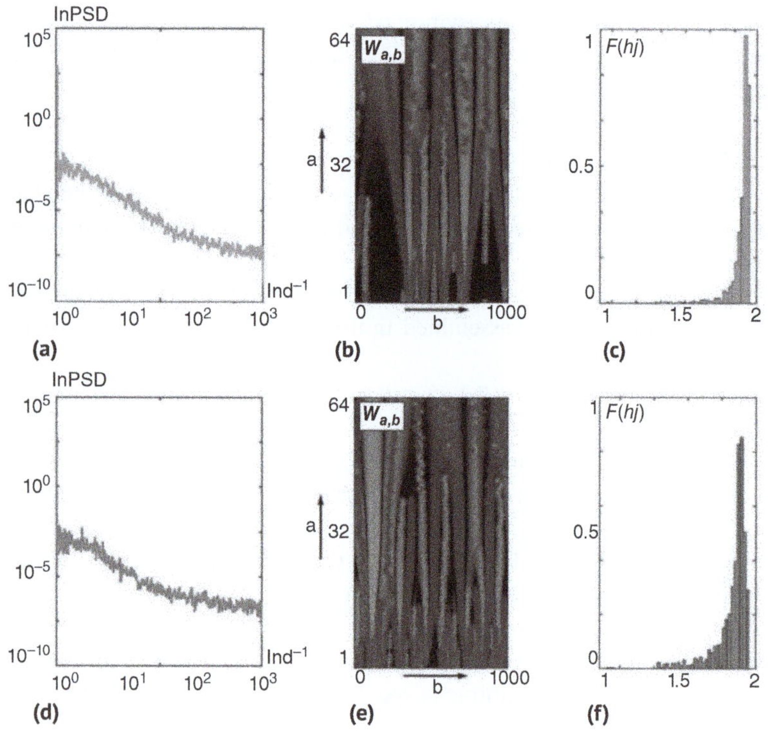

Figure 13.9 Power spectra logarithmic dependences (a), (d); wavelet transform modules maxima skeletons (b), (e) and multifractal spectra of BPF1 phase anisotropy maps $JT^{LB;CB}(\varphi_k = \pi/4)$ (c), (f).

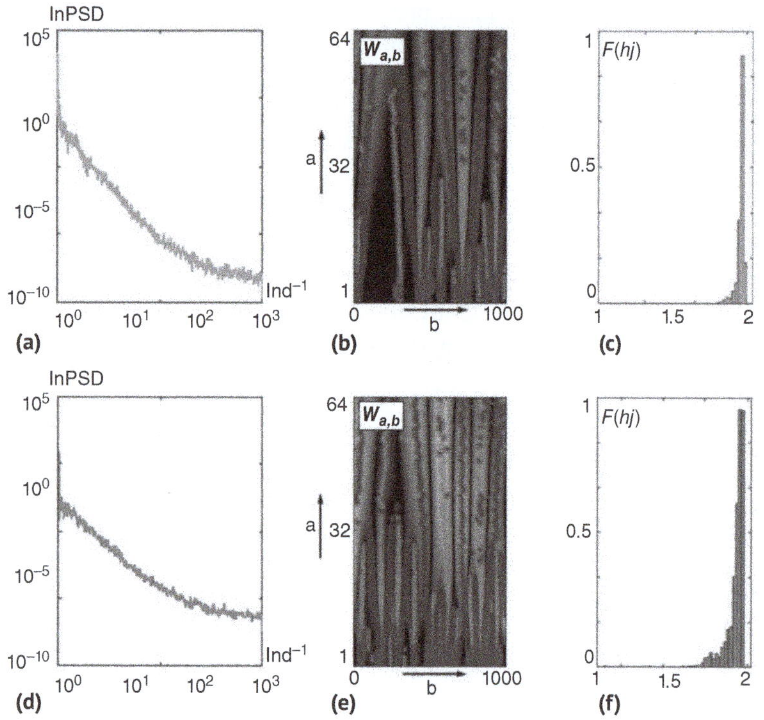

Figure 13.10 Power spectra logarithmic dependencies (a), (d); maxima skeletons of wavelet transform modules (b), (e); and multifractal spectra of BPF1 phase anisotropy maps. $JT^{LB;CB}(\varphi_k = \pi/8)$ (c), (f).

Table 13.10 Statistical characteristics of $JT^{LB;CB}$ multifractal spectra $F^{LB;CB}(h_r, \varphi_k = \pi/4)$ – BPF1.

JT	LB	CB
M_1	1.31 ± 0.065	1.65 ± 0.076
M_2	0.09 ± 0.005	0.18 ± 0.009
M_3	2.64 ± 0.14	1.63 ± 0.093
M_4	3.35 ± 0.19	2.24 ± 0.13

Table 13.11 Statistical characteristics of $JT^{LB;CB}$ multifractal spectra $F^{LB;CB}(h_r, \varphi_k = \pi/8)$ – BPF2.

JT	LB	CB
M_1	1.23 ± 0.065	1.34 ± 0.076
M_2	0.07 ± 0.004	0.16 ± 0.008
M_3	3.24 ± 0.19	1.98 ± 0.11
M_4	4.41 ± 0.28	2.91 ± 0.16

The third- fourth-order statistical moments, which characterize the distributions $M_{3;4}(F^{LB;CB}(h_r, \varphi_k \downarrow))$, were the most sensitive and, accordingly, diagnostically promising to changes in the matrix reproduced maps $JT^{LB;CB}$ multifractal structure:

The change $\Delta M_{3;4}^{LB}(\varphi_k \downarrow)$ range is from 2 to 2.5 times;
The change $\Delta M_{3;4}^{CB}(\varphi_k \downarrow)$ range is from 1.6 to 2.1 times.

From a physical perspective, the results observed are due to the fact that during phase scanning $(\varphi_k \downarrow)$, the BPF1 geometric depths decrease. Consequently, the Jones matrix more accurately captures the large-scale, self-similar (monofractal) optical anisotropy of the polycrystalline supramolecular network. This is primarily due to its birefringence parameter' $JT^{LB;CB}(h_r, \varphi_k)$ distribution is approaching a fractal to form a fairly narrow multifractal spectrum $F^{LB;CB}(h_r, \varphi_k)$.

For multifractal measures $JT^{CB}(h_r, \varphi_k)$, optically active crystals exhibit a similar but less pronounced trend across a broader range of fractal dimensions in the spectrum $F^{CB}(h_r, \varphi_k)$, Figures 13.7, 13.9, and 13.11, panels b, c, e, f.

Figures 13.11 and 13.12 series fragments a–f, along with Tables 13.12 and 13.13, display the results of the multifractal and statistical analysis of BPF2 linear $(JT^{LB}(\varphi_k))$ and circular $(JT^{LB}(\varphi_k))$ birefringence in the phase planes $\varphi_k = \pi/4$ and $\varphi_k = \pi/8$, respectively.

Within the BPF2 maps (Figures 13.8 and 13.9, panels a, d) in various phase sections ($\varphi_k = \pi/4$ and $\varphi_k = \pi/8$) fractal analysis framework, it was found (similar as for BPF1) that with a decrease in the phase scanning parameter $\varphi_k \downarrow$ value, number (L_i) decreases and the linear sections in dependencies $\ln\left(PD(JT^{LB,CB}\right) - \ln L^{-1}$ geometric extent (d_i) increases. Simultaneously, the oscillation amplitude decreases, indicating that the large-scale self-similar organization of the supramolecular BPF2 network's crystallites is increasing.

The BPF2 WTMM method analysis showed generally similar trends across all phase-scanning planes for these films, though they were less clearly defined.

A similar difference is observed in the values of the statistical moments that define the multifractal spectra across the phase planes, $\begin{pmatrix} M_{i=1;2}(F^{CB}(h_r, \varphi_k)) < M_{i=1;2}(F^{LB}(h_r, \varphi_k)) \\ M_{i=3;4}F^{CB}(h_r, \varphi_k) > M_{i=3;4}(F^{LB}(h_r, \varphi_k)) \end{pmatrix}$ – Tables 13.12 and 13.13.

When a phase parameter φ_k decrease implemented previously discovered statistical scenario – $M_{i=1;2;3;4}(F^{LB;CB}(h_r, \varphi_k \downarrow)) \Rightarrow \begin{pmatrix} M_{i=1;2}(F^{LB;CB}(h_r, \varphi_k)) \downarrow \\ M_{i=3;4}(F^{LB;CB}(h_r, \varphi_k)) \uparrow \end{pmatrix}$, Tables 13.12 and 13.13.

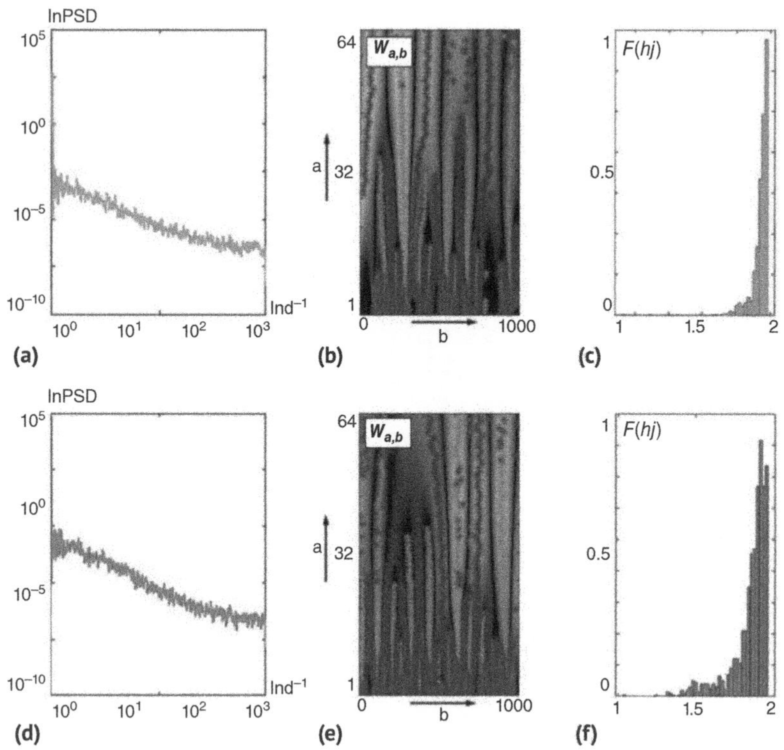

Figure 13.11 Power spectra logarithmic dependences (a), (d); maxima skeletons of wavelet transform modules (b), (e); and multifractal spectra of BPF2 phase anisotropy maps. $JT^{LB;CB}(\varphi_k = \pi/4)$ (c), (f).

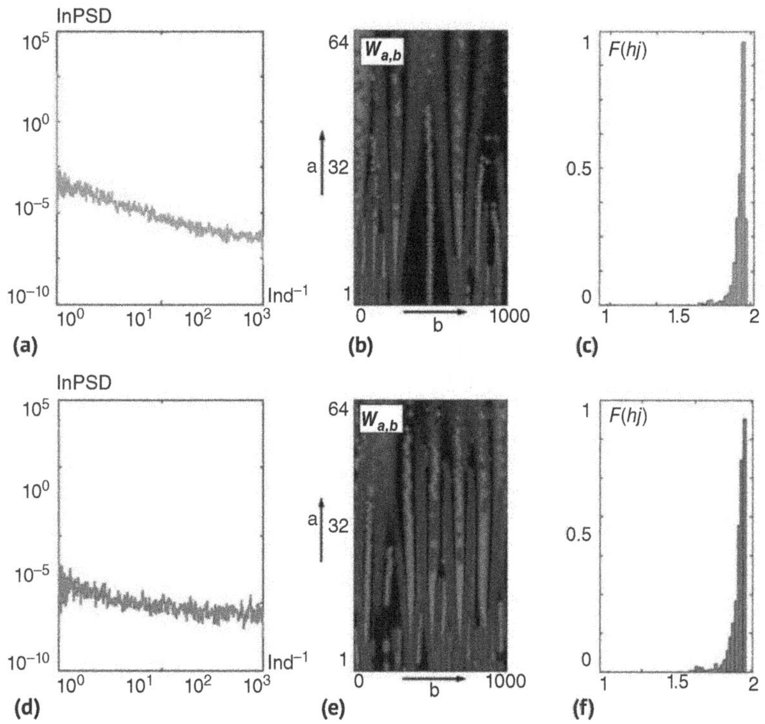

Figure 13.12 Power spectra with logarithmic dependencies (a), (d); wavelet transform module maxima skeletons (b), (e); and multifractal spectra of BPF2 phase anisotropy maps. $JT^{LB;CB}(\varphi_k = \pi/8)$ (c), (f).

Table 13.12 Statistical characteristics of BPF2 maps $F^{LB;CB}(h_r, \varphi_k = \pi/4)$.

JT	LB	CB
M_1	1.41 ± 0.075	1.67 ± 0.086
M_2	0.13 ± 0.008	0.16 ± 0.009
M_3	1.49 ± 0.074	1.18 ± 0.053
M_4	1.81 ± 0.082	1.41 ± 0.076

Table 13.13 Statistical characteristics of BPF2 maps $F^{LB;CB}(h_r, \varphi_k = \pi/8)$.

JT	LB	CB
M_1	1.34 ± 0.079	1.51 ± 0.098
M_2	0.07 ± 0.004	0.11 ± 0.011
M_3	1.83 ± 0.079	1.48 ± 0.065
M_4	2.44 ± 0.099	2.02 ± 0.086

The third- to fourth-order statistical moments, which characterize the distributions $F^{LB;CB}(h_r, \varphi_k \downarrow)$, were also the most sensitive:

The change $\Delta M_{3;4}^{LB}(\varphi_k \downarrow)$ range is from 1.3 to 1.7 times.
The change $\Delta M_{3;4}^{CB}(\varphi_k \downarrow)$ range is from 1.4 to 1.8 times.

Therefore, in the BPF2 phase scanning $(\varphi_k \downarrow)$ case demonstrated the process of "approximation" $JT^{LB;CB}(h_r, \varphi_k)$ large-scale self-similar measures monofractal to which they correspond fairly narrow multifractal spectra $F^{LB;CB}(h_r, \varphi_k)$.

However, the specificity of polycrystalline BPF2 scale-self-similar architectonics networks differs from that of BPF1 fractal measures in their construction $F^{LB;CB}(h_r, \varphi_k)$, Tables 13.12 and 13.13 [74–84].

13.2.2 Conclusions

A method for layer-by-layer Jones-matrix introscopy was created and experimentally validated to examine the optically anisotropic structure of supramolecular networks in blood plasma films.

Maps showing LB and CB across various phase regions of the complex amplitude field of blood plasma films were measured and analyzed.

The accuracy of differential diagnosis for late COVID-19 consequences based on statistical markers $M_{3;4}(JT^{LB;CB}(h_r, \varphi_k))$ does not exceed satisfactory $(Ac(M_{3;4}, \varphi_k = \pi/8) = 83\% - 85\%)$ levels.

Maps of blood plasma films $JT^{LB;CB}$ power spectra logarithmic dependences $\ln(PD(JT^{LB;CB})) - \ln L^{-1}$ analysis revealed supramolecular polycrystalline network's multifractal structure signs – approximating curve $H(\vartheta)$ linear sections (monofractal measures) for different geometric dimension d_i presence.

For the first time, the wavelet transform was applied to the maxima of the modulus in LB and CB distributions to evaluate the multifractal structure of blood plasma films, focusing on optically anisotropic supramolecular networks via layer-by-layer maps.

The distributions of fractal dimension spectra in the layered optical anisotropy maps were derived using algorithms and analyzed through statistical methods.

Multifractal analysis identified statistically significant markers that offer differential diagnostic insights into the architectonics of supramolecular networks in blood plasma during COVID-19 transformation. These multifractal statistical markers allow for very good $(Ac(M_{3;4}) = 92\%)$ and excellent $(Ac(M_{3;4}, \varphi_k = \pi/8) = 97\%)$ accuracy in diagnosing the long-term effects of COVID-19 over periods of one to three years.

References

1 Shankaran V, Walsh Jr. JT, Maitland DJ. Comparative study of polarized light propagation in biological tissues. J Biomed Opt. 2002;7(3):300–306.

2 Daubechies I, Ten Lectures on Wavelets. Society for Industrial and Applied Mathematics; 1992.

3 Cohen A, Daubechies I, Vial P. Wavelets on the interval and fast wavelet transforms. Appl Comput Harmon Anal. 1993;1(1):54–81.

4 Meyer Y. Wavelets: Algorithms & Applications. Philadelphia: Society for Industrial and Applied Mathematics (SIAM); 1993.

5 Mallat S. A theory for multiresolution signal decomposition: the wavelet representation. IEEE Trans Pattern Anal Mach Intell. 1989;11(7):674–693.

6 Daubechies I. Wavelets on the interval. In Progress in Wavelet Analysis and Applications: Proceedings of the International Conference "Wavelets and Applications", Toulouse, France. 1992.

7 Farge M. Wavelet transforms and their applications to turbulence. Annu Rev Fluid Mech. 1992;24(1):395–458.

8 Ma H, He H, Ramella-Roman JC. Mueller matrix microscopy. In Ramella-Roman JC, Novikova T, editors. Polarized Light in Biomedical Imaging and Sensing. Cham: Springer; 2023. p. 281–321.

9 He H, Liao R, Zeng N, Li P, Chen Z, Liu X, et al. Mueller matrix polarimetry—an emerging new tool for characterizing the microstructural feature of complex biological specimen. J Lightwave Technol. 2019;37:2534–2548.

10 Wang Y, He H, Chang J, Zeng N, Liu S, Li M, et al. Differentiating characteristic microstructural features of cancerous tissues using Mueller matrix microscope. Micron 2015;79:8–15.

11 Abbasian V, Moradi A-R. Microsphere-assisted super-resolved Mueller matrix microscopy. Opt Lett. 2020;45:4336–4339.

12 Zhang Y, Lee SY, Zhang Y, Furst D, Fitzgerald J, Ozcan A. Wide-field imaging of birefringent synovial fluid crystals using lens-free polarized microscopy for gout diagnosis. Sci Rep 2016;6:28793.

13 de Boer JF, Milner TE. Review of polarization sensitive optical coherence tomography and Stokes vector determination. J Biomed Opt. 2002;7(3):359–371. https://doi.org/10.1117/1.1483879.

14 Ku S, Mahato KK, Mazumder N. Polarization-resolved Stokes-Mueller imaging: a review of technology and applications. Lasers Med Sci. 2019;34(7):1283–1293. https://doi.org/10.1007/s10103-019-02752-1.

15 He Y, Li K, Li W, Qiu Y, Li D, Wang C, et al. Polarization coherency matrix tomography. J Biophotonics. 2023;16(9):e202300093. https://doi.org/10.1002/jbio.202300093.

16 Yang D, Yuan Z, Hu M, Liang Y. Zebrafish brain and skull imaging based on polarization-sensitive optical coherence tomography. J Biophotonics. 2022;15(12):e202200112. https://doi.org/10.1002/jbio.202200112.

17 Park JE, Xin Z, Kwon DY, Kim SW, Lee H et al. Application of polarization sensitive-optical coherence tomography to the assessment of phase retardation in subpleural cancer in rabbits. Tissue Eng Regen Med. 2021;18(1):61–69. https://doi.org/10.1007/s13770-020-00318-9.

18 Willemse J, Gräfe MGO, Verbraak FD, de Boer JF. In vivo 3D determination of peripapillary scleral and retinal layer architecture using polarization-sensitive optical coherence tomography. Transl Vis Sci Technol. 2020;9(11):21. https://doi.org/10.1167/tvst.9.11.21.

19 Killeen AA, Ossina N, McGlennen RC, Minnerath S, Borgos J, Alexandrov V, et al. Protein self-organization patterns in dried serum reveal changes in B-cell disorders. Mol Diag Ther. 2006;10:371–380.

20 Brutin D, Sobac B, Loquet B, Sampol J. Pattern formation in drying drops of blood. J Fluid Mech. 2011;667:85–95.

21 Heckbert SR, Vedder NB, Hoffman W, Winn RK, Hudson LD, Jurkovich GJ, et al. Outcome after hemorrhagic shock in trauma patients. J Trauma Acute Care Surg. 1998;45(3):545–549.

22 Johar RS, Smith RP. Assessing gravimetric estimation of intraoperative blood loss. J Gynecol Surg. 1993;9(3):151–154. https://doi.org/10.1089/gyn.1993.9.151.

23 Convertino VA, Lye KR, Koons NJ, Joyner MJ. Physiological comparison of hemorrhagic shock and V˙ O2max: a conceptual framework for defining the limitation of oxygen delivery. Exp Biol Med (Maywood). 2019;244(8):690–701. https://doi.org/10.1177/1535370219846425.

24 Lemmer Jr. JH, Despotis GJ. Antithrombin III concentrate to treat heparin resistance in patients undergoing cardiac surgery. J Thorac Cardiovasc Surg. 2002;123:213–217. https://doi.org/10.1067/mtc.2002.119060.

25 Verma AK, Xu D, Garg A, Blaber AP, Tavakolian K. Effect of aging on muscle-pump baroreflex of individual leg muscles during standing. Front Physiol. 2017;8:767. https://doi.org/10.3389/fphys.2017.00767.

26 Swan HJ, Ganz W, Forrester J, Marcus H, Diamond G, Chonette D. Catheterization of the heart in man with use of a flow-directed balloon-tipped catheter. N Engl J Med. 1970;283(9):447–451. https://doi.org/10.1056/NEJM197008272830902.

27 Bercovitz RS, Shewmake AC, Newman DK, Niebler RA, Scott JP, Stuth E, et al. Validation of a definition of excessive postoperative bleeding in infants undergoing cardiac surgery with cardiopulmonary bypass. J Thorac Cardiovasc Surg. 2018;155(5):2112–2124. https://doi.org/10.1016/j.jtcvs.2017.12.038.
28 Hunter R. Foundations of Colloid Science. New York: Oxford University Press; 2004. p. 806.
29 Chen X, Chen PG, Ouazzani J, Liu Q. Numerical simulations of sessile droplet evaporating on heated substrate. Eur Phys J Spec Top. 2017;226(6):1325–1335.
30 Dash S, Garimella SV. Droplet evaporation on heated hydrophobic and superhydrophobic surfaces. Phys Rev E. 2014;89:042402.
31 Gleason K, Putnam SA. Microdroplet evaporation with a forced pinned contact line. Langmuir. 2014;30(34):10548–10555.
32 Hu H, Larson RG. Evaporation of a sessile droplet on a substrate. J Phys Chem B. 2020;106:1334–1344.
33 Vladmir Z, Wang JB, Yan XH. Human blood plasma crystal and molecular biocolloid textures – dismetabolism and genetic breaches. Nat Sci J Xiangtan Univ. 2001;23:18–127.
34 Abe S, Tanaka J, Kojima M, Kanamaru S, Hirata K, Yamashita K, et al. Cell-free protein crystallization for nanocrystal structure determination. *Sci Rep*. 2022;12:16031.
35 Salla GR, Kumar V, Miyamoto Y, Singh RP. Scattering of Poincare beams: polarization speckles. Opt Express. 2017;25(17): 19886–19893. https://doi.org/10.1364/OE.25.019886.
36 Li J, Yao G, Wang LV. Degree of polarization in laser speckles from turbid media: implications in tissue optics. J Biomed Opt. 2002;7(3):307–312. https://doi.org/10.1117/1.1483313.
37 Jiao S, Wang LV. Jones-matrix imaging of biological tissues with quadruple-channel optical coherence tomography. J Biomed Opt. 2002;7(3):350–358. https://doi.org/10.1117/1.1483878.
38 Miyazawa A, Makita S, Li E, Yamazaki K, Kobayashi M, Sakai S, et al. Polarization-sensitive optical coherence elastography. Biomed Opt Express. 2019;10(10):5162–5181. https://doi.org/10.1364/BOE.10.005162.
39 Zhu L, Makita S, Tamaoki J, Zhu Y, Mukherjee P, Lim Y, et al. Polarization-artifact reduction and accuracy improvement of Jones-matrix polarization-sensitive optical coherence tomography by multi-focus-averaging based multiple scattering reduction. Biomed Opt Express. 2023;15(1):256–276. https://doi.org/10.1364/BOE.509763.
40 Mandelbrot BB. Fractals and Multifractals: Noise, Turbulence and Galaxies. New York, NY, USA: Springer-Verlag; 1989.
41 Angelsky OV, Ushenko A, Ushenko YA, Pishak VP, Peresunko AP. Statistical, correlation, and topological approaches in diagnostics of the structure and physiological state of birefringent biological tissues. In Tuchin VV, editor, Handbook of Photonics for Biomedical Science. London, UK: CRC Press, Taylor & Francis Publishing; 2010. pp. 283–322.
42 Muzy, JF, Bacry E, Arneodo A. Multifractal formalism for fractal signals: the structure-function approach versus the wavelet-transform modulus-maxima method. Int J Bifurc Chaos. 1994;4:245–302.
43 Grassberger P. Estimating the Kolmogorov entropy from a chaotic signal. Phys Lett A. 1983;97:227–232.
44 Grassberger P, Procaccia I. Measuring the strangeness of strange attractors. Phys D Nonlinear Phenom. 1983;9:189–208.
45 Hentschel HGE, Procaccia I. The infinite number of generalized dimensions of fractals and strange attractors. Phys D Nonlinear Phenom. 1983;8:435–444.
46 Oldenbourg R. A new view on polarization microscopy. Nature. 1996;381:811–812.
47 Gurjar RS, Backman V, Perelman LT, Georgakoudi I, Badizadegan K, Itzkan I, et al. Imaging human epithelial properties with polarized lightscattering spectroscopy. Nat Med. 2001;7:1245–1248.
48 Ramella-Roman JC, Saytashev I, Piccini M. A review of polarization-based imaging technologies for clinical and preclinical applications. J Opt. 2020;22:123001.
49 Qi J, Elson DS. Mueller polarimetric imaging for surgical and diagnostic applications: a review. J Biophotonics. 2017;10:950–982.
50 Fujiwara H. Spectroscopic Ellipsometry: Principles and Applications. John Wiley & Sons; 2007.
51 Mendoza-Galván A, Muñoz-Pineda E, Ribeiro SJ, Santos MV, Järrendahl K, Arwin H. Mueller matrix spectroscopic ellipsometry study of chiral nanocrystalline cellulose films. J Opt. 2018;20:024001.
52 Azzam RMA. Stokes-vector and Mueller-matrix polarimetry. J Opt Soc Am A. 2016;33:1396–1408.
53 Azzam RMA. Mueller-matrix ellipsometry: a review. Proc SPIE. 1997;3121:369–405.
54 Song BK, Gu H, Zhu S, Jiang H, Chen X, Zhang C, et al. Broadband optical properties of graphene and HOPG investigated by spectroscopic Mueller matrix ellipsometry. Appl Surf Sci. 2018;439:1079–1087.
55 Jiang H, Ma Z, Gu H, Chen X, Liu S. Characterization of volume gratings based on distributed dielectric constant model using Mueller matrix ellipsometry. Appl Sci. 2019;9:698.

56 Chang J, He H, Wang Y, Huang Y, Li X, He C, et al. Division of focal plane polarimeter-based 3×4 Mueller matrix microscope: a potential tool for quick diagnosis of human carcinoma tissues. J Biomed Opt. 2016;21:056002.

57 Angelsky PO, Ushenko AG, Dubolazov AV, Sidor MI, Bodnar GB, Koval G, et al. The singular approach for processing polarization-inhomogeneous laser images of blood plasma layers. J Opt. 2013;15(4):044030.

58 Ushenko AG, Dubolazov AV, Ushenko VA, Novakovskaya OY. Statistical analysis of polarization-inhomogeneous Fourier spectra of laser radiation scattered by human skin in the tasks of differentiation of benign and malignant formations. J Biomed Opt. 2016;21(7):071110.

59 Garazdyuk MS, Bachinskyi VT, Vanchulyak OY, Ushenko AG, Dubolazov OV, Gorsky MP. Polarization-phase images of liquor polycrystalline films in determining time of death. Appl Opt 2016;55(12):B67–B71.

60 Angelskii OV, Ushenko OG, Burkovets DN, Arkhelyuk OD, Ushenko YA. Polarization-correlation studies of multifractal structures in biotissues and diagnostics of their pathologic changes. Laser Phys. 2000; 10(e):1136–1142.

61 Yermolenko S, Ushenko A, Ivashko P, Goudail F, Gruia I, Gavrilă C, Zimnyakov D, Mikhailova A. Spectropolarimetry of cancer change of biotissues. In Ninth International Conference on Correlation Optics. SPIE. 2009;7388:404–410.

62 Jóźwicki R, Patorski K, Angelsky OV, Ushenko AG, Burkovets DN, Ushenko YA. Automatic polarimetric system for early medical diagnosis by biotissue testing. Opt Appl. 2002;32(4):603–612.

63 Ushenko AG, Dubolazov OV, Ushenko VA, Novakovskaya OY, Olar OV. Fourier polarimetry of human skin in the tasks of differentiation of benign and malignant formations. Appl Opt. 2016;55(12):B56–B60.

64 Ushenko AG, Pishak VP. Coherent-domain optical methods: biomedical diagnostics, environmental and material science. In Laser Polarimetry of Biological Tissue: Principles and Applications. Kluwer Academic Publisher; 2004.

65 Ushenko AG, Fediv AI, Marchuk YF. Correlation and fractal structure of Jones matrices of human bile secret. In European Conference on Biomedical Optics. 2009;7368:73681Q.

66 Ushenko AG. Depolarization of a laser emission field as the correlation development of its polarization structure. In International Conference on Correlation Optics. 1997;3317:331–339.

67 Bachinskyi V, Boychuk T, Ushenko A. Laser Polarimetry of Biological Tissues and Fluids: Chapter 4. Methods of Spatial-frequency, Singular and Correlation Analysis of Biological Layers Object Fields. LAP Lambert Academic Publishing; 2018.

68 Ushenko AG, Burkovets DN, Yermolenko SB, Arkhelyuk AD, Pishak VP, Yuzko AM, et al. Stokes polarimetry of biotissues. In Fourth International Conference on Correlation Optics. SPIE. 1999;3904:527–533.

69 Ushenko VA, Dubolazov AV, Pidkamin LY, Sakchnovsky MY, Bodnar AB, Ushenko YA, et al. Mapping of polycrystalline films of biological fluids utilizing the Jones-matrix formalism. Laser Phys. 2018;28(2):025602.

70 Ushenko AG, Angelsky PO, Sidor M, Marchuk YF, Andreychuk DR, Pashkovskaya NV. Spatial-frequency selection of complex degree of coherence of laser images of blood plasma in diagnostics and differentiation of pathological states of human organism of various nosology. Appl Opt. 2014;53(10):B172–B180.

71 Olar EI, Ushenko AG, Ushenko YA. Polarization correlation measurements of the phase tomograms of optically anisotropic biofractals. Laser Phys. 2004;14(8):1115–1121.

72 Guminetskiy SH, Ushenko OG, Polyanskiy IP, Motrych AV, Grynchuk FV. The optical method for investigation of the peritonitis progressing process. In Eighth International Conference on Correlation Optics. 2008;7008:626–632.

73 Ushenko AG, Fediv AI, Marchuk YF. Singular structure of polarization images of bile secret in diagnostics of human physiological state. In: European Conference on Biomedical Optics. Optica Publishing Group; 2009.

74 Ushenko Y, Ushenko A, Dubolazov A, Soltys I, Litvinenko O, Wanchuliak O, et al. Mueller-matrix interferometric multifractal scaling of optically anisotropic architectonics of diffuse blood facies: fundamental and applied aspects. J Biophotonics. 2025;18(3):e202400412. https://doi.org/10.1002/jbio.202400412.

75 Ushenko A, Soltys I, Dubolazov A, Ushenko Y, Bilookyi V, Bilookyi O, et al. 3D Jones-matrix thesiography of biological fluid facies. J Innov Opt Health Sci. 2025;18(2):2443002. https://doi.org/10.1142/S1793545824430028.

76 Ushenko A, Pavlyukovich N, Khukhlina O, Pavlyukovich O, Gorsky M, Soltys I, et al. Layer-by-layer multifractal scanning of optically anisotropic architectonics of blood plasma films: fundamental and applied aspects. Photonics. 2025;12:215. https://doi.org/10.3390/photonics12030215.

77 Ushenko A, Pavlyukovich N, Khukhlina O, Pavlyukovich O, Soltys I, Dubolazov A, et al. Blood plasma film multifractal scanning in COVID-19 consequences diagnostics. J Biophotonics. 2024;17(11):e202400356. https://doi.org/10.1002/jbio.202400356.

78 Ushenko A, Dubolazov A, Zheng J, Litvinenko A, Gorsky M, Ushenko Y, et al. 3D polarization-interference holographic histology for wavelet-based differentiation of biological tissues with different necrotic states. J Biomed Opt. 2024;29(5):052920. https://doi.org/10.1117/1.JBO.29.5.052920.

79 Ushenko O, Bilookyi O, Zheng J, Dubolazov A, Olar O, Ushenko Y, et al. 3D digital holographic polarimetry of laser speckle fields formed by polycrystalline blood films: a tool for differential diagnosis of thyroid pathology. Front Phys. 2024;12: 1426469. https://doi.org/10.3389/fphy.2024.1426469.

80 Wójcik W, Hu Z, Ushenko Y, Smolarz A, Soltys I, Dubolazov O, et al. Optical sensor system for 3D Jones matrix reconstruction of optical anisotropy maps of self-assembled polycrystalline soft matter films. Sensors. 2024;24(5):1589. https://doi.org/10.3390/s24051589.

81 Ushenko A, Zheng J, Gorsky M, Dubolazov A, Ushenko Y, Soltys I, et al. 3D digital holographic polarimetry of diffuse optically anisotropic biological tissue object fields. Front Phys. 2023;11:1288935. https://doi.org/10.3389/fphy.2023.1288935.

82 Dubolazov A, Ushenko V, Trifonyuk L, Stashkevich A, Soltys I, Ushenko Y, et al. Polarization-singular approach to imaging Mueller-matrix polarimetry in differential diagnosis of histological sections of biopsy of tumors of the uterus and prostate. Front Phys. 2021;9:711212. https://doi.org/10.3389/fphy.2021.711212.

83 Ushenko VO, Trifonyuk L, Ushenko YA, Dubolazov OV, Gorsky MP, Ushenko AG. Polarization singularity analysis of Mueller-matrix invariants of optical anisotropy of biological tissue samples in cancer diagnostics. J Opt. 2021;23(6):064004. https://doi.org/10.1088/2040-8986/abf97a.

84 Ushenko YA, Koval GD, Ushenko AG, Dubolazov OV, Ushenko VA, Novakovskaia OY. Mueller-matrix of laser-induced autofluorescence of polycrystalline films of dried peritoneal fluid in diagnostics of endometriosis. J Biomed Opt. 2016;21(7):071116. https://doi.org/10.1117/1.JBO.21.7.071116.

14

Scanning of Blood Films for Organ Pathology

Yurii Ushenko[1,2], Weidong Dou[1], Oleksandr Ushenko[2,3], Oleksandr Dubolazov[2], Iryna Soltys[2], Vitaliy Rozhko[4], Oleksandr Olar[2], and Vladyslava Sholota[5]

[1] *Shaoxing University, Shaoxing, People's Republic of China*
[2] *Yurii Fedkovych Chernivtsi National University, Chernivtsi, Ukraine*
[3] *Taizhou Institute of Zhejiang University, Taizhou, People's Republic of China*
[4] *Bukovinian State Medical University, Chernivtsi, Ukraine*
[5] *Vinnytsia National Technical University, Vinnytsia, Ukraine*

14.1 Objects of Investigations

As part of our study in experimental medicine, two groups of blood film samples from laboratory rats were examined: healthy (control group 1) and those subjected to septic inflammatory processes (experimental group 2) [1–5].

Blood film samples were obtained by applying a liquid drop to a glass substrate at room temperature ($t = 22\,°\mathrm{C}$) and drying for 24 hours. We reduced the volume of a liquid drop to $0.035 - 0.004\,\mu\mathrm{L}$. As a result of blood dehydration, blood films were formed with an average height of up to $60\,\mu\mathrm{m}$ and a diameter of up to 3–4 mm.

Figure 14.1 presents microscopic images (×4) of the polycrystalline structure of blood films from rats in group 1 (Figure 14.1a) and group 2 (Figure 14.1b). A comparative visual analysis of both images reveals no significant differences in the geometric structure of the supramolecular networks in the blood film samples from healthy and inflamed blood. This fact indicates that septic changes do not involve visible macroscopic transformations in the optically anisotropic polycrystalline structure of dehydrated blood films. This further underscores the importance of obtaining and statistically analyzing data from a set of maps of optical anisotropy at various geometric scales of supramolecular networks.

14.2 Jones Matrix Maps of Linear and Circular Birefringence of Blood Films

From Figures 14.2 to 14.3, maps $T(\mathrm{LB}; \mathrm{CB}_{\otimes;\oplus}(\mathrm{Im}\, w_{ik}))$ and histograms $G(\mathrm{LB}, \mathrm{CB}_{\otimes;\oplus})$ of the phase anisotropy parameter distributions $(m \times n)$ for blood films of rats from the control group 1 ((a), (c), (e), (g)) and experimental group 2 ((b), (d), (f), (h)) are presented.

The analysis of the obtained data in different phase planes φ_k revealed the individual topographic $(m \times n)$ and statistical structure $(G(\mathrm{LB}, \mathrm{CB}_{\otimes;\oplus}))$ of the maps $T(\mathrm{LB}; \mathrm{CB}_{\otimes;\oplus}(\mathrm{Im}\, w_{ik}))$ of blood films from rats in different groups [6–11].

Quantitatively, the scenarios of phase transformation (φ_k) of layer-by-layer distributions of linear $\mathrm{LB}(m \times n)$ and circular $\mathrm{CB}_{\otimes;\oplus}(m \times n)$ birefringence of polycrystalline supramolecular networks of blood films are illustrated by the results of statistical analysis presented in Table 14.1.

In the phase plane $\varphi_k = \pi/8$, the most sensitive parameters (markers) to changes in the topographic structure of the maps $T(\varphi_k, \mathrm{LB}; \mathrm{CB}_{\otimes;\oplus}(\mathrm{Im}\, w_{ik}))$ were identified – statistical moments of the third and fourth orders $M_{i=3;4}(\mathrm{LB}, \mathrm{CB}_{\otimes;\oplus}(m \times n))$, which characterize the skewness and kurtosis of the distributions of the phase anisotropy parameter values [9–19].

For the maps of linear birefringence $T(\varphi_k = \pi/8, \mathrm{LB}(\mathrm{Im}\, w_{ik}))$, the differences $\Delta M_{i=3;4}$ between the values of diagnostic markers for the control and experimental groups of samples do not exceed 30%. For the maps of circular birefringence

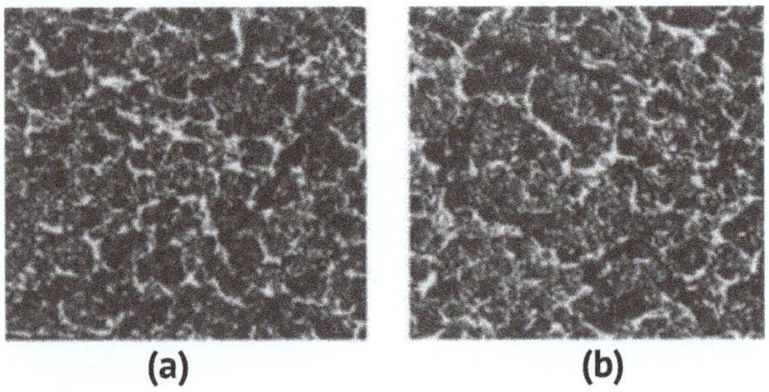

Figure 14.1 Microscopic images (×4) of blood films from healthy (a) and septic (b) rats.

$T(\varphi_k = \pi/8, \mathrm{CB}_{\otimes;\oplus}(\mathrm{Im}\, w_{ik}))$, the differences $\Delta M_{i=3;4}$ between the values of diagnostic markers significantly increase and reach 45–55%.

From a physical perspective, the obtained results can be explained by the following considerations. The magnitude of linear birefringence (LB) is primarily determined by the degree of spatial orientation of the optical axis directions (γ) of large-scale ($l \sim 40\,\mu\mathrm{m} \div 80\,\mu\mathrm{m}$) needle-like crystals of albumin, fibrin, and elastin. The orientation variations of the optical axis directions within the dendritic network are significant and nearly equiprobable within the range $0 \leq \gamma \leq 2\pi$. As a result, the values of the first- and second-order statistical moments, which characterize the mean and variance of the $\mathrm{LB}(m \times n)$ distributions, are relatively large. Conversely, the values of the higher-order statistical moments $M_{i=3;4}$ are minimal due to the Equations $M_{i=3} \sim 1/(M_{i=2}{}^3)$; $M_{i=4} \sim 1/(M_{i=2}{}^4)$ [20–33].

In a septic process, the concentration of needle-like crystals slightly decreases. As a result, during crystallization, there is a slight decline (↓) in the degree of structural anisotropy of the supramolecular network in the films of septic inflamed blood $\begin{pmatrix} M_{i=1;2}(\mathrm{LB}) \downarrow; \\ M_{i=3;4}(\mathrm{LB}) \uparrow \end{pmatrix}$.

A different pattern is observed in the formation of a finer-scale spherulitic component of the supramolecular networks in the blood films from experimental group 2. During inflammation, the concentration of small-scale ($l \sim 10\,\mu\mathrm{m} \div 20\,\mu\mathrm{m}$) optically active molecular globulin complexes increase.

As a result, the value of circular birefringence $\mathrm{CB}_{\otimes;\oplus}$ of globulin crystals and formed elements (erythrocytes, platelets, leukocytes, and monocytes) also increases. Consequently, the statistical moments of the first to fourth order, which characterize the circular birefringence maps $T(\mathrm{CB}_{\otimes;\oplus}(\mathrm{Im}\, w_{ik}))$, undergo significantly greater changes – $M_{i=1;2}(\mathrm{CB}_{\otimes;\oplus}) \uparrow$ and $M_{i=3;4}(\mathrm{CB}_{\otimes;\oplus}) \downarrow$ – compared to the analogous statistical parameters of the linear birefringence maps $T(\mathrm{LB}(\mathrm{Im}\, w_{ik}))$.

It should be noted that the Jones matrix reconstructed maps $T\begin{pmatrix} \mathrm{LB}(\mathrm{Im}\, w_{ik}); \\ \mathrm{CB}_{\otimes;\oplus}(\mathrm{Im}\, w_{ik}) \end{pmatrix}$ correspond to the conditions of multiple light scattering within the volume of polycrystalline blood films. As a result, the coordinate structure features of the linear $\mathrm{LB}(m \times n)$ and circular $\mathrm{CB}_{\otimes;\oplus}(m \times n)$ birefringence parameters of the supramolecular networks are statistically averaged to a significant extent [34–41].

Consequently, the statistical differences between the integral maps $\Delta M_{i=1;2;3;4}\left(T\begin{pmatrix} \mathrm{LB}(\mathrm{Im}\, w_{ik}); \\ \mathrm{CB}_{\otimes;\oplus}(\mathrm{Im}\, w_{ik}) \end{pmatrix}\right)$ of the blood films in both groups do not reach maximal values (Table 14.1).

At the same time, with a decrease in the phase scanning shift $\varphi_k \downarrow$ in the object field, components with minimal scattering multiplicity are distinguished in the diffuse samples of blood films. This enables the reconstruction of Jones matrix maps $T\begin{pmatrix} \varphi_k = \pi/8, \mathrm{LB}(\mathrm{Im}\, w_{ik}); \\ \varphi_k = \pi/8, \mathrm{CB}_{\otimes;\oplus}(\mathrm{Im}\, w_{ik}) \end{pmatrix}$, which are almost unambiguously associated with the specific topographic structure of the supramolecular networks in the blood films of healthy and diseased rats.

Against this background, septic differences in the fine-scale polycrystalline structure of the optically anisotropic networks in the blood samples of rats from experimental group 2 become most clearly pronounced – $\Delta M_{i=1;2;3;4}(T(\varphi_k = \pi/8, \mathrm{CB}_{\otimes;\oplus}(\mathrm{Im}\, w_{ik}))) \to \max$.

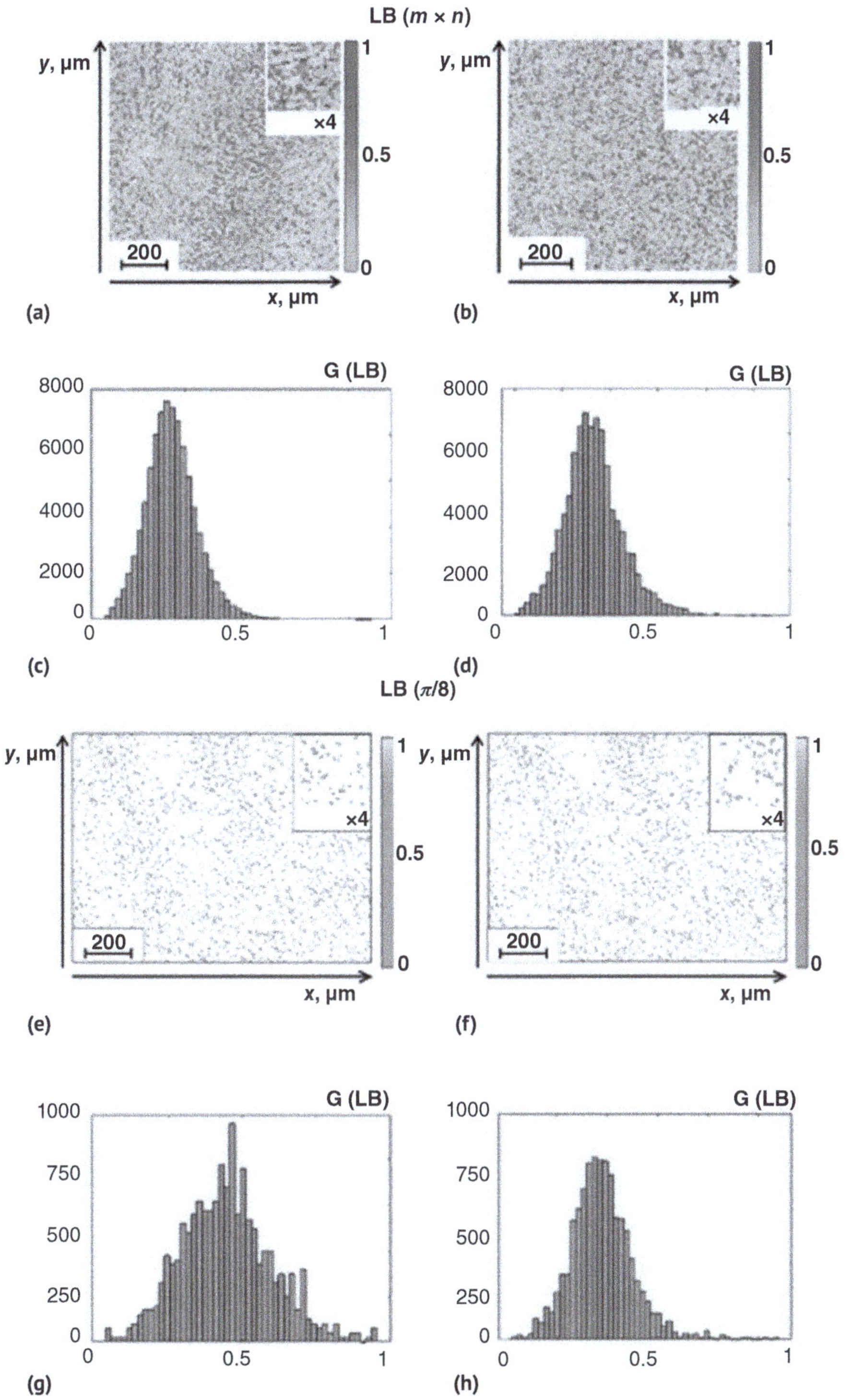

Figure 14.2 Maps $T(\text{LB}(\text{Im } w_{ik}))$ (a), (b); $T(\text{LB}(\varphi_k = \pi/8))$ (e), (f) and histograms $G(\text{LB})$ (c), (d); $G(\text{LB}, \varphi_k = \pi/8)$ (g), (h) of the distributions of the linear birefringence magnitude of blood film from healthy (a), (c), (e), (g) and septic (b), (d), (f), (h) rats.

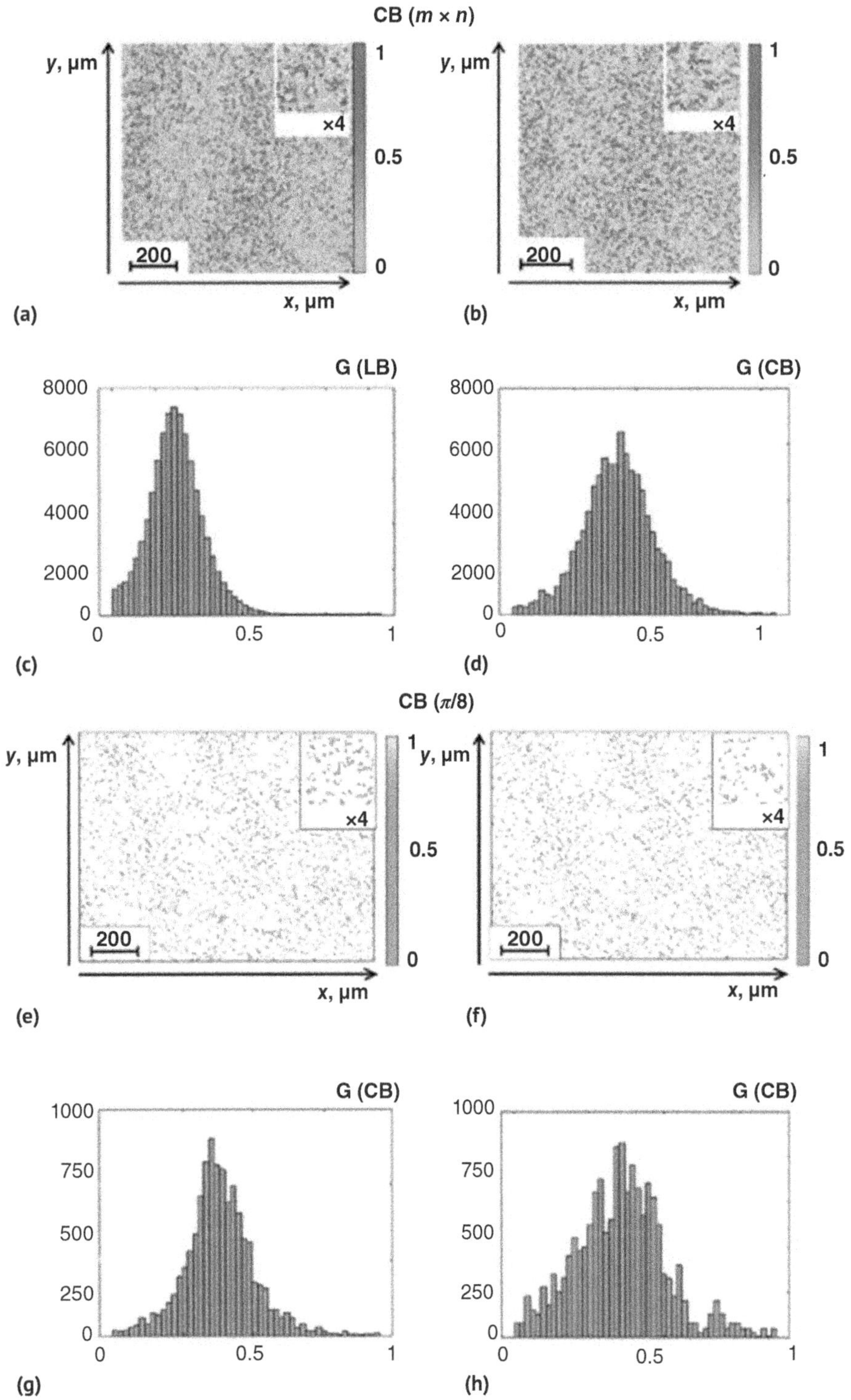

Figure 14.3 Maps $\text{CB}_{\otimes;\oplus}(\text{Im}\, w_{ik})$ (a), (b); $T(\text{CB}_{\otimes;\oplus}(\varphi_k = \pi/8))$ (e), (f) and histograms $G(\text{CB}_{\otimes;\oplus})$ (c), (d); $G(\text{CB}_{\otimes;\oplus}, \varphi_k = \pi/8)$ (g), (h) of the distributions of the circular birefringence magnitude of blood films from healthy (a), (c), (e), (g) and septic (b), (d), (f), (h) rats.

Table 14.1 Statistical parameters of maps $T(\mathrm{LB})$; $T(\mathrm{LB}(\varphi_k = \pi/8))$ and $T(\mathrm{CB}_{\otimes;\oplus})$; $T(\mathrm{CB}_{\otimes;\oplus}(\varphi_k = \pi/8))$ of blood films from healthy (group 1) and septic (group 2) rats.

M_i	LB		$\mathrm{LD}(\varphi_k = \pi/8)$	
	Group 1	Group 2	Group 1	Group 2
M_3	0.41 ± 0.021	0.54 ± 0.027	0.89 ± 0.044	1.21 ± 0.054
M_4	0.56 ± 0.032	0.76 ± 0.034	1.18 ± 0.052	1.44 ± 0.077
M_i	CB		$\mathrm{CB}_{\otimes;\oplus}(\varphi_k = \pi/8)$	
	Group 1	Group 2	Group 1	Group 2
M_3	0.87 ± 0.046	0.73 ± 0.039	1.33 ± 0.028	0.98 ± 0.057
M_4	0.99 ± 0.057	0.78 ± 0.041	1.34 ± 0.081	1.07 ± 0.059

14.3 Jones Matrix Maps of Linear and Circular Dichroism of Blood Films

Figures 14.4 and 14.5 present the integral (a), (b) and layer-by-layer (e), (f) maps $T(\mathrm{LD};\ \mathrm{CD}_{\otimes;\oplus}(Re\ w_{ik}))$ and the histograms $G(\mathrm{LD},\ \mathrm{CD}_{\otimes;\oplus})$ of the distributions of optical dichroism parameters of the polycrystalline supramolecular networks of blood films from group 1 ((a), (c), (e), (g)) and group 2 ((b), (d), (f), (h)) rats [42–53].

The statistical scenarios of the phase transformation of the map set $T(\mathrm{LD};\ \mathrm{CD}_{\otimes;\oplus}(Re\ w_{ik}))$ are illustrated by the values of the first- to fourth-order statistical moments, which are presented in Table 14.2.

The analysis of the obtained data revealed a dependence of the statistical distributions $M_{i=1;2;3;4}(\mathrm{LD},\ \mathrm{CD}_{\otimes;\oplus})$ on the phase scanning depth φ_k of the volume of the polycrystalline architectonics of the blood film samples. The most sensitive parameters were found to be the third- and fourth-order statistical moments, which characterize the skewness and kurtosis of the distributions of the linear and circular dichroism magnitudes of the optically anisotropic network of the blood samples in the phase plane $\varphi_k = \pi/8$.

For the maps of linear dichroism $T(\varphi_k = \pi/8,\ \mathrm{LD}(Re\ w_{ik}))$, the differences $\Delta M_{i=3;4}$ between the diagnostic marker values of the control and experimental groups of blood drop samples amounted to 25–30%. For the maps of circular dichroism $T(\varphi_k = \pi/8,\ \mathrm{CD}_{\otimes;\oplus}(Re\ w_{ik}))$, the differences $\Delta M_{i=3;4}$ between the diagnostic marker values increased and reached 35–45%.

The obtained results can be attributed to the fact that the magnitude of linear dichroism LD, similarly to linear birefringence LB, is determined by the degree of spatial orientation of the optical axis directions γ of the large-scale component of the supramolecular network of blood films. Due to minor concentration changes of needle-shaped protein crystals, the statistical differences between the maps $T(\varphi_k = \pi/8,\ \mathrm{LD}(Re\ w_{ik}))$ are relatively small. In contrast, due to the septic increase in the concentration of globulins, there is an increase in $M_{i=1;2}(\mathrm{CD}_{\otimes;\oplus}) \uparrow$ alongside a sharper decrease in $M_{i=3;4}(\mathrm{CD}_{\otimes;\oplus}) \downarrow$.

Thus, the statistical analysis of the algorithmically reconstructed layer-by-layer Jones matrix maps $T\begin{pmatrix} \mathrm{LB}(\varphi_k,\ \mathrm{Im}\ w_{ik}); \\ \mathrm{CB}_{\otimes;\oplus}(\varphi_k,\ \mathrm{Im}\ w_{ik}); \\ \mathrm{LD}(\varphi_k,\ Re\ w_{ik}); \\ \mathrm{CD}_{\otimes;\oplus}(\varphi_k,\ Re\ w_{ik}) \end{pmatrix}$ revealed different "scale-dependent" scenarios of septic alterations in the polycrystalline supramolecular networks of blood films.

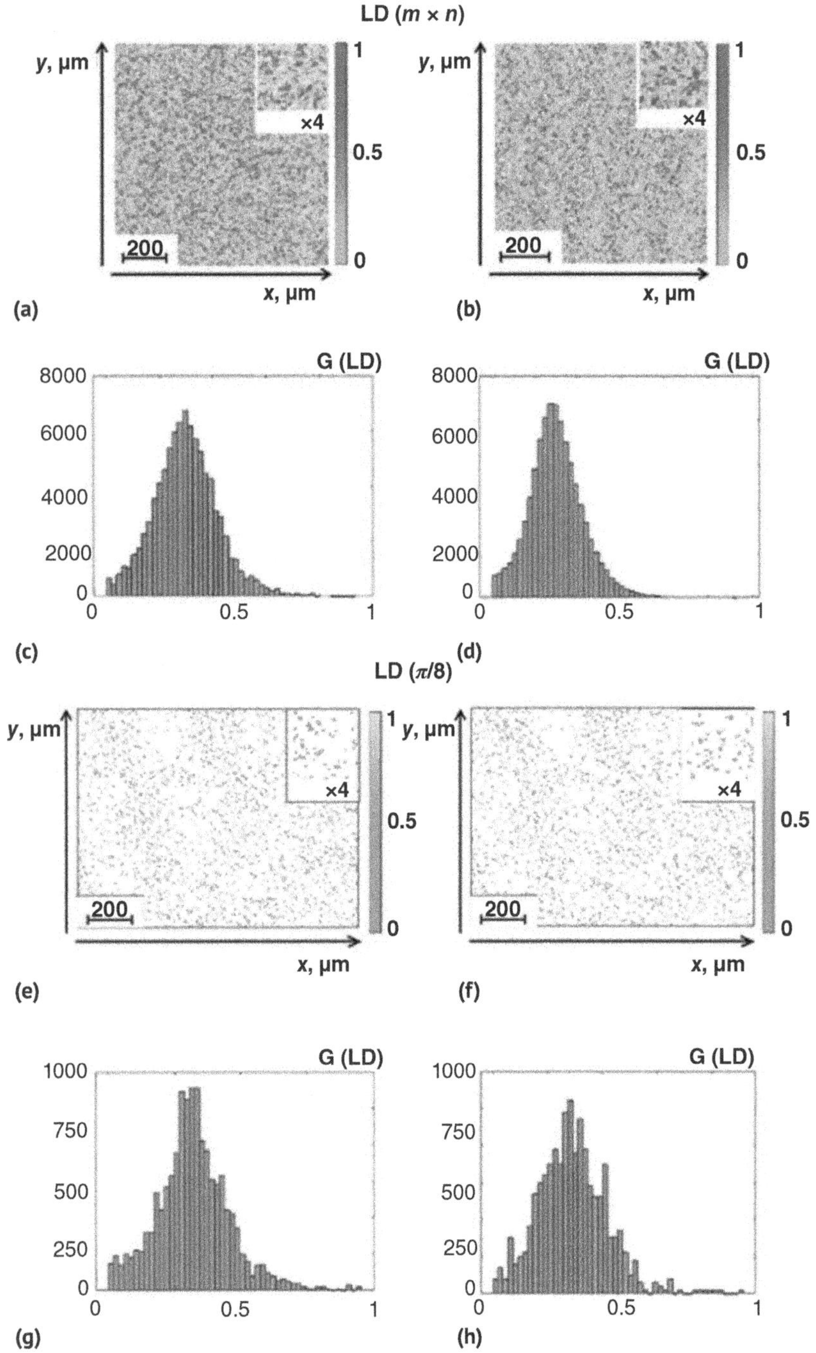

Figure 14.4 Maps $T(\text{LD}(\text{Im}\, w_{ik}))$ (a), (b); $T(\text{LD}(\varphi_k = \pi/8))$ (e), (f) and histograms $G(\text{LD})$ (c), (d); $G(\text{LD}, \varphi_k = \pi/8)$ (g), (h) of the distributions of the linear dichroism magnitude of blood films from healthy (a), (c), (e), (g) and septic (b), (d), (f), (h) rats.

To enable a more detailed quantitative analysis of these pathological transformations, we applied a scale-selective wavelet analysis to the set of maps $T\begin{pmatrix} \text{LB}(\varphi_k, \text{Im}\, w_{ik}); \\ \text{CB}_{\otimes;\oplus}(\varphi_k, \text{Im}\, w_{ik}); \\ \text{LD}(\varphi_k, Re\, w_{ik}); \\ \text{CD}_{\otimes;\oplus}(\varphi_k, Re\, w_{ik}) \end{pmatrix}$ in the phase plane $\varphi_k = \pi/8$.

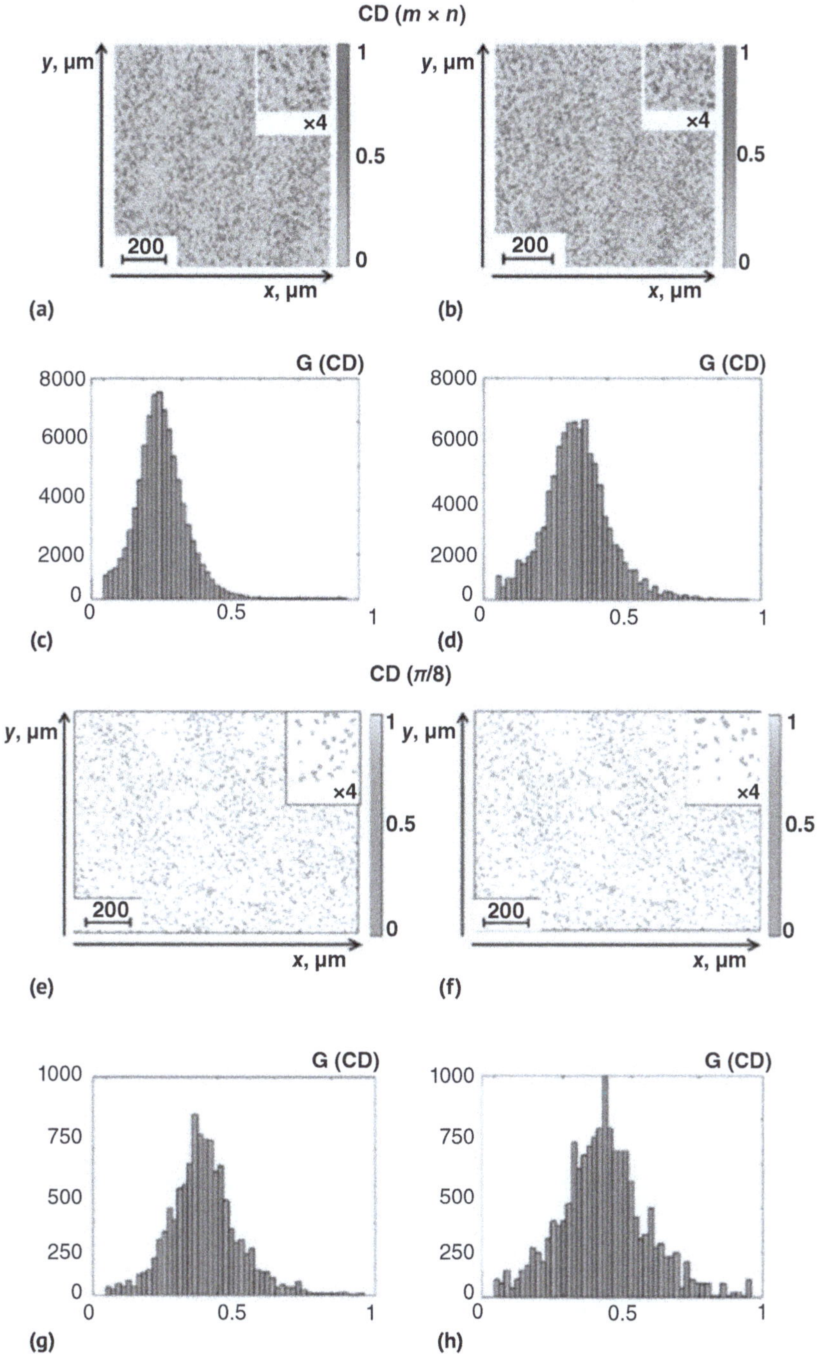

Figure 14.5 Maps $CD_{\otimes;\oplus}(\text{Im}\, w_{ik})$ (a), (b); $T(CD_{\otimes;\oplus}(\varphi_k = \pi/8))$ (e), (f) and histograms $G(CD_{\otimes;\oplus})$ (c), (d); $G(CD_{\otimes;\oplus}, \varphi_k = \pi/8)$ (g), (h) of the distributions of the circular dichroism magnitude of blood films from healthy (a), (c), (e), (g) and septic (b), (d), (f), (h) rats.

14.4 Wavelet Analysis of Phase Anisotropy Maps of Blood Films

Figures 14.6 and 14.7 present the results of wavelet analysis at two scales ($a_{min} = 15$ – "light gray lines" and $a_{max} = 55$ – "dark gray lines") of the MHAT function, which scans the Jones matrix reproduced maps $T\begin{pmatrix} LB(\varphi_k = \pi/8, \text{Im}\, w_{ik}); \\ CB_{\otimes;\oplus}(\varphi_k = \pi/8, \text{Im}\, w_{ik}) \end{pmatrix}$ of blood films from rats in the control and experimental groups.

Table 14.2 Statistical parameters of maps $T(\text{LD})$; $T(\text{LD}(\varphi_k = \pi/8))$ and $T(\text{CD}_{\otimes;\oplus})$; $T(\text{CD}_{\otimes;\oplus}(\varphi_k = \pi/8))$ of blood films from healthy (group 1) and septic (group 2) rats.

M_i	LD		$\text{LD}(\varphi_k = \pi/8)$	
	Group 1	Group 2	Group 1	Group 2
M_3	0.51 ± 0.023	0.43 ± 0.023	0.79 ± 0.038	0.91 ± 0.045
M_4	0.76 ± 0.032	0.64 ± 0.034	0.98 ± 0.052	1.21 ± 0.062
M_i	$\text{CD}_{\otimes;\oplus}$		$\text{CD}_{\otimes;\oplus}(\varphi_k = \pi/8)$	
	Group 1	Group 2	Group 1	Group 2
M_3	0.76 ± 0.036	0.63 ± 0.036	1.19 ± 0.068	0.91 ± 0.067
M_4	0.93 ± 0.047	0.81 ± 0.044	1.44 ± 0.089	1.15 ± 0.089

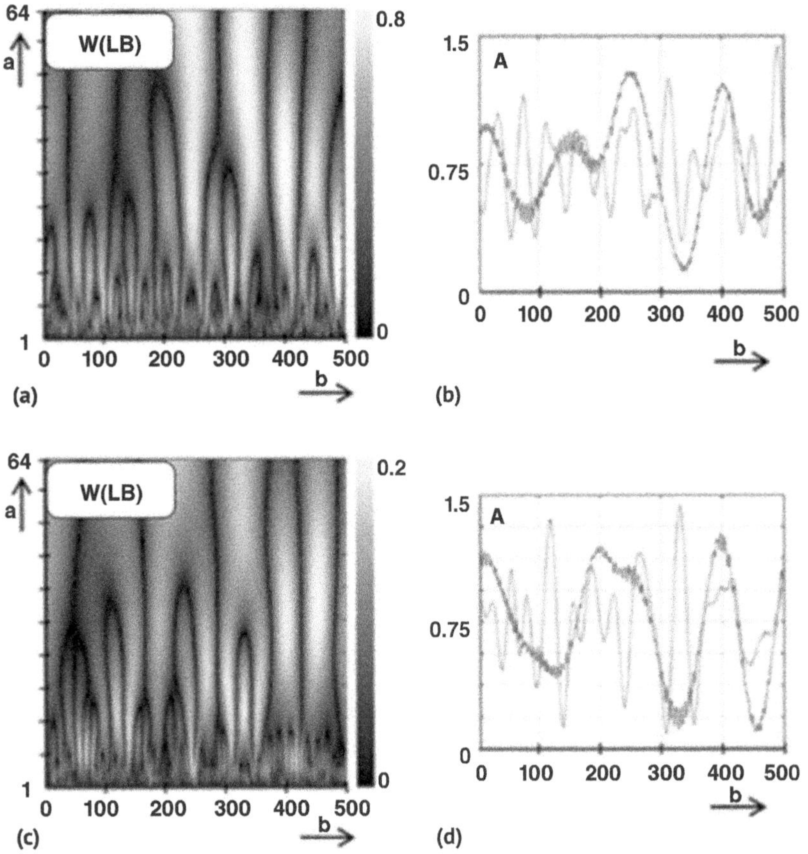

Figure 14.6 Wavelet coefficient maps $W(\text{LB}, a, b)$ (a), (c) and multiscale ($a_{\min} = 15$, "light gray lines" and $a_{\max} = 55$, "dark gray lines") dependencies of the wavelet coefficient amplitude $A(\mu_k)$ of maps $T(\text{LB}(\varphi_k = \pi/8))$ for the linear birefringence LB of the supramolecular network of blood films from healthy (a), (b) and septic (c), (d) rats.

The comparative analysis of the obtained maps $W(\text{LB}, \text{CB}_{\otimes;\oplus}, \mu_k, b)$ (panels (a), (c)) revealed individual modulation of the amplitude values $\mu_k(a_{min} = 15)$ and $\mu_k(a_{\max} = 55)$ of the wavelet coefficients $W(\mu_k, b)$ of maps $T\begin{pmatrix} \text{LB}(\varphi_k = \pi/8, \text{Im}\, w_{ik}); \\ \text{CB}_{\otimes;\oplus}(\varphi_k = \pi/8, \text{Im}\, w_{ik}) \end{pmatrix}$ of blood films from rats in groups 1 and 2, – (b), (d).

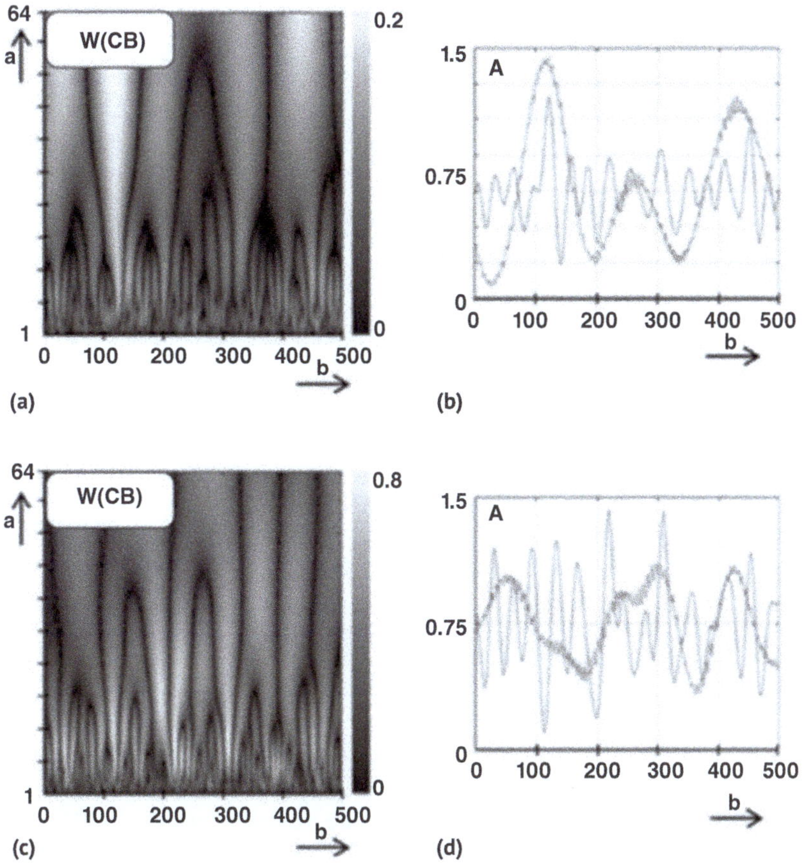

Figure 14.7 Wavelet coefficient maps $W(CB_{\otimes;\oplus}, a, b)$ (a), (c) and multiscale ($a_{min} = 15$, "light gray lines" and $a_{max} = 55$, "dark gray lines") dependencies of the wavelet coefficient amplitude $A(\mu_k)$ of maps $T(CB_{\otimes;\oplus}(\varphi_k = \pi/8))$ for the circular birefringence $CB_{\otimes;\oplus}$ of the supramolecular network of blood films from healthy (a), (b) and septic (c), (d) rats.

For structural birefringence $LB(\varphi_k = \pi/8, \mathrm{Im}\, w_{ik})$, the maximum amplitude and variations in the wavelet coefficients μ_k occur at all scales of the scanning soliton-like MHAT function of the maps of blood film samples from healthy and septic rats (Figure 14.6a, d).

For optical activity $CB_{\otimes;\oplus}(\varphi_k = \pi/8, \mathrm{Im}\, w_{ik})$ of blood components and molecular domains of globulin, the maximum amplitude and variations in the wavelet coefficients occur at smaller scales ($a_{min} = 15$) of the scanning soliton-like MHAT function of the maps of blood film samples from septic rats (Figure 14.7b, d).

Comparison of the results of statistical analysis of the amplitude distributions of the multiscale wavelet coefficients $A(\mu_k)$ of the maps $T\begin{pmatrix} LB(\varphi_k = \pi/8, \mathrm{Im}\, w_{ik}); \\ CB_{\otimes;\oplus}(\varphi_k = \pi/8, \mathrm{Im}\, w_{ik}) \end{pmatrix}$ of blood films from rats in groups 1 and 2 revealed the following patterns.

For the blood samples of healthy rats, the statistical moments of the first- and second-order $M_{1;2}(W(LB(\varphi_k = \pi/8, a_{min}, a_{max})))$ LB were 1.45 to 1.55 times larger than the corresponding statistical parameters that characterize the scale-selective structure $W(LB, \mu_i, b)$ of maps of the structural anisotropy of the blood drop films of septic rats (Table 14.3).

This fact can be associated with the narrowing of the orientation range of the optical axes of needle-like crystals ($0 \leq \gamma \leq \pi$) during the formation of supramolecular networks of films in septic blood ($\gamma_{min} \leq \gamma \leq \gamma_{max}$).

An inverted and more pronounced pattern is observed for the correlation between the statistical parameters $M_{1;2}(W(CB_{\otimes;\oplus}(\varphi_k = \pi/8, a_{min}, a_{max})))$, which characterize the films of blood from rats in the control and experimental groups.

The corresponding difference between the first- and second-order statistical moments is 1.6 to 1.8 times.

Table 14.3 Statistical parameters of the amplitude distribution of wavelet coefficients $A(\mu_k)$ of maps $T(LB(\varphi_k = \pi/8))$ and $T(CB_{\otimes;\oplus}(\varphi_k = \pi/8))$ of blood films from healthy (group 1) and septic (group 2) rats.

Linear Birefringence, LB						
a_i	$a_{min} = 15$		Ac, %	$a_{max} = 55$		Ac, %
M_i	Group 1	Group 2		Group 1	Group 2	
M_1	0.57 ± 0.031	0.39 ± 0.016	91.2	0.46 ± 0.024	0.36 ± 0.019	88.2
M_2	0.68 ± 0.035	0.45 ± 0.024	94.1	0.54 ± 0.029	0.41 ± 0.024	91.2
Circular birefringence, $CB_{\otimes;\oplus}$						
a_i	$a_{min} = 15$		Ac, %	$a_{max} = 55$		Ac, %
M_i	Group 1	Group 2		Group 1	Group 2	
M_1	0.31 ± 0.016	0.48 ± 0.021	94.1	0.24 ± 0.013	0.39 ± 0.018	91.2
M_2	0.42 ± 0.021	0.67 ± 0.031	94.1	0.35 ± 0.021	0.54 ± 0.027	94.1

14.5 Wavelet Analysis of the Amplitude Anisotropy in Blood Films Maps

The results of the wavelet analysis of the amplitude anisotropy of the polycrystalline network of blood films are presented in Figures 14.8 and 14.9.

The analysis of the obtained data revealed individual modulation of the amplitude values $A(\mu_k(a_{min} = 15))$ and $A(\mu_k(a_{max} = 55))$ of the wavelet coefficients μ_k of linear birefringence of wavelet coefficients $W(\mu_k, b)$ for the maps of linear LD and circular CD dichroism at all scales of the scanning soliton-like MHAT function, as shown in Figures 14.8 and 14.9 (right columns).

The results of the statistical analysis of the amplitude distribution of wavelet coefficients $A(\mu_k)$ for the amplitude anisotropy maps $T\begin{pmatrix} LD(\varphi_k = \pi/8, Re\, w_{ik}); \\ CD_{\otimes;\oplus}(\varphi_k = \pi/8, Re\, w_{ik}) \end{pmatrix}$ for blood films of healthy and septic rats are presented in Table 14.4.

It was found that for the blood facie samples of healthy rats, the statistical moments of the first and second orders, defined for different scales a of the MHAT function, are 1.35–1.4 times greater than the corresponding statistical parameters characterizing the wavelet maps $W(LD, \mu_i, b)$ for the blood facies of septic rats, as shown in Table 14.4.

The opposite pattern is observed for the relationship between the statistical parameters $M_{i=1;2}$, which characterize the multiscale distributions of the wavelet coefficients for the maps of circular dichroism CD in the blood facies of rats from the control and experimental groups. The corresponding difference between the first- and second-order statistical moments reaches 1.5 times.

Thus, it can be stated that the application of wavelet analysis of algorithmically reconstructed multilayered Jones matrix maps $T\begin{pmatrix} LB(\varphi_k = \pi/8, \mathrm{Im}\, w_{ik}); \\ CB_{\otimes;\oplus}(\varphi_k = \pi/8, \mathrm{Im}\, w_{ik}); \\ LD(\varphi_k = \pi/8, Re\, w_{ik}); \\ CD_{\otimes;\oplus}(\varphi_k = \pi/8, Re\, w_{ik}) \end{pmatrix}$ significantly increased the sensitivity of statistical analysis in differentiating between blood facie samples from healthy and septic rats.

The next step involved conducting a comparative study of the effectiveness of differential diagnosis of sepsis by determining the balanced accuracy of two methods:

1. Layered polarimetric-interference mapping of statistical and wavelet structures of azimuth and ellipticity maps of polarization $P\begin{pmatrix} \alpha(\varphi_k = \pi/8, (m \times n)); \\ \beta(\varphi_k = \pi/8, (m \times n)) \end{pmatrix}$ of the object fields of blood facies.

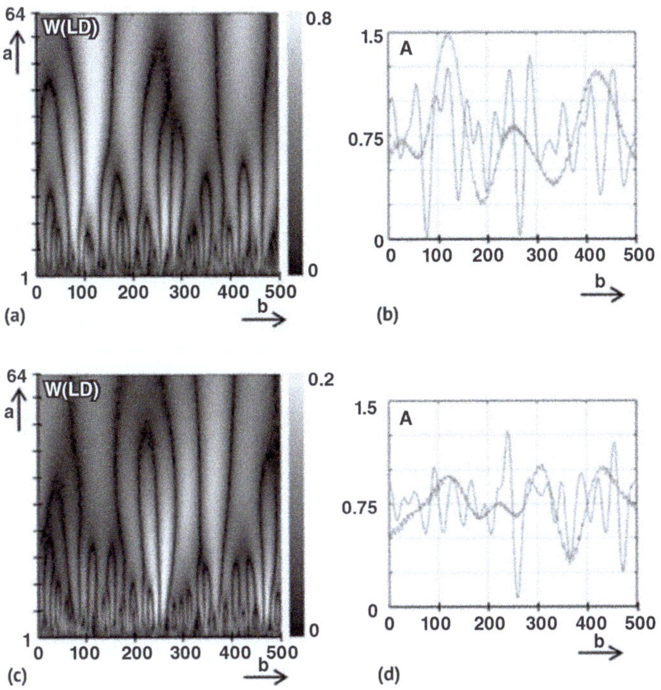

Figure 14.8 Wavelet coefficient maps $W(\mathrm{LD}, a, b)$ (a), (c) and multiscale ($a_{\min} = 15$, "light gray lines" and $a_{\max} = 55$, "dark gray lines") dependencies of the wavelet coefficient amplitude $A(\mu_k)$ of maps $T(\mathrm{LD}(\varphi_k = \pi/8))$ for the linear dichroism LD of the supramolecular network of blood films from healthy (a), (b) and septic (c), (d) rats.

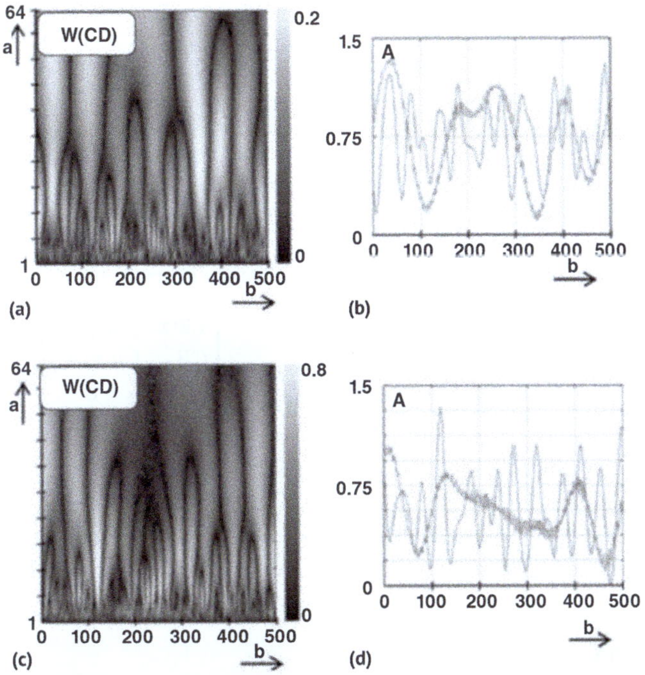

Figure 14.9 Wavelet coefficient maps $W(\mathrm{CD}_{\otimes;\oplus}, a, b)$ (a), (c) and multiscale ($a_{\min} = 15$, "light gray lines" and $a_{\max} = 55$, "dark gray lines") dependencies of the wavelet coefficient amplitude $A(\mu_k)$ of maps $T(\mathrm{CD}_{\otimes;\oplus}(\varphi_k = \pi/8))$ for the circular dichroism $\mathrm{CD}_{\otimes;\oplus}$ of the supramolecular network of blood films from healthy (a), (b) and septic (c), (d) rats.

Table 14.4 Statistical parameters of the amplitude distribution of wavelet coefficients $A(\mu_k)$ of maps $T(\mathrm{LD}(\varphi_k = \pi/8))$ and $T(\mathrm{CD}_{\otimes;\oplus}(\varphi_k = \pi/8))$ of blood films from healthy (group 1) and septic (group 2) rats.

Linear dichroism, LD						
a_i	$a_{min} = 15$		Ac, %	$a_{max} = 55$		Ac, %
M_i	Group 1	Group 2		Group 1	Group 2	
M_1	0.56 ± 0.031	0.46 ± 0.023	88.2	0.45 ± 0.023	0.36 ± 0.019	85.3
M_2	0.67 ± 0.034	0.51 ± 0.024	91.2	0.59 ± 0.032	0.45 ± 0.024	88.2
Circular dichroism, $CB_{\otimes;\oplus}$						
a_i	$a_{min} = 15$		Ac, %	$a_{max} = 55$		Ac, %
M_i	Group 1	Group 2		Group 1	Group 2	
M_1	0.31 ± 0.017	0.44 ± 0.027	91.2	0.24 ± 0.014	0.35 ± 0.018	88.2
M_2	0.42 ± 0.025	0.64 ± 0.035	94.2	0.33 ± 0.017	0.49 ± 0.026	91.2

2. Layered Jones matrix maps of statistical and wavelet structures of maps of optical anisotropy $T\begin{pmatrix} \mathrm{LB}(\varphi_k = \pi/8, (m \times n)); \\ \mathrm{CB}_{\otimes;\oplus}(\varphi_k = \pi/8, (m \times n)); \\ \mathrm{LD}(\varphi_k = \pi/8, (m \times n)); \\ \mathrm{CD}_{\otimes;\oplus}(\varphi_k = \pi/8, (m \times n)) \end{pmatrix}$ of blood facie supramolecular networks.

14.6 Comparative Analysis of the Diagnostic Effectiveness of Layered Polarization–Interference Mapping and Jones Matrix Mapping Methods for Blood Films

The results of the informational analysis of data obtained by the methods of layered polarization–interference mapping of azimuth maps $\alpha(m \times n)$ and ellipticity maps $\beta(m \times n)$ of the object field, and Jones matrix mapping $T\begin{pmatrix} \mathrm{LB}(m \times n); \\ \mathrm{CB}_{\otimes;\oplus}(m \times n); \\ \mathrm{LD}(m \times n); \\ \mathrm{CD}_{\otimes;\oplus}(m \times n) \end{pmatrix}$ of the optical anisotropy of supramolecular networks of blood films are presented for the phase plane $\varphi_k = \pi/8$ in Tables 14.5 and 14.6.

Based on the results of the statistical analysis of the data obtained by the method of polarization–interference mapping of scattered object fields of blood films (diagnostic markers $M_{i=3;4}(\alpha)$ and $M_{i=3;4}(\beta)$), the following maximum levels of balanced accuracy Ac, 100% for the differential diagnosis of septic inflammation were established:

- $M_{i=3;4}(\alpha)$ and $Z_{i=3;4}(\beta)$ – a satisfactory level of 80% < Ac < 85%.
- $M_{i=1;2}(W(\alpha; (a_{min} = 15; b)))$ – good level Ac = 87%.

Wavelet analysis $W(a_{min} = 15; a_{max} = 55; b)$ of the polarization maps of azimuth $\alpha(m \times n)$ and ellipticity $\beta(m \times n)$ of the object field (statistical markers $M_{i=1;2}(W(a_{max} = 55; b))$ and $M_{i=1;2}(W(a_{min} = 15; b))$ revealed the following maximum levels of balanced accuracy:

- $M_{i=1;2}(W(\alpha; (a_{max} = 55; b)))$ and $M_{i=1;2}(W(\beta; (a_{min} = 15; a_{max} = 55; b)))$ – satisfactory level 80% < Ac < 85%;
- $M_{i=1;2}(W(\alpha; (a_{min} = 15; b)))$ – good level Ac = 87%.

Table 14.5 Balanced accuracy Ac, % of differential diagnosis of sepsis.

M_i	$\alpha(m \times n)$	$\beta(m \times n)$	$LB(m \times n)$	$CB_{\otimes;\oplus}(m \times n)$	$LD(m \times n)$	$CD_{\otimes;\oplus}(m \times n)$
$M_{i=3}$	80.4 c = 4 d = 5	80.4 c = 3 d = 6	89.1 c = 3 d = 2	91.3 c = 1 d = 3	87 c = 2 d = 4	89.1 c = 2 d = 3
$M_{i=4}$	82.6 c = 3 d = 5	80.4 c = 4 d = 5	89.1 c = 2 d = 3	93.5 c = 1 d = 2	87 c = 3 d = 3	91.3 c = 1 d = 3

Table 14.6 Balanced accuracy Ac, % of differential diagnosis of sepsis using wavelet analysis.

a_i			$a_{max} = 55$			
M_i	$\alpha(m \times n)$	$\beta(m \times n)$	$LB(m \times n)$	$CB_{\otimes;\oplus}(m \times n)$	$LD(m \times n)$	$CD_{\otimes;\oplus}(m \times n)$
$M_{i=3}$	82.6 c = 4 d = 4	82.6 c = 3 d = 5	91.3 c = d =	93.5 c = 2 d = 1	89.1 c = 2 d = 3	91.3 c = 1 d = 3
$M_{i=4}$	84.8 c = 3 d = 4	82.6 c = 3 d = 5	93.5 c = 1 d = 2	95.7 c = 1 d = 1	91.3 c = 2 d = 2	93.5 c = 1 d = 2
a_i			$a_{min} = 15$			
M_i	$\alpha(m \times n)$	$\beta(m \times n)$	$LB(m \times n)$	$CB_{\otimes;\oplus}(m \times n)$	$LD(m \times n)$	$CD_{\otimes;\oplus}(m \times n)$
$M_{i=3}$	84.8 c = 3 d = 4	84.8 c = 3 d = 4	89.1 c = 2 d = 3	97.8 c = 0 d = 1	87 c = 3 d = 3	93.5 c = 1 d = 2
$M_{i=4}$	87 c = 2 d = 4	84.8 c = 3 d = 4	91.3 c = 2 d = 2	97.8 c = 0 d = 1	89.1 c = 2 d = 3	95.7 c = 1 d = 1

Based on the results of statistical analysis of the Jones matrix mapping method $T(m \times n)$ for the optical anisotropy parameters of the supramolecular networks of blood films (diagnostic markers $M_{i=3;4}(LB; CB_{\otimes;\oplus}; LD; CD_{\otimes;\oplus})$, the following maximum levels of balanced accuracy Ac, 100% for the differential diagnosis of septic inflammation were established:

- Maps of linear birefringence $M_{i=3;4}(LB(m \times n))$ – good level $Ac = 89.1\%$;
- Maps of circular birefringence $M_{i=3;4}(CB_{\otimes;\oplus}(m \times n))$ – very good level $Ac = 93.5\%$;
- Maps of linear dichroism $M_{i=3;4}(LD(m \times n))$ – good level $Ac = 87\%$;
- Maps of circular dichroism $M_{i=1;2}(W(LB; (a_{min} = 15; b))$ very good level $Ac = 91.3\%$.

Wavelet analysis $W(a_{min} = 15; b)$ of the set of maps $T(m \times n)$ (statistical markers $M_{i=1;2}(W(T, a_{max} = 55; b))$ revealed the following maximum levels of balanced accuracy:

- Amplitude distributions of wavelet coefficients of linear birefringence maps $M_{i=1;2}(W(LB; (a_{max} = 55; b))$ – very good level Ac = 93.5%;
- Amplitude distributions of wavelet coefficients of circular birefringence maps $M_{i=1;2}(W(CB_{\otimes;\oplus}; (a_{max} = 55; b))$ – excellent level $Ac = 95.7\%$;
- Amplitude distributions of wavelet coefficients of linear dichroism maps $M_{i=1;2}(W(LD; (a_{max} = 55; b))$ – very good level $Ac = 91.3\%$;
- Amplitude distributions of wavelet coefficients of circular dichroism maps $M_{i=1;2}(W(CD_{\otimes;\oplus}; (a_{max} = 55; b))$ – very good level $Ac = 93.5\%$.

Wavelet analysis $W(a_{min} = 15; b)$ of the set of maps $T(m \times n)$ (statistical markers $M_{i=1;2}(W(T, a_{max} = 55; b))$ revealed the following maximum levels of balanced accuracy:

- Amplitude distributions of wavelet coefficients of linear birefringence maps $M_{i=1;2}(W(\text{LB}; (a_{min} = 15; b))$ – very good level Ac = 91.3%;
- Amplitude distributions of wavelet coefficients of circular birefringence maps $M_{i=1;2}\left(W\left(\text{CB}_{\otimes;\oplus}; (a_{min} = 15; b)\right)\right.$ – excellent level $Ac = 97.8\%$;
- Amplitude distributions of wavelet coefficients of linear dichroism maps $M_{i=1;2}(W(\text{LD}; (a_{min} = 15; b))$ – good level $Ac = 89.1\%$;
- Amplitude distributions of wavelet coefficients of circular dichroism maps $M_{i=1;2}\left(W\left(\text{CD}_{\otimes;\oplus}; (a_{min} = 15; b)\right)\right.$ – excellent level $Ac = 95.7\%$.

Comparative analysis of the diagnostic effectiveness of both methods revealed a significantly higher maximum level of balanced accuracy for the Jones matrix mapping technique of optical anisotropy parameters of polycrystalline diffusely scattering blood films compared to layered polarization–interference mapping of laser object fields:

- $\text{Ac}\left(\text{CB}_{\otimes;\oplus}(m \times n)\right) = 93.5\%$ and $\text{Ac}(\alpha(m \times n)) = 84.8\%$.
- $\text{Ac}\left(W\left(\text{CB}_{\otimes;\oplus}; (a_{min} = 15; b)\right)\right) = 97.8\%$ and $\text{Ac}(W(\alpha; (a_{min} = 15; b))) = 87\%$.

14.7 Conclusions

1. Algorithms for laser layer-by-layer Jones matrix mapping of optical anisotropy in supramolecular networks of blood films have been developed and experimentally tested.
2. Experimental layer-by-layer maps of linear and circular birefringence and dichroism of supramolecular networks of blood films have been presented and physically analyzed.
3. A set of diagnostic relationships between statistical (central statistical moments of the first to fourth orders) and wavelet (coordinate and linear distributions of wavelet coefficients' amplitudes) parameters of Jones matrix maps and orientation-phase structure (distributions of optical axis directions and phase shifts) of blood films has been identified.
4. Within the framework of statistical analysis of the data from layer-by-layer Jones matrix mapping, the most sensitive markers to septic changes in the structure of optical anisotropy maps were determined – skewness and kurtosis of the distributions of linear and circular birefringence and dichroism parameters of supramolecular networks in blood films.
5. Within the framework of wavelet analysis, the physical laws of modulation formation on different scales of the soliton-like MHAT function of wavelet coefficient amplitudes μ_k for phase and amplitude anisotropy maps have been established, and the most sensitive parameters to septic changes in supramolecular networks of blood films have been determined – central statistical moments of the first and second orders, which characterize the distributions of fluctuations in the wavelet coefficient amplitudes of Jones matrix reconstructed maps of optical anisotropy.
6. Information analysis of the wavelet decomposition data of Jones matrix maps of circular double refraction of polycrystalline supramolecular networks demonstrated excellent accuracy (97.4%) in the differential diagnosis of sepsis, which is 10% higher than the accuracy of current methods of layer-by-layer polarization–interference mapping of azimuth and ellipticity of polarization in object fields of blood films.

References

1 Pircher M, Hitzenberger CK, Schmidt-Erfurth U. Polarization sensitive optical coherence tomography in the human eye. Prog. Retin. Eye Res. 2011;30(6):431–451.

2 Stifter D. Beyond biomedicine: a review of alternative applications and developments for optical coherence tomography. Appl Phys B. 2007;88(3):337–357.

3 Jones RC. A new calculus for the treatment of optical systems I. Description and discussion of the calculus. J Opt Soc Am. 1941;31(7):488–493.

4 Gil JJ, Bernabeu E. Obtainment of the polarizing and retardation parameters of a non-depolarizing optical system from the polar decomposition of its Mueller matrix. Optik (Stuttg.). 1987;76:67–71.

5 Bohren CF, Huffman DR. Absorption and Scattering of Light by Small Particles. New York, NY: Wiley Interscience; 1983.

6 Shurcliff WA, Ballard SS. Polarized Light. New York: Van Nostrand; 1964.

7 Park BH, Saxer C, Srinivas SM, Nelson JS, de Boer JF. In vivo burn depth determination by highspeed fiber-based polarization sensitive optical coherence tomography. J Biomed Opt. 2001;6(4):474–479.

8 Park BH, Pierce MC, Cense B, de Boer JF. Real-time multi-functional optical coherence tomography. Opt Express. 2003;11(7):782–793.

9 Saxer CE, de Boer JF, Park BH, Zhao Y, Chen Z, Nelson JS. High-speed fiber based polarizationsensitive optical coherence tomography of in vivo human skin. Opt Lett. 2000;25(18):1355–1357.

10 Park BH, Pierce MC, Cense B, de Boer JF. Jones matrix analysis for a polarization-sensitive optical coherence tomography system using fiber-optic components. Opt Lett. 2004;29(21):2512–2514.

11 Park BH, Pierce MC, Cense B, de Boer JF. Optic axis determination accuracy for fiber-based polarization-sensitive optical coherence tomography. Opt Lett. 2005;30(19):2587–2589.

12 Kemp NJ, Park J, Zaatari HN, Rylander HG, Milner TE. High-sensitivity determination of birefringence in turbid media with enhanced polarization-sensitive optical coherence tomography. J Opt Soc Am A. 2005;22(3):552–560.

13 Kemp NJ, Zaatari HN, Park J, Rylander III HG, Milner TE. Form-biattenuance in fibrous tissues measured with polarization-sensitive optical coherence tomography (PS-OCT). Opt Express. 2005;13(12):4611–4628.

14 Götzinger E, Pircher M, Geitzenauer W, Ahlers C, Baumann B, Michels S, et al. Retinal pigment epithelium segmentation by polarization sensitive optical coherence tomography. Opt Express. 2008;16(21):16410–16422.

15 Haskell RC, Carlson FD, Blank PS. Form birefringence of muscle. Biophys J. 1989;56(2):401–413.

16 van Blokland GJ. Ellipsometry of the human retina in vivo: preservation of polarization. J Opt Soc Am A. 1985;2(1):72–75.

17 Klein Brink HB, van Blokland GJ. Birefringence of the human foveal area assessed in vivo with Mueller-matrix ellipsometry. J Opt Soc Am A. 1988;5(1):49–57.

18 Todorović M, Jiao S, Wang LV, Stoica G. Determination of local polarization properties of biological samples in the presence of diattenuation by use of Mueller optical coherence tomography. Opt Lett. 2004;29(20):2402–2404.

19 Schmitt JM, Xiang SH. Cross-polarized backscatter in optical coherence tomography of biological tissue. Opt Lett. 1998;23(13):1060–1062.

20 Daubechies I. Ten Lectures on Wavelets. Society for Industrial and Applied Mathematics; 1992.

21 Cohen A, Daubechies I, Vial P. Wavelets on the interval and fast wavelet transforms. Appl Comput Harmon Anal. 1993;1(1):54–81.

22 Meyer Y. Wavelets: Algorithms & Applications. Philadelphia: Society for Industrial and Applied Mathematics (SIAM); 1993.

23 Mallat S. A theory for multiresolution signal decomposition: the wavelet representation. IEEE Trans Pattern Anal Mach Intell. 1989;11(7):674–693.

24 Daubechies I. Wavelets on the interval. In Progress in Wavelet Analysis and Applications: Proceedings of the International Conference "Wavelets and Applications", Toulouse, France. 1992.

25 Farge M. Wavelet transforms and their applications to turbulence. Annu Rev Fluid Mech. 1992;24(1):395–458.

26 Angelsky PO, Ushenko AG, Dubolazov AV, Sidor MI, Bodnar GB, Koval G, et al. The singular approach for processing polarization-inhomogeneous laser images of blood plasma layers. J Opt. 2013;15(4):044030.

27 Ushenko AG, Dubolazov AV, Ushenko VA, Novakovskaya OY. Statistical analysis of polarization-inhomogeneous Fourier spectra of laser radiation scattered by human skin in the tasks of differentiation of benign and malignant formations. J Biomed Opt. 2016;21(7):071110.

28 Garazdyuk MS, Bachinskyi VT, Vanchulyak OY, Ushenko AG, Dubolazov OV, Gorsky MP. Polarization-phase images of liquor polycrystalline films in determining time of death. Appl Opt. 2016;55(12):B67–B71.

29 Angelskii OV, Ushenko OG, Burkovets DN, Arkhelyuk OD, Ushenko YA. Polarization-correlation studies of multifractal structures in biotissues and diagnostics of their pathologic changes. Laser Phys. 2000;10(5):1136–1142.

30 Yermolenko S, Ushenko A, Ivashko P, Goudail F, Gruia I, Gavrilă C. Spectropolarimetry of cancer change of biotissues. In Ninth International Conference on Correlation Optics. SPIE. 2009;7388:404–410.

31 Jóźwicki R, Patorski K, Angelsky OV, Ushenko AG, Burkovets DN, Ushenko YA. Automatic polarimetric system for early medical diagnosis by biotissue testing. Opt Appl. 2002;32(4):603–612.

32 Ushenko AG, Dubolazov OV, Ushenko VA, Novakovskaya OY, Olar OV. Fourier polarimetry of human skin in the tasks of differentiation of benign and malignant formations. Appl Opt. 2016;55(12):B56–B60.

33 Ushenko AG, Pishak VP. Coherent-domain optical methods: biomedical diagnostics, environmental and material science. In Laser Polarimetry of Biological Tissue: Principles and Applications. 2004.

34 Ushenko AG, Fediv AI, Marchuk YF. Correlation and fractal structure of Jones matrices of human bile secret. In European Conference on Biomedical Optics. 2009;7368:73681Q.

35 Ushenko AG. Depolarization of a laser emission field as the correlation development of its polarization structure. In International Conference on Correlation Optics. 1997;3317:331–339.

36 Bachinskyi V, Boychuk T, Ushenko A. Laser Polarimetry of Biological Tissues and Fluids: Chapter 4. Methods of Spatial-frequency, Singular and Correlation Analysis of Biological Layers Object Fields. LAP Lambert Academic Publishing; 2018.

37 Ushenko AG, Burkovets DN, Yermolenko SB, Arkhelyuk AD, Pishak VP, Yuzko AM, et al. Stokes polarimetry of biotissues. In Fourth International Conference on Correlation Optics. SPIE. 1999;3904:527–533.

38 Ushenko VA, Dubolazov AV, Pidkamin LY, Sakchnovsky MY, Bodnar AB, Ushenko YA, et al. Mapping of polycrystalline films of biological fluids utilizing the Jones-matrix formalism. Laser Phys. 2018;28(2):025602.

39 Ushenko AG, Angelsky PO, Sidor M, Marchuk YF, Andreychuk DR, Pashkovskaya NV. Spatial-frequency selection of complex degree of coherence of laser images of blood plasma in diagnostics and differentiation of pathological states of human organism of various nosology. Appl Opt. 2014;53(10):B172–B180.

40 Olar EI, Ushenko AG, Ushenko YA. Polarization correlation measurements of the phase tomograms of optically anisotropic biofractals. Laser Phys. 2004;14(8):1115–1121.

41 Guminetskiy SH, Ushenko OG, Polyanskiy IP, Motrych AV, Grynchuk FV. The optical method for investigation of the peritonitis progressing process. In Eighth International Conference on Correlation Optics. 2008;7008:626–632.

42 Ushenko AG, Fediv AI, Marchuk YF. Singular structure of polarization images of bile secret in diagnostics of human physiological state. In: European Conference on Biomedical Optics. Optica Publishing Group; 2009.

43 Ushenko Y, Ushenko A, Dubolazov A, Soltys I, Litvinenko O, Wanchuliak O, et al. Mueller-matrix interferometric multifractal scaling of optically anisotropic architectonics of diffuse blood facies: fundamental and applied aspects. J Biophotonics. 2025;18(3):e202400412. https://doi.org/10.1002/jbio.202400412.

44 Ushenko A, Soltys I, Dubolazov A, Ushenko Y, Bilookyi V, Bilookyi O, et al. 3D Jones-matrix thesiography of biological fluid facies. J Innov Opt Health Sci. 2025;18(2):2443002. https://doi.org/10.1142/S1793545824430028.

45 Ushenko A, Pavlyukovich N, Khukhlina O, Pavlyukovich O, Gorsky M, Soltys I, et al. Layer-by-layer multifractal scanning of optically anisotropic architectonics of blood plasma films: fundamental and applied aspects. Photonics. 2025;12:215. https://doi.org/10.3390/photonics12030215.

46 Ushenko A, Pavlyukovich N, Khukhlina O, Pavlyukovich O, Soltys I, Dubolazov A, et al. Blood plasma film multifractal scanning in COVID-19 consequences diagnostics. J Biophotonics. 2024;17(11):e202400356. https://doi.org/10.1002/jbio.202400356.

47 Ushenko A, Dubolazov A, Zheng J, Litvinenko A, Gorsky M, Ushenko Y, et al. 3D polarization-interference holographic histology for wavelet-based differentiation of biological tissues with different necrotic states. J Biomed Opt. 2024;29(5):052920. https://doi.org/10.1117/1.JBO.29.5.052920.

48 Ushenko O, Bilookyi O, Zheng J, Dubolazov A, Olar O, Ushenko Y, et al. 3D digital holographic polarimetry of laser speckle fields formed by polycrystalline blood films: a tool for differential diagnosis of thyroid pathology. Front Phys. 2024;12:1426469. https://doi.org/10.3389/fphy.2024.1426469.

49 Wójcik W, Hu Z, Ushenko Y, Smolarz A, Soltys I, Dubolazov O, et al. Optical sensor system for 3D Jones matrix reconstruction of optical anisotropy maps of self-assembled polycrystalline soft matter films. Sensors. 2024;24(5):1589. https://doi.org/10.3390/s24051589.

50 Ushenko A, Zheng J, Gorsky M, Dubolazov A, Ushenko Y, Soltys I, et al. 3D digital holographic polarimetry of diffuse optically anisotropic biological tissue object fields. Front Phys. 2023;11:1288935. https://doi.org/10.3389/fphy.2023.1288935.

51 Dubolazov A, Ushenko V, Trifonyuk L, Stashkevich A, Soltys I, Ushenko Y, et al. Polarization-singular approach to imaging Mueller-matrix polarimetry in differential diagnosis of histological sections of biopsy of tumors of the uterus and prostate. Front Phys. 2021;9:711212. https://doi.org/10.3389/fphy.2021.711212.

52 Ushenko VO, Trifonyuk L, Ushenko YA, Dubolazov OV, Gorsky MP, Ushenko AG. Polarization singularity analysis of Mueller-matrix invariants of optical anisotropy of biological tissue samples in cancer diagnostics. J Opt. 2021;23(6):064004. https://doi.org/10.1088/2040-8986/abf97a.

53 Ushenko YA, Koval GD, Ushenko AG, Dubolazov OV, Ushenko VA, Novakovskaia OY. Mueller-matrix of laser-induced autofluorescence of polycrystalline films of dried peritoneal fluid in diagnostics of endometriosis. J Biomed Opt. 2016;21(7):071116. https://doi.org/10.1117/1.JBO.21.7.071116.

15

Polarimetry of Blood Films for Thyroid Pathology

Yurii Ushenko[1,2], Weidong Dou[1], Oleksandr Ushenko[2,3], Oleksandr Dubolazov[2], Iryna Soltys[2], Vitaliy Rozhko[4], Oleksandr Olar[2], and Vladyslava Sholota[5]

[1] *Shaoxing University, Shaoxing, People's Republic of China*
[2] *Yurii Fedkovych Chernivtsi National University, Chernivtsi, Ukraine*
[3] *Taizhou Institute of Zhejiang University, Taizhou, People's Republic of China*
[4] *Bukovinian State Medical University, Chernivtsi, Ukraine*
[5] *Vinnytsia National Technical University, Vinnytsia, Ukraine*

15.1 Three-dimensional Digital Holographic Polarimetry of Laser Speckle Fields from Polycrystalline Blood Films for the Identification of Thyroid Diseases

15.1.1 Introduction

Timely identification of thyroid disorders remains a significant global healthcare challenge. According to World Health Organization data, thyroid diseases affect approximately 10–30% of the adult population worldwide and are associated with both functional impairments and structural alterations of the gland. Moreover, the prevalence of these conditions continues to rise by about 5% annually [1–12].

At present, histopathological examination based on biopsy or surgical specimens stained with hematoxylin and eosin is considered the reference standard for diagnosing thyroid pathology. However, reliable early-stage detection is often complicated by substantial variability in morphological interpretation, while malignant alterations may be subtle and difficult to recognize at initial stages.

In recent years, these limitations have driven the development of digital pathology approaches that incorporate quantitative image analysis techniques. Using such methods, the feasibility of differential diagnosis of thyroid nodules – distinguishing benign lesions such as follicular adenoma from malignant entities including papillary carcinoma – has been demonstrated. The integration of objective, high-resolution quantitative tools enables pathologists to enhance diagnostic accuracy in thyroid disease assessment.

The advancement of PSHG microscopy has been closely associated with the development of noninvasive polarization-sensitive optical coherence tomography (PSOCT). The application of direct interferometric recording of object speckle-field structures enables depth-resolved measurements of spatial distributions of Mueller-matrix elements, resulting in the formation of Mueller-matrix images at different depths within biological samples. This approach provides a quantitative, layer-by-layer evaluation of tissue optical anisotropy. Using such methods, accurate detection of fibrotic regions and differentiation of tumor areas characterized by low fibrosis have been demonstrated. However, the diagnostic sensitivity of these systems is often constrained by a strong depolarized laser speckle background [13–25].

Phase-scanning polarization–interferometric techniques applied to biological object fields with complex amplitude distributions, combined with digital layer-by-layer reconstruction, offer an effective strategy to suppress the depolarized background component. This significantly enhances the diagnostic sensitivity of both PSHG microscopy and PSOCT-based imaging methods.

These approaches enable phase-selective isolation of the single-scattering, non-depolarized component of laser speckle fields. The non-depolarized fraction of the complex-amplitude field formed by a biological layer is uniquely associated with the parameters of its optical anisotropy. Consequently, within a statistical analysis framework, this makes it possible to extract highly reliable information on pathological alterations in human biopsy samples, including changes in polycrystalline structural organization. Using this methodology, successful diagnosis of endometriosis, discrimination between benign and malignant prostate tumors, and identification of necrotic and traumatic myocardial damage have been demonstrated.

Further extension of polarization–interference phase-scanning techniques may involve their application to alternative object types, particularly biological fluids. Such media are readily obtainable and do not require invasive biopsy procedures. Biological fluids play a key role in sustaining vital physiological functions and participate actively in metabolic regulation. Their molecular composition and intermolecular interactions undergo rapid modifications under both normal physiological conditions and in response to pathological or extreme states.

Any physiological or pathological process within a living organism is accompanied by specific structural alterations of proteins and other biomolecules. These molecular-level transformations are highly informative for assessing homeostasis and form a robust basis for early-stage disease diagnosis.

Blood is among the most informative biological fluids reflecting the physiological state of the human body, as it contains nearly 100 distinct proteins. Traditionally, based on differences in solubility, these proteins are classified into five principal fractions: albumins (55–65%), $\alpha 1$ – (2–4%), $\alpha 2$ – (6–12%), β – (8–12%), and γ – (2–4%) globulins [26–36].

Albumins are present in blood at high concentrations and are characterized by relatively small molecular dimensions, resulting in a large total surface area of the albumin molecular ensemble. This structural feature underlies their high transport capacity, enabling the binding and transfer of substances such as bilirubin, urobilin, fatty acids, bile salts, and lipid hormones. In addition to albumin itself, this fraction contains transthyretin (prealbumin), which, together with thyroxine-binding globulin (TBG) and albumin, plays a key role in the transport of the thyroid hormone thyroxine and its metabolite iodothyronine.

Globulins constitute a class of blood proteins whose polypeptide chains are folded into compact, nearly spherical conformations corresponding to tertiary protein structures. One representative of this group is thyroglobulin (Tg), a protein synthesized by thyroid follicular cells and stored as a colloidal component within the follicle, the gland's basic structural and functional unit. In patients with Hashimoto's thyroiditis or Graves' disease, elevated levels of antithyroglobulin antibodies are frequently observed and are widely used as diagnostic indicators. Moreover, Tg serves as a reliable postoperative prognostic marker in differentiated thyroid cancers, including papillary and follicular carcinoma, and its concentration may also be increased in Graves' disease.

Despite the clinical relevance of these biomolecular systems, their optically anisotropic organization and polarization-related properties remain insufficiently characterized. To date, only a limited number of studies have addressed the application of polarization interferometry to biological fluids for the diagnosis of pathological changes in human organs.

This section is therefore devoted to the identification of novel objective markers for minimally invasive differential diagnosis of thyroid disorders by employing comprehensive polarization–interference mapping combined with digital holographic reconstruction of depth-resolved maps of speckle-field ellipticity in blood films. To maximize diagnostic performance, a statistical analysis of polarization-ellipticity maps corresponding to the singly scattered component of the speckle field in blood films from healthy donors and patients with nodular goiter, autoimmune thyroiditis, and papillary carcinoma is applied here for the first time [37–44].

15.1.2 Biological Samples

Four representative groups of dehydrated blood films were formed for the study.

Group 1: Control group. This group included 51 healthy donors aged 30–60 years (48 women and 3 men).

Inclusion criteria: thyroid fine-needle aspiration biopsy without evidence of pathological changes; normal thyroid hormone levels and absence of elevated autoantibodies.

Exclusion criteria: Biopsy-confirmed thyroid pathology.

Group 2: *Experimental group (nodular goiter)*. The group comprised 51 patients aged 30–60 years (46 women and 5 men) diagnosed with nodular goiter. Inclusion criteria: Biopsy-confirmed thyroid pathology; presence of one or more colloid nodules in the thyroid gland; normal thyroid hormone concentrations and autoantibody levels.

Exclusion criteria: Hypo- or hyperthyroidism; elevated autoantibody levels; concomitant diseases including diabetes mellitus, hypertension, atherosclerosis, chronic adrenal insufficiency, rheumatoid arthritis, chronic active hepatitis, and systemic lupus erythematosus.

This group is characterized by disturbances in the negative feedback mechanisms regulating endocrine function.

Group 3: Experimental group (autoimmune thyroiditis). This group consisted of 51 patients aged 30–60 years (49 women and 2 men) with autoimmune thyroiditis.

Inclusion criteria: Biopsy-confirmed thyroid pathology corresponding to stages 1–3 (stage 1 – hypertrophic thyroiditis/Hashimoto's goiter; stage 2 – atrophic thyroiditis; stage 3 – fibrous invasive Riedel's goiter); autoantibody levels exceeding 1000 IU/mL; presence of hypothyroidism.

Exclusion criteria: Elevated thyroid hormone levels; normal levels of antibodies to thyroperoxidase; associated systemic diseases, including diabetes, hypertension, atherosclerosis, chronic adrenal insufficiency, rheumatoid arthritis, chronic active hepatitis, and systemic lupus erythematosus.

This group is characterized by impairment of the hematothyroid barrier.

Group 4: Experimental group (papillary thyroid carcinoma). The group included 51 patients aged 30–60 years (44 women and 7 men) with papillary thyroid cancer.

Inclusion criteria: Biopsy-confirmed malignant process with the presence of papillary epithelial cancer cells; absence of extracapsular extension and lymph node invasion; normal autoantibody levels; euthyroid status.

Exclusion criteria: Abnormal thyroid hormone levels; elevated antibodies to thyroperoxidase; concomitant systemic diseases such as diabetes, hypertension, atherosclerosis, chronic adrenal insufficiency, rheumatoid arthritis, chronic active hepatitis, and systemic lupus erythematosus.

This group is characterized by uncontrolled atypical proliferation of thyroid cells with papillary structure formation, nuclear inclusions, tumor progression, and a history of frequent exposure to the consequences of the Chernobyl accident.

The optical parameters of polycrystalline blood films corresponding to these groups are summarized in Table 15.1.

15.1.3 Experimental Results

To date, experimentally established trends in pathological alterations of polarization maps corresponding to optically anisotropic components of thyroid biopsy specimens can be summarized as follows.

Nodular goiter is characterized by an increase in the optical activity of thyroglobulin protein complexes, which arises from follicular enlargement and the development of structural birefringence associated with nodular proliferation of connective tissue.

Autoimmune thyroiditis exhibits proportional enhancement of both optical activity of thyroglobulin complexes and birefringence of the fibrous stromal network, primarily caused by squamous metaplasia of follicular epithelial cells accompanied by an increase in their size.

Papillary carcinoma is dominated by the effects of structural birefringence resulting from hyperplastic processes manifested as elongated papillary formations.

The interpretation and diagnostic application of these findings are based on a two-component amorphous–anisotropic model of the optical properties of the thyroid gland. Within this framework, the amorphous component, represented by the epithelial parenchyma, is considered polarization-inactive and does not modify the polarization state of incident laser radiation. The anisotropic component comprises optically active elements, namely globulin protein complexes corresponding to follicles filled with thyroglobulin-rich colloid, as well as structural elements formed by supramolecular fibrillar protein networks of connective tissue within the parenchyma. The optically active component predominantly contributes to the formation of polarization azimuth maps $\alpha(m, n)$. The structural or polycrystalline component predominantly forms maps of polarization ellipticity.

Table 15.1 Optical parameters of blood films.

Parameter	Group 1	Group 2	Group 3	Group 4
Attenuation coefficient τ, cm^{-1}	0.64 ± 0.04	0.61 ± 0.04	0.63 ± 0.04	0.62 ± 0.04
Depolarization degree Λ, %	41 ± 0.9	38 ± 0.8	42 ± 0.8	40 ± 0.7

During dehydration of a blood droplet, spontaneous self-organization of protein molecules leads to the formation of a supramolecular protein network. The structural features of these proteins govern the architecture of the resulting blood films. Protein structure is determined by the spatial arrangement of atoms within the molecule, which in three-dimensional space gives rise to polymers composed of polypeptide chains. Each polypeptide is formed by a specific sequence of α-amino acids. With the exception of glycine, all α-amino acids present in living organisms contain at least one asymmetric carbon atom (with threonine and isoleucine containing two), which gives rise to optical activity manifested as circular birefringence. Protein organization is commonly described in terms of hierarchical structural levels.

The primary structure corresponds to the amino acid sequence of the polypeptide chain. A key characteristic of this level is the quasi-random distribution of amino acid residues, which, from a biophysical standpoint, is responsible for the emergence of circular birefringence. The secondary structure arises from the spatial arrangement of the main polypeptide backbone, largely independent of side-chain conformations, and at this level the orientation of the linear birefringence (LB) optical axis is established [43–55].

The tertiary structure defines the three-dimensional conformation of a single polypeptide chain. Depending on the mode of spatial organization, proteins are classified as globular or fibrillar. Globular proteins typically exhibit compact, ellipsoidal shapes and predominantly demonstrate circular birefringence, whereas fibrillar proteins possess elongated geometries and are mainly associated with LB. The quaternary structure results from the spatial association of multiple polypeptide chains, usually sharing similar primary, secondary, and tertiary structures.

As a consequence of these hierarchical interactions, a structurally and functionally integrated supramolecular network is formed in dehydrated blood films, characterized by phase anisotropy that includes both circular and LB. The specific polycrystalline organization of this network, reflected in the relationships between optical anisotropy parameters, is determined by conformationally autonomous polypeptide fragments incorporated into the protein ensemble.

Based on this framework, alterations in the circular and LB characteristics of blood film (BF) supramolecular networks are expected in the presence of thyroid pathology. In particular, pathological increases in thyroglobulin concentration, along with albumin acting as its transport protein, may lead to enhanced phase anisotropy of the supramolecular protein network, as observed in affected patients.

In this study, primary attention is devoted to polarization maps $\beta(x, y)$ corresponding to the singly scattered component of the object field formed by blood films from healthy donors and patients with various thyroid disorders. It has been demonstrated that this polarization parameter is governed by the phase-shifting properties, or phase anisotropy, of the underlying supramolecular networks.

15.1.4 Statistical Analysis of Polarization Ellipticity Maps

Figures 15.1–15.4 present a series of results illustrating the integral maps (panels (a)) and the corresponding histograms of ellipticity value distributions (panels (b)) obtained for healthy donors (control group 1) and for patients with various thyroid gland pathologies (experimental groups 2–4).

Analysis of the obtained results demonstrates a clear correspondence between the experimental data (Figures 15.1–15.4) and the proposed model describing the formation of the polarization structure of speckle fields $H_{\text{blood}}(x, y)$ in blood films from both control subjects and patients. It was established that, for all examined thyroid pathologies – nodular goiter, autoimmune thyroiditis, and papillary carcinoma – the object speckle fields of BF samples exhibit comparable polarization ellipticity maps $\beta_{\text{blood}}(\vartheta_{xy})$ (panels (a) in Figures 15.1–15.4).

This similarity is further confirmed by the corresponding histograms $N(\beta)$ representing the distributions of polarization ellipticity values (panels (b) in Figures 15.1–15.4). For each BF sample, the polarization ellipticity map $\beta_{\text{blood}}(\vartheta_{xy})$ spans a relatively broad range of values ($0 \leq \beta \leq 0.5\pi$). In addition, the probabilities associated with the occurrence of specific ellipticity values in the speckle fields of blood films from healthy donors and patients with thyroid disorders are of comparable magnitude. At first inspection, the experimental findings appear inconsistent with the scenario predicted by the proposed model describing the formation and pathological modification of the polarization structure $\beta_{\text{object}}(\delta_{xy})$ of the object field $E_{\text{object}}(x, y) \equiv \begin{pmatrix} E_x \\ E_y \\ \delta_{xy} \end{pmatrix}$ generated by optically anisotropic supramolecular networks in blood films. As discussed above, the architectonics of these networks, characterized by $\left(\rho(x, y); \delta_{xy}(x, y)\right)$, exhibit a multilevel hierarchical organization, encompassing nucleotide monomers, polypeptide chains, and the tertiary and quaternary structures of globular and fibrillar proteins.

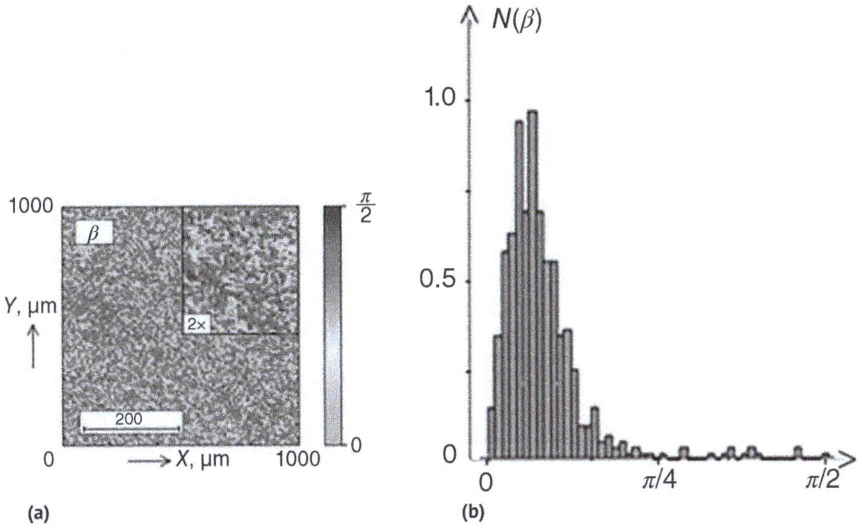

Figure 15.1 BF polarization map (a) and the corresponding histogram (b) for a representative healthy donor (control group 1).

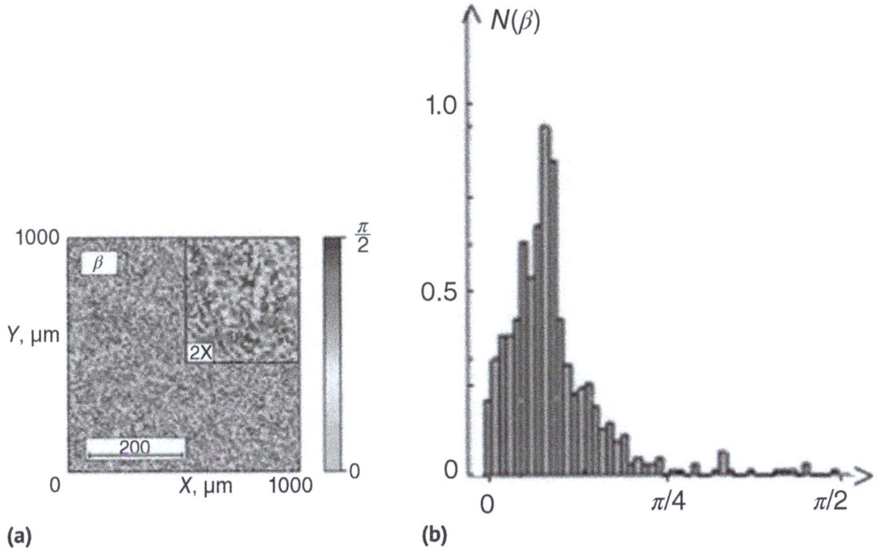

Figure 15.2 Polarization map of the BF (a) and the corresponding histogram (b) obtained for a representative patient from experimental group 2 (nodular goiter).

The development of pathological conditions is accompanied by an increase in optical anisotropy of blood films arising from both biochemical factors, such as elevated thyroglobulin concentration, and conformational factors associated with the enhanced structuring of globular and fibrillar protein networks. From a polarization standpoint, these changes would be expected to result in an increase in both the magnitude of ellipticity and the breadth of its variation range $\beta_{\text{object}}(\delta_{xy})$.

In the presence of multiple scattering within the volume of blood films, an additional "speckle" component $U_{\text{speckle}}(x, y) \equiv \begin{pmatrix} U_x \\ U_y \\ \varphi_{xy} \end{pmatrix}$ is formed alongside the "object" field $E_{\text{object}}(x, y) \equiv \begin{pmatrix} E_x \\ E_y \\ \delta_{xy} \end{pmatrix}$, irrespective of the presence or type of pathology. The interference mechanisms responsible for the formation of this speckle component lead to transformation and averaging of the polarization ellipticity $\beta_{\text{object}}(\delta_{xy})$ distributions, which are otherwise specific to individual pathological states. As a consequence of this averaging process, diagnostically relevant differences between the polarization ellipticity maps $\beta_{\text{object}}(x, y, \delta_{xy})$ are substantially reduced.

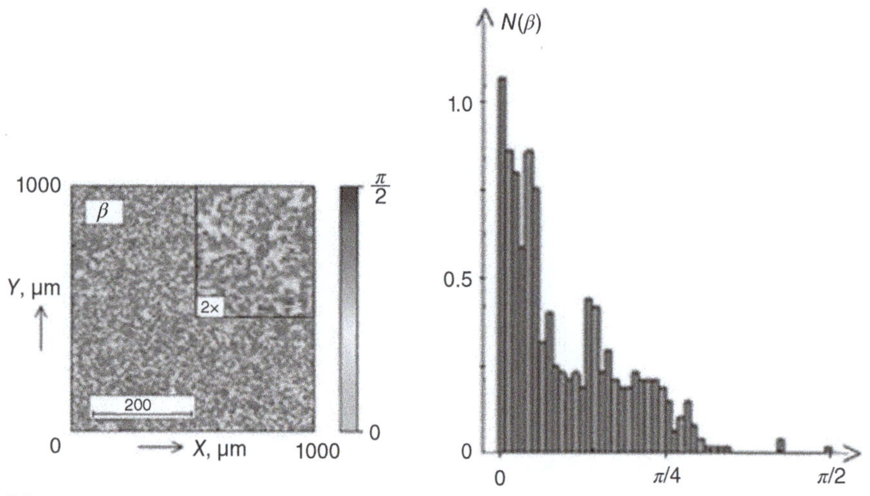

Figure 15.3 Polarization map of the BF (a) and the corresponding histogram (b) obtained for a representative patient from experimental group 3 diagnosed with autoimmune thyroiditis.

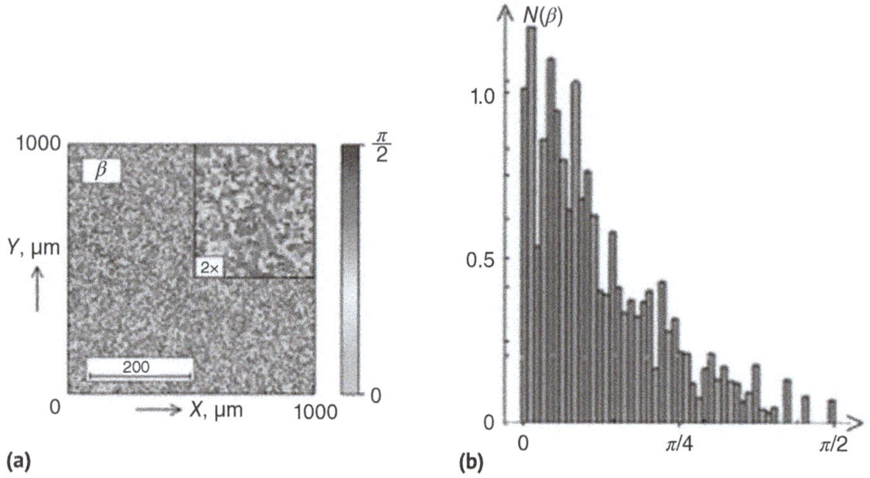

Figure 15.4 Polarization map of the BF (a) and the corresponding histogram (b) obtained for a representative patient from experimental group 4 diagnosed with papillary thyroid carcinoma.

Each act of interference involving the scattered wave components $\widetilde{E}_{xj}(r)$ and $\widetilde{E}_{yj}(r)$ modifies the resulting polarization ellipticity value $\widetilde{\beta}(x, y, \varphi_{x,y})$. Moreover, the interference-induced phase shifts φxy can significantly exceed the corresponding object-related phase parameters δxy. Consequently, the secondary influence of interference becomes the dominant factor governing the formation of polarization ellipticity maps $\beta_{\text{speckle}}(x, y, \varphi_{xy})$ in blood films from both healthy donors and patients with thyroid disorders.

The results of statistical analysis of the random ellipticity value distributions confirm the quantitative similarity of the speckle-field polarization structures across all studied groups, as summarized in Table 15.2.

A nonparametric statistical comparison of polarization ellipticity random values obtained from independent samples was performed using the Mann–Whitney U test. Pairwise comparisons were conducted between the following groups: group 1 versus group 2, group 1 versus group 3, group 1 versus group 4, group 2 versus group 3, group 2 versus group 4, and group 3 versus group 4.

Table 15.2 Statistical analysis of polarization and ellipticity maps of blood films in donors and patients with thyroid diseases.

Groups	Group 1 (51 samples)	Group 2 (51 samples)	Group 3 (51 samples)	Group 4 (51 samples)
Average, M_1	0.187 ± 0.01	0.205 ± 0.01	0.212 ± 0.01	0.22 ± 0.01
p_{ik}		$p_{12;13;14} > 0.05$ $p_{23;24;34} > 0.05$		
Variance, M_2	0.082 ± 0.005	0.093 ± 0.006	0.114 ± 0.007	0.123 ± 0.008
p_{ik}		$p_{12;13;14} > 0.05$ $p_{23;24;34} > 0.05$		
Skewness, M_3	0.89 ± 0.05	0.78 ± 0.04	0.71 ± 0.04	0.66 ± 0.04
p_{ik}		$p_{12;13;14} > 0.05$ $p_{23;24;34} > 0.05$		
Kurtosis, M_4	1.47 ± 0.09	1.18 ± 0.06	0.94 ± 0.06	0.83 ± 0.05
p_{ik}		$p_{12;13;14} < 0.05$ $p_{23;24;34} > 0.05$		

The statistical analysis demonstrated the following trends:

- The mean value M_1 of the ellipticity distribution $\beta(m, n)$ exhibited a slight increase from 0.187 to 0.22 across the studied samples; however, this change was not statistically significant $(p_{12;13;14} > 0.05; p_{23;24;34} > 0.05)$.
- The variance M_2 of the $\beta(m, n)$ distributions ranged from 0.082 to 0.123 and likewise did not show statistically significant intergroup differences $(p_{12;13;14} > 0.05; p_{23;24;34} > 0.05)$.
- The third-order statistical moment M_3, describing the skewness of the ellipticity distributions and varying between 0.66 and 0.89, was also found to be statistically insignificant for differential diagnostic purposes.
- In contrast, the sensitivity of BF polarization maps to intergroup variations was primarily associated with the kurtosis M_4 of the ellipticity random value distributions.
- The kurtosis M_4 demonstrated statistically significant differences $(p_{12;13;14} \leq 0.05)$, increasing from 0.83 to 1.47, and can therefore be considered a reliable marker for detecting the presence of thyroid gland pathology.

Overall, for the sequence of pathological states ("normal – nodular goiter – autoimmune thyroiditis – papillary cancer"), the mean and variance of the polarization ellipticity distributions exhibit only minor changes. Although the third- and fourth-order statistical moments vary by approximately 30–40% between groups, only the kurtosis M_4 provides sufficient sensitivity for effective differentiation of thyroid pathologies [56–65].

The results of differential diagnosis of thyroid gland pathologies, expressed in terms of balanced accuracy, are summarized in Table 15.3.

Here, a denotes the total number of correctly classified samples across all groups, including the control group 1 and, sequentially, the experimental groups 2 (nodular goiter), 3 (autoimmune thyroiditis), and 4 (papillary carcinoma), whereas b represents the total number of false-positive classifications aggregated over all groups.

The information analysis indicates a limited diagnostic capability for differentiating thyroid pathology when comparing the control group (group 1) with the combined set of experimental groups (groups 2–4). Specifically, none of the evaluated statistical markers achieved a balanced accuracy exceeding 80%.

A markedly different outcome is obtained when polarization–interference mapping of BF speckle fields $\begin{pmatrix} H_x; \\ H_y; \\ \vartheta_{xy} \end{pmatrix}$ is combined with phase scanning and digital layer-by-layer reconstruction of polarization ellipticity maps $\beta_{\text{blood}}(\Delta\theta_{xy} \to \vartheta_{xy})$.

Table 15.3 Balanced accuracy of thyroid pathology differential diagnosis.

	Average, M_1	
$a = 70$	$a = 71$	$a = 72$
$b = 32$	$b = 31$	$b = 30$
$Ac_{12} = 68.6\%$	$Ac_{13} = 69.6\%$	$Ac_{14} = 70.6\%$
$a = 70$	$a = 70$	$a = 71$
$b = 32$	$b = 32$	$b = 31$
$Ac_{23} = 68.6\%$	$Ac_{24} = 68.6\%$	$Ac_{34} = 69.6\%$
	Variance, M_2	
$a = 72$	$a = 73$	$a = 74$
$b = 30$	$b = 29$	$b = 28$
$Ac_{12} = 70.6\%$	$Ac_{13} = 71.5\%$	$Ac_{14} = 72.5\%$
$a = 71$	$a = 72$	$a = 73$
$b = 31$	$b = 30$	$b = 29$
$Ac_{23} = 69.6\%$	$Ac_{24} = 70.6\%$	$Ac_{34} = 71.6\%$
	Skewness, M_3	
$a = 75$	$a = 76$	$a = 77$
$b = 27$	$b = 26$	$b = 25$
$Ac_{12} = 73.5\%$	$Ac_{13} = 74.5\%$	$Ac_{14} = 75.5\%$
$a = 73$	$a = 74$	$a = 75$
$b = 29$	$b = 28$	$b = 27$
$Ac_{23} = 71.5\%$	$Ac_{24} = 72.5\%$	$Ac_{34} = 73.5\%$
	Kurtosis, M_4	
$a = 76$	$a = 77$	$a = 78$
$b = 26$	$b = 25$	$b = 24$
$Ac_{12} = 74.5\%$	$Ac_{13} = 75.5\%$	$Ac_{14} = 76.5\%$
$a = 74$	$a = 75$	$a = 76$
$b = 28$	$b = 27$	$b = 26$
$Ac_{23} = 72.5\%$	$Ac_{24} = 73.5\%$	$Ac_{34} = 74.5\%$

Figures 15.5–15.8 present algorithmically reconstructed polarization ellipticity maps of the "object" component $\begin{pmatrix} E_x; \\ E_y; \\ \delta_{xy} \end{pmatrix}$, obtained in the phase plane ($\theta_k = \pi/8$), for BF speckle fields $\begin{pmatrix} H_x; \\ H_y; \\ \vartheta_{xy} \end{pmatrix}$ from healthy donors and patients with thyroid pathologies.

The phase value ($\theta_k = \pi/8$) was determined through stepwise phase scanning $(\Delta\theta_{xy})$ combined with statistical analysis of the algorithmically reconstructed polarization ellipticity distributions $\beta_{\text{speckle}}(\Delta\theta_{xy}, x, y)$. This phase plane corresponds to the statistical criterion $SM_{g=1;2;3;4}(\beta(\Delta\theta_{xy} \leq \theta_k)) \approx \text{const}$, which enables the selection of the singly scattered "object" component $\begin{pmatrix} E_x; \\ E_y; \\ \delta_{xy} \end{pmatrix}$.

Comparative analysis of the "object" polarization maps $\beta_{\text{object}}(\theta_k, x, y)$ (panels (a) in Figures 15.5–15.8) revealed a pronounced reduction in the range of ellipticity variation relative to the corresponding "speckle" distributions $\beta_{\text{speckle}}(\vartheta_{xy}, x, y)$ shown in panels (a) of Figures 15.1–15.4. This observation confirms the effective algorithmic suppression of the distorting influence of the scattered speckle component $\beta_{\text{speckle}}(\vartheta_{xy}, x, y)$ on the spatial distributions of "object" ellipticity.

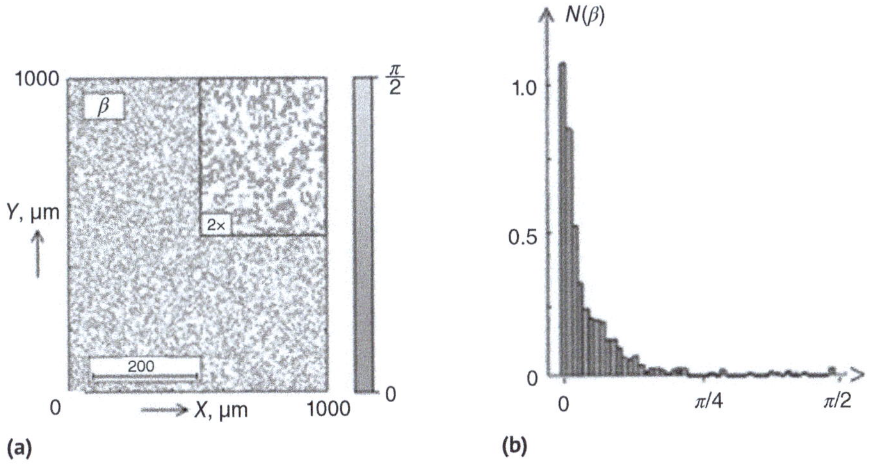

Figure 15.5 Phase-resolved ($\theta_k = \pi/8$) polarization map of the BF (a) and the corresponding histogram (b) for a representative subject from control group 1 (healthy donors).

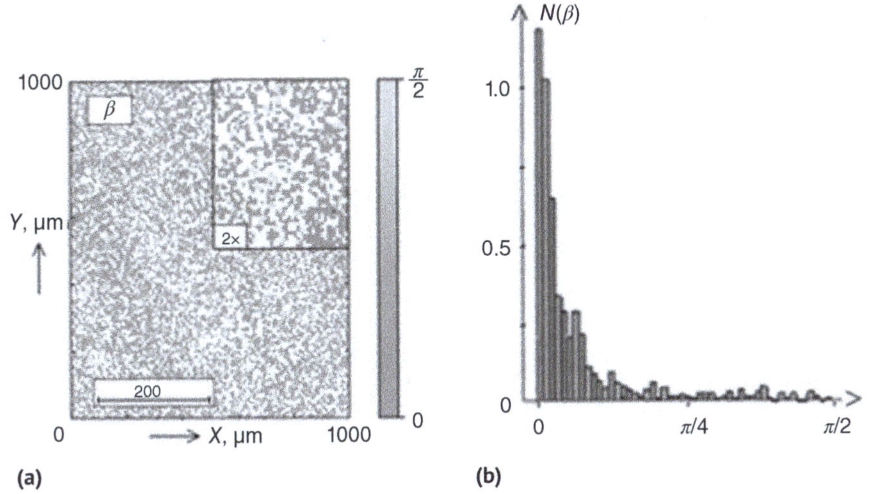

Figure 15.6 Phase-resolved ($\theta_k = \pi/8$) polarization map of the BF (a) and the corresponding histogram (b) obtained for a representative patient from experimental group 2 diagnosed with nodular goiter.

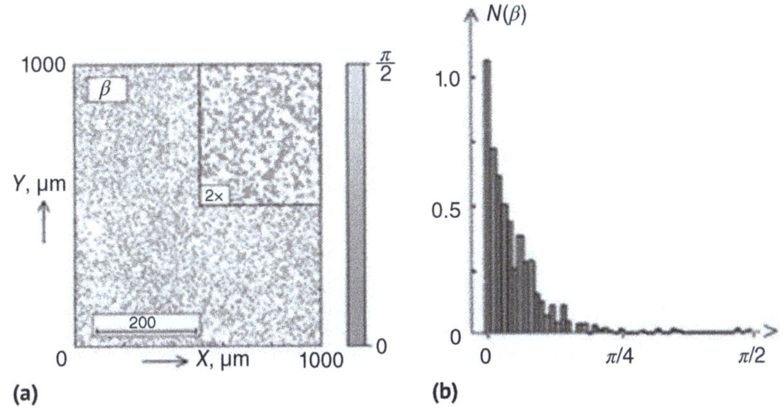

Figure 15.7 Phase-resolved ($\theta_k = \pi/8$) polarization map of the BF (a) and the corresponding histogram (b) obtained for a representative patient from experimental group 3 diagnosed with autoimmune thyroiditis.

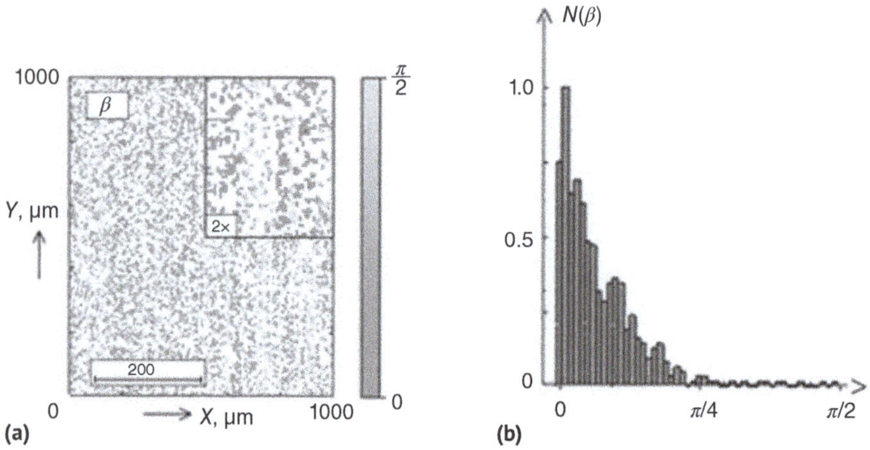

Figure 15.8 Phase-resolved ($\theta_k = \pi/8$) polarization map of the BF (a) and the corresponding histogram (b) obtained for a representative patient from experimental group 4 diagnosed with papillary thyroid carcinoma.

Within the framework of the proposed optical anisotropy model of blood films, it becomes possible to estimate the most probable variations of the polarization ellipticity associated with the singly scattered component $\begin{pmatrix} E_x; \\ E_y; \\ \delta_{xy} \end{pmatrix}$. The birefringence of protein structures is characterized by ($\Delta n \sim 1.5 \times 10^{-3}$, while the geometric thickness of blood films varies from approximately 30 to 100 μm, and the laser wavelength is $\lambda = 0.6328$ μm. Under the assumption that the entire BF volume exhibits birefringence, the corresponding "object" phase shift δ_{xy} would be expected to lie within the range from $\pi/6$ to $\pi/2$. Consequently, considering the formation conditions of singly scattered laser waves $\begin{pmatrix} E_x; \\ E_y; \\ \delta_{xy} \end{pmatrix}$, a maximum polarization ellipticity of about $\pi/4$ may be anticipated.

However, experimental data obtained from real BF samples indicate that polarization states with substantially lower ellipticity values are more probable, as illustrated by the histograms in panels (b) of Figures 15.5–15.8. This discrepancy can be attributed to the fact that optically anisotropic supramolecular networks are spatially heterogeneous and do not uniformly occupy the entire volume of the blood films.

As a consequence, the effective geometric dimensions $l_{\text{effective}}$ of birefringent globular and fibrillar protein chains are reduced. This leads to a corresponding narrowing of the "object" phase shift δ_{xy} and polarization ellipticity β_{object} ranges, which become confined to the interval from 0 to $\pi/4$.

Comparative analysis of the histograms $N(\beta)$ of the ellipticity distributions $\beta_{\text{object}}(\theta_k, x, y)$ in the phase plane $\theta_k = \pi/8$ reveals a consistent trend toward increasing ellipticity values for blood films obtained from healthy donors (Figure 15.5b) and from patients with nodular goiter (Figure 15.6b), autoimmune thyroiditis (Figure 15.7b), and papillary carcinoma (Figure 15.8b).

This behavior is reflected by a decrease in histogram kurtosis M_4, which is associated with a reduction in the minimum polarization ellipticity values β_{object}. Simultaneously, the probability distributions $N(\beta)$ become more symmetric, as evidenced by a decrease in the skewness parameter M_3. The observed pattern of pathological transformation of the "object" component $\begin{pmatrix} E_x; \\ E_y; \\ \delta_{xy} \end{pmatrix}$ of BF speckle fields $\begin{pmatrix} H_x; \\ H_y; \\ \vartheta_{xy} \end{pmatrix}$, manifested in the ellipticity maps $\beta_{\text{object}}(\theta_k, x, y)$, is consistent with the proposed optical anisotropy model of supramolecular protein networks.

According to this model, pathological states – from nodular goiter and autoimmune thyroiditis to papillary carcinoma – are characterized by a progressive increase in ellipticity magnitude and an expansion of its variation range, driven by both biochemical factors (elevated thyroglobulin concentration) and conformational changes related to the structuring of globular and fibrillar protein networks.

Within the statistical framework, such pathological transformations are expected to be accompanied by an increase in the mean value M_1 and variance M_2 of the polarization maps $\beta_{object}(\theta_k, x, y)$. In contrast, higher-order statistical moments $M_{3;4}$, which scale approximately as $M_3 \sim 1/(M_2)^3$ and $M_4 \sim 1/(M_2)^4$, should decrease.

This mechanism underlying the formation of differences between the polarization "object" maps $\beta_{object}(\theta_k, x, y)$ for all studied groups is confirmed by the calculated first- to fourth-order central statistical moments $M_{g=1;2;3;4}$, as summarized in Table 15.4.

Statistical analysis of the polarization ellipticity maps $\beta_{object}(\theta_k, x, y)$ revealed the following key findings:

- The analyzed statistical parameters demonstrated statistically significant sensitivity for both diagnosis $(p_{12;13;14})$ and differentiation $(p_{23;24;34})$ across the entire set of thyroid pathologies.
- The mean value M_1 and variance M_2 exhibited statistically significant differences ($p_{ik} < 0.05$) and showed a monotonic increase from the control group (group 1) to the experimental groups (groups 2–4), with $M_1 \sim$ from 0.15 to 0.194 and $M_2 \sim$ from 0.069 to 0.11.
- The higher-order statistical moments, skewness M_3 and kurtosis M_4, displayed high diagnostic $(p_{12;13;14} < 0.001)$ and differentiation $(p_{23;24;34} < 0.05)$ sensitivity to structural changes in the layer-resolved ellipticity maps $\beta_{object}(\theta_k, x, y)$.
- The skewness parameter M_3, characterizing asymmetry of the polarization ellipticity distributions, decreased from 1.21 to 0.81 across the studied groups.
- The kurtosis M_4, reflecting the sharpness of the distribution peak for the singly scattered component $\begin{pmatrix} E_x; \\ E_y; \\ \delta_{xy} \end{pmatrix}$ of the BF speckle field $\begin{pmatrix} H_x; \\ H_y; \\ \vartheta_{xy} \end{pmatrix}$, exhibited a consistent reduction from 2.07 to 1.22.

Information analysis of these results (Table 15.5) indicates a substantial improvement in diagnostic performance, with sensitivity levels increasing from *unsatisfactory* to *very good* and *excellent* when compared with conventional integral polarization ellipticity mapping.

Table 15.4 Statistical parameters of BF polarization, ellipticity, and phase maps.

Groups	Group 1 (51 samples)	Group 2 (51 samples)	Group 3 (51 samples)	Group 4 (51 samples)
Average, M_1	0.15 ± 0.008	0.17 ± 0.009	0.185 ± 0.01	0.194 ± 0.01
p_{ik}		$p_{12;13;14} \leq 0.05$; $p_{23;24;34} \leq 0.05$		
Variance, M_2	0.069 ± 0.004	0.081 ± 0.005	0.096 ± 0.006	0.11 ± 0.007
p_{ik}		$p_{12;13;14} \leq 0.05$; $p_{23;24;34} \leq 0.05$		
Skewness, M_3	1.21 ± 0.07	0.99 ± 0.05	0.88 ± 0.05	0.81 ± 0.04
p_{ik}		$p_{12;13;14} \leq 0.001$; $p_{23;24;34} \leq 0.05$		
Kurtosis, M_4	2.07 ± 0.1	1.79 ± 0.1	1.54 ± 0.08	1.22 ± 0.07
p_{ik}		$p_{12;13;14} \leq 0.001$; $p_{23;24;34} \leq 0.05$		

Table 15.5 Balanced accuracy of differential diagnosis of thyroid pathology.

	Average, M_1	
$a = 85$ $b = 17$ $Ac_{12} = 84.3\%$ $a = 80$ $b = 22$ $Ac_{23} = 78.4\%$	$a = 88$ $b = 14$ $Ac_{13} = 86.3\%$ $a = 82$ $b = 20$ $Ac_{24} = 80.4\%$	$a = 89$ $b = 13$ $Ac_{14} = 87.3\%$ $a = 83$ $b = 19$ $Ac_{34} = 81.4\%$
	Variance, M_2	
$a = 85$ $b = 17$ $Ac_{12} = 84.3\%$ $a = 80$ $b = 22$ $Ac_{23} = 78.4\%$	$a = 88$ $b = 14$ $Ac_{13} = 86.3\%$ $a = 82$ $b = 20$ $Ac_{24} = 80.4\%$	$a = 89$ $b = 13$ $Ac_{14} = 87.3\%$ $a = 83$ $b = 19$ $Ac_{34} = 81.4\%$
	Skewness, M_3	
$a = 93$ $b = 9$ $Ac_{12} = 91.2\%$ $a = 89$ $b = 13$ $Ac_{23} = 87.3\%$	$a = 94$ $b = 8$ $Ac_{13} = 92.1\%$ $a = 90$ $b = 12$ $Ac_{24} = 88.2\%$	$a = 95$ $b = 7$ $Ac_{14} = 93.1\%$ $a = 91$ $b = 11$ $Ac_{34} = 89.2\%$
	Kurtosis, M_4	
$a = 93$ $b = 9$ $Ac_{12} = 91.2\%$ $a = 89$ $b = 13$ $Ac_{23} = 87.3\%$	$a = 94$ $b = 8$ $Ac_{13} = 92.1\%$ $a = 90$ $b = 12$ $Ac_{24} = 88.2\%$	$a = 95$ $b = 7$ $Ac_{14} = 93.1\%$ $a = 91$ $b = 11$ $Ac_{34} = 89.2\%$

The following balanced accuracy values were obtained for the differential diagnosis of thyroid gland pathologies:

1. Detection of thyroid pathology (comparison between the control group 1 and the combined experimental groups 2–4):
   - *Normal vs. nodular goiter*: $Ac_{12}(M_3; M_4) \sim 91.2\%$, corresponding to an excellent diagnostic level;
   - *Normal vs. autoimmune thyroiditis*: $Ac_{13} \sim 92.1\%$, classified as an excellent diagnostic level;
   - *Normal vs. papillary thyroid carcinoma*: $Ac_{14} \sim 93.1\%$, also indicating an excellent diagnostic level.
2. Differential diagnosis between thyroid pathologies:
   - *Nodular goiter vs. autoimmune thyroiditis*: $Ac_{23} \sim 87\%$, corresponding to a very good differentiation level;
   - *Nodular goiter vs. papillary thyroid carcinoma*: $Ac_{24} \sim 88\%$, indicating a very good differentiation level;
   - *Autoimmune thyroiditis vs. papillary thyroid carcinoma*: $Ac_{34} \sim 89\%$, also classified as a very good differentiation level.

15.1.5 Conclusions

1. For the first time, a minimally invasive laser-based approach employing polarization–interference, layer-by-layer mapping of dehydrated blood films was developed and experimentally validated for the differential diagnosis of thyroid disorders.
2. To provide a physical interpretation and predictive assessment of speckle-field polarization structures, a theoretical model of polycrystalline supramolecular networks in blood films was proposed for the first time, based on the approximation of LB and CB.

3. It was shown that direct polarization–interference mapping and statistical analysis of integral ellipticity maps of the diffusely scattered speckle-field component do not yield statistically significant diagnostic markers. Consequently, the balanced accuracy of thyroid pathology differentiation using this approach does not exceed a satisfactory level ($\leq$80%).
4. The application of digital phase scanning to algorithmically reconstructed complex-amplitude fields enabled the isolation and quantitative analysis of polarization ellipticity maps corresponding to the singly scattered component of BF speckle fields. This made it possible to identify and physically justify new statistically significant markers for minimally invasive differential diagnosis of thyroid pathology.
5. Using the proposed methodology, the following diagnostic performance was achieved:
   5.1. Detection of thyroid pathology (comparison between control group 1 and experimental groups 2–4):
   - *Normal vs. nodular goiter*: $Ac_{12}(M_3; M_4) \sim 91.2\%$, corresponding to an excellent diagnostic level;
   - *Normal vs. autoimmune thyroiditis*: $Ac_{13} \sim 92.1\%$, classified as an excellent diagnostic level;
   - *Normal vs. papillary thyroid carcinoma*: $Ac_{14} \sim 93.1\%$, also indicating an excellent diagnostic level.

   5.2. Differential diagnosis between thyroid pathologies:
   - *Nodular goiter vs. autoimmune thyroiditis*: $Ac_{23} \sim 87\%$, corresponding to a very good differentiation level;
   - *Nodular goiter vs. papillary thyroid carcinoma*: $Ac_{24} \sim 88\%$, indicating a very good differentiation level;
   - *Autoimmune thyroiditis vs. papillary thyroid carcinoma*: $Ac_{34} \sim 89\%$, also classified as a very good differentiation level.

15.2 Polarization–phase Holographic Layer-by-layer Scanning of Polycrystalline Architectonics in Blood Films

15.2.1 Maps of BF Phase Anisotropy

Figure 15.9 shows histograms of $\text{LB}(\varphi_k, (m, n))$ (a) and $\text{CB}(\varphi_k, (m, n))$ (b) birefringence, as well as histograms of phase anisotropy parameters $G(\text{LB, CB})$ (c), (d).

The analysis of the obtained data revealed:

- each mechanism of phase anisotropy corresponds to the individual topographic structure of the maps $\text{LB}(\varphi_k, (m, n))$ and $\text{CB}(\varphi_k, (m, n))$ (Figure 15.9a, b, e, f);
- asymmetric structure of $G(\text{LB, CB})$ histograms (Figure 15.9c, d, g, h);
- the dependence of the histogram structure $G(\text{LB, CB})$ on the magnitude of the phase shift φ_k.

The experimentally revealed scenarios of $\text{LB}(m, n)$ and $\text{CB}(m, n)$ histograms phase transformation for various light scattering conditions in the BF volume illustrate the data presented in Tables 15.6 and 15.7.

It was established that:

- the holographically reconstructed BF maps of $\text{LB}(m, n)$ $\text{CB}(m, n)$ and $\text{CB}(m, n)$ exhibit individual, non-Gaussian statistical behavior, as indicated by nonzero statistical moments ($M_{i=1;2;3;4}(\text{LB, CB}) \neq 0$), in contrast to the Gaussian case where $M_{3;4}(\text{LB, CB}) = 0$;
- the third- and fourth-order statistical moments $M_{i=3;4}(\text{LB, CB}(m, n))$ are the most sensitive to variations in the topographic structure of BF phase anisotropy maps;
- a dynamic variation range of the higher-order moments $M_{i=3;4}(\text{LB}(m, n))$ and $M_{i=3;4}(\text{CB}(m, n))$, reaching approximately a twofold change, is observed with variation of the phase-scanning parameter φ_k under different light-scattering conditions within the BF volume.

The obtained results can be interpreted based on physical considerations derived from the model analysis of the optical anisotropy of blood films. As noted earlier, the biochemical composition of blood may be approximated as a two-component system consisting of plasma and formed elements. The magnitude of LB is predominantly governed by the spatial orientation of the optical axes of needle-like albumin crystals forming the marginal layer, characterized by a wide angular distribution – $0 \leq \gamma \leq 2\pi$. As a consequence, the first- and second-order statistical moments, describing the mean value and variance of the $\text{LB}(m, n)$ distributions, remain relatively small. In contrast, the higher-order moments $M_{i=3;4}(\text{LB}(m, n))$, which quantify the skewness and kurtosis of the polycrystalline dendritic network, attain comparatively large values (Table 15.7).

The magnitude of circular birefringence in BF, arising from globulin crystals and shaped cellular elements such as erythrocytes, platelets, and leukocytes, is also substantial. In dehydrated blood drops, the high concentration of spherulitic

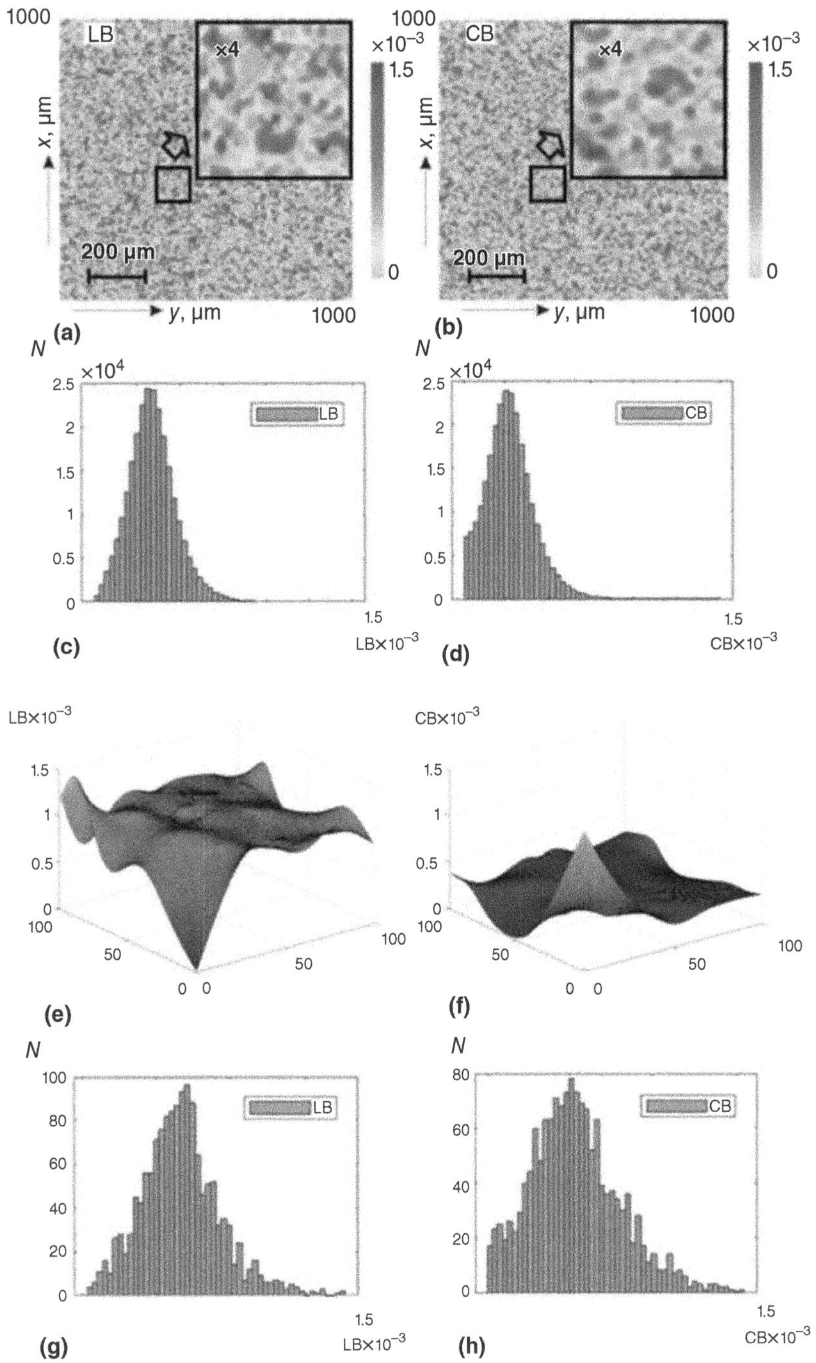

Figure 15.9 Distributions and histograms of integral (a), (b) and layer-by-layer ($\varphi = \pi/8$) (e), (f) LB (c), (g) and CB (d), (h) birefringence BF maps.

Table 15.6 Optical characteristics of BF samples.

Parameters	BF
Attenuation coefficient τ, cm^{-1}	0.38 ± 0.021
Depolarization degree Λ, %	29 ± 0.18

Table 15.7 First to fourth-orders statistical moments, that characterize the LB(m, n) and CB(m, n) maps.

M_i	LB	LB ($\varphi = \pi/2$)	LB ($\varphi = \pi/4$)	LB ($\varphi = \pi/8$)
$M_1 \times 10^{-3}$	0.42 ± 0.082	0.27 ± 0.071	0.16 ± 0.053	0.093 ± 0.0048
$M_2 \times 10^{-3}$	0.31 ± 0.016	0.28 ± 0.015	0.24 ± 0.014	0.21 ± 0.012
M_3	0.35 ± 0.018	0.41 ± 0.023	0.53 ± 0.027	0.65 ± 0.036
M_4	0.51 ± 0.028	0.67 ± 0.037	0.78 ± 0.042	0.91 ± 0.046
M_i	CB	CB ($\varphi = \pi/2$)	CB ($\varphi = \pi/4$)	CB ($\varphi = \pi/8$)
$M_1 \times 10^{-3}$	0.32 ± 0.067	0.13 ± 0.055	0.094 ± 0.0046	0.089 ± 0.0045
$M_2 \times 10^{-3}$	0.29 ± 0.017	0.23 ± 0.013	0.19 ± 0.011	0.17 ± 0.009
M_3	0.56 ± 0.029	0.71 ± 0.037	0.82 ± 0.043	0.99 ± 0.047
M_4	0.71 ± 0.038	0.94 ± 0.046	0.17 ± 0.064	0.38 ± 0.073

structures leads to circular birefringence values that are comparable, on average, to the LB of the albumin needle-crystal network. Accordingly, the statistical moments satisfy the relations $M_{i=1;2}(\mathrm{CB}) \approx M_{i=1;2}(\mathrm{LB})$ and $M_{i=3;4}(\mathrm{CB}) \approx M_{i=3;4}(\mathrm{LB})$, as summarized in Table 15.7.

During digital holographic phase scanning of the BF volume along depth, distinct realizations of polycrystalline networks with spatially varying optical anisotropy parameters – such as optical axis orientations and phase shifts between orthogonal field components – are sequentially resolved. Each phase plane is characterized by its own orientation distribution $\gamma_{\min} \leq \gamma \leq \gamma_{\max}$) with an inherent random dispersion. As the scattering regime approaches single or weak scattering, the range of random phase shifts becomes progressively narrower. In the limit of an optically thin layer, the statistics of BF phase anisotropy maps LB(m, n) and CB(m, n)) evolve according to the following trend:

$$\varphi_k \downarrow \Longrightarrow \begin{pmatrix} M_{1;2}(\mathrm{LB},\,\mathrm{CB}) \downarrow; \\ M_{3;4}(\mathrm{LB},\,\mathrm{CB}) \uparrow \end{pmatrix}.$$

15.2.2 BF Amplitude Anisotropy Maps

Figure 15.10 presents the results of the BF amplitude anisotropy studies – integral (m, n) and layer-by-layer (φ_k, (m, n)) linear LD(φ_k, (m, n)) and circular CD(φ_k, (m, n)) dichroism maps and histograms.

Established:

- individual (for different films volume depth scattering conditions) statistical distributions of the BF linear (LD(φ_k, (m, n))) and circular (CD(φ_k, (m, n))) dichroism (Figure 15.10a, b, e, f);
- asymmetric structure (as for the phase anisotropy parameters G(LB, CB) distributions) BF linear and circular dichroism histograms G(LD, CD), – Figure 15.10c, d, f, h.

Phase transformation of the LD(m, n) and CD(m, n) maps illustrate the first- to fourth-orders central statistical moments (Table 15.8).

It was established that:

- for all phase-resolved depths of blood films, the central statistical moments of first to fourth order satisfy $M_{i=1;2;3;4}(\mathrm{LD},\,\mathrm{CD}) \neq 0$, indicating pronounced non-Gaussian statistical behavior of the linear and circular dichroism maps;
- the magnitudes of the statistical parameters $M_{i=1;2;3;4}(\mathrm{LD},\,\mathrm{CD}) \leftrightarrow q(\varphi_k)$ exhibit a clear dependence on the phase-scanning depth φ_k;
- the third- and fourth-order statistical moments, $M_{i=3;4}(\mathrm{LD},\,\mathrm{CD}(m,\,n))$, demonstrate the highest sensitivity to structural variations in amplitude anisotropy maps;
- with increasing values of the phase-scanning parameter $\varphi_k \uparrow$ (corresponding to conditions approaching single scattering), the higher-order moments increase markedly: $M_{i=3;4}(\mathrm{LD}(m,\,n))$ rise by up to ~85%, while $M_{i=3;4}(\mathrm{CD}(m,\,n))$ increase by approximately 60–80%.

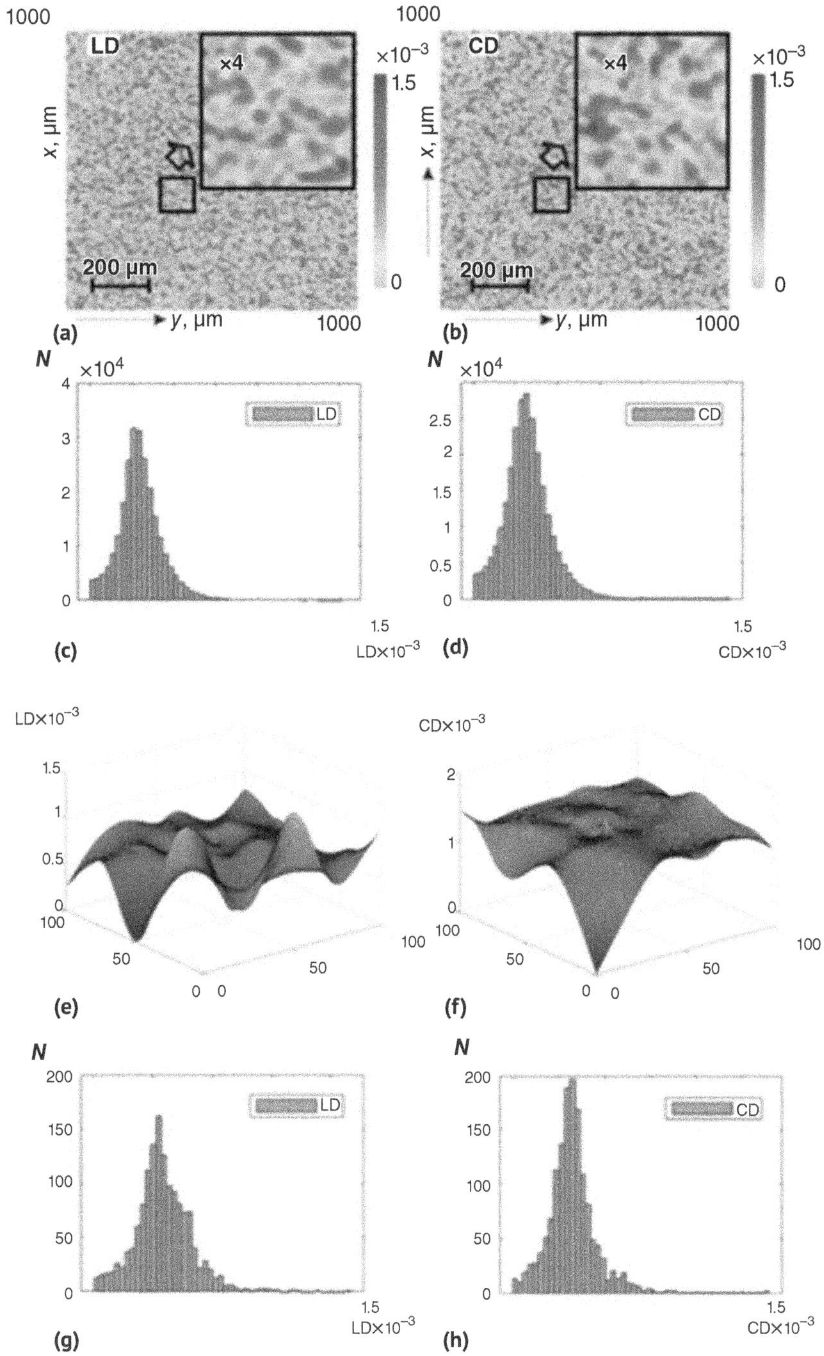

Figure 15.10 Distributions and histograms of integral (a), (b) and layer-by-layer ($\varphi = \pi/8$) (e), (f) linear LD (c), (g) and circular CD (d), (h) dichroism BF maps.

From a physical standpoint, these observations can be explained by the fact that the magnitude of LD, similarly to LB, at all phase depths is governed by the spatial orientation of the optical axes of needle-like protein crystals (albumin, fibrin, elastin). As the phase-scanning process approaches the regime of an optically thin layer, both the dispersion of optical axis orientations and the magnitude of phase retardations are modified. This results in the following characteristic evolution of statistical parameters:

$\varphi_k \downarrow \Longrightarrow \begin{pmatrix} M_{1;2}(\text{LD, CD}) \downarrow; \\ M_{3;4}(\text{LD, CD}) \uparrow \end{pmatrix}$, as summarized in Table 15.8.

At the same time, the magnitude of circular dichroism (CD) associated with the combined polycrystalline network formed by shaped cellular elements, albumin, and globulin grids (globulin crystals, erythrocytes, platelets, and leukocytes) is comparable to, and not lower than, that of LD. Consequently, the statistical moments satisfy the approximate relations $M_{i=1;2}(\mathrm{LD}) \approx M_{i=1;2}(\mathrm{CD})$ and $M_{i=3;4}(\mathrm{LD}) \approx M_{i=3;4}(\mathrm{CD})$.

Thus, the performed cycle of fundamental polarization–interference investigations of blood films – encompassing their optical anisotropy, polycrystalline organization, and layer-resolved mapping – forms a solid physical foundation for the applied implementation of Jones-matrix-based mapping techniques.

15.2.3 Applied Aspect

In this section, layer-by-layer Jones-matrix mapping of blood films was performed for healthy donors (control group 1) and patients diagnosed with papillary thyroid carcinoma (experimental group 2). The diagnostic performance was evaluated using the accuracy scale summarized in Table 15.9.

Two groups of BF samples were formed for the study (Table 15.10):

- Control group 1: healthy donors, 26 samples;
- Experimental group 2: patients diagnosed with papillary thyroid carcinoma, 26 samples.

15.2.4 Maps of BF Phase Anisotropy

Figures 15.11 and 15.12 present integral LB and CB maps (panels (a) and (b)), as well as their layer-resolved representations $\mathrm{LB}(\varphi_k, (m, n))$ and $\mathrm{CB}(\varphi_k, (m, n))$ obtained by phase scanning (panels (e) and (f)). The corresponding histograms of the birefringence value distributions $G(\mathrm{LB}, \mathrm{CB})$ are shown in panels (c), (d), (g), and (h). Results for the control group (group 1) are illustrated in panels (a), (c), (e), and (g), while data for the experimental group (group 2) are presented in panels (b), (d), (f), and (h).

Analysis of the experimental data demonstrates that, at all investigated phase depths, the Jones-matrix–reconstructed birefringence maps and their associated histograms exhibit distinct structural features for blood films from different groups, as shown in Figures 15.11 and 15.12.

Table 15.8 First- to fourth-order statistical moments of BF linear and circular dichroism maps.

$M_1 \times 10^{-3}$	LD	LD ($\varphi = \pi/2$)	LD ($\varphi = \pi/4$)	LD ($\varphi = \pi/8$)
$M_2 \times 10^{-3}$	0.24 ± 0.069	0.19 ± 0.064	0.07 ± 0.058	0.96 ± 0.054
M_3	0.28 ± 0.015	0.24 ± 0.013	0.22 ± 0.013	0.18 ± 0.009
M_4	0.41 ± 0.022	0.56 ± 0.029	0.68 ± 0.036	0.77 ± 0.039
M_i	0.62 ± 0.033	0.79 ± 0.041	0.92 ± 0.046	0.09 ± 0.051
$M_1 \times 10^{-3}$	**CD**	**CD ($\varphi = \pi/2$)**	**CD ($\varphi = \pi/4$)**	**CD ($\varphi = \pi/8$)**
$M_2 \times 10^{-3}$	0.95 ± 0.055	0.93 ± 0.049	0.88 ± 0.046	0.82 ± 0.045
M_3	0.21 ± 0.011	0.19 ± 0.011	0.17 ± 0.009	0.15 ± 0.008
M_4	0.65 ± 0.034	0.81 ± 0.043	0.97 ± 0.051	0.16 ± 0.066
$M_1 \times 10^{-3}$	0.83 ± 0.043	0.05 ± 0.054	0.31 ± 0.072	0.54 ± 0.087

Table 15.9 Balanced accuracy threshold criteria.

Diagnostic Accuracy Assessment	Accuracy, Ac, %
Unsatisfactory	≤ 80
Satisfactory	81–85
Good	86–90
Very good	91–95
Excellent	> 95

Table 15.10 Optical and geometric parameters of BF samples.

Parameters	Values
Geometrical thickness, h, μm	60–65
Attenuation coefficient τ, cm^{-1}	0.21–0.23
Depolarization degree, Δ, %	26–29

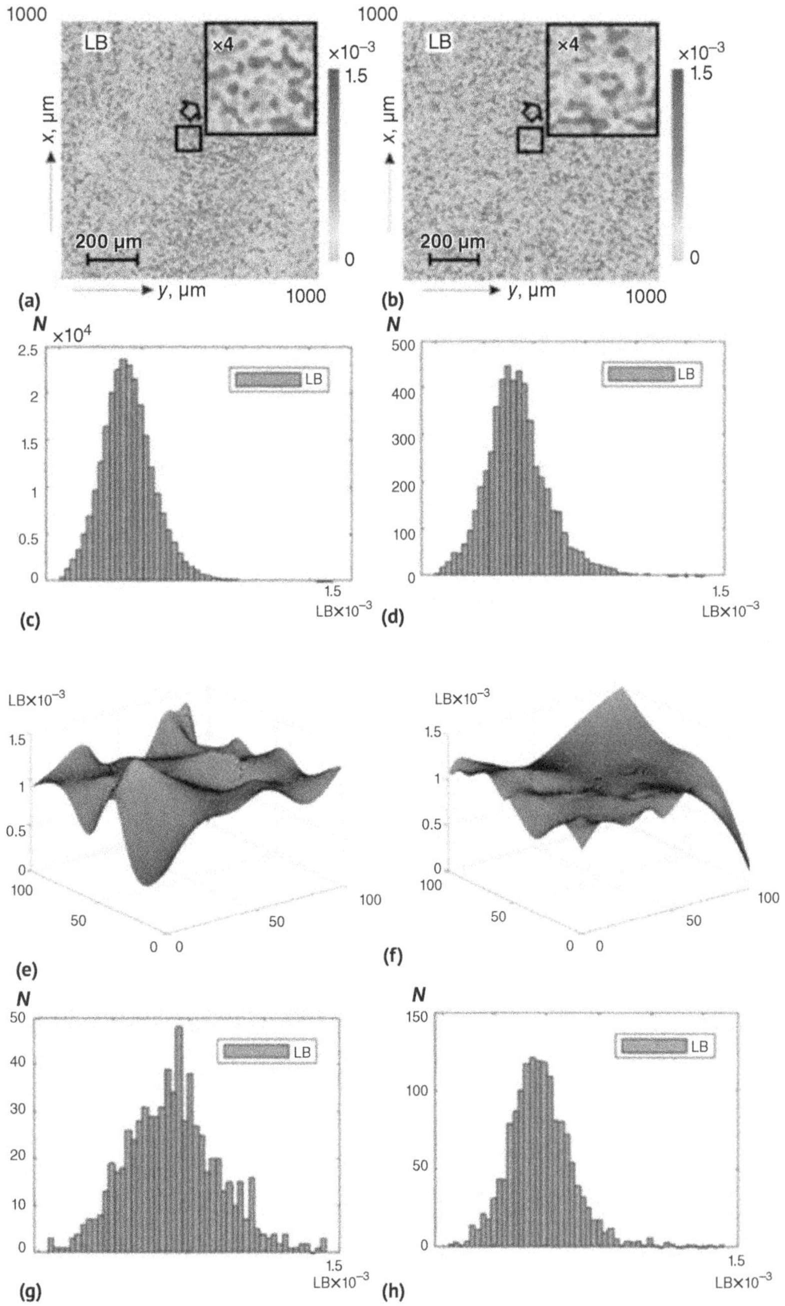

Figure 15.11 BF maps $\{\mathrm{LB}(m, n)\}$ and histograms $G\{\mathrm{LB}(m, n)\}$ of healthy donors (left column) and patients with papillary thyroid carcinoma (right column).

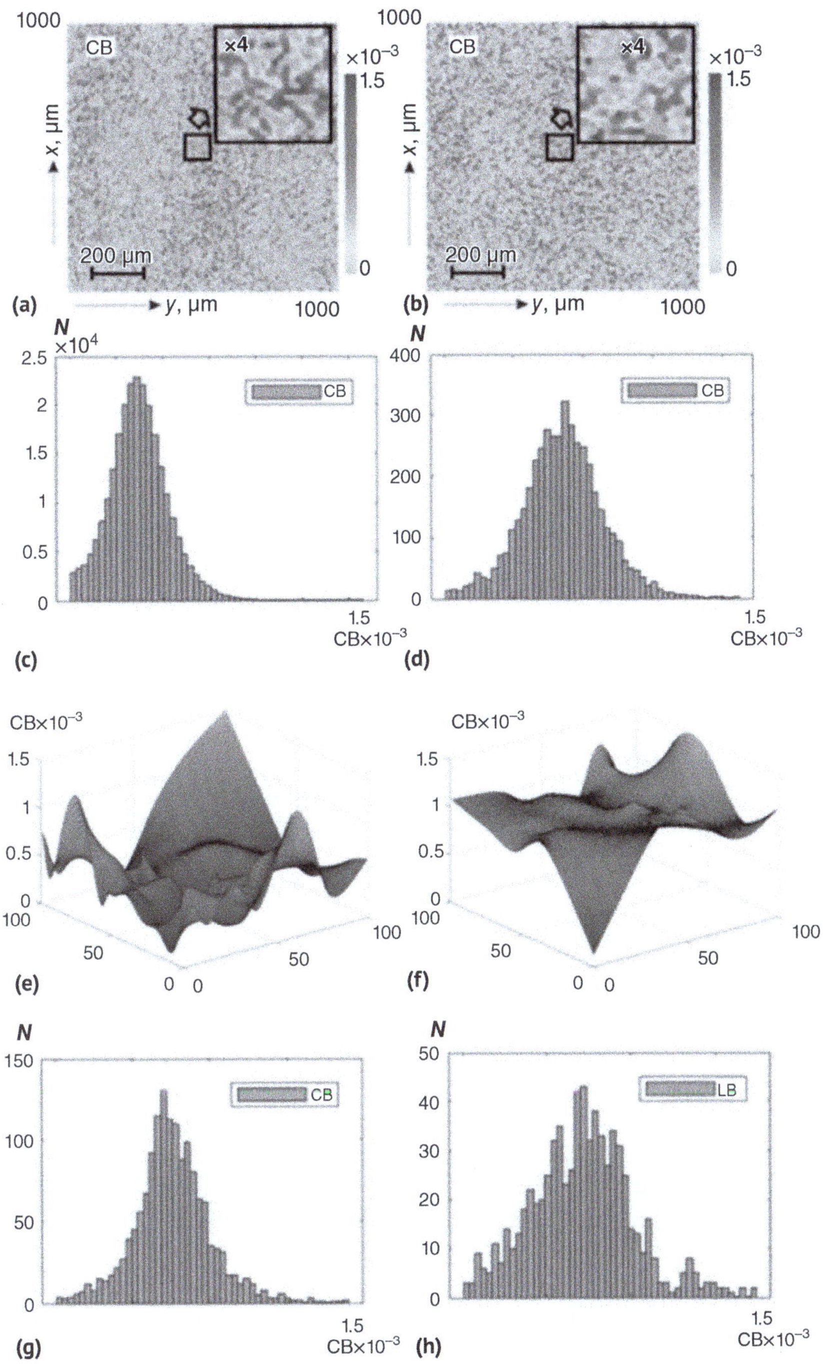

Figure 15.12 BF maps $\{\text{CB}(m,n)\}$ and histograms $G\{\text{CB}(m,n)\}$ of healthy donors (left column) and patients with papillary thyroid carcinoma (right column).

Quantitative characterization of phase-dependent transformations and intergroup differences in the maps – $\text{LB}(\varphi_k,(m,n))$ and $\text{LB}(\varphi_k,(m,n))$ is provided by the results of statistical analysis summarized in Tables 15.11 and 15.12.

Statistical analysis of the phase anisotropy maps $\text{LB}(m,n)$ and $\text{CB}(m,n)$ yielded the following results:

- For all investigated phase depths, the central statistical moments of the first through fourth order are nonzero, i.e., $M_{i=1;2;3;4}(\text{LB},\text{CB}) \neq 0$, indicating non-Gaussian behavior of the birefringence distributions.
- The third- and fourth-order statistical moments, $M_{i=3;4}(\text{LB}(m,n),\text{CB}(m,n))$, exhibit the highest sensitivity to variations in BF birefringence, with maximal discriminative power observed in the phase plane $\varphi_k = \pi/8$.

Table 15.11 Statistical parameters of $LB(\varphi_k, (m, n))$.

M_i	LB		Ac, %	$\{LB(\pi/8)\}$		Ac, %
	Group 1	Group 2		Group 1	Group 2	
$M_1 \times 10^{-3}$	0.41 ± 0.081	0.34 ± 0.081	73.5	0.89 ± 0.046	0.94 ± 0.049	76.5
$M_2 \times 10^{-3}$	0.33 ± 0.015	0.28 ± 0.015	76.5	0.21 ± 0.011	0.24 ± 0.013	79.4
M_3	0.31 ± 0.017	0.42 ± 0.021	82.4	0.68 ± 0.035	0.57 ± 0.037	88.2
M_4	0.45 ± 0.026	0.53 ± 0.027	82.4	0.89 ± 0.044	0.81 ± 0.048	88.2

Table 15.12 Statistical parameters of $CB(\varphi_k, (m, n))$.

M_i	CB		Ac, %	$\{CB(\pi/8)\}$		Ac, %
	Group 1	Group 2		Group 1	Group 2	
$M_1 \times 10^{-3}$	0.02 ± 0.062	0.11 ± 0.064	76.5	0.88 ± 0.049	0.95 ± 0.046	79.4
$M_2 \times 10^{-3}$	0.27 ± 0.015	0.32 ± 0.016	79.4	0.16 ± 0.008	0.21 ± 0.011	82.4
M_3	0.59 ± 0.025	0.49 ± 0.027	85.3	0.29 ± 0.064	0.87 ± 0.048	90.2

- For the LB maps, $LB(\varphi_k = \pi/8; (m, n))$, the relative differences $\Delta M_{i=3;4}(LB(a, b))$ between the control and experimental groups range from approximately 30% to 40%.
- For the circular birefringence maps $CB(\varphi_k = \pi/8(m, n))$, the corresponding intergroup differences $\Delta M_{i=3;4}(CB(m, n))$ are more pronounced, increasing to about 45–55%.

The maximum septic inflammation differential diagnosis accuracy is the following levels:

- maps $LB(\varphi_k = \pi/8; (m, n)) - Ac(M_{i=3;4}(\varphi = \pi/8)) = 88.2\%$ is a good level;
- maps $CB(\varphi_k = \pi/8; (m, n)) - Ac(M_{i=3;4}(\varphi = \pi/8)) = 90.2\%$ is a very good level.

From a physical perspective, the obtained results can be explained by specific features of pathological alterations in the polycrystalline architectonics of BF supramolecular networks.

First, earlier classical vector–parametric investigations of optical anisotropy in blood films and blood plasma demonstrated that the magnitude of LB is primarily governed by the spatial orientation of the optical axes of needle-like protein crystals, such as albumin, fibrin, and elastin. The orientations of these optical axes within the dendritic network are highly dispersed and nearly uniformly distributed over the full angular range $0 \leq \gamma \leq 2\pi$. As a consequence, the first- and second-order statistical moments, which describe the mean value and variance of the $LB(m, n)$ distributions, attain relatively large values, whereas the higher-order moments $M_{i=3;4}(LB(m, n))$ remain comparatively small.

During oncological processes, blood crystallization becomes more pronounced. For dehydrated blood films obtained from patients in experimental group 2, this leads to a narrowing of the orientation distribution range. As a result, the lower-order statistical moments $M_{i=1;2}(LB(m, n)) \downarrow$ decrease, while the third- and fourth-order moments $M_{i=3;4}(LB(m, n)) \uparrow$ increase, reflecting enhanced asymmetry and peakedness of the distributions.

Second, the CB of the spherulitic optically anisotropic network formed by globulin crystals and shaped cellular elements (erythrocytes, platelets, leukocytes, and monocytes) is comparable in magnitude to the LB. Under cancer-associated inflammatory conditions, the concentrations of globulins and white blood cells increase, which enhances the optical activity of chiral molecular domains in dehydrated blood films from patients with papillary thyroid carcinoma. Accordingly, the statistical moments characterizing the circular birefringence distributions exhibit an increase in the first- and second-order moments $M_{i=1;2}(CB) \uparrow$, accompanied by a reduction in the higher-order moments $M_{i=3;4}(CB) \downarrow$.

Third, as the phase-scanning parameter $\varphi_k \downarrow$ decreases, progressively thinner optically active layers of the dehydrated BF are effectively selected. Under these conditions, the distribution of optical axis orientations of albumin, fibrin, and elastin needle crystals transforms into an asymmetric form ($\gamma_{min} \leq \gamma \leq \gamma_{max}$) with a certain degree of randomness. Simultaneously, the range of random phase shifts between orthogonal components of the laser radiation becomes narrower. These effects lead to systematic changes in the statistical properties of the phase anisotropy maps LB(m, n) and CB(m, n), described by the trend $\begin{pmatrix} M_{1;2}(\text{LB, CB}) \downarrow; \\ M_{3;4}(\text{LB, CB}) \uparrow \end{pmatrix}$.

Against this background, pathological differences in the polycrystalline structure of optically anisotropic networks in blood samples from patients of experimental group 2 become more clearly pronounced.

15.2.5 Maps of BF Amplitude Anisotropy

Figures 15.13 and 15.14 show integral (a, b) (panels (a), (b)) and layer-by-layer $(\varphi_k, (m, n))$ (panels (e), (f)) BF amplitude anisotropy linear LD$(\varphi_k, (m, n))$ and a circular CD$(\varphi_k, (m, n))$ dichroism maps, as well as histograms of G(LD, CD) (panels (c), (d), (g),(h)). Control group 1 (panels (a), (c), (e), (g)) and experimental group 2 (fragments (b), (d), (f), (h)).

Phase transformations of maps LD(m, n) and CD(m, n) are shown in Tables 15.13 and 15.14.

Analysis of BF amplitude anisotropy maps over all phase depths yielded the following results:

- for all investigated phase planes, the central statistical moments of the first through fourth order differ significantly from zero, i.e., $M_{i=1;2;3;4}(\text{LD, CD}) \neq 0$, indicating non-Gaussian statistical behavior of the linear and circular dichroism maps;
- the statistical characteristics $M_{i=1;2;3;4}(\text{LD, CD}) \leftrightarrow q(\varphi_k)$ exhibit a pronounced dependence on the phase-scanning parameter φk, reflecting variations in the polycrystalline architectonics of blood films;
- the third- and fourth-order statistical moments, $M_{i=3;4}(\text{LD}(m, n), \text{CD}(m, n))$;, were identified as the most sensitive diagnostic markers;
- for LD maps, LD(a, b), the relative differences $\Delta M_{i=3;4}(\text{LD}(m, n))$ between the control and experimental BF sample groups amount to approximately 25–30%;
- for CD maps, CD(m, n), the corresponding intergroup differences $\Delta M_{i=3;4}(\text{CD}(m, n))$ are more pronounced, reaching values of about 35–45%.

The maximum diagnostic accuracy for differentiation of cancer-associated inflammatory changes was determined as follows:

- for LD(m, n) – $\text{Ac}(M_{i=3;4}(\varphi = \pi/8)) = 82.8\%$, corresponding to a satisfactory diagnostic level;
- for CD(m, n) – $\text{Ac}(M_{i=3;4}(\varphi = \pi/8)) = 88.2\%$, corresponding to a good diagnostic level.

From a physical standpoint, these findings can be explained by the fact that decreasing the phase-scanning parameter $\varphi_k \downarrow$ enables the effective selection of a more optically thin region of the blood film. Under these conditions, the topographic structure of the BF maps increasingly reflects the intrinsic statistics of needle-like crystal optical axis orientations (governing LD) and the associated phase retardations (governing CD). As a result, the maximum achievable accuracy for identifying polycrystalline structural differences between BF samples from the two groups is attained.

Table 15.13 Statistical parameters of LD$(\varphi_k, (m, n))$.

M_i	LD		Ac, %	{LD$(\pi/8)$}		Ac, %
	Group 1	Group 2		Group 1	Group 2	
$M_1 \times 10^{-3}$	0.21 ± 0.063	0.29 ± 0.068	73.5	0.94 ± 0.054	0.89 ± 0.051	76.5
$M_2 \times 10^{-3}$	0.27 ± 0.014	0.33 ± 0.016	76.5	0.18 ± 0.009	0.15 ± 0.008	76.5
M_3	0.51 ± 0.023	0.43 ± 0.023	79.4	0.79 ± 0.038	0.91 ± 0.045	82.4
M_4	0.76 ± 0.032	0.64 ± 0.034	79.4	0.98 ± 0.052	0.21 ± 0.062	82.4

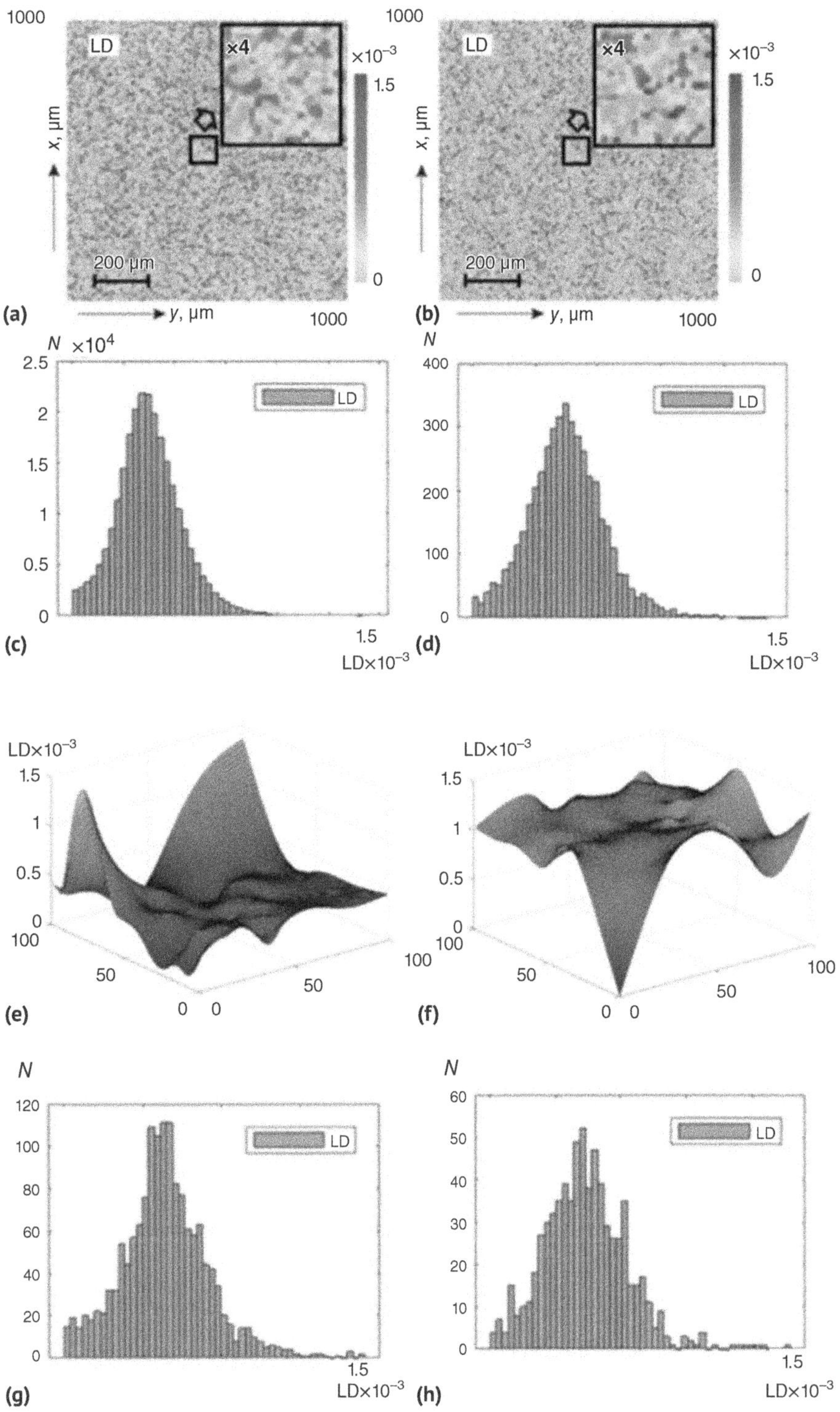

Figure 15.13 BF maps $\{LD(m,n)\}$ and histograms $G\{LD(m,n)\}$ of healthy donors (left column) and patients with papillary thyroid carcinoma (right column).

The slightly lower diagnostic accuracy observed for amplitude anisotropy mapping, compared with phase anisotropy analysis (Tables 15.11 and 15.12), may be attributed to the weaker contribution of anisotropic absorption relative to laser scattering in blood films. Consequently, the sensitivity of Jones- or dichroism-based matrix mapping of the polycrystalline BF component is reduced.

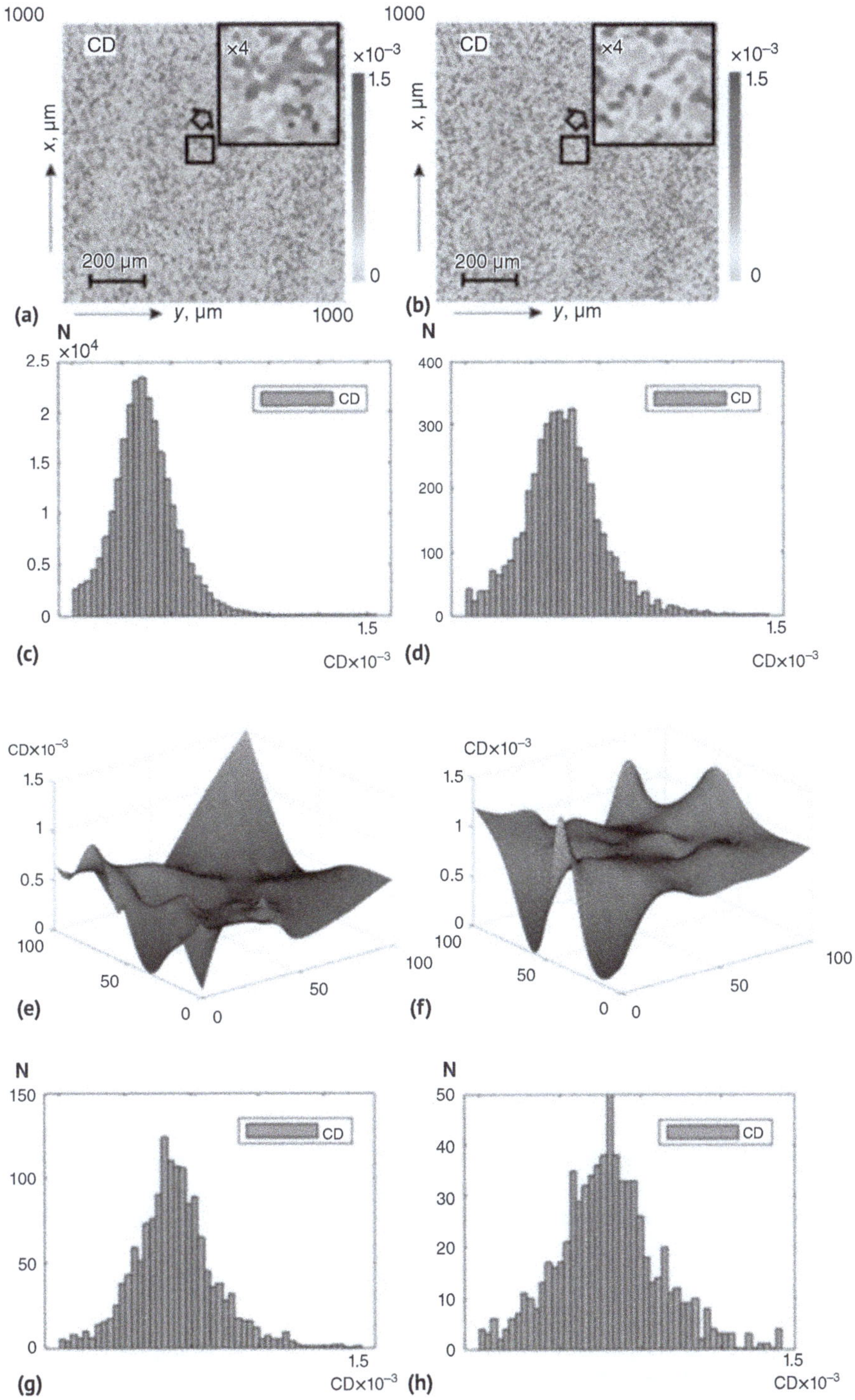

Figure 15.14 BF maps $\{\mathrm{CD}(m, n)\}$ and histograms $G\{\mathrm{CD}(m, n)\}$ of healthy donors (left column) and patients with papillary thyroid carcinoma (right column).

15.2.6 Conclusions

1. Jones-matrix mapping algorithms for polycrystalline blood films are developed and experimentally tested.
2. A comprehensive set of integral and layer-resolved maps (a, b) describing the phase anisotropy (birefringence: LB, CB) and amplitude anisotropy (dichroism: LD, CD) of the polycrystalline components of blood films was obtained and subjected to physical analysis.

Table 15.14 Statistical parameters of $\mathrm{CD}(\varphi_k, (a, b))$.

M_i	CD		Ac, %	$\{\mathrm{CD}(\pi/8)\}$		Ac, %
	Group 1	Group 2		Group 1	Group 2	
$M_1 \times 10^{-3}$	0.89 ± 0.051	0.94 ± 0.054	76.5	0.81 ± 0.046	0.88 ± 0.043	79.4
$M_2 \times 10^{-3}$	0.18 ± 0.011	0.21 ± 0.011	76.5	0.14 ± 0.007	0.18 ± 0.009	79.4
M_3	0.76 ± 0.036	0.63 ± 0.036	82.4	0.19 ± 0.068	0.91 ± 0.067	88.2
M_4	0.93 ± 0.047	0.81 ± 0.044	82.4	0.44 ± 0.089	0.15 ± 0.089	88.2

3. Characteristic regularities and statistical scenarios of phase-dependent transformations in the holographically reconstructed optical anisotropy maps of blood films were identified. In particular, with increasing phase-scanning parameter, the following trend is observed:
4. $(\varphi_k \uparrow) \Longrightarrow \begin{pmatrix} M_{i=1;2}(\mathrm{LB, CB, LD, CD}(a, b)) \downarrow; \\ M_{i=3;4}(\mathrm{LB, CB, LD, CD}(a, b)) \uparrow \end{pmatrix}$, reflecting a redistribution of statistical weight toward higher-order moments.
5. The third- and fourth-order statistical moments $M_{i=3;4}(\mathrm{LB, CB, LD, CD}(a, b))$ were identified as the most sensitive markers of changes in the layer-by-layer topographic organization of both phase and amplitude anisotropy in blood films.
6. Within the information analysis framework, the following maximum levels cancer differential diagnosis balanced accuracy are established:
7. Maps $\mathrm{LB}(\varphi_k = \pi/8; (a, b)) - \mathrm{Ac}(M_{i=3;4}(\varphi = \pi/8)) = 88.2\%$ – good level;
8. Maps $\mathrm{CB}(\varphi_k = \pi/8; (a, b)) - \mathrm{Ac}(M_{i=3;4}(\varphi = \pi/8)) = 90.2\%$ – very good level;
9. Maps $\mathrm{LD}(m, n) - \mathrm{Ac}(M_{i=3;4}(\varphi = \pi/8)) = 82.8\%$ – satisfactory level; and
10. Maps $\mathrm{CD}(m, n) - \mathrm{Ac}(M_{i=3;4}(\varphi = \pi/8)) = 88.2\%$ – good level.

References

1 Layden D, Ghosh N, Vitkin IA. Quantitative polarimetry for tissue characterization and diagnosis. In Wang RK, Tuchin VV, editors. Advanced Biophotonics: Tissue Optical Sectioning. Boca Raton: CRC Press; 2013. pp. 73–108.

2 Vitkin A, Ghosh N, de Martino A. Tissue polarimetry. In Andrews DL, editor. Photonics: Scientific Foundations, Technology and Applications. John Wiley & Sons, Ltd.; 2015. p. 239–321.

3 Tuchin VV. Tissue Optics and Photonics: Biological Tissue Structures. J Biomed Photonics Eng. 2015;1:3–21.

4 Bai X, Liang Z, Zhu Z, Schwing A, Forsyth D, Gruev V. Polarization-based underwater geolocalization with deep learning. eLight. 2023;3:15.

5 Lee HR, Lotz C, Groeber-Becker FK, Dembski S, Novikova T. Digital histology with Mueller polarimetry and fast DBSCAN. Appl Opt. 2022;61(32):9616–9624.

6 Kim M, Lee HR, Ossikovski R, Malfait-Jobart A, Lamarque D, Novikova T. Optical diagnosis of gastric tissue biopsies with Mueller microscopy and statistical analysis. J Europ Opt Soc Rapid Publ. 2022;18(2):10.

7 Lee HR, Li P, Yoo TS, Lotz C, Groeber-Becker FK, Dembski S, et al. Digital histology with Mueller microscopy: how to mitigate an impact of tissue cut thickness fluctuations. J Biomed Opt. 2019;24(7):076004.

8 Li P, Lee HR, Chandel S, Lotz C, Groeber-Becker FK, Dembski S, et al. Analysis of tissue microstructure with Mueller microscopy: logarithmic decomposition and Monte Carlo modelling. J Biomed Opt. 2020;25(1):015002.

9 Lee HR, Yoo TS, Li P, Lotz C, Groeber-Becker FK, Dembski S, et al. Mueller microscopy of anisotropic scattering media: theory and experiments. In Proc SPIE 10677 Unconventional Optical Imaging. 2018;10677: 222–229.

10 Ma H, He H, Ramella-Roman JC. Mueller matrix microscopy. In Ramella-Roman JC, Novikova T, editors. Polarized Light in Biomedical Imaging and Sensing. Cham: Springer; 2023. pp. 281–321.

11 Eftimie LG, Glogojeanu RR, Tejaswee A, Gheorghita P, Stanciu SG, Chirila A, et al. Differential diagnosis of thyroid nodule capsules using random forest guided selection of image features. Sci Rep. 2022;12:21636.

12 Yang WT, Ma BY, Chen Y. A narrative review of deep learning in thyroid imaging: current progress and future prospects. Quant Imaging Med Surg. 2024;14(2):2069–2088. https://doi.org/10.21037/qims-23-908.

13 Tokarz D, Cisek R, Joseph A, Asa SL, Wilson BC, Barzda V. Characterization of pathological thyroid tissue using polarization-sensitive second harmonic generation microscopy. Lab Invest. 2020;100(10):1280–1287. https://doi.org/10.1038/s41374-020-0475-7.

14 Cisek R, Joseph A, Harvey M, Tokarz D. Polarization-sensitive second harmonic generation microscopy for investigations of diseased collagenous tissues. Front Phys Sec Opt Photonics. 2021;9:726996. https://doi.org/10.3389/fphy.2021.726996.

15 Hristu R, Eftimie LG, Stanciu SG, Tranca DE, Paun B, Sajin M, et al. Quantitative second harmonic generation microscopy for the structural characterization of capsular collagen in thyroid neoplasms. Biomed Opt Express. 2018;9(8):3923–3936. https://doi.org/10.1364/BOE.9.003923.

16 de Boer JF, Milner TE. Review of polarization sensitive optical coherence tomography and Stokes vector determination. J Biomed Opt. 2002;7(3):359–371. https://doi.org/10.1117/1.1483879.

17 Ku S, Mahato KK, Mazumder N. Polarization-resolved Stokes-Mueller imaging: a review of technology and applications. Lasers Med Sci. 2019;34(7):1283–1293. https://doi.org/10.1007/s10103-019-02752.

18 He Y, Li K, Li W, Qiu Y, Li D, Wang C, Tang Q, Li Z. Polarization coherency matrix tomography. J Biophotonics. 2023;16(9):e202300093. https://doi.org/10.1002/jbio.202300093.

19 Yang D, Yuan Z, Hu M, Liang Y. Zebrafish brain and skull imaging based on polarization-sensitive optical coherence tomography. J Biophotonics. 2022;15(12):e202200112. https://doi.org/10.1002/jbio.202200112.

20 Park JE, Xin Z, Kwon DY, Kim SW, Lee H, Jung MJ, et al. Application of polarization sensitive-optical coherence tomography to the assessment of phase retardation in subpleural cancer in rabbits. Tissue Eng Regen Med. 2021;18(1):61–69. https://doi.org/10.1007/s13770-020-00318-9.

21 Willemse J, Gräfe MGO, Verbraak FD, de Boer JF. In vivo 3D determination of peripapillary scleral and retinal layer architecture using polarization-sensitive optical coherence tomography. Transl Vis Sci Technol. 2020;9(11):21. https://doi.org/10.1167/tvst.9.11.21.

22 Baumann B, Augustin M, Lichtenegger A, Harper D, Muck M, Eugui P, et al. Polarization-sensitive optical coherence tomography imaging of the anterior mouse eye. J Biomed Opt. 2018;23(8):1–12. https://doi.org/10.1117/1.JBO.23.8.086005.

23 Hariri LP, Adams DC, Applegate MB, Miller AJ, Roop BW, Villiger M, et al. Distinguishing tumor from associated fibrosis to increase diagnostic biopsy yield with polarization-sensitive optical coherence tomography. Clin Cancer Res. 2019;25(17):5242–5249. https://doi.org/10.1158/1078-0432.CCR-19-0566.

24 Tao K, Sun K, Ding Z, Ma Y, Kuang H, Zhao H, et al. Catheter-based polarization sensitive optical coherence tomography using similar Mueller matrix method. IEEE Trans Biomed Eng. 2020;67(1):60–68. https://doi.org/10.1109/TBME.2019.2908031.

25 De Boer JF, Hitzenberger CK, Yasuno Y. Polarization sensitive optical coherence tomography. Biomed Opt Express. 2017;8(3):1838–1873.

26 Gleason K, Putnam SA. Microdroplet evaporation with a forced pinned contact line. Langmuir. 2014;30(34):10548–10555.

27 Hu H, Larson RG. Evaporation of a sessile droplet on a substrate. J Phys Chem B. 2020;106:1334–1344.

28 Vladmir Z, Wang JB, Yan XH. Human blood plasma crystal and molecular biocolloid textures – dismetabolism and genetic breaches. Nat Sci J Xiangtan Univ. 2001;23:118–127.

29 Abe S, Tanaka J, Kojima M, Kanamaru S, Hirata K, Yamashita K, et al. Cell-free protein crystallization for nanocrystal structure determination. *Sci Rep*. 2022;12:16031.

30 Van den Beld AW, Kaufman JM, Zillikens MC, Lamberts SWJ, Egan JM, van der Lely AJ. The physiology of endocrine systems with ageing. Lancet Diabetes Endocrinol. 2018;6(8):647–658. https://doi.org/10.1016/S2213-8587(18)30026-3.

31 Asa SL, Erickson LA, Rindi G. The spectrum of endocrine pathology. Endocr Pathol. 2023;34(4):368–381. https://doi.org/10.1007/s12022-023-09758-0.

32 Brix K, Szumska J, Weber J, Qatato M, Venugopalan V, Al-Hashimi A, et al. Auto-regulation of the thyroid gland beyond classical pathways. Exp Clin Endocrinol Diabetes. 2020;128(6-7):437–445. https://doi.org/10.1055/a-1080-2969.

33 Grieco G, Wang T, Delcorte O, et al. Class III PI3K Vps34 controls thyroid hormone production by regulating thyroglobulin iodination, lysosomal proteolysis, and tissue homeostasis. Thyroid. 2020;30(1):133–146. https://doi.org/10.1089/thy.2019.0182.

34 Boron WF. Medical Physiology: A Cellular and Molecular Approach. Elsevier/Saunders; 2003. p. 1300. ISBN 1-4160-2328-3.

35 Delom F, Mallet B, Carayon P, Lejeune PJ. Role of extracellular molecular chaperones in the folding of oxidized proteins. Refolding of colloidal thyroglobulin by protein disulfide isomerase and immunoglobulin heavy chain-binding protein. J Biol Chem. 2001;276(24):21337–21342. https://doi.org/10.1074/jbc.M101086200.

36 Wagner I, Musso H. New Naturally Occurring Amino Acids. Angew Chem Int Ed Engl. 1983;22(11):816–828. https://doi.org/10.1002/anie.198308161.

37 Stoker HS. Organic and Biological Chemistry. Cengage Learning; 2011. 371. ISBN 978-1-305-68645-8.

38 Angelsky PO, Ushenko AG, Dubolazov AV, Sidor MI, Bodnar GB, Koval G, et al. The singular approach for processing polarization-inhomogeneous laser images of blood plasma layers. J Opt. 2013;15(4):044030.

39 Ushenko AG, Dubolazov AV, Ushenko VA, Novakovskaya OY. Statistical analysis of polarization-inhomogeneous Fourier spectra of laser radiation scattered by human skin in the tasks of differentiation of benign and malignant formations. J Biomed Opt. 2016;21(7):071110.

40 Garazdyuk MS, Bachinskyi VT, Vanchulyak OY, Ushenko AG, Dubolazov OV, Gorsky MP. Polarization-phase images of liquor polycrystalline films in determining time of death. Appl Opt. 2016;55(12):B67–B71.

41 Angelskii OV, Ushenko OG, Burkovets DN, Arkhelyuk OD, Ushenko YA. Polarization-correlation studies of multifractal structures in biotissues and diagnostics of their pathologic changes. Laser Phys. 2000;10(5):1136–1142.

42 Yermolenko S, Ushenko A, Ivashko P, Goudail F, Gruia I, Gavrilă C, et al. Spectropolarimetry of cancer change of biotissues. In Ninth International Conference on Correlation Optics. SPIE. 2009;7388:404–410.

43 Jóźwicki R, Patorski K, Angelsky OV, Ushenko AG, Burkovets DN, Ushenko YA. Automatic polarimetric system for early medical diagnosis by biotissue testing. Opt Appl. 2002;32(4):603–612.

44 Ushenko AG, Dubolazov OV, Ushenko VA, Novakovskaya OY, Olar OV. Fourier polarimetry of human skin in the tasks of differentiation of benign and malignant formations. Appl Opt. 2016;55(12):B56–B60.

45 Ushenko AG, Pishak VP. Coherent-domain optical methods: biomedical diagnostics, environmental and material science. In Laser Polarimetry of Biological Tissue: Principles and Applications. Kluwer Academic Publisher; 2004.

46 Ushenko AG, Fediv AI, Marchuk YF. Correlation and fractal structure of Jones matrices of human bile secret. In European Conference on Biomedical Optics. 2009;7368:73681Q.

47 Ushenko AG. Depolarization of a laser emission field as the correlation development of its polarization structure. In International Conference on Correlation Optics. 1997;3317:331–339.

48 Bachinskyi V, Boychuk T, Ushenko A. Laser Polarimetry of Biological Tissues and Fluids: Chapter 4. Methods of Spatial-frequency, Singular and Correlation Analysis of Biological Layers Object Fields. LAP Lambert Academic Publishing; 2018.

49 Ushenko AG, Burkovets DN, Yermolenko SB, Arkhelyuk AD, Pishak VP, Yuzko AM, et al. Stokes polarimetry of biotissues. In Fourth International Conference on Correlation Optics. SPIE. 1999;3904:527–533.

50 Ushenko VA, Dubolazov AV, Pidkamin LY, Sakchnovsky MY, Bodnar AB, Ushenko YA, et al. Mapping of polycrystalline films of biological fluids utilizing the Jones-matrix formalism. Laser Phys. 2018;28(2):025602.

51 Ushenko AG, Angelsky PO, Sidor M, Marchuk YF, Andreychuk DR, Pashkovskaya NV. Spatial-frequency selection of complex degree of coherence of laser images of blood plasma in diagnostics and differentiation of pathological states of human organism of various nosology. Appl Opt. 2014;53(10):B172–B180.

52 Olar EI, Ushenko AG, Ushenko YA. Polarization correlation measurements of the phase tomograms of optically anisotropic biofractals. Laser Phys. 2004;14(8):1115–1121.

53 Guminetskiy SH, Ushenko OG, Polyanskiy IP, Motrych AV, Grynchuk FV. The optical method for investigation of the peritonitis progressing process. In Eighth International Conference on Correlation Optics. 2008;7008:626–632.

54 Ushenko AG, Fediv AI, Marchuk YF. Singular structure of polarization images of bile secret in diagnostics of human physiological state. In: European Conference on Biomedical Optics. Optica Publishing Group; 2009.

55 Ushenko Y, Ushenko A, Dubolazov A, Soltys I, Litvinenko O, Wanchuliak O, et al. Mueller-matrix interferometric multifractal scaling of optically anisotropic architectonics of diffuse blood facies: fundamental and applied aspects. J Biophotonics. 2025;18(3):e202400412. https://doi.org/10.1002/jbio.202400412.

56 Ushenko A, Soltys I, Dubolazov A, Ushenko Y, Bilookyi V, Bilookyi O, et al. 3D Jones-matrix thesiography of biological fluid facies. J Innov Opt Health Sci. 2025;18(2):2443002. https://doi.org/10.1142/S1793545824430028.

57 Ushenko A, Pavlyukovich N, Khukhlina O, Pavlyukovich O, Gorsky M, Soltys I, et al. Layer-by-layer multifractal scanning of optically anisotropic architectonics of blood plasma films: fundamental and applied aspects. Photonics. 2025;12:215. https://doi.org/10.3390/photonics12030215.

58 Ushenko A, Pavlyukovich N, Khukhlina O, Pavlyukovich O, Soltys I, Dubolazov A, et al. Blood plasma film multifractal scanning in COVID-19 consequences diagnostics. J Biophotonics. 2024;17(11):e202400356. https://doi.org/10.1002/jbio.202400356.

59 Ushenko A, Dubolazov A, Zheng J, Litvinenko A, Gorsky M, Ushenko Y, et al. 3D polarization-interference holographic histology for wavelet-based differentiation of biological tissues with different necrotic states. J Biomed Opt. 2024;29(5):052920. https://doi.org/10.1117/1.JBO.29.5.052920.

60 Ushenko O, Bilookyi O, Zheng J, Dubolazov A, Olar O, Ushenko Y, et al. 3D digital holographic polarimetry of laser speckle fields formed by polycrystalline blood films: a tool for differential diagnosis of thyroid pathology. Front Phys. 2024;12: 1426469. https://doi.org/10.3389/fphy.2024.1426469.

61 Wójcik W, Hu Z, Ushenko Y, Smolarz A, Soltys I, Dubolazov O, et al. Optical sensor system for 3D Jones matrix reconstruction of optical anisotropy maps of self-assembled polycrystalline soft matter films. Sensors. 2024;24(5):1589. https://doi.org/10.3390/s24051589.

62 Ushenko A, Zheng J, Gorsky M, Dubolazov A, Ushenko Y, Soltys I, et al. 3D digital holographic polarimetry of diffuse optically anisotropic biological tissue object fields. Front Phys. 2023;11:1288935. https://doi.org/10.3389/fphy.2023.1288935.

63 Dubolazov A, Ushenko V, Trifonyuk L, Stashkevich A, Soltys I, Ushenko Y, et al. Polarization-singular approach to imaging Mueller-matrix polarimetry in differential diagnosis of histological sections of biopsy of tumors of the uterus and prostate. Front Phys. 2021;9:711212. https://doi.org/10.3389/fphy.2021.711212.

64 Ushenko VO, Trifonyuk L, Ushenko YA, Dubolazov OV, Gorsky MP, Ushenko AG. Polarization singularity analysis of Mueller-matrix invariants of optical anisotropy of biological tissue samples in cancer diagnostics. J Opt. 2021;23(6):064004. https://doi.org/10.1088/2040-8986/abf97a.

65 Ushenko YA, Koval GD, Ushenko AG, Dubolazov OV, Ushenko VA, Novakovskaia OY. Mueller-matrix of laser-induced autofluorescence of polycrystalline films of dried peritoneal fluid in diagnostics of endometriosis. J Biomed Opt. 2016;21(7):071116. https://doi.org/10.1117/1.JBO.21.7.071116.

A1

Programs for 3D Jones–Mueller Holographic Reconstruction

```
set(0,'DefaultAxesFontSize',7,'DefaultAxesFontName','Arial Cyr');

filename = 'e';

path = '';
rezpath = '';

data_name = {'0';'45';'90';'135';'prav';'liv'};
rez = struct('s0',[],'s45',[],'s90',[],'s135',[],'sprav',[],'sliv',[]);
corr_air = struct('c0', 0, 'c45', 0, 'c90', 0, 'c135', 0, 'cprav', pi/2, 'cliv', -pi/2);
corr_en = 0;
no_int_change = 0;
num_of_slices = 200;
disp_slices = 1;
save_data = 1;
new3d = 0;

corrx = 1100;
corry = 700;
corrx = 100;
corry = 100;
sigsmallx = 100;
sigsmally = 100;
sigsmallst = 400;
corrn = 0;

for i = 1:numel(data_name)
    slug = data_name{i};
    B = imread(strcat(path, filename, slug, 'x.bmp'));
    c1 = double(B(:,:,1));
    B = imread(strcat(path, filename, slug, 'y.bmp'));
    c2 = double(B(:,:,1));
    try
        %[p1,cs1,n2] = FTmethod(c1);
        [p1,cs1] = FTmethod_o(c1);
    catch ME
        p1 = zeros(size(c1));
        cs1 = zeros(size(c1));
    end
    try
        %[p2,cs2,n2] = FTmethod(c2);
        [p2,cs2] = FTmethod_o(c2);
    catch ME
        p2 = zeros(size(c2));
        cs2 = zeros(size(c2));
    end
    for rowc=1:size(p2,1)
```

```
        for colc=1:size(p2,2)
            if isnan(p2(rowc,colc)) && ~isnan(p1(rowc,colc))
                p2(rowc,colc) = p1(rowc,colc);
                cs2(rowc,colc) = 0;
            end
            if isnan(p1(rowc,colc)) && ~isnan(p2(rowc,colc))
                p1(rowc,colc) = p2(rowc,colc);
                cs1(rowc,colc) = 0;
            end
        end
    end

    if (corr_en == 1)
        corr1 = p1(corry:corry+100, corrx:corrx+100);
        corr2 = p2(corry:corry+100, corrx:corrx+100);
        %figure;imshow(mat2gray(corr1));
        %figure;imshow(mat2gray(corr2));
        corrn = mean2(corr2 - corr1);
        %corrn = max(max(corr2 - corr1));
        disp(corrn);
        corrn = corrn - corr_air.(strcat('c', slug));
        p2 = p2 - corrn;
    end


      diff = p2 - p1;
      intx = abs(cs1).^(2);
      inty = abs(cs2).^(2);
      int = intx + inty;

    absx = abs(cs1);
    absy = abs(cs2);
    %figure;imshow(mat2gray(int));
    %error('stop');
    [re, im] = pol2cart(diff, int);
    sig = re + im*1i;
    diff = mod(diff, pi);
    %[s1, s2, s3, s4] = calculateS2(intx, inty, diff);
    [s1, s2, s3, s4] = calculateS(cs1, cs2);

     s2 = s2./s1;
     s3 = s3./s1*(-1);
     s4 = (s4./s1)*(-1);
     s1 = s1./s1;

    disp(strcat(slug, ' - ', ' diff:' , num2str(mean2(diff)), ...
        '; absx:', num2str(mean2(absx)), '; absy:', num2str(mean2(absy)), ...
        '; absx/y:', num2str(mean2(absx./absy)),...
        '; s1:', num2str(mean2(s1)), '; s2:', num2str(mean2(s2)), ...
        '; s3:', num2str(mean2(s3)), '; s4:', num2str(mean2(s4)), ';'));



    sigsmall = sig(sigsmally:sigsmally+sigsmallst, sigsmallx:sigsmallx+sigsmallst);
    s1small = s1(sigsmally:sigsmally+sigsmallst, sigsmallx:sigsmallx+sigsmallst);
    s2small = s2(sigsmally:sigsmally+sigsmallst, sigsmallx:sigsmallx+sigsmallst);
    s3small = s3(sigsmally:sigsmally+sigsmallst, sigsmallx:sigsmallx+sigsmallst);
    s4small = s4(sigsmally:sigsmally+sigsmallst, sigsmallx:sigsmallx+sigsmallst);



    s1sl = SliceFazeAdd(sigsmall, num_of_slices, s1small);
    s2sl = SliceFazeAdd(sigsmall, num_of_slices, s2small);
    s3sl = SliceFazeAdd(sigsmall, num_of_slices, s3small);
    s4sl = SliceFazeAdd(sigsmall, num_of_slices, s4small);
```

```
    rez.(strcat('s',slug)) = struct('s1', s1sl, 's2', s2sl, 's3', s3sl, 's4', s4sl);
    rez.(strcat('ss',slug)) = struct('s1', s1small, 's2', s2small, 's3', s3small,
  's4', s4small);
end;
%memory
clear
('B','p1','cs1','n1','p2','cs2','n2','diff','int','s1','s2','s3','s4','sigsmall',
's1small','s2small','s3small','s4small','s1sl','s2sl','s3sl','s4sl');
clear('c1','c2','im','re','sig');
%memory
im = 0.5 * atan2(rez.(strcat('s','0')).s3, rez.(strcat('s','0')).s2);
save(strcat(rezpath, 'mat/', 'azimuth'), 'im');
return;


formulas = struct(...
'm11', '(rez.(''s0'').s1 + rez.(''s90'').s1)./2',...
'm12', '(rez.(''s0'').s2 + rez.(''s90'').s2)./2',...
'm13', '(rez.(''s0'').s3 + rez.(''s90'').s3)./2',...
'm14', '(rez.(''s0'').s4 + rez.(''s90'').s4)./2',...
'm21', '(rez.(''s0'').s1 - rez.(''s90'').s1)./2',...
'm22', '(rez.(''s0'').s2 - rez.(''s90'').s2)./2',...
'm23', '(rez.(''s0'').s3 - rez.(''s90'').s3)./2',...
'm24', '(rez.(''s0'').s4 - rez.(''s90'').s4)./2',...
'm31', 'rez.(''s45'').s1 - (rez.(''s0'').s1 + rez.(''s90'').s1)./2',...
'm32', 'rez.(''s45'').s2 - (rez.(''s0'').s2 + rez.(''s90'').s2)./2',...
'm33', 'rez.(''s45'').s3 - (rez.(''s0'').s3 + rez.(''s90'').s3)./2',...
'm34', 'rez.(''s45'').s4 - (rez.(''s0'').s4 + rez.(''s90'').s4)./2',...
'm41', 'rez.(''sprav'').s1 - (rez.(''s0'').s1 + rez.(''s90'').s1)./2',...
'm42', 'rez.(''sprav'').s2 - (rez.(''s0'').s2 + rez.(''s90'').s2)./2',...
'm43', 'rez.(''sprav'').s3 - (rez.(''s0'').s3 + rez.(''s90'').s3)./2',...
'm44', 'rez.(''sprav'').s4 - (rez.(''s0'').s4 + rez.(''s90'').s4)./2'...
);
formulas2 = struct(...
'm11', '(rez.(''ss0'').s1 + rez.(''ss90'').s1)./2',...
'm12', '(rez.(''ss0'').s2 + rez.(''ss90'').s2)./2',...
'm13', '(rez.(''ss0'').s3 + rez.(''ss90'').s3)./2',...
'm14', '(rez.(''ss0'').s4 + rez.(''ss90'').s4)./2',...
'm21', '(rez.(''ss0'').s1 - rez.(''ss90'').s1)./2',...
'm22', '(rez.(''ss0'').s2 - rez.(''ss90'').s2)./2',...
'm23', '(rez.(''ss0'').s3 - rez.(''ss90'').s3)./2',...
'm24', '(rez.(''ss0'').s4 - rez.(''ss90'').s4)./2',...
'm31', 'rez.(''ss45'').s1 - (rez.(''ss0'').s1 + rez.(''ss90'').s1)./2',...
'm32', 'rez.(''ss45'').s2 - (rez.(''ss0'').s2 + rez.(''ss90'').s2)./2',...
'm33', 'rez.(''ss45'').s3 - (rez.(''ss0'').s3 + rez.(''ss90'').s3)./2',...
'm34', 'rez.(''ss45'').s4 - (rez.(''ss0'').s4 + rez.(''ss90'').s4)./2',...
'm41', 'rez.(''ssprav'').s1 - (rez.(''ss0'').s1 + rez.(''ss90'').s1)./2',...
'm42', 'rez.(''ssprav'').s2 - (rez.(''ss0'').s2 + rez.(''ss90'').s2)./2',...
'm43', 'rez.(''ssprav'').s3 - (rez.(''ss0'').s3 + rez.(''ss90'').s3)./2',...
'm44', 'rez.(''ssprav'').s4 - (rez.(''ss0'').s4 + rez.(''ss90'').s4)./2'...
);


fields = fieldnames(formulas);
h = figure;
sh = gobjects(numel(fields), 1);
for i = 1:numel(fields)
    str = fields{i};
    posx = str2double(str(2:2));
    posy = str2double(str(3:3));

    sh(i) = subplot(4,5,(posy-1)*5 + posx);
end;
```

```
m11 = eval(formulas2.('m11'));
map = colormap('jet');
figure;
for i = 1:numel(fields)
    str = fields{i};
    im = eval(formulas2.(str));
    posx = str2double(str(2:2));
    posy = str2double(str(3:3));
    imshow(mat2gray(im).*2 - 1, 'Parent', sh(i)); colormap jet;

end

print(strcat(rezpath, 'matrix'),'-dpng');

if (disp_slices == 1)
if (save_data == 1)
    m11 = eval(formulas.('m11'));
    for i = 1:numel(fields)
        tosave = zeros(sigsmallst+1, sigsmallst+1, num_of_slices);
        str = fields{i};
        tmp = eval(formulas.(str));
        for p=1:num_of_slices
            im = tmp(:,:,p);
            %hack for correct values
            if (contains(str, 'm23') || contains(str, 'm32'))
               im = (eval(formulas.('m23')) + eval(formulas.('m32')))./2;
               im = im(:,:,p);
            end
            if (contains(str, '1') && ~contains(str, 'm11'))
                mxim = max(max(abs(im)));
                mxm11 = max(max(abs(m11(:,:,p))));
                if ((mxim/mxm11) > 0.1)
                   im = im .* 0.1;
                end
            end
            im(isnan(im)) = 0.0;
            im(isinf(im)) = 0.0;
            posx = str2double(str(2:2));
            posy = str2double(str(3:3));
            tosave(:,:,p) = im;
        end
        save(strcat(rezpath, 'mat/', filename, str), 'tosave');
    end;
    save(strcat(rezpath, 'mat/', filename, '_stocks'), 'rez')
else
     m11 = eval(formulas.('m11'));
     for p=1:num_of_slices
        fields = fieldnames(formulas);
        h = figure;
        sh = gobjects(numel(fields), 1);
        for i = 1:numel(fields)
            str = fields{i};
            posx = str2double(str(2:2));
            posy = str2double(str(3:3));

            sh(i) = subplot(4,5,(posy-1)*5 + posx);
        end;
        map = colormap('jet');
        for i = 1:numel(fields)
            str = fields{i};
            tmp = eval(formulas.(str));
            im = tmp(:,:,p);

            im(isnan(im)) = 0;
            im(isinf(im)) = 0;
```

```
            posx = str2double(str(2:2));
            posy = str2double(str(3:3));

            tmp2(:,:,p).(str) = im;
            imshow(im, 'Parent', sh(i));
            pos = get(sh(i), 'Position');
            colormap(sh(i), map);
            cbh = colorbar(sh(i));
            range = [min(min(im)), max(max(im))];
                if (range(1) < range(2) && ~contains(str, '1'))
                    caxis(sh(i), range)
                else
                    caxis(sh(i), [-1, 1])
                end;
            set(sh(i), 'Position', [1.3 * pos(1) - 0.1 pos(2) pos(3) pos(4)]);
            cbt=get(cbh,'Limits');
            maxlim = str2double(sprintf('%.1f',cbt(1)));
            minlim = str2double(sprintf('%.1f',cbt(2)));
            %set(cbh,'Ticks',linspace(maxlim,minlim,5));
        end;

        print(strcat(rezpath, 'matrix',int2str(p)),'-dpng','-r600');
        set(h, 'paperunits', 'centimeters', 'paperposition', [0 0 30 20]);
     end
end
else
    for i = 1:numel(fields)
        to3d = eval(formulas.(fields{i}));
        %hack for correct values
        if (contains(str, 'm23') || contains(str, 'm32'))
           to3d = (eval(formulas.('m23')) + eval(formulas.('m32')))./2;
        end
        if (contains(str, '1') && ~contains(str, 'm11'))
            mxim = max(max(to3d));
            mxm11 = max(max(max(m11)));
            if ((mxim/mxm11) > 0.1)
               to3d = to3d .* 0.1;
            end
        end

        if (new3d == 1)
            im = zeros(sigsmallst+1,sigsmallst+1);
            for p=1:num_of_slices
                imtmp = to3d(:,:,p);
                imtmp(imtmp>=0.5) = 1;
                im(imtmp>=0.5) = (num_of_slices - p)/num_of_slices;
            end
            h = figure;
            h1 = surf(im);
            set(h1,'LineStyle','none');%contour3(im);
            print(strcat(rezpath, '3d/',
    filename,'_',num2str(num_of_slices),'_contour', fields{i}),'-dpng')

            else
                im = zeros(sigsmallst+1,sigsmallst+1);
            for p=1:num_of_slices
              im(:,:,1,p) = uint8((1-abs(to3d(:,:,p))).*255);
            end

            h = figure; vol3d('cdata', squeeze(im), 'xdata', [0 1],
    'ydata', [0 1], 'zdata', [0 0.7]);
            colormap(bone(256));
            alphamap([0 linspace(0.1, 0, 255)]);
            axis equal off
```

```
            set(h, 'color', 'w');
            view(3);
            view(-55, 50);


            print(strcat(rezpath, '3d/', filename, fields{i}),'-dpng')

            view(-90, 82);
            print(strcat(rezpath, '3d/', filename,'_top', fields{i}),'-dpng')
            view(-90, 0);
            print(strcat(rezpath, '3d/', filename,'_front', fields{i}),'-dpng')
            view(0, 0);
            print(strcat(rezpath, '3d/', filename,'_left', fields{i}),'-dpng')
            close(h);
        end
    end;
end
close all;
```

A2

Programs for Layer-by-layer Reconstruction of Polycrystalline Structures

```
set(0,'DefaultAxesFontSize',7,'DefaultAxesFontName','Arial Cyr');

filename = 'e';

path = '';
rezpath = '';

data_name = {'0';'45';'90';'135';'prav';'liv'};
rez = struct('s0',[],'s45',[],'s90',[],'s135',[],'sprav',[],'sliv',[]);
corr_air = struct('c0', 0, 'c45', 0, 'c90', 0, 'c135', 0, 'cprav', pi/2, 'cliv', -pi/2);
corr_en = 0;
no_int_change = 0;
num_of_slices = 200;
disp_slices = 1;
save_data = 1;
new3d = 0;

corrx = 1100;
corry = 700;
corrx = 100;
corry = 100;
sigsmallx = 100;
sigsmally = 100;
sigsmallst = 400;
corrn = 0;

for i = 1:numel(data_name)
    slug = data_name{i};
    B = imread(strcat(path, filename, slug, 'x.bmp'));
    c1 = double(B(:,:,1));
    B = imread(strcat(path, filename, slug, 'y.bmp'));
    c2 = double(B(:,:,1));
    try
        %[p1,cs1,n2] = FTmethod(c1);
        [p1,cs1] = FTmethod_o(c1);
    catch ME
        p1 = zeros(size(c1));
        cs1 = zeros(size(c1));
    end
    try
        %[p2,cs2,n2] = FTmethod(c2);
        [p2,cs2] = FTmethod_o(c2);
    catch ME
        p2 = zeros(size(c2));
        cs2 = zeros(size(c2));
    end
    for rowc=1:size(p2,1)
```

```
        for colc=1:size(p2,2)
            if isnan(p2(rowc,colc)) && ~isnan(p1(rowc,colc))
                p2(rowc,colc) = p1(rowc,colc);
                cs2(rowc,colc) = 0;
            end
            if isnan(p1(rowc,colc)) && ~isnan(p2(rowc,colc))
                p1(rowc,colc) = p2(rowc,colc);
                cs1(rowc,colc) = 0;
            end
        end
    end

    if (corr_en == 1)
        corr1 = p1(corry:corry+100, corrx:corrx+100);
        corr2 = p2(corry:corry+100, corrx:corrx+100);
        %figure;imshow(mat2gray(corr1));
        %figure;imshow(mat2gray(corr2));
        corrn = mean2(corr2 - corr1);
        %corrn = max(max(corr2 - corr1));
        disp(corrn);
        corrn = corrn - corr_air.(strcat('c', slug));
        p2 = p2 - corrn;
    end

      diff = p2 - p1;
      intx = abs(cs1).^(2);
      inty = abs(cs2).^(2);
      int = intx + inty;

    absx = abs(cs1);
    absy = abs(cs2);
    %figure;imshow(mat2gray(int));
    %error('stop');
    [re, im] = pol2cart(diff, int);
    sig = re + im*1i;
    diff = mod(diff, pi);
    %[s1, s2, s3, s4] = calculateS2(intx, inty, diff);
    [s1, s2, s3, s4] = calculateS(cs1, cs2);

     s2 = s2./s1;
     s3 = s3./s1*(-1);
     s4 = (s4./s1)*(-1);
     s1 = s1./s1;

    disp(strcat(slug, ' - ', ' diff:' , num2str(mean2(diff)), ...
        '; absx:', num2str(mean2(absx)), '; absy:', num2str(mean2(absy)), ...
        '; absx/y:', num2str(mean2(absx./absy)),...
        '; s1:', num2str(mean2(s1)), '; s2:', num2str(mean2(s2)), ...
        '; s3:', num2str(mean2(s3)), '; s4:', num2str(mean2(s4)), ';'));

function [Phase, ComplexSig] = FTmethod_o(Im)
CarrSigma = 200;

Spectr = fftshift(fft2(Im)); %find spectrum of the image
[rowPeak,colPeak] = find(Spectr == max(Spectr(:)));
[rowS, colS] = size(Spectr);

[ColInd, RowInd] = meshgrid(linspace(1,colS,colS),linspace(1,rowS,rowS));
SpectrLoc = abs(Spectr(rowPeak-2:rowPeak+2,colPeak-2:colPeak+2));
CoordLocCol = ColInd(rowPeak-2:rowPeak+2,colPeak-2:colPeak+2);
CoordLocRow = RowInd(rowPeak-2:rowPeak+2,colPeak-2:colPeak+2);
rowPeakSub = sum(SpectrLoc(:).*CoordLocRow(:))./sum(SpectrLoc(:));
colPeakSub = sum(SpectrLoc(:).*CoordLocCol(:))./sum(SpectrLoc(:));

Spectr = Spectr.*exp(-((RowInd-rowPeakSub).^2+(ColInd-colPeakSub).^2)/2/(CarrSigma/2)^2);
```

```
NewSpectr = zeros(size(Spectr));
NewSpectr(round(rowS/2)+1-CarrSigma:round(rowS/2)+1+CarrSigma,
round(colS/2)+1-CarrSigma:round(colS/2)+1+CarrSigma)=
Spectr(rowPeak-CarrSigma:rowPeak+CarrSigma,
colPeak-CarrSigma:colPeak+CarrSigma);

[ColF, RowF] = meshgrid(linspace(-colS/2,colS/2-1,colS),linspace(-rowS/2,rowS/2-1,rowS));
MaskSpatial = exp(1i*(ColF.*(colPeakSub-(round(colS/2)+1))*2*pi/rowS + RowF.*
(rowPeakSub-(round(rowS/2)+1))*2*pi/colS));
NewSpatial = ifft2(fftshift(Spectr))./MaskSpatial;

ComplexSig = ifft2(fftshift(NewSpectr));
Phase = angle(ComplexSig);
```

A3

Programs for Statistical and Correlation Data Analysis

```
[dim1,dim2]=size(M14n);
for i=1:dim1
    AKFx(i,:)=xcorr(M14n(i,:)-mean(M14n(i,:)));
end

XtotalAKF=sum(AKFx);


if min(min(XtotalAKF))>0;
   XtotalAKF=XtotalAKF-abs(min(min(XtotalAKF)));
end
if min(min(XtotalAKF))<0;
   XtotalAKF=XtotalAKF+abs(min(min(XtotalAKF)));
end

XtotalAKF1=XtotalAKF./max(max(XtotalAKF));

XPSD1=periodogram(XtotalAKF1);


[dim1,dim2]=size(H1);
STD2=std(H1);
Seredne=mean(H1);
Mx=mean(H1);
Dx=STD2^2;
MEDx - median(H1);
MAX = max(H1);
MIN = min(H1);
Ax=skewness(H1);
Ex=kurtosis(H1)-3;

fid=fopen('f14(x,y) - stat.txt','w')
fprintf(fid,'Stat f14(x,y)\n')
fprintf(fid,'\n')
fprintf(fid,'X-dimension of the matrix dim1=%14.7f\n',dim1);
fprintf(fid,'Y-dimension of the matrix dim2=%14.7f\n',dim2);
fprintf(fid,'Min. element of the matrix    =%14.7f\n',MIN);
fprintf(fid,'Max. element of the matrix    =%14.7f\n',MAX);
fprintf(fid,'Mean                          =%14.7f\n',Seredne);
fprintf(fid,'Math. waiting                 =%14.7f\n',Mx);
fprintf(fid,'S. K. V.                      =%14.7f\n',STD2);
fprintf(fid,'Dispersion                    =%14.7f\n',Dx);
fprintf(fid,'Assimetrie                    =%14.7f\n',Ax);
fprintf(fid,'Extcess                       =%14.7f\n',Ex);
fprintf(fid,'Median                        =%14.7f\n',MEDx);
fclose(fid);
movefile('f14(x,y) - stat.TXT','C:/MATLAB7/work');
```

```
[dim1,dim2]=size(M41n);
for i=1:dim1
    AKFx(i,:)=xcorr(M41n(i,:)-mean(M41n(i,:)));
end

%Сумарна АКФ у напрямку X
XtotalAKF=sum(AKFx);

if min(min(XtotalAKF))>0;
   XtotalAKF=XtotalAKF-abs(min(min(XtotalAKF)));
end
if min(min(XtotalAKF))<0;
   XtotalAKF=XtotalAKF+abs(min(min(XtotalAKF)));
end

XtotalAKF2=XtotalAKF./max(max(XtotalAKF));

XPSD2=periodogram(XtotalAKF2);

[dim1,dim2]=size(H2);
STD2=std(H2);
Seredne=mean(H2);
Mx=mean(H2);
Dx=STD2^2;
MEDx = median(H2);
MAX = max(H2);
MIN = min(H2);
Ax=skewness(H2);
Ex=kurtosis(H2)-3;

fid=fopen('f41(x,y) - stat.txt','w')
fprintf(fid,'Stat f41(x,y)\n')
fprintf(fid,'\n')
fprintf(fid,'X-dimension of the matrix dim1=%14.7f\n',dim1);
fprintf(fid,'Y-dimension of the matrix dim2=%14.7f\n',dim2);
fprintf(fid,'Min. element of the matrix    =%14.7f\n',MIN);
fprintf(fid,'Max. element of the matrix    =%14.7f\n',MAX);
fprintf(fid,'Mean                          =%14.7f\n',Seredne);
fprintf(fid,'Math. waiting                 =%14.7f\n',Mx);
fprintf(fid,'S. K. V.                      =%14.7f\n',STD2);
fprintf(fid,'Dispersion                    =%14.7f\n',Dx);
fprintf(fid,'Assimetrie                    =%14.7f\n',Ax);
fprintf(fid,'Extcess                       =%14.7f\n',Ex);
fprintf(fid,'Median                        =%14.7f\n',MEDx);
fclose(fid);
movefile('f41(x,y) - stat.TXT','C:/MATLAB7/work');


[dim1,dim2]=size(M44n);
for i=1:dim1
    AKFx(i,:)=xcorr(M44n(i,:)-mean(M44n(i,:)));
end

XtotalAKF=sum(AKFx);

if min(min(XtotalAKF))>0;
   XtotalAKF=XtotalAKF-abs(min(min(XtotalAKF)));
end
if min(min(XtotalAKF))<0;
   XtotalAKF=XtotalAKF+abs(min(min(XtotalAKF)));
end

XtotalAKF3=XtotalAKF./max(max(XtotalAKF));
```

```
XPSD3=periodogram(XtotalAKF3);

[dim1,dim2]=size(H3);
STD2=std(H3);
Seredne=mean(H3);
Mx=mean(H3);
Dx=STD2^2;
MEDx = median(H3);
MAX = max(H3);
MIN = min(H3);
Ax=skewness(H3);
Ex=kurtosis(H3)-3;

fid=fopen('f44(x,y) - stat.txt','w')
fprintf(fid,'Stat f44(x,y)\n')
fprintf(fid,'\n')
fprintf(fid,'X-dimension of the matrix dim1=%14.7f\n',dim1);
fprintf(fid,'Y-dimension of the matrix dim2=%14.7f\n',dim2);
fprintf(fid,'Min. element of the matrix    =%14.7f\n',MIN);
fprintf(fid,'Max. element of the matrix    =%14.7f\n',MAX);
fprintf(fid,'Mean                          =%14.7f\n',Seredne);
fprintf(fid,'Math. waiting                 =%14.7f\n',Mx);
fprintf(fid,'S. K. V.                      =%14.7f\n',STD2);
fprintf(fid,'Dispersion                    =%14.7f\n',Dx);
fprintf(fid,'Assimetrie                    =%14.7f\n',Ax);
fprintf(fid,'Extcess                       =%14.7f\n',Ex);
fprintf(fid,'Median                        =%14.7f\n',MEDx);
fclose(fid);
movefile('f44(x,y) - stat.TXT','C:/MATLAB7/work');


[dim1,dim2]=size(F1n);
for i=1:dim1
    AKFx(i,:)=xcorr(F1n(i,:)-mean(F1n(i,:)));
end

XtotalAKF=sum(AKFx);

if min(min(XtotalAKF))>0;
   XtotalAKF=XtotalAKF-abs(min(min(XtotalAKF)));
end
if min(min(XtotalAKF))<0;
   XtotalAKF=XtotalAKF+abs(min(min(XtotalAKF)));
end

XtotalAKF4=XtotalAKF./max(max(XtotalAKF));

XPSD4=periodogram(XtotalAKF4);

[dim1,dim2]=size(H4);
STD2=std(H4);
Seredne=mean(H4);
Mx=mean(H4);
Dx=STD2^2;
MEDx = median(H4);
MAX = max(H4);
MIN = min(H4);
Ax=skewness(H4);
Ex=kurtosis(H4)-3;

fid=fopen('f23-f32(x,y) - stat.txt','w')
fprintf(fid,'Stat f23-f32(x,y)\n')
fprintf(fid,'\n')
fprintf(fid,'X-dimension of the matrix dim1=%14.7f\n',dim1);
```

```
fprintf(fid,'Y-dimension of the matrix dim2=%14.7f\n',dim2);
fprintf(fid,'Min. element of the matrix    =%14.7f\n',MIN);
fprintf(fid,'Max. element of the matrix    =%14.7f\n',MAX);
fprintf(fid,'Mean                          =%14.7f\n',Seredne);
fprintf(fid,'Math. waiting                 =%14.7f\n',Mx);
fprintf(fid,'S. K. V.                      =%14.7f\n',STD2);
fprintf(fid,'Dispersion                    =%14.7f\n',Dx);
fprintf(fid,'Assimetrie                    =%14.7f\n',Ax);
fprintf(fid,'Extcess                       =%14.7f\n',Ex);
fprintf(fid,'Median                        =%14.7f\n',MEDx);
fclose(fid);
movefile('f23-f32(x,y) - stat.TXT','C:/MATLAB7/work');
```

A4

Album of Mueller-matrix Images

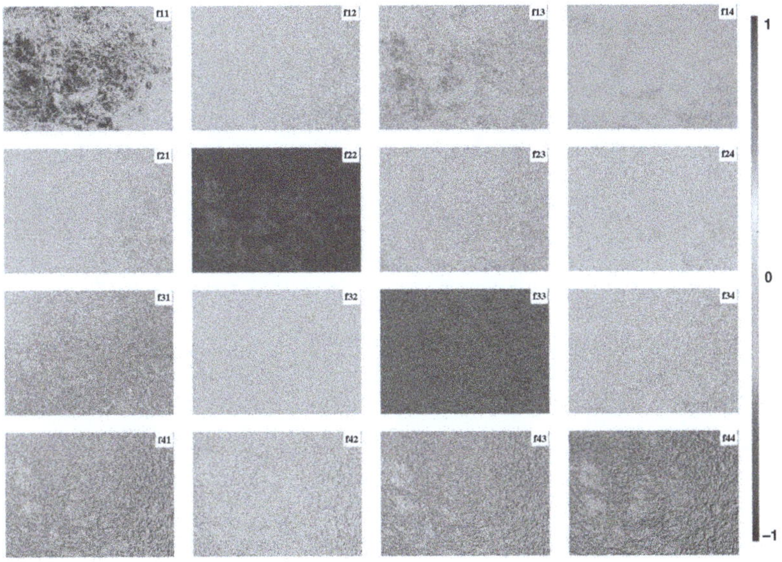

Figure A4.1 Mueller-matrix images of a native histological section of the endometrium (normal).

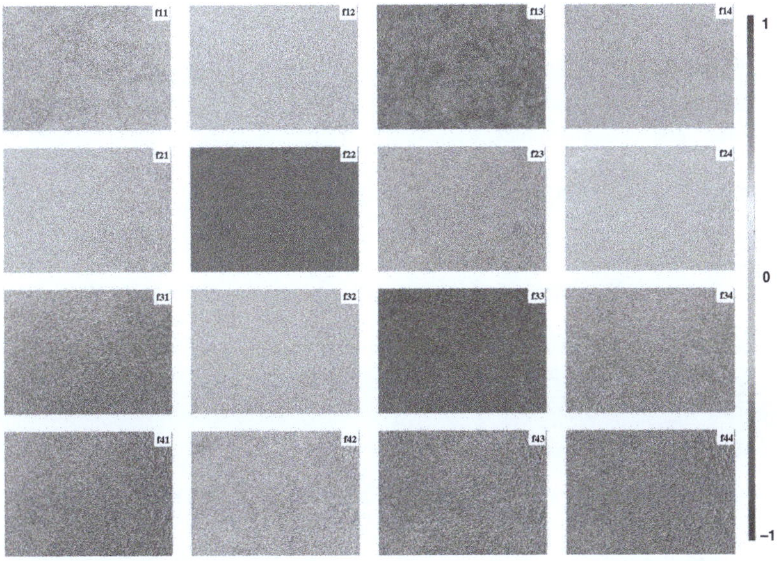

Figure A4.2 Mueller-matrix images of a native histological section of the endometrium (endometriosis).

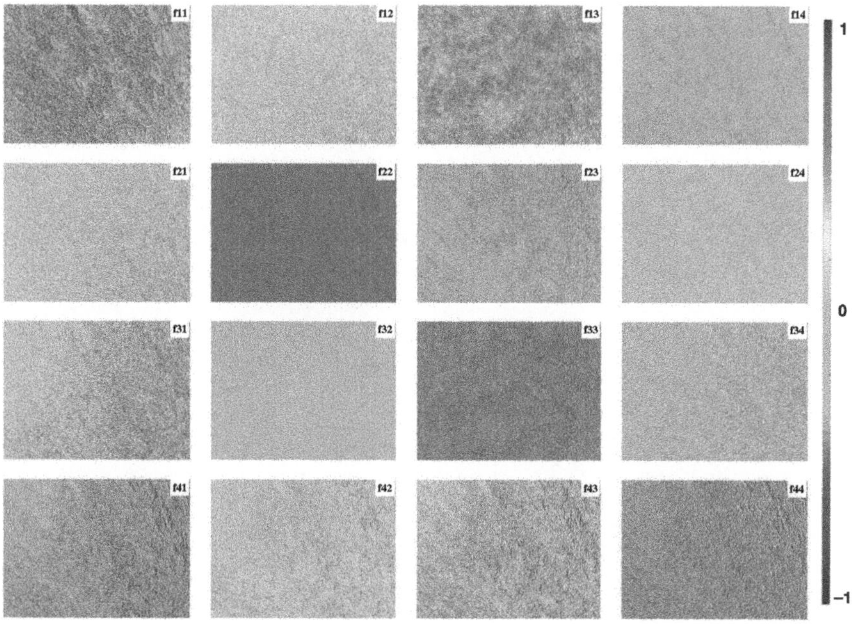

Figure A4.3 Mueller-matrix images of a native histological section of well-differentiated prostate tissue (Gleason score 3 + 3).

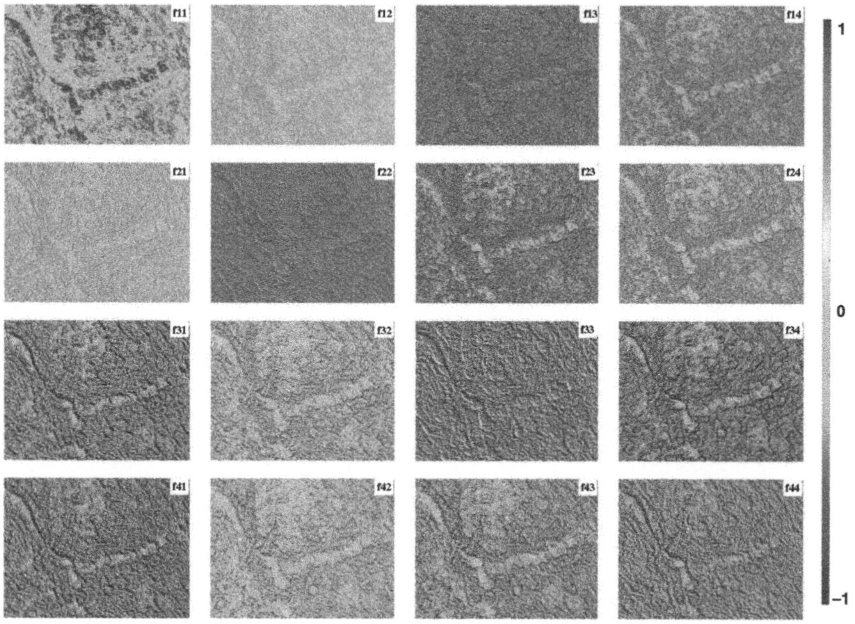

Figure A4.4 Mueller-matrix images of a native histological section of poorly differentiated prostate tissue (Gleason score 4 + 4).

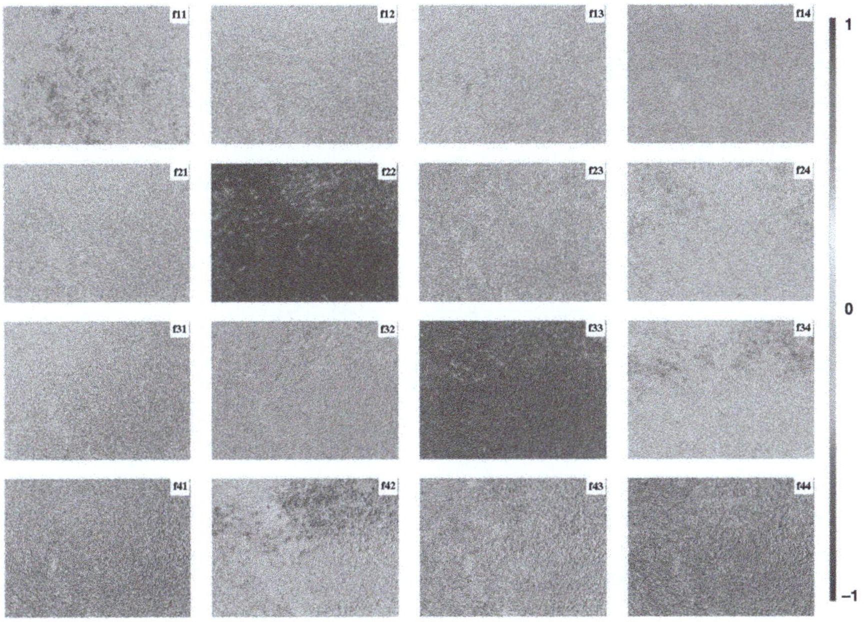

Figure A4.5 Mueller-matrix images of a native histological section of brain tissue (ischemic infarction).

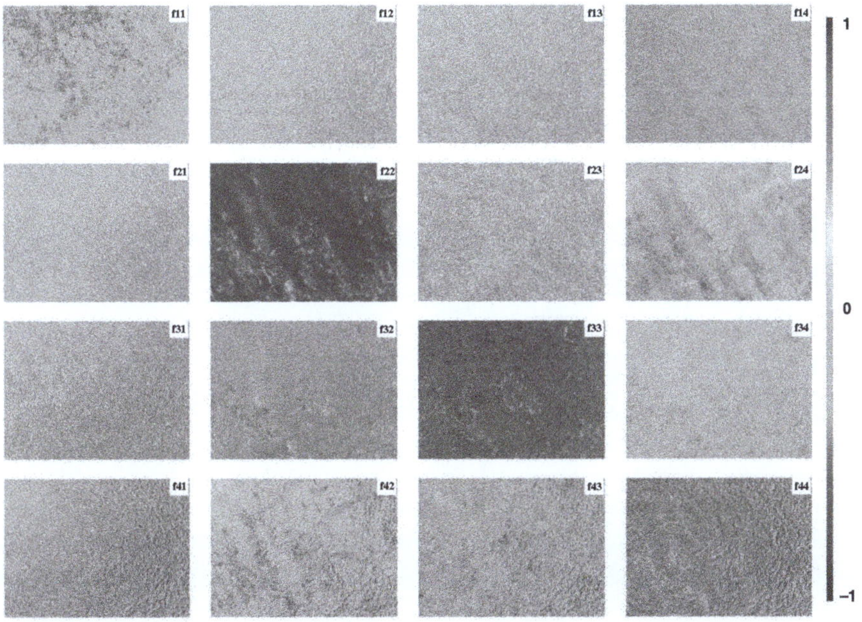

Figure A4.6 Mueller-matrix images of a native histological section of brain tissue (hemorrhagic infarction).

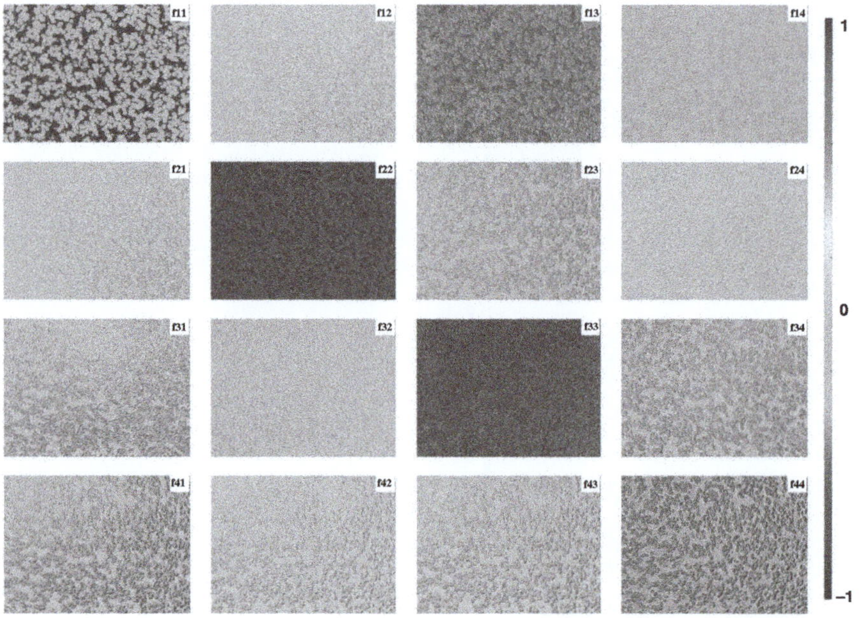

Figure A4.7 Mueller-matrix images of a dehydrated film of blood plasma (healthy donor).

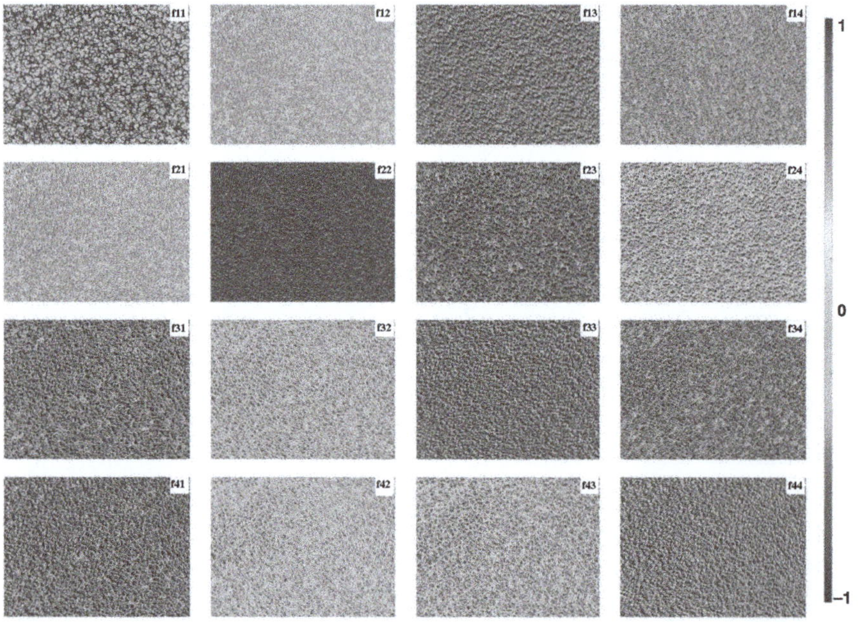

Figure A4.8 Mueller-matrix images of a dehydrated film of blood plasma (stage 1 papillary thyroid cancer).

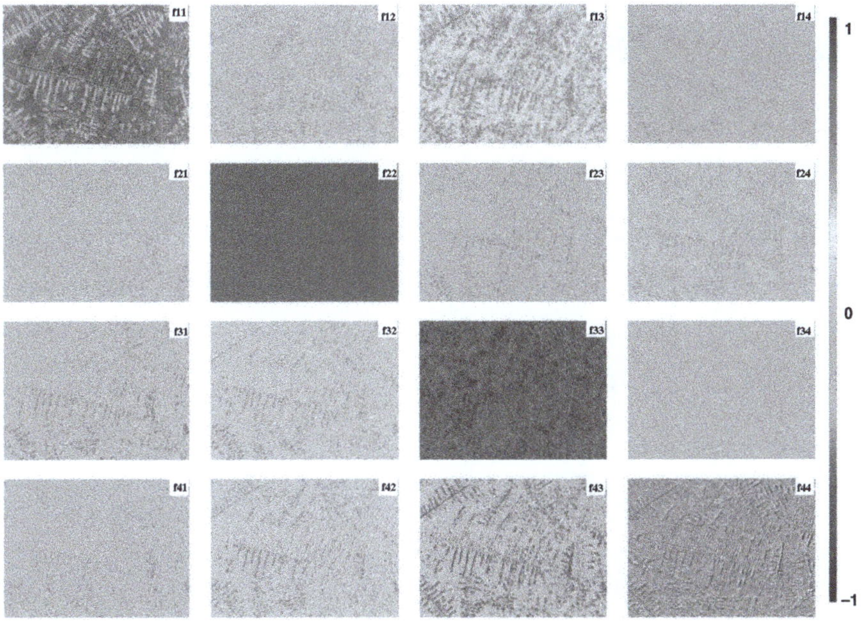

Figure A4.9 Mueller-matrix images of dehydrated synovial fluid film in the knee joint (reactive arthrosis).

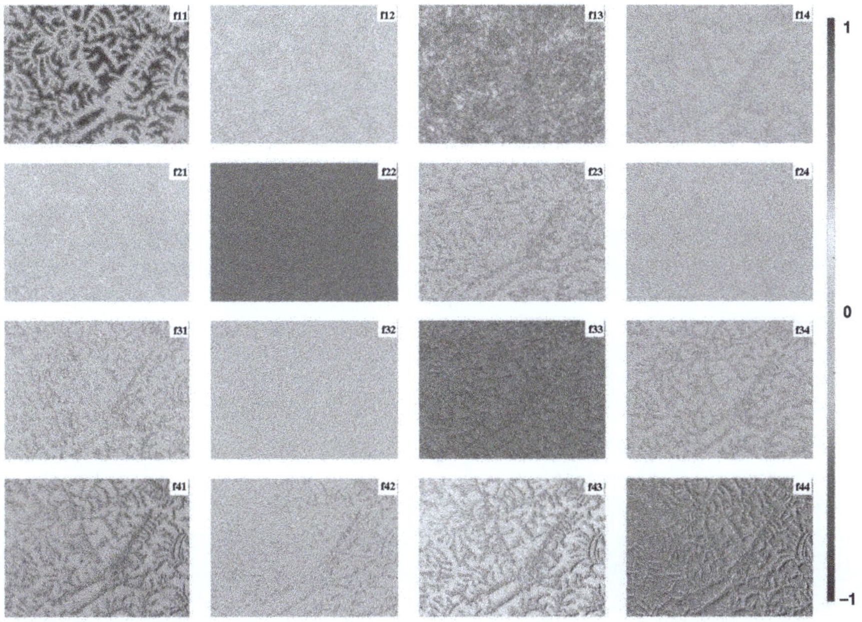

Figure A4.10 Mueller-matrix images of dehydrated synovial fluid film in the knee joint (septic arthritis).

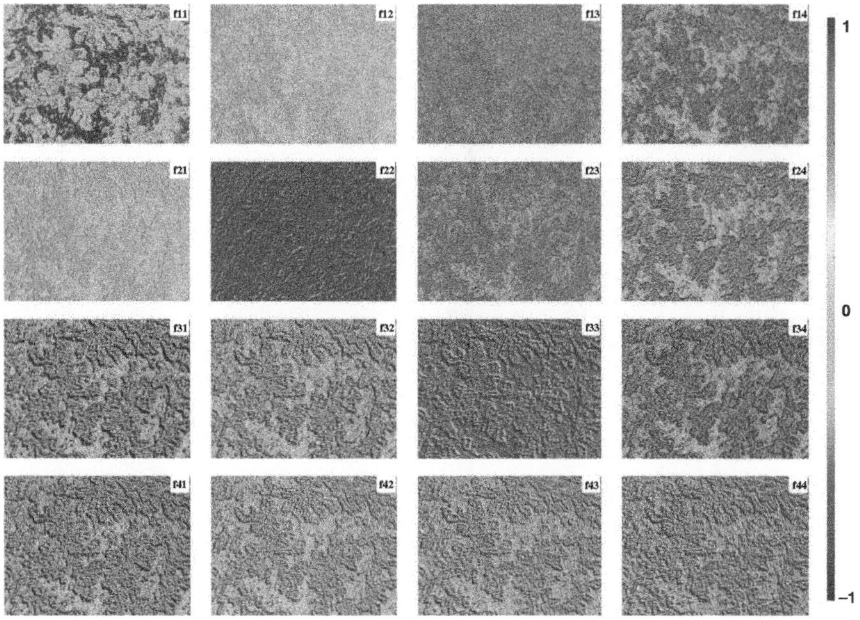

Figure A4.11 Mueller-matrix images of dehydrated gallbladder bile film (healthy patient).

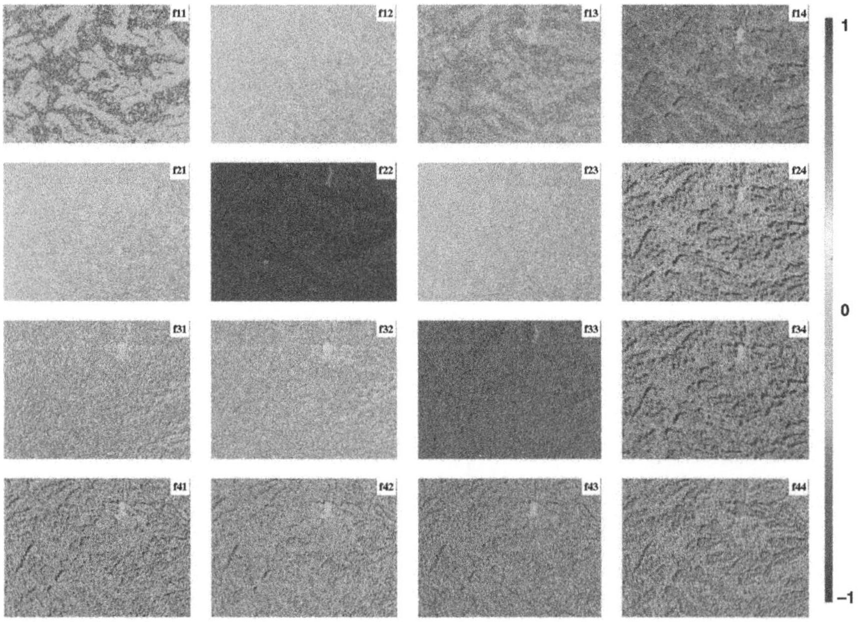

Figure A4.12 Mueller-matrix images of dehydrated gallbladder bile film (calculous cholecystitis).

A5

Album of Optical Anisotropy Maps

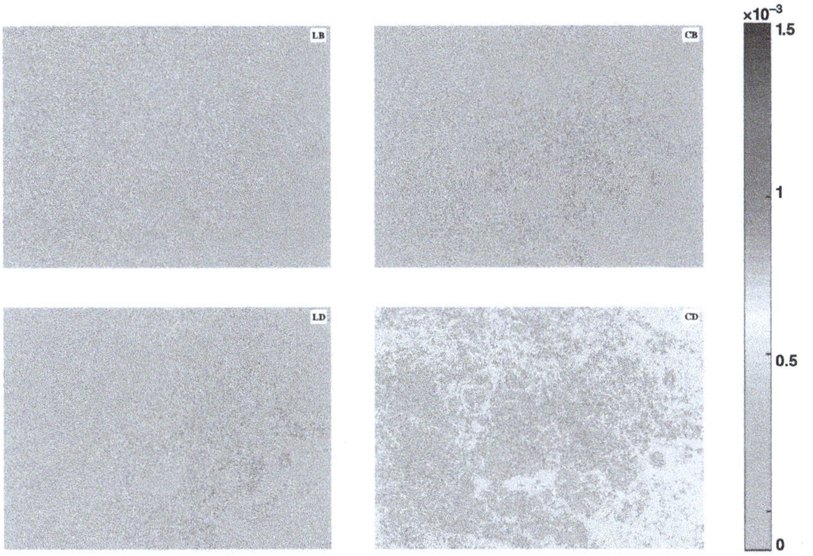

Figure A5.1 Maps of the phase (linear [LB] and circular [CB] birefringence) and amplitude (linear [LD] and circular [CD] dichroism) of the polycrystalline component of a native histological section of the endometrium (normal).

Figure A5.2 Maps of the phase (LB and CB) and amplitude (LD and CD) of the polycrystalline component of a native histological section of the endometrium (endometriosis).

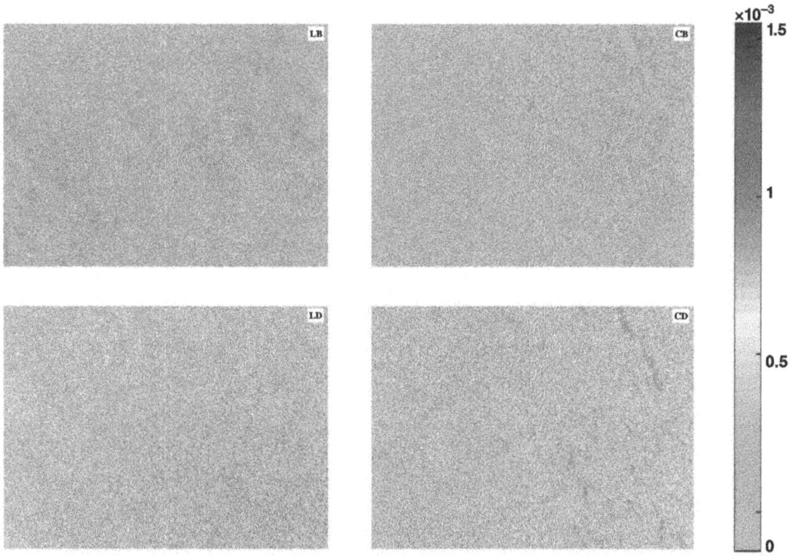

Figure A5.3 Maps of the phase (LB and CB) and amplitude (LD and CD) of the polycrystalline component of a native histological section of highly differentiated prostate tissue (Gleason score 3 + 3).

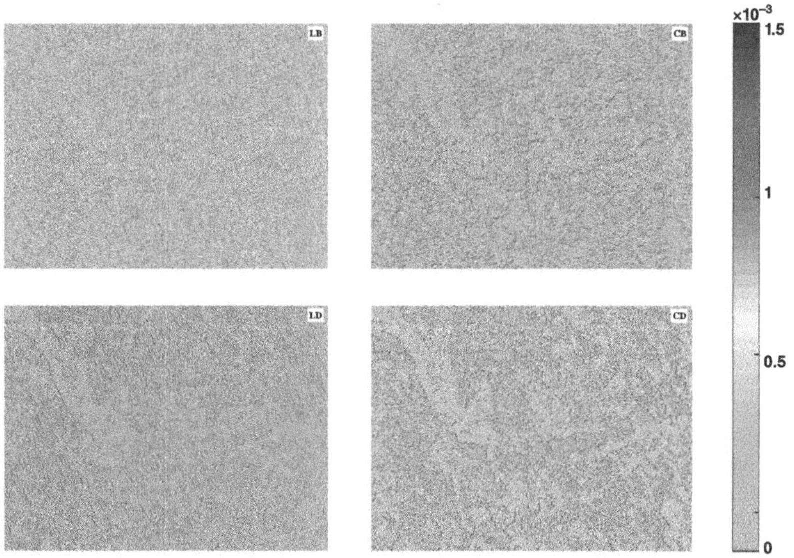

Figure A5.4 Maps of the phase (LB and CB) and amplitude (LD and CD) of the polycrystalline component of a native histological section of poorly differentiated prostate tissue (Gleason score 4 + 4).

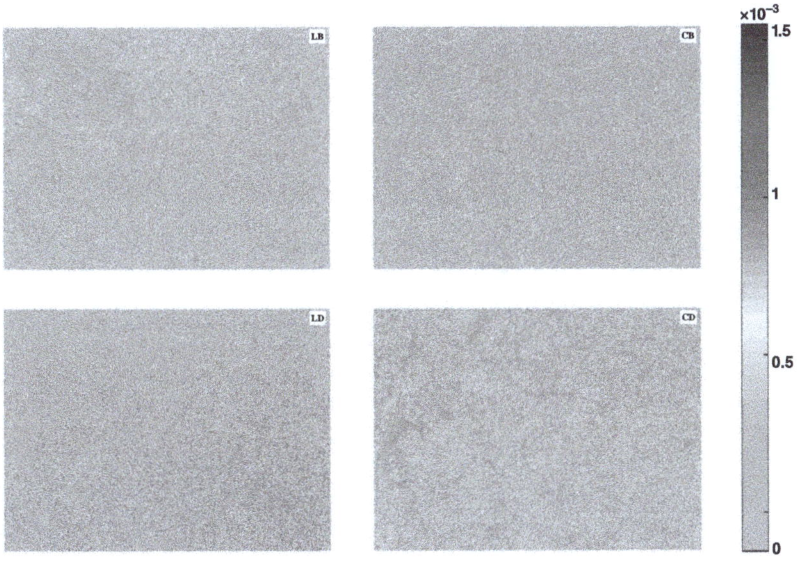

Figure A5.5 Maps of the phase (LB and CB) and amplitude (LD and CD) of the polycrystalline component of a native histological section of brain tissue (ischemic infarction).

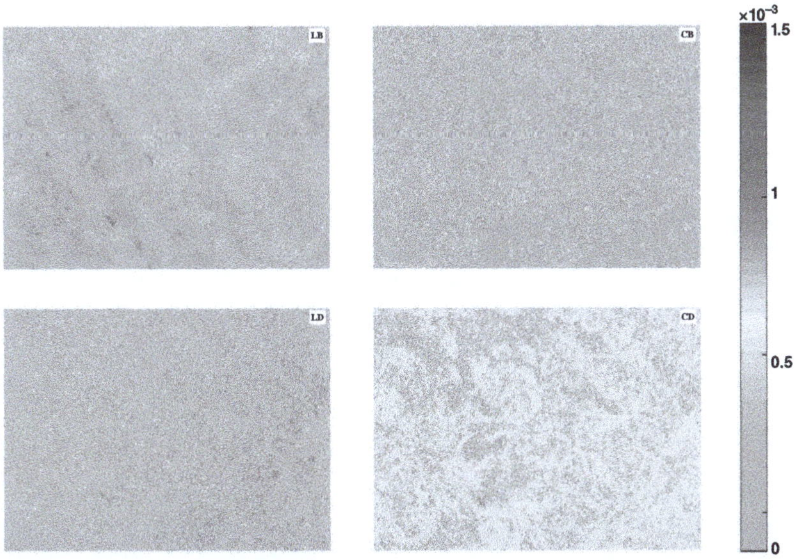

Figure A5.6 Maps of the phase (LB and CB) and amplitude (LD and CD) of the polycrystalline component of a native histological section of brain tissue (hemorrhagic infarction).

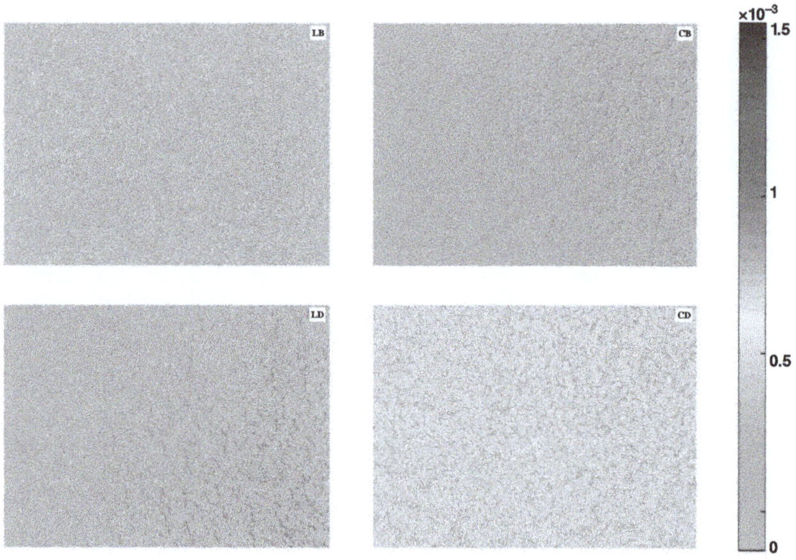

Figure A5.7 Maps of the phase (LB and CB) and amplitude (LD and CD) of the polycrystalline component of a dehydrated blood plasma film (healthy donor).

Figure A5.8 Maps of the phase (LB and CB) and amplitude (LD and CD) of the polycrystalline component of a dehydrated blood plasma film (stage 1 papillary thyroid cancer).

Figure A5.9 Maps of the phase (LB and CB) and amplitude (LD and CD) of the polycrystalline component of a dehydrated synovial fluid film from the knee joint (reactive arthrosis).

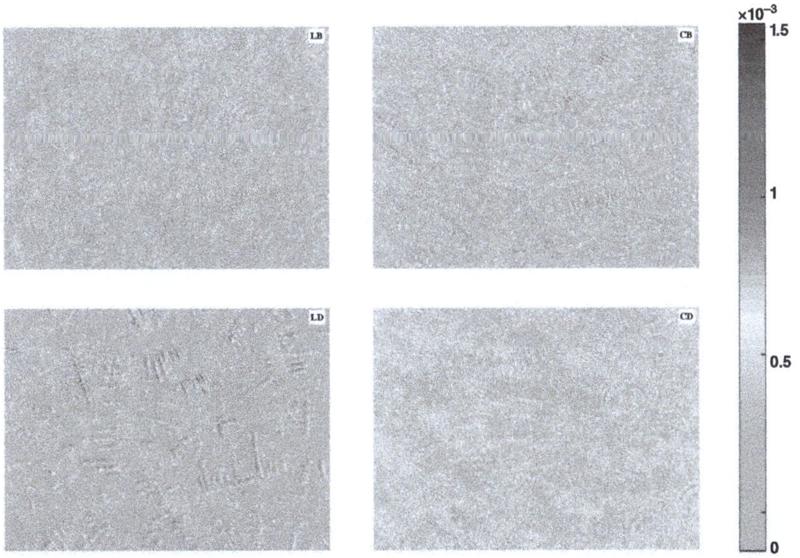

Figure A5.10 Maps of the phase (LB and CB) and amplitude (LD and CD) of the polycrystalline component of the dehydrated film of synovial fluid from the knee joint (septic arthritis).

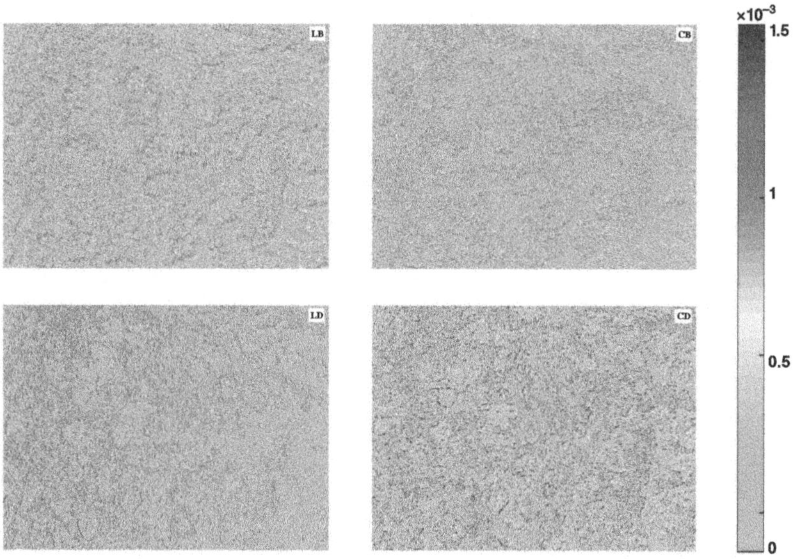

Figure A5.11 Maps of the phase (LB and CB) and amplitude (LD and CD) of the polycrystalline component of the dehydrated film of gallbladder bile (healthy patient).

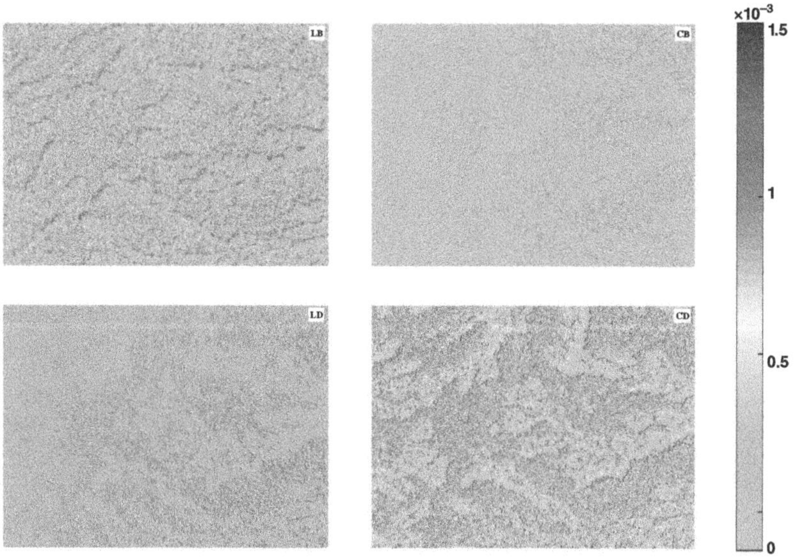

Figure A5.12 Maps of the phase (LB and CB) and amplitude (LD and CD) of the polycrystalline component of the dehydrated film of gallbladder bile (calculous cholecystitis).

Conclusion

This book organizes and examines original materials on the development of a new multifunctional area in biomedical laser imaging, utilizing advanced principles, digital techniques, and systems for rapid, highly sensitive laser digital holographic reconstruction of the layer-by-layer micro- and nanostructures of optically anisotropic polycrystalline components in biological tissues and human organ fluids.

The presented experimental data can serve as a solid foundation for further developing principles for early (preclinical) detection and monitoring of the most common systemic inflammatory, pathological, and necrotic conditions in the human body. This book introduces a core concept in the field of three-dimensional digital holographic Jones–Mueller mapping based on a single principle: the search for and physical validation of a set of diagnostically important relationships between objective (statistical, correlation, fractal, singular, wavelet, Fourier, etc.) digital markers that characterize multilayer Jones–Mueller matrix images and reconstructed optical anisotropy maps, as well as a broad spectrum of inflammatory, pathological, and necrotic changes in the polycrystalline structure of biological tissues and fluid samples.

The main focus is on reconstructing and performing multifunctional digital analysis of the polycrystalline, optically anisotropic part of biological samples.

The analysis of experimental methods and their instrumentations is detailed. They are built on a synthesis of the principles of Stokes polarimetry, polarization interferometry, digital holography, phase scanning, and the formalism of Mueller and Jones matrices. This approach allowed for a comparative study of multiparameter information regarding the polarization effects of combined linear and circular birefringence and optical anisotropy dichroism in the polycrystalline structure of biological samples. Consequently, new digital markers were discovered, and their diagnostic effectiveness (sensitivity, specificity, and accuracy) was assessed within an evidence-based medicine framework across various clinical conditions.

We hope that promoting and sharing our book will establish a modern, unified methodological foundation for a theoretical, algorithmic, and engineering platform. This platform will support the practical use of a wide range of highly sensitive, fast, multiparameter laser biophysical optical mapping methods in the early (preclinical) diagnosis of many common and severe human pathologies.

Index

Note: Page numbers in *italics* and **bold** refers to figures and tables respectively.

n

o

r

s

t

W